QUANTUM DISSIPATIVE SYSTEMS

Fourth Edition

QUANTUM DISSIPATIVE SYSTEMS

Fourth Edition

Ulrich Weiss
University of Stuttgart, Germany

World Scientific

NEW JERSEY · LONDON · SINGAPORE · BEIJING · SHANGHAI · HONG KONG · TAIPEI · CHENNAI

Published by

World Scientific Publishing Co. Pte. Ltd.
5 Toh Tuck Link, Singapore 596224
USA office: 27 Warren Street, Suite 401-402, Hackensack, NJ 07601
UK office: 57 Shelton Street, Covent Garden, London WC2H 9HE

British Library Cataloguing-in-Publication Data
A catalogue record for this book is available from the British Library.

QUANTUM DISSIPATIVE SYSTEMS
Fourth Edition

ISBN-13 978-981-4374-91-0 (pbk)
ISBN-10 981-4374-91-1 (pbk)

Printed in Singapore by Mainland Press Pte Ltd.

Preface

The prospect of a fourth edition of *Quantum Dissipative Systems*, almost twenty years after initial publication, and four years after publication of the third edition, shows on the one hand that the field is rapidly developing and on the other hand that the book has been paid the compliment of growing use as a text since it appeared for the first time.

In preparing the fourth edition I have tried to sustain as much as possible the advantages of the previous editions while taking into account developments in the subject itself and its applications to other fields. It was a persistent intention of mine to escalate the pedagogical objectives. What has emerged is a thorough revision and a very considerable enlargement of the third edition to which about 60 pages have been added. The bibliography has undergone similar expansion, which reflects the enormous advance in this field in the four years since the third edition. Hardly a passage of the text has been left untouched with the aim of enhancing physical insights and clarity. The advanced formal techniques retained are in my opinion essential as to avoid that the reader must take anything on trust. At the same time I hope that the book is relatively easily readable by, for example, beginning graduate students in theory, or by a general readership with a background in quantum statistical physics. The major changes are of various kinds.

In Part I new sections have been added. They deal with ergodicity and treat a scheme in which stochastic non-Markovian quantum dynamics based on the unraveled influence functional is combined with semiclassical propagation within the frozen Gaussian approximation. The section on the harmonic oscillator bath with linear coupling has been augmented by subsections on the fractional Langevin equation and on the problem of a charged particle interacting with the radiation field.

The major additions in Part II are as follows. The chapter on the damped linear oscillator has been extended by a section on quantum mechanical master equations for the reduced density matrix. Further subsections dealing with radiation damping and with thermodynamic properties (e.g., internal energy, free energy and entropy) of the damped quantum oscillator have been added. The chapter on quantum Brownian free motion has been augmented by a section on the partition function and on thermodynamic properties. Further, a section on electron decoherence in a disordered conductor has been subjoined.

In Part IV, the chapter on the dissipative two-state dynamics has been rearranged and extended considerably in order to improve clarity. Now, there is a chapter on the basics and methods, and a chapter dealing with explicit results in various regimes of the parameter space. New sections on the pure dephasing regime and on decoherence resulting from $1/f$ noise have been appended.

Part V now contains a chapter on twisted partition functions for the dissipative multi-state system in the field theory limit. This includes derivation of the exact scaling solution for the partition function using properties of Jack symmetric functions and presentation of a conjecture by which the nonlinear conductance can be extracted from the the twisted partition function. This chapter complements the chapter on duality symmetry by giving a different perspective.

Errors that I have caught, or which have been pointed out to me, have been corrected. I hope that not too many new ones have crept into the revised material. I shall be very grateful to any reader who brings such deficiencies to my attention.

The selection made here represents something of a personal compromise. During recent years several topics, in particular those connected with quantum state engineering and quantum computing, have undergone enormous expansion both on the theoretical and experimental side. I am very conscious that the coverage of most recent developments in these fields is far from being complete. One area omitted deserves special mention – decoherence and relaxation in quantum gases. The importance of the field is without question, but I felt that an adequate treatment deserves a separate book.

An unforeseen event deferred the delivery by several months. I am very grateful to the responsible editor Kim Tan of World Scientific for her continuous advice and patience.

Finally, I would like to thank many colleagues and readers who made valuable remarks on previous editions. Particular thanks are due to Doron Cohen, Giuseppe Falci, Hermann Grabert, Milena Grifoni, Frank Großmann, Gert-Ludwig Ingold, Elisabetta Paladino, Gerd Schön, Udo Seifert, Jürgen Stockburger, and Andrei Zaikin for their comments and suggestions.

Stuttgart
November 2011   *Ulrich Weiss*

Preface to the Third Edition

In the fourteen years since the appearance of the first edition, the subject kept freshness. There have been interesting theoretical progress, important new applications and lots of stunning new experiments in the field.

The present edition of *Quantum Dissipative Systems* reflects two endeavors on my part: the improvement and refinement of material contained already in the second edition; the addition of new topics (and the omission of few).

The emphasis and major intensions are still the same, but there are changes, augmentations and additions. The major extensions, altogether about 60 pages, are: Chapter 3 contains a more detailed discussion of the quasiclassical Langevin equation and a subsection on Josephson flux and charge qubits. Chapter 4 gives wider space to the basics of path integration and to the treatment of an electromagnetic environment. Chapter 5 discusses the stochastic unraveling of path integrals for the reduced density matrix. Chaper 6 gives an extended discussion of the damped quantum harmonic oscillator. It includes discussion of internal energy, purity, and uncertainty. Chapter 15 presents a generalization of the Smoluchowski diffusion equation which includes quantum effects. Chapter 20 offers a broader discussion of single-charge tunneling in the weak-tunneling or Coulomb blockade regime. Chapter 21 discusses relaxation and decoherence in the spin-boson model at zero temperature. It presents analytical results for the relaxation and decoherence rate at general damping strength which cover the entire regime extending from weak to strong tunneling. Chapter 24 includes a discussion of the full counting statistics for Poissonian quantum transport and presents many analytical results available in special cases. Chapter 25 presents the scaling-invariant solution of the full counting statistics in diverse limits and discusses application to charge transport in Josephson junctions. Chapter 26 gives an extended discussion of charge transport in quantum impurity systems, including full counting statistics. It points out an intimate connection of these systems with models for coherent conductors and with others discussed in the preceding Chapters 23 − 25. The bibliography is updated.

This new edition has benefited from comments, suggestions and criticisms from many students and colleagues. Among those to whom I owe specific debt of gratitude are Holger Baur, Pino Falci, Hermann Grabert, Milena Grifoni, Yuli Nazarov, Elisabetta Paladino, Jürgen Stockburger, and Ruggero Vaia.

It is a pleasure to thank the students P. Diemand and A. Herzog for proofreading and tracking down misprints. I also wish to thank the responsible editor Kim Tan for advice and patience until completion of the third edition. Finally, I am grateful to my wife Christel for her sympathy and constant encouragement.

Stuttgart
October 2007   *Ulrich Weiss*

Preface to the Second Edition

Since the first publication of this book in 1993, there have been enormous research activities in quantum dissipative mechanics both experimentally and theoretically. For this reason, it has been highly desirable after the book has been sold out almost three years ago to undergo a number of extensions and improvements. I have been encouraged by the positive reception of this book by a large community and by many colleagues to write not simply an updated second edition. What came out now after all is almost a new book of roughly double content.

In an extensive rewriting, the 19 Chapters of the First Edition have been expanded by about one third to better meet the desires of both the newcomers to the field and the advanced readership, and I have added 7 new chapters. The most relevant extensions are as follows. In the first part, I have added a section on stochastic dynamics in Hilbert space and I have extended the discussion of relevant microscopic global models considerably. Now, there are also treated acoustic phonons with two-phonon coupling, a microscopic model for tunneling between surfaces, charging and environmental effects in normally conducting and superconducting tunnel junctions, and nonlinear quantum environments. Part II now contains an extended discussion of the damped harmonic oscillator (e.g., a study of the density of states is added), and new chapters on the thermodynamic variational approach and variational perturbation expansion method, and on the quantum decoherence problem. Part III, which deals with quantum-statistical decay, is extended by two chapters. In the new edition, the turnover theory to the energy-diffusion limited regime is discussed, and the treatment of dissipative quantum tunneling has been extended and improved. Ample space is now provided in Part IV to a thorough discussion of the dissipative two-state system. A number of new results on the thermodynamics and dynamics of this archetypal system are presented. An extensive discussion of electron transfer in a solvent, incoherent tunneling in the nonadiabatic regime, and single-charge tunneling is provided in a unified framework. Regarding dynamics, new sections on exact master equations, improved approximation schemes, and recent results on correlation functions have been written, and a new chapter on the driven dissipative two-state is included. Part V, which deals with the dissipative multi-state system, is completely rewritten. It now contains four chapters on quantum Brownian motion in a cosine potential, multi-state dynamics, duality symmetry, and tunneling of charge through an impurity in a quantum wire. Many new results available only very recently are presented. The about 460 references are suggestions for additional reading on particular subjects and are not intended as a comprehensive bibliography.

Stuttgart
December 1998   *Ulrich Weiss*

Preface to the First Edition

This book is an outgrowth of a series of lectures which I taught at the ICTP at Trieste and at the University of Stuttgart during the spring and summer of 1991. The purpose of my lectures was to present the approaches and techniques that accurately treat quantum processes in the presence of frictional influences.

The problem of *open* quantum systems has been around since the beginnings of quantum mechanics. Important contributions to this general area have been made by researchers working in fields as diverse as solid-state physics, chemical physics, biophysics, quantum measurement theory, quantum optics, nuclear and particle physics. Often, there has been used, and still is used, a language well known in one context or one field, yet sufficiently different from others that it is not altogether easy to make out the connection. Here, I offer a collection of ideas and examples rather than a comprehensive review of the topic and the history.

The central theme is the space-time functional integral or path integral formulation of quantum theory. This approach is particularly well suited for treating the quantum generalization of friction. Here we are faced to understand the behavior of a system with few quantal degrees of freedom coupled to a thermal reservoir. After integrating out the bath while keeping the system's coordinates fixed we get the influence functional describing the influence of the many bath degrees of freedom on the few relevant ones. This leads to an effective action weighting the paths of the open system in the functional integral description. Indeed, if one wishes to perform numerical computations on a rigorous level, there are no alternatives to this approach at present.

Path integration in condensed matter and chemical physics has become a growth industry in the last one or two decades. A newcomer to this thriving field may not yet be very familiar with the path integral method. Here, I do assume a knowledge of standard text book quantum mechanics and statistical mechanics augmented by a knowledge of Feynman's approach on a first introductory level. The books by Baym [1], Chandler [2], Feynman and Hibbs [3], and by Feynman [4] provide the elementary material in this regard. Further background and supplementary material on the path integral method are contained in the books by Schulman [5] and by Kleinert [6]. However, advanced mastery of these subjects is not necessary.

Some of the more sophisticated concepts, such as preparation functions, propagating functions, and correlation functions, are basic to the development as it is presented here. To cover this material at an introductory level, I make frequent use of simplified models. In this way, I can keep the mathematics relatively simple.

Many of the problems, methods, and ideas which are discussed here have become essential to the current understanding of quantum statistical mechanics. I have made a considerable effort to make the material largely self-contained. Thus, although the theoretical tools are not developed systematically and in its full beauty, the material may be useful to many graduate students to become familiar with the field and learn the methods. For the most part, I refrain from just quoting results without explaining where they come from. With regard to citations, I have tried to give references to the historical development and also to provide a selection of the very recent important ones. But the list is surely not a comprehensive bibliography.

This book exists because of the physics I learned and enjoyed from the fertile collaboration with Hermann Grabert, Peter Hänggi, Gert-Ludwig Ingold, Peter Riseborough, and Maura Sassetti. I am particularly indebted to Maura who took time off her research to weed out points of confusion and who persistently encouraged me to finish this venture. I am also grateful to my students Reinhold Egger, Manfried Milch, Jürgen Stockburger, and Dietmar Weinmann for helpful comments concerning the presentation of many subjects discussed here and for preparing the figures.

In writing this book, I have benefited from the discussion with many companions; in particular Uli Eckern, Enrico Galleani d'Agliano, Anthony J. Leggett, Hajo Leschke, Franco Napoli, Albert Schmid, Gerd Schön, Larry Schulman, Peter Talkner, Valerio Tognetti, Andrei Zaikin, and Wilhelm Zwerger, who helped me in increasing my understanding of many of the subjects which are discussed here.

It is a pleasure to thank my teacher, colleague, and friend Wolfgang Weidlich for many fruitful discussions over the years.

Finally, and most importantly, I am deeply grateful to my wife Christel and my children Ulrike, Jan, and Meike for their infinite patience and omnipresent sympathy.

Stuttgart
October 1992   *Ulrich Weiss*

Acknowledgements

A number of colleagues and friends have made valuable remarks on the first edition. In this regard, I am especially grateful to Theo Costi, Thorsten Dröse, Reinhold Egger, Hermann Grabert, Peter Hänggi, Gert-Ludwig Ingold, Chi Mak, Maura Sassetti, Rolf Schilling, Herbert Spohn, and Wilhelm Zwerger. In writing the new edition, discussions with Pino Falci, Igor Goychuk, Milena Grifoni, Gunther Lang, Elisabetta Paladino, and Manfred Winterstetter have been extremely profitable, and I wish to thank them for their suggestions and for tracking down misprints. I also would like to acknowledge the preparation of a number of figures by Jochen Bauer, Gunther Lang, Jörg Rollbühler, and Manfred Winterstetter. Finally, I would like to thank Mrs. Karen Yeo as the responsible Editor of World Scientific Publishing Co. for her useful advice.

Contents

1. Introduction

Quantum-statistical mechanics is a very rich and checkered field. It is the theory dealing with the dynamical behavior of spontaneous quantal fluctuations.

When probing dynamical processes in complex many-body systems, one usually employs an external force which drives the system slightly or far away from equilibrium, and then measures the time-dependent response to this force. The standard experimental methods are quasielastic and inelastic scattering of light, electrons, and neutrons off a sample, and the system's dynamics is analyzed from the line shapes of the corresponding spectra. Other experimental tools are, e. g., spin relaxation experiments, study of the absorptive and dispersive acoustic behaviors, and investigation of transport properties. In such experiments, the system's response gives information about the dynamical behavior of the spontaneous fluctuations. Theoretically, the response is rigorously described in terms of time correlation functions. Therefore, time correlation functions are the center of interest in theoretical studies of the relaxation dynamics of nonequilibrium systems.

This book deals with the theories of open quantum systems with emphasis on phenomena in condensed matter physics. A preliminary consideration for any particular real dissipative quantum system is the separation of the related global quantum system into a relevant subsystem - the dissipative system of interest - and the irrelevant environment, of which the detailed dynamics is insignificant. In most cases of practical interest, the environment is thought to be in thermal equilibrium. The quantum statistical environment manifests itself in a fluctuating force acting on the relevant system and carrying the characteristics of the heat reservoir. It is the very nature of the fluctuating force to cause decoherence and damping, and to drive everything to disorder.

While quantum mechanics was conceived as a theory for the microcosm, there is apparently no contradiction with this theory in the mesoscopic and macroscopic world. The understanding of the appearance of classical behavior within quantum mechanics is of fundamental importance. This issue is intimately connected with the understanding of decoherence. Despite the stunning success of quantum theory, there is still no general agreement on the interpretation. The main disputes circle around "measurement" and "observation".

Decoherence is the phenomenon that the superposition of macroscopically distinct states decays on a short time scale. It is omnipresent because information about quantum interference is carried away in some physical form into the surroundings. In a sense, the environmental coupling acts as a continuous measuring apparatus, leading to an incessant destruction of phase correlations. The relevance of this coupling for macroscopic systems is nowadays generally accepted by the respectable community.

This is a book of methods, techniques, and applications. The level is such that anyone with a first course in quantum mechanics and rudimentary knowledge of path integration should not find difficulties. An attempt is made to present the problem of dissipation in quantum mechanics in a unified form. A general framework is developed which can deal with weak and strong dissipation, and with all kinds of memory effects. The reader will find a presentation of the relevant ideas and theoretical concepts, and a discussion of a wide collection of microscopic models. In the models and applications, emphasis is put on condensed matter physics. I have tried to use vocabulary and notation which should be fairly familiar to scientists working in chemical and condensed matter physics.

The book is divided into five parts. The following sequence of topics is adopted. The first part of the book is devoted to the general theory of open quantum systems. In Chapter 2, I review traditional approaches, such as formulations by master equations for weak coupling, Lindblad theory, operator-valued and quasiclassical Langevin equations. I also discuss attempts to interpret the dynamics of an open quantum system in terms of a stochastic process in the Hilbert space of state vectors pertaining to the reduced system. In Chapter 3, a variety of system-plus-reservoir models are introduced. They are partly motivated by phenomenological reasoning and partly connected with microscopic models which are of relevance in condensed matter physics. The models represent both continuous and discrete quantum systems, and they cover the diverse regimes of frequency-dependent friction and noise. The criteria for ergodic behavior of the open system are discussed. Emphasis is put on models in which dissipation is linear. Also scenarios are treated in which the quantum environment is intrinsically nonlinear.

Chapter 4 is devoted to the equilibrium statistical mechanics for the relevant subsystems of these models using the imaginary-time path integral approach. Chapter 5 concerns dynamics – quantum-mechanical motion, decoherence and relaxation of macroscopic systems that are far from or close to equilibrium. I discuss the Feynman-Vernon real-time influence functional approach and the relation with the Keldysh method. The concepts of preparation functions, propagating functions, and correlation functions are treated. Exact formal expressions for these quantities are derived using path sum techniques. This chapter also deals with more recent developments in which the two-time nonlocal influence functional is unravelled into a single-time stochastic influence functional. The appealing aspect is that one may derive stochastic Schrödinger equations and equivalent stochastic Liouville–von Neumann equations which are free of quantum memory effects.

Part II with Chapters 6 – 9 covers a discussion of exactly solvable damped linear quantum systems (damped harmonic oscillator and free Brownian particle), the useful thermodynamic variational approach with extension to open nonlinear quantum systems, and different scenarios for loss of quantum coherence. I have also tried to provide a balanced survey of electron dephasing in a diffusive conductor at low temperature.

Part III deals with quantum-statistical metastability: a problem of fundamental importance in chemical physics and reaction theory. After an introduction into the problem in Chapter 10, the relevant theoretical concepts and the characteristic features of the decay are discussed in Chapters 11 to 17. The treatment mainly relies on a thermodynamic method in which the decay rate is related to the imaginary part of the analytically continued free energy of the damped system. This allows for a uniform theoretical description in the entire temperature range. The discussion extends from high temperatures where thermal activation prevails down to zero temperature where the system can only decay by quantum-mechanical tunneling out of the ground state in the metastable well. Results in analytic form are presented where available.

In Part IV, I have tried to consider the thermodynamics and dynamics of the dissipative two-state or spin-boson system on a uniform theoretical basis. The spin-boson system is the simplest nonlinear system featuring the interplay between quantum coherence, quantal and thermal fluctuations, and friction, and it is the premaster of the archetypal qubit. After an introduction into the model in Chapter 18, the discussion in Chapter 19 is focused on equilibrium properties for a general form of the system-bath coupling. In particular, the partition function is discussed and the specific heat and static susceptibility are studied. The relationship with Kondo and Ising models is explained. Chapter 20 is devoted to the electron transfer problem in a solvent, nonadiabatic tunneling under exchange of energy, and single charge tunneling in the presence of an electromagnetic environment. Chapter 21 deals with the basics and suitable methods relevant to the dynamics of the dissipative two-state system. Different kinds of initial preparations of the system-plus-bath complex are treated and exact formal expressions for the system's dynamics in the form of series expressions and generalized master equations are derived. Various approximation schemes useful in different regimes of the parameter space are introduced and discussed. Chapter 22 is devoted to miscellaneous dynamical aspects of the dissipative two-state system. Ample space is given to the discussion of non-equilibrium and equilibrium correlation functions, and to adequate approximation schemes in the various regions of the parameter space. The effects of $1/f$ noise, sub-Ohmic, Ohmic and super-Ohmic friction and noise are discussed. Part IV closes with a chapter on the dynamics of the dissipative two-state system driven by time-dependent external forces.

The last part reviews dissipative quantum transport of a quantum Brownian particle in a tilted cosine potential. In Chapter 24, I introduce the weak- and strong-tunneling representations of the respective global model. Chapter 25 provides an outline of the nonequilibrium quantum transport formalism and derivation of exact formal expressions describing the system's dynamics for factorizing and thermal initial states including the full counting statistics. Explicit solutions in analytic form are given in various limits. Chapter 26 contains a discussion of the duality symmetry between the weak- and strong-tunneling representations which becomes an exact self-duality in the so-called Ohmic scaling limit. I show that self-duality offers the possibility to construct the exact scaling function at zero temperature for the nonlinear mobility, and in addition, with utilization of a hidden symmetry, the full counting statistics. In Chapter 27 I discuss a route in which (i) the partition function on a ring with a magnetic charge at the origin is solved with the aid of Jack polynomials, and (ii) the nonlinear mobility is found via a conjectured relation with the partition function and by analytic continuation of the winding number to the physical bias.

In Chapter 28 I am concerned with quantum transport of charge in quantum impurity models. Both the weak- and strong-tunneling representations are discussed, and the close relationship of the quantum impurity model with the Brownian particle model, and with charge transport in a coherent conductor and in Josephson systems is pointed out. This allows to to translate results obtained for one of these system in corresponding results for other systems which are related by a map of the parameters. The book closes with a chapter on quantum transport of a particle subject to sub- and super-Ohmic friction and associated quantum noise.

Throughout the book I have tried to concentrate on models which are simple enough to be largely tractable by means of analytical methods. There are, however, important examples where numerical computations have given clues to the analytical solution of a problem. If one wishes to calculate the full dynamics of the global system, one is faced with the problem that the number of basis states is growing exponentially. Therefore, even on supercomputers, the number of reservoir modes which can be treated numerically exactly, is rather limited. When the number of bath modes is above ten or even tends to infinity, an inclusive description of the environmental effects, e.g., in terms of the influence functional method (cf. Chapters 4 and 5) is indispensable. Various numerical schemes developed within the framework of the influence functional approach are available. The most valuable numerical tool in many-body quantum theory is probably the path integral Monte Carlo simulation method. Unfortunately, in simulations of the real-time quantum dynamics, the numerical stability of long-time propagation is spoilt by the destructive interference of different paths contributing to the path sum. This so-called *dynamical sign problem* is intrinsic in real-time quantum mechanics, and is characterized by an exponential drop of the signal-to-noise ratio with increasing propagation time.

In recent years, considerable progress in reducing the sign problem has been achieved. Promising routes have been, e.g., blocking algorithms in quantum Monte Carlo simulations, iterative procedures such as the tensor-propagator approach, which relies on a maximal memory time of the bath correlation function, and techniques for stochastic unraveling of the influence functional. I have refrained from adding sections which deal with numerical methods in detail. Where appropriate, I have tried to provide relevant informations and literature.

After all, the reader may not find a comprehensive account of what interests him most. Since the number of articles in this general field has become enormous in recent years, a somewhat arbitrary choice among the various efforts is inevitable. My choice of topics is just one possibility. It reflects, to some extent, the author's personal valuation of an active and rapidly developing area in science.

PART I

GENERAL THEORY OF OPEN QUANTUM SYSTEMS

2. Diverse limited approaches: a brief survey

Often in condensed phases, a rather complex physical situation can adequately be described by a global model system consisting of only one or few relevant dynamical variables in contact with a huge environment, of which the number of degrees of freedom is very large or even infinity. If we are interested in the physical properties of the *small* relevant system alone, we have to handle this system as an *open system* which exchanges energy with its surroundings in a random manner. In the last fifty years, a great variety of different theoretical methods for open quantum systems has been developed and employed. In this book, emphasis is put on the functional integral approach to open quantum systems. Over the years, this method has turned out to be very powerful and has found broad application. Nevertheless, I find it appropriate to begin with a survey of various other formalisms. Clearly, the discussion given subsequently can not do justice to all of them. However, I hope that the interested reader will be able to get a line along the given references for deeper studies. I find it appropriate to begin with a brief discussion of the classical regime.

2.1 Langevin equation for a damped classical system

It is our everyday experience that the motion of any macroscopic physical system comes to a stop when supply of energy is cut off. The reason for this is energy dissipation: in fact, there is no physical system that can completely be isolated from the surroundings. Therefore, energy accumulated in the system is inevitably given away to the environment. For instance, a particle in a fluid collides with the surrounding molecules. Thereby, momentum and energy of the particle is transferred to the fluid and the velocity of the particle declines. Eventually, the particle is trapped in a local minimum of the potential. The loss of energy is phenomenologically described in terms of a smooth frictional force. Since the energy exchange goes on randomly, there is in addition a fluctuating force which is zero on average. Now, a fundamental principle of classical and quantum equilibrium thermodynamics asserts that absorption of energy and fluctuations of the noise force are closely related.

Consider for simplicity an open system with a single degree of freedom, which we may associate with the coordinate $q(t)$ of a particle with mass M. The simplest assumptions about the dissipative process one can make is that dissipation is state-independent. Then the frictional force is a *linear* functional of the history of the velocity $\dot{q}(t)$, and the fluctuating force $\xi(t)$ obeys stationary Gaussian statistics. It is fully characterized by the classical ensemble averages

$$\langle \xi(t) \rangle_{\mathrm{cl}} = 0 \,, \qquad \langle \xi(t)\xi(0) \rangle_{\mathrm{cl}} \equiv \mathcal{X}_{\mathrm{cl}}(t) \,. \tag{2.1}$$

A classical heat reservoir at temperature T with zero memory time constitutes a white noise source. Then the frictional force is local in time, $F_{\mathrm{fric}}(t) = -M\gamma\dot{q}(t)$, where γ is the damping rate, and the stochastic force $\xi(t)$ is δ-correlated according to

$$\mathcal{X}_{\mathrm{cl}}(t) = 2M\gamma k_{\mathrm{B}}T\,\delta(t) \,. \tag{2.2}$$

The dynamics of the damped particle is described by the Langevin equation

$$M\ddot{q}(t) + M\gamma\,\dot{q}(t) + V'(q) = \xi(t) \,. \tag{2.3}$$

Time-local friction proportional to the velocity is usually called Ohmic because of the correspondence with a series resistor in an electrical circuit. The dynamical equation (2.3) describes, for example, a heavy Brownian particle with mass M immersed in a fluid of light particles and driven by a systematic force $-V'(q)$, where $V(q)$ is an externally applied potential. Equation (2.3) along with the relations (2.1) and (2.2) forms the basis of the theory of Brownian motion since the seminal studies by Einstein, Langevin and Smoluchowski. The early theoretical work on Brownian motion was reviewed in an excellent article by Chandrasekhar [7].

In many cases of practical interest, the heat reservoir exhibits retardation. Then the noise is colored, and friction depends on the velocity in the past. The appropriate dynamical equation is the generalized classical Langevin equation

$$M\ddot{q}(t) + M\int_{-\infty}^{t} dt'\,\gamma(t-t')\,\dot{q}(t') + V'(q) = \xi(t) \,. \tag{2.4}$$

Since the random force $\xi(t)$ has zero mean, the effect of the reservoir on average is in the memory-friction kernel $\gamma(t)$. The frequency-dependent damping function[1]

$$\tilde{\gamma}(\omega) \equiv \tilde{\gamma}'(\omega) + i\,\tilde{\gamma}''(\omega) = \int_{-\infty}^{\infty} dt\,\gamma(t)\,\mathrm{e}^{i\omega t} \tag{2.5}$$

has three important properties which are directly connected with three fundamental physical principles. The first of these is that $\tilde{\gamma}(\omega)$ is analytic in the upper half-plane $\mathrm{Im}(\omega) > 0$ and therefore satisfies Kramers-Kronig relations. This is an implication of causality, $\gamma(t) = 0$ for $t < 0$. The second property is that $\tilde{\gamma}(\omega)$ has a positive real part everywhere on the real axis, $\tilde{\gamma}'(\omega) > 0$ in the range $-\infty < \omega < \infty$. This is a consequence of the second law of thermodynamics. The third property is $\tilde{\gamma}(\omega) = \tilde{\gamma}^*(-\omega)$. This is a consequence of the physical requirement that the memory-friction kernel

---

[1]Throughout this book, I use for the Fourier transform the normalization and sign convention in the exponent as given in Eq. (2.5). Where appropriate, I mark Fourier transforms with a tilde and Laplace transforms with a hat.

$\gamma(t)$ is a real function. The damping function $\tilde{\gamma}(\omega)$ is connected with the force autocorrelation function $\langle \xi(t)\xi(t') \rangle_{\mathrm{cl}} - \langle \zeta(t - t')\xi(0) \rangle_{\mathrm{cl}}$ by the Green-Kubo formula [8]

$$\tilde{\gamma}(\omega) = \frac{1}{Mk_{\mathrm{B}}T} \int_0^\infty dt\, \langle \xi(t)\xi(0) \rangle_{\mathrm{cl}}\, e^{i\omega t} . \tag{2.6}$$

From this we infer that the power spectrum of the classical stochastic force

$$\tilde{\mathcal{X}}_{\mathrm{cl}}(\omega) = \int_{-\infty}^\infty dt\, \mathcal{X}_{\mathrm{cl}}(t)\cos(\omega t) \tag{2.7}$$

is related to the real part $\tilde{\gamma}'(\omega)$ of the damping function $\tilde{\gamma}(\omega)$ by the relation

$$\tilde{\mathcal{X}}_{\mathrm{cl}}(\omega) = 2Mk_{\mathrm{B}}T\,\tilde{\gamma}'(\omega) . \tag{2.8}$$

This is a version of the classical fluctuation-dissipation or Nyquist theorem.

The Langevin equation (2.4) will be derived from a global system-plus-reservoir Hamiltonian in Subsection 3.1.2.

Evidently, the equation (2.4) with (2.8) is limited to the classical domain. However, we should expect that at sufficiently low temperatures all types of quantum effects will occur. Since the standard procedure of quantization relies upon the existence of a Lagrangian or a Hamiltonian function for the system, the question arises: how can one reconcile dissipation with the canonical scheme of quantization?

2.2 New schemes of quantization

The equation of damped motion (2.3) can not be obtained from the application of Hamilton's principle unless the Lagrangian has an explicit time dependence. The use of time-dependent Lagrangians or Hamiltonians would permit us to use the standard schemes of quantization directly. Historically, the first researchers taking this path were Caldirola [9] and Kanai [10] who employed a time-dependent mass chosen in such a way that a friction term appears in the classical equation of motion. However, the commutator of the corresponding momentum and position operator decays exponentially with time, and hence the uncertainty principle is violated [11]. In a study by Schuch, explicitly time-dependent Hamiltonians were related to nonlinear Schrödinger equations, and violation of the uncertainty principle was obliterated by a noncanonical transformation of variables q and p and a nonunitary transformation of the wave function [12]. Nevertheless, it is generally accepted meanwhile that dissipation cannot be described adequately by simply employing a time-dependent mass.

Many approaches to open quantum systems were introduced over the last forty years. The variety of historical attempts falls into three main categories. One either modifies the procedure of quantization, or one trusts in heuristic stochastic Schrödinger equations for state vectors, or one starts out from the system-plus-reservoir approach.

Among the first group, Dekker [13] proposed a theory with a canonical quantization procedure for complex variables, thereby reproducing the Fokker-Planck equation for the Wigner distribution function. However, some *ad hoc* assumptions in the theory seem questionable, such as the introduction of noise sources in the canonical equations for position and momentum. Kostin [14] introduced a theory with a nonlinear Schrödinger equation. The same equation was found later by Yasue [15] using Nelson's stochastic quantization procedure [16]. However, this theory violates the superposition principle. It also yields some dubious results such as stationary undamped states. Apart from the fact that the theoretical foundations are completely unclear, these approaches can reproduce, at best, known results only for very limited cases, such as weakly damped linear systems. Therefore, all attempts of the first group can be assessed to have failed. I shall not consider them further here. The second category is discussed in some detail in Section 2.4.

The more natural, and also more successful approach has been to view the dissipative system as a relevant system with a single or few significant degrees of freedom, which is in contact with (infinitely) many degrees of freedom. These additional degrees of freedom are commonly referred to as bath, or reservoir or environment. Both the relevant system and the reservoir are the constituents of an energy-conserving global system which obeys the standard rules of quantization. In this picture, friction comes about by the transfer of energy from the "small" system to the "huge" environment. The energy, once transferred, dissipates into the environment and is not given back within any physically relevant period of time.

2.3 Traditional system-plus-reservoir methods

Common approaches to open quantum systems based on system-plus-reservoir models are generally divided into two classes. Working in the Schrödinger picture, the dynamics is conventionally described in terms of generalized quantum master equations for the reduced density matrix or density operator [17] – [19]. Working in the Heisenberg picture, the description is given in terms of generalized Langevin equations for the relevant set of operators of the reduced system [20] – [22].

2.3.1 Quantum-mechanical master equations for weak coupling

The starting point of this method is the familiar Liouville equation of motion for the density operator $W(t)$ of the global system,

$$\dot{W} = -\frac{i}{\hbar}[H, W(t)] \equiv \mathcal{L}W(t), \qquad (2.9)$$

where H is the Hamiltonian of the total system, and where the second equality defines the Liouville operator $\mathcal{L}$. Next, assume that the Hamiltonian H and the Liouvillian $\mathcal{L}$ of the total system are decomposed as

$$H = H_{\mathrm{S}} + H_{\mathrm{R}} + H_{\mathrm{I}}; \qquad \mathcal{L} = \mathcal{L}_{\mathrm{S}} + \mathcal{L}_{\mathrm{R}} + \mathcal{L}_{\mathrm{I}}. \qquad (2.10)$$

The individual parts refer to the free motion of the relevant system and of the reservoir, and to the interaction term, respectively. Employing a certain projection operator P, chosen as to project on the relevant part of the density matrix, the full density operator is reduced to an operator acting only in the space of the relevant variables,

$$\rho(t) = PW(t). \tag{2.11}$$

The operator $\rho(t)$ is usually called *reduced density* operator. For systems with the Hamiltonian form (2.10), the projection operator contains a trace operation over the reservoir coordinates. By means of the projection operator P, the density operator can be decomposed into the relevant part $\rho(t)$ and the irrelevant part $(1 - P)W(t)$,

$$W(t) = \rho(t) + (1 - P)W(t); \qquad P^2 = P. \tag{2.12}$$

Upon substituting the decomposition (2.12) into Eq. (2.9), and acting on the resulting equation from the left with the operator P and with the operator $1 - P$, respectively, we obtain two coupled equations for the relevant part $\rho(t)$ and the irrelevant part $(1 - P)W(t)$. A closed equation for $\rho(t)$ is obtained by inserting the formal integral for $(1 - P)W(t)$ into the first equation. We then finally arrive at the formally exact generalized quantum master equation, the *Nakajima-Zwanzig equation* [17, 18]

$$\dot{\rho}(t) = P\mathcal{L}\rho(t) + \int_0^t dt'\, P\mathcal{L}\, e^{(1-P)\mathcal{L}t'} (1 - P)\mathcal{L}\,\rho(t - t') \\ + P\mathcal{L}\, e^{(1-P)\mathcal{L}t} (1 - P)W(0). \tag{2.13}$$

The generalized master equation is an inhomogeneous integro-differential equation in time. It describes the dynamics of the open (damped) system in contact with the reservoir $\mathcal{R}$. Observe that the inhomogeneity in Eq. (2.13) still depends on the initial value of the irrelevant part $(1 - P)W(0)$. In applications, it is attempted to choose the projection operator in such a way that the irrelevant part of the initial state $(1 - P)W(0)$ can be disregarded. Assuming further that P commutes with $\mathcal{L}_S$ one then finds the homogeneous time-retarded quantum master equation

$$\dot{\rho}(t) = P(\mathcal{L}_S + \mathcal{L}_I)\rho(t) + \int_0^t dt'\, P\mathcal{L}_I\, e^{(1-P)\mathcal{L}t'} (1 - P)\mathcal{L}_I\rho(t - t'). \tag{2.14}$$

The first (instantaneous) term describes the reversible motion of the relevant system while the second (time-retarded) term brings on irreversibility. It includes all effects the reservoir may exert on the system, such as relaxation, decoherence and energy shifts. Equation (2.14) is still too complicated for explicit evaluation. First, the kernel of (2.14) contains any power of $\mathcal{L}_I$. Secondly, the dynamics of ρ at time t depends on the whole history of the density matrix. In order to surmount these difficulties, one usually considers the kernel of Eq. (2.14) only to second order in $\mathcal{L}_I$. Disregarding also retardation effects, one finally arrives at the quantum master equation in Born-Markov approximation

$$\dot{\rho}(t) \; = \; P(\mathcal{L}_\mathrm{S} + \mathcal{L}_\mathrm{I})\rho(t) \; + \; \int_0^t dt' \, P\mathcal{L}_\mathrm{I} \, \mathrm{e}^{(1-P)(\mathcal{L}_\mathrm{S} + \mathcal{L}_\mathrm{R})\,t'} \, (1 - P)\mathcal{L}_\mathrm{I}\,\rho(t) \,. \tag{2.15}$$

Master equations of this form were successfully used to describe weak-damping phenomena, for instance in quantum optics or spin dynamics. Various excellent reviews
of this sort of approach including many applications are available in Refs. [23] – [31].
While the Markov assumption can easily be dropped, the more severe limitation of
this method is the Born approximation for the kernel. The truncation of the Born
series at second order in the interaction $\mathcal{L}_\mathrm{I}$ effectively restricts the application of the
master equation (2.15) or of its non-Markovian generalization to weakly damped sys-
tems with relaxation times that are large compared to the relevant time scales of the
reversible dynamics.

When the Born-Markov quantum master equation (2.15) is given in the energy
eigenstate basis of H_S, it is usually referred to as *Redfield* equation [32, 28, 33]

$$\dot{\rho}_{nm}(t) \; = \; -i\,\omega_{nm}\,\rho_{nm}(t) \; - \; \sum_{k,l} R_{nmkl}\,\rho_{kl}(t) \,. \tag{2.16}$$

The first term represents the reversible motion in terms of the transition frequencies
ω_{nm}, and the second term describes relaxation. The Redfield relaxation tensor reads

$$R_{nmkl} \; = \; \delta_{lm} \sum_r \Gamma^{(+)}_{nrrk} + \delta_{nk} \sum_r \Gamma^{(-)}_{lrrm} - \Gamma^{(+)}_{lmnk} - \Gamma^{(-)}_{lmnk} \,. \tag{2.17}$$

The rates are given by the Golden Rule expressions

$$\begin{aligned}
\Gamma^{(+)}_{lmnk} &= \hbar^{-2} \int_0^\infty dt \, \mathrm{e}^{-i\omega_{nk}t} \big\langle \tilde{H}_{\mathrm{I},\,lm}(t)\tilde{H}_{\mathrm{I},\,nk}(0) \big\rangle_\mathrm{R} \,, \\
\Gamma^{(-)}_{lmnk} &= \hbar^{-2} \int_0^\infty dt \, \mathrm{e}^{-i\omega_{lm}t} \big\langle \tilde{H}_{\mathrm{I},\,lm}(0)\tilde{H}_{\mathrm{I},\,nk}(t) \big\rangle_\mathrm{R} \,.
\end{aligned} \tag{2.18}$$

Here, $\tilde{H}_\mathrm{I}(t) = \mathrm{e}^{iH_\mathrm{R}t/\hbar} H_\mathrm{I} \, \mathrm{e}^{-iH_\mathrm{R}t/\hbar}$ is the interaction in the interaction picture, and the
angular bracket denotes thermal average of the bath degrees of freedom.

The Redfield equations (2.16) are well-established in wide areas of physics and
chemistry, e.g., in nuclear magnetic resonance (NMR), in optical spectroscopy, and in
laser physics. In NMR, one deals with the externally driven dynamics of the density
matrix for the nuclear spin [34, 35, 36]. In optical spectroscopy, a variant of the
Redfield equations are the optical Bloch equations [37, 38].

Multilevel Redfield theory has been applied to the electron transfer dynamics in
condensed phase reactions by several authors [39].

Markovian reduced density matrix (RDM) theory has been also utilized in the diabatic state representation [40]. Denoting electronic-vibrational direct-product states
in the diabatic representation[2] by $\alpha, \beta, \cdots$, the RDM equations of motion read

---

[2]The concepts of diabatic states are discussed in recent reviews [41, 42, 43].

$$\dot{\rho}_{\alpha\beta}(t) \;=\; -i\,\omega_{\alpha\beta}\,\rho_{\alpha\beta}(t) \;-\; \frac{i}{\hbar}\sum_{\nu}\left[V_{\alpha\nu}\rho_{\nu\beta}(t)-V_{\nu\beta}\rho_{\alpha\nu}(t)\right] \;-\; \sum_{\nu,\sigma}R_{\alpha\beta\nu\sigma}\,\rho_{\nu\sigma}(t)\,. \quad (2.19)$$

The $\omega_{\alpha\beta}$ are the transition frequencies between unperturbed diabatic surfaces, the $V_{\alpha\beta}$ are the matrix elements of the diabatic interstate coupling, and the relaxation tensor $R_{\alpha\beta\nu\sigma}$ describes relaxation of the diabatic electronic-vibrational states. For electron transfer processes, the diabatic surfaces can often be taken as harmonic, which simplifies the calculation of the relaxation tensor drastically. The master equation (2.19) in the diabatic basis is especially useful when the interstate coupling is weak or moderate. In contrast to Redfield theory in the system's eigenstate basis, the computation of the Redfield tensor in the diabatic basis, Eq. (2.19), is without difficulties even for multimode vibronic-coupling systems [43]. When the RDM approach is applied to complex systems, the dimension N of the Hilbert space of the relevant system H_S is possibly very large. Since the density matrix scales with N^2 and the relaxation tensor with N^4, the computational problem may become easily nontrivial. The so-called Monte-Carlo wave function propagation or quantum jump method discussed below in Section 2.4, which is equivalent to the Born-Markov RDM approach, provides a considerably more favorable scaling of the computational costs, i.e. *linear* scaling with the number of states, than direct integration of the Redfield equations.

2.3.2 Lindblad theory

Within the conception of quantum mechanics, the time evolution of the density operator of a closed system is a unitary map. If the system is open, the possible transformations are thought to be "completely positive" [44, 30], $\rho \to \sum_n O_n \rho O_n^{\dagger}$. Here $\{O_n\}$ is a set of linear operators on the reduced state space, restricted only by $\sum_n O_n^{\dagger} O_n = 1$, which guarantees that $\operatorname{tr}\rho$ does not change. The most general form of generators $\mathcal{L}$, $\dot{\rho}(t) = \mathcal{L}\rho(t)$, preserving complete positivity of density operators and conveying time-directed irreversibility is established by the Lindblad theory [44]. The Lindblad form of the quantum master equation for an open system reads

$$\frac{d\rho(t)}{dt} \;=\; -\frac{i}{\hbar}\left[H_S,\rho(t)\right] \;+\; \frac{1}{2}\sum_{j}\left\{\left[L_j\rho(t),L_j^{\dagger}\right] + \left[L_j,\rho(t)L_j^{\dagger}\right]\right\}. \quad (2.20)$$

The first term represents the reversible dynamics of the relevant system. The Lindblad operators L_j describe the effect of the environment on the system in Born-Markov approximation. In concrete applications of Eq. (2.20), the L_j transmit emission and absorption processes. For linear dissipation, the simplest form of a Lindblad operator is a linear combination of a raising and lowering operator, or equivalently of a coordinate and a momentum operator [45],

$$L \;=\; \mu q + i\nu p\,, \qquad\qquad L^{\dagger} \;=\; \mu q - i\nu p\,, \quad (2.21)$$

where μ and ν are c-numbers. Explicit temperature-dependent expressions for μ and ν are obtained for the damped harmonic quantum oscillator by matching the dissipation terms of the Lindblad master equation with corresponding expressions of this exactly solvable model (cf. Chapter 6) in the Born-Markov limit [46].

Complete positivity of reduced density operators is considered as a strict guideline by researchers working on stochastic Schrödinger equations (see Section 2.4). However, as argued by Pechukas [47, a], complete positivity of the RDM is an artefact of product initial conditions $W^{(i)} = \rho^{(i)} \otimes \rho_R^{(i)}$ for the global density matrix. Product initial conditions are only appropriate in the weak-coupling limit. In general, reduced dynamics need not be completely positive. It is known that the Redfield-Bloch master equation may break the positivity of the density matrix. Violation of positivity may occur if memory effects are not adequately taken into account at the initial stage of the time evolution. Then some "slippage" of the initial conditions must occur before the reduced dynamics looks Markovian [47, b]. The slippage captures the effects of the actual non-Markovian evolution in a short transient regime of the order of the reservoir's memory time. (see also [47, c]). The Markovian master equation holds only in a subspace in which rapidly decaying components of the density matrix are disregarded, whereas Lindblad theory requires validity for *any* reduced density matrix. These findings have been confirmed in a study of the exactly solvable damped harmonic oscillator [48]. Using the exact path integral solution, it has been shown that in general there is no exact dissipative Liouville operator describing the dynamics of the oscillator in terms of an exact master equation that is independent of the initial preparation. Exact non-stationary Liouville operators can be found only for particular preparations. Time-independent Liouville operators which are valid for arbitrary preparation of the initial state can be extracted only in the sub-space left after the fast transient components have died out. However, the Liouville operators are still not of the Lindblad form. The Lindblad master equation is gained only after the weak-coupling limit is performed and a coarse graining in time is carried out. Discussion of this issue is given in Section 6.8. It is perfectly obvious from the study of this exactly solvable linear model that the reduced dynamics in general does not constitute a dynamical semigroup. This shows that the Lindblad master equation has no fundamental significance.

Besides the Markov approximation and the weak-coupling limit, there is another severe limitation of the applicability of the above quantum master equations. These equations do not describe the relaxation contributions with time scales of the inverse Matsubara frequencies, $\sum_n a_n \mathrm{e}^{-|\nu_n|t}$ ($\nu_n = 2\pi n/\hbar\beta$), which originate from the fluctuation dissipation theorem (see, e.g., the discussion in Chapters 6 and 17). Hence the relaxation dynamics towards the equilibrium state ρ_β is correctly described by master equations of Lindblad form only when the system's relaxation times are large compared to the thermal time $\hbar\beta$.

In conclusion, the Born-Markov quantum master equation approach provides a reasonable description in many cases, e. g., in NMR, in quantum optics, and in a variety of chemical reactions. However, this method is not applicable in many problems in solid state physics at low temperature in which neither the Born approximation is valid nor the Markov assumption holds.

2.3.3 Operator Langevin equations for weak coupling

Just as we projected the density matrix $W(t)$ of the global system onto the relevant part $\rho(t)$, we may proceed in the Heisenberg picture by projecting the operators of the global system on the set of macroscopically relevant operators. The various efforts in studying the dynamics of these operators have been described by Gardiner [50].

Let us denote the set of operators of the global system by $\{X\}$ and the set of macroscopically relevant operators governing the open system by $\{Y\}$. Now consider the operators X_i and Y_μ as elements $|X_i)$ and $|Y_\mu)$ in the Liouville space Λ.[3] At time $t = 0$, the operators $\{Y\}$ span a subspace Λ_Y of the Liouville space Λ. I use the convention that an operator in the Heisenberg representation without time argument denotes the operator at time zero. Next, it is convenient to define a time-independent projection operator which projects onto the subspace Λ_Y,

$$\mathcal{P} = \sum_{\mu,\nu} |Y_\mu)\, g_{\mu\nu}\, (Y_\nu| \; ; \qquad \mathcal{P}^2 = \mathcal{P} \,. \qquad (2.22)$$

The metric $g_{\mu\nu}$ is the inverse of the scalar product $(Y_\mu|Y_\nu)$ which has to be chosen appropriately in practical calculations. For quantum statistical linear response and relaxation problems, a suitable form is the Mori scalar product [22, 49]

$$(Y_\mu|Y_\nu) \equiv \frac{1}{\beta} \int_0^\beta d\lambda \left\langle e^{-\lambda H} Y_\mu^\dagger e^{\lambda H} Y_\nu \right\rangle \,. \qquad (2.23)$$

The angular brackets denote average with respect to the canonical ensemble of the global system $W_\beta = Z_\beta^{-1} e^{-\beta H}$, and $\beta = 1/k_B T$. The superoperator $\mathcal{P}$ projects onto the subspace Λ_Y according to

$$\mathcal{P}|X_i) = \sum_{\mu,\nu} |Y_\mu)\, g_{\mu\nu}\, (Y_\nu|X_i) \,. \qquad (2.24)$$

Acting now from the left with $\mathcal{P}$ and with $1 - \mathcal{P}$ on the Heisenberg equation of motion

$$|\dot{X}_i) = \mathcal{L}|X_i) \,, \qquad (2.25)$$

where $\mathcal{L}$ is the Liouville superoperator, and eliminating $(1 - \mathcal{P})|X_i)$ with the aid of the exact formal solution, it is straightforward to derive for $Y_\mu(t)$ a system of coupled integro-differential equations, which have been popularized as the Mori equations.

$$\dot{Y}_\mu(t) = i\sum_\nu Y_\nu(t)\, \Omega_{\nu\mu}(t) - \sum_\nu \int_0^t ds\, \gamma_{\nu\mu}(s) Y_\nu(t-s) + \xi_\mu(t) \,. \qquad (2.26)$$

The generally temperature-dependent drift matrix $\Omega_{\nu\mu}(t)$ is given by

$$i\,\Omega_{\nu\mu}(t) = \sum_\rho g_{\nu\rho}\, (Y_\rho|\dot{Y}_\mu(t)) \,. \qquad (2.27)$$

---

[3]See Ref. [49] for a review of the formulation of quantum mechanics in Liouville space.

The stochastic force $\xi_\nu(t)$ is a functional of the operators of the *irrelevant* part,

$$\xi_\nu(t) = \exp[(1 - \mathcal{P})\mathcal{L}\,t]\,(1 - \mathcal{P})\dot{X}_\nu(t) \,. \tag{2.28}$$

Finally, the memory matrix is expressed in terms of the correlation function of the stochastic force as

$$\gamma_{\nu\mu}(t) = \sum_\rho g_{\nu\rho}(\xi_\rho|\xi_\mu(t)) \,. \tag{2.29}$$

The actual computation of the fluctuating force and of the memory matrix is again restricted to weak coupling. Altogether, the operator Langevin method is subject to exactly the same limitations we encountered above in the master equation approach.

In conclusion, it is important in the Mori formalism that the complete set of macro variables spans the subspace. Otherwise, the fluctuating force contains slowly varying components, and the separation of time scales is incomplete.

2.3.4   Generalized quantum Langevin equation

One further attempt consists in a quantum mechanical generalization of the classical Langevin equation (2.4) [21, 51, 52, 80]. For a particle moving in one dimension, the quantum Langevin equation (QLE) is the Heisenberg equation of motion for the coordinate operator $\hat{q}(t)$. It may be written in the form

$$M\frac{d^2\hat{q}(t)}{dt^2} + M\int_{t_0}^t dt'\,\gamma(t - t')\frac{d\hat{q}(t')}{dt'} + V'[\hat{q}(t)] = \hat{\xi}(t) \,, \tag{2.30}$$

where $\hat{\xi}(t)$ is the Gaussian random force operator with symmetric autocorrelation

$$\frac{1}{2}\langle\hat{\xi}(t)\hat{\xi}(0) + \hat{\xi}(0)\hat{\xi}(t)\rangle_\beta = \frac{M}{\pi}\int_0^\infty d\omega\,\hbar\omega\,\tilde{\gamma}'(\omega)\coth(\tfrac{1}{2}\beta\hbar\omega)\cos(\omega t) \,. \tag{2.31}$$

and the nonequal-time commutator

$$[\hat{\xi}(t), \hat{\xi}(0)] = -i\frac{2M}{\pi}\int_0^\infty d\omega\,\hbar\omega\,\tilde{\gamma}'(\omega)\sin(\omega t) \,. \tag{2.32}$$

Piecing Eqs. (2.31) and (2.32) together, the quantum-statistical force correlation is

$$\mathcal{X}(t) \equiv \langle\hat{\xi}(t)\hat{\xi}(0)\rangle_\beta = \frac{\hbar M}{2\pi}\int_{-\infty}^\infty d\omega\,\omega\tilde{\gamma}'(\omega)[\coth(\tfrac{1}{2}\beta\hbar\omega)\cos(\omega t) - i\sin(\omega t)] \,, \tag{2.33}$$

and the power spectrum of the random force is

$$\tilde{\mathcal{X}}(\omega) \equiv \int_{-\infty}^\infty dt\,\langle\hat{\xi}(t)\hat{\xi}(0)\rangle\cos(\omega t) = M\hbar\omega\coth\left(\frac{\hbar\omega}{2k_\mathrm{B}T}\right)\tilde{\gamma}'(\omega) \,. \tag{2.34}$$

The dynamical equation (2.30) is a Langevin equation with a linear memory-friction force and a random force operator representing additive stationary Gaussian noise. The expression (2.31) for the symmetrized force auto-correlation function and concomitant with this the expression (2.34) for the power spectrum of the stochastic force are general implications of the fluctuation-dissipation theorem (cf. Section 6.1). They are therefore independent of the model. On the other hand, the Gaussian property of the stochastic force operator $\hat{\xi}(t)$ strictly holds only when the reservoir is harmonic.

The QLE (2.30) can be derived for a system which is bilinearly coupled to a bath of harmonic oscillators. The derivation is given in Section 3.1. The dependence of the force operator $\hat{\xi}(t)$ on the initial conditions of the system $\hat{q}(0)$ and of the set of bath positions $\{\hat{x}_\alpha(0)\}$ and bath momenta $\{\hat{p}_\alpha(0)\}$, and the subtleties of the thermal average $\langle\cdots\rangle_\beta$ in Eq. (2.31) are discussed in Subsection 3.1.4.

Benguria and Kac [53], and Ford and Kac [51] argued that a system which obeys the QLE (2.30) with Gaussian noise characteristics (2.31) and commutator (2.32) approaches the proper equilibrium state.

When friction is strictly Ohmic, $\tilde{\gamma}(\omega) = \gamma$, the damping function is memoryless, $\gamma(t) = 2\gamma\Theta(t)\delta(t)$, while the symmetrized force auto-correlation function (2.31) is not delta-correlated. Hence the quantum-mechanical process is non-Markovian. The white-noise (Markovian) limit is only reached in the classical limit $\hbar \to 0$ in which the commutator (2.32) vanishes and the stochastic force is delta-correlated as in Eq. (2.2).

2.3.5   Generalized quasiclassical Langevin equation

The practical application of the generally nonlinear QLE is severely limited because it is an operator equation in the Hilbert space of system and reservoir. A manifest approximation now is (i) to abandon the operator character of the QLE and (ii) to replace the quantum noise by a classical colored noise source. In the so-called *quasiclassical* Langevin equation [20, 54], the power spectrum of the classical noise source is matched with that of the quantum noise,[4]

$$\tilde{\mathcal{X}}_{\text{qucl}}(\omega) \equiv \int_{-\infty}^{\infty} dt\, \langle\xi(t)\xi(0)\rangle_\beta \cos(\omega t) = M\hbar\omega \coth\left(\frac{\hbar\omega}{2k_B T}\right) \tilde{\gamma}'(\omega) . \qquad (2.35)$$

While in the QLE the operators $\hat{q}(t)$ and $\hat{p}(t)$ obey the Heisenberg uncertainty relations for all times, this feature is abandoned in the quasiclassical Langevin equation. The quasiclassical Langevin equation is an exact dynamical equation for *linear* damped quantum systems. It yields a reasonable description for systems which are nearly harmonic [55, 56]. However, the predictions of the quasiclassical Langevin equation are unreliable when the anharmonicity of the potential is crucial like, for instance, in quantum tunneling processes. The conclusion of a detailed study is that use of the colored noise (2.35) in a classical Langevin equation is insufficient to render a proper description of the quantum statistical decay of a metastable state [55].

When the anharmonicity of the potential is pivotal, the most successful approach is the functional integral method. Like in the classical regime, the dissipative system is considered to interact with a complex environment, and the "complete universe"

---

[4]The derivation of the QLE within the path integral method following the approach by Schmid [54] is presented in Section 5.6.

formed by the system plus environment is assumed to be energy-conserving so that it can be quantized in the standard way. For equations of motion which are linear in the bath coordinates, the environment can easily be eliminated. Thus one obtains closed equations for the damped system alone. In the path integral representation, the environment reveals itself through an influence functional which depends on the spectral properties of the environmental coupling and on temperature. A general discussion of the influence functional method is presented in Chapters 4 and 5.

2.3.6 Phenomenological methods

Often, a physical or chemical system cannot be characterized by a simple model Hamiltonian of the form (2.10), or the Hamiltonian is even unknown. In such situations, it is sometimes useful to describe dissipative quantum dynamics on a phenomenological level. For instance, one may introduce a dynamical description of the system in terms of occupation probabilities $p_n(t)$ of energy levels or of spatially localized states, rather than in terms of complex probability amplitudes or wave functions, or the full reduced density matrix ρ_{nm}. Then, the relaxation dynamics of a macroscopic system is described by a *Pauli master equation* for $p_n(t) \equiv \rho_{nn}(t)$,

$$\dot{p}_n(t) = \sum_m \left[A_{nm} p_m(t) - A_{mn} p_n(t) \right] . \tag{2.36}$$

The first term on the r.h.s. describes the gain and the second term the loss of probability to occupy the state n. In this formulation, knowledge of the full set of transition rates $\{A_{nm}\}$ is required in order to have a complete description of the relaxation process. The transition rates may be inferred, e. g., from standard quantum mechanical perturbation theory, or from experimental data, or they may be chosen by a phenomenological ansatz. The Pauli master equation (2.36) has found widespread application to the study of rate dynamics in physics, chemical kinetics, and biology.

   In the quantum coherence regime, off-diagonal matrix elements of the density matrix become relevant. Nevertheless, the full coherent dynamics of the populations can still be formulated in terms of dynamical equations for diagonal matrix elements. However, the corresponding master equation is time-nonlocal (see Section 21.2.6).

2.4 Stochastic dynamics in Hilbert space

In recent years, there have been made numerous attempts to postulate non-Hamiltonian dynamics as a fundamental modification of the Schrödinger equation in order to explain the spontaneous stochastic collapse of the wave function and the appearance of a "classical world" (cf. for a survey the article by I.-O. Stamatescu in Ref. [57]). In these approaches, the non-unitary dynamics of open quantum systems is interpreted in terms of a fundamental stochastic process in the Hilbert space of state vectors pertaining to the system. To retain the standard probability rules, the respective dynamical equations become inevitably nonlinear. The evolution of state vectors is considered as a stochastic Markov process, and the covariance matrix of the state vector is taken as the density operator. The stochastic process is usually constructed in such a way that the equation of motion of the density operator is the familiar Markovian quantum master equation in Lindblad form (2.20), e.g., the optical Bloch equations [37, 38]. First of all, approaches of this type were introduced on phenomenological grounds. Later on, a fundamental significance has been allocated to the "stochastic Schrödinger equations" by several authors, in particular to propose explanation of the omnipresent decoherence phenomena observed in real quantum systems. The above scheme does not lead to a definite stochastic representation of the dynamics of the reduced system in Hilbert space, even though the Markov approximation is made, since the stochastic process is not unambiguously determined by merely fixing the covariance. An infinity of different realizations is possible. Basically, one may distinguish two classes of stochastic models for the evolution of state vectors in the Schrödinger picture. In the first class of stochastic Schrödinger equations, the stochastic increment is a diffusion process, the so-called Wiener process. In the second class, the evolution of the state vector is represented as a stochastic process of which the realizations are piecewise deterministic paths, and the smooth segments are interrupted by stochastic sudden jump processes [58].

The *quantum-state diffusion* (QSD) method proposed by Gisin [59] and developed further by Gisin, Percival and coworkers [60, 61] belongs to the first class. The QSD method is based upon a correspondence between the solutions of the master equation for the ensemble density operator ρ and the solutions of a Langevin-Itô diffusion equation for the normalized pure state vector $|\psi>$ of an individual system of the ensemble. If the master equation has the Lindblad form (2.20), then the corresponding QSD equation is the nonlinear stochastic differential equation

$$|d\psi> = -\frac{i}{\hbar}H|\psi> dt + \sum_j \left(<L_j^\dagger> L_j - \tfrac{1}{2}L_j^\dagger L_j - \tfrac{1}{2} <L_j^\dagger><L_j> \right)|\psi> dt$$
$$+ \sum_j \left(L_j - <L_j> \right)|\psi> d\xi_j , \qquad (2.37)$$

where $<L_j> \equiv <\psi|L_j|\psi>$ is the quantum expectation.[5] The first sum describes the nonlinear drift of the state vector in the state space and the second sum the random fluctuations. The $d\xi_j$ are complex differential variables of a Wiener process satisfying

$$\langle d\xi_j \rangle = 0 , \qquad \langle d\xi_j\, d\xi_k \rangle = 0 , \qquad \langle d\xi_j^*\, d\xi_k \rangle = \delta_{j,k}\, dt . \qquad (2.38)$$

Here $\langle \cdots \rangle$ denotes the ensemble average. The respective density operator is the ensemble average of the projector onto the quantum state,

$$\rho = \langle\, |\psi><\psi|\, \rangle . \qquad (2.39)$$

A relativistic quantum state diffusion model has been proposed in Ref. [62].

Attempts have also been made to describe the stochastic evolution of the state vector in terms of a stochastic differential equation with a linear drift [63].

---

[5]I constantly employ the ket symbol $|\cdots>$ for a pure state, and $\langle \cdots \rangle$ for the ensemble average.

The Monte-Carlo wave function simulation or *quantum jump methods* proposed by Diósi [64], by Dalibard, Castin, and Mølmer [65], by Zoller and coworkers [66], and by Carmichael *et al.* [67] belong to the second class. In these related methods, the Schrödinger equation is supplemented by a non-Hermitian term and by a stochastic term undergoing a Poisson jump process. Because of the non-unitary time evolution of the state vector under a non-Hermitian Hamiltonian, the trace of the density operator is no more conserved. Conservation of probability is restored again and again by imposing stochastically chosen quantum jumps (see Refs. [65, 66, 68]). In the Monte Carlo algorithm by Mølmer *et al.* [69], the deviation of the norm δp of the wave function from unity after a certain time step is compared with a number ε, which is randomly chosen from the interval [0, 1]. If $\delta p > \varepsilon$, a quantum jump occurs by which the wave function is renormalized to unity. A comparison of some of the quantum jump and state diffusion models is given in Ref. [70].

Breuer and Petruccione have shown that a unique stochastic process in Hilbert space for the dynamics of the open system may be derived directly from the underlying microscopic system-plus-reservoir model [71, 72]. They employed a description of quantum mechanical ensembles in terms of probability distributions on a projective Hilbert space. In the elimination of the reservoir, they made the Markov approximation, and they employed second-order perturbation theory in the system-reservoir coupling. They then obtained a Liouville-master functional equation for the reduced probability distribution. The Liouville part of this equation corresponds to a deterministic Schrödinger-type equation with a non-Hermitian Hamiltonian which is intrinsically nonlinear in order to preserve the norm. The master part of this equation describes gain and loss of the probabilities for individual states due to discontinuous quantum jumps. In this description, the realization of the stochastic process is very similar to those generated by the piecewise deterministic quantum jump method [65] – [67]. Therefore, the stochastic simulation algorithms of all these approaches are very similar likewise. The equation of motion for the reduced density matrix derived from the Liouville-master equation is in the Lindblad form (2.20).

In the first place, the stochastic wave function methods are computational tools with which the solution of the Born-Markov master equation is simulated by using Monte Carlo importance-sampling techniques [68]. The stochastic methods are numerically superior to the conventional integration of the master equation when the rank of the reduced density matrix is large. Over and above the computational advantage of stochastic wave function methods for Born-Markov processes, some groups are presuming to claim that the instantaneous discontinuous processes are *real* and provide a natural description of individual quantum jump events (and not only of their statistics) as observed, e.g., in experiments with single ions in radio-frequency traps [73, 74]. Against that, I wish to point out that the assignment of definite states to a subsystem is *incompatible* with standard quantum theory, and has been proven wrong, e.g., in Einstein-Rosen-Podolsky experiments. Moreover, there is no experimental indication for non-standard phenomena (e.g., spontaneous collapse) in connection with the explanation of classical properties. Hence there is no phenomenological necessity for the introduction of a stochastic equation for state vectors. Besides computational advantages in the simulation of Born-Markov processes, the quantum-state diffusion method provides an alternative approach to measurement theory. In this method, a continuous measurement process, by which the system is steadily reduced within a certain time period to an eigenstate, is an integral part of the dynamical description. In conclusion, the stochastic wave function approaches provide efficient numerical simulation schemes for quantum master equations which are of the Lindblad form. However, since the Markov and the Born approximation are made, the application of these methods to solid state physics problems is as limited as the Born-Markov quantum master equation approach.

In the sequel we move on firm ground by taking the conservative view that the Hamiltonian dynamics of a global system induces a non-unitary dynamics for a subsystem. Upon performing a reduction of the global system to the relevant subsystem, all effects of the environmental coupling are put in an influence functional. Non-Markovian generalizations of quantum state diffusion and related stochastic Schrödinger equations can be found by a stochastic unraveling of the influence functional. The related discussion is given in Section 5.9.

3. System-plus-reservoir models

For many complex quantum systems we do not have a clear understanding of the microscopic origin of damping. In some systems, however, it is possible to track down the power spectrum of the stochastic force in the classical regime, and hence the spectral damping function $\tilde{\gamma}(\omega)$. Therefore, it is very important to have phenomenological system-plus-reservoir models which on the one hand open up the full quantum mechanical treatment, and on the other hand reduce in the classical limit to a description of the stochastic process in terms of a Langevin equation of the form (2.4).

The simplest model of a dissipative quantum mechanical system that one can envisage is a damped quantum mechanical linear oscillator: a central harmonic oscillator is coupled linearly via its displacement coordinate q to a fluctuating collective coordinate of a dynamical reservoir or bath. If the equilibrium state of the reservoir is only weakly perturbed by the central oscillator, its classical dynamics can be represented by linear equations. Therefore, it can be described in terms of a boson field or a (infinite) set of harmonic oscillators. Then the noise statistics of the stochastic process induced by the reservoir is strictly Gaussian. This simple system-plus-reservoir model has been introduced and discussed in a series of four papers by Ullersma [75]. Zwanzig generalized the model to the case in which the central particle moves in an anharmonic potential and studied the classical regime [76]. Caldeira and Leggett [77] were among the first who applied this model to a study of quantum mechanical tunneling of a macroscopic variable. The relevant model is considered in this chapter.

In the first section of this chapter I introduce the model and I seek the relations between the parameters of the model and the quantities appearing on the classical phenomenological level. In the next section I study the conditions that the open system is ergodic. In the subsequent sections I introduce the versatile spin-boson model, and I discuss a number of physically important particle-plus-reservoir systems for which we can base the description to some extent on a microscopic footing. In this chapter, I cannot deal with these systems in any detail. I shall outline the underlying Hamiltonians and postpone the path integral formulation of the quantum statistical mechanics until the next chapter. In the last section, I touch on the discussion of nonlinear quantum environments.

3.1 Harmonic oscillator bath with linear coupling

In this section, I first introduce the most general Hamiltonian underlying a dissipative system obeying Eq. (2.4) in the classical limit, and I explain the important generalization to the case where the viscosity is state-dependent. After that, I determine the relation between the parameters of the global model and the phenomenological frequency-dependent friction coefficient $\tilde{\gamma}(\omega)$.

3.1.1 The Hamiltonian of the global system

Consider a system with one or few degrees of freedom which is coupled to a huge environment and imagine that the environment is represented by a bath of harmonic excitations above a stable ground state. The interaction of the system with each individual degree of freedom of the reservoir is proportional to the inverse of the volume of the reservoir. Hence, *the coupling to an individual bath mode is weak for a geometrically macroscopic environment.* Therefore, it is physically very reasonable for macroscopic global systems to assume that the system-reservoir coupling is a *linear* function of the bath coordinates. This property is favorable since it allows to eliminate the environment exactly. Importantly, the weak perturbation of any individual bath mode does not necessarily mean that the dissipative influence of the reservoir on the system is weak as well since the couplings of the individual bath modes add up and the number of modes can be very large.

The general form of the Hamiltonian for the global system complying with these properties (barring pathological cases) is (see Ref. [77], Appendix C)

$$H \;=\; H_S + H_R + H_I \,. \tag{3.1}$$

Here, H_S is the Hamiltonian of the relevant system. For simplicity, let us imagine a particle of mass M moving in a potential $V(q)$,

$$H_S \;=\; \frac{p^2}{2M} + V(q) \,. \tag{3.2}$$

The reservoir consists of a set of harmonic oscillators,

$$H_{\mathrm{R}} = \sum_{\alpha=1}^{N} \left(\frac{p_\alpha^2}{2m_\alpha} + \frac{1}{2} m_\alpha \omega_\alpha^2 x_\alpha^2 \right) , \tag{3.3}$$

and the system-bath interaction H_{I} is assumed to be linear in the bath coordinates,

$$H_{\mathrm{I}} = -\sum_{\alpha=1}^{N} F_\alpha(q) x_\alpha + \Delta V(q) . \tag{3.4}$$

For specific purpose I have added a counter-term $\Delta V(q)$ which depends on $F_\alpha(q)$, and on the parameters m_α, ω_α of the reservoir, but not on its dynamical variables $x_\alpha(t)$. The additional potential term $\Delta V(q)$ is introduced in order to compensate a renormalization of the potential $V(q)$ which is caused by the coupling term linear in x_α in the interaction H_{I}. In the absence of $\Delta V(q)$, the minimum of the potential surface of the global system for fixed q in x_α-direction is at $x_\alpha = F_\alpha(q)/m_\alpha \omega_\alpha^2$. Thus, the "effective" potential renormalized by the coupling is given by

$$V_{\mathrm{eff}}(q) = V(q) - \sum_{\alpha=1}^{N} \frac{F_\alpha^2(q)}{2m_\alpha \omega_\alpha^2} . \tag{3.5}$$

In the special case $F_\alpha(q) = c_\alpha q$, the second term in Eq. (3.5) causes a negative shift $(\Delta \omega)^2 = - \sum_\alpha c_\alpha^2 / M m_\alpha \omega_\alpha^2$ in the squared circular frequency ω_0^2 of small oscillations about the minimum. This coupling-induced renormalization of the potential can be very large, and, if $\omega_{\mathrm{eff}}^2 = \omega_0^2 + (\Delta \omega)^2 < 0$, it changes the potential even qualitatively.

If we wish that the coupling of the relevant system to the reservoir solely introduces dissipation – and not in addition a renormalization of the potential $V(q)$ – we must compensate the second term in Eq. (3.5) by a suitable choice of $\Delta V(q)$. Full compensation of the coupling-induced potential deformation is achieved if we put

$$\Delta V(q) = \sum_{\alpha=1}^{N} \frac{F_\alpha^2(q)}{2m_\alpha \omega_\alpha^2} . \tag{3.6}$$

The specific choice of *separable* interaction

$$F_\alpha(q) = c_\alpha F(q) , \tag{3.7}$$

where $F(q)$ is independent of α, is of particular interest. Thus, under the assumptions specified above, the most general translational invariant Hamiltonian with a separable interaction is

$$H = \frac{p^2}{2M} + V(q) + \frac{1}{2} \sum_{\alpha=1}^{N} \left[\frac{p_\alpha^2}{m_\alpha} + m_\alpha \omega_\alpha^2 \left(x_\alpha - \frac{c_\alpha}{m_\alpha \omega_\alpha^2} F(q) \right)^2 \right] . \tag{3.8}$$

The case of a nonlinear function $F(q)$ occurs e.g. in rotational tunneling systems, in polaron systems, and in Josephson systems.

It has been argued [78] that the periodicity of a hindering potential in a *rotational tunneling system* is an exact symmetry which cannot be destroyed whatever the external influences are. The argument applies when several identical particles are tunneling at the same time, as e.g., in a rotating molecule complex. If there are N identical particles coherently tunneling [e.g., $N = 3$ for a methyl-(CH_3-)group], both the potential $V(\varphi)$ and the coupling function $F(\varphi)$, where φ is the dynamical angular variable, belong to the same symmetry group C_N, i.e., $V(\varphi) = V(\varphi + 2\pi/N)$ and $F(\varphi) = F(\varphi + 2\pi/N)$.

In a polaron system the particle's interaction energy due to a linear lattice distortion is nonlinear in the particle's coordinate according to $F_{\mathbf{k}}(\mathbf{q}) = e^{i\mathbf{k}\cdot\mathbf{q}}$ [cf. Sec. 3.4].

Quasiparticle tunneling in systems is another important case. Then the coordinate q is again identified with a phase variable φ. In a phenomenological modeling of charge tunneling between superconductors the interaction term is [79]

$$H_{\mathrm{I}} = \sin(\varphi/2) \sum_\alpha c_\alpha^{(1)} x_\alpha^{(1)} + \cos(\varphi/2) \sum_\alpha c_\alpha^{(2)} x_\alpha^{(2)} , \qquad (3.9)$$

where $\{x_\alpha^{(1)}\}$ and $\{x_\alpha^{(2)}\}$ represent two independent sets of oscillators. For a discussion of the microscopic theory, see Subsection 4.2.10.

If we require that the dissipation be *strictly* linear, we must constrain $F_\alpha(q)$ as

$$F_\alpha(q) = c_\alpha q . \qquad (3.10)$$

This is the case of state-independent dissipation. The form (3.10) describes in von Neumann's sense an ideal measurement of the particle's position by the reservoir.

Substituting Eq. (3.10) into Eq. (3.1), the Hamiltonian takes the form

$$H = \frac{p^2}{2M} + V(q) + \frac{1}{2} \sum_{\alpha=1}^{N} \left[\frac{p_\alpha^2}{m_\alpha} + m_\alpha \omega_\alpha^2 \left(x_\alpha - \frac{c_\alpha}{m_\alpha \omega_\alpha^2} q \right)^2 \right] . \qquad (3.11)$$

For later convenience, I rewrite the Hamiltonian (3.11) as

$$H = \frac{p^2}{2M} + \sum_{\alpha=1}^{N} \frac{p_\alpha^2}{2m_\alpha} + V(q, \boldsymbol{x}) , \qquad (3.12)$$

$$V(q, \boldsymbol{x}) = V(q) + \frac{1}{2} \sum_{\alpha=1}^{N} m_\alpha \omega_\alpha^2 \left(x_\alpha - \frac{c_\alpha}{m_\alpha \omega_\alpha^2} q \right)^2 . \qquad (3.13)$$

Here, $V(q, \boldsymbol{x})$ is the potential of the global system, and $\boldsymbol{x}$ represents the set of bath coordinates $\{x_\alpha\}$. The Hamiltonian (3.11) has been used to model dissipation for about thirty years. Early studies were limited to a harmonic potential $V(q)$. Probably the first who showed that Eq. (3.11) leads to dissipation was Magalinskiĭ [80]. Shortly later studies include the work by Rubin [81] for classical systems, and Senitzky [20], Ford et al. [21], and Ullersma [75] for quantum systems. In the more recent literature, the model described by Eq. (3.11) is usually referred to as the Caldeira-Leggett model.

3.1.2   The road to generalized Langevin equations

The equations of motion of a global system described by the Hamiltonian (3.11) read

$$M\ddot{q} + V'(q) + \sum_\alpha \left(c_\alpha^2 / m_\alpha \omega_\alpha^2 \right) q \;=\; \sum_\alpha c_\alpha x_\alpha \,, \qquad (3.14)$$

$$m_\alpha \ddot{x}_\alpha + m_\alpha \omega_\alpha^2 x_\alpha \;=\; c_\alpha q \,, \qquad (3.15)$$

where $V'(q) = \partial V / \partial q$. The dynamical equation for the oscillator position $x_\alpha(t)$ is an ordinary second order linear differential equation with inhomogeneity $c_\alpha q(t)$. This equation is solved by standard Green function techniques. In Fourier space, an individual solution of the inhomogeneous equation is

$$\tilde{x}_\alpha(\omega) \;=\; \tilde{\chi}_\alpha(\omega) c_\alpha \tilde{q}(\omega) \qquad \text{with} \qquad \tilde{\chi}_\alpha(\omega) \;=\; \lim_{\varepsilon \to 0^+} \frac{1}{m_\alpha} \frac{1}{\omega_\alpha^2 - \omega^2 - i\omega\varepsilon} \,. \qquad (3.16)$$

The function $\tilde{\chi}_\alpha(\omega)$ is the dynamical susceptibility of an individual bath oscillator. Of particular later interest is the absorptive part $\tilde{\chi}_\alpha''(\omega) = (\pi/m_\alpha)\,\mathrm{sgn}(\omega)\,\delta(\omega_\alpha^2 - \omega^2)$. Without loss of generality, I fix the time of preparation of the initial state to $t_0 = 0$. The solution evolving from the initial values $x_\alpha^{(0)}$ and $p_\alpha^{(0)}$ reads

$$x_\alpha(t) \;=\; x_\alpha^{(0)} \cos(\omega_\alpha t) + \frac{p_\alpha^{(0)}}{m_\alpha \omega_\alpha} \sin(\omega_\alpha t) + \frac{c_\alpha}{m_\alpha \omega_\alpha} \int_0^t dt' \sin\left[\omega_\alpha(t - t')\right] q(t') \,. \qquad (3.17)$$

It is convenient to rewrite Eq. (3.17) as a functional of the particle's velocity. Integrating the last term by parts, we get

$$\begin{aligned} x_\alpha(t) \;=\;& x_\alpha^{(0)} \cos(\omega_\alpha t) + \frac{p_\alpha^{(0)}}{m_\alpha \omega_\alpha} \sin(\omega_\alpha t) \\[2mm] & + \frac{c_\alpha}{m_\alpha \omega_\alpha^2} \left(q(t) - \cos(\omega_\alpha t) q(0) - \int_0^t dt' \cos[\omega_\alpha(t - t')]\, \dot{q}(t') \right) . \end{aligned} \qquad (3.18)$$

Next, we may eliminate the bath degrees of freedom by inserting Eq. (3.18) into Eq. (3.14). Then, the last term of the l.h.s. in Eq. (3.14) originating from the counter term in Eq. (3.4) cancels out. The dynamical equation for $q(t)$ alone is found to read

$$M\ddot{q}(t) + M \int_0^t dt'\, \gamma(t - t')\dot{q}(t') + V'(q) \;=\; \zeta(t) - M\gamma(t)q(0) \,. \qquad (3.19)$$

Here we have introduced the memory-friction kernel

$$\gamma(t) \;=\; \Theta(t)\, k(t) \,, \qquad \text{where} \qquad k(t) \;=\; \frac{1}{M} \sum_\alpha \frac{c_\alpha^2}{m_\alpha \omega_\alpha^2} \cos(\omega_\alpha t) \,, \qquad (3.20)$$

and the force

$$\zeta(t) \;=\; \sum_\alpha c_\alpha \left(x_\alpha^{(0)} \cos(\omega_\alpha t) + \frac{p_\alpha^{(0)}}{m_\alpha \omega_\alpha} \sin(\omega_\alpha t) \right) . \qquad (3.21)$$

The dynamical equation of motion (3.19) is a generalized Langevin equation extended with an inhomogeneous term which is the initial slip $-M\gamma(t)q(0)$. The random force $\zeta(t)$ depends explicitly on the preparation at time zero of the set of bath variables $\{q_\alpha^{(0)}\}$ and $\{p_\alpha^{(0)}\}$. The statistical properties of these variables decide on the statistical properties of the stochastic force $\zeta(t)$. Further processing of the Langevin equation (3.19) is given in Subsection 3.1.4.

Let us next consider the reduced dynamics when the system-reservoir coupling is nonlinear in the system's coordinate. If we had started at the Hamiltonian (3.8) and had eliminated the bath dynamics, we would have arrived at the equation

$$M\ddot{q}(t) + MF'[q(t)]\int_0^t dt'\, \gamma(t-t')F'[q(t')]\,\dot{q}(t') + V'(q)$$
$$= F'[q(t)]\left\{\zeta(t) - M\gamma(t)q(0)\right\}\ . \tag{3.22}$$

In this modified Langevin equation the damping term describes nonlinear (state-dependent) memory friction and the random force represents multiplicative noise.

As important examples for state-dependent dissipation, I mention quasiparticle tunneling through a barrier between superconductors (contact) and strong electron tunneling through a mesoscopic junction [cf. Subsections 4.2.10 and 5.8.2].

Before we embark in the subsection after next on the statistical properties of the stochastic force, we deal with the phenomenological modeling of the friction kernel.

3.1.3 Phenomenological modeling of friction

If the number N of bath oscillators is small, the period of time, in which the energy transferred to the bath is fed back to the central system, is of the order of other relevant time scales. However, when N is about 20 or larger, the Poincaré recurrence time is found to be practically infinity. In such cases it is appropriate to replace the sum over the discrete bath modes by a frequency integral with a continuous spectral density of the reservoir coupling.

With these preliminary remarks it is now straightforward to establish the connection of the frequency-dependent damping function $\tilde{\gamma}(\omega)$ defined in Eq. (2.5) with the parameters of the Hamiltonian (3.8) or (3.11). The Fourier transform of the retarded memory-friction kernel (3.20) is

$$\tilde{\gamma}(\omega) = \lim_{\varepsilon \to 0^+} \frac{-i\,\omega}{M} \sum_{\alpha=1}^N \frac{c_\alpha^2}{m_\alpha \omega_\alpha^2} \frac{1}{\omega_\alpha^2 - \omega^2 - i\,\omega\varepsilon}\ . \tag{3.23}$$

Next we introduce the spectral density of the environmental coupling

$$J(\omega) = \frac{\pi}{2} \sum_\alpha \frac{c_\alpha^2}{m_\alpha \omega_\alpha} \delta(\omega - \omega_\alpha)\ . \tag{3.24}$$

For a set of discrete modes, the spectral density consists of a sequence of δ-peaks. In order to work on a genuine heat bath, we assume that the eigenfrequencies ω_α are so dense as to form a continuous spectrum. In the continuum limit, $J(\omega)$ becomes a smooth function of ω, and the sum in Eq. (3.23) is replaced by the integral

$$\tilde{\gamma}(\omega) = \lim_{\varepsilon \to 0^+} \frac{-i\omega}{M} \frac{2}{\pi} \int_0^\infty d\omega' \frac{J(\omega')}{\omega'} \frac{1}{\omega'^2 - \omega^2 - i\omega\varepsilon} . \tag{3.25}$$

By this subtle modification, the function $\tilde{\gamma}(\omega)$ acquires a smooth real part,

$$\tilde{\gamma}'(\omega) = \tilde{\gamma}'(-\omega) = J(\omega)/M\omega . \tag{3.26}$$

Upon choosing for $J(\omega)$ a positive function which drops to zero both in the IR and UV limit, the resulting spectral damping function $\tilde{\gamma}(\omega)$ owns all of the essential properties specified after Eq. (2.5).

Often it is convenient to work in Laplace space. The Laplace transform $\hat{\gamma}(z)$ of the damping function $\gamma(t)$ is related to the Fourier transform $\tilde{\gamma}(\omega)$ by analytic continuation,

$$\hat{\gamma}(z) = \tilde{\gamma}(\omega = iz) ; \qquad \tilde{\gamma}(\omega) = \lim_{\varepsilon \to 0^+} \hat{\gamma}(z = -i\omega + \varepsilon) . \tag{3.27}$$

With use of Eqs. (3.23) and (3.25) we obtain the relations

$$\hat{\gamma}(z) = \frac{z}{M} \sum_{\alpha=1}^{N} \frac{c_\alpha^2}{m_\alpha \omega_\alpha^2} \frac{1}{(\omega_\alpha^2 + z^2)} = \frac{2}{\pi} \frac{z}{M} \int_0^\infty d\omega' \frac{J(\omega')}{\omega'} \frac{1}{\omega'^2 + z^2} , \tag{3.28}$$

where the latter form applies in the continuum limit.

We may express $\tilde{\gamma}(\omega)$ and $J(\omega)$ in terms of the dynamical susceptibilities of the individual reservoir modes given in Eq. (3.16). The resulting expressions are

$$\tilde{\gamma}(\omega) = \frac{1}{-i\omega M} \sum_{\alpha=1}^{N} c_\alpha^2 \Big(\tilde{\chi}_\alpha(0) - \tilde{\chi}_\alpha(\omega) \Big) , \tag{3.29}$$

and

$$J(\omega) = \Theta(\omega) \sum_{\alpha=1}^{N} c_\alpha^2 \, \tilde{\chi}_\alpha''(\omega) . \tag{3.30}$$

We see from Eq. (3.29) that the aforementioned counter term serves to eliminate the contribution of the static susceptibility. These forms pave the way for situations in which the bath modes are effectively nonlinear, as discussed below in Section 3.6.

As far as we are interested in properties of the particle alone, the dynamics is fully determined by the mass M, the potential $V(q)$, and the spectral density $J(\omega)$. We see from Eq. (3.26) that $J(\omega)$ is uniquely determined by the classical frequency-dependent damping coefficient $\tilde{\gamma}(\omega)$. With the roles reversed, we can also express the microscopic characteristics in terms of the phenomenological real-time damping kernel $\gamma(t)$. Using Eq. (3.25), we can give the damping kernel $\gamma(t)$ in terms of the spectral density $J(\omega)$ of the environmental coupling,

$$\gamma(t) = \Theta(t) \frac{1}{M} \frac{2}{\pi} \int_0^\infty d\omega \frac{J(\omega)}{\omega} \cos(\omega t) . \tag{3.31}$$

The inversion of the Fourier integral (3.31) gives

$$J(\omega) \; = \; M\omega \int_0^\infty dt\, \gamma(t) \cos(\omega t) \, . \tag{3.32}$$

The spectral density $J(\omega)$ may also be expressed in terms of the Laplace transform of the damping kernel. From Eq. (3.32) we find

$$J(\omega) \; = \; \lim_{\varepsilon \to 0+} M\omega [\, \hat{\gamma}(\varepsilon + i\omega) + \hat{\gamma}(\varepsilon - i\omega)\,]/2 \, . \tag{3.33}$$

We conclude this subsection with an important remark. The spectral density $J(\omega)$, and after all dissipative quantum mechanics, is determined by quantities that appear already in the classical phenomenological equation of motion [82, 83]. This property holds exactly in the case of strict linear (i.e. state-independent) dissipation. For instance, a conventional molecular dynamics simulation may be used to compute $\tilde{\gamma}'(\omega)$, which then determines $J(\omega)$ by Eq. (3.26). Therefore, this relation plays a fundamental role in the phenomenological modeling of dissipative quantum systems.

3.1.4   Quantum statistical properties of the stochastic force

In Subsection 3.1.2 we have studied the reduced dynamics on the classical level. Evidently, the quantum mechanical equations of motion in the Heisenberg picture are the same. In the quantum regime, the random force $\zeta(z)$ given in Eq. (3.21) bears operator character via the bath operators $\hat{x}_\alpha^{(0)}$ and $\hat{p}_\alpha^{(0)}$. It is pertinent to write these operators in terms of annihilation and creation operators b_α and $b_\alpha^\dagger$, respectively,

$$\hat{x}_\alpha^{(0)} \; = \; \left(\frac{\hbar}{2m_\alpha \omega_\alpha}\right)^{1/2} (b_\alpha + b_\alpha^\dagger) \, , \qquad \hat{p}_\alpha^{(0)} \; = \; i\left(\frac{m_\alpha \omega_\alpha \hbar}{2}\right)^{1/2} (b_\alpha^\dagger - b_\alpha) \, , \tag{3.34}$$

which obey the commutation relations

$$\left[\, b_\alpha, b_\gamma \,\right] \; = \; \left[\, b_\alpha^\dagger, b_\gamma^\dagger \,\right] \; = \; 0 \, , \qquad \text{and} \qquad \left[\, b_\alpha, b_\gamma^\dagger \,\right] = \delta_{\alpha\gamma} \, . \tag{3.35}$$

In the language of these operators, the random force operator (3.21) reads

$$\hat{\zeta}(t) \; = \; \sum_\alpha \left(\frac{\hbar c_\alpha^2}{2m_\alpha \omega_\alpha}\right)^{1/2} \left(e^{i\omega_\alpha t} b_\alpha^\dagger + e^{-i\omega_\alpha t} b_\alpha\right) \, . \tag{3.36}$$

It obeys the commutator relation

$$[\, \hat{\zeta}(t), \, \hat{\zeta}(0) \,] \; = \; -i\hbar \sum_\alpha \frac{c_\alpha^2}{m_\alpha \omega_\alpha} \sin(\omega_\alpha t) \, . \tag{3.37}$$

Taking operator averages with the canonical density operator of the unperturbed bath

$$
\begin{aligned}
\hat{\rho}_{\mathrm{R}}^{(0)} \; &= \; \frac{1}{Z} \exp\left[-\beta \sum_\alpha \left(\frac{\hat{p}_\alpha^{(0)\,2}}{2m_\alpha} + \frac{m_\alpha \omega_\alpha^2}{2} \hat{x}_\alpha^{(0)\,2}\right)\right] \\
&= \; \frac{1}{Z'} \exp\left[-\beta\hbar \sum_\alpha \omega_\alpha b_\alpha^\dagger b_\alpha\right] \, ,
\end{aligned}
\tag{3.38}
$$

where Z (Z') is the partition functions with (without) the zero-point energy, we obtain

$$\langle b_\alpha \rangle_{\hat{\rho}_R^{(0)}} = \langle b_\alpha^\dagger \rangle_{\hat{\rho}_R^{(0)}} = \langle b_\alpha b_\gamma \rangle_{\hat{\rho}_R^{(0)}} = \langle b_\alpha^\dagger b_\gamma^\dagger \rangle_{\hat{\rho}_R^{(0)}} = 0 \,,$$

$$\langle b_\alpha^\dagger b_\gamma \rangle_{\hat{\rho}_R^{(0)}} = \delta_{\alpha\gamma} \, n(\omega_\alpha) \,, \qquad \langle b_\alpha b_\gamma^\dagger \rangle_{\hat{\rho}_R^{(0)}} = \delta_{\alpha\gamma} \left[1 + n(\omega_\alpha) \right] \,, \tag{3.39}$$

where $n(\omega)$ is the single-particle Bose distribution

$$n(\omega) = \frac{1}{e^{\beta\hbar\omega} - 1} \,. \tag{3.40}$$

Thus, $\zeta(t)$ is a stationary Gaussian noise operator with ensemble averages

$$\langle \hat{\zeta}(t) \rangle_{\hat{\rho}_R^{(0)}} = 0 \,, \tag{3.41}$$

$$\frac{1}{2} \langle \hat{\zeta}(t)\hat{\zeta}(0) + \hat{\zeta}(0)\hat{\zeta}(t) \rangle_{\hat{\rho}_R^{(0)}} = \hbar \sum_\alpha \frac{c_\alpha^2}{2m_\alpha \omega_\alpha} \coth(\tfrac{1}{2}\beta\hbar\omega_\alpha) \cos(\omega_\alpha t) \,. \tag{3.42}$$

The stochastic force conveys absorption into and emission from the reservoir of a single quantum of energy $\hbar\omega_\alpha$. Since we have

$$\frac{n(\omega)}{1 + n(\omega)} = e^{-\beta\hbar\omega} \,, \tag{3.43}$$

emission and absorption are related by detailed balance. The form (3.42) is a version of the quantum-mechanical fluctuation-dissipation theorem [cf. Section 6.1].

The operator form of the stochastic equation of motion (3.19) still contrasts with the usual form of the Langevin equation by the spurious term $-M\gamma(t)\hat{q}(0)$. This term is a transient which depends on the initial value $\hat{q}(0)$. It is an artifact of the the decoupled thermal initial state (3.38) [84]. The annoying term $-M\gamma(t)\hat{q}(0)$ may be removed (i) by a shift of the random force operator $\hat{\zeta}(t)$,

$$\hat{\zeta}(t) = \hat{\xi}(t) + M\gamma(t)\hat{q}(0) \,, \tag{3.44}$$

and (ii) by taking quantum-statistical average with the canonical distribution of shifted bath oscillators,

$$\hat{\rho}_R = Z^{-1} \exp\left\{ -\beta \sum_\alpha \left[\frac{\hat{p}_\alpha^{(0)\,2}}{2m_\alpha} + \frac{m_\alpha\omega_\alpha^2}{2}\left(\hat{x}_\alpha^{(0)} - \frac{c_\alpha}{m_\alpha\omega_\alpha^2}\hat{q}(0) \right)^2 \right] \right\} \,. \tag{3.45}$$

With these modifications the generalized quantum Langevin equation is put into the standard form

$$M\frac{d^2\hat{q}(t)}{dt^2} + M\int_0^t dt' \, \gamma(t-t')\frac{d\hat{q}(t')}{dt'} + V'[\hat{q}(t)] = \hat{\xi}(t) \,. \tag{3.46}$$

The centered random force operator $\hat{\xi}(t)$ has the commutator relation

$$[\hat{\xi}(t), \hat{\xi}(0)] = -i\hbar \sum_\alpha \frac{c_\alpha^2}{m_\alpha\omega_\alpha} \sin(\omega_\alpha t) \,. \tag{3.47}$$

and obeys stationary Gaussian statistics with ensemble averages

$$\langle \hat{\xi}(t) \rangle_{\hat{\rho}_R} = 0 , \tag{3.48}$$

$$\frac{1}{2} \langle \hat{\xi}(t)\hat{\xi}(0) + \hat{\xi}(0)\hat{\xi}(t) \rangle_{\hat{\rho}_R} = \hbar \sum_\alpha \frac{c_\alpha^2}{2 m_\alpha \omega_\alpha} \coth(\tfrac{1}{2}\beta\hbar\omega_\alpha) \cos(\omega_\alpha t) . \tag{3.49}$$

In the continuum limit (3.24) with (3.26), the expressions (3.47) and (3.49) take the forms (2.32) and (2.31), respectively. The QLE (3.46) is an operator equation that acts in the full Hilbert space of system and bath. The coupling induces entanglement with time even when the density matrix of the global system is initially in factorized form.

From the above we see that care has to be taken if one refers to statistical properties of a random force [84]. We have been able to remove the initial transient slippage by taking the thermal average in the initial state with respect to bath modes which are shifted by the coupling to the particle.

3.1.5 Displacement correlation function

Consider as a little exercise of the antecedent findings the displacement correlation function $D^+(t) \equiv \langle [\hat{q}(0) - \hat{q}(t)]\hat{q}(0) \rangle_\beta$, where the position satisfies the operator Langevin equation (3.46). The linear response to the fluctuating force $\xi(t)$ is expressed in terms of the response function $\chi(t)$ as

$$\hat{q}(t) = \int_{-\infty}^{t} dt' \, \chi(t - t')\hat{\xi}(t') . \tag{3.50}$$

With use of Eq. (3.50) the displacement correlation function takes the form

$$D^+(t) = \int_{-\infty}^{\infty} dt' \, [\chi(-t') - \chi(t - t')] \int_{-\infty}^{\infty} dt'' \, \chi(-t'') \, \langle \hat{\xi}(t')\hat{\xi}(t'') \rangle_\beta . \tag{3.51}$$

Observing that $\langle \hat{\xi}(t')\hat{\xi}(t'') \rangle_\beta = \tfrac{1}{2}\{ [\hat{\xi}(t'), \hat{\xi}(t'')] + \langle \hat{\xi}(t')\hat{\xi}(t'') + \hat{\xi}(t'')\hat{\xi}(t') \rangle_\beta \}$ and using the relations (2.32) and (2.31) we obtain

$$D^+(t) = \frac{\hbar}{\pi} \int_0^t d\omega \, \tilde{\chi}''(\omega) \left\{ \coth(\tfrac{1}{2}\beta\hbar\omega)[1 - \cos(\omega t)] + i\sin(\omega t) \right\} . \tag{3.52}$$

where $\tilde{\chi}''(\omega)$ is the absorptive part of the dynamical susceptibility,

$$\tilde{\chi}''(\omega) \equiv \mathrm{Im} \int_0^{\infty} dt \, \chi(t) \, e^{i\omega t} = M\omega\tilde{\gamma}'(\omega)\tilde{\chi}(\omega)\tilde{\chi}^*(\omega) . \tag{3.53}$$

For linear quantum systems the quantum mechanical response function coincides with the classical response function (see Section 6.2). Putting $V'(q) = M\omega_0^2 q$, we then have

$$\tilde{\chi}(\omega) = \frac{1}{M} \frac{1}{\omega_0^2 - \omega^2 - i\omega\tilde{\gamma}(\omega)} . \tag{3.54}$$

The correlation function (3.52) is studied for the damped quantum oscillator in Section 6 and for the free Brownian particle in Chapter 7.

3.1.6 Thermal propagator and imaginary-time correlations

As an expedient groundwork we now analytically continue the force auto-correlation function $\mathcal{X}(t) = \langle \hat{\xi}(t)\hat{\xi}(0)\rangle_\beta$ to imaginary time $t = -i\tau$. It is immediately clear that the square bracket in the spectral representation (2.33) takes in the principal interval $0 \le \tau < \hbar\beta$ the form

$$D_\omega(\tau) = [1 + n(\omega)]\,e^{-\omega\tau} + n(\omega)\,e^{\omega\tau} = \frac{\cosh[\omega(\frac{1}{2}\hbar\beta - \tau)]}{\sinh(\frac{1}{2}\omega\hbar\beta)}. \tag{3.55}$$

Outside the principal interval $0 \le \tau < \hbar\beta$, the function $D_\omega(\tau)$ may be periodically continued with cusps at the points $\tau = n\hbar\beta$ $(n = 0, \pm1, \pm2, \cdots)$ by writing it as Fourier series in terms of the bosonic Matsubara frequency $\nu_n = 2\pi n/\hbar\beta$,

$$D_\omega(\tau) = \frac{1}{\hbar\beta}\sum_{n=-\infty}^{+\infty}\frac{2\omega}{\nu_n^2 + \omega^2}\,e^{i\nu_n\tau}. \tag{3.56}$$

The function $D_\omega(\tau)$ solves the inhomogeneous differential equation

$$\left[-\partial^2/\partial\tau^2 + \omega^2\right]D_\omega(\tau) = 2\omega\!:\!\delta(\tau)\!:\,, \tag{3.57}$$

where $:\!\delta(\tau)\!:$ is the periodically continued δ-function

$$:\!\delta(\tau)\!: \equiv \frac{1}{\hbar\beta}\sum_{n=-\infty}^{+\infty}e^{i\nu_n\tau} = \sum_{n=-\infty}^{+\infty}\delta(\tau - n\hbar\beta). \tag{3.58}$$

This points up that $D_\omega(\tau)$ is the imaginary-time propagator in thermal equilibrium or thermal Green function of a boson.

In terms of the thermal propagator, the imaginary-time force autocorrelation function $\mathcal{X}^{(\mathrm{E})}(\tau) \equiv \langle\xi(\tau)\xi(0)\rangle_\beta = \mathcal{X}(t = -i\tau)$ takes the form

$$\mathcal{X}^{(\mathrm{E})}(\tau) = \frac{\hbar}{\pi}\int_0^\infty d\omega\, J(\omega)D_\omega(\tau) = \frac{\hbar M}{2\pi}\int_{-\infty}^\infty d\omega\,\omega\tilde{\gamma}'(\omega)D_\omega(\tau). \tag{3.59}$$

Alternatively, it may be written as Fourier series. Use of Eq. (3.28) gives

$$\mathcal{X}^{(\mathrm{E})}(\tau) = \frac{M}{\beta}\sum_{n=-\infty}^{\infty}|\nu_n|\,\hat{\gamma}(|\nu_n|)\,e^{i\nu_n\tau}. \tag{3.60}$$

In a similar manner, the imaginary-time displacement correlation function $D^{(\mathrm{E})}(\tau) = D^+(t = -i\tau)$ may be written in the form

$$D^{(\mathrm{E})}(\tau) = \frac{\hbar}{\pi}\int_0^\infty d\omega\,\tilde{\chi}''(\omega)[D_\omega(0) - D_\omega(\tau)]. \tag{3.61}$$

Alternatively, we may write $D^{(\mathrm{E})}(\tau)$ with the Fourier series representation (3.56) as a Matsubara sum and employ the dispersion relation

$$\mathrm{Im}\,\frac{1}{\pi}\int_{-\infty}^\infty d\omega\,\frac{\omega\tilde{\chi}(\omega)}{\omega^2 + z^2} = \frac{2}{\pi}\int_0^\infty d\omega\,\frac{\omega\tilde{\chi}''(\omega)}{\omega^2 + z^2} = \hat{\chi}(z). \tag{3.62}$$

We then obtain

$$D^{(\mathrm{E})}(\tau) = \frac{1}{\beta}\sum_{n=-\infty}^{\infty}\hat{\chi}(|\nu_n|)\left[1 - e^{i\nu_n\tau}\right] = \frac{1}{\beta}\sum_{n=-\infty}^{\infty}\frac{1 - e^{i\nu_n\tau}}{M[\nu_n^2 + |\nu_n|\hat{\gamma}(|\nu_n|)]}. \tag{3.63}$$

3.1.7 Ohmic and frequency-dependent damping

In the strict *Ohmic* (instant response) limit, damping is frequency-independent,

$$\tilde{\gamma}(\omega) = \gamma . \tag{3.64}$$

We see from Eq. (3.32) or from Eq. (3.33) that strict Ohmic damping is described by the model (3.11) in which the spectral density $J(\omega)$ is chosen as [77]

$$J(\omega) = \eta\omega = M\gamma\omega \tag{3.65}$$

for all frequencies ω. In the first equality, we have introduced the familiar viscosity coefficient η. The relation (3.65) implies memoryless friction $\gamma(t) = 2\gamma\Theta(t)\delta(t)$.[1] This, of course, is an idealized situation. In reality, any particular spectral density $J(\omega)$ of physical origin falls off in the limit $\omega \to \infty$, because there is always a microscopic memory time setting the time scale for inertia effects in the reservoir. If $J(\omega)$ would grow steadily with ω, certain physical quantities, e.g. the momentum dispersion, would be divergent (cf. the discussion in Sections 6.6 and 7.3).

In the simplest form, the damping kernel $\gamma(t)$ in the classical equation of motion (2.4) is regularized with a Drude memory time $\tau_D = 1/\omega_D$,

$$\gamma(t) = \gamma\omega_D\Theta(t)\exp(-\omega_D t) , \tag{3.66}$$

which is known as *Drude regularization*. We then have in the frequency domain

$$\tilde{\gamma}(\omega) = \gamma/[1 - i\omega/\omega_D] , \quad \text{and} \quad \hat{\gamma}(z) = \gamma/[1 + z/\omega_D] . \tag{3.67}$$

Since the imaginary part of $\tilde{\gamma}(\omega)$ differs from the real part by a factor ω/ω_D, it is small well below the Drude frequency.

The Drude form (3.67) emerges from a spectral density with algebraic cutoff,

$$J(\omega) = M\gamma\omega/[1 + \omega^2/\omega_D^2] . \tag{3.68}$$

The damping kernel carries memory-friction on the time scale $\tau_D = \omega_D^{-1}$. When the relevant frequencies of the system's dynamics are much lower than the Drude cutoff frequency ω_D, the spectral density of the reservoir coupling (3.68) induces Ohmic damping with effective damping strength $\gamma_{\text{eff}} = \int_0^\infty dt\,\gamma(t) = \gamma$.

Next, consider the extension to general frequency-dependent damping. It is convenient to assume (though it is not strictly necessary) that the function $J(\omega)$ has a power law form at low frequencies, $J(\omega) \propto \omega^s$. It will become clear later on that the dissipative influences can be classified by the power s. Negative values of s are excluded since otherwise the frequency integral in Eq. (3.31) would be infrared-divergent, which is pathological. The power-law form is assumed to hold in the frequency range $0 \leq \omega \lesssim \omega_c$, where ω_c is much less than a characteristic cutoff frequency ω_{ch}, which is of the order of the Drude, Debye, or Fermi frequency etc., depending on

---

[1]In the integral in Eq. (2.4) the δ-function counts only half, and thus the damping term reduces to the Ohmic form $M\gamma\dot{q}(t)$ in Eq. (2.3).

the model. For frequencies of the order of or greater than ω_{ch}, the behavior of $J(\omega)$ may be complicated and may not easily be inferable from the classical motion or from microscopic considerations. The important point, however, is that, as long as we are interested in times $t \gg \omega_c^{-1}$, the effect of the environmental modes with $\omega \gtrsim \omega_c$ can be absorbed into a renormalization of parameters appearing in the Hamiltonian H_S of the system. The only additional property we need to postulate is that $J(\omega)$ falls off at least with some negative power of ω in the limit $\omega \to \infty$. The minimal decrease of $J(\omega)$ required depends on the physical quantity under consideration.

To make the discussion quantitative, we decompose $J(\omega)$ into a low-frequency and a high-frequency contribution,

$$J(\omega) = J_{\mathrm{lf}}(\omega) + J_{\mathrm{hf}}(\omega) , \qquad (3.69)$$

$$J_{\mathrm{lf}}(\omega) = J(\omega) f(\omega/\omega_c) , \qquad (3.70)$$

$$J_{\mathrm{hf}}(\omega) = J(\omega) [1 - f(\omega/\omega_c)] . \qquad (3.71)$$

Here, $f(\omega/\omega_c)$ is a cutoff function defined in such a way that $J_{\mathrm{hf}}(\omega)$ is negligibly small for $\omega \ll \omega_c$ and $J_{\mathrm{lf}}(\omega)$ is negligibly small for $\omega \gg \omega_c$. According to convenience, we may choose either a sharp cutoff $f(\omega/\omega_c) = \Theta(1 - \omega/\omega_c)$, or a smooth cutoff. Expedient forms are a Gaussian, an exponential, or the rational Drude form $f(\omega/\omega_c) = 1/[1 + (\omega/\omega_c)^2]$.

We shall consider in most applications the case in which $J_{\mathrm{lf}}(\omega)$ has a power-law form with an exponential cutoff,

$$J_{\mathrm{lf}}(\omega) = \eta_s \omega_{\mathrm{ph}}^{1-s} \omega^s \, e^{-\omega/\omega_c} , \qquad \eta_s = M\gamma_s . \qquad (3.72)$$

Here I have introduced for $s \neq 1$ a "phononic" reference frequency ω_{ph}, so that the coupling constant η_s has dimension of viscosity for all s. I find it useful to distinguish the frequency ω_{ph} from the cutoff frequency ω_c. The latter is of different relevance in the regimes $s < 1$ and $s > 1$, as we shall see. Where convenient, I shall use the temperature scales

$$T_{\mathrm{ph}} \equiv \hbar\omega_{\mathrm{ph}}/k_B \qquad \text{and} \qquad T_c \equiv \hbar\omega_c/k_B . \qquad (3.73)$$

In many physical situations, the frequency ω_c at which the power law $J_{\mathrm{lf}}(\omega) \propto \omega^s$ is cut off is very large compared with all other relevant frequencies of the system. A number of analytic results will be available in this limit.

The case $s = 1$ in Eq. (3.72) describes Ohmic friction, as we have discussed already. The cases $0 < s < 1$ and $s > 1$ have been dubbed *sub-Ohmic* and *super-Ohmic*, respectively [85]. I shall present in Sections 3.4 and 4.2 various microscopic models and consider the respective spectral densities of the coupling. The Ohmic case is important, e.g., for tunneling systems in a metallic environment. Importantly, Ohmic friction is virtually ubiquitous in real physical systems at low temperature, as we shall see in Subsection 4.2.5. A phonon bath in d spatial dimensions corresponds to the case $s = d$ or $s = d + 2$, depending on the underlying symmetry of the strain field (see Subsection 4.2.4). Irrational values of s may occur for complex environments with effectively fractal dimension, such as porous and viscoelastic media, and reservoirs with chaotic dynamics.

In the limit $s \to 0$ in the sub-Ohmic regime, the spectral density $J(\omega)$ takes the constant value $\eta_0 \omega_{\mathrm{ph}}$. Then we have $\tilde{\gamma}'(\omega) = \eta_0 \omega_{\mathrm{ph}} / M|\omega|$, and the power spectrum (2.34) of the random force takes the form $\tilde{\mathcal{X}}(\omega) = \eta_0 \hbar \omega_{\mathrm{ph}} \coth(\beta \hbar |\omega|/2)$. Thus, in the regime $\omega \ll k_{\mathrm{B}} T/\hbar$ the power spectrum is $1/f$ or flicker noise

$$\tilde{\mathcal{X}}(\omega) = 2k_{\mathrm{B}} T \eta_0 \omega_{\mathrm{ph}}/|\omega| . \tag{3.74}$$

$1/f$ noise may be attributed, e.g., to a set of bistable (two-level) systems making random transitions between the two states. Discussion of this and of possible other origin of $1/f$ noise, and of the implications on decoherence is given in Section 22.5.

With a destructive combination of two Drude kernels of the form (3.66) super-Ohmic friction $J(\omega) \propto \omega^3$ can be generated [86]. Upon putting

$$\gamma(t) = \Theta(t) \gamma \left(\omega_1 e^{-\omega_1 t} - \omega_2 e^{-\omega_2 t} \right) \tag{3.75}$$

we find with use of Eq. (3.32)

$$J(\omega) = M \gamma \omega^3 \frac{\omega_1^2 - \omega_2^2}{(\omega^2 + \omega_1^2)(\omega^2 + \omega_2^2)} . \tag{3.76}$$

The damping function $\hat{\gamma}(z)$ acquires in the low-frequency regime $z \ll \omega_{\mathrm{c}}$ from the partial spectral density $J_{\mathrm{hf}}(\omega)$ the contribution

$$\hat{\gamma}_{\mathrm{hf}}(z) = z \Delta M_{\mathrm{hf}}/M , \tag{3.77}$$

$$\Delta M_{\mathrm{hf}} = \frac{2}{\pi} \int_0^\infty d\omega \, \frac{J_{\mathrm{hf}}(\omega)}{\omega^3} , \tag{3.78}$$

as follows from Eq. (3.28). With the form (3.77), the Laplace transform of the memory-friction force $M z \hat{\gamma}_{\mathrm{hf}}(z) \hat{q}(z)$ reduces at low frequency to the expression $\Delta M_{\mathrm{hf}} z^2 \hat{q}(z)$. This term, however, may simply be added to the kinetic term $M z^2 \hat{q}(z)$. Thus for asymptotic time $t \gg \omega_{\mathrm{c}}^{-1}$, which in Laplace space is $z \ll \omega_{\mathrm{c}}$, the partial spectral density $J_{\mathrm{hf}}(\omega)$ manifests itself just as mass renormalization,

$$M[z^2 + z \hat{\gamma}_{\mathrm{hf}}(z)] \hat{q}(z) = (M + \Delta M_{\mathrm{hf}}) z^2 \hat{q}(z) . \tag{3.79}$$

From this we conclude that if we are not interested in the time regime $t \lesssim \omega_{\mathrm{c}}^{-1}$, we may treat $J_{\mathrm{hf}}(\omega)$ in the adiabatic approximation. If not stated differently, I will take into account $J_{\mathrm{hf}}(\omega)$ only globally by regarding the mass M as the renormalized mass which is already dressed by the reservoir's high-frequency modes.

The integral (3.78) is infra-red convergent also for the spectral density $J_{\mathrm{lf}}(\omega)$ when the power s in Eq. (3.72) exceeds 2. Upon extracting this term from the integral (3.78) we then obtain

$$\hat{\gamma}(z) = \frac{\Delta M_s}{M} z - \frac{2}{\pi} \frac{z^3}{M} \int_0^\infty d\omega \, \frac{J(\omega)}{\omega^3} \frac{1}{\omega^2 + z^2} , \tag{3.80}$$

where

$$\Delta M_s = \frac{2}{\pi} \int_0^\infty d\omega \, \frac{J_{\mathrm{lf}}(\omega)}{\omega^3} = \frac{2}{\pi} \Gamma(s-2) \left(\frac{\omega_c}{\omega_{\mathrm{ph}}} \right)^{s-1} \frac{\gamma_s}{\omega_c} M . \tag{3.81}$$

In the range $4 < s < 6$, an additional term with powers z^3 can be extracted from the integral (3.78). The extension to higher values of s is evident.

Choosing for the spectral density the form (3.72) with a sharp high-frequency cutoff, the spectral damping function (3.28) takes the analytic form [87]

$$\hat{\gamma}(z) = \frac{2\gamma_s}{\pi s} \left(\frac{\omega_c}{\omega_{\mathrm{ph}}} \right)^{s-1} \frac{\omega_c}{z} \, {}_2F_1(1, \tfrac{1}{2}s; 1+\tfrac{1}{2}s; -\omega_c^2/z^2) , \tag{3.82}$$

where ${}_2F_1(a, b; c; x)$ is a hypergeometric function [88, 89]. The asymptotic expansion of the hypergeometric function yields the low-frequency expansion of the spectral damping function $\hat{\gamma}(z)$. It may be calculated upon using standard transformation formulas of the hypergeometric function. Putting $\lambda_s \equiv \gamma_s / \sin(\pi s/2)$, we find

$$\hat{\gamma}(z) = \begin{cases} \lambda_s \left(\dfrac{z}{\omega_{\mathrm{ph}}} \right)^{s-1} \left\{ 1 + \mathcal{O}\!\left(\dfrac{z^{2-s}}{\omega_c^{2-s}} \right) \right\}, & 0 < s < 2, \\[2ex] \dfrac{2\gamma_2}{\pi} \dfrac{z}{\omega_{\mathrm{ph}}} \left\{ \ln\left(\dfrac{\omega_c}{z} \right) + \mathcal{O}\!\left(\dfrac{z^2}{\omega_c^2} \right) \right\}, & s = 2, \\[2ex] z \dfrac{\Delta M_s}{M} \left\{ 1 + \mathcal{O}\!\left(\dfrac{z^2}{\omega_c^2} \right) \right\} + \lambda_s \left(\dfrac{z}{\omega_{\mathrm{ph}}} \right)^{s-1}, & 2 < s < 4, \\[2ex] z \dfrac{\Delta M_4}{M} - \dfrac{2\gamma_4}{\pi} \left(\dfrac{z}{\omega_{\mathrm{ph}}} \right)^3 \left\{ \ln\left(\dfrac{\omega_c}{z} \right) + \mathcal{O}\!\left(\dfrac{z^2}{\omega_c^2} \right) \right\}, & s = 4, \\[2ex] z \dfrac{\Delta M_s}{M} \left\{ 1 - \dfrac{s-2}{s-4} \dfrac{z^2}{\omega_c^2} + \mathcal{O}\!\left(\dfrac{z^4}{\omega_c^4} \right) \right\} + \lambda_s \left(\dfrac{z}{\omega_{\mathrm{ph}}} \right)^{s-1}, & 4 < s < 6. \end{cases} \tag{3.83}$$

In the most interesting regime $0 < s < 2$ the damping function $\hat{\gamma}(z)$ varies as z^{s-1}. For a free particle, this leads to sub-diffusive behavior in the sub-Ohmic regime $0 < s < 1$, and to super-diffusive behavior in the super-Ohmic regime $1 < s < 2$, as we shall see in Section 7.4. In the range $s > 2$, the leading term at low z induces renormalization of the mass of the free particle. Sub-leading terms are relevant in the transient time regime (cf. Section 7.4).

The case of Ohmic friction is particularly important since it is frequently encountered in nature. Over and above it is responsible for a quantum phase transition at a critical damping strength, and it shows interesting scaling properties, as we shall see. Here I anticipate that in the context of quantum state engineering the weak coupling regime for a general spectral density is of interest.

3.1.8 Fractional Langevin equation

For a spectral coupling with non-integer power s with the associated spectral damping function $\hat{\gamma}(z) \propto z^{s-1}$, the generalized Langevin equation (3.46) can be cast into a fractional Langevin equation. Fractional calculus is the branch of mathematical analysis that generalizes the derivative of a function to non-integer order. We start off with Cauchy's formula for the calculation of an n-fold iterated integral,

$$D^{-n}g(t) \equiv \int_0^t d\tau_n \int_0^{\tau_n} d\tau_{n-1} \cdots \int_0^{\tau_2} d\tau_1 \, g(\tau_1) \; = \; \frac{1}{\Gamma(n)} \int_0^t d\tau \, g(\tau)(t-\tau)^{n-1} \,. \quad (3.84)$$

Evidently, the resulting integral can be analytically continued to non-integer values of n. The choice $n = -\alpha$ is the Riemann-Liouville derivative

$$D^\alpha g(t) \equiv \frac{\partial^\alpha g(t)}{\partial t^\alpha} = \frac{1}{\Gamma(-\alpha)} \int_0^t d\tau \, \frac{g(\tau)}{(t-\tau)^{1+\alpha}} \,, \qquad -1 < \alpha < 0 \,. \quad (3.85)$$

Since the integral is constrained to the regime $-1 < \alpha < 0$, it is actually a fractional antiderivative. In the fractional derivative of order α, where α is assumed in the range $n - 1 < \alpha < n$, the integer derivative of order n has to be performed before the fractional antiderivative. Thus we have

$$D^\alpha g(t) = \frac{1}{\Gamma(n-\alpha)} \left(\frac{\partial}{\partial t}\right)^n \int_0^t d\tau \, \frac{g(\tau)}{(t-\tau)^{1+\alpha-n}} \,, \qquad n-1 < \alpha < n \,. \quad (3.86)$$

The key point now is that the representation (3.86) for the fractional derivative is a convolution. Taking the Laplace transform, and observing that

$$\int_0^\infty dt \, e^{-zt} \, t^{n-\alpha-1} = \Gamma(n-\alpha) \, z^{\alpha-n} \,, \qquad n-1 < \alpha < n \,, \quad (3.87)$$

we have

$$\int_0^\infty dt \, e^{-zt} \, D^\alpha g(t) = z^\alpha \, \hat{g}(z) \,. \quad (3.88)$$

Being targeted to the Langevin equation (3.46) with the friction kernel

$$M\hat{\gamma}(z) = M\lambda_s \omega_{\mathrm{ph}}^{1-s} z^{s-1} \,, \qquad 0 < s < 2 \,, \quad (3.89)$$

the fractional Langevin equation reads

$$M\frac{d^2q(t)}{dt^2} + M\lambda_s \omega_{\mathrm{ph}}^{1-s} \, D^{s-1} \, \dot{q}(t) + V'[q(t)] = \xi(t) \,. \quad (3.90)$$

The operator-valued form of the fractional Langevin equation (3.90) is equivalent to the generalized quantum Langevin equation (3.46) with the damping kernel (3.89).

Fractional kinetic equations were used to describe chaotic dynamics in low-dimensional systems [90], and anomalous diffusion in biology and physics [91, 92].

3.1.9   Rubin model

In Rubin's model, a heavy particle of mass M and coordinate q is bi-linearly coupled to a semi-infinite chain of harmonic oscillators with masses m and spring constants $f = m\omega_{\mathrm{R}}^2/4$ [93], as sketched in Fig. 3.1. The Hamiltonian of the global model is

$$M \qquad f \qquad m$$

Figure 3.1: Pictorial sketch of the mechanical analogue of the Rubin model.

$$H = \frac{p^2}{2M} + V(q) + \sum_{n=1}^{\infty} \left(\frac{p_n^2}{2m} + \frac{f}{2}(x_{n+1} - x_n)^2 \right) + \frac{f}{2}(q - x_1)^2 . \qquad (3.91)$$

The model is not yet in the standard form (3.11) since the harmonic bath modes are coupled with each other. The harmonic chain is diagonalized with the transformation

$$x_n = \sqrt{2/\pi} \int_0^\pi dk \, \sin(kn) \, X(k) . \qquad (3.92)$$

In the normal mode representation of the reservoir, the Hamiltonian takes the form

$$H = \frac{p^2}{2M} + V(q) + \frac{f}{2}q^2 + \int_0^\pi dk \left(\frac{P^2(k)}{2m} + \frac{m}{2}\omega^2(k)X^2(k) - c(k)X(k)\, q \right) , \qquad (3.93)$$

where the eigenfrequencies $\omega(k)$ and the coupling function $c(k)$ are given by

$$\omega(k) = \omega_R \sin(\tfrac{1}{2}k) , \qquad c(k) = \frac{1}{2\sqrt{2\pi}} m\omega_R^2 \sin(k) . \qquad (3.94)$$

The frequency ω_R is the highest frequency of the reservoir modes. The spectral density (3.24) of the coupling to the semi-infinite chain is found with Eq. (3.94) as

$$J_{\mathrm{si}}(\omega) = \frac{m\omega_R^3}{16} \int_0^\pi dk \, \frac{\sin^2(k)}{\sin(\tfrac{1}{2}k)} \, \delta\left[\omega - \omega(k)\right] = \frac{m\omega_R}{2} \omega \left(1 - \frac{\omega^2}{\omega_R^2}\right)^{1/2} \Theta(\omega_R - \omega) .$$

If the particle is embedded in an infinite chain, the sum in Eq. (3.91) is also over negative integers n, and the resulting spectral density is $J_{\mathrm{inf}}(\omega) = 2J_{\mathrm{si}}(\omega)$. With this form, the damping kernel (3.31) in Laplace and real space emerges as [94, 95]

$$\hat{\gamma}_{\mathrm{inf}}(z) = \frac{m}{M}\left(\sqrt{z^2 + \omega_R^2} - z\right) , \quad \text{and} \quad \gamma_{\mathrm{inf}}(t) = \frac{m}{M}\omega_R \frac{J_1(\omega_R t)}{t} , \qquad (3.95)$$

respectively, where $J_1(z)$ is a Bessel function of the first kind. The damping kernel of the Rubin model oscillates with time, and the envelope function decays algebraically with power $\frac{3}{2}$. This particular behavior significantly differs from the exponential decay of the Drude kernel (3.66). The memory time of the Rubin kernel is of the order of ω_R^{-1}.

Finally, consider the velocity response $\mathcal{R}(t)$ to an impulsive force in the absence of the potential $V(q)$. In Laplace space we have with $\kappa = m/M$

$$\hat{\mathcal{R}}(z) \equiv \frac{1}{z + \hat{\gamma}_{\mathrm{inf}}(z)} = \frac{1}{(1 - \kappa)z + \kappa\sqrt{z^2 + \omega_R^2}} , \qquad (3.96)$$

There are three interesting cases:

(1) In the limit $m/M \to 0$ with $\gamma_0 \equiv \omega_R m/M$ held fixed, the case of *memory-less* Ohmic friction is reached, $\hat{\gamma}_{\rm inf}(z) = \gamma_0$. The response function in Laplace space then is $\hat{R}(z) = 1/[z + \gamma_0]$, and thus in the time regime

$$R(t) = e^{-\gamma_0 t} . \tag{3.97}$$

(2) When $M = m$, we have $\hat{R}(z) = 1/\sqrt{z^2 + \omega_R^2}$. In the time regime, the velocity response then displays oscillatory behavior of a Bessel function of order zero,

$$R(t) = J_0(\omega_R t) . \tag{3.98}$$

The memory time is of the scale $1/\omega_R$, and the envelope function of $R(t)$ decays asymptotically as $1/\sqrt{t}$.

(3) When $M = 2m$, we have $\hat{R}(z) = 2/[z + \sqrt{z^2 + \omega_R^2}]$. This yields in the time domain

$$R(t) = \frac{J_1(\omega_R t)}{\omega_R t} . \tag{3.99}$$

Interestingly, in this particular case the time-dependence of $R(t)$ coincides, except for an overall factor, with that of $\gamma_{\rm inf}(t)$.

3.1.10 Interaction of a charged particle with the radiation field

The dynamics of a one-dimensional particle of mass M and charge e in a potential $V(q)$ and interacting with the 3-d radiation field is described by the Hamiltonian

$$H = \frac{1}{2M}\left(p - \frac{e}{c}A(q)\right)^2 + V(q) + \sum_{\mathbf{k},\lambda} \hbar \omega_k a_{\mathbf{k},\lambda}^\dagger a_{\mathbf{k},\lambda} . \tag{3.100}$$

The last term is the Hamiltonian of the radiation field. In the dipole approximation, the vector potential is spatially constant [96]. We then have the mode decomposition

$$A(t) = \sqrt{\frac{2\pi\hbar c^2}{L^3}} \sum_{\mathbf{k},\lambda}\left(\mathbf{e}_q\!\cdot\!\mathbf{e}_{\mathbf{k},\lambda}\frac{f_k^*}{\sqrt{\omega_k}}\, a_{\mathbf{k},\lambda}(t) + {\rm h.c.}\right), \tag{3.101}$$

where $\mathbf{e}_q\!\cdot\!\mathbf{e}_{\mathbf{k},\lambda}$ is the projection of the polarization vector $\mathbf{e}_{\mathbf{k},\lambda}$ on the particle's straight trajectory, L^3 is the black-body volume, and f_k is the form factor of the charged particle. The form factor is unity at small k and falls to zero above a cutoff frequency ω_D.[2] The Heisenberg equations of motion are found from Eq. (3.100) as

$$M\dot{q}(t) = p(t) - (e/c)A(t) , \qquad \dot{p}(t) = -V'[q(t)] , \tag{3.102}$$

$$\dot{a}_{\mathbf{k},\lambda}(t) = -i\,\omega_k a_{\mathbf{k},\lambda}(t) + i\sqrt{2\pi e^2/\hbar L^3}\,\mathbf{e}_q\!\cdot\!\mathbf{e}_{\mathbf{k},\lambda}^*(f_k/\sqrt{\omega_k})\dot{q}(t) . \tag{3.103}$$

Eq. (3.103) is easily solved with the result

---

[2]For convenience in Sect. 6.3 the cutoff frequency is denoted, as the Drude frequency, by ω_D.

$$a_{\mathbf{k},\lambda}(t) \;=\; a_{\mathbf{k},\lambda}^{(0)}\, e^{-i\omega_k t} + i\sqrt{\frac{2\pi e^2}{\hbar L^3}}\,\frac{f_k}{\sqrt{\omega_k}}\; \mathbf{e}_q\!\cdot\!\mathbf{e}_{\mathbf{k},\lambda}^* \int_0^t dt'\, e^{-i\omega_k(t-t')}\dot{q}(t')\,, \qquad (3.104)$$

where $a_{\mathbf{k},\lambda}^{(0)}$ is the free-field annihilation operator. The expansion (3.101) with (3.104) yields an explicit form for $A(t)$. When this expression is put into Eq. (3.102) we readily obtain the generalized Langevin equation

$$M\ddot{q}(t) + M\int_0^t dt'\,\gamma(t-t')\dot{q}(t') + V'[q(t)] \;=\; \xi(t)\,, \qquad (3.105)$$

where the friction kernel $\gamma(t)$ and the stochastic force $\xi(t)$ are given by [97]

$$\gamma(t) \;=\; \frac{4\pi e^2}{ML^3}\sum_{\mathbf{k},\lambda}|\mathbf{e}_q\!\cdot\!\mathbf{e}_{\mathbf{k},\lambda}|^2\,|f_k|^2\,\cos(\omega_k t)\,, \qquad (3.106)$$

$$\xi(t) \;=\; \sqrt{\frac{2\pi\hbar e^2}{L^3}}\sum_{\mathbf{k},\lambda}\sqrt{\omega_k}\Big(i\,f_k^*\,\mathbf{e}_q\!\cdot\!\mathbf{e}_{\mathbf{k},\lambda}^*\,a_{\mathbf{k},\lambda}^{(0)}\,e^{-i\omega_k t} + \text{h.c.}\Big)\,. \qquad (3.107)$$

The relation (3.32) with the form (3.106) gives the spectral density of the coupling as

$$J(\omega) \;=\; \frac{2\pi^2 e^2}{L^3}\sum_{\mathbf{k},\lambda}|\mathbf{e}_x\!\cdot\!\mathbf{e}_{\mathbf{k},\lambda}|^2\,|f_k|^2\,\omega_k\,\delta(\omega-\omega_k) - \frac{2e^2}{3c^3}\frac{\omega^3}{1+\omega^2/\omega_D^2}\,. \qquad (3.108)$$

The second form holds for the convenient choice $|f_k|^2 = 1/(1+\omega_k^2/\omega_D^2)$ and after execution of the continuum limit. Thus, radiation damping is super-Ohmic with power $s = 3$. In addition, we find with use of the thermal averages discussed in Subsection 3.1.4 that black body radiation equips the stochastic force (3.107) with the correlations (2.31) and (2.32), where $\tilde{\gamma}'(\omega) = J(\omega)/M\omega$.

3.2 Ergodicity

A quantum statistical system is called ergodic when the ensemble average of observables coincides with the time average at asymptotic time. Expressed differently, the system is ergodic when it forgets over time its initial state and therefore relaxes to a unique equilibrium state. In order that the system is not caught in a particular state, it must be able to exchange an arbitrarily small amount of energy with the environment. This implies that the spectral density of the environmental coupling $J(\omega)$ must have nonzero weight at the smallest frequencies. In turn, an open system is *nonergodic* when the spectral density has a low-frequency cutoff.

Whether an open quantum system of interest is ergodic can be found, e.g., with a study of the asymptotic dynamics of the response function $\chi(t)$ pertinent to the Langevin equation (3.46). The two opposite cases are that the Brownian particle is confined in a potential and that it is free. In the first case, the static susceptibility χ_0 is finite, and in the second case it is infinity.

The paradigm of the first case is the damped linear oscillator of which the susceptibility in Laplace space is (cf. Section 6.2)

$$\hat{\chi}(z) = \frac{1}{M[\omega_0^2 + z^2 + z\hat{\gamma}(z)]} \,. \tag{3.109}$$

In general, the response function $\chi(t)$ consists of a coherent part $\chi_{\mathrm{coh}}(t)$, which is due to a complex conjugate pair of simple poles of $\hat{\chi}(z)$ in the cut z-plane located at z_{p} and z_{p}^*, and a incoherent cut contribution $\chi_{\mathrm{cut}}(t)$,

$$\chi(t) = \chi_{\mathrm{coh}}(t) + \chi_{\mathrm{cut}}(t) \,. \tag{3.110}$$

Since $\mathrm{Re}(z_{\mathrm{p}}) < 0$, the contribution $\chi_{\mathrm{coh}}(t)$ describes damped oscillations. Thus we have asymptotically $\lim_{t\to\infty} \chi_{\mathrm{coh}}(t) = 0$.

Next we turn to the leading cut contribution at asymptotic time. The expansion of $\hat{\chi}(z)$ at $z = 0$ is

$$\hat{\chi}(z) = \chi_0 - M\chi_0^2 z^2 - M\chi_0^2 z\hat{\gamma}(z) + \mathcal{O}[(z^2 + z\hat{\gamma}(z))^2/\omega_0^4] \,, \tag{3.111}$$

where $\chi_0 = 1/M\omega_0^2$ is the static susceptibility. At first we observe that integer powers of z do not contribute to $\chi(t)$. Hence the terms of order z^{2n+1} with $n = 0,\ 1,\ 2,\cdots$ in the expressions for $\hat{\gamma}(z)$ in Eq. (3.83) are irrelevant, and the leading relevant term of $\hat{\gamma}(z)$ for all s is the term $\lambda_s(z/\omega_{\mathrm{ph}})^{s-1}$. Insertion of this expression in the third term of Eq. (3.111) yields the leading cut contribution as

$$\hat{\chi}_{\mathrm{cut}}(z) = - M\chi_0^2 \lambda_s \omega_{\mathrm{ph}}^{1-s} z^s \,, \tag{3.112}$$

and in the time regime[3]

$$\chi_{\mathrm{cut}}(t) = \frac{2\Gamma(1+s)}{\pi} \cos\left(\frac{\pi s}{2}\right) \frac{M\gamma_s}{\omega_{\mathrm{ph}}^{s-1}} \frac{\chi_0^2}{t^{1+s}} \,. \tag{3.113}$$

We see that there is no cut contribution when s is an odd integer.

The important result of this analysis is that the response function drops to zero for all s. We shall see in Chapter 6 that also the equilibrium correlation functions of the damped linear quantum oscillator, e.g., the equilibrium autocorrelation function of the position, approach asymptotically a unique equilibrium state for all s.

The dissipative two-state system is another open quantum system with a finite static susceptibility. We shall find below in Chapter 21 that this nonlinear system shows similar behavior. Like the damped linear oscillator, it relaxes for general s to a unique equilibrium state.

Consider next as open system with divergent static susceptibility a free Brownian particle. The Langevin equation (3.46) yields for the velocity operator $\hat{v}(t)$ in the absence of deterministic forces the solution

$$\hat{v}(t) = \mathcal{R}(t)\hat{v}(0) + \frac{1}{M}\int_0^t dt'\, \mathcal{R}(t-t')\hat{\xi}(t') \,, \tag{3.114}$$

---

[3]While $\hat{\gamma}(z)$ has logarithmic terms for $s = 2n$, the form (3.113) also holds for these values of s.

where $\mathcal{R}(t) = M\dot{\chi}(t)$ describes the velocity response to an impulsive force,

$$\hat{\mathcal{R}}(z) = \frac{1}{z + \hat{\gamma}(z)} \, . \tag{3.115}$$

In the regime $0 < s < 2$, we find with use of Eq. (3.83) $\lim_{z \to 0} \hat{\mathcal{R}}(z) z^{s-1} = \omega_{\rm ph}^{s-1}/\lambda_s$ and hence the asymptotic time dependence

$$\mathcal{R}(t) = \frac{\sin(\pi s/2)}{\Gamma(s-1)} \frac{\omega_{\rm ph}}{\gamma_s} \frac{1}{(\omega_{\rm ph} t)^{2-s}} \, , \qquad 0 < s < 2 \, . \tag{3.116}$$

When the parameter s exceeds 2, the dominant term in $\hat{\mathcal{R}}(z)$ is the kinetic one with the polaronic mass term added, $\lim_{z \to 0} \hat{\mathcal{R}}(z) z = 1/(1 + \Delta M_s/M)$. Hence the frictional effects of the damping kernel fade away in the course of time, and $\mathcal{R}(t)$ approaches asymptotically a constant, like for a free undamped particle but with change from bare to renormalized mass,

$$\mathcal{R}(t \to \infty) = \frac{M}{M + \Delta M_s} \, , \qquad s > 2 \, . \tag{3.117}$$

Hence the mean velocity of the particle depends asymptotically on the initial velocity v_0 as $\langle v(t \to \infty) \rangle = [M/(M + \Delta M_s)] v_0$ when $s > 2$. Further discussion of free Brownian motion is given in Chapter 7.

From these little thoughts we deduce the following conjecture. Every open system with a finite static susceptibility is ergodic in the entire regime $s > 0$. On the other hand, an open system with divergent static susceptibility, like a free Brownian particle, is ergodic in the range $0 < s < 2$ and nonergodic for super-Ohmic friction with power $s > 2$. The findings given in this section are in correspondence with the analysis in Ref. [98].

To corroborate these findings, consider two spectral densities $\propto \omega$, the one with a high-frequency cutoff, $J_{\rm h}(\omega) = M\gamma\omega \, \Theta(\omega_{\rm c} - \omega)$, and the other with a low frequency cutoff, $J_\ell(\omega) = M\gamma\omega \, \Theta(\omega - \omega_\ell)$. The respective spectral damping functions are found from Eq. (3.28) as

$$\begin{aligned} \hat{\gamma}_{\rm h}(z) &= (2/\pi)\gamma \arctan(\omega_{\rm c}/z) \, , \\ \hat{\gamma}_\ell(z) &= (2/\pi)\gamma \arctan(z/\omega_\ell) \, . \end{aligned} \tag{3.118}$$

We have $\lim_{z \to 0} \hat{\gamma}_{\rm h}(z) = \gamma$ and $\lim_{z \to 0} \hat{\gamma}_\ell(z)/z = (2/\pi)\gamma/\omega_\ell$. In the first case, there is Ohmic friction down to the lowest frequency. In the second case, the spectral damping function ends up in mass renormalization at time $t \gg 1/\omega_\ell$, $\Delta M/M = (2/\pi)\gamma/\omega_\ell$. Hence the system is ergodic in the first and nonergodic in the second case.

A discussion of susceptibility, correlation functions, and transport properties of linear open systems is given in Section 6.3, and in Subsections 6.4.2 and 7.4.1. We conclude with the remark that the relation (3.112) is the essence of the Shiba relation. This relation is studied in Subsection 6.4.3 for the damped linear oscillator, and in Subsections 22.6.5, and 22.7.2 for the dissipative two-state system.

3.3   The spin-boson model

Many physical and chemical systems can be described by a generalized coordinate
with which is associated an effective potential energy function with two separate minima at roughly the same energy. At thermal energies which are very small compared
with the level spacing of the low-lying states in the individual wells, only the ground
states of the two wells are involved in the dynamics. Then the time evolution goes off
in a two-dimensional Hilbert space. For a high barrier, the two eigen states are clearly
localized in the left and right well, respectively, and they are spatially well-separated.
The localized states are weakly coupled through a transfer or tunneling matrix element. The two-state system (TSS) is the simplest system showing constructive and
destructive quantum interference effects, e.g., clockwise oscillations of the occupation
of the left and right well. There is a great amount of interest in the thermodynamical
and dynamical behavior of this model due to the diversity of physical realizations, as
well as to theoretical advances in recent years.

   The basic element in a quantum computer is a *qubit*. Principally, it can be formed
by any physical system, whose motion is effectively restricted to a two-dimensional
Hilbert space. It is often convenient to regard such system as a spin-$\frac{1}{2}$ system. In any
real physical situation, the spin is in contact with the environment. If the surroundings
is imagined as a reservoir of bosons, the global system consisting of the spin and the
boson bath is represented by the *spin-boson model*. Most interesting in this system,
in particular in the context of quantum state engineering and quantum computing,
is the extent to which the phase relation between the "spin-up" and "spin-down"
components of the wave function is preserved.

   Other examples of the situation just described include the motion of defects in
crystalline solids (e.g., impurity ions in alkali halides), tunneling of light particles (e.g.,
hydrogen, muon, proton) in metals [99, 100, 101], the tunneling entities believed to
be responsible for the anomalous low-temperature thermal and acoustic properties
of oxide glasses and amorphous metals [102] and for the anomalous conductance
in mesoscopic wires [103, 104], and also some types of chemical reactions involving
electron transfer processes [105, 106].

3.3.1   The model Hamiltonian

I begin with considering the reduction of a spatially expanded double-well system
to a two-state system. I characterize the asymmetric double well potential by a "detuning" energy $\hbar\epsilon$ between the two potential minima and by a coupling energy $\hbar\Delta_0$
representing the tunnel splitting of the symmetric system, as sketched in Fig. 3.2.
Throughout the book, I shall use the convention that the right well is the lower one
for positive bias, $\epsilon > 0$. The tunneling matrix element Δ_0 may be calculated using
standard WKB or instanton methods (see Refs. [107, 108]).

   Consider a symmetric double well potential $V(q)$ with minima at $q = \pm\frac{1}{2}q_0$ and
potential energy $V(\pm\frac{1}{2}q_0) = 0$. The frequency of small oscillations in the wells is

$\omega_0 = \sqrt{V''(\pm\frac{1}{2}q_0)/M}$. In the semiclassical limit, the tunnel splitting for a particle of mass M is determined by properties of the so-called instanton path $q_{inst}(\tau)$. This path describes motion in the *upside-down* potential $-V(q)$ with zero total energy,

$$\tfrac{1}{2}M\dot{q}_{inst}^2(\tau) = V[q_{inst}(\tau)], \tag{3.119}$$

and satisfies the boundary conditions $q_{inst}(\tau \to \pm\infty) = \pm\frac{1}{2}q_0$. The instanton is a kink-like path centered at the time where the path rushes with maximal velocity through the minimum of the upside-down potential $-V(q)$. The action of this path is

$$S_{inst} = M \int_{-\infty}^{\infty} d\tau\, \dot{q}_{inst}^2(\tau). \tag{3.120}$$

The instanton with center at $\tau = 0$ approaches the boundary values $\pm\frac{1}{2}q_0$ as

$$q_{inst}(\tau \to \pm\infty) = \pm\frac{q_0}{2} \mp C_0 \left(\frac{S_{inst}}{2M\omega_0}\right)^{1/2} e^{-\omega_0|\tau|}, \tag{3.121}$$

where I have chosen the pre-exponential factor conveniently. The factor C_0 is a numerical constant of order unity which depends on the shape of the potential barrier. In the semiclassical limit, the tunneling matrix element Δ_0 for any particular symmetric double well is determined by the instanton action S_{inst}, the frequency ω_0, and by the numerical constant C_0 occurring in the asymptotic behavior (3.121). The semiclassical expression for the tunneling matrix element expressed in terms of instanton parameters reads [107, 108]

$$\Delta_0 = 2\omega_0 C_0 (S_{inst}/2\pi\hbar)^{1/2} e^{-S_{inst}/\hbar}, \tag{3.122}$$

For the archetypal double-well potential of quartic form

$$V(q) = \frac{M\omega_0^2}{2}\left(\frac{q^2 - \frac{1}{4}q_0^2}{q_0}\right)^2, \tag{3.123}$$

which has barrier height $V(q = 0) \equiv V_b = M\omega_0^2 q_0^2/32$, the instanton trajectory reads

$$q_{inst}(\tau) = \tfrac{1}{2}q_0 \tanh\left(\tfrac{1}{2}\omega_0\tau\right). \tag{3.124}$$

For this particular path, the action S_{inst} and the numerical factor C_0 are given by

$$S_{inst} = M\omega_0 q_0^2/6 = 16V_b/3\omega_0, \quad \text{and} \quad C_0 = 2\sqrt{3}. \tag{3.125}$$

Thus we find from Eq. (3.122) for the tunneling matrix element the expression

$$\Delta_0 = 8(2V_b/\pi\hbar\omega_0)^{1/2}\omega_0 \exp\left(-16V_b/3\hbar\omega_0\right). \tag{3.126}$$

For later convenience, we employ in the sequel instead of the bare Δ_0 the renormalized tunnel matrix element Δ, which is dressed by a Franck-Condon factor representing the polarization cloud of the high-frequency environmental modes in adiabatic approximation (see the discussion in Subsection 18.1.2). In the parameter regime

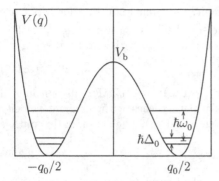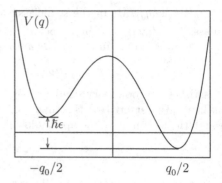

Figure 3.2: Symmetric double well potential (left) with barrier height V_b. The spacing between the first excited state and the ground state in each well is $\hbar\omega_0$, and the tunnel splitting is $\hbar\Delta_0$. The biased double well with detuning energy $\hbar\epsilon$ is sketched on the right.

$$V_b \gg \hbar\omega_0 \gg \hbar\Delta, \hbar|\epsilon|, k_B T , \qquad (3.127)$$

where V_b is the barrier height and $\hbar\omega_0$ is the separation of the first excited state from the ground state in either well, the system will be effectively restricted to a two-dimensional Hilbert space spanned, e.g., by the states $|R> \,\hat{=}\, |\uparrow>$ (right) and $|L> \,\hat{=}\, |\downarrow>$ (left) localized in the right and left well, respectively. The two-state Hamiltonian may be expressed in terms of the Pauli matrices in the pseudo-spin form[4]

$$H_{\text{TSS}} = -\tfrac{1}{2}\hbar\Delta\,\sigma_x - \tfrac{1}{2}\hbar\epsilon\,\sigma_z = -\frac{\hbar}{2}\begin{pmatrix} \epsilon & \Delta \\ \Delta & -\epsilon \end{pmatrix} . \qquad (3.128)$$

The basis is formed by the localized states $|R>$ and $|L>$ which are eigenstates of σ_z with eigenvalues $+1$ and -1, respectively. The position operator is $\hat{q} = \tfrac{1}{2}q_0\sigma_z$ and the eigenvalues $\pm\tfrac{1}{2}q_0$ of $\hat{q}$ are the positions of the localized states. In this representation, the Pauli operators may equivalently be written in the form

$$\begin{aligned} \sigma_z &= |R><R| - |L><L| , \\ \sigma_x &= |R><L| + |L><R| , \qquad (3.129) \\ \sigma_y &= i\,(|L><R| - |R><L|) . \end{aligned}$$

The localized states $|R>$ and $|L>$ are related to the eigenstates $|g>$ (ground) and $|e>$ (excited) of the Hamiltonian (3.128) by the orthogonal transformation

---

[4]I choose spin matrix representations $\{\sigma_j\}$ and $\{\tau_j\}$ in which σ_z and τ_z are diagonal,

$$\sigma_x = \begin{pmatrix} 0 & 1 \\ 1 & 0 \end{pmatrix} , \qquad \sigma_y = \begin{pmatrix} 0 & -i \\ i & 0 \end{pmatrix} , \qquad \sigma_z = \begin{pmatrix} 1 & 0 \\ 0 & -1 \end{pmatrix} .$$

$$\begin{pmatrix} |R> \\ |L> \end{pmatrix} = \mathcal{R}(\varphi) \begin{pmatrix} |g> \\ |e> \end{pmatrix} \equiv \begin{pmatrix} \cos\frac{1}{2}\varphi & -\sin\frac{1}{2}\varphi \\ \sin\frac{1}{2}\varphi & \cos\frac{1}{2}\varphi \end{pmatrix} \begin{pmatrix} |g> \\ |e> \end{pmatrix}, \qquad (3.130)$$

where

$$\sin\varphi = \frac{\Delta}{\Delta_b}, \qquad \cos\varphi = \frac{\epsilon}{\Delta_b}, \qquad \tan\varphi = \frac{\Delta}{\epsilon}. \qquad (3.131)$$

The tunnel splitting energy of the biased TSS is

$$E_e - E_g = \hbar\Delta_b = \hbar\sqrt{\Delta^2 + \epsilon^2}. \qquad (3.132)$$

With the orthogonal transformation

$$\mathcal{R}(-\varphi)[\sin\varphi\,\sigma_x + \cos\varphi\,\sigma_z]\mathcal{R}(\varphi) = \tau_z, \qquad (3.133)$$

the TSS Hamiltonian is converted into diagonal form

$$H'_{\text{TSS}} \equiv \mathcal{R}(-\varphi)H_{\text{TSS}}\mathcal{R}(\varphi) = -\tfrac{1}{2}\hbar\Delta_b\,\tau_z. \qquad (3.134)$$

The expectation values of the operators $\tau_{\pm} \equiv \frac{1}{2}(\tau_x \pm i\tau_y)$ are the off-diagonal elements or coherences of the TSS density matrix in energy representation. These are expressed in terms of the matrices σ_x, σ_y and σ_z as

$$\tau_{\pm} = \mathcal{R}(-\varphi)\tfrac{1}{2}[\cos\varphi\,\sigma_x \pm i\,\sigma_y - \sin\varphi\,\sigma_z]\mathcal{R}(\varphi). \qquad (3.135)$$

Consider the coupling to a heat bath which is sensitive to the position of the TSS and which is represented by a collective bath mode, the polarization energy $\mathfrak{E}(t)$,

$$H_{\text{I}} = -\tfrac{1}{2}\sigma_z\mathfrak{E}(t), \qquad \mathfrak{E}(t) \equiv q_0\sum_{\alpha=1}^{N}c_\alpha x_\alpha(t). \qquad (3.136)$$

The $x_\alpha(t)$ are the individual modes of the reservoir. A dipole-local-field interaction provides a simple physical model for this kind of coupling. As already discussed in some detail in Subsection 3.1.2, the coupling to the heat bath introduces a fluctuating force $\xi(t) = \sum_\alpha c_\alpha x_\alpha(t)$ which causes a fluctuating bias or polarization energy $\mathfrak{E}(t)$. Assuming strict Gaussian statistics, the heat bath can again be modeled exactly by a harmonic bath of bosons. Thus the essential physics is captured by a model Hamiltonian which has become known in the literature as the "spin-boson" Hamiltonian,

$$H_{\text{SB}} = -\tfrac{1}{2}\hbar\Delta\,\sigma_x - \tfrac{1}{2}\hbar\epsilon\,\sigma_z - \tfrac{1}{2}\sigma_z\mathfrak{E}(t) + H_{\text{R}}. \qquad (3.137)$$

In the eigenbasis of H_{TSS} the spin-boson Hamiltonian takes the form

$$\begin{aligned} H'_{\text{SB}} &\equiv \mathcal{R}(-\varphi)H_{\text{SB}}\mathcal{R}(\varphi) \\ &= -\tfrac{1}{2}\hbar\Delta_b\,\tau_z - \tfrac{1}{2}(\cos\varphi\,\tau_z - \sin\varphi\,\tau_x)\,\mathfrak{E}(t) + H_{\text{R}}, \end{aligned} \qquad (3.138)$$

where τ_z and τ_x are the Pauli matrices in the transformed basis. In the energy representation of H_{TSS} the reservoir coupling consists of a transverse ($\propto \sin\varphi$) and a longitudinal ($\propto \cos\varphi$) coupling. Clearly, only the transverse part can flip the spin.

The Hamiltonian (3.137) can equivalently be expressed in terms of the annihilation and creation operators b_α and $b_\alpha^\dagger$ introduced in Eq. (3.34). We then have

$$H_{\mathrm{SB}} = H_{\mathrm{TSS}} - \frac{1}{2}\sigma_z \sum_{\alpha=1}^{N} \hbar\lambda_\alpha \left(b_\alpha + b_\alpha^\dagger \right) + \sum_{\alpha=1}^{N} \hbar\omega_\alpha b_\alpha^\dagger b_\alpha . \qquad (3.139)$$

Here I have dropped the irrelevant zero-point energy. The environmental effects are again administered by a spectral density of the coupling,

$$G(\omega) = \sum_{\alpha=1}^{N} \lambda_\alpha^2 \, \delta(\omega - \omega_\alpha) = \frac{q_0^2}{2\hbar} \sum_{\alpha=1}^{N} \frac{c_\alpha^2}{m_\alpha \omega_\alpha} \, \delta(\omega - \omega_\alpha) . \qquad (3.140)$$

The second equality relates λ_α to c_α. The spin-boson spectral density (3.140) is related to the spectral density of the continuous model $J(\omega)$ defined in Eq. (3.24) by

$$G(\omega) = (q_0^2/\pi\hbar) \, J(\omega) . \qquad (3.141)$$

The spectral density $J(\omega)$ of the continuous model (3.8) or (3.11) has dimension mass times frequency squared whereas $G(\omega)$ has dimension frequency.

3.3.2 Flux and charge qubits: reduction to the spin-boson model

Superconducting circuits based on tunnel junctions are considered as candidates for quantum computing devices. Making qubits from such electrical elements might be advantageous because coupling of qubits, electrical control, and scaling to large numbers should be feasible using integrated-circuit fabrication technology.

The simplest design is an rf SQUID which is formed by a superconducting ring interrupted by a junction. The appropriate macroscopic degree of freedom is the magnetic flux ϕ threading the ring. It is related to the phase difference ψ of the Cooper pair wave function[5] across the junction (cf. Subsection 4.2.10) by the relation

$$\psi = 2\pi(n - \phi/\phi_0) , \qquad (3.142)$$

where $2\pi n$ is the change of the phase of the pair wave function per cycle around the ring, and where $\phi_0 = h/2e$ is the flux quantum. The total flux ϕ through the ring consists of the externally applied flux ϕ_{ext} and the flux induced by the super-current I resulting from Cooper pair tunneling, $\phi = \phi_{\mathrm{ext}} + LI$. Here L is the self-inductance of the ring and the super-current is $I = I_c \sin\psi$. The critical super-current I_c depends on the junction properties (see Subsection 4.2.10). The relation between the coupling energy E_{J} and the critical current I_c is $E_{\mathrm{J}} = I_c\phi_0/2\pi = I_c\hbar/2e$. In the regime $E_{\mathrm{J}} \gg E_{\mathrm{C}}$, the flux ϕ is the relevant quantum degree of freedom.

According to the phenomenological resistively shunted junction (RSJ) model [109] (see Subsection 3.4.4), the Josephson junction is shunted by an Ohmic resistance R

---

[5]We repeatedly use for the phase difference across a junction the variable ψ and across a normal junction the variable φ.

and by a capacitance C, which bears charging energy $Q^2/2C$. Correspondingly, a current through the loop splits into three pieces,

$$I = I_c \sin \psi + U/R + C\dot{U} . \tag{3.143}$$

Josephson's second relation

$$\dot{\psi} = 2eU/\hbar \tag{3.144}$$

relates the change of ψ per unit time with the voltage drop U across the junction. Insertion of Eqs. (3.142) and (3.144) into Eq. (3.143) yields the deterministic equation of motion for the total flux ϕ threading the ring,

$$C\ddot{\phi} + \dot{\phi}/R + \partial V(\phi)/\partial \phi = 0 , \tag{3.145}$$

in which the potential $V(\phi)$ is a sinusoidally modulated parabola,

$$V(\phi) = \frac{1}{2L} \left(\phi - \phi_{\text{ext}} \right)^2 - E_J \cos \left(\frac{2\pi\phi}{\phi_0} \right) . \tag{3.146}$$

Apart from the missing stochastic force, the equation of motion (3.145) is homologous with the Langevin equation of a damped particle, Eq. (2.3).

If the self-inductance L is so large that $\beta_L \equiv 2\pi L I_c/\phi_0 = 4\pi^2 L E_J/\phi_0^2 > 1$, then the SQUID is hysteretic, i.e., the potential $V(\phi)$ has one or several relative minima in which the flux may be trapped. When the flux is trapped in a well, the flux state is metastable with respect to a neighboring flux state of lower energy and fluxoid macroscopic transitions to the lower state may occur. At high temperature, the transition comes about by thermal activation. At sufficiently low temperature, the transition predominantly occurs via quantum tunneling.

Josephson systems have been employed since the mid 1980s in order to study new specific quantum effects, such as macroscopic quantum tunneling (MQT) of the phase (or flux) [see Part III], as well as indirect (spectroscopic) evidence for quantum superpositions of macroscopic states [110, 111].

Flux qubit

Upon adjusting the external flux ϕ_{ext} to the value $\frac{1}{2}\phi_0$, the potential (3.146) becomes symmetric in the shifted phase variable $\tilde{\psi} = 2\pi\phi/\phi_0 - \pi$. We have

$$V(\tilde{\psi}) = \left(\frac{1}{2}\tilde{\psi}^2/\beta_L + \cos \tilde{\psi} \right) E_J , \tag{3.147}$$

and in quartic approximation

$$V_{\text{quart}}(\tilde{\psi}) = \left(1 - \frac{1}{2}(1 - 1/\beta_L)\tilde{\psi}^2 + \frac{1}{24}\tilde{\psi}^4 \right) E_J . \tag{3.148}$$

For $\beta_L > 1$, the potential $V_{\text{quart}}(\tilde{\psi})$ is a symmetric double well. The two degenerate minima are at $\tilde{\psi} = \tilde{\psi}_\pm$, where $\tilde{\psi}_\pm = \pm\sqrt{6(1 - 1/\beta_L)}$, and they correspond to the two different senses of rotation of the supercurrent in the ring. The barrier height is

$$V_b = \frac{3}{2} (1 - 1/\beta_L)^2 E_J . \tag{3.149}$$

When the flux is put out of tune, $\phi_{\text{ext}} \neq \frac{1}{2}\phi_0$, the double well becomes asymmetric. In the regime $\beta_{\text{L}} - 1 \ll 1$, the bias energy $\hbar\epsilon \equiv V_{\text{quart}}(\tilde{\psi}_+) - V_{\text{quart}}(\tilde{\psi}_-)$ depends linearly on the externally applied flux ϕ_{ext} as $\hbar\epsilon = 4\pi\sqrt{6(\beta_{\text{L}} - 1)}(\phi_{\text{ext}}/\phi_0 - \frac{1}{2})E_{\text{J}}$. In an extended setting, in which the single junction is replaced by a dc SQUID ring, the trapped flux $\tilde{\phi}_{\text{ext}}$ in the auxiliary ring serves as an independent parameter which controls the strength of the coupling energy between the two localized states.

The condition $\beta_{\text{L}} > 1$ requires a relatively large loop, which makes the system quite vulnerable by external noise. To overcome this shortcoming, one may use smaller superconducting loops with three or four junctions [112, 113].

The environment of flux-qubit systems usually consists of resistive elements in the circuits needed for manipulations and measurements. These elements produce voltage and current noise. In the usual case the noise is Gaussian. Friction and noise can be introduced in the relevant model by coupling the flux ϕ dynamically to a harmonic heat bath, as discussed above in Section 3.1. Under these conditions, the dynamics of the extended flux qubit setting is well described by the spin-boson model with tunable bias and tunable coupling energy.

Charge qubit

The simplest form of a charge qubit is a superconducting charge box populated by excess Cooper pair charges. It consists of a small superconducting island ("box") connected to a superconducting electrode by a tunnel junction with capacitance C_{J} and coupling energy E_{J} [114]. A control gate voltage V_{g} is coupled to the system via a gate capacitor C_{g}. When the gap energy $\hbar\Delta_{\text{g}}$ is the largest relevant energy of the system, quasiparticle tunneling is suppressed so that quasiparticle excitations on the island are negligible. Then the system is described by the Hamiltonian

$$H = 4E_{\text{C}}(N - N_{\text{g}})^2 - E_{\text{J}}\cos\psi \,, \tag{3.150}$$

where $E_{\text{C}} = e^2/2(C_{\text{g}} + C_{\text{J}})$ is the single-electron charging energy and N is the number operator of excess Cooper pair charges on the island, which is conjugate to the phase of the order parameter, $N = -i\partial/\partial\psi$. The gate voltage V_{g} controls the gate charge $2eN_{\text{g}}$ according to the relation $2eN_{\text{g}} = C_{\text{g}}V_{\text{g}}$. In the regime $E_{\text{C}} \gg E_{\text{J}}$, the charge on the island is almost sharp. It is then convenient to introduce a basis of charge states, parametrized by the number N of Cooper pairs on the island,

$$Q = 2e\sum_N N|N\rangle\langle N| \,, \qquad e^{i\psi} = \sum_N |N+1\rangle\langle N| \,. \tag{3.151}$$

In this basis, the Hamiltonian (3.150) reads

$$H = \sum_N \left[4E_{\text{C}}(N - N_{\text{g}})^2 |N\rangle\langle N| - \tfrac{1}{2}E_{\text{J}}\left(|N+1\rangle\langle N| + |N\rangle\langle N+1|\right)\right] \,. \tag{3.152}$$

When the gate voltage is tuned to the symmetry point, $V_{\text{g}} = e/C_{\text{g}}$, the gate charge N_{g} is $\frac{1}{2}$. Hence the states with $N = 0$ and $N = 1$ are degenerate. When the charging energy E_{C} is large enough, the charge states $N > 1$ are elevated to fairly high energy, and they can be disregarded in the thermal regime $k_B T \ll 9 E_C$. The charge box then reduces to a two-state or spin-$\frac{1}{2}$ quantum system (qubit) of the form (3.128). The charge states with $N = 0$ and $N = 1$ correspond to the localized states $|R\rangle$ and $|L\rangle$ of the TSS. The coupling of the states is described by E_J, and the bias energy is

$$\delta E_{ch}(V_g) = 4 E_C [1 - 2 N_g(V_g)] . \tag{3.153}$$

When the single junction is replaced by two identical junctions in a loop, each with coupling energy E_J^0, and when the loop is penetrated by an external flux ϕ_{ext}, the effective coupling energy of the modified device is

$$E_J(\phi_{ext}) = E_J^0 \cos(\pi \phi_{ext}/\phi_0) . \tag{3.154}$$

The SQUID-controlled charge qubit is thus a TSS with tunable coupling energy and tunable bias energy,

$$H = -\frac{1}{2} E_J(\phi_{ext}) \, \sigma_x - \frac{1}{2} \delta E_{ch}(V_g) \, \sigma_z . \tag{3.155}$$

Quantum manipulations of Josephson charge qubits described by the Hamiltonian (3.155) have been performed [115].

In charge qubits the most serious source of decoherence is the noise of the biasing voltage ("charge noise"). Fluctuations of the voltage entail fluctuations of the phase ψ, which are described by the phase autocorrelation function given below in Eq. (3.234) with (3.235). In charge-qubit experiments, the major source of decoherence was found to be low-frequency $1/f$ noise which originates from background charge fluctuators [116, 117, 118]. When the system is affected by $1/f$ noise in the σ_z or longitudinal ($\propto \cos\varphi$) coupling, this source of decoherence exceeds Ohmic noise at very low temperature. There are experimental indications for a connection between high-frequency Ohmic noise, which is responsible for relaxation, and low frequency $1/f$ noise, which accounts for dephasing. This can be explained with a distribution of coherent two-level background charges which is log-uniform in the tunnel splitting and linear in the bias, like the distribution of two-level tunneling systems in amorphous materials, which has been introduced in order to understand anomalous low-temperature properties [119, 120, 121]. A discussion of $1/f$ noise is given in Section 22.5.

The possibility of tuning the coupling energy and the biasing energy offers twofold advantage. First, one can adjust the gate voltage V_g to the degeneracy point $V_{g,0}$ at which the biasing charging energy vanishes, $\delta E_{ch}(V_{g,0}) = 0$. Secondly, one can tune the applied flux ϕ_{ext} such that the Josephson coupling energy has an extremum, $\partial E_J(\phi_{ext})/\partial \phi_{ext} = 0$. At this optimal point the qubit Hamiltonian (3.155) depends on voltage fluctuations δV and flux fluctuations $\delta\phi$ as

$$H_{opt} = -\frac{1}{2}\left(E_J(\phi_{ext,0}) + \frac{1}{2}\frac{\partial^2 E_J}{\partial \phi_{ext}^2}\Big|_{\phi_{ext,0}} \delta\phi^2 \right) \sigma_x - \frac{1}{2}\frac{\partial E_{ch}}{\partial V_g}\Big|_{V_{g,0}} \delta V \, \sigma_z . \tag{3.156}$$

Hence the energy splitting of the two states depends only quadratically on the fluctuations δV and $\delta\phi$ [122]. Operation of the charge qubit at this optimal point can increase the coherence time by 2-3 orders of magnitude as shown for a charge-phase qubit ($E_c \approx E_J$) [123]. Meanwhile several types of superconducting circuits based on Josephson junctions have given evidence that control of coherent quantum state evolution is feasible. Overall, they have proven to be promising candidates for qubit implementation [115], [123] - [126]. There has been made also substantial progress in the control of the interactions between individual flux qubits operating at the optimal point while retaining quantum coherence [127].

3.4 Microscopic models

There are numerous physical systems for which the Hamiltonian of the global system can be determined from microscopic considerations.

In unpolar crystals like semiconductors and metals, a charged particle, which may be an electron, a heavy interstitial particle, or a tunneling defect particle, distorts the lattice in its neighborhood, and when the particle moves, the lattice distortion moves with it. In the underlying microscopic description, the particle interacts with acoustic phonons. In the conceptually simplest case, the interaction is described by a scalar deformation potential. The particle together with its attached vibrating environment is called *acoustic polaron*.

An electron in an *ionic* crystal interacts with surrounding ions and thus polarizes the lattice. The interaction lowers the energy of the electron, and when the electron moves, the polarization cloud moves with it. The accompanying cloud of vibrating displacements of the ionic sub-lattices originates from the electron's coupling to longitudinal optical phonons. The electron together with its polarized environment is called *optical polaron*.

Even without lattice vibrations, there is a rigid periodic potential $H_{rig}(q)$ acting on the particle at position q. The Hamiltonian in the absence of lattice vibrations is

$$H_S = \frac{p^2}{2M} + H_{rig}(q) . \qquad (3.157)$$

One distinguishes between *small* and *large* polaron, depending on whether it is represented by a localized state in a discrete lattice (small polaron) or by an extended state in a continuous medium (large polaron).

In the *large-polaron* system, the rigid potential $H_{rig}(q)$ provides the electron with an effective mass M_{eff}. This is the usual case for ionic crystals, semiconductors and metals. For a moving electron, the lattice must adjust the local distortions. Therefore, the actual mass of the polaron is even higher than the effective mass M_{eff} for a rigid lattice. This dynamical effect is the subject of Feynman's polaron problem [4].

In the opposite *small-polaron* limit, the particle moves in a discrete tight-binding lattice. The rigid lattice potential leads to spatially localized states. The overlap of these states is represented by a transfer matrix describing transitions of the system between the localized states. The dynamics of the particle in the absence of lattice vibrations is then described by the tight-binding Hamiltonian

$$H_{\mathrm{S}} = -\frac{1}{2}\hbar\Delta \sum_n [a_{n+1}^\dagger a_n + a_n^\dagger a_{n+1}] \,, \tag{3.158}$$

where I have used the language of creation and annihilation operators. To this class belong motion of a single excess electron (or hole) in a molecular crystal [128, 129], reorientation processes of dipolar defects [130], and phonon assisted transport [131] and quantum diffusion of light interstitials and defects in crystals [132] – [134].

The particle's interaction energy due to linear lattice distortions is described by the Fröhlich interaction [135]. The Hamiltonian of the particle-phonon system reads

$$H = H_{\mathrm{S}} + \sum_{\boldsymbol{k},\lambda} \hbar\omega_{\boldsymbol{k},\lambda} b_{\boldsymbol{k},\lambda}^\dagger b_{\boldsymbol{k},\lambda} + \sum_{\boldsymbol{k},\lambda} W_{\boldsymbol{k},\lambda}\,\mathrm{e}^{i\boldsymbol{k}\cdot\boldsymbol{q}} \left(b_{\boldsymbol{k},\lambda} + b_{-\boldsymbol{k},\lambda}^\dagger \right) \,, \tag{3.159}$$

where the $b_{\boldsymbol{k},\lambda}$ and $b_{\boldsymbol{k},\lambda}^\dagger$ are the annihilation and creation operators for the normal mode with wave vector $\boldsymbol{k}$ and branch index λ. In order that the interaction is Hermitian, we need to have $W_{\boldsymbol{k},\lambda}^* = W_{-\boldsymbol{k},\lambda}$. The linear coupling to the phonon modes leads to a shift of the equilibrium position and to a shift of the potential energy, as we have already seen in Subsection 3.1.1 [cf. Eq. (3.5)],

$$\Delta V_{\mathrm{relax}} = -\sum_{\boldsymbol{k},\lambda} |W_{\boldsymbol{k},\lambda}|^2 / \hbar\omega_{\boldsymbol{k},\lambda} \,. \tag{3.160}$$

In connection with optical polarons, $\Delta V_{\mathrm{relax}}$ is called the *relaxation energy*. The relaxation energy has no effect on the dynamics of the polaron. It is a matter of convenience whether $\Delta V_{\mathrm{relax}}$ is subtracted from the Hamiltonian (3.159).

Below, the form factor $W_{\boldsymbol{k},\lambda}$ is calculated from a microscopic consideration both for acoustic and optical polarons. In many particular cases, the form factor is practically independent of the lattice coordinates. In this case, the Hamiltonian is translational invariant. The Fröhlich Hamiltonian has been of basic importance for numerous branches of solid state physics over the last five decades.

In metals, the polaron is also interacting with the conduction electrons. The electromagnetic interaction of the polaron distorts the electron gas or Fermi liquid. This results in a screening cloud of virtual electron-hole excitations around the particle which is dragged along as the particle moves. The high density of these excitations leads to various singular behaviors, as we shall see. These phenomena are known as Fermi surface effects [136].

3.4.1 Acoustic polaron: one-phonon and two-phonon coupling

The interaction of the charged particle with the lattice shifts the position of the atom n from the equilibrium value $\boldsymbol{R}^{(n)}$ to the actual position $\boldsymbol{X}^{(n)}$ by the displacement vector $\boldsymbol{u}^{(n)}$, $\boldsymbol{X}^{(n)} = \boldsymbol{R}^{(n)} + \boldsymbol{u}^{(n)}$. Since the displacement $\boldsymbol{u}^{(n)}$ is usually small, it is convenient to expand the particle-lattice potential in a power series in the displacement. We assume that the interaction potential depends only on the distance of the

particle from the host atoms, and we express this property by the short-hand notation
$H_{\text{latt}}(q - \{X\})$. We then have [137]

$$H_{\text{latt}}(q - \{X\}) \;=\; H_{\text{rig}}(q) + H_{\text{lin}}(q) + H_{\text{quadr}}(q) + \cdots , \qquad (3.161)$$

where

$$H_{\text{rig}}(q) \;=\; H_{\text{latt}}(q - \{R\}) ,$$

$$H_{\text{lin}}(q) \;=\; \sum_{n} \left(\nabla_{X^{(n)}} H_{\text{latt}}(q - \{X\}) \Big|_{u^{(n)}=0} \right) \cdot u^{(n)}(q) , \qquad (3.162)$$

$$H_{\text{quadr}}(q) \;=\; \frac{1}{2} \sum_{n,m} u^{(n)}(q) \cdot \left(\nabla_{X^{(n)}} \nabla_{X^{(m)}} H_{\text{latt}}(q - \{X\}) \Big|_{u^{(n/m)}=0} \right) \cdot u^{(m)}(q) .$$

The first term is the rigid lattice potential introduced already in Eq. (3.157). The
displacement is expressed in terms of the usual bosonic annihilation and creation
operators. We find it convenient to choose the form [138]

$$u^{(n)}(q) \;=\; i \sum_{k,\lambda} \left(\frac{\hbar}{2V\varrho\omega_{k,\lambda}} \right)^{1/2} e(k,\lambda) \exp\left(ik\cdot(q - R^{(n)}) \right) \left(b_{k,\lambda} + b^{\dagger}_{-k,\lambda} \right) , \quad (3.163)$$

where V is the volume of the lattice, ϱ is the mass density, and $e(k,\lambda)$ is the polarization vector. Hermiticity of the Hamiltonian requires that $-i\,e^{*}(k,\lambda) = i\,e(-k,\lambda)$.
We choose the polarization vectors $e(k,\lambda)$ real and to change sign with k direction,

$$e(-k,\lambda) \;=\; -e(k,\lambda) . \qquad (3.164)$$

For notational clearness, we use in the sequel the short form

$$k := (k,\lambda) \qquad \text{and} \qquad -k := (-k,\lambda) . \qquad (3.165)$$

Next we define the scalar coupling functions

$$\kappa_{k}^{(n)}(q) \;=\; -i\nabla_{X^{(n)}} H_{\text{latt}}(q - \{X\}) \Big|_{u^{(n)}=0} \cdot e(k) \;\equiv\; g^{(n)}(q)\cdot e(k) , \quad (3.166)$$

$$\gamma_{k,k'}^{(n,m)}(q) \;=\; -e(k) \cdot \left(\nabla_{X^{(n)}} \nabla_{X^{(m)}} H_{\text{latt}}(q - \{X\}) \Big|_{u^{(n/m)}=0} \right) \cdot e(k') . \quad (3.167)$$

The interaction which is linear in the lattice diplacement reads

$$H_{\text{lin}}(q) \;=\; -\left(\frac{\hbar}{2V\varrho} \right)^{1/2} \sum_{n} \sum_{k} \frac{\kappa_{k}^{(n)}(q)}{\sqrt{\omega_k}} \left[b_k + b^{\dagger}_{-k} \right] \exp\left(ik\cdot(q - R^{(n)}) \right) , \quad (3.168)$$

and the interaction which is quadratic in the displacement is given by

$$H_{\text{quadr}}(q) \;=\; \frac{\hbar}{4V\varrho} \sum_{n,m} \sum_{k,k'} \frac{\gamma_{k,k'}^{(n,m)}(q)}{\sqrt{\omega_k \omega_{k'}}} \left[b_k + b^{\dagger}_{-k} \right] \left[b_{k'} + b^{\dagger}_{-k'} \right]$$

$$\times \exp\left(ik\cdot(q - R^{(n)}) \right) \exp\left(ik'\cdot(q - R^{(m)}) \right) . \qquad (3.169)$$

The elementary process of the interaction $H_{\text{lin}}(q)$ is absorption or emission of a single acoustic phonon. In the nth order of the perturbation series in $H_{\text{lin}}(q)$, the polaron absorbs and emits altogether n *uncorrelated* phonons. In contrast, the elementary process of the interaction $H_{\text{quadr}}(q)$, which is of second order in the lattice displacement, describes simultaneous absorption and emission of two phonons. In different terms, the contribution which is quadratic in $H_{\text{lin}}(q)$ is a single-phonon process of second order while the first order in $H_{\text{quadr}}(q)$ describes a genuine two-phonon process. We shall see in Section 4.2 that the effects of acoustic phonons on the tunneling of atoms between surfaces, as occurs in a scanning-tunneling microscope, are qualitatively different from atom tunneling in the bulk. The first case implies Ohmic dissipation while dissipation in the second case belongs to the super-Ohmic variety. The correlated two-phonon contribution (3.169) is disregarded in most studies because it is usually small compared to the contributions of $H_{\text{lin}}(q)$. Interestingly, as we shall see in Subsection 4.2.5, the two-phonon coupling $H_{\text{quadr}}(q)$ brings about Ohmic dissipation for tunneling in the bulk. However, the viscosity coefficient turns out to be strongly temperature dependent.

3.4.2 Optical polaron

In *polar* crystals, such as sodium chloride, the ions of positive charge oscillate in anti-phase with the ions of negative charge. Due to the relative displacement of the ionic sub-lattices, polarization charges are set up. The latter induce a polarization field which scatters the electrons. The displacement mode which causes strong polarization is the *longitudinal optical* (LO) mode [$\lambda = \ell$ in Eq. (3.159)]. The relevant polarization vector is a unit vector in k direction,

$$e(k, \ell) = \hat{k} . \tag{3.170}$$

Next we assume that the LO mode is independent of the positions of the individual ions and that the frequency of the optical phonon is constant, $\omega_{k,\ell} = \omega_{\text{LO}}$. The relative displacement of the ionic sub-lattices by the longitudinal optical mode is then obtained from Eq. (3.163) as

$$u(q) = i \left(\frac{\hbar}{2V \varrho \, \omega_{\text{LO}}} \right)^{1/2} \sum_k \hat{k} \, e^{ik \cdot q} \left(b_k + b^\dagger_{-k} \right) . \tag{3.171}$$

The optical mode induces a polarization field $P(q)$ which is proportional to the displacement $u(q)$. We may write

$$P(q) = U u(q) . \tag{3.172}$$

The constant U has dimension charge density and is to be determined yet. The polarization charge density

$$\rho_{\text{pol}}(q) = -\nabla \cdot P(q) \tag{3.173}$$

is a source for an electrical field which itself is a gradient field of a scattering potential for the electron. Thus, the scattering potential $V_{\text{scatt}}(\boldsymbol{q})$ satisfies the Poisson equation

$$\nabla^2 V_{\text{scatt}}(\boldsymbol{q}) = 4\pi e \rho_{\text{pol}}(\boldsymbol{q}) = -4\pi e U \nabla \cdot \boldsymbol{u}(\boldsymbol{q}) , \qquad (3.174)$$

where the electron has charge $-e$. The solution of Eq. (3.174) reads

$$V_{\text{scatt}}(\boldsymbol{q}) = -4\pi e U \left(\frac{\hbar}{2V \varrho \omega_{\text{LO}}}\right)^{1/2} \sum_{\boldsymbol{k}} \frac{e^{i \boldsymbol{k} \cdot \boldsymbol{q}}}{|\boldsymbol{k}|} \left(b_{\boldsymbol{k}} + b^{\dagger}_{-\boldsymbol{k}}\right) . \qquad (3.175)$$

The interaction of the electron with the modes of the polar crystal, Eq. (3.175), is exactly of the Fröhlich form (3.159) [4, 135, 138]. Upon comparing Eq. (3.175) with the coupling term in Eq. (3.159), the form factor of the longitudinal optical polaron $W_{\boldsymbol{k},\ell}$ reads

$$W_{\boldsymbol{k},\ell} = -4\pi e U \left(\frac{\hbar}{2V \varrho \omega_{\text{LO}}}\right)^{1/2} \frac{1}{|\boldsymbol{k}|} . \qquad (3.176)$$

Finally, the charge density parameter U is found from the consideration of the interaction potential between two electrons at fixed distance $|\boldsymbol{q}|$. The interaction energy originates from the relaxation of the ionic environment in which the charges are embedded. For a finite distance $\boldsymbol{q}$ of the two charges, we find from Eq. (3.160) the relaxation contribution

$$V_{\text{relax}}(\boldsymbol{q}) = -2 \sum_{\boldsymbol{k}} \frac{|W_{\boldsymbol{k},\ell}|^2}{\hbar \omega_{\text{LO}}} e^{i \boldsymbol{k} \cdot \boldsymbol{q}} . \qquad (3.177)$$

With insertion of Eq. (3.176) we obtain

$$V_{\text{relax}}(\boldsymbol{q}) = -\frac{(4\pi e U)^2}{\varrho \omega_{\text{LO}}^2} \int \frac{d^3 k}{(2\pi)^3} \frac{e^{i \boldsymbol{k} \cdot \boldsymbol{q}}}{|\boldsymbol{k}|^2} , \qquad (3.178)$$

which in real space is the Coulomb potential

$$V_{\text{relax}}(\boldsymbol{q}) = -\gamma \frac{e^2}{|\boldsymbol{q}|} \qquad \text{with} \qquad \gamma = \frac{4\pi U^2}{\varrho \omega_{\text{LO}}^2} . \qquad (3.179)$$

The relaxation potential $V_{\text{relax}}(\boldsymbol{q})$ represents the contribution of the optical phonons to the dielectric screening of the Coulomb potential. This is the difference in screening between the cases of low and high frequencies. Thus we have

$$\frac{e^2}{\epsilon_0 |\boldsymbol{q}|} = \frac{e^2}{|\boldsymbol{q}|} \left(\frac{1}{\epsilon_\infty} - \gamma\right) \qquad \text{with} \qquad \gamma = \frac{1}{\epsilon_\infty} - \frac{1}{\epsilon_0} . \qquad (3.180)$$

Here ϵ_∞ is the high-frequency (optical) and ϵ_0 the static dielectric constant. The charge density U is gathered from Eqs. (3.179) and (3.180) as

$$U^2 = \frac{\varrho \omega_{\text{LO}}^2}{4\pi} \left(\frac{1}{\epsilon_\infty} - \frac{1}{\epsilon_0}\right) . \qquad (3.181)$$

Thus, we have found that U is given entirely in terms of measurable quantities. In the literature, it is customary to use the dimensionless polaron coupling constant

$$\alpha \equiv \left(\frac{1}{\epsilon_\infty} - \frac{1}{\epsilon_0}\right) \frac{e^2}{\hbar\omega_{LO}} \left(\frac{M\omega_{LO}}{2\hbar}\right)^{1/2}. \tag{3.182}$$

Expressed in terms of the parameter α, the form factor for optical polarons (3.176) takes the form

$$W_{\bm{k},\ell} = -\left(\frac{4\pi\alpha}{V}\right)^{1/2} \left(\frac{\hbar}{2M\omega_{LO}}\right)^{1/4} \frac{\hbar\omega_{LO}}{|\bm{k}|}. \tag{3.183}$$

With the coupling parameter α given by Eq. (3.182), the self-energy of the polaron can be put in a simple form [4]. Typical values of α run from about 1 to 20.

3.4.3   Interaction with fermions (normal and superconducting)

As an implication of the Pauli principle, the response of the noninteracting electron gas to a time-dependent local potential shows interesting characteristic features. The transient perturbations of the Fermi gas or Fermi liquid lead to a screening cloud of virtual electron-hole excitations which turns out to be very singular in character. The singular behavior can be seen, o. g., in the energy dependence of soft X-ray absorption or emission of metals (see, e.g., the review by Mahan [139], and by Ohtaka and Tanabe [140]). The singular response of the electron gas is also reflected in the electrical resistivity when dilute magnetic impurities are dissolved in a non-magnetic metallic host crystal. The understanding of these peculiar properties constitutes the crux of the Kondo problem (cf. the review by Tsvelik and Wiegmann [141]). Closely related to the above phenomena, quantum diffusion of charged interstitials in normally conducting metals also exhibits singular behavior, as shown first by Kondo.

The singular transient response of the conduction electrons results from the high density of electron-hole excitations around the Fermi surface.[6] When the particle moves, it drags behind it a screening cloud of electron-hole pairs. These excitations have bosonic character. As we shall see, they can adequately be represented by an Ohmic spectral density of a virtual coupling to a bosonic reservoir. This particular coupling leads, e.g., to self-trapping at zero temperature above a critical value of the coupling strength [142] – [147], and to anomalous temperature dependence such as the increase of electron transfer rates and of the diffusion coefficient with decreasing temperature [136]). This issue is studied in Subsections 20.2.5 and 25.5.1. Ohmic friction fades away when the fermionic reservoir changes from the normally conducting to the superconducting state. We shall discuss the respective properties of the coupling in some detail in Subsection 4.2.8.

In this subsection I introduce the underlying Hamiltonians for the normal and the superconducting state of the fermions. I postpone the derivation of the effective

---

[6]The Fermi surface effects are reviewed by Kondo in Ref. [136].

action of the particle until Subsection 4.2.8. The description of the fermionic effects
on the charged interstitial is based on the Hamiltonian

$$H = H_S(\boldsymbol{q}) + H_R + H_I(\boldsymbol{q}) \, , \tag{3.184}$$

where H_S is the Hamiltonian of the system in the absence of the environment, and
H_R and H_I are the terms due to the reservoir and the interaction, respectively. It is
convenient to formulate H_R and H_I in the scheme of second quantization.

For fermions in the *normally conducting* state, the reservoir Hamiltonian has the
standard free fermion form

$$H_R^{(\mathrm{nc})} = \sum_{k,\sigma} \hbar\omega_k \, c_{k\sigma}^\dagger c_{k\sigma} \, , \tag{3.185}$$

where $c_{k\sigma}^\dagger$ is the creation operator for a conduction electron with wave vector $\boldsymbol{k}$ and
spin polarization σ. The energy $\hbar\omega_k$ is measured relative to the Fermi energy E_F.

The charged interstitial couples to local fluctuations of the electronic density via
a potential $U(\boldsymbol{r})$. For a point particle at position $\boldsymbol{q}$, the interaction term reads

$$H_I^{(\mathrm{nc})}(\boldsymbol{q}) = \int d^3r \, U(\boldsymbol{r} - \boldsymbol{q}) \Psi^\dagger(\boldsymbol{r}) \Psi(\boldsymbol{r}) \, . \tag{3.186}$$

Upon using the normal mode expansion $\Psi^\dagger(\boldsymbol{r}) = \sum_{k,\sigma} \mathrm{e}^{-i \boldsymbol{k} \cdot \boldsymbol{r}} \, c_{k,\sigma}^\dagger$, we obtain

$$H_I^{(\mathrm{nc})}(\boldsymbol{q}) = \sum_{k,k',\sigma,\sigma'} < k,\sigma| U |k',\sigma' > \mathrm{e}^{i\,(k'-k)\cdot q} \, c_{k\sigma}^\dagger c_{k'\sigma'} \, . \tag{3.187}$$

The potential $U(\boldsymbol{r})$ depends on the particular case. For a charged particle in a metal,
it is the Coulomb potential, while in a superconductor, it is a screened Coulomb
potential. Other choices, such as dipolar or multipolar couplings for neutral particles,
are also pertinent.

When the fermions are in the *superconducting* state, the reservoir is made up of
BCS quasiparticles,

$$H_R^{(\mathrm{sc})} = \sum_{k,\alpha} \hbar\Omega_k \, \gamma_{k\alpha}^\dagger \gamma_{k\alpha} \, . \tag{3.188}$$

Here, $\gamma_{k\alpha}^\dagger$ is the creation operator for a quasi-particle with wave vector $\boldsymbol{k}$, spin polarization α, and excitation energy $\hbar\Omega_k$, where

$$\hbar\Omega_k = \hbar(\omega_k^2 + \Delta_g^2)^{1/2} \, , \tag{3.189}$$

and $\hbar\Delta_g$ is the temperature dependent gap energy of the superconducting state. The
operators $\gamma_{k\alpha}^\dagger$ are connected with $c_{k\sigma}$ and $c_{k\sigma}^\dagger$ by Bogoliubov relations [148]

$$\begin{aligned}
\gamma_{k\uparrow}^\dagger &= u_k c_{k\uparrow}^\dagger - v_k c_{-k\downarrow} \, , \\
\gamma_{k\downarrow}^\dagger &= u_k c_{k\downarrow}^\dagger + v_k c_{-k\uparrow} \, .
\end{aligned} \tag{3.190}$$

Since the screening of the electrons in the superconducting state is the same as in the normal state [149], the interaction is the same as in Eq. (3.187). Next I wish to express the interaction as a function of the quasi-particle operators. The inversion of Eq. (3.190) gives

$$c_{k\alpha}^\dagger = u_k \gamma_{k\alpha}^\dagger + \sum_\beta \rho_{\alpha\beta} v_k \gamma_{-k\beta} , \qquad (3.191)$$

where

$$\rho = \begin{pmatrix} 0 & -1 \\ 1 & 0 \end{pmatrix} . \qquad (3.192)$$

Insertion of Eq. (3.191) into Eq. (3.187) yields

$$
\begin{aligned}
H_I(q) = &\sum_{k,k',\alpha,\alpha'} <k,\alpha|\,U\,|k',\alpha'> e^{i(k'-k)\cdot q} \\
&\times \Bigg\{ u_k u_{k'} \gamma_{k\alpha}^\dagger \gamma_{k'\alpha'} + v_k v_{k'} \sum_{\sigma,\sigma'} \rho_{\alpha\sigma} \rho_{\alpha'\sigma'} \gamma_{-k\sigma} \gamma_{-k'\sigma'}^\dagger \\
&+ u_k v_{k'} \sum_{\sigma'} \rho_{\alpha'\sigma'} \gamma_{k\alpha}^\dagger \gamma_{-k'\sigma'}^\dagger + u_{k'} v_k \sum_\sigma \rho_{\alpha\sigma} \gamma_{-k\sigma} \gamma_{k'\alpha'} \Bigg\} .
\end{aligned}
\qquad (3.193)
$$

We are interested in the first two terms of the interaction since they describe scattering of quasiparticles. The other terms are irrelevant at low temperature because they create and annihilate two quasiparticles, respectively. Thus we have

$$H_I^{(\mathrm{sc})}(q) = \sum_{k,k',\alpha,\alpha'} M(k,\alpha|k',\alpha') e^{i(k'-k)\cdot q} \gamma_{k\alpha}^\dagger \gamma_{k'\alpha'} , \qquad (3.194)$$

where $M(k,\alpha|k',\alpha')$ is determined by the first two terms in Eq. (3.193). Upon using $\gamma_i \gamma_j^\dagger = -\gamma_j^\dagger \gamma_i$ for $i \neq j$, we find [148]

$$M(k,\alpha|k',\alpha') = u_k u_{k'} <k,\alpha|\,U\,|k',\alpha'> - v_k v_{k'} \sum_{\sigma,\sigma'} \rho_{\sigma'\alpha'} \rho_{\sigma\alpha} <-k',\sigma'|\,U\,|-k,\sigma> .$$

The term $\sum_{\sigma\sigma'} \rho_{\sigma'\alpha'} \rho_{\sigma\alpha} <-k',\sigma'|U|-k,\sigma>$ is essentially the matrix element of U in which the polarizations and momenta of the electrons have the opposite sign. Accordingly, the sign of this term depends on the behavior of the interaction under time-reversal. Thus we may write

$$M(k,\alpha|k',\alpha') = \left\{ u_k u_{k'} - \zeta v_k v_{k'} \right\} <k,\alpha|\,U\,|k',\alpha'> , \qquad (3.195)$$

where $\zeta = \pm 1$. The curly bracket in Eq. (3.195) is called the coherence factor of the transition. With use of the definitions

$$u_k v_k = \frac{\Delta_g}{2\Omega_k} ; \quad u_k^2 = \frac{1}{2}\left(1 + \frac{\omega_k}{\Omega_k}\right) ; \quad v_k^2 = \frac{1}{2}\left(1 - \frac{\omega_k}{\Omega_k}\right) , \qquad (3.196)$$

we find

$$|M(\boldsymbol{k},\sigma|\boldsymbol{k}',\sigma')|^2 \;=\; \frac{1}{2}\left\{1+\frac{\omega_{\boldsymbol{k}}\omega_{\boldsymbol{k}'}}{\Omega_{\boldsymbol{k}}\Omega_{\boldsymbol{k}'}}-\zeta\frac{\Delta_{\mathrm{g}}^2}{\Omega_{\boldsymbol{k}}\Omega_{\boldsymbol{k}'}}\right\}\;|<\boldsymbol{k},\sigma|U|\boldsymbol{k}',\sigma'>|^2 \;. \qquad (3.197)$$

If the interaction does not break time-reversal symmetry, which is the case of most interest, we have $\zeta = +1$. Since to each value of $\Omega_{\boldsymbol{k}}$ there belong the values $\pm\omega_{\boldsymbol{k}}$, the term $\omega_{\boldsymbol{k}}\omega_{\boldsymbol{k}'}/\Omega_{\boldsymbol{k}}\Omega_{\boldsymbol{k}'}$ cancels out when Eq. (3.197) is summed over these two values.

Our subsequent study of the dissipative influences of the fermionic environment, which is given in in Subsection 4.2.8, will be based on the above microscopic Hamiltonians. For normally conducting electrons, these are Eq. (3.184) with Eq. (3.185) and Eq. (3.187), whereas in the superconducting case the underlying model is defined by Eq. (3.184) with Eq. (3.188) and Eq. (3.194).

3.4.4 Superconducting tunnel junction

A junction consists of two superconducting electrodes which are weakly coupled through a normally conducting barrier. The phase difference of the superconducting state across the barrier represents a collective variable which determines the dynamics of the device. The electronic degrees of freedom constitute a heat reservoir which causes phase fluctuations. A well-known simple phenomenological description of the reservoir's effects is rendered by the *resistively shunted junction* (RSJ) model in which the barrier acts as an Ohmic resistor shunted in parallel to the displacement current and to the super-current [109].

Alternatively, one may start from the microscopic theory, as shown by Ambegaokar *et al.* [150], and by Larkin and Ovchinnikov [151]. We shall see that the microscopic theory leads to a model Hamiltonian equivalent to the form (3.8) with linear or nonlinear Ohmic friction, depending on the magnitude of the phase difference across the junction. The description of the bulk superconductor is based on the grand canonical Hamiltonian [152]

$$H_{\mathrm{bulk}} \;=\; \int d^3\boldsymbol{r}\,\Psi_\sigma^\dagger(\boldsymbol{r})\left[-\frac{\hbar^2}{2m}\left(\boldsymbol{\nabla}-\frac{ie}{\hbar}\boldsymbol{A}\right)^2-\mu+U(\boldsymbol{r})\right]\Psi_\sigma(\boldsymbol{r}) \qquad (3.198)$$

$$-\frac{1}{2}\int d^3\boldsymbol{r}\,\Psi_\sigma^\dagger(\boldsymbol{r})\,\Psi_{-\sigma}^\dagger(\boldsymbol{r})\,g(\boldsymbol{r})\,\Psi_{-\sigma}(\boldsymbol{r})\,\Psi_\sigma(\boldsymbol{r})\;+\;\frac{1}{8\pi}\int d^3\boldsymbol{r}\,(\boldsymbol{h}-\boldsymbol{h}_{\mathrm{ext}})^2\;.$$

Here a summation over spin polarizations is implied. The vector field $\boldsymbol{A}$ represents the vector potential, and μ is the chemical potential. The Hamiltonian describes conduction electrons in a potential $U(\boldsymbol{r})$. This may also account for impurities and boundaries. The second line describes an effective attractive BCS interaction of strength $g(\boldsymbol{r})$, and in the last line we have added the magnetic field contribution, where $\boldsymbol{h}$ and $\boldsymbol{h}_{\mathrm{ext}}$ are the magnetic field and the externally applied magnetic field, respectively.

A tunnel junction with superconducting electrodes to the left (L) and to the right (R) of the barrier is conveniently described by the Hamiltonian

$$H \;=\; H_{\mathrm{bulk,L}}+H_{\mathrm{bulk,R}}+H_{\mathrm{T}}+H_{\mathrm{Q}}\;. \qquad (3.199)$$

The coupling of the leads is due to the transfer of electrons through the barrier and due to the Coulomb interaction. The former is described by the tunneling term

$$H_{\mathrm{T}} = \int_{r \in \mathrm{R}} d^3r \int_{r' \in \mathrm{L}} d^3r' \left(T_{rr'} \, \Psi_{\mathrm{R},\sigma}^{\dagger}(r) \, \Psi_{\mathrm{L},\sigma}(r') + \text{h.c.} \right) . \tag{3.200}$$

The range of the tunneling matrix element $T_{rr'}$ is limited to the vicinity of the barrier. As long as we are interested in frequencies much smaller than the plasma frequency, the Coulomb interaction across the barrier effectively behaves as a capacitive interaction depending on the Cooper pair charges Q_{L} and Q_{R} stored on the electrodes and on the capacitor with capacitance C. The latter is determined by the geometry and by properties of the insulating barrier. Then the Coulomb term reads

$$H_{\mathrm{Q}} = (Q_{\mathrm{L}} - Q_{\mathrm{R}})^2 / 8C , \tag{3.201}$$

$$Q_{\mathrm{L(R)}} = e \int_{r \in \mathrm{ins}} d^3r \, \Psi_{\mathrm{L(R)},\sigma}^{\dagger}(r) \, \Psi_{\mathrm{L(R)},\sigma}(r) . \tag{3.202}$$

The model also describes inhomogeneous systems, such as normally conducting domains $[g(r) = 0]$ and constrictions, as well as superconducting rings with a weak link in which the electrodes L and R are joined up in a loop.

The relevant quantum statistical properties of the system can be extracted from the generating functional

$$Z(\xi) = \mathrm{tr}_{A,\Psi} \left\{ T_\tau \exp \left(-\frac{1}{\hbar} \int_0^{\hbar\beta} d\tau \left[H(\tau) - \hbar\xi(\tau)B(\tau) \right] \right) \right\} , \tag{3.203}$$

where $B(\tau)$ is the quantity of interest and $\xi(\tau)$ is the usual source term. Here, T_τ is the time ordering operator along the imaginary-time axis τ (cf. Chapter 4). The trace is taken over the fermion fields Ψ and the vector potential A. The respective imaginary-time effective action is discussed in Subsection 4.2.10.

3.5 Charging and environmental effects in tunnel junctions

With the enormous progress in micro-fabrication techniques, it has become possible to fabricate metallic tunnel junctions with capacitances C of 10^{-15} F or less. The corresponding charging energy of an electron, $E_C = e^2/2C$, is larger than the thermal energy for temperatures of 1 K or smaller. Therefore, it is nowadays possible to study charge transport through barriers in a regime where charging effects play an important role. Here we are interested in a quantum mechanical treatment of these effects.

Charging effects are also relevant in semiconductor nanostructures, e.g., quantum dots in a 2-D electron gas with typical capacitances of 10^{-15} F. In the still smaller molecular electron transfer systems, the charging energy can be so large that single-electron tunneling becomes observable even at room temperatures. On the one hand, the systems are small, on the other hand, they are still large enough that they can be connected to macroscopic current and voltage sources or probes. Just for this reason, the systems are sensitive to the electromagnetic environment.

The concepts introduced here and in Chapter 20.3 are generally important for metallic, semiconductor, and molecular systems. Here we concentrate the attention on metallic systems with a high electron density of states.

A collection of research papers in the field of single-charge tunneling has been published as a special issue of Zeitschrift für Physik B [153]. A series of nine tutorial articles by renowned experts which deal with particular aspects of the field are compiled in the book *Single Charge Tunneling* [154]. We now introduce the global system for single-electron tunneling and present various examples of electromagnetic environments. For a further study, we refer the reader to the review by Ingold and Nazarov [155].

For an ideal current-biased junction at zero temperature, a charge can only tunnel if the balance of the electrostatic charging energy before and after the charge transfer is positive,

$$\Delta E_C \equiv \frac{Q^2}{2C} - \frac{(Q - e)^2}{2C} = e[V_a - e/2C] > 0 \,. \tag{3.204}$$

This condition is satisfied if $Q > e/2$, which in turn implies that the voltage V_a across the junction must be larger than the critical voltage $V_{crit} = e/2C$. When the junction is current-biased with current I, the charge Q on the junction capacitor increases until the threshold charge $e/2$ is reached. At $Q = e/2$, the charge e tunnels through the junction thereby leaving a charge $Q = -e/2$ on the capacitor. After this a new charging cycle begins. Altogether, the junction voltage performs sawtooth single electron tunneling (SET) oscillations [156, 152] with a frequency $\nu_{SET} = I/e$. [7]

Alternatively, we may drive the current through the junction with an ideal voltage source. According to the above argument, there is no current in the regime $-e^*/2C < V_a < e^*/2C$, where e^* is the charge e in the normal state and is $2e$ in the superconducting state. This is the *Coulomb blockade* phenomenon for a single junction. For a voltage larger than $e/2C$ one finds for a normal state junction at $T = 0$ an Ohmic (linear) current-voltage characteristics which is shifted by the Coulomb gap $e/2C$, $I(V_a) = (V_a - e/2C)/R_T$ [cf. Subsection 20.3.2].

3.5.1 The global system for single electron tunneling

In reality, the single-electron device is influenced by stray capacitances, inductances, and resistances, as well as by quantum and thermal fluctuations. The simplest model in which we can study these influences on charge tunneling is a single junction in series with a general impedance $Z(\omega)$ and connected to an ideal voltage source V_x as sketched in Fig. 3.3. The impedance $Z(\omega)$ represents the frequency response of the electromagnetic environment. Generally, the dc voltage drop V_a across the junction differs from the externally applied voltage V_{ext} by the dc voltage drop at the

---

[7]By a similar argument, analogous so-called Bloch oscillations occur for a superconducting junction device with fundamental frequency $\nu_{Bloch} = I/2e$.

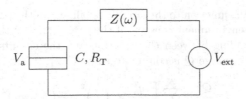

Figure 3.3: A voltage-biased tunnel junction with tunnel resistance R_T and capacitance C. The impedance $Z(\omega)$ models the frequency response of the electromagnetic environment. This is dominated by the effects of the leads attached to the junction.

Table I. Correspondence between mechanical and electrical quantities

Mechanical quantity	Electrical quantity
mass M	capacitance C
momentum p	charge Q
velocity $v = p/M$	voltage $V = Q/C$
coordinate x	phase φ
$[x, p] = i\hbar$	$[\varphi, Q] = ie$
spring constant f	inverse inductance $1/L$
harmonic oscillator	LC-circuit

impedance. Granted that we have calculated the $I(V_a)$ characteristics of the junction, the applied voltage V_{ext} can be reconstructed from the relation $V_{ext} = V_a + Z(0)\, I(V_a)$.

To derive a phenomenological "electromagnetic" Hamiltonian for the environment, we resort to the well-known correspondence between electrical and mechanical quantities assembled in Table I [155]. We introduce the phase difference at the junction

$$\varphi_V(t) = \frac{e}{\hbar} \int_{-\infty}^{t} dt'\, V(t') , \qquad (3.205)$$

where $V(t) = V_a + \delta V(t)$ is the voltage drop at the junction. The impedance $Z(\omega)$ introduces voltage fluctuations $\delta V(t)$ and charge fluctuations $\mathcal{Q}(t)$ which vanish on average. The total charge $Q(t)$ on the junction is the canonical conjugate variable to $(\hbar/e)\varphi_V(t)$. In equilibrium, the average charge on the capacitor is CV_a. Next, we introduce the fluctuations of the charge $\mathcal{Q}(t)$ and of the phase $\varphi(t)$ around the mean values determined by the voltage V_a,

$$\mathcal{Q} = Q - CV_a , \qquad \varphi(t) = \varphi_V(t) - eV_a t/\hbar , \qquad \text{with} \qquad [\varphi, \mathcal{Q}] = i\,e . \quad (3.206)$$

In a phenomenological modeling analogous to a harmonic oscillator bath, the electromagnetic reservoir is formed by LC-circuits with eigenfrequencies $\omega_\alpha = 1/\sqrt{L_\alpha C_\alpha}$, and the coupling of the junction to the circuits is bilinear in the phase variables. Thus, the electromagnetic environment consists of the capacitance of the junction, the LC-circuits, and the coupling between the junction and the circuits. It is described by the Hamiltonian [the charge Q is conjugate to $(\hbar/e)\varphi$]

$$H_{\text{env}} \;=\; \frac{Q^2}{2C} + \sum_{\alpha=1}^{N} \left[\frac{q_\alpha^2}{2C_\alpha} + \left(\frac{\hbar}{e}\right)^2 \frac{1}{2L_\alpha}(\varphi - \varphi_\alpha)^2\right]. \tag{3.207}$$

This model is of the type (3.11) with particular choice $c_\alpha = m_\alpha \omega_\alpha^2 = 1/L_\alpha$. With use of the relations $Q = (\hbar/e)C\dot\varphi$ and $q_\alpha = (\hbar/e)C_\alpha\dot\varphi_\alpha$, the equations of motion read

$$C\ddot\varphi + \sum_\alpha \frac{1}{L_\alpha}\varphi \;=\; \sum_\alpha \frac{1}{L_\alpha}\varphi_\alpha, \tag{3.208}$$

$$C_\alpha \ddot\varphi_\alpha + C_\alpha \omega_\alpha^2 \varphi_\alpha \;=\; \frac{1}{L_\alpha}\varphi, \tag{3.209}$$

Next we determine the reduced dynamics analogously to the treatment given in Subsection 3.1.2. This yields the Langevin equation for $\varphi(t)$ in the form of the equation of motion of a Brownian particle with frequency-dependent damping, Eq. (3.46),

$$C\ddot\varphi(t) + \int_0^t dt'\, Y(t-t')\dot\varphi(t') \;=\; \frac{e}{\hbar} I_{\text{noise}}(t). \tag{3.210}$$

The damping kernel is given by

$$Y(t) \;=\; \sum_\alpha \frac{1}{L_\alpha}\cos(\omega_\alpha t), \qquad \tilde Y^*(\omega) \;=\; \lim_{\varepsilon\to 0^+} \sum_\alpha \frac{1}{L_\alpha} \frac{-i\,\omega}{\omega_\alpha^2 - (\omega + i\,\varepsilon)^2}. \tag{3.211}$$

The Fourier transform of $Y(t)$ is the admittance $\tilde Y^*(\omega)$, and $\tilde Z^*(\omega) = 1/\tilde Y^*(\omega)$ is the impedance function.[8] The quantum mechanical noise current $I_{\text{noise}}(t)$ obeys stationary Gaussian statistics with equilibrium averages analogous to (2.33). The correspondence relation is $M\tilde\gamma'(\omega) \leftrightarrow \tilde Y'(\omega)$ with $\tilde Y'(\omega) = \operatorname{Re}\tilde Y(\omega)$. It is expedient to pass on by integration from the current-current to the charge-charge correlation function $Q_Q(t) = \langle[Q(0) - Q(t)]Q(0)\rangle_\beta$. We then get

$$Q_Q(t) \;=\; \int_0^\infty d\omega\, \frac{G_Q(\omega)}{\omega^2}\Big(\coth(\tfrac{1}{2}\beta\hbar\omega)[\,1 - \cos(\omega t)\,] + i\sin(\omega t)\Big) \tag{3.212}$$

with the spectral density

$$G_Q(\omega) \;=\; 2\left(\frac{e}{2\pi}\right)^2 R_{\text{K}} \tilde Y'(\omega)\,\omega. \tag{3.213}$$

It is also interesting to consider the phase displacement correlation function $Q_\varphi(t) = \langle[\,\varphi(0) - \varphi(t)\,]\varphi(0)\rangle_\beta$. In analogy to the displacement correlation function

---

[8]We use the quantum mechanical convention $e^{-i\omega t}$ for the kernels, whereas impedances are expressed in terms of the engineering convention $e^{+i\omega t}$. This is the reason for complex conjugation.

(3.52), the function $Q_\varphi(t)$ is governed by the absorptive part of the dynamical susceptibility, which describes the response of the phase to the current $I_{\text{noise}}(t)$. Fourier-transformation of Eq. (3.210) yields $\tilde{\varphi}(\omega) = \tilde{\chi}(\omega)\, e\tilde{I}_{\text{noise}}(\omega)/\hbar$, where

$$\tilde{\chi}(\omega) = \frac{1}{-\omega^2 C - i\,\omega\tilde{Y}^*(\omega)} = \frac{\tilde{Z}_t^*(\omega)}{-i\,\omega}\,, \qquad \tilde{Z}_t(\omega) = \frac{1}{i\,\omega C + \tilde{Z}^{-1}(\omega)}\,. \tag{3.214}$$

Hence the total impedance of the circuit seen by the junction consists of the external impedance $\tilde{Z}(\omega)$ in parallel with the capacitance C of the junction. The resulting expression is

$$Q_\varphi(t) = \int_0^\infty d\omega\, \frac{G_\varphi(\omega)}{\omega^2}\Big(\coth(\tfrac{1}{2}\beta\hbar\omega)[\,1 - \cos(\omega t)\,] + i\sin(\omega t)\Big) \tag{3.215}$$

with the spectral density

$$G_\varphi(\omega) = (e^2/\pi\hbar)\,\omega^2\tilde{\chi}''(\omega) = 2\omega\,\tilde{Z}_t'(\omega)/R_K = 2R_K\bar{Y}'(\omega)\,\omega\,, \tag{3.216}$$

where $R_K = 2\pi\hbar/e^2$ is the resistance quantum. In the third form, we have introduced for later convenience the admittance $\bar{Y}(\omega) = \tilde{Z}_t(\omega)/R_K^2$ [cf. Subsection 26.3]. The spectral coupling $G_\varphi(\omega)$ directly corresponds to the spectral density $G(\omega)$ introduced in Eq. (3.140). This will be advantageous in the study of single-charge tunneling, which is given in Section 20.3.

An alternative approach is presented in Subsection 4.2.11. There we calculate the imaginary-time correlation function $W_\varphi(\tau) = Q_\varphi(t = -i\tau)$ upon performing average of the phase correlator $e^{i[\varphi(0)-\varphi(\tau)]}$ with the Gaussian weight $\exp[-S_{\text{eff},0}^{(E)}(\varphi)/\hbar]$, where $S_{\text{eff},0}^{(E)}(\varphi)$ is the effective Euclidean action of a LC-circuit environment.

Apparently, any particular harmonic electromagnetic environment can be modeled by a suitable choice of the parameters L_α and C_α in the expression (3.207). Eventually, the sum in Eq. (3.211) turns into an integral over a continuous distribution, as discussed in Subsection 3.1.3.

So far we have treated the junction merely as a capacitor. Now we admit tunneling of charge through the junction by adding the tunneling Hamiltonian

$$H_T' = \sum_{k,k',\sigma} T_{k,k'} c_{k',\sigma}^\dagger c_{k,\sigma}\, e^{-i\varphi_V} + \text{h.c.}\,. \tag{3.217}$$

The operators $c_{k,\sigma}$ and $c_{k,\sigma}^\dagger$ are annihilation and creation operators for quasiparticles with wave vector k, energy E_k, and spin polarization σ. The wave vectors k and k' correspond to quasiparticles in the left and right electrode of the tunnel barrier, respectively. Finally, the model is completed by a Hamiltonian describing the quasiparticles in the left and right electrode, respectively,

$$H_{\text{qp}}' = \sum_{k,\sigma} E_k c_{k,\sigma}^\dagger c_{k,\sigma} + \sum_{k',\sigma} E_{k'} c_{k',\sigma}^\dagger c_{k',\sigma}\,. \tag{3.218}$$

In the next step we replace in the tunneling term the phase φ_V by the phase φ. This can be achieved by a time-dependent transformation with the unitary operator acting on quasiparticle operators in the left electrode

$$U = \prod_{k,\sigma} \exp\left(i \frac{e}{\hbar} V_a t\, c^\dagger_{k,\sigma} c_{k,\sigma} \right) . \tag{3.219}$$

We have

$$H_T = U^\dagger H'_T U = \sum_{k,k',\sigma} T_{k,k'} c^\dagger_{k',\sigma} c_{k,\sigma}\, \mathrm{e}^{-i\varphi} + \text{h.c.} , \tag{3.220}$$

where φ is given in Eq. (3.206). This transformation has no effect on H_{env}, whereas it shifts the quasiparticle energies in the left electrode by eV_a,

$$H_{\mathrm{qp}} = U^\dagger H'_{\mathrm{qp}} U - i\hbar U^\dagger \partial U/\partial t = \sum_{k,\sigma}[E_k + eV_a] c^\dagger_{k,\sigma} c_{k,\sigma} + \sum_{k',\sigma} E_{k'} c^\dagger_{k',\sigma} c_{k',\sigma} . \tag{3.221}$$

The Hamiltonian of the global system takes the form

$$H = H_T + H_{\mathrm{qp}} + H_{\mathrm{env}} , \tag{3.222}$$

in which the tunneling Hamiltonian (3.220) couples the quasiparticle Hamiltonian (3.221) to the reservoir part (3.207).

For weak tunneling, the charge on the junction is nearly sharp. It is then natural to choose as basis the discrete charge representation

$$Q = e \sum_N N |N\rangle\langle N| , \qquad \mathrm{e}^{i\varphi} = \sum_N |N+1\rangle\langle N| . \tag{3.223}$$

In this limit, the phase fluctuations given by the correlation function (3.215) are large. In Section 20.3, single-electron tunneling is considered to lowest order in the tunneling coupling. There we shall see that the energy-dependence of the Golden Rule tunneling rate is governed by the probability density $P(E)$ for exchange of energy E with the environment. The function $P(E)$ will turn out as the Fourier transform of $\mathrm{e}^{-Q_\varphi(t)}$, where $Q_\varphi(t)$ is the phase correlation function given in Eq. (3.215).

On the other hand, for strong tunneling, the jump of the phase at the junction is nearly sharp. Then, it is favorable to switch over to the discrete phase representation

$$\varphi = 2\pi \sum_n n |n\rangle\langle n| , \qquad \mathrm{e}^{i2\pi Q/e} = \sum_n |n\rangle\langle n+1| . \tag{3.224}$$

In this representation, the probability for exchange of energy with the environment will turn out as the Fourier transform of the function $\mathrm{e}^{-Q_Q(t)}$.

These two different representation are dual to each other. We shall discuss the charge-phase duality in the context of Cooper pair tunneling in Subsection 26.3.

The full weak- and strong-tunneling series expansions of charge transport across a weak link and a weak constriction, respectively, will be discussed in Section 28.1.

3.5.2 Resistor, inductor, and transmission lines

A significant case is when the impedance of the electromagnetic environment may be represented by an ideal Ohmic resistor, $Z(\omega) = R$. Then we obtain from Eqs. (3.214) and (3.216)

$$G_\varphi(\omega) = \frac{2\alpha\,\omega}{1 + (\omega/\omega_R)^2}\,. \tag{3.225}$$

Here we have introduced a dimensionless resistance α and a cutoff frequency ω_R,

$$\alpha = R/R_K = R\,e^2/2\pi\hbar\,, \qquad \omega_R = 1/RC = E_C/\pi\alpha\hbar\,, \tag{3.226}$$

and $E_C = e^2/2C$ is the charging energy. The spectral density is Ohmic at low frequencies, and the junction capacitance provides a Drude form with cutoff ω_R.

Consider next the case where the impedance $Z(\omega)$ is inductive, $Z(\omega) = i\omega L$. Now, the total impedance is given by $Z_t(\omega) = (i\omega/C)/[\omega_L^2 - (\omega - i0^+)^2]$. Here, $\omega_L = 1/\sqrt{LC}$ is the resonance frequency of the LC circuit. Thus, the junction is effectively coupled to a single environmental mode,

$$G_\varphi(\omega) = (E_C/\hbar)\,\omega_L\,\delta(\omega - \omega_L)\,. \tag{3.227}$$

When a resistance is put in series with the inductance, $Z(\omega) = R + i\omega L$, the resonance at frequency ω_L is broadened and the quality factor $Q_{qual} \equiv \omega_R/\omega_L$ becomes finite. Upon employing Eqs. (3.214) and (3.216), we get the spectral density

$$G_\varphi(\omega) = \frac{2\alpha\,\omega}{[1 - (\omega/\omega_R)^2 Q_{qual}^2]^2 + (\omega/\omega_R)^2}\,. \tag{3.228}$$

This form comprises the Ohmic case (3.225) in the limit $Q_{qual} \to 0$, and the single mode case (3.227) in the limit $Q_{qual} \to \infty$. Thus, upon tuning the quality factor we can vary the characteristics of the environment between resonant and resistive.

So far, we have considered impedances which are built by one or two lumped circuit elements. A more realistic model, which describes e.g. the coaxial leads attached to a junction in a real experiment, is a resistive transmission line model with distributed resistors, inductors, and capacitors. A resistive transmission line model for $Z(\omega)$ can be thought of as a ladder of discrete building blocks, in which the inductor and the resistor are in series along one stringboard of the ladder, and the capacitor is on the rung. Putting the resistance, inductance, and capacitance per building block as R_0, L_0, and C_0, one finds for an infinitely long transmission line the impedance [155]

$$Z(\omega) = \sqrt{\frac{R_0 + i\omega L_0}{i\omega C_0}}\,. \tag{3.229}$$

A LC transmission line is characterized by a resistive impedance $Z_{LC}(0) \approx \sqrt{L_0/C_0}$. This form results from Eq. (3.229) by disregarding R_0. The corresponding spectral density $G_\varphi(\omega)$ is of the Ohmic form (3.225) in which R is replaced with $\sqrt{L_0/C_0}$.

On the other hand, a RC transmission line, which corresponds to the limit $L_0 \to 0$ in Eq. (3.229), has impedance $\propto \omega^{-1/2}$. Insertion of $Z_{\mathrm{RC}}(\omega) = \sqrt{R_0/(i\,\omega C_0)}$ into Eq. (3.214), puts the spectral density (3.216) into the form

$$G_\varphi(\omega) = \frac{2\delta\,\omega_{\mathrm{ph}}^{1/2}\,\omega^{1/2}}{1 + (2\omega/\omega_{\mathrm{c}})^{1/2} + \omega/\omega_{\mathrm{c}}}\,, \tag{3.230}$$

where

$$\delta = \frac{R_0}{R_{\mathrm{K}}}\,, \qquad \omega_{\mathrm{ph}} = \frac{1}{2R_0 C_0}\,, \qquad \omega_{\mathrm{c}} = \frac{C_0}{C^2 R_0}\,. \tag{3.231}$$

Thus in the low-frequency range $\omega \ll \omega_{\mathrm{c}}$, the spectral density modeling the coupling to a RC transmission line is sub-Ohmic with the power $s = \frac{1}{2}$. The corresponding spectral density does not depend on properties of the junction.

I shall study the effect of these diverse electromagnetic environments on single-electron tunneling in Section 20.3.

3.5.3 Charging effects in junctions

There are two kinds of charge carriers traversing the barrier in tunnel junctions: Cooper pairs and fermionic quasiparticles.[9] At temperature well below the critical temperature of the superconductor and applied voltage well below the gap voltage, the tunneling of quasiparticles may be disregarded. Tunneling of Cooper pairs through the barrier gives rise to the potential [cf. Eq. (3.146)]

$$H_{\mathrm{J}} = -E_{\mathrm{J}} \cos\psi\,. \tag{3.232}$$

Here, ψ is the phase leap of the Cooper pair wave function across the junction. The coupling energy depends linearly on the critical current, $E_{\mathrm{J}} = (\hbar/2e)I_{\mathrm{c}}$. The operator relation $e^{i\psi}Q\,e^{-i\psi} = Q - 2e$ points up that tunneling of a Cooper pair changes the charge on the junction by $2e$. Furthermore, we assume that the junction has capacitance C and that it is coupled to an ideal voltage source V_{x} through a general impedance $Z(\omega)$, as sketched in Fig. 3.3.

As the voltage drop V_{R} at the resistor is $V_{\mathrm{R}} = V_{\mathrm{x}} - V_{\mathrm{a}} = V_{\mathrm{x}} - \hbar\dot\psi/2e$ [cf. Eq. (3.144)], we may associate with V_{R} the phase $\psi_{\mathrm{R}} = 2eV_{\mathrm{x}}t/\hbar - \psi$. Hence the junction's coupling to the electromagnetic environment is in the form (3.206) with (3.207) with substitutions $e \to 2e$ and $V_{\mathrm{a}} \to V_{\mathrm{x}}$. As a result, the Hamiltonian of the junction-plus-environment system is found to read

$$H = \frac{Q^2}{2C} + \sum_{\alpha=1}^{N}\left[\frac{q_\alpha^2}{2C_\alpha} + \left(\frac{\hbar}{2e}\right)^2\frac{1}{2L_\alpha}\left(\psi - 2eV_{\mathrm{x}}t/\hbar - \psi_\alpha\right)^2\right] - E_{\mathrm{J}}\cos\psi\,. \tag{3.233}$$

The charge q_α is conjugate to $(\hbar/2e)\psi_\alpha$, and the charging energy for Cooper pairs is $E_{\mathrm{C}} = 2e^2/C$.

---

[9]The effective action governing the equilibrium properties of the junction including quasiparticle and Cooper pair tunneling is considered below in Subsection 4.2.10.

In the regime $E_J \ll E_C$, the charge on the junction is quite sharp. Hence the charge representation (3.151) is appropriate, and the coupling term may be treated as a perturbation. The environmental coupling leads to phase fluctuations. The phase autocorrelation function $Q_\psi(t) = \langle [\,\psi(0) - \psi(t)\,]\psi(0)\rangle_\beta$ has the spectral representation

$$Q_\psi(t) = \int_0^\infty d\omega \, \frac{G_\psi(\omega)}{\omega^2} \Big(\coth(\tfrac{1}{2}\beta\hbar\omega)[\,1 - \cos(\omega t)\,] + i\sin(\omega t) \Big) \qquad (3.234)$$

with the spectral density of the coupling

$$G_\psi(\omega) = 2\omega \, Z_t'(\omega)/R_Q . \qquad (3.235)$$

Here, $Z_t(\omega)$ is the total impedance of the circuit. It is specified in Eq. (3.214). The reference value is the resistance quantum for Cooper pairs, $R_Q = R_K/4 = 2\pi\hbar/4e^2$. Observe the close resemblance with the corresponding expression for a normal junction, Eq.(3.215) with (3.216). The case of weak coupling is studied below in Subsection 20.3.4.

In the opposite limit of large Josephson coupling, $E_J \gg E_C$, the phase is rather sharp, and the environmental influences produce charge fluctuations. We shall discuss the limits of weak and strong coupling in some more detail below in Subsection 26.3. There I point out a striking duality symmetry between charge and phase representation of the device.

3.6 Nonlinear quantum environments

Generally, the low-energy physics of a complex system which is coupled to complex surroundings can be studied upon using an effective Hamiltonian. The reduction usually leads to one of only few canonical forms. For instance, a double well may be reduced to a two-state system, and a reservoir mode with a whole ladder of roughly equidistant excited states may be represented by a harmonic oscillator. The opposite extreme case is a nonlinear bath mode with only one or few accessible excited states. This mode behaves like a spin. Evidently, a distribution of such modes forms a bath of spins. As opposed to a harmonic bath, a spin bath features saturation at elevated temperature. Every nonlinear bath is situated between these two idealized cases.

Consider a global Hamiltonian analogous to Eq. (3.1) consisting of a system part, a nonlinear reservoir part, and an interaction part with a counter term,

$$H = \frac{p^2}{2M} + V(q) + H_R(\{\boldsymbol{x}\}) - F(q) \sum_\alpha c_\alpha x_\alpha + \frac{1}{2} F^2(q) \sum_\alpha c_\alpha^2 \tilde{\chi}_\alpha(0) . \qquad (3.236)$$

The counter-term is defined in terms of the static susceptibility $\tilde{\chi}_\alpha(\omega = 0)$ of the nonlinear mode x_α. It would reduce to the previous form (3.6) with (3.7) for a linear bath. It is again easy to write down the equations of motion and to eliminate the reservoir coordinates. This is most conveniently performed in frequency space upon introducing the dynamical susceptibility of the reservoir mode x_α. We then find an equation of motion of the form (3.19) in which the Fourier transform of the damping kernel has the form (3.29). The corresponding expression for the spectral density of the coupling is as in Eq. (3.30),

$$J(\omega) = \Theta(\omega) \sum_\alpha c_\alpha^2 \, \tilde{\chi}_\alpha''(\omega) \,. \tag{3.237}$$

The derivation reveals that the form (3.237) is generally valid for any global system in which the coupling is *linear* in the bath coordinate. For a nonlinear bath, the dynamical susceptibility may drastically differ from the harmonic form (3.16). In the general case, it depends on $\hbar$ and on temperature. Hence the spectral density of the coupling for a nonlinear environment is temperature-dependent as well.

A "nanomagnet", i.e. a mono-domain particle, may contain as many as 10^8 magnetically ordered spins. Such a giant spin, denoted by $\boldsymbol{S}$, is usually coupled to surrounding nuclear spins and to paramagnetic impurity spins. The global Hamiltonian for a central spin bi-linearly coupled to a spin bath in a magnetic field and with internal interactions has the canonical form

$$H(\boldsymbol{S}, \{\sigma^\alpha\}) = H_{\mathrm{S}} + \frac{1}{S} \sum_\alpha E_\alpha \boldsymbol{S} \cdot \sigma^\alpha + \sum_\alpha h^\alpha \cdot \sigma^\alpha + \tfrac{1}{2} \sum_{\alpha,\alpha';i,j} V_{ij}^{\alpha,\alpha'} \sigma_i^\alpha \sigma_j^{\alpha'} \,.$$

Many different forms for the giant spin Hamiltonian H_{S} have been discussed in the literature (cf. the book [157] and the review [158], and references therein). A simple example of the giant spin Hamiltonian H_{S} is the biaxial (easy axis/easy plane) form

$$H_{\mathrm{S}} = \left(- K_\parallel S_z^2 + K_\perp S_x^2 \right)/S + g\mu_{\mathrm{B}} H_z S_z \,, \tag{3.238}$$

in which tunneling between the classical minima at $S_z = \pm S$ is accomplished by the symmetry-breaking transverse term $K_\perp$ and we have added a magnetic field term. The low-energy physics of the Hamiltonian (3.238) can be understood in terms of the spin-$\frac{1}{2}$ Hamiltonian (3.128). The passage from the form (3.238) to the form (3.128) for $S \gg 1$ has been studied using WKB methods [160] and instanton methods [161].

Single molecule magnets like $Mn_{12}ac$ are conveniently described by similar spin Hamiltonians [162].

For giant spin systems, there are two scenarios. In the usual case, the coupling between the giant spin and the nuclear spins is very large. Then the nuclear spins are slaved to the giant spin, and the spectrum of the spin environment is drastically modified. The coupling to the nuclear spins may introduce three different effects [158]. Transitions of the giant spin may cause a change of the phase in the nuclear bath state, which then reacts back upon the giant spin and induces phase randomization ("topological decoherence"). Secondly, transitions of the giant spin may be hindered by a mismatch between the initial and final bath state ("orthogonality blocking"). This would lead to a Franck-Condon type dressing factor (cf. Subsection 18.1.4). Thirdly, the spread of coupling energies of the nuclear spins leads to a spread of bias energies which may bring the initial state of the tunneling giant spin out of resonance with the final state. This would result in a Landau-Zener type suppression of tunneling transitions ("degeneracy blocking").

The strong-coupling case contrasts with the usual weakly-coupled harmonic oscillator environment. It is impossible to integrate out the spin bath using the functional averaging method described in the next chapter, and one has to study the global system directly, e.g., the instanton trajectory in the full system-bath space [163].

In the other scenario, the coupling to an individual bath spin is weak. In this case, the treatment is like that of a boson bath. The resulting spectral density $J_{\text{spin}}(\omega)$ has the form (3.237), in which $\tilde{\chi}_\alpha(\omega)$ is the dynamical susceptibility for a spin degree. Consider as example a global system with a system part $H_S(p, q)$, a spin-$\frac{1}{2}$ bath with excitation frequencies $\{\omega_\alpha\}$, a bilinear coupling term and a counter term,

$$ H = H_S(p, q) - \tfrac{1}{2} \sum_\alpha \hbar \omega_\alpha \sigma_x^\alpha - \tfrac{1}{2} \sum_\alpha c_\alpha q_0^{(\alpha)} \sigma_z^\alpha q + \tfrac{1}{2} \sum_\alpha c_\alpha^2 \tilde{\chi}_\alpha(0) q^2 . \qquad (3.239) $$

The coupling terms in Eq. (3.11) and (3.239) correspond to each other at $T = 0$ if we identify the length $\tfrac{1}{2} q_0^{(\alpha)}$ with the position spread of the equivalent bath oscillator with eigenfrequency ω_α in the ground state, $\tfrac{1}{2} q_0^{(\alpha)} = \sqrt{\hbar/2m_\alpha \omega_\alpha}$ (cf. Subsection 6.4.3).

The spin correlation function is Re $\langle \sigma_z^\alpha(t) \sigma_z^\alpha(0) \rangle = \cos(\omega_\alpha t)$. Taking the Fourier transform and anticipating the fluctuation dissipation theorem (6.24), we obtain

$$ \tilde{\chi}_{\alpha, \text{spin}}''(\omega) = (\pi/\hbar) \tanh(\beta \hbar \omega_\alpha/2) \delta(\omega - \omega_\alpha) . \qquad (3.240) $$

With the equivalent oscillator susceptibility, $\tilde{\chi}_\alpha(\omega) = (\tfrac{1}{2} q_0^{(\alpha)})^2 \tilde{\chi}_{\alpha, \text{spin}}(\omega)$, we find

$$ J_{\text{spin}}(\omega) = \Theta(\omega) \sum_\alpha c_\alpha^2 \tilde{\chi}_\alpha''(\omega) = \frac{\pi}{2} \sum_\alpha \frac{c_\alpha^2}{m_\alpha \omega_\alpha} \tanh(\beta \hbar \omega_\alpha/2) \delta(\omega - \omega_\alpha) . \qquad (3.241) $$

The same form is found by calculation of the influence function (see Section 4.2) of a spin bath up to second order in c_α [164]. Because of the non-Gaussian nature of the spin-bath, the cumulant expansion does not break off at this order. Cumulants of higher order in c_α are not considered here. We see from Eq. (3.241) that, to order c_α^2, the spectral densities of a spin-$\frac{1}{2}$ bath and a boson bath are related by

$$ J_{\text{spin}}(\omega) = J_{\text{boson}}(\omega) \tanh(\beta \hbar \omega/2) . \qquad (3.242) $$

The two reservoirs lead to the same spectral coupling function at zero temperature. At finite temperature, the spin bath has a smaller effect on the system owing to the possibility for saturation of the populations of excited bath states.

The striking counter-example to a spin bath is when the ladder of excited states of a nonlinear bath mode is narrowing with increasing energy. In this case, we find enhancement of $J(\omega)$ compared with the spectral density of a harmonic bath.

The findings of this section can be summarized as follows. For weak coupling of the individual reservoir degrees of freedom to the system, the elimination of a nonlinear bath is similar to that of a linear bath, and the effects of a nonlinear environment are captured, to second order in the coupling, by a temperature-dependent spectral density of the coupling.

4. Imaginary-time approach and equilibrium dynamics

In a canonical ensemble – the assembly of all microstates with fixed particle number and volume – the system is kept at equilibrium by being in contact with a heat reservoir at temperature T, and only the energy can fluctuate. The equilibrium characteristics of a canonical ensemble of quantum systems governed by the Hamiltonian H is determined by properties of the canonical density operator $W_\beta = Z^{-1}\,\mathrm{e}^{-\beta H}$, where $Z = \mathrm{tr}\,\mathrm{e}^{-\beta H} = \sum_k \mathrm{e}^{-\beta E_k}$ is the partition function. The source of the imaginary-time path integral approach to equilibrium thermodynamics is the fact that the canonical density operator $\mathrm{e}^{-\beta H}$ is related to the time-evolution operator $\mathrm{e}^{-iHt/\hbar}$ by analytic continuation of time, $t = -i\hbar\beta$, known in field theory as *Wick rotation* from Minkowskian to Euclidean field theory. Feynman taught us that the coordinate matrix elements of $\mathrm{e}^{-iHt/\hbar}$ and $\mathrm{e}^{-\beta H}$ can be written as a sum over histories of real- and imaginary-time paths in configuration space, respectively. From a contemporary point of view, path integrals represent not only an approach alternative to canonical quantization of classical mechanics, but are basic to the foundation and interpretation of quantum mechanics. Besides that, they form a perfect basis for numerical simulations of quantum thermodynamics and dynamics employing Monte-Carlo methods.

4.1   General concepts

4.1.1   Density matrix and reduced density matrix

In quantum-statistical mechanics, the state of a system at a particular time is not perfectly known. When one has incomplete information about a system, one usually appeals to the concept of probability. Typically, the incomplete information about a system presents itself, in quantum mechanics, as a *statistical mixture*: the state of the system may be either the state $|\psi_1>$ with a probability p_1, or the state $|\psi_2>$ with a probability p_2, etc. There holds $0 \le p_1, p_2, \cdots, p_k, \cdots \le 1$, and $\sum_k p_k = 1$.

A system in a mixed state is specified by the density operator

$$W(t) \;=\; \sum_k p_k\, |\psi_k(t)><\psi_k(t)| \,. \qquad (4.1)$$

We have $\mathrm{tr}\,W = \sum_k p_k = 1$, whereas $\mathrm{tr}\,W^2 = \sum_k p_k^2 \le 1$. The equality holds for a pure state. An important quantity is the equilibrium density operator in which the p_k are canonically distributed, and the $|\psi_k><\psi_k|$ are projectors on energy eigenstates,

$$W^{(\mathrm{eq})} \;\equiv\; W_\beta \;=\; \frac{1}{Z}\,\mathrm{e}^{-\beta H} \;=\; \frac{1}{Z}\sum_k \mathrm{e}^{-\beta E_k}\, |\psi_k><\psi_k| \,. \qquad (4.2)$$

In the orthonormal basis $\{|\varphi_n>\}$, the density matrix reads

$$W_{n,m} = \sum_k p_k <\varphi_n|\psi_k><\psi_k|\varphi_m> . \tag{4.3}$$

The diagonal matrix elements $W_{n,n}$ of W in the $\{|\varphi_n>\}$ basis are

$$W_{n,n} = \sum_k p_k |<\varphi_n|\psi_k>|^2 , \tag{4.4}$$

where $| <\varphi_n|\psi_k> |^2$ is a real positive number with the following physical interpretation: if the state of the system is $|\psi_k>$, it is the probability of finding, in a measurement, this system in the state $|\varphi_n>$. Because of the uncertainty of the state $|\psi_k>$ before the measurement, $W_{n,n}$ represents the average probability of finding the system in the state $|\varphi_n>$. For this reason, the diagonal matrix element $W_{n,n}$ is called the *population* of the state $|\varphi_n>$. The off-diagonal element

$$W_{n,m} = \sum_k p_k <\varphi_n|\psi_k><\psi_k|\varphi_m> \tag{4.5}$$

is the average of the cross terms. While $W_{n,n}$ is a sum of real positive numbers, $W_{n,m}$ is a sum of complex numbers. When $W_{n,m}$ is nonzero, coherence between the states $|\varphi_n >$ and $|\varphi_m >$ persists. For this reason, the off-diagonal elements of the density matrix are dubbed *coherences*. When $W_{n,m}$ is zero, there is no interference between the states $|\varphi_n>$ and $|\varphi_m>$. It is easy to prove that

$$W_{n,n}W_{m,m} \geq |W_{n,m}|^2 . \tag{4.6}$$

We remark that the distinction between populations and coherences depends on the basis $\{|\varphi_n >\}$ chosen in the state space. The density matrix concept for coupled few-state systems in $SU(n)$ representation is neatly presented in Ref. [165].

The density operator of the global system acts in the full state space $\mathcal{H} = \mathcal{H}_S \otimes \mathcal{H}_R$. Within the density matrix representation, expectation values are expressed as

$$\langle A \rangle = \mathrm{tr}\{WA\} , \qquad \langle A \rangle^{(\mathrm{eq})} = \mathrm{tr}\{W^{(\mathrm{eq})}A\} . \tag{4.7}$$

For observables of which the respective operators act only in the system's space $\mathcal{H}_S$, the trace with respect to the reservoir in Eq. (4.7) can be separated. This suggests to introduce a *reduced* density operator ρ acting only in the system's space $\mathcal{H}_S$,

$$\rho = \mathrm{tr}_R W . \tag{4.8}$$

The operation in Eq. (4.8) is the *partial trace* with respect to the reservoir's degrees of freedom. The essence of the reduced density operator ρ is that all physical predictions about measurements bearing only on the system S are captured by this quantity,

$$\langle A^{(\mathrm{S})} \rangle = \mathrm{tr}_R \mathrm{tr}_S\{W A^{(\mathrm{S})}\} = \mathrm{tr}_S\{\rho A^{(\mathrm{S})}\} . \tag{4.9}$$

For a normalized reduced density matrix, $\mathrm{tr}\,\rho = 1$, we have

$$0 < \mathrm{tr}\,\rho^2 \leq 1 . \tag{4.10}$$

The maximum value $\mathrm{tr}\,\rho^2 = 1$ corresponds to a pure quantum state, whereas small values of $\mathrm{tr}\,\rho^2$ describe almost classical situations.

4.1.2   Imaginary-time path integral

To begin with we sketch the derivation of the path integral for the imaginary time propagator $\langle q''|\,e^{-\tau H/\hbar}|q'\rangle$ for the simple Hamiltonian $H = p^2/2M+V(q)$. The starting point of the derivation is a group property of the imaginary-time evolution,

$$\langle q''|\,e^{-\tau H/\hbar}|q'\rangle \;=\; \langle q''|\,e^{-\tau H/N\hbar}\,\mathbf{1}\,e^{-\tau H/N\hbar}\,\mathbf{1}\,e^{-\tau H/N\hbar}\,\cdots\,e^{-\tau H/N\hbar}|q'\rangle\;. \tag{4.11}$$

Upon insertion of the completeness relation of the orthonormal position eigenstates

$$\mathbf{1} \;=\; \int_{-\infty}^{\infty} dq\,|q\rangle\langle q|\;, \tag{4.12}$$

the thermal propagator $\langle q''|\,e^{-\tau H/\hbar}|q'\rangle$ becomes a multiple integral representing succession of N short-time propagators over imaginary time intervals of length $\varepsilon = \tau/N$,

$$\langle q''|\,e^{-\tau H/\hbar}|q'\rangle \;=\; \lim_{N\to\infty} \int_{-\infty}^{\infty} dq_{N-1}\cdots dq_1 \prod_{n=1}^{N} \langle q_n|\,e^{-\varepsilon H/\hbar}|q_{n-1}\rangle\;, \tag{4.13}$$

where $q_N = q''$ and $q_0 = q'$, and where we eventually take the limit $N \to \infty$. The short-time propagator with initial condition $\langle q_n|q_{n-1}\rangle = \delta(q_n - q_{n-1})$ reads

$$\langle q_n|\,e^{-\varepsilon H/\hbar}|q_{n-1}\rangle \;=\; \sqrt{\frac{M}{2\pi\hbar\varepsilon}}\;e^{-S_n/\hbar}\;, \tag{4.14}$$

where S_n is the Euclidean short-time action,

$$S_n \;=\; \varepsilon\left[\frac{M}{2}\left(\frac{q_n - q_{n-1}}{\varepsilon}\right)^2 + V(q_n)\right]\;. \tag{4.15}$$

With the normalization in Eq. (4.14) the limit $N \to \infty$ exists and we may write

$$\langle q''|\,e^{-\tau H/\hbar}|q'\rangle \;=\; \lim_{N\to\infty} \int_{-\infty}^{\infty} \frac{d^{N-1}q}{(2\pi\hbar\varepsilon/M)^{N/2}}\;e^{-S_N/\hbar}\;, \tag{4.16}$$

where $S_N = \sum_{n=1}^{N} S_n$. As $N \to \infty$, S_N turns into the Euclidean action integral[1]

$$\lim_{N\to\infty} S_N \;\to\; S^{(\mathrm{E})}[q(\cdot)] \;=\; \int_0^{\tau} d\tau'\left(\frac{M}{2}\dot{q}^2(\tau') + V[q(\tau')]\right)\;. \tag{4.17}$$

With the short-hand notation for the well-behaved multiple-integral measure

$$\int_{q'}^{q''} \mathcal{D}q(\cdot)\,\cdots \;=\; \lim_{N\to\infty} \int_{-\infty}^{\infty} \frac{d^{N-1}q}{(2\pi\hbar\varepsilon/M)^{N/2}}\,\cdots\;, \tag{4.18}$$

the propagator takes the suggestive form of a path sum or path integral

---

[1]Throughout the book, the center dot in "$q(\cdot)$" is used to express that the path q is a functional of time (here imaginary-time), while $q(\tau)$ means the value the path takes at a particular time τ. The notation $q(\tau)$ for the path – though standard in the literature – is not used here since the path $q(\cdot)$ and the value $q(\tau)$ it takes at a particular time will often appear near by.

$$\langle q''|\,e^{-\tau H/\hbar}|q'\rangle \;=\; \int_{q'}^{q''}\mathcal{D}q(\cdot)\,e^{-S^{(\mathrm{E})}[q(\cdot)]/\hbar}\,. \tag{4.19}$$

As already emphasized by Feynman [3], the concept of the sum over all paths, like the concept of an ordinary integral, is independent of a special definition of the path sum. It may be chosen according to convenience. In the representation (4.16), the path is a zigzag of straight line segments. Evaluation of path integrals by Fourier series is discussed in Subsection 4.3.4. The notion of a path integral can be extended to discrete systems for which no classical analogue exists. The tight-binding (TB) position operator has a discrete spectrum. Hence the paths piecewise stay for some time at localized TB positions with a countable number of sudden jumps between these states. The respective path sum is discussed in Chapters 19, 21 and 25.

This concludes the brief introduction to path integrals. The reader interested in getting familiarized with this field is referred to Refs. [3] – [6].

Let us first consider a one-dimensional system described by a coordinate q and a Lagrangian (in view of correlation functions, we add a source term)

$$\mathcal{L}_{\mathrm{S}}[q(t),\dot{q}(t)] \;=\; \tfrac{1}{2}M\dot{q}^2(t) - V[q(t)] - \mathcal{J}(t)q(t)\,. \tag{4.20}$$

For simplicity, we assume that there is no velocity-dependent force. The canonical density matrix $W_\beta(q'',q')$ of the system may be expressed in terms of the eigenstates $\psi_n(q)$ and eigenvalues E_n of the Hamiltonian H_{S}. We then have

$$< q''|\,W_\beta\,|q' > \equiv\; W_\beta(q'',q') \;=\; Z_{\mathrm{S}}^{-1}\sum_n \psi_n(q'')\,\psi_n^*(q')\,\exp(-\beta E_n)\,, \tag{4.21}$$

where Z_{S} is the partition function which normalizes W_β as $\operatorname{tr} W_\beta = 1$. The canonical density matrix can be represented as an imaginary-time path integral (see e.g. [3, 4])

$$W_\beta(q'',q') \;=\; Z_{\mathrm{S}}^{-1}\int_{q(0)=q'}^{q(\hbar\beta)=q''}\mathcal{D}q(\cdot)\,\exp\left(-\int_0^{\hbar\beta}d\tau\,\mathcal{L}_{\mathrm{S}}^{(\mathrm{E})}[q(\tau),\dot{q}(\tau)]/\hbar\right)\,. \tag{4.22}$$

Here, $\hbar\beta$ is the *imaginary* or *thermal* time. The path integral in Eq. (4.22) runs over all paths $q(\tau)$ in imaginary time τ which leave position q' at time zero and arrive at position q'' at imaginary time $\hbar\beta$. In the absence of velocity-dependent forces the "classical dynamics" in imaginary time is described by the *Euclidean* Lagrangian $\mathcal{L}_{\mathrm{S}}^{(\mathrm{E})}(q,\dot{q})$. As a result of the analytical continuation, the Euclidean Lagrangian differs from the real-time Lagrangian (4.20) by the reverse sign in the potential term,

$$\mathcal{L}_{\mathrm{S}}^{(\mathrm{E})}[q(\tau),\dot{q}(\tau)] \;=\; \tfrac{1}{2}M\dot{q}^2(\tau) + V[q(\tau)] + \mathcal{J}(\tau)q(\tau)\,. \tag{4.23}$$

Here we use the convention that in real-time quantities, as e. g. in Eq. (4.20), the overdot denotes differentiation with respect to t while in *Euclidean* quantities the overdot denotes differentiation with respect to the imaginary time τ.

The reduced density operator in thermal equilibrium is defined as the partial trace of the canonical density operator W_β of the total system with respect to the reservoir coordinates, $\rho_\beta \equiv \mathrm{tr_R} W_\beta$. Our main goal in this chapter is to derive for the global systems discussed in the preceding chapter the corresponding path integral expressions for the reduced density matrix in thermal equilibrium. The essential element of the imaginary-time path integral for the open system is the effective Euclidean action. This is found by integrating out the reservoir coordinates in the path integral for the equilibrium density matrix W_β of the global system.

In the next section of this chapter, we shall calculate the effective Euclidean actions for the various system-plus-reservoir models introduced in the preceding section. We then study the partition function of these systems and discuss several methods and concepts for the approximate evaluation of imaginary-time path integrals.

4.2   Effective action and equilibrium density matrix

We now generalize the discussion to the case in which the system is coupled to a thermal reservoir. The equilibrium density matrix of the "universe" (system plus environment) may be written in the form

$$W_\beta(q'', \boldsymbol{x}''; q', \boldsymbol{x}') = Z_{\text{tot}}^{-1} \sum_{\{n\}} \Psi_{\{n\}}(q'', \boldsymbol{x}'') \, \Psi_{\{n\}}^*(q', \boldsymbol{x}') \, \exp(-\beta E_{\{n\}}) \, . \qquad (4.24)$$

The N-component vector $\boldsymbol{x}$ stands for the coordinates $x_1, x_2, \ldots, x_N$ of the reservoir, $\Psi_{\{n\}}$ denotes the eigenstates of the Hamiltonian of the global system, and $\{n\}$ is the full set of quantum numbers. Again, Z_{tot} is a normalization factor so that $\mathrm{tr}\, W_\beta = 1$. We shall assume that the forces in the Hamiltonian of the global system are velocity-independent. Then the canonical density matrix (4.24) can be written as a standard path integral,

$$W_\beta(q'', \boldsymbol{x}''; q', \boldsymbol{x}') = Z_{\text{tot}}^{-1} \int_{q(0)=q'}^{q(\hbar\beta)=q''} \mathcal{D}q(\cdot) \int_{\boldsymbol{x}(0)=\boldsymbol{x}'}^{\boldsymbol{x}(\hbar\beta)=\boldsymbol{x}''} \mathcal{D}\boldsymbol{x}(\cdot) \, \exp\left(-S^{(\mathrm{E})}[q(\cdot), \boldsymbol{x}(\cdot)]/\hbar\right) \, . \qquad (4.25)$$

As is explicitly indicated, the functional integrations run over all paths taken by the coordinates $q(\tau)$ and $\boldsymbol{x}(\tau)$ with endpoints $q(0) = q'$, $\boldsymbol{x}(0) = \boldsymbol{x}'$, $q(\hbar\beta) = q''$ and $\boldsymbol{x}(\hbar\beta) = \boldsymbol{x}''$. The Euclidean action receives contributions from the system (S), the reservoir (R), and the interaction (I),

$$S^{(\mathrm{E})} = S_{\mathrm{S}}^{(\mathrm{E})} + S_{\mathrm{R}}^{(\mathrm{E})} + S_{\mathrm{I}}^{(\mathrm{E})} = \int_0^{\hbar\beta} d\tau \left(\mathcal{L}_{\mathrm{S}}^{(\mathrm{E})} + \mathcal{L}_{\mathrm{R}}^{(\mathrm{E})} + \mathcal{L}_{\mathrm{I}}^{(\mathrm{E})} \right) \, . \qquad (4.26)$$

From here on we focus our interest on a reduced description in which the reservoir coordinates are eliminated. To this end, we introduce the *reduced* equilibrium density matrix describing the open (damped) system

$$\rho_\beta(q'', q') = \mathrm{tr}\, W_\beta = \int_{-\infty}^{+\infty} d\boldsymbol{x}' \, W_\beta(q'', \boldsymbol{x}'; q', \boldsymbol{x}') \, , \qquad (4.27)$$

and the *reduced* partition function of the damped system,

$$Z(\hbar\beta) = \frac{Z_{\text{tot}}(\hbar\beta)}{Z_{\text{R}}(\hbar\beta)} , \tag{4.28}$$

where Z_{tot} is the partition function of the global system, and Z_{R} is the partition function of the unperturbed reservoir. For a bath of N harmonic oscillators we have

$$Z_{\text{R}} \equiv \prod_{\alpha=1}^{N} Z_{\text{R}}^{(\alpha)} = \prod_{\alpha=1}^{N} \frac{1}{2\sinh(\beta\hbar\omega_\alpha/2)} . \tag{4.29}$$

With use of Eqs. (4.25) – (4.28), the reduced equilibrium density matrix can be written in the form

$$\rho_\beta(q'', q') = \frac{1}{Z} \int_{q(0)=q'}^{q(\hbar\beta)=q''} \mathcal{D}q(\cdot) \exp\left\{-S_{\text{S}}^{(\text{E})}[q(\cdot)]/\hbar\right\} \mathcal{F}^{(\text{E})}[q(\cdot)] . \tag{4.30}$$

The *influence functional*

$$\mathcal{F}^{(\text{E})}[q(\cdot)] \equiv \exp\left\{-S_{\text{infl}}^{(\text{E})}[q(\cdot)]/\hbar\right\} = \frac{1}{Z_{\text{R}}} \oint \mathcal{D}\boldsymbol{x}(\cdot) \exp\left\{-S_{\text{R,I}}^{(\text{E})}[q(\cdot), \boldsymbol{x}(\cdot)]/\hbar\right\} \tag{4.31}$$

with

$$S_{\text{R,I}}^{(\text{E})}[q, \boldsymbol{x}] = S_{\text{R}}^{(\text{E})}[\boldsymbol{x}] + S_{\text{I}}^{(\text{E})}[q, \boldsymbol{x}] . \tag{4.32}$$

captures the effects of the environmental coupling on the equilibrium properties of the open system. The trace operation in Eq. (4.27) appears in the path integral (4.31) as the sum over all *periodic* paths with period $\hbar\beta$ taken by the coordinate $\boldsymbol{x}$. The factor Z_{R}^{-1} normalizes $\mathcal{F}^{(\text{E})}[q]$, so that $\mathcal{F}^{(\text{E})} = 1$ when the coupling $S_{\text{I}}^{(\text{E})}[q, \boldsymbol{x}]$ is switched off, and the factor Z^{-1} normalizes ρ_β as $\text{tr}\,\rho_\beta = 1$.

In the following subsections, we calculate the Euclidean influence action $S_{\text{infl}}^{(\text{E})}[q(\cdot)]$ for particular systems.

4.2.1 Open system with bilinear coupling to a harmonic reservoir

For the model described by the Hamiltonian (3.11), the Euclidean Lagrangian $\mathcal{L}_{\text{S}}^{(\text{E})}$ is given in Eq. (4.23) with $\mathcal{J} = 0$, and

$$\begin{aligned}
\mathcal{L}_{\text{R}}^{(\text{E})} &= \frac{1}{2} \sum_{\alpha=1}^{N} m_\alpha \left(\dot{x}_\alpha^2 + \omega_\alpha^2 x_\alpha^2\right) , \\
\mathcal{L}_{\text{I}}^{(\text{E})} &= \sum_{\alpha=1}^{N} \left(-c_\alpha x_\alpha q + \frac{1}{2} \frac{c_\alpha^2 q^2}{m_\alpha \omega_\alpha^2}\right) .
\end{aligned} \tag{4.33}$$

The stationary paths of the total action (4.26), which we denote by $\bar{q}$ and by $\bar{x}_\alpha$, obey the Euclidean classical equations of motion

$$M\ddot{\bar{q}} - \frac{\partial V(\bar{q})}{\partial \bar{q}} + \sum_{\alpha=1}^{N} c_\alpha \left(\bar{x}_\alpha - \frac{c_\alpha}{m_\alpha \omega_\alpha^2} \bar{q} \right) = 0 \,, \qquad (4.34)$$

$$m_\alpha \ddot{\bar{x}}_\alpha - m_\alpha \omega_\alpha^2 \bar{x}_\alpha + c_\alpha \bar{q} = 0 \,. \qquad (4.35)$$

Since the Lagrangian is a quadratic form in $\boldsymbol{x}$ and $\dot{\boldsymbol{x}}$, the multiple functional integration of the periodic path $\boldsymbol{x}(\tau)$ can be performed in closed form. To proceed, we choose periodic continuation of the paths $q(\tau)$ and $\boldsymbol{x}(\tau)$ outside the range $0 \leq \tau < \hbar\beta$ by writing them as Fourier series

$$x_\alpha(\tau) = \frac{1}{\hbar\beta} \sum_{n=-\infty}^{+\infty} x_{\alpha,n} \, e^{i\nu_n \tau} \,,$$

$$q(\tau) = \frac{1}{\hbar\beta} \sum_{n=-\infty}^{+\infty} q_n \, e^{i\nu_n \tau} \,,$$

where $x_{\alpha,n} = x_{\alpha,-n}^*$, $q_n = q_{-n}^*$, and $\nu_n = 2\pi n/\hbar\beta$ is a bosonic Matsubara frequency. Insertion of the series expressions (4.36) into Eq. (4.32) with Eq. (4.33) yields

$$S_{\mathrm{R,I}}^{(\mathrm{E})}[q, \boldsymbol{x}] = \sum_{\alpha=1}^{N} \frac{1}{\hbar\beta} \sum_{n=-\infty}^{+\infty} \frac{m_\alpha}{2} \left(\nu_n^2 |x_{\alpha,n}|^2 + \omega_\alpha^2 \left| x_{\alpha,n} - \frac{c_\alpha}{m_\alpha \omega_\alpha^2} q_n \right|^2 \right) . \qquad (4.36)$$

Next, we split $x_{\alpha,n}$ into the classical part $\bar{x}_{\alpha,n}$ and the quantum fluctuation part $y_{\alpha,n}$,

$$x_{\alpha,n} = \bar{x}_{\alpha,n} + y_{\alpha,n} = \frac{c_\alpha}{m_\alpha(\nu_n^2 + \omega_\alpha^2)} q_n + y_{\alpha,n} \,. \qquad (4.37)$$

In the second form, we have used the solution of Eq. (4.35) in Fourier space. Since $\bar{x}_\alpha(\tau)$ is a stationary point of the action $S_{\mathrm{R,I}}^{(\mathrm{E})}[q, \boldsymbol{x}]$, the term linear in the deviation $y_{\alpha,n}$ is absent, and we get pure quadratic forms in $\boldsymbol{y}$ and q,

$$S_{\mathrm{R,I}}^{(\mathrm{E})}[q, \bar{\boldsymbol{x}} + \boldsymbol{y}] = S_{\mathrm{R}}^{(\mathrm{E})}[\boldsymbol{y}] + S_{\mathrm{infl}}^{(\mathrm{E})}[q] \,, \qquad (4.38)$$

$$S_{\mathrm{R}}^{(\mathrm{E})}[\boldsymbol{y}] = \sum_{\alpha=1}^{N} \frac{1}{\hbar\beta} \sum_{n=-\infty}^{+\infty} \frac{m_\alpha}{2}(\nu_n^2 + \omega_\alpha^2)|y_{\alpha,n}|^2 = \sum_{\alpha=1}^{N} \int_0^{\hbar\beta} d\tau \, \frac{m_\alpha}{2}(\dot{y}_\alpha^2 + \omega_\alpha^2 \, y_\alpha^2) \,,$$

$$S_{\mathrm{infl}}^{(\mathrm{E})}[q] = \sum_{\alpha=1}^{N} \frac{c_\alpha^2}{2m_\alpha} \frac{1}{\hbar\beta} \sum_{n=-\infty}^{+\infty} \left(\frac{|q_n|^2}{\omega_\alpha^2} - \frac{|q_n|^2}{\nu_n^2 + \omega_\alpha^2} \right) . \qquad (4.39)$$

The first term in $S_{\mathrm{infl}}^{(\mathrm{E})}[q]$ originates from the potential counter term $\propto q^2$ in $\mathcal{L}_{\mathrm{I}}^{(\mathrm{E})}$. The path sum of all $\hbar\beta$-periodic quantum fluctuations $\boldsymbol{y}(\tau)$ directly yields the partition function of the reservoir,

$$Z_{\mathrm{R}} = \oint \mathcal{D}\boldsymbol{y}(\cdot) \exp\left\{ -S_{\mathrm{R}}^{(\mathrm{E})}[\boldsymbol{y}(\cdot)]/\hbar \right\} . \qquad (4.40)$$

Hence this term cancels the factor Z_R^{-1} in Eq. (4.31). The influence action (4.39) may be concisely written as

$$S_{\text{infl}}^{(E)}[q(\cdot)] = \frac{M}{2} \frac{1}{\hbar\beta} \sum_{n=-\infty}^{+\infty} \xi_n |q_n|^2 , \qquad (4.41)$$

$$\xi_n = \frac{1}{M} \sum_{\alpha=1}^{N} \frac{c_\alpha^2}{m_\alpha \omega_\alpha^2} \frac{\nu_n^2}{(\nu_n^2 + \omega_\alpha^2)} = \frac{2}{M\pi} \int_0^\infty d\omega \frac{J(\omega)}{\omega} \frac{\nu_n^2}{\nu_n^2 + \omega^2} . \qquad (4.42)$$

In the second form, we have introduced the spectral density (3.24). The Fourier coefficient ξ_n of the influence action may be expressed in terms of the spectral damping functions $\tilde{\gamma}(\omega)$ and $\hat{\gamma}(z)$ defined in Eq. (3.25) and (3.27), respectively. We have

$$\xi_n = |\nu_n| \tilde{\gamma}(i|\nu_n|) = |\nu_n| \hat{\gamma}(|\nu_n|) . \qquad (4.43)$$

The expression (4.41) is the bosonic Matsubara action for general spectral density $J(\omega)$. In imaginary-time representation, the influence action (4.41) is a convolution,

$$S_{\text{infl}}^{(E)}[q(\cdot)] = \int_0^{\hbar\beta} d\tau \int_0^\tau d\tau' \, k(\tau - \tau') \, q(\tau) \, q(\tau') , \qquad (4.44)$$

$$k(\tau) = \frac{M}{\hbar\beta} \sum_{n=-\infty}^{+\infty} \xi_n \, e^{i\nu_n \tau} . \qquad (4.45)$$

The kernel $k(\tau)$ satisfies the symmetry relations

$$k(\tau) = k(\hbar\beta - \tau) = k(\tau + \hbar\beta) . \qquad (4.46)$$

With use of the property

$$\int_0^{\hbar\beta} d\tau \, k(\tau) = M\xi_0 = 0 \qquad (4.47)$$

the influence action (4.44) may be cast into the form

$$S_{\text{infl}}^{(E)}[q(\cdot)] = -\frac{1}{2} \int_0^{\hbar\beta} d\tau \int_0^\tau d\tau' \, k(\tau - \tau') \left(q(\tau) - q(\tau') \right)^2 . \qquad (4.48)$$

Evidently, the influence action $S_{\text{infl}}^{(E)}[q]$ is fully *nonlocal* and therefore does not cause potential renormalization.

Alternatively, we may express the influence action (4.48) in terms of the kernel

$$K(\tau) = \mu :\delta(\tau): - k(\tau) , \qquad (4.49)$$

where $:\delta(\tau):$ is the periodically continued δ-function (3.58). Hence we may write

$$S_{\text{infl}}^{(E)}[q(\cdot)] = \frac{1}{2} \int_0^{\hbar\beta} d\tau \int_0^\tau d\tau' \, K(\tau - \tau') \left(q(\tau) - q(\tau') \right)^2 . \qquad (4.50)$$

The local part of the kernel does not contribute. With the choice

$$\mu = \sum_{\alpha=1}^{N} \frac{c_\alpha^2}{m_\alpha \omega_\alpha^2} = \frac{2}{\pi} \int_0^\infty d\omega \, \frac{J(\omega)}{\omega} = \lim_{t \to 0^+} M\gamma(t) = M\gamma(0^+) \tag{4.51}$$

the Fourier coefficient of the kernel

$$K(\tau) = \frac{M}{\hbar\beta} \sum_{n=-\infty}^{+\infty} \zeta_n \, e^{i\nu_n \tau} \tag{4.52}$$

may be written as

$$\zeta_n = \frac{\mu}{M} - \xi_n = \frac{1}{M} \sum_{\alpha=1}^{N} \frac{c_\alpha^2}{m_\alpha} \frac{1}{\nu_n^2 + \omega_\alpha^2} = \frac{2}{\pi M} \int_0^\infty d\omega \, J(\omega) \frac{\omega}{\nu_n^2 + \omega^2} \, . \tag{4.53}$$

Expressing the Fourier sum in (4.52) in terms of the boson propagator (3.56), we find

$$K(\tau) = \frac{1}{\pi} \int_0^\infty d\omega \, J(\omega) \, D_\omega(\tau) = \frac{1}{\pi} \int_0^\infty d\omega \, J(\omega) \frac{\cosh(\omega(\hbar\beta/2 - \tau)}{\sinh(\omega\hbar\beta\tau)} \, . \tag{4.54}$$

The second form is obtained with Eq. (3.55) and holds in the regime $0 \le \tau < \hbar\beta$. The kernel $K(\tau)$ is related to the autocorrelation function $\mathcal{X}^{(\mathrm{E})}(\tau)$ of the stochastic force $\xi(\tau) = \sum_\alpha c_\alpha x_\alpha(\tau)$ given in Eq. (3.59) by

$$K(\tau) = \mathcal{X}^{(\mathrm{E})}(\tau)/\hbar \, . \tag{4.55}$$

The function $K(\tau)$ may also be expressed in terms of the real-time memory-friction kernel $\gamma(t)$. Upon substituting the expression (3.32) for $J(\omega)$ into Eq. (4.54), reversing the order of integrations, and performing the ω-integral with use of the form (3.55) for $D_\omega(\tau)$, we get

$$K(\tau) = \frac{M}{\hbar\beta} \int_0^\infty dt \, \gamma(t) \frac{\partial}{\partial t} \left(\frac{\sinh(\nu t)}{\cosh(\nu t) - \cos(\nu \tau)} \right) , \tag{4.56}$$

where $\nu \equiv \nu_1 = 2\pi/\hbar\beta$ is the lowest Matsubara frequency.

With these forms, the kernel $K(\tau)$ may be calculated for general spectral density $J(\omega)$. In the important Ohmic case $J(\omega) = \eta\omega$, the kernel takes the explicit form

$$K(\tau) = \frac{\eta}{\pi} \frac{(\pi k_\mathrm{B} T/\hbar)^2}{\sinh^2(\pi k_\mathrm{B} T\tau/\hbar)} \, . \tag{4.57}$$

Another form of the influence action $S_{\mathrm{infl}}^{(\mathrm{E})}[q]$ is found if we agree to periodically continue the path $q(\tau)$ outside the range $0 \le \tau < \hbar\beta$ according to $q(\tau + n\hbar\beta) = q(\tau)$. By extension of the integration regime, the expression (4.50) is transformed into [77]

$$S_{\mathrm{infl}}^{(\mathrm{E})}[q(\cdot)] = \frac{1}{4} \int_0^{\hbar\beta} d\tau \int_{-\infty}^\infty d\tau' \, K_0(\tau - \tau') \left(q(\tau) - q(\tau') \right)^2 , \tag{4.58}$$

where $K_0(\tau)$ is the kernel $K(\tau)$ at $T = 0$,

$$K_0(\tau) = \sum_{\alpha=1}^{N} \frac{c_\alpha^2}{2m_\alpha\omega_\alpha}\, e^{-\omega_\alpha|\tau|} = \frac{1}{\pi}\int_0^\infty d\omega\, J(\omega)\, e^{-\omega|\tau|}. \qquad (4.59)$$

While $K(\tau - \tau')$ in Eq. (4.50) is acting only within the principal interval of length $\hbar\beta$, the function $K_0(\tau - \tau')$ in Eq. (4.58) arranges also correlations between the path in the principal interval and the periodically continued path.

Finally, we may insert Eq. (4.31) with Eq. (4.50) into Eq. (4.30). Then the reduced equilibrium density matrix of the open system takes the concise form

$$< q''|\rho_\beta|q' > \equiv \rho_\beta(q'', q') = Z^{-1} \int_{q(0)=q'}^{q(\hbar\beta)=q''} \mathcal{D}q(\cdot)\, \exp\left\{-S_{\mathrm{eff}}^{(\mathrm{E})}[q(\cdot)]/\hbar\right\}. \qquad (4.60)$$

The effective action $S_{\mathrm{eff}}^{(\mathrm{E})}[q(\cdot)] = S_S^{(\mathrm{E})}[q(\cdot)] + S_{\mathrm{infl}}^{(\mathrm{E})}[q(\cdot)]$ may be written as

$$\begin{aligned}
S_{\mathrm{eff}}^{(\mathrm{E})}[q(\cdot)] &= \int_0^{\hbar\beta} d\tau \left(\tfrac{1}{2}M\dot{q}^2 + V(q)\right) + \frac{1}{2}\int_0^{\hbar\beta} d\tau \int_0^{\tau} d\tau'\, K(\tau - \tau')\left(q(\tau) - q(\tau')\right)^2 \\
&= \int_0^{\hbar\beta} d\tau \left(\tfrac{1}{2}M\dot{q}^2 + V(q)\right) + \int_0^{\hbar\beta} d\tau \int_0^{\tau} d\tau'\, k(\tau - \tau')q(\tau)q(\tau'). \qquad (4.61)
\end{aligned}$$

The action (4.61) and the trivial generalization to many dimensions describe any system exposed to a linear dissipative process. It turns out to be the essential component in the path integral approach towards thermodynamics of open quantum systems.

In a perturbative treatment of the potential $V(q)$, it is pertinent to introduce the effective action in absence of the potential term,

$$S_{\mathrm{eff},0}^{(\mathrm{E})}[q(\cdot)] = \frac{M}{2}\frac{1}{\hbar\beta}\sum_{n=-\infty}^{\infty} \lambda_n|q_n|^2, \qquad (4.62)$$

where

$$\lambda_n = \nu_n^2 + \xi_n = \nu_n^2 + |\nu_n|\hat{\gamma}(|\nu_n|). \qquad (4.63)$$

Consider now the displacement correlation function $D^{(\mathrm{E})}(\tau) = \langle[q(0)-q(\tau)]q(0)\rangle_\beta$. Here $\langle\cdots\rangle_\beta$ denotes average with the weight function $\exp[-S_{\mathrm{eff},0}^{(\mathrm{E})}[q]/\hbar]$. With the pure quadratic form (4.62) we get

$$D^{(\mathrm{E})}(\tau) = \frac{1}{\beta}\sum_{n\neq 0}\hat{\chi}(|\nu_n|)\left[1 - e^{i\nu_n\tau}\right] \quad \text{with} \quad \hat{\chi}(z) = \frac{1}{M[z^2 + z\hat{\gamma}(z)]}. \qquad (4.64)$$

The function $\hat{\chi}(z)$ is the Laplace transform of the susceptibility $\chi(t)$ of a free Brownian particle. Importantly, the expression (4.64) coincides with the earlier result (3.63) found along different line. Free Brownian motion will be discussed in Chapter 7.

4.2.2 State-dependent memory friction

The analysis given in the previous subsection can easily be generalized to nonlinear state-dependent friction with all types of memory effects provided that the system-bath coupling is linear in the bath degrees of freedom. For the global system described by the Hamiltonian (3.8), the reduced density matrix in thermal equilibrium adopts again the form (4.60), but now the influence action reads

$$
\begin{aligned}
S_{\text{infl}}^{(\text{E})}[q(\cdot)] \; &= \; \frac{1}{2} \int_0^{\hbar\beta} d\tau \int_0^{\tau} d\tau' \, K(\tau - \tau') \Big(F[q(\tau)] - F[q(\tau')] \Big)^2 \\
&= \; \int_0^{\hbar\beta} d\tau \int_0^{\tau} d\tau' \, k(\tau - \tau') F[q(\tau)] F[q(\tau')] \,.
\end{aligned}
\tag{4.65}
$$

A nonlinear form for $F(q)$ is relevant, e.g., in rotational tunneling systems, as the function $F(q)$ must obey the symmetries of the hindering potential [78] [cf. the discussion following Eq. (3.8)], and for quasiparticle tunneling between superconductors. For the phenomenological coupling (3.9), the influence action reads

$$
S_{\text{infl}}^{(\text{E})}[\varphi(\cdot)] \; = \; \sum_{m=1}^{2} \int_0^{\hbar\beta} d\tau \int_0^{\tau} d\tau' \, k^{(m)}(\tau - \tau') F^{(m)}[\varphi(\tau)] \, F^{(m)}[\varphi(\tau')] \,,
\tag{4.66}
$$

where $F^{(1)}[\varphi(\tau)] = \sin[\varphi(\tau)/2]$ and $F^{(2)}[\varphi(\tau)] = \cos[\varphi(\tau)/2]$. By straightforward manipulation, the action (4.66) can be brought into the form (4.187) of the microscopic model for quasiparticle tunneling in a Josephson junction [79] (cf. Subsection 4.2.10).

An important example of state-dependent friction is when the relevant system interacts with individual bath modes within a spatially restricted region [166, 167]. The case in which the system interacts with localized modes is discussed in the context of decoherence in Section 9.3. The reasoning of this subsection applies accordingly to a weakly coupled nonlinear bath considered in Sec. 3.6.

4.2.3 Spin-boson model

In the dissipative two-state or spin-boson model (3.137) or (3.139), the transitions between the two states are sudden. Correspondingly, in the path sum representation of the TSS, the path $q(\tau)$ is piecewise constant and it occasionally jumps between the positions $+\frac{1}{2}q_0$ and $-\frac{1}{2}q_0$. It is convenient to put

$$
q(\tau) \; = \; \tfrac{1}{2} q_0 \, \sigma(\tau) \,.
\tag{4.67}
$$

Here $\sigma(\tau)$ is a spin path in the model (3.139). The path $\sigma(\tau)$ jumps forth and back between the values $+1$ and -1. A spin path with 8 moves (spin flips, or kinks, or instantons of width zero) is sketched in Fig. 4.1. A path of 2m alternating flips with centers $\{s_j\}$ in chronological order (m sequential anti-kink–kink pairs) reads

$$
\sigma^{(m)}(\tau) \; = \; 1 + 2 \sum_{j=1}^{2m} (-1)^j \Theta(\tau - s_j) \,.
\tag{4.68}
$$

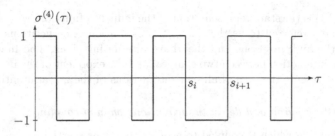

Figure 4.1: The spin path (4.68) composed of eight jumps. s_j.

For the shifted canonical distribution (3.45) with initial state $q(0) = \frac{1}{2}q_0\sigma_i$, the mean value of the polarization energy is $\frac{1}{2}\mathfrak{E}_0(\sigma_i) = \frac{1}{4}q_0^2 M\gamma(0)\sigma_i$ as follows with use of Eqs. (3.136) and (3.44). The energy $\Lambda_{cl} \equiv \frac{1}{2}[\mathfrak{E}_0(+1) - \mathfrak{E}_0(-1)] = \frac{1}{2}q_0^2 M\gamma(0_+)$ is called the classical *reorganization energy* of the bath in a transition of the spin from state $\sigma_i = +1$ to state $\sigma_i = -1$. With use of Eqs. (3.20) and (3.140) we get

$$\Lambda_{cl} = \frac{q_0^2}{2}\sum_\alpha \frac{c_\alpha^2}{m_\alpha\omega_\alpha^2} = \hbar\int_0^\infty d\omega \frac{G(\omega)}{\omega} \ . \tag{4.69}$$

The reorganization energy of the bath plays a crucial role in the Marcus theory of electron transfer in a solvent, which is discussed in Section 20.1.

Consider next the polarization fluctuations $\delta\mathfrak{E}(\tau) \equiv \mathfrak{E}(\tau) - \mathfrak{E}_0 = q_0\sum_\alpha c_\alpha x_\alpha(\tau)$. We obtain with use of Eqs. (4.55) and (4.54)

$$\frac{1}{\hbar^2}\left\langle \delta\mathfrak{E}(\tau)\delta\mathfrak{E}(0)\right\rangle_\beta \equiv \mathcal{K}(\tau) = \frac{q_0^2}{\hbar}K(\tau) = \int_0^\infty d\omega\, G(\omega)D_\omega(\tau) \ . \tag{4.70}$$

In the last form, we have employed the spin spectral density $G(\omega)$ introduced in Eq. (3.140). Upon writing the action (4.50) in terms of the spin path $\sigma(\tau)$, the Euclidean influence functional (4.31) takes the form

$$\mathcal{F}^{(E)}[\sigma(\cdot)] = \exp\left\{-\frac{1}{8}\int_0^{\hbar\beta} d\tau \int_0^\tau d\tau'\, \mathcal{K}(\tau - \tau')\big(\sigma(\tau) - \sigma(\tau')\big)^2\right\} \ . \tag{4.71}$$

With the form (4.68) for $\sigma(\tau)$, the time integrations can be done. We then get a representation in terms of the kink-pair interaction $W(\tau)$, which is a second integral of the kernel $\mathcal{K}(\tau)$. We have $\ddot{W}(\tau) = -\mathcal{K}(\tau)$. For a path with m consecutive kink-anti-kink pairs, the influence functional takes the form

$$\mathcal{F}_m^{(E)}[\{s_j\}] = \exp\left\{\sum_{j=2}^{2m}\sum_{i=1}^{j-1}(-1)^{i+j}W(s_j - s_i)\right\} \ ,$$

$$W(\tau) = \int_0^\infty d\omega \frac{G(\omega)}{\omega^2}\big(D_\omega(0) - D_\omega(\tau)\big) \ . \tag{4.72}$$

The form (4.71) is the state representation of the influence function, whereas the form (4.72) is usually referred to as the charge representation. The fictitious charges are situated at the kink positions and they have alternating signs. The function $W(\tau)$ describes the interaction between two charges, and the exponent of the influence function (4.72) represents the sum of all pair interactions of the $2m$ alternating charges.

4.2.4 Acoustic polaron and defect tunneling: one-phonon coupling

Consider a particle which is coupled to acoustic phonons via the interaction $H_{\text{lin}}(\boldsymbol{q})$ given in Eq. (3.168). Since the global Hamiltonian is a mixed quadratic form in the phonon variables, the thermal average of the phonon degrees of freedom can be done exactly. Rather than perform the average in the path sum representation, here we choose, for a change, to proceed directly in the operator formulation. The influence functional which describes the effects of the coupling H_{lin} is formally defined by

$$F_{\text{lin}}^{(\text{E})}[\boldsymbol{q}(\cdot)] \equiv \exp\left(-S_{\text{lin}}^{(\text{E})}[\boldsymbol{q}(\cdot)]/\hbar\right) = \left\langle T_{\tau}\exp\left(-\int_{0}^{\hbar\beta} d\tau\, \tilde{H}_{\text{lin}}[\boldsymbol{q}(\tau)]/\hbar\right)\right\rangle_{\beta} . \quad (4.73)$$

Here T_{τ} is the imaginary-time ordering operator, $\langle\cdots\rangle_{\beta}$ denotes the thermal average with respect to $\exp(-\sum_{k}\beta\hbar\omega_{k}b_{k}^{\dagger}b_{k})$, where again $k = (\boldsymbol{k},\lambda)$, and $\tilde{H}_{\text{lin}}[\boldsymbol{q}(\tau)]$ is the electron-phonon interaction (3.168) in the interaction picture (denoted by the tilde),

$$\tilde{H}_{\text{lin}}[\boldsymbol{q}(\tau)] = \exp(H_{\text{R}}\tau/\hbar)H_{\text{lin}}[\boldsymbol{q}(\tau)]\exp(-H_{\text{R}}\tau/\hbar) . \quad (4.74)$$

In imaginary time we have

$$\tilde{b}_{k}(\tau) = b_{k}\, e^{-\omega_{k}\tau} , \qquad \tilde{b}_{k}^{\dagger}(\tau) = b_{k}^{\dagger}\, e^{\omega_{k}\tau} . \quad (4.75)$$

Next, the formal expression (4.73) is expanded into a power series in $\tilde{H}_{\text{lin}}[\boldsymbol{q}]$,

$$\begin{aligned}
\mathcal{F}_{\text{lin}}^{(\text{E})}[\boldsymbol{q}(\cdot)] = {}& 1 + \sum_{n=1}^{\infty}\left(\frac{-1}{\hbar}\right)^{n}\int_{0}^{\hbar\beta} d\tau_{n}\cdots\int_{0}^{\tau_{3}} d\tau_{2}\int_{0}^{\tau_{2}} d\tau_{1} \\
& \times \left\langle \tilde{H}_{\text{lin}}[\boldsymbol{q}(\tau_{n})]\cdots\tilde{H}_{\text{lin}}[\boldsymbol{q}(\tau_{1})]\right\rangle_{\beta} ,
\end{aligned} \quad (4.76)$$

and *Wick's theorem* for thermodynamic averages (see, e.g., Ref. [168]) is employed. The resulting cumulant expansion then yields the influence action. Since the interaction (3.168) is linear in the phonon field, the averages of all odd powers of $\tilde{H}_{\text{lin}}[\boldsymbol{q}]$ vanish. Because of the Gaussian statistics, also the cumulants of fourth and of higher even order in $\tilde{H}_{\text{lin}}[\boldsymbol{q}]$ are zero. Hence the only non-vanishing cumulant contraction is the second order term $\langle\tilde{H}_{\text{lin}}[\boldsymbol{q}(\tau)]\tilde{H}_{\text{lin}}[\boldsymbol{q}(\tau')]\rangle_{\beta}$, and thus the influence action is

$$S_{\text{lin}}^{(\text{E})}[\boldsymbol{q}(\cdot)] = -\frac{1}{\hbar}\int_{0}^{\hbar\beta} d\tau\int_{0}^{\tau} d\tau'\, \left\langle\tilde{H}_{\text{lin}}[\boldsymbol{q}(\tau)]\tilde{H}_{\text{lin}}[\boldsymbol{q}(\tau')]\right\rangle_{\beta} . \quad (4.77)$$

With use of the form (3.168) for $H_{\text{lin}}[\boldsymbol{q}(\tau)]$, the contraction may be written in terms of the boson imaginary-time propagator (3.55) as

$$\left\langle \tilde{H}_{\text{lin}}[\boldsymbol{q}(\tau)] \tilde{H}_{\text{lin}}[\boldsymbol{q}(\tau')] \right\rangle_{\beta} = \sum_{n,m} \sum_{\boldsymbol{k},\lambda} \frac{\hbar \kappa_k^{(n)}[\boldsymbol{q}(\tau)] \kappa_k^{(m)}[\boldsymbol{q}(\tau')]}{2V \varrho \, \omega_k} \, e^{-i\boldsymbol{k} \cdot \left[\boldsymbol{R}^{(n)} - \boldsymbol{R}^{(m)} \right]}$$

$$\times \exp\left(i\boldsymbol{k} \cdot \left[\boldsymbol{q}(\tau) - \boldsymbol{q}(\tau') \right] \right) D_{\omega_k}(\tau - \tau') . \tag{4.78}$$

Quantum diffusion of light interstitials and defect tunneling in solids usually take place by a succession of incoherent tunneling transitions between localized states in the rigid multi-well potential of the host crystal. At thermal energy small compared with the level spacing of the low-lying states in the individual wells, only the lowest state in each local minimum is relevant to the dynamics [169].[2]

Consider the case of a defect particle which tunnels between two spatially separated interstitial positions located at $\boldsymbol{q}_1 = \frac{1}{2}\boldsymbol{q}_0$ and $\boldsymbol{q}_2 = -\frac{1}{2}\boldsymbol{q}_0$. We then have

$$\boldsymbol{q}(\tau) = \tfrac{1}{2}\boldsymbol{q}_0 \, \sigma(\tau) , \tag{4.79}$$

where $\sigma(\tau) = \pm 1$, as sketched in Fig. 4.1. The values $\pm\frac{1}{2}\boldsymbol{q}_0$ are the two energetically accessible interstitial positions of the defect. For this formation, the contraction (4.78) takes the form

$$\left\langle \tilde{H}_{\text{lin}}[\sigma(\tau)] \tilde{H}_{\text{lin}}[\sigma(\tau')] \right\rangle_{\beta} = \sum_{n,m} \sum_{\boldsymbol{k},\lambda} \frac{\hbar \kappa_k^{(n)}(\sigma) \kappa_k^{(m)}(\sigma')}{2V \varrho \, \omega_k} \, e^{-i\boldsymbol{k} \cdot [\boldsymbol{R}^{(n)} - \boldsymbol{R}^{(m)}]} D_{\omega_k}(\tau - \tau')$$

$$\times \left[\cos^2\left(\frac{\boldsymbol{k} \cdot \boldsymbol{q}_0}{2} \right) + \sin^2\left(\frac{\boldsymbol{k} \cdot \boldsymbol{q}_0}{2} \right) \sigma(\tau)\sigma(\tau') \right] . \tag{4.80}$$

Here we have used the notation

$$\kappa_k^{(n)}(\sigma) \equiv \kappa_k^{(n)}(\boldsymbol{q} = \tfrac{1}{2}\sigma\boldsymbol{q}_0) . \tag{4.81}$$

In defect tunneling, the energy shift which is caused by the linear coupling $H_{\text{lin}}(\boldsymbol{q})$ plays an ancillary role [cf. the discussion below Eq. (3.4)]. Therefore we may restrict the attention to the change of the action when the particle hops. With a suitable subtraction of the contribution resulting from a constant path σ we have

$$S_{\text{lin}}^{(E)}[\sigma(\cdot)] = \frac{1}{\hbar} \int_0^{\hbar\beta} d\tau \int_0^\tau d\tau' \left\langle \tilde{H}_{\text{lin}}[\sigma] \tilde{H}_{\text{lin}}[\sigma] - \tilde{H}_{\text{lin}}[\sigma(\tau)] \tilde{H}_{\text{lin}}[\sigma(\tau')] \right\rangle_{\beta} . \tag{4.82}$$

Putting $[\kappa_k^{(n)}(+)]^2 = [\kappa_k^{(n)}(-)]^2$, the influence action may be written in the form (4.71),

$$S_{\text{lin}}^{(E)}[\sigma(\cdot)]/\hbar = \frac{1}{8} \int_0^{\hbar\beta} d\tau \int_0^\tau d\tau' \, \mathcal{K}_{\text{lin}}(\tau - \tau') \left(\sigma(\tau) - \sigma(\tau') \right)^2 , \tag{4.83}$$

where the kernel $\mathcal{K}_{\text{lin}}(\tau)$ is defined in terms of the integral representation

$$\mathcal{K}_{\text{lin}}(\tau) = \int_0^\infty d\omega \, G_{\text{lin}}(\omega) D_\omega(\tau) . \tag{4.84}$$

---

[2]The dissipative two- and multi-state system are discussed in Part IV and Part V, respectively.

Here, $D_\omega(\tau)$ is the phonon propagator (3.55), and the spectral density reads

$$
G_{\rm lin}(\omega) = \sum_{\boldsymbol{k},\lambda} \frac{\delta(\omega - \omega_{\boldsymbol{k},\lambda})}{2V\varrho\,\hbar\omega_{\boldsymbol{k},\lambda}} \left\{ \cos^2\left(\frac{\boldsymbol{k}\cdot\boldsymbol{q}_0}{2}\right) \left| \sum_n e^{i\boldsymbol{k}\cdot\boldsymbol{R}^{(n)}} \left(\kappa_{\boldsymbol{k},\lambda}^{(n)}(+) - \kappa_{\boldsymbol{k},\lambda}^{(n)}(-) \right) \right|^2 \right.
$$

$$
\left. + \sin^2\left(\frac{\boldsymbol{k}\cdot\boldsymbol{q}_0}{2}\right) \left| \sum_n e^{i\boldsymbol{k}\cdot\boldsymbol{R}^{(n)}} \left(\kappa_{\boldsymbol{k},\lambda}^{(n)}(+) + \kappa_{\boldsymbol{k},\lambda}^{(n)}(-) \right) \right|^2 \right\} . \tag{4.85}
$$

Consider next the Debye model for the phonon degrees of freedom. For three space dimension, linear dispersion $\omega_{\boldsymbol{k},\lambda} = v_\lambda|\boldsymbol{k}|$, and with the same Debye frequency $\omega_{\rm D}$ for the longitudinal and transversal phonon branches, we have

$$
\frac{1}{V}\sum_{\boldsymbol{k},\lambda} \cdots \rightarrow \frac{1}{2\pi^2}\sum_\lambda \frac{1}{v_\lambda^3} \int \frac{d\Omega}{4\pi} \int_0^\infty d\omega_{\boldsymbol{k},\lambda}\, \omega_{\boldsymbol{k},\lambda}^2 \Theta(\omega_{\rm D} - \omega_{\boldsymbol{k},\lambda}) \cdots . \tag{4.86}
$$

Let N denote the number of atoms per volume V. Then the summation (4.86) yields

$$
\omega_{\rm D} = \left(\frac{2\pi^2 N}{V}\right)^{1/3} \bar{v} \qquad \text{with} \qquad \frac{1}{\bar{v}^3} = \frac{1}{3}\left(\frac{1}{v_{\rm l}^3} + \frac{2}{v_{\rm t}^3}\right) . \tag{4.87}
$$

The low-frequency expansion of the spectral density $G_{\rm lin}(\omega)$ is found upon Taylor expansion of the exponential and trigonometric functions in Eq. (4.85),

$$
G_{\rm lin}(\omega) = \left(2\alpha_1\omega + 2\alpha_3\frac{\omega^3}{\omega_{\rm D}^2} + 2\alpha_5\frac{\omega^5}{\omega_{\rm D}^4} + \mathcal{O}\left(\omega^7\right) \right) \Theta(\omega_{\rm D} - \omega) . \tag{4.88}
$$

The coupling parameter of the Ohmic contribution $G_{\rm lin}(\omega) \propto \omega$ is given by

$$
\alpha_1 = \frac{1}{8\pi^2\varrho\hbar}\sum_\lambda \frac{1}{v_\lambda^3} \int \frac{d\Omega}{4\pi} \left| \sum_n \left(\kappa_{\boldsymbol{k},\lambda}^{(n)}(+) - \kappa_{\boldsymbol{k},\lambda}^{(n)}(-) \right) \right|^2 . \tag{4.89}
$$

The dimensionless coupling coefficients α_3 and α_5 are conveniently expressed in terms of the symmetrized and anti-symmetrized components

$$
\mathbb{P}_{\rm s} = \frac{1}{2}\Big(\mathbb{P}(+) + \mathbb{P}(-)\Big) \qquad \text{and} \qquad \mathbb{P}_{\rm a} = \frac{1}{2}\Big(\mathbb{P}(+) - \mathbb{P}(-)\Big) \tag{4.90}
$$

of the deformation potential tensor ($\mathbb{P}$ has dimension energy, κ has dimension force)

$$
\mathbb{P}(\pm) = \sum_n \boldsymbol{R}^{(n)} \otimes \boldsymbol{g}^{(n)}(\pm) , \tag{4.91}
$$

where $\boldsymbol{g}^{(n)}(\sigma)$ is the Kanzaki force defined in Eq. (3.166). Equation (4.85) yields [137]

$$
\alpha_3 = \frac{\omega_{\rm D}^2}{2\pi^2\varrho\hbar}\sum_\lambda \frac{1}{v_\lambda^5} \int \frac{d\Omega}{4\pi} \left| \hat{\boldsymbol{k}}\cdot\mathbb{P}_{\rm a}\cdot\boldsymbol{e}(\hat{\boldsymbol{k}},\lambda) \right|^2 , \tag{4.92}
$$

and

$$\alpha_5 = \frac{\omega_D^4}{32\pi^2 \varrho \hbar} \sum_\lambda \frac{1}{v_\lambda^7} \int \frac{d\Omega}{4\pi} \left\{ \left| \sum_n \left(\hat{k} \cdot R^{(n)} \right)^2 \left(\kappa_{\hat{k},\lambda}^{(n)}(+) - \kappa_{\hat{k},\lambda}^{(n)}(-) \right) \right|^2 \right.$$
$$\left. + 4 \left(\hat{k} \cdot q_0 \right)^2 \left(\left| \hat{k} \cdot \mathbb{P}_s \cdot e(\hat{k},\lambda) \right|^2 - \left| \hat{k} \cdot \mathbb{P}_a \cdot e(\hat{k},\lambda) \right|^2 \right) \right\}, \tag{4.93}$$

where $\hat{k}$ is a unit vector. The amplitude factors $|\sum_n (R^{(n)})^{(\nu-1)/2} \cdots|$ in the coefficients α_ν are moments of the positions of the atoms. Therefore, we may regard α_1 as the coefficient of a monopole force, α_3 as the coefficient of a dipole force, etc. [170].

For a crystalline environment, the *monopole* force is absent because of the symmetry of the interstitial positions. Therefore, the coupling coefficient α_1 is zero. Hence dissipation by acoustic phonons is super-Ohmic instead of Ohmic.

The tensor character of $\mathbb{P}$ in Eqs. (4.92) and (4.93) may lead to strong selection rules for the coefficients α_3 and α_5. The particular case depends on the symmetry of the interstitial sites [132, 134, 137].

In bcc lattices, such as Nb and Fe, the octahedral and tetrahedral interstitial sites have tetragonal symmetry. Hence there is no selection rule. Denoting the eigenvalues of the deformation potential tensor (4.91) by p_1 and p_3, where p_1 is double and p_3 belongs to the tetragonal axis, we obtain for an isotropic elastic medium

$$\alpha_3 = \frac{\omega_D^2 (p_1 - p_3)^2}{2\pi^2 \varrho \hbar} \left(\frac{1}{10} \frac{1}{v_t^5} + \frac{1}{15} \frac{1}{v_l^5} \right). \tag{4.94}$$

For niobium, we have $\alpha_3 \approx 1.3 \, (p_1 - p_3)^2/[eV]^2$ (Ref. [137]), and tabulated values of p_1 and p_3 can be found in Ref. [171]. In bcc metals, the coefficient α_5 is estimated to be smaller than or at most of the same order of magnitude as the coefficient α_3.

In fcc lattices, such as Cu and Al, the octahedral and tetrahedral sites have cubic symmetry. Therefore, the deformation potential tensor is degenerate, $p_1 = p_2 = p_3 \equiv p$. As a result of this symmetry, the dipole force is absent. Hence the coefficient α_3 is zero. Furthermore, there follows from Eq. (4.93) that only the longitudinal phonon branch contributes to the coefficient α_5. One finds [137]

$$\alpha_5 = \frac{\omega_D^4 q_0^2}{24\pi^2 \varrho \hbar v_l^7} p^2. \tag{4.95}$$

For tetrahedral sites in Al, Eq. (4.95) gives the numerical value $\alpha_5 = 0.1 \, p^2/[eV]^2$, and for octahedral sites the value $\alpha_5 = 0.2 \, p^2/[eV]^2$.

For tunneling centers in amorphous solids, the monopole force term is again missing for symmetry reasons, i.e., the coupling parameter α_1 of the Ohmic spectral density vanishes again. However, there are generally no selection rules for the dipole force. It is convenient to introduce the coupling strength as in Eq. (4.88),

$$G_{\text{lin}}(\omega) = 2\alpha_3 \frac{\omega^3}{\omega_D^2}, \quad \text{where} \quad \alpha_3 = \frac{\omega_D^2}{2\pi^2 \varrho \hbar} \left(\frac{p_l^2}{v_l^5} + \frac{2p_t^2}{v_t^5} \right). \tag{4.96}$$

The elastic energies or deformation potentials p_l and p_t describe again the coupling to longitudinal and transversal acoustic phonons, respectively.

For acoustic phonons in d dimensions, the spectral density of states per unit volume at low frequencies is

$$\sum_{k,\lambda} \delta(\omega - \omega_{k,\lambda}) \equiv D_{ac}(\omega) = 2D_{ac,0} \left(\frac{\omega}{\omega_D}\right)^{d-1}, \tag{4.97}$$

where $D_{ac,0}$ is a normalization constant. Next, we assume a frequency-dependent coupling of the power-law form [$\lambda_\alpha \to \lambda(\omega)$ in Eq. (3.140)],

$$\lambda(\omega) = \lambda_0 (\omega/\omega_D)^\mu \quad \text{for} \quad \omega \lesssim \omega_D. \tag{4.98}$$

Here, the parameter λ_0 is proportional to the deformation potential. Then the spectral density $G_{lin}(\omega)$ in Eq. (3.140) takes the power-law form [cf. Eq. (3.72)]

$$G_{lin}(\omega) \equiv D_{ac}(\omega)\lambda^2(\omega) = 2\alpha_s \omega_D^{1-s}\omega^s, \tag{4.99}$$

where $\alpha_s = D_{ac,0}\lambda_0^2/\omega_D$. The power s of the spectral algebraic law is

$$s = d - 1 + 2\mu. \tag{4.100}$$

The relation (4.100) connects the power s of the power-law form (4.99) for $G_{lin}(\omega)$ with the dimension d of the lattice and with the power μ of the frequency-dependent coupling function $\lambda(\omega)$ in Eq. (4.98). In the absence or presence of a cubic symmetry, we have $\mu = 1/2$ and $\mu = 3/2$, respectively.

A concluding thought is appropriate. When the polaron moves coherently in the crystal, the two-state assumption breaks down. In this case, it is reasonable to eliminate the dependence of the expression (4.78) on the individual positions $R^{(n)}$ of the atoms by averaging over many atoms. Then the angular integrals in Eq. (4.78) are easily carried out. We then find

$$S_{lin}^{(E)}[q(\cdot)] = \frac{3}{\pi}\int_0^\infty d\omega \, \frac{J(\omega)}{k^2(\omega)}\int_0^{\hbar\beta} d\tau \int_0^\tau d\tau' \, D_\omega(\tau - \tau')$$
$$\times \left\{1 - \frac{\sin[k(\omega)|q(\tau) - q(\tau')|]}{k(\omega)|q(\tau) - q(\tau')|}\right\}, \tag{4.101}$$

where $D_\omega(\tau)$ is the thermal propagator introduced in Subsection 3.1.6. The function $k(\omega)$ is the dispersion relation solved for $k = |k|$, and the spectral density $J(\omega)$ is related to the function $G(\omega)$ given in Eq. (4.88) by Eq. (3.141), where q_0 is a characteristic length. For $k|q(\tau) - q(\tau')| \ll 1$ in the relevant integration regime, we may expand the sine in Eq. (4.101). Then, the expression (4.101) reduces to the standard form given in Eq. (4.50).

Generally, the influence action (4.50) overestimates the dissipative influences of the environment as compared with the more accurate expression (4.101). The influence action (4.101) has been used repeatedly in studies of the ground-state energy and the effective mass of the acoustic polaron.[3]

---

[3]Cf., e.g., Refs. [172, 173]. See also the review in Ref. [174].

Particular interest has been devoted over about forty years to the investigation of the possibility of a phonon-induced self-trapping phase transition. In recent years, this question has been answered in the negative. For a clarification of this issue we refer to a review by Gerlach and Löwen [174].

4.2.5 Acoustic polaron: two-phonon coupling

We now turn to the discussion of the two-phonon coupling (3.169). The influence functional of the two-phonon process $\mathcal{F}_{\text{quadr}}^{(E)}[\boldsymbol{q}]$ is formally given by the series expansion (4.76) in which $\tilde{H}_{\text{lin}}[\boldsymbol{q}(\tau)]$ is replaced by $\tilde{H}_{\text{quadr}}[\boldsymbol{q}(\tau)]$. With use of Eqs. (4.75), (3.39), and (3.55), the thermal average of the first order takes the form

$$\left\langle \tilde{H}_{\text{quadr}}[\boldsymbol{q}(\tau)] \right\rangle_{\beta} = \frac{\hbar}{4V\varrho} \sum_{n,m} \sum_{k,\lambda} \frac{\gamma_{k,-k}^{(n,m)}[\boldsymbol{q}(\tau)]}{\omega_k} e^{i\boldsymbol{k}\cdot[\boldsymbol{R}^{(m)}-\boldsymbol{R}^{(n)}]} \coth(\beta\hbar\omega_k/2) . \quad (4.102)$$

Now assume that the particle-lattice potential in Eq. (3.161) or Eq. (3.167) is a sum of pair potentials. Furthermore, at low temperature umklapp processes may be disregarded. Then one finds [137]

$$\sum_{n,m} \gamma_{k,-k}^{(n,m)}[\boldsymbol{q}(\tau)] e^{i\boldsymbol{k}\cdot[\boldsymbol{R}^{(m)}-\boldsymbol{R}^{(n)}]} = 0 , \quad (4.103)$$

and thus

$$\langle \tilde{H}_{\text{quadr}}[\boldsymbol{q}(\tau)]\rangle_{\beta} = 0 . \quad (4.104)$$

The computation of the second cumulant is straightforward. With use of relations for Gaussian thermal averages involving four phonon operators , e.g.,

$$\left\langle b_k b_{-k'}^{\dagger} b_{-k''}^{\dagger} b_{k'''} \right\rangle_{\beta} = \left\langle b_k b_{-k'}^{\dagger} \right\rangle_{\beta} \left\langle b_{-k''}^{\dagger} b_{k'''} \right\rangle_{\beta} + \left\langle b_k b_{-k''}^{\dagger} \right\rangle_{\beta} \left\langle b_{-k'}^{\dagger} b_{k'''} \right\rangle_{\beta} , \quad (4.105)$$

and of Eqs. (4.104), (3.39), (3.55), and with the relation $\gamma_{k,k'}^{(n,m)} = \gamma_{-k,-k'}^{(n,m)}$, we find

$$\left\langle \tilde{H}_{\text{quadr}}[\boldsymbol{q}(\tau)]\tilde{H}_{\text{quadr}}[\boldsymbol{q}(\tau')] \right\rangle_{\beta} = \sum_{n,m} \sum_{n',m'} \sum_{k,k'} \frac{\hbar^2 \gamma_{k,k'}^{(n,m)}[\boldsymbol{q}(\tau)]\gamma_{k,k'}^{(n',m')}[\boldsymbol{q}(\tau')]}{2(2V\varrho)^2 \omega_k \omega_{k'}} \quad (4.106)$$

$$\times \quad e^{i\boldsymbol{k}\cdot[\boldsymbol{R}^{(n')}-\boldsymbol{R}^{(n)}]} e^{i\boldsymbol{k'}\cdot[\boldsymbol{R}^{(m')}-\boldsymbol{R}^{(m)}]} e^{i(\boldsymbol{k}+\boldsymbol{k'})\cdot[\boldsymbol{q}(\tau)-\boldsymbol{q}(\tau')]}$$

$$\times \quad \left\{ 2\Theta\left(\omega_k - \omega_{k'}\right)\left(n(\omega_{k'}) - n(\omega_k)\right) D_{\omega_k - \omega_{k'}}(\tau - \tau') \right.$$

$$\left. + \left(1 + n(\omega_k) + n(\omega_{k'})\right) D_{\omega_k + \omega_{k'}}(\tau - \tau') \right\} .$$

In the sequel, we restrict again the attention to defect tunneling between two interstitial sites. With use of the parametrization (4.79) we may write

$$\mathrm{Re}\, e^{i(\boldsymbol{k}+\boldsymbol{k}')\cdot[\boldsymbol{q}(\tau)-\boldsymbol{q}(\tau')]} = \cos^2\left(\frac{(\boldsymbol{k}+\boldsymbol{k}')\cdot\boldsymbol{q}_0}{2}\right) + \sin^2\left(\frac{(\boldsymbol{k}+\boldsymbol{k}')\cdot\boldsymbol{q}_0}{2}\right)\sigma(\tau)\sigma(\tau') . \quad (4.107)$$

As in the preceding analysis of linear phonon coupling, we are not interested in the constant energy shift induced by the coupling for $\sigma = \sigma'$. Therefore, as before in Eq.(4.82), we subtract this term from the expression (4.106). With use of Eq. (4.107) we then obtain the influence action in the form

$$S_{\mathrm{quadr}}^{(\mathrm{E})}[\sigma(\cdot)]/\hbar = \frac{1}{8}\int_0^{\hbar\beta} d\tau \int_0^{\tau} d\tau'\, \mathcal{K}_{\mathrm{quadr}}(\tau - \tau')\Big(\sigma(\tau) - \sigma(\tau')\Big)^2 \quad (4.108)$$

with the kernel

$$\mathcal{K}_{\mathrm{quadr}}(\tau) = \int_0^{\infty} d\omega\, G_{\mathrm{quadr}}(\omega) D_\omega(\tau) . \quad (4.109)$$

The spectral density resulting from two-phonon coupling is given by

$$G_{\mathrm{quadr}}(\omega) = \frac{1}{2(2V\varrho)^2}\sum_{k,k'}\left\{2\delta\big(\omega - \omega_k + \omega_{k'}\big)\frac{n(\omega_{k'}) - n(\omega_k)}{\omega_k\omega_{k'}}\right. \quad (4.110)$$

$$\left. + \delta\big(\omega - \omega_k - \omega_{k'}\big)\frac{1 + n(\omega_k) + n(\omega_{k'})}{\omega_k\omega_{k'}}\right\} f(k, k') ,$$

where

$$f(k, k') = \cos^2\left(\frac{(\boldsymbol{k}+\boldsymbol{k}')\cdot\boldsymbol{q}_0}{2}\right)\left|\sum_{n,m} e^{i\boldsymbol{k}\cdot\boldsymbol{R}^{(n)}} e^{i\boldsymbol{k}'\cdot\boldsymbol{R}^{(m)}}\left(\gamma_{k,k'}^{(n,m)}(+) - \gamma_{k,k'}^{(n,m)}(-)\right)\right|^2$$

$$+ \sin^2\left(\frac{(\boldsymbol{k}+\boldsymbol{k}')\cdot\boldsymbol{q}_0}{2}\right)\left|\sum_{n,m} e^{i\boldsymbol{k}\cdot\boldsymbol{R}^{(n)}} e^{i\boldsymbol{k}'\cdot\boldsymbol{R}^{(m)}}\left(\gamma_{k,k'}^{(n,m)}(+) + \gamma_{k,k'}^{(n,m)}(-)\right)\right|^2 ,$$

and where we used the notation

$$\gamma_{k,k'}^{(n,m)}(\sigma) \equiv \gamma_{k,k'}^{(n,m)}[\boldsymbol{q} = \sigma\boldsymbol{q}_0/2] . \quad (4.111)$$

Next, we expand the expression (4.110) in powers of ω. To linear order in ω, we find

$$G_{\mathrm{quadr}}(\omega) = 2\alpha_{\mathrm{quadr}}\,\omega + \mathcal{O}\big((\omega/\omega_{\mathrm{D}})^3\big) , \quad (4.112)$$

where the prefactor α_{quadr} is the dimensionless expression

$$\alpha_{\mathrm{quadr}} = \frac{\hbar\beta}{2(2V\varrho)^2}\sum_{k,k'}\delta\big(\omega_k - \omega_{k'}\big)\frac{n(\omega_k)[1 + n(\omega_{k'})]}{\omega_k\omega_{k'}}f(k, k') . \quad (4.113)$$

The contributions $\propto \omega^3, \omega^5, \cdots$ are of minor interest since they can be merged with the corresponding terms of $G_{\mathrm{lin}}(\omega)$ in the series (4.88). This would result in small temperature dependent contributions to the coefficients α_3, α_5, etc.

Turning to the Debye model (4.86) and expanding the trigonometric functions in $\boldsymbol{k}$ and $\boldsymbol{k}'$, we obtain the low temperature expansion for α_{quadr}. We find

$$\alpha_{\text{quadr}} = \kappa_1 \left(\frac{k_{\text{B}}T}{\hbar\omega_{\text{D}}}\right)^6 + \kappa_2 \left(\frac{k_{\text{B}}T}{\hbar\omega_{\text{D}}}\right)^8 + \mathcal{O}\left(\left(\frac{k_{\text{B}}T}{\hbar\omega_{\text{D}}}\right)^{10}\right). \tag{4.114}$$

The coupling parameter κ_1 is given by

$$\kappa_1 = \frac{9\pi^2\omega_{\text{D}}^6}{370\varrho^2} \sum_{\lambda,\lambda'} \frac{1}{v_\lambda^5 v_{\lambda'}^5} \int \frac{d\Omega}{4\pi} \int \frac{d\Omega'}{4\pi}$$
$$\times \left| \sum_{n,m} \left(\gamma_{\hat{\boldsymbol{k}},\lambda}^{(n,m)}(+) - \gamma_{\hat{\boldsymbol{k}},\lambda}^{(n,m)}(-)\right) \left(\hat{\boldsymbol{k}} \cdot \boldsymbol{R}^{(n)}\right) \left(\hat{\boldsymbol{k}}' \cdot \boldsymbol{R}^{(m)}\right) \right|^2. \tag{4.115}$$

It is also straightforward to calculate the coefficient κ_2. Since the resulting expression is quite lengthy, it is not given here, and we refer to Ref. [137].

For tetrahedral interstitial sites in bcc crystals, the coefficient κ_1 is estimated to be about 10^4. On the other hand, for cubic symmetry of the interstitial sites, there holds $\gamma^{(n)}(+) = \gamma^{(n)}(-)$, and therefore the coefficient κ_1 vanishes. The coefficient κ_2 is estimated to be about 10^6.

In summary, the spectral density due to the two-phonon process (4.112) is Ohmic. However, the viscosity coefficient varies strongly with T as T^6 or T^8, depending on the particular symmetry of the interstitial sites. We anticipate that coupling to electron-hole excitations in metals leads also to an Ohmic form of the spectral density, but with a temperature-independent viscosity (see the discussion in Subsection 4.2.8).

Correlated n-phonon processes of higher order, $n \geq 3$, which arise from the terms of order $\boldsymbol{u}^n$ in the expansion (3.161), yield again an Ohmic contribution to the spectral density. The respective viscosity coefficient behaves as T^{2n+2} in the absence of cubic symmetry, and as T^{2n+4} for cubic symmetry. Therefore, at low temperatures, multi-phonon processes with $n > 2$ merely result in a weak temperature-dependent renormalization of the parameters κ_1 and κ_2. The two-phonon process has been studied in Refs. [133, 175, 177, 178, 137].

In conclusion, even if one has, for some model, in leading order of the coupling a super-Ohmic form for $G(\omega)$, in most (if not all) cases an Ohmic spectral density contribution is found if one considers higher orders of the coupling. However, the respective viscosity will turn out to be strongly *temperature-dependent*.

4.2.6 Tunneling between surfaces: one-phonon coupling

One of the most interesting developments in surface science in the last twenty years is the possibility to manipulate atoms and molecules at a surface on an atomic scale by a scanning tunneling microscope (STM) [179]. In the atomic switch realized by Eigler *et al.* [180], a Xe atom has been reversibly transferred between a Ni surface and a tungsten tip. Thus, scanning-tunneling microscopy can be used not only for imaging, but also to test fundamental aspects of quantum mechanics [181]. The crystalline environment can have strong influence on the transfer of the atom. Louis and Sethna have emphasized that linear coupling of acoustic phonons to atoms tunneling between surfaces results in Ohmic dissipation, as opposed to the bulk case where dissipation is super-Ohmic [170]. With the groundwork already done in Subsection 4.2.4, the case of surface tunneling is easily understood.

The potential for the surface-tip system of a STM or atomic-force microscope (AFM) has a double well shape. Consider the parameter regime (3.127) which allows us to truncate the Hilbert space to two states. The tunneling atom exerts a force on the surface which changes by an amount $\Delta \boldsymbol{F}$ when the atom hops from the surface to the tip. In response to the atom being on the surface or on the tip, the surface atoms will switch their equilibrium positions. Since the displacement is small, we may consider the interaction of the tunneling atom with the surface in linear order in the displacement (one-phonon coupling). The atom on the surface or tip together with the relaxation of the atoms on the surface can thus be modeled by the spin-boson Hamiltonian (3.139). The possibility of tuning many parameters is very appealing. By application of an electrical field and by shift of the tip position, the bias energy and the tunneling coupling can be varied in a wide range.

All we need to connect tunneling between surfaces with defect tunneling in the bulk is to identify the force $\Delta \boldsymbol{F}$ with the change of the sum of the Kanzaki forces $\boldsymbol{g}^{(n)}$ when the atom moves from the surface $(+)$ to the tip $(-)$. We have

$$\Delta \boldsymbol{F} \;=\; \sum_n \Big(\boldsymbol{g}^{(n)}(+) - \boldsymbol{g}^{(n)}(-) \Big) , \tag{4.116}$$

where the Kanzaki forces are defined in Eq. (3.166). The link missing yet is easily gathered from the findings in Subsection 4.2.4, which for the present purpose are Eq. (4.88) with Eq. (4.89). Thus, the tip-plus-atom system coupled to acoustic phonons is represented by an Ohmic spectral density[4]

$$G(\omega) \;=\; \Big(2K\omega + \mathcal{O}(\omega^3) \Big)\Theta(\omega_{\mathrm{D}} - \omega) \tag{4.117}$$

with the dimensionless coupling parameter

$$K \;=\; \frac{1}{8\pi^2 \varrho \hbar} \sum_\lambda \frac{1}{v_\lambda^3} \int \frac{d\Omega}{4\pi} \left| \Delta \boldsymbol{F} \cdot \boldsymbol{e}(\hat{\boldsymbol{k}}, \lambda) \right|^2 . \tag{4.118}$$

While for defect tunneling in the bulk of a solid the monopole force is absent and hence super-Ohmic dissipation prevails, atom tunneling between a tip and a surface has lower symmetry, which results in a nonzero monopole force $\Delta \boldsymbol{F}$ on the surface and hence in Ohmic dissipation. Assuming a point force $\Delta \boldsymbol{F}/2$ on a semi-infinite isotropic medium, the Ohmic coupling parameter K is estimated as [170]

$$K \;=\; \frac{\delta}{8\pi^2 \varrho \hbar v_{\mathrm{t}}^3} (\Delta \boldsymbol{F})^2 , \tag{4.119}$$

---

[4]We deliberately denote the coupling parameter by K in order to emphasize the similarity with the Kondo parameter K occuring for a fermionic environment [see Eqs. (4.154) and (18.20)].

where v_t is the tranversal sound velocity and the numerical constant δ varies between 0.3 and 0.8, depending on the Lamé constants of the medium. By variation of the distance between the tip and the surface, the force ΔF (and thus the coupling parameter K) can be tuned besides the tunneling coupling Δ and the bias ϵ. Hence, atomic tunneling in a STM or AFM seems to be a prominent test ground for studying *macroscopic quantum tunneling* (MQT) and *macroscopic quantum coherence* (MQC) phenomena in different parameter regimes. In conclusion, we have studied the effects of phonons on the tunneling of an atom between two surfaces, and we have found, because of the lower symmetry compared with tunneling in the bulk, Ohmic dissipation.

4.2.7 Optical polaron

Consider an electron interacting with longitudinal optical phonons in a polar crystal as described by the Hamiltonian (3.159) with Eq. (3.183). In the interaction picture, the interaction term reads

$$\tilde{H}_{\mathrm{I}}[\boldsymbol{q}(\tau)] = \sum_k W_{\boldsymbol{k},\ell} \, e^{i\boldsymbol{k}\cdot\boldsymbol{q}(\tau)} \left(e^{-\omega_{\mathrm{LO}}\tau} b_{\boldsymbol{k}} + e^{\omega_{\mathrm{LO}}\tau} b_{\boldsymbol{k}}^\dagger \right). \tag{4.120}$$

The influence functional $\mathcal{F}_{\mathrm{LO}}^{(\mathrm{E})}[\boldsymbol{q}(\cdot)]$ is calculated following the lines presented in the two preceding subsections. The formal expression for $\mathcal{F}_{\mathrm{LO}}^{(\mathrm{E})}[\boldsymbol{q}(\cdot)]$ is expanded into a power series in $\tilde{H}_{\mathrm{I}}$, as in Eq. (4.76), and subsequently *Wick's theorem* for thermodynamic Gaussian averages is applied. For the simple form (4.120) the only nonvanishing cumulant contraction is again the two-point function $\langle \tilde{H}_{\mathrm{I}}[\boldsymbol{q}(\tau)]\tilde{H}_{\mathrm{I}}[\boldsymbol{q}(\tau')]\rangle_\beta$, and thus the influence action is again given by [cf. Eq. (4.77)]

$$S_{\mathrm{LO}}^{(\mathrm{E})}[\boldsymbol{q}(\cdot)] = -\frac{1}{\hbar} \int_0^{\hbar\beta} d\tau \int_0^\tau d\tau' \, \langle \tilde{H}_{\mathrm{I}}[\boldsymbol{q}(\tau)]\tilde{H}_{\mathrm{I}}[\boldsymbol{q}(\tau')]\rangle_\beta. \tag{4.121}$$

The contraction takes the form

$$\langle \tilde{H}_{\mathrm{I}}[\boldsymbol{q}(\tau)]\tilde{H}_{\mathrm{I}}[\boldsymbol{q}(\tau')]\rangle_\beta = \sum_k |W_{\boldsymbol{k},\ell}|^2 \, e^{i\boldsymbol{k}\cdot[\boldsymbol{q}(\tau)-\boldsymbol{q}(\tau')]} \tag{4.122}$$

$$\times \left[e^{-\omega_{\mathrm{LO}}(\tau-\tau')} \langle b_{\boldsymbol{k}} b_{\boldsymbol{k}}^\dagger \rangle_\beta + e^{\omega_{\mathrm{LO}}(\tau-\tau')} \langle b_{\boldsymbol{k}}^\dagger b_{\boldsymbol{k}} \rangle_\beta \right].$$

Next, we observe that the square bracket in Eq. (4.122) is just the boson propagator $D_{\omega_{\mathrm{LO}}}(\tau - \tau')$, as follows from inspection of the expression (3.55). Thus we have

$$\langle \tilde{H}_{\mathrm{I}}[\boldsymbol{q}(\tau)]\tilde{H}_{\mathrm{I}}[\boldsymbol{q}(\tau')]\rangle_\beta = \sum_k |W_{\boldsymbol{k},\ell}|^2 \, e^{i\boldsymbol{k}\cdot[\boldsymbol{q}(\tau)-\boldsymbol{q}(\tau')]} D_{\omega_{\mathrm{LO}}}(\tau - \tau'). \tag{4.123}$$

The effective action of a longitudinal optical polaron has a kinetic and an influence contribution,

$$S_{\mathrm{eff}}^{(\mathrm{E})}[\boldsymbol{q}(\cdot)] = \frac{M}{2} \int_0^{\hbar\beta} d\tau \, \dot{\boldsymbol{q}}^2(\tau) + S_{\mathrm{LO}}^{(\mathrm{E})}[\boldsymbol{q}(\cdot)]. \tag{4.124}$$

With use of the form (3.183) for $W_{k,l}$ the influence action is found to read

$$S_{LO}^{(E)}[q(\cdot)] \;=\; -\frac{\hbar}{V}\frac{4\pi\alpha\omega_{LO}^2\sqrt{\hbar}}{(2M\omega_{LO})^{1/2}}\int_0^{\hbar\beta}d\tau\int_0^{\tau}d\tau'\,D_{\omega_{LO}}(\tau-\tau')\sum_k\frac{e^{ik\cdot[q(\tau)-q(\tau')]}}{|k|^2}\,. \qquad (4.125)$$

Finally, in the continuum limit of a three-dimensional isotropic medium, the influence action takes the form

$$S_{LO}^{(E)}[q(\cdot)] \;=\; -\alpha\hbar\left(\frac{\hbar}{2M\omega_{LO}}\right)^{1/2}\omega_{LO}^2\int_0^{\hbar\beta}d\tau\int_0^{\tau}d\tau'\,\frac{D_{\omega_{LO}}(\tau-\tau')}{|q(\tau)-q(\tau')|}\,. \qquad (4.126)$$

The polaron problem represents one of the simplest examples of the interaction of a particle with a field. With the influence action (4.126) however, we are left with a considerably complicated non-Gaussian path integral. Fortunately, one may tackle the polaron path integral with a variational approach (cf. Subsection 8.2). The variational method has been successfully utilized in diverse applications and gives reliable results for arbitrary strength of the coupling.

4.2.8   Heavy particle in a metal

We now turn our attention to a study of the response of a noninteracting Fermi liquid to a local time-dependent perturbation. The above elimination procedure adequate for bosons is generally not suitable for fermions. The anti-commutation relations for fermions involve difficulties which are absent in the case of bosons [182]. Fortunately, the low-energy excitations of the Fermi liquid have bosonic character. Therefore, it is possible to circumvent these difficulties and give a consistent formulation following previous lines. To determine the effective action for a heavy particle interacting with conduction electrons, we have to evaluate the influence functional (4.31) for the Hamiltonian (3.184). It is convenient to work out $\mathcal{F}^{(E)}[q(\cdot)]$ using the operator formulation for the fermionic degrees of freedom. We have

$$\begin{aligned}
\mathcal{F}^{(E)}[q(\cdot)] \;&=\; Z_R^{-1}\mathrm{tr}_R\left\{e^{-\beta H_R}\,T_\tau\exp\left(-\int_0^{\hbar\beta}d\tau\,\tilde{H}_I[q(\tau)]/\hbar\right)\right\}\\
&\equiv\; \left\langle T_\tau\exp\left(-\int_0^{\hbar\beta}d\tau\,\tilde{H}_I[q(\tau)]/\hbar\right)\right\rangle_\beta,
\end{aligned} \qquad (4.127)$$

where T_τ is the imaginary-time ordering operator for fermion fields. In the second form, $\langle\cdots\rangle_\beta$ denotes thermal average with respect to $\exp(-\beta H_R)$. The tilde indicates again that we are in the interaction representation [cf. Eq. (4.74)]. We proceed by expanding $\mathcal{F}^{(E)}[q]$ into a power series in $\tilde{H}_I$ as given in Eq. (4.76).

Consider first *normally conducting electrons*. We find from Eq. (3.187)

$$\tilde{H}_I \;=\; \sum_{k,k',\sigma,\sigma'}<k,\sigma|U|k',\sigma'>\exp[i(k'-k)\cdot q(\tau)]\exp[-(\omega_{k'}-\omega_k)\tau]\,c_{k\sigma}^\dagger c_{k'\sigma'}\,. \qquad (4.128)$$

Working in perturbation theory, all terms with odd powers of $\tilde{H}_I$ vanish because $\langle \tilde{H}_I[q(\tau)] \rangle_\beta = 0$. The lowest-order term $(n = 2)$ in the series expression analogous to Eq. (4.76) is written in terms of the two-time electron-hole contraction

$$\mathcal{H}_2[q(\tau), q(\tau')] \equiv \langle \tilde{H}_I[q(\tau)] \, \tilde{H}_I[q(\tau')] \rangle_\beta \tag{4.129}$$

as

$$\mathcal{F}_2^{(E)}[q(\cdot)] = \int_0^{\hbar\beta} d\tau \int_0^\tau d\tau' \, \mathcal{H}_2[q(\tau), q(\tau')]/\hbar^2 . \tag{4.130}$$

Next, we use the expression (4.128) and employ for the thermal averages the relations

$$\langle c_{k\sigma}^\dagger c_{k'\sigma'} \rangle_\beta = \delta_{\sigma\sigma'}\delta_{kk'} f_k , \qquad \langle c_{k\sigma} c_{k'\sigma'}^\dagger \rangle_\beta = \delta_{\sigma\sigma'}\delta_{kk'}(1 - f_k) . \tag{4.131}$$

Here, f_k is the Fermi distribution function,

$$f_k \equiv f(\omega_k) = \frac{1}{\exp(\beta\hbar\omega_k) + 1} , \tag{4.132}$$

and the energy $\hbar\omega_k$ is measured relatively to the Fermi energy. We then get

$$\mathcal{H}_2[q(\tau), q(\tau')] = \sum_{k,k',\sigma,\sigma'} |<k,\sigma|U|k',\sigma'>|^2 e^{i(k'-k)\cdot[q(\tau)-q(\tau')]}$$
$$\times \exp\left((\omega_{k'} \quad \omega_k)(\tau - \tau')\right) f_k(1 - f_{k'}) . \tag{4.133}$$

The expression (4.130) with Eq. (4.133) is a density response function of the non-interacting Fermi gas. The function $\mathcal{H}_2[q(\tau), q(\tau')]$ may be interpreted in terms of an electron-hole pair injected at imaginary time τ' and removed at a later imaginary time τ. This is indicated by diagram (a) in Fig. 4.2. All those diagrams representing multiple self-energy insertions arising from uncorrelated electron-hole loops, e. g., the two shown in (b) of Fig. 4.2, can be summed up and expressed as an exponential of the simple diagram (a). As we shall see immediately, the spectral density of the particle's coupling to the boson-like electron-hole excitations implicit in Eq. (4.130) is of the Ohmic form $J(\omega) \propto \omega$, this being due to the constant density of states around the Fermi surface. Contractions of higher order in the cumulant expansion for the action resulting from Eq. (4.127) [cf. Eq. (4.76)], e.g., diagram (c) in Fig. 4.2, correspond to coherent excitation of two or more electron-hole pairs. Since they represent convolutions of the densities of pairs, they come with additional powers of ω at low frequency in comparison to diagram (a). Therefore, contributions of type (c) to the response of the electron gas can be disregarded at low temperature.

Within the approximation of uncorrelated electron-hole pairs, we thus obtain the influence functional in the form

$$\mathcal{F}^{(E)}[q(\cdot)] = \exp\left(\int_0^{\hbar\beta} d\tau \int_0^\tau d\tau' \, \mathcal{H}_2[q(\tau), q(\tau')]\Big/\hbar^2\right) . \tag{4.134}$$

The evaluation of the expression (4.133) is simplified when the impurity potential is a contact potential, $<k,\sigma|U|k',\sigma'> = \delta_{\sigma\sigma'}U_0$. The coupling strength U_0 is called

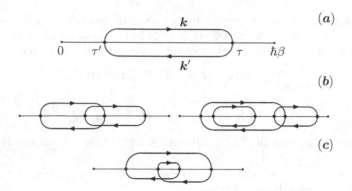

Figure 4.2: Some virtual electron-hole diagrams appearing in the expansion in $\tilde{H}_{\mathrm{I}}[\boldsymbol{q}(\tau)]$ are shown. The oval (thick) lines represent electron and hole propagators, respectively. Diagrams (a) and (b) are included in the expression (4.134), while diagram (c) is neglected.

deformation potential. This form is also an appropriate approximation for reasonably extended potentials, since the Fermi functions effectively restrict $\hbar|\omega_{\boldsymbol{k}}|$ and $\hbar|\omega_{\boldsymbol{k}'}|$ in Eq. (4.133) to small values compared to the Fermi energy E_{F}.

Computation of the angular integrals for an isotropic medium ($d = 3$) gives

$$
\mathcal{H}_2[\Delta q(\tau, \tau')] = 2 U_0^2 \sum_{\boldsymbol{k}, \boldsymbol{k}'} \frac{\sin[k \Delta q(\tau, \tau')]}{k \Delta q(\tau, \tau')} \frac{\sin[k' \Delta q(\tau, \tau')]}{k' \Delta q(\tau, \tau')}
$$
$$
\times\ \mathrm{e}^{(\omega_k - \omega_{k'})(\tau - \tau')} f_k (1 - f_{k'}) ,
$$

(4.135)

with

$$
\Delta q(\tau, \tau') \equiv |\boldsymbol{q}(\tau) - \boldsymbol{q}(\tau')| .
$$

(4.136)

Next, we perform the continuum limit

$$
\sum_{\boldsymbol{k}} \cdots \ \to \ \hbar \varrho \int d\omega\, \mathcal{N}(\omega) \cdots ,
$$

(4.137)

and count the energy relative to the Fermi energy E_{F}. The quantity $\varrho \mathcal{N}(\omega)$ is the number of states per energy interval $\hbar\, d\omega$, where ϱ is the density of states at the Fermi energy and has dimension inverse energy. Thus by definition

$$
\mathcal{N}(\omega = 0) = 1 .
$$

(4.138)

As we are interested in the low energy excitations around the Fermi surface, we may replace k and k' in the fractions of Eq. (4.135) by the Fermi wave number k_{F}. At this point it is convenient to absorb various factors into the "viscosity" coefficient

$$
\eta = 4\pi U_0^2 \varrho^2 \hbar k_{\mathrm{F}}^2 / 3 ,
$$

(4.139)

which has the usual dimension mass times frequency. Thus we may write

$$\mathcal{H}_2[\Delta q(\tau, \tau')] = \frac{3\hbar}{2k_F^2} \frac{\sin^2[k_F \Delta q(\tau, \tau')]}{[k_F \Delta q(\tau, \tau')]^2} K(\tau - \tau') , \qquad (4.140)$$

$$K(\tau) = \frac{\eta}{\pi} \int_{-\infty}^{\infty} d\omega' \int_{-\infty}^{\infty} d\omega'' \, \mathcal{N}(\omega') \mathcal{N}(\omega'') \exp[(\omega'' - \omega')\tau] f(\omega'') \, f(-\omega') . \qquad (4.141)$$

With the substitution $\omega'' = \omega' - \omega$, the kernel $K(\tau)$ takes the form

$$K(\tau) = \frac{1}{\pi} \int_{-\infty}^{\infty} d\omega \, \frac{\exp(-\omega\tau)}{1 - \exp(-\omega\hbar\beta)} J(\omega) \qquad (4.142)$$

with

$$J(\omega) = \eta \left[1 - e^{-\omega\hbar\beta} \right] \int_{-\infty}^{\infty} d\omega' \, \mathcal{N}(\omega') \mathcal{N}(\omega' - \omega) f(\omega' - \omega) f(-\omega') . \qquad (4.143)$$

Use of the explicit form of the Fermi function yields

$$J(\omega) = \frac{\eta}{2} \int_{-\infty}^{\infty} d\omega' \, \mathcal{N}(\omega') \mathcal{N}(\omega' - \omega) \left\{ \tanh(\tfrac{1}{2}\beta\hbar\omega') - \tanh[\tfrac{1}{2}\beta\hbar(\omega' - \omega)] \right\} , \qquad (4.144)$$

which shows that $J(\omega)$ is antisymmetric in ω. The antisymmetric part of the fraction in Eq. (4.142) is just the phonon propagator (3.55). Thus we may write

$$K(\tau) = \frac{1}{\pi} \int_0^{\infty} d\omega \, J(\omega) D_\omega(\tau) . \qquad (4.145)$$

This form coincides with the familiar bosonic heat bath kernel (4.54). Next, we require again that the influence functional $\mathcal{F}^{(E)}[q(\cdot)]$ is unity for a constant path. Therefore, we subtract from $\mathcal{H}_2[\Delta q(\tau, \tau')]$ in Eq. (4.140) the expression $\mathcal{H}_2[\Delta q = 0]$. We then find $\mathcal{F}^{(E)}[q(\cdot)]$ in the form (4.31) with the influence action

$$S_{\text{infl}}^{(E)}[q(\cdot)] = \frac{3}{2k_F^2} \int_0^{\hbar\beta} d\tau \int_0^{\tau} d\tau' \, K(\tau - \tau') \left(1 - \frac{\sin^2[k_F|q(\tau) - q(\tau')|]}{[k_F|q(\tau) - q(\tau')|]^2} \right) . \qquad (4.146)$$

The influence action (4.146) holds without restriction on the numerical value of $k_F \Delta q$.

For defect tunneling in metals between two interstitial sites of distance q_0, we may use again the spin parametrization (4.67). Then the influence action may be rewritten in the spin-boson form [cf. Eq. (4.71) with Eq. (4.70)]

$$S_{\text{infl}}^{(E)}[\sigma(\cdot)] = \frac{\hbar}{8} \int_0^{\hbar\beta} d\tau \int_0^{\tau} d\tau' \, \mathcal{K}(\tau - \tau') [\sigma(\tau) - \sigma(\tau')]^2 \qquad (4.147)$$

with

$$\mathcal{K}(\tau) = \int_0^{\infty} d\omega \, G(\omega) D_\omega(\tau) . \qquad (4.148)$$

The influence action (4.146) takes the form (4.147) wherein the spectral density $G(\omega)$ is related to $J(\omega)$ as

$$G(\omega) = \frac{3}{(k_F q_0)^2} \left(1 - \frac{\sin^2(k_F q_0)}{(k_F q_0)^2} \right) \frac{q_0^2}{\pi\hbar} J(\omega) . \qquad (4.149)$$

For conduction electrons, the density of states around the Fermi level is constant,

$$\mathcal{N}(\omega) = 1 . \tag{4.150}$$

If we disregard, in addition, the finite width of the conduction band, the ω'-integral in Eq. (4.144) can easily be performed. We then get

$$J(\omega) = \eta\omega , \tag{4.151}$$

which is just the Ohmic form (3.65).

The expression (4.139) is the viscosity in Born approximation, $\eta \propto U_0^2$, resulting from the sea of electron-hole excitations. In a non-perturbative treatment, Sassetti *et al.* [183] showed by calculation of the single-particle propagator for a contact potential that the friction coefficient may be expressed in terms of the s-wave scattering phase shift δ_0 at the Fermi energy,

$$\eta = 4\hbar\, k_{\mathrm{F}}^2 \sin^2\delta_0 / 3\pi , \qquad \text{where} \qquad \tan\delta_0 = -\pi U_0 \varrho . \tag{4.152}$$

The same form is also found by calculation of the force auto-correlation function of a heavy particle in a free electron gas. For a spin-independent spherically symmetric finite-range potential, the partial wave analysis yields [184]

$$\eta = \frac{4\hbar}{3\pi} k_{\mathrm{F}}^2 \sum_{l=0}^{\infty}(l+1)\sin^2(\delta_{l+1}-\delta_l) , \tag{4.153}$$

where δ_l represents the scattering phase shift of the lth partial wave.

As intermediate result, we emphasize that the electronic excitations near to the Fermi energy collectively behave as if they were bosons with an Ohmic spectral density of the coupling, regardless of whether the impurity interaction is a contact or a finite-range potential. Similar conclusions were drawn by Chang and Chakravarty [185] who studied Dyson's equation for the fermion propagator in presence of the impurity within a real-time formulation. They solved the problem by employing the long-time approximation of the free fermion propagator, a method originally proposed by Nozières and De Dominicis [186]. Furthermore, it has been shown via bosonization of the fermion operators [187] that the low-energy excitations of a fermionic bath can be mapped onto a bosonic bath. Importantly, the scattering phase of a contact potential depends on the regularization scheme employed. Only the leading Born term is found to be universal, i.e., does not depend on the particular regularization prescription. For a discussion of this point we refer to Ref. [187].

With the Ohmic form (4.151, the spin-boson spectral density $G(\omega)$ is Ohmic too,

$$G(\omega) = 2K\omega . \tag{4.154}$$

The dimensionless damping strength K is usually called Kondo parameter for reasons which become clear shortly. For interstitials in metals with a tunneling distance q_0 of the order of the lattice constant, we have $k_{\mathrm{F}} q_0 \ll 1$. In this important limit we then find from (4.149) with (4.151)

$$K = \frac{\eta q_0^2}{2\pi\hbar} + \mathcal{O}\left((k_F q_0)^4\right) . \tag{4.155}$$

The expression (4.155) coincides with a corresponding result found by Sols and Guinea [188] in a linear response calculation. Schönhammer [189] has shown by taking all the scattering phase shifts into account that the relation (4.155) is generally valid in the limit $k_F q_0 \ll 1$ for any spherically symmetric potential independent of the strength of the coupling constant.

Use of Eq. (4.151) with (4.152) in Eq. (4.149) yields

$$K = (2/\pi^2) \sin^2 \delta_0 \left[1 - \sin^2(k_F q_0)/(k_F q_0)^2 \right] . \tag{4.156}$$

This relation has the limiting forms

$$K = 2(U_0\rho)^2 [1 - \sin^2(k_F q_0)/(k_F q_0)^2] , \qquad \delta_0 \ll \pi/2 , \tag{4.157}$$

$$K = \frac{2}{3\pi^2} (k_F q_0)^2 \sin^2 \delta_0 + \mathcal{O}[(k_F q_0)^4] , \qquad k_F q_0 \ll 1 . \tag{4.158}$$

The nature of the electronic screening cloud around an impurity in a metal has been the subject of intense studies. Anderson [190] considered the overlap of the ground state $|\phi>$ of the fermions in the absence of the impurity with the corresponding state $|\psi>$ in the presence of the impurity and found that the overlap integral for a state with N conduction electrons behaves as

$$<\psi|\phi> \propto N^{-K_+} , \tag{4.159}$$

where K_+ is a positive number. This relation, known as Anderson's *orthogonality theorem*, indicates that the overlap tends to zero as the system size tends to infinity. For a contact potential, one finds

$$K_+ = (\delta_0/\pi)^2 , \tag{4.160}$$

where δ_0 is the s-wave phase for electron scattering off the impurity.

Consider next a version of the orthogonality theorem which concerns the overlap of the ground state of the fermions with the impurity at position q_1, denoted by $|\psi_1>$, and the ground state of the fermions with the impurity at position q_2, denoted by $|\psi_2>$. Kondo has found that these states obey a similar orthogonality theorem [136],

$$<\psi_1|\psi_2> \propto N^{-K} . \tag{4.161}$$

Interestingly, the overlap parameter K introduced by Kondo coincides with the dimensionless damping strength K in the spectral density (4.154). It is is a function of the phase shifts and the distance $q_0 = |q_1 - q_2|$. A general calculation of K as a function of the distance is rather difficult [191, 192, 189], since each partial wave centered at q_1 mixes with *all* partial waves centered at q_2. Even when the impurity potential is spherically symmetric, the problem lacks spherical symmetry.

For s-wave scattering and spin degeneracy, one finds [191]

$$K = \frac{2}{\pi^2} \left\{ \arctan\left(\frac{\sqrt{1 - x(q_0)} \tan\delta_0}{\sqrt{1 + x(q_0)\tan^2\delta_0}} \right) \right\}^2 . \tag{4.162}$$

The phase shift δ_0 is given in Eq. (4.152). Equation (4.162) gives the bound $K \leq 1/2$. The function $x(q_0)$ depends on the space dimension d,

$$
\begin{aligned}
x(q_0) &= \sin^2(k_F q_0)/(k_F q_0)^2 &&\text{for} &&d = 3, &&\tag{4.163}\\
x(q_0) &= \sin^2(k_F q_0) &&\text{for} &&d = 1. &&\tag{4.164}
\end{aligned}
$$

For $d = 3$, the expression (4.162) coincides in the limits $\delta_0 \ll \pi/2$ and $k_F q_0 \ll 1$ with the expressions (4.157) and (4.158), respectively.

The *short-distance* form of K, Eq. (4.155), does not depend on the spatial dimension d [189], as distinguished from the large distance behavior which depends on d. In the one-dimensional case, the function $x(q_0)$ is oscillating around the mean value $\frac{1}{2}$. For $d \geq 2$, the function $x(q_0)$ goes to zero as $k_F q_0 \to \infty$. Putting $x = 0$ in Eq. (4.162), we obtain $K = 2(\delta_0/\pi)^2$, and thus $K = 2K^+$.

For s-wave scattering, the overlap parameter K has the upper bound $1/2$ for arbitrary distances as follows from Eq. (4.162). In the general case of an extended potential, K is *not* bounded in principle by this value in two or three dimensions, since the exponent K_+ is not bounded [189]. Experimentally, K turns out to be small, typically about 0.1 for charged interstitials in metals. The parameter K is small because surrounding charges are effectively screening the impurity potential.

For the Ohmic spectral density (4.154), the kernel $\mathcal{K}(\tau)$ in the retarded action (4.147) can be calculated in analytic form. The integral expression (4.148) yields

$$\mathcal{K}(\tau) = 2K \frac{(\pi/\hbar\beta)^2}{\sin^2(\pi\tau/\hbar\beta)} . \tag{4.165}$$

The influence action (4.147) with the kernel (4.165) conveys the effects of a normal-state metallic environment on interstitial tunneling at thermal energies well below the Fermi energy.

4.2.9 Heavy particle in a superconductor

It is straightforward to generalize the discussion of the preceding subsection to the *superconducting* state. The underlying Hamiltonian is given by Eqs. (3.188) and (3.194) with (3.197). According to the BCS theory, the electronic spectrum is modified and the coherence factor is as specified in Eq. (3.197). The influence action is found again in the form (4.146), and with the TSS path (4.67) in the form (4.147), respectively. Furthermore, the kernel $\mathcal{K}(\tau)$ takes again the form (4.148), but now the spectral density of the coupling (4.144) is adjusted to BCS-quasiparticle excitations,

$$
\begin{aligned}
G_{qp}(\omega) = K \int_{-\infty}^{\infty} d\omega'\, \mathcal{N}_{qp}(\omega') \mathcal{N}_{qp}(\omega' - \omega) \left\{ 1 - \frac{\Delta_g^2}{\omega'(\omega' - \omega)} \right\} \\
\times \left\{ \tanh(\tfrac{1}{2}\beta\hbar\omega') - \tanh[\tfrac{1}{2}\beta\hbar(\omega' - \omega)] \right\} .
\end{aligned}
\tag{4.166}
$$

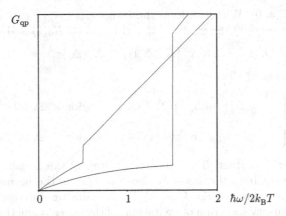

Figure 4.3: Quasiparticle spectral density G_{qp} as a function of $\hbar\omega/2k_BT$. The solid curves correspond to $\hbar\Delta_g/k_BT = 0.5$ and $\hbar\Delta_g/k_BT = 1.5$, respectively. The dotted straight line represents the spectral density in the normally conducting state.

Here $\mathcal{N}_{qp}(\omega)$ is the density of states of BCS quasiparticles, and the coherence factor is $1 - \Delta_g^2/[\omega'(\omega' - \omega)]$, which holds for time-reversal-invariant interaction. Within the BCS theory, the density of states is given by

$$\mathcal{N}_{qp}(\omega) = \Theta(|\omega| - \Delta_g)\frac{|\omega|}{\sqrt{\omega^2 - \Delta_g^2}} \, . \tag{4.167}$$

In the low- frequency range $\omega \ll \Delta_g$, the function $G_{qp}(\omega)$ takes the Ohmic form

$$G_{qp}(\omega) = 2K_{qp}\,\omega \tag{4.168}$$

with the temperature-dependent dimensionless Ohmic coupling constant

$$K_{qp} = 2Kf(\Delta_g) = 2K/[1 + \exp(\beta\hbar\Delta_g)] \, . \tag{4.169}$$

Thus, K_{qp} is downsized by the gap and even drops to zero at $T = 0$. The function $G_{qp}(\omega)$ is discontinuous at the excitation threshold $\omega = 2\Delta_g$, with a jump of height

$$\Delta G_{qp} = 2K\pi\Delta_g \tanh\left(\frac{\hbar\Delta_g}{2k_BT}\right) \, . \tag{4.170}$$

Above the threshold, $G_{qp}(\omega)$ increases with increasing frequency and approaches in the regime $\omega \gg \Delta_g$ the Ohmic form $G_{qp}(\omega) = 2K\omega$. Fig. 4.3 shows plots of $G_{qp}(\omega)$ for different temperatures.

At zero temperature, the integral in Eq. (4.166) can be expressed in terms of the complete elliptic integral of the second kind $E(\pi/2, k)$ [88]. We then get

$$G_{qp}(\omega) = \Theta\left(\omega - 2\Delta_g\right) 2K\omega\, E\left(\pi/2, \sqrt{1 - 4\Delta_g^2/\omega^2}\right) \, , \tag{4.171}$$

where $\Delta_g = \Delta_g(T = 0)$. With this form the kernel $\mathcal{K}(\tau) = \int_0^\infty d\omega \, G_{qp}(\omega) D_\omega(\tau)|_{T=0}$ may be written in terms of the modified Bessel functions $K_0(z)$ and $K_1(z)$ as

$$\mathcal{K}(\tau) = 2K\Delta_g^2 \left\{ K_1^2(\Delta_g|\tau|) + K_0^2(\Delta_g|\tau|) \right\} . \tag{4.172}$$

The limiting behavior of $\mathcal{K}(\tau)$ is

$$\mathcal{K}(\tau) = 2K \begin{cases} \dfrac{1}{\tau^2} + \Delta_g^2 \Big(\ln(\Delta_g|\tau|/2) - C_E \Big)^2 , & \text{for} \quad \Delta_g|\tau| \ll 1 , \\[2mm] \dfrac{\pi}{|\tau|}\Delta_g \, e^{-2\Delta_g|\tau|} , & \text{for} \quad \Delta_g|\tau| \gg 1 , \end{cases} \tag{4.173}$$

where C_E is Euler's constant. The exponential drop of the kernel at long time is qualitatively different from the algebraic decay $\propto 1/\tau^2$ in the normally conducting state at zero temperature [cf. Eq. (4.165)]. The qualitative difference is due to the gap in the low frequency spectrum of excitations of the superconducting environment.

4.2.10   Effective action of a junction

To find the effective action of a junction, we start out from the microscopic Hamiltonian (3.199) put down in Subsection 3.4.4 and follow the procedure outlined by Ambegaokar *et al.* [150]. In the first step, the quartic interactions in Eqs. (3.198) and (3.201) are eliminated in the functional integral representation of the expression (3.203) by means of a Gaussian identity which has become known as Hubbard-Stratonovich transformation. In this method, new fields are introduced in exchange for the quartic pair interaction and for the quartic Coulomb interaction, respectively. The former can be identified with the complex order parameter fields $\Delta_L(\boldsymbol{r}, \tau)$ and $\Delta_R(\boldsymbol{r}, \tau)$, the latter with the (real) voltage field $U(\tau)$. As an outcome of the Hubbard-Stratonovich identity, the resulting action is quadratic in the fermion field. Hence the trace with respect to this field can readily be performed. The partition function takes the form

$$Z = \int \mathcal{D}^2 \Delta_L(\boldsymbol{r}, \tau) \, \mathcal{D}^2 \Delta_R(\boldsymbol{r}, \tau) \, \mathcal{D}U(\tau) \, \mathcal{D}\boldsymbol{A}(\boldsymbol{r}, \tau) \exp(-\mathcal{A}[\Delta_L, \Delta_R, U, \boldsymbol{A}]/\hbar) , \tag{4.174}$$

with the effective action

$$\mathcal{A} = \hbar \, \mathrm{tr} \ln \underline{\hat{G}}^{-1} + \frac{1}{2} \int_0^{\hbar\beta} d\tau \, CU^2(\tau) + \frac{1}{8\pi} \int_0^{\hbar\beta} d\tau \int d^3r \left(\boldsymbol{h}(\boldsymbol{r}, \tau) - \boldsymbol{h}_{\mathrm{ext}} \right)^2$$

$$+ \int_0^{\hbar\beta} d\tau \left\{ \int_{r\in L} d^3r \, g_L^{-1}(\boldsymbol{r}) \, |\Delta_L(\boldsymbol{r}, \tau)|^2 + \int_{r\in R} d^3r \, g_R^{-1}(\boldsymbol{r}) \, |\Delta_R(\boldsymbol{r}, \tau)|^2 \right\} . \tag{4.175}$$

Here, $\underline{\hat{G}}$ is a 4×4 matrix Green function in the space spanned by (L) and (R) and in the Nambu pseudo-spin space.[5] In the sequel, we indicate matrices in Nambu space by carets, and matrices describing the (L) and (R) superconductors by underlines. The inverse of the Green function is given by

---

[5]The two-dimensional Nambu space facilitates a compact formulation of BCS superconductivity.

$$\underline{\hat{G}}^{-1}(\boldsymbol{r},\tau;\boldsymbol{r}',\tau') = \begin{pmatrix} \hat{G}_{\rm L}^{-1}(\boldsymbol{r},\tau;\boldsymbol{r}',\tau') & \hat{T}_{\boldsymbol{r}\boldsymbol{r}'}\,\delta(\tau-\tau') \\ \hat{T}_{\boldsymbol{r}\boldsymbol{r}'}^{\dagger}\,\delta(\tau-\tau') & \hat{G}_{\rm R}^{-1}(\boldsymbol{r},\tau;\boldsymbol{r}',\tau') \end{pmatrix}. \tag{4.176}$$

The diagonal elements in the L-R space are

$$\hat{G}_{\rm L/R}^{-1}(\boldsymbol{r},\tau;\boldsymbol{r}',\tau') = \left\{ \hbar\frac{\partial}{\partial\tau}\hat{1} + \left[-\frac{\hbar^2}{2m}\left(\boldsymbol{\nabla} - \frac{ie}{\hbar c}\boldsymbol{A}\hat{\tau}_3 \right)^2 - \mu + ie\,U_{\rm L/R}(\tau) \right]\hat{\tau}_3 \right.$$

$$\left. + \hat{\Delta}_{\rm L/R}(\boldsymbol{r},\tau) \right\}\delta(\boldsymbol{r}-\boldsymbol{r}')\,\delta(\tau-\tau'), \tag{4.177}$$

where $\hat{\tau}_3$ is the Pauli matrix in the Nambu space, and where $U(\tau) = U_{\rm L}(\tau) - U_{\rm R}(\tau)$. The order parameter field in Nambu space reads

$$\hat{\Delta}_{\rm L/R}(\boldsymbol{r},\tau) = \begin{pmatrix} 0 & \Delta_{\rm L/R}(\boldsymbol{r},\tau) \\ \Delta_{\rm L/R}^{*}(\boldsymbol{r},\tau) & 0 \end{pmatrix} = |\Delta_{\rm L/R}|\exp(-i\,\varphi_{\rm L/R}\hat{\tau}_3)\,\hat{\tau}_1. \tag{4.178}$$

The two superconductors (L) and (R) are coupled by the transfer matrix

$$\hat{T}_{\boldsymbol{r}\boldsymbol{r}'} - \begin{pmatrix} T_{\boldsymbol{r}\boldsymbol{r}'} & 0 \\ 0 & -T_{\boldsymbol{r}\boldsymbol{r}'}^{*} \end{pmatrix}, \tag{4.179}$$

It is useful to perform a gauge transformation with the aim of making the off-diagonal elements in Eq. (4.178) real. This is achieved with a transformation induced by the unitary matrix

$$\underline{\hat{U}} = \begin{pmatrix} e^{i\varphi_{\rm L}(\boldsymbol{r},\tau)\hat{\tau}_3/2} & 0 \\ 0 & e^{i\varphi_{\rm R}(\boldsymbol{r},\tau)\hat{\tau}_3/2} \end{pmatrix}. \tag{4.180}$$

Then the diagonal element of $\underline{\hat{G}}'^{-1} = \underline{\hat{U}}\,\underline{\hat{G}}^{-1}\underline{\hat{U}}^{-1}$ in the L-space becomes

$$\hat{G}_{\rm L}'^{-1}(\boldsymbol{r},\tau;\boldsymbol{r}',\tau') = \left\{ \hbar\frac{\partial}{\partial\tau}\hat{1} - i\hbar\left(\boldsymbol{v}_{\rm s,L} \cdot \boldsymbol{\nabla} \right)\hat{1} + \hat{\Delta}_{\rm L}(\boldsymbol{r},\tau) \right. \tag{4.181}$$

$$\left. + \left[-\frac{\hbar^2\boldsymbol{\nabla}^2}{2m} - \mu + \frac{m}{2}v_{\rm s,L}^2 - i\left(\frac{\hbar}{2}\frac{\partial\varphi_{\rm L}}{\partial\tau} - eU_{\rm L}(\tau) \right) \right]\hat{\tau}_3 \right\}\delta(\boldsymbol{r}-\boldsymbol{r}')\,\delta(\tau-\tau').$$

In the transformation, the left-right off-diagonal elements pick up a phase,

$$\hat{T}_{\boldsymbol{r}\boldsymbol{r}'}' = \begin{pmatrix} T_{\boldsymbol{r}\boldsymbol{r}'}\,e^{i\psi(\tau)/2} & 0 \\ 0 & -T_{\boldsymbol{r}\boldsymbol{r}'}^{*}\,e^{-i\psi(\tau)/2} \end{pmatrix}, \tag{4.182}$$

where $\psi(\tau)$ is the phase difference across the junction plane. Here we assumed that $T_{\boldsymbol{r}\boldsymbol{r}'}$ differs essentially from zero only for $\boldsymbol{r}$, $\boldsymbol{r}'$ near to the right and left of the barrier, respectively. Above we have introduced the gauge-invariant superfluid velocity

$$v_{s,\mathrm{L/R}} = -\frac{\hbar}{2m}\left(\boldsymbol{\nabla}\varphi_{\mathrm{L/R}} + \frac{2e}{\hbar c}\boldsymbol{A}\right).\tag{4.183}$$

In Eq. (4.181), the order parameter matrix in Nambu space $\hat{\Delta}_{\mathrm{L}}$ has *real* matrix elements instead of the complex ones in Eq. (4.178).

The action $\mathcal{A}$ depends on the independent collective variables $|\Delta_{\mathrm{L/R}}|$, $\boldsymbol{A}$, U and ψ. The partition function is a functional integral over all these variables. For bulk superconductors, the quantities $|\Delta_{\mathrm{L/R}}|$, $\boldsymbol{A}$, and U have insignificant uncertainties, so that we may disregard fluctuations of these variables. This simplifies the further discussion of Z considerably. We proceed by observing the following three points:

(i) For uncoupled homogeneous bulk superconductors, the extremal condition $\delta\mathcal{A}/\delta|\Delta_{\mathrm{L/R}}| = 0$ reproduces the gap equation of the BCS theory, which yields the mean field value $|\Delta_{\mathrm{L/R}}^{\mathrm{BCS}}|$. Since the second variation $\delta^2\mathcal{A}/(\delta|\Delta_{\mathrm{L/R}}|)^2$ is proportional to the density N_0 of free electron states at E_{F} and proportional to the third power of the BCS coherence length, fluctuations of $|\Delta_{\mathrm{L/R}}|$ about $|\Delta_{\mathrm{L/R}}^{\mathrm{BCS}}|$ are suppressed except in a narrow temperature range just below the transition temperature T_{c}.

(ii) Expansion of the action to second order in the superfluid velocity yields

$$\mathcal{A} = \mathcal{A}_0 + \int_0^{\hbar\beta} d\tau \int d^3r \left(\frac{m}{2}\rho_{\mathrm{S}}v_{\mathrm{S}}^2 + \frac{1}{8\pi}\Big(\boldsymbol{h}(\boldsymbol{r},\tau) - \boldsymbol{h}_{\mathrm{ext}}\Big)^2\right),\tag{4.184}$$

where ρ_{S} is the superfluid density [150]. From this one finds that the least action condition $\delta\mathcal{A}/\delta\boldsymbol{A}(\boldsymbol{r},\tau) = 0$ reproduces the familiar law $\boldsymbol{j} = \boldsymbol{j}_{\mathrm{ext}} + \boldsymbol{j}_{\mathrm{S}}$, where $\boldsymbol{j}_{\mathrm{ext}}$ is the current density related to the external field, and where $\boldsymbol{j}_{\mathrm{S}}$ is the supercurrent

$$\boldsymbol{j}_{\mathrm{S}} = \frac{e\hbar}{2m}\rho_{\mathrm{S}}\left(\boldsymbol{\nabla}\psi + \frac{2e}{\hbar c}\boldsymbol{A}\right).\tag{4.185}$$

In a bulk superconductor, fluctuations of $\boldsymbol{j}$ are suppressed, as follows from the study of the second variation of $\mathcal{A}$ with respect to the vector potential $\boldsymbol{A}$.

(iii) The action is extremal when the phase difference across the junction satisfies the relation (3.144). Again, we may disregard deviations from the mean field relation (3.144) since fluctuations are suppressed by bulk energies.

Evidently, the only remaining fluctuating variable is the phase difference ψ. In a SQUID geometry, in which the two superconductors are joined in a loop, the relevant fluctuating variable is the total flux ϕ through the loop. The flux ϕ is related to the phase variable ψ as given in Eq. (3.142). When the quantities $|\Delta_{\mathrm{L/R}}|$, U and $\boldsymbol{j}$ are fixed to their mean values, the multiple functional integral (4.174) is reduced to a single one over the phase variable ψ. The partition function of a junction then reads

$$Z = \oint \mathcal{D}\psi(\cdot)\exp\left\{-\frac{C}{2\hbar}\int_0^{\hbar\beta}d\tau\left(\frac{\hbar}{2e}\frac{\partial\psi}{\partial\tau}\right)^2 - \frac{1}{\hbar}\mathcal{A}_{\mathrm{T}}[\psi(\cdot)]\right\},\tag{4.186}$$

where $\mathcal{A}_{\mathrm{T}}[\psi(\cdot)]$ represents the tunneling contribution to the action. Working out $\mathcal{A}_{\mathrm{T}}[\psi(\cdot)]$ to second order in the (averaged) tunnel matrix element from the first term on the right hand side of (4.175), one finds[6] [150] – [152]

$$
\mathcal{A}_{\mathrm{T}} = 2 \int_0^{\hbar\beta} d\tau \int_0^{\hbar\beta} d\tau' \left\{ \alpha(\tau - \tau') \left(1 - \cos\frac{\psi(\tau) - \psi(\tau')}{2} \right) \right.
$$
$$
\left. + \beta(\tau - \tau') \cos\frac{\psi(\tau) + \psi(\tau')}{2} \right\}.
$$
(4.187)

For later convenience, we have extracted a factor 2 in front of the integrals, and we have added a ψ-independent term in order that the first contribution to $\mathcal{A}_{\mathrm{T}}$ vanishes for a constant path, $\psi(\tau) = \psi(\tau')$. The kernels $\alpha(\tau)$ and $\beta(\tau)$ are given in terms of the diagonal $(1,1)$ and off-diagonal $(1,2)$ components of the Green function $\hat{G}$ in Nambu space, usually denoted by $G(\tau, \boldsymbol{p})$ and $F(\tau, \boldsymbol{p})$,

$$
\alpha(\tau) = -\frac{|T_{\mathrm{t}}|^2}{\hbar} \int \frac{d^3 \boldsymbol{p}_{\mathrm{L}}}{(2\pi\hbar)^3} \int \frac{d^3 \boldsymbol{p}_{\mathrm{R}}}{(2\pi\hbar)^3} G_{\mathrm{L}}(\tau, \boldsymbol{p}_{\mathrm{L}}) G_{\mathrm{R}}(-\tau, \boldsymbol{p}_{\mathrm{R}}),
$$
$$
\beta(\tau) = -\frac{|T_{\mathrm{t}}|^2}{\hbar} \int \frac{d^3 \boldsymbol{p}_{\mathrm{L}}}{(2\pi\hbar)^3} \int \frac{d^3 \boldsymbol{p}_{\mathrm{R}}}{(2\pi\hbar)^3} F_{\mathrm{L}}(\tau, \boldsymbol{p}_{\mathrm{L}}) F_{\mathrm{R}}(-\tau, \boldsymbol{p}_{\mathrm{R}}).
$$
(4.188)

Here we have assumed that the tunneling coupling T_{t} is independent of momentum near to the Fermi momentum p_{F}. The Green functions are given by

$$
\int \frac{d^3 \boldsymbol{p}}{(2\pi\hbar)^3} \begin{pmatrix} G \\ F \end{pmatrix} (\tau, \boldsymbol{p}) = -N_0 \int_{-\infty}^{\infty} d\omega\, \mathcal{N}_{\mathrm{qp}}(\omega) \begin{pmatrix} 1 \\ \Delta_{\mathrm{g}}/\omega \end{pmatrix}
$$
$$
\times e^{-\omega\tau} \left(f(-\omega)\Theta(\tau) - f(\omega)\Theta(-\tau) \right).
$$
(4.189)

Here $\mathcal{N}_{\mathrm{qp}}(\omega)$ is the density of states of BCS quasiparticles (4.167), $f(\omega)$ denotes the Fermi distribution function and Δ_{g} is the mean field gap frequency $\Delta_{\mathrm{g}} = \Delta^{\mathrm{BCS}}/\hbar$. For notational simplicity, we now assume that $\mathcal{N}_{\mathrm{qp}}^{(\mathrm{L})}(\omega) = \mathcal{N}_{\mathrm{qp}}^{(\mathrm{R})}(\omega) = \mathcal{N}_{\mathrm{qp}}$ and $\Delta_{\mathrm{g,L}} = \Delta_{\mathrm{g,R}} = \Delta_{\mathrm{g}}$. Next, we insert the expression (4.189) into Eq. (4.188) and assume that both superconductors are described by the same mean field parameters. We then get

$$
\begin{pmatrix} \alpha(\tau) \\ \beta(\tau) \end{pmatrix} = \frac{|T_{\mathrm{t}}|^2}{\hbar} N_0^2 \int_{-\infty}^{\infty} d\omega' \int_{-\infty}^{\infty} d\omega''\, \mathcal{N}_{\mathrm{qp}}(\omega') \mathcal{N}_{\mathrm{qp}}(\omega'') \begin{pmatrix} 1 \\ \Delta_{\mathrm{g}}^2/\omega'\omega'' \end{pmatrix}
$$
$$
\times e^{(\omega''-\omega')|\tau|} f(\omega'') f(-\omega').
$$
(4.190)

These expressions are similar to the previous form (4.141). Next, we follow the steps described below Eq. (4.141). We then find in analogy to the expression (4.54)

$$
\begin{pmatrix} \alpha(\tau) \\ \beta(\tau) \end{pmatrix} = \frac{1}{\pi} \int_0^{\infty} d\omega \begin{pmatrix} J_\alpha(\omega) \\ J_\beta(\omega) \end{pmatrix} D_\omega(\tau).
$$
(4.191)

---

[6]This expression is the leading term in the cumulant expansion.

Here, $D_\omega(\tau)$ is the boson propagator (3.55), and $J_{\alpha(\beta)}(\omega)$ is the spectral density

$$
\begin{pmatrix} J_\alpha(\omega) \\ J_\beta(\omega) \end{pmatrix} = \frac{\pi}{2} \frac{|T_t|^2}{\hbar} N_0^2 \int_{-\infty}^{\infty} d\omega' \, \mathcal{N}_{\rm qp}(\omega') \mathcal{N}_{\rm qp}(\omega' - \omega) \begin{pmatrix} 1 \\ \frac{\Delta_g^2}{\omega'(\omega'-\omega)} \end{pmatrix}
$$

$$
\times \left\{ \tanh(\tfrac{1}{2}\beta\hbar\omega') - \tanh[\tfrac{1}{2}\beta\hbar(\omega' - \omega)] \right\} . \tag{4.192}
$$

The physics of the two terms in Eq. (4.187) is quite different. The α-term describes dissipation because of quasiparticle tunneling. The β-term describes tunneling of Cooper pairs (tunneling). In this term, the kernel $\beta(\tau - \tau')$ appears in combination with $\cos[(\psi(\tau) + \psi(\tau'))/2]$. Assuming that $\psi(\tau)$ varies slowly on the time scale on which the kernel $\beta(\tau)$ decays, the β-term reduces to an ordinary potential term, the "washboard potential" action

$$
\mathcal{A}_{\rm T}^{(\beta)}[\psi(\cdot)] = -I_{\rm c}\frac{\hbar}{2e} \int_0^{\hbar\beta} d\tau \, \cos\psi(\tau) , \tag{4.193}
$$

where

$$
I_{\rm c} = -\frac{4e}{\hbar} \int_{-\hbar\beta/2}^{\hbar\beta/2} d\tau \, \beta(\tau) = -\frac{8e}{\pi\hbar} \int_0^{\infty} d\omega \, \frac{J_\beta(\omega)}{\omega} \tag{4.194}
$$

is the critical current of the junction. The nonlocal correction in the β-term represents the nonlocal supercurrent found by Wertheimer [193]. This contribution is usually referred to as the "$\cos\psi$" term or quasi-particle pair interference current [150, 151]. In order to keep the discussion simple, we shall ignore time-nonlocal current corrections, apart from a contribution to the effective capacitance (see below).

When the superconducting tunnel junction is biased by an externally applied current $I_{\rm ext}$, there is an additional potential contribution $\propto \psi$ which tilts the washboard. At this level of approximation, the effective action of a current biased tunnel junction with capacitance C is

$$
\mathcal{A}_{\rm eff}[\psi(\cdot)] = \int_0^{\hbar\beta} d\tau \left[\frac{C}{2} \left(\frac{\hbar}{2e} \frac{\partial\psi}{\partial\tau} \right)^2 + V(\psi) \right] + \mathcal{A}_{\rm T}^{(\alpha)}[\psi] ,
$$

$$
\mathcal{A}_{\rm T}^{(\alpha)}[\psi(\cdot)] = 2 \int_0^{\hbar\beta} d\tau \int_0^{\hbar\beta} d\tau' \, \alpha(\tau - \tau') \left(1 - \cos\frac{\psi(\tau) - \psi(\tau')}{2} \right) , \tag{4.195}
$$

where $V(\psi)$ is the tilted washboard potential

$$
V(\psi) = -\frac{\hbar}{2e} I_{\rm c} \cos\psi - \frac{\hbar}{2e} I_{\rm ext}\psi . \tag{4.196}
$$

The potential energy of the total flux ϕ in a SQUID geometry is given in Eq. (3.146).

The first term in $\mathcal{A}_{\rm eff}$ accounts for the charging energy, and the potential $V(\psi)$ for the coupling and a bias term. The term $\mathcal{A}_{\rm T}^{(\alpha)}$ describes state-dependent damping through quasiparticle tunneling in superconducting junctions across the barrier, or through single electrons in normal junctions. The trigonometric dependence on the phase difference describes discrete quasiparticle or single electron tunneling rather than continuous flow of charge. When the phase fluctuations in thermal equilibrium are weak, we may expand the trigonometric function in the α-term. Then we recover the standard quadratic form (4.50),

$$A_T^{(\alpha)}[\psi(\cdot)] = \frac{1}{2} \int_0^{\hbar\beta} d\tau \int_0^\tau d\tau' \, \alpha(\tau - \tau') \left[\psi(\tau) - \psi(\tau') \right]^2 . \qquad (4.197)$$

Observing the analogy $\psi(\tau)/2\pi \,\hat{=}\, q(\tau)/q_0$ we see that the correspondence of the dissipation kernel for quasiparticle tunneling with that of the phenomenological model discussed in Subsection 4.2.1 is $4\pi^2\alpha(\tau) \,\hat{=}\, q_0^2 K(\tau)$. Accordingly, the correspondence of the spectral densities is $4\pi^2 J_\alpha(\omega) \,\hat{=}\, q_0^2 J(\omega)$.

Important limiting cases are:

(i) When the junction is formed by ideal BCS superconductors, the excitation spectrum has a gap, and thus $J_\alpha(\omega)$ has a behavior similar to the one of $G_{qp}(\omega)$ sketched in Fig. 4.3. At zero temperature, we find [cf. Eq. (4.172)]

$$\alpha(\tau) = \frac{\hbar^2 \Delta_{g,L} \Delta_{g,R}}{4\pi e^2 R_N} K_1(\Delta_{g,L}|\tau|) \, K_1(\Delta_{g,R}|\tau|) , \qquad (4.198)$$

where

$$R_N^{-1} = 4\pi N_{0,L} N_{0,R} \frac{|T|^2}{\hbar^2} \frac{e^2}{\hbar} \qquad (4.199)$$

denotes the normal state conductance of the tunnel junction. For equal gap frequencies, $\Delta_{g,L} = \Delta_{g,R} = \Delta_g$, the limiting cases are

$$\alpha(\tau) = \frac{\hbar^2}{4\pi e^2 R_N} \begin{cases} \dfrac{1}{\tau^2} & \text{for} \quad \Delta_g|\tau| \ll 1 , \\[2mm] \dfrac{\pi\Delta_g}{2|\tau|} e^{-2\Delta_g|\tau|} & \text{for} \quad \Delta_g|\tau| \gg 1 . \end{cases} \qquad (4.200)$$

If the phase varies slowly with time on the time scale $1/\Delta_g$, we may expand the term $\psi(\tau) - \psi(\tau')$ in (4.197) in the relative variable $\xi = \tau - \tau'$. We then get

$$A_T^{(\alpha)}[\psi(\cdot)] = \frac{1}{2} \int_0^{\hbar\beta} d\tau \left(\frac{\partial\psi}{\partial\tau} \right)^2 \int_0^{\hbar\beta/2} d\xi \, \xi^2 \, \alpha(\xi) . \qquad (4.201)$$

Hence on this condition, the α-term yields a kinetic contribution which leads to an increase of the effective capacitance. With use of the form (4.198), the additional contribution is found to read

$$\delta C^{(\alpha)} = \frac{3\pi}{32} \frac{1}{\Delta_g R_N} . \qquad (4.202)$$

The renormalization of the capacitance is similar to the mass renormalization in a mechanical analogue for super-Ohmic dissipation $(s > 2)$ discussed in Section 3.1.7.

(ii) In a normal junction ($\Delta_g = 0$), the kernel $\alpha(\tau)$ is in Ohmic form (4.165),

$$\alpha(\tau) = \frac{\hbar^2}{4\pi e^2 R_N} \frac{(\pi/\hbar\beta)^2}{\sin^2(\pi\tau/\hbar\beta)} . \tag{4.203}$$

However even in this case, quasiparticle tunneling differs from an Ohmic resistor by the trigonometric dependence of the action $\mathcal{A}_T^{(\alpha)}[\psi(\cdot)]$ on the phase in Eq. (4.195), instead of a quadratic dependence for a resistor, Eq. (4.197). The difference reflects different physics [194]. In the former case we have a discrete transfer of charge in units of e, while in the latter case charge is flowing continuously.

(iii) In a non-ideal junction we may have pair-breaking effects, spatial variation of the order parameter, or locally non-perfect energy gaps which lead to a finite subgap conductance even at $T = 0$ [152]. These effects may be accounted for either by smearing the density of states $\mathcal{N}_{qp}(\omega)$ in Eq. (4.192), or by writing the kernel $\alpha(\tau)$ as a linear combination of the expressions (4.198) and (4.203).

Given the action $\mathcal{A}_{eff}[\psi(\cdot)]$, one may ask whether ψ is an extended variable defined in the range $-\infty \leq \psi \leq \infty$, or a compact variable restricted either to the interval $0 < \psi \leq 2\pi$ or to $0 < \psi \leq 4\pi$. In the first case, the minima of the potential $U(\psi)$ are distinguishable, and $Q = (2e/i)\partial/\partial\psi$ is the usual charge operator with continuous eigenvalues, while in the other two cases Q is a quasi-charge operator with discrete eigenvalues similar to quasi-momentum of a particle in a periodic potential [152]. All these cases are possible, and which case applies depends on the experimental conditions. If charge transport is dominated by Cooper pair tunneling, the 2π-periodic potential $U(\psi)$ determines the symmetry of the problem and it is convenient to choose a basis of corresponding Bloch states. If quasiparticle tunneling is relevant, the symmetry is reduced to 4π periodicity, which is the symmetry of $\mathcal{A}_T^{(\alpha)}[\psi]$. Finally, if there is a continuous flow of charge, as in the Ohmic case, or in a SQUID, or when the junction is coupled to an external circuit, Bloch states are inappropriate and we have to choose a basis of continuous charge states. When ψ is defined on a ring, the path sum of the partition function generally includes a summation over winding numbers. In many cases it is convenient to work with the restricted space of discrete charges and allow for a continuous change of charges perturbatively [152].

When quasiparticle tunneling is hampered by the superconducting gap, higher-order processes may become relevant, e.g., correlated tunneling of two electrons across a junction with a normal and a superconducting electrode. A diagram taking into account two diagonal (G) propagators on the normal metal side and two off-diagonal (F) propagators on the superconductor side describes Andreev scattering across the interface. The resulting action is similar in form to the quasiparticle contribution $\mathcal{A}_T^{(\alpha)}$ in Eq. (4.195) except that it is of order $|T|^4$ and that the factor $\frac{1}{2}$ in the argument of the cosine function is missing because, instead of the charge e, the charge $2e$ is transferred in the Andreev scattering process. This implies that dissipation by Andreev reflection is very similar to dissipation by quasiparticle or single-electron tunneling [195, 196]. If the normal electrode is made of a dirty metal, the average over many impurities must be performed which results in a Cooperon propagator. The respective analysis for different junction geometries is given in Ref. [197].

4.2.11 Electromagnetic environment

In analogy to the derivation of the influence action (4.41) with (4.43) for the mechanical oscillator model Hamiltonian (3.11), it is straightforward to calculate the effective action connected with the Hamiltonian (3.207), which represents besides the capacitive energy an electromagnetic environment formed by LC-circuits. The resulting action, which corresponds to the expression (4.62) with (4.63), is

$$S_{\text{eff},0}^{(E)}[\varphi] = \frac{R_K}{4\pi} \frac{1}{\beta} \sum_n \left[|\nu_n|^2 C + |\nu_n| \hat{Y}(|\nu_n|) \right] |\varphi_n|^2 , \tag{4.204}$$

where $\hat{Y}(z)$ is the Laplace transform of the admittance function $Y(\tau)$ and φ_n is the Fourier coefficient of the phase function in imaginary time, $\varphi(\tau) = \sum_n e^{i\nu_n\tau}\varphi_n/\hbar\beta$.

Consider next the phase correlator $\langle e^{i[\varphi(0)-\varphi(\tau)]}\rangle_\beta$ in imaginary time $0 \leq \tau < \hbar\beta$. Here, $\langle\cdots\rangle_\beta$ denotes average with the weight function $\exp[-S_{\text{eff},0}^{(E)}[\varphi]/\hbar]$. With the form (4.204), we get

$$\langle e^{i[\varphi(0)-\varphi(\tau)]}\rangle_\beta = e^{-\langle[\varphi(0)-\varphi(\tau)]\varphi(0)\rangle_\beta} = e^{-W_\varphi(\tau)} . \tag{4.205}$$

The Gaussian average (4.205) yields for $W_\varphi(\tau)$ the expression

$$W_\varphi(\tau) = \frac{2\pi}{\hbar\beta} \sum_{n\neq 0} \frac{\hat{Z}_t(|\nu_n|)}{|\nu_n|R_K} \left[1 - e^{i\nu_n\tau} \right] \quad \text{with} \quad \hat{Z}_t(\nu_n) = \frac{1}{\nu_n C + \hat{Y}(\nu_n)} . \tag{4.206}$$

Observing the equivalence of the expression (3.63) to the expression (3.61), the Matsubara sum in Eq. (4.206) may be transformed into the frequency integral

$$W_\varphi(\tau) = \int_0^\infty d\omega \frac{G_\varphi(\omega)}{\omega^2} \left[D_\omega(0) - D_\omega(\tau) \right] , \tag{4.207}$$

where [cf. Eq. (3.216) with (3.214)]

$$G_\varphi(\omega) = 2\omega \operatorname{Re} \frac{1}{R_K[i\omega C + Z^{-1}(\omega)]} . \tag{4.208}$$

and where $D_\omega(\tau)$ is the thermal propagator (3.55).

For a resistive environment described by an Ohmic impedance R, the phase correlation function $W_\varphi(\tau)$ takes in the regime $\tau \gg RC$ the form

$$W_\varphi(\tau) = \frac{R}{R_K} \frac{2\pi}{\hbar\beta} \sum_{n\neq 0} \frac{1}{|\nu_n|} \left[1 - e^{i\nu_n\tau} \right] . \tag{4.209}$$

We shall employ these expressions in Section 20.3 and in Subsection 28.1.4. There we discuss quantum transport of charge through a weak link and a weak constriction under influence of an electromagnetic environment.

4.3   Partition function of the open system

4.3.1   General path integral expression

Quantum statistical equilibrium properties can be calculated directly if the partition function Z is known as a function of temperature, volume, external field, etc. The standard thermodynamic quantities are internal energy, entropy, pressure, specific heat, and susceptibility. They are obtained by differentiation of Z with respect to the control parameters. Since the partition function is the Laplace transform of the density of states, it also carries information about the energy spectrum (see the discussion of the density of states of the damped harmonic oscillator in Sect. 6.5).

The thermodynamical key quantity of open quantum systems is the reduced partition function. The subsequent discussion is based on the respective Euclidean path integral or path sum representation

$$Z(\hbar\beta) \;=\; \oint \mathcal{D}q(\cdot) \, \exp\left(-\, S_{\text{eff}}^{(\text{E})}[q(\cdot)]/\hbar\right). \tag{4.210}$$

Here the symbol $\oint$ indicates that the path sum is over all periodic paths of period $\hbar\beta$, and $S_{\text{eff}}^{(\text{E})}[q]$ is the effective action of the open system under consideration.

If not stated otherwise, we shall restrict our attention in the remainder of this chapter to the simple phenomenological model discussed previously in Subsection 4.2.1. The exact formal expression (4.210) with the effective action (4.61) finds application in diverse open quantum systems. We may use it as a starting point, e. g., for the calculation of thermodynamic properties, or, in problems connected with tunneling, for the calculation of the quantum statistical decay of metastable states. The respective discussion is postponed until Part III.

The "time-retarded" Euclidean action $S_{\text{eff}}^{(\text{E})}[q(\cdot)]$ determines the path probability $\propto \exp[-S_{\text{eff}}^{(\text{E})}[q(\cdot)]/\hbar]$ in the functional integral expression of the reduced partition function. Unfortunately, the path sum can be carried out exactly only when the exponent $S_{\text{eff}}^{(\text{E})}[q]$ is a quadratic form in the dynamical variable q. Therefore, I find it useful to expound a variety of different approaches which yield reasonable approximations over the whole temperature range. Among them are the semiclassical approximation and the variational approach to quantum-statistical mechanics. I find it convenient to defer the discussion of the latter approach until Chapter 8, which is after the discussion of the damped harmonic oscillator.

4.3.2   Semiclassical approximation

In the semiclassical regime, $S_{\text{eff}}^{(\text{E})}[q(\cdot)]/\hbar \gg 1$, the path sum (4.60) or (4.210) is dominated by paths which are in function space close to the stationary points of the action. The first variation of the action vanishes for the *extremal* path $\bar{q}(\tau)$ obeying $\partial S_{\text{eff}}^{(\text{E})}[q(\cdot)]/\partial q|_{\bar{q}(\cdot)} = 0$. This path obeys the equation of motion

$$-M\,\ddot{\bar{q}}(\tau) + V'[\bar{q}(\tau)] + \int_0^{\hbar\beta} d\tau'\, k(\tau - \tau')\,\bar{q}(\tau') \;=\; 0\,. \tag{4.211}$$

The sign of the inertia term is negative because the motion is in imaginary time. The time-nonlocal third term originates from the projection of the original energy-conserving path $\{\bar{q}(\tau), \bar{x}(\tau)\}$ in the $(N+1)$-dimensional $\{q, x\}$ coordinate space [see Eqs. (4.34) and (4.35)] onto the q-axis. As regards the partition function, the solution of Eq. (4.211) must be $\hbar\beta$-periodic, $\bar{q}(\tau + \hbar\beta) = \bar{q}(\tau)$. To account for the contributions of paths in the vicinity of $\bar{q}(\tau)$, we expand the action (4.61) at the stationary action till second order in the deviations from the stationary path. Putting

$$q(\tau) \;=\; \bar{q}(\tau) + y(\tau)\,, \tag{4.212}$$

we find

$$S_{\mathrm{eff}}^{(\mathrm{E})}[q(\cdot)] \;=\; S_{\mathrm{eff}}^{(\mathrm{E})}[\bar{q}(\cdot)] + \frac{1}{2}M\int_0^{\hbar\beta} d\tau\, y(\tau)\Lambda[\bar{q}(\tau)]\,y(\tau) + \mathcal{O}(y^3)\,. \tag{4.213}$$

Here, $\Lambda[\bar{q}(\tau)]$ is a linear operator acting in the space of $\hbar\beta$-periodic functions,

$$\Lambda[\bar{q}(\tau)]y(\tau) \;\equiv\; \left(-\frac{\partial^2}{\partial\tau^2} + \frac{1}{M}V''[\bar{q}(\tau)]\right)y(\tau) + \frac{1}{M}\int_0^{\hbar\beta} d\tau'\, k(\tau - \tau')\,y(\tau')\,. \tag{4.214}$$

The functional integral for Z may now be written as a path sum of all deviations $y(\tau)$ with Gaussian weight,

$$Z \;=\; \exp\left(-S_{\mathrm{eff}}^{(\mathrm{E})}[\bar{q}(\cdot)]/\hbar\right)\oint \mathcal{D}y(\cdot)\,\exp\left(-\frac{M}{2\hbar}\int_0^{\hbar\beta} d\tau\, y(\tau)\Lambda[\bar{q}(\tau)]y(\tau)\right)\,. \tag{4.215}$$

Execution of all the Gaussian integrals in a suitable representation yields

$$Z \;=\; \frac{N}{\sqrt{D[\bar{q}(\cdot)]}}\,\exp\left(-S_{\mathrm{eff}}^{(\mathrm{E})}[\bar{q}(\cdot)]/\hbar\right)\,, \tag{4.216}$$

where N is a normalization factor, and where $D[\bar{q}(\cdot)]$ is the determinant of the fluctuation operator $\Lambda[\bar{q}(\cdot)]$ on the space of $\hbar\beta$-periodic functions,

$$D[\bar{q}(\cdot)] \;\equiv\; \det(\Lambda[\bar{q}(\cdot)])\,. \tag{4.217}$$

The yet undetermined normalization factor N is universal in the sense that it neither depends on the specific form of the potential nor on the specific form of the dissipative mechanism. This is because the normalization factor N is fixed by the kinetic part $-\partial^2/\partial\tau^2$ of $\Lambda[\bar{q}(\cdot)]$ (see Subsection 4.3.4). Therefore, in ratios of partition functions the normalization factor N drops out. Thus we may write

$$Z \;=\; Z_0\left(\frac{D_0}{D[\bar{q}(\cdot)]}\right)^{1/2}\exp\left(-S_{\mathrm{eff}}^{(\mathrm{E})}[\bar{q}(\cdot)]/\hbar\right)\,. \tag{4.218}$$

We may choose, for convenience, $Z_0 = N/\sqrt{D_0}$ as the partition function of the un-damped harmonic oscillator. Here D_0 is the determinant of the respective Gaussian fluctuation operator $\mathbf{\Lambda}_0$. With the form (4.218), the remaining computational problem is reduced to the calculation of a ratio of determinants.

The semiclassical form (4.218) is a valid expression when all of the fluctuation modes can be treated in Gaussian approximation and when all eigenvalues of $\mathbf{\Lambda}[\bar{q}(\cdot)]$ are positive. The appropriate treatment of modes with negative eigenvalues and with zero or quasi-zero eigenvalues is given in Part III.

4.3.3 Partition function of the damped harmonic oscillator

Consider a harmonic potential with the minimum located at $q = 0$, $V(q) = \frac{1}{2}M\omega_0^2 q^2$, where ω_0 denotes the frequency of oscillations about the minimum of the well. For this potential, the imaginary-time equation of motion (4.211) admits only the trivial periodic solution in which the particle sits for ever at the barrier top of the upside-down potential $-V(q)$. Hence we have $\bar{q}(\tau) = 0$ and $S_{\mathrm{eff}}^{(\mathrm{E})}[\bar{q} = 0] = 0$. The eigenvalues of the related fluctuation operator

$$\mathbf{\Lambda}^{(0)} y(\tau) \equiv \left(-\frac{\partial^2}{\partial\tau^2} + \omega_0^2\right) y(\tau) + \frac{1}{M} \int_0^{\hbar\beta} d\tau' \, k(\tau - \tau') y(\tau') \qquad (4.219)$$

on the space of $\hbar\beta$-periodic functions are found with use of Eqs. (4.45) and (4.43) as

$$\Lambda_n^{(0)} = \nu_n^2 + \omega_0^2 + |\nu_n|\hat{\gamma}(|\nu_n|) \,, \qquad (4.220)$$

where $\nu_n = 2\pi n/\hbar\beta$ $(n = 0, \pm 1, \pm 2, \cdots)$ is a bosonic Matsubara frequency, and where $\hat{\gamma}(z)$ is the Laplace transform of the friction kernel $\gamma(t)$. With the expression (4.220) the determinant of the fluctuation operator $\mathbf{\Lambda}^{(0)}$ takes the form

$$D^{(0)} = \prod_{n=-\infty}^{\infty} \Lambda_n^{(0)} = \omega_0^2 \prod_{n=1}^{\infty} \left(\Lambda_n^{(0)}\right)^2 \,. \qquad (4.221)$$

For the undamped oscillator, this expression reduces to

$$D_0 = \quad = \omega_0^2 \prod_{n=1}^{\infty} \left(\nu_n^2 + \omega_0^2\right)^2 \,, \qquad (4.222)$$

and the Matsubara representation of the corresponding partition function is

$$Z_0 = \frac{1}{2\sinh(\beta\hbar\omega_0/2)} = \frac{1}{\beta\hbar\omega_0} \prod_{n=1}^{\infty} \frac{\nu_n^2}{\omega_0^2 + \nu_n^2} \,. \qquad (4.223)$$

With use of Eqs. (4.221), (4.222) and (4.223), and with $S_{\mathrm{eff}}^{(\mathrm{E})}[\bar{q} = 0] = 0$, the semiclassical partition function (4.218) of the damped oscillator may be written as

$$Z^{(0)} = \frac{N}{\sqrt{D^{(0)}}} = Z_0 \sqrt{\frac{D_0}{D^{(0)}}} = \frac{1}{\beta\hbar\omega_0} \prod_{n=1}^{\infty} \frac{\nu_n^2}{\Lambda_n^{(0)}} \,. \qquad (4.224)$$

Semiclassics is exact for a harmonic system, since the action is of second order in the fluctuation modes. Thus we finally ended up at the exact expression of the partition function of the damped harmonic oscillator. The resulting expression in Matsubara representation

$$Z^{(0)} \; = \; \frac{1}{\beta \hbar \omega_0} \prod_{n=1}^{\infty} \frac{\nu_n^2}{\omega_0^2 + \nu_n^2 + \nu_n \hat{\gamma}(\nu_n)} \tag{4.225}$$

holds for arbitrary linear dissipation.

4.3.4   Functional measure in Fourier space

Lagrangians which consist of a kinetic term $\frac{1}{2} M \dot{q}^2$ and a general potential $V(q)$ are usually referred to as standard Lagrangians. Open systems described by Eqs. (4.26), (4.23) and (4.33) also belong to this class, whereas magnetic systems are non-standard. Consider now the normalized functional measure for standard Lagrangians in Fourier space. Upon expanding the $\hbar\beta$-periodic fluctuation $y(\tau)$ into Fourier series

$$y(\tau) \; = \; \sum_{n=-\infty}^{\infty} y_n \, e^{i\nu_n \tau} \,, \tag{4.226}$$

the exponent of the Gaussian weight function becomes

$$\frac{M}{2\hbar} \int_0^{\hbar\beta} d\tau \, y(\tau) \Lambda^{(0)} \, y(\tau) \; = \; \frac{M\beta}{2} \sum_{n=-\infty}^{\infty} \Lambda_n^{(0)} y_n y_{-n} \,. \tag{4.227}$$

The fluctuation operator $\Lambda^{(0)}$ is defined in Eq. (4.219). Apparently, the expression (4.225) is directly found from (4.215) if we define the functional measure as

$$\oint \mathcal{D}y(\cdot) \cdots \; \equiv \; \int_{-\infty}^{\infty} \frac{dy_0}{\sqrt{2\pi\hbar^2\beta/M}} \prod_{n=1}^{\infty} \left[\int_{-\infty}^{\infty} \int_{-\infty}^{\infty} \frac{dy_n \, dy_{-n}}{2\pi/(iM\beta\nu_n^2)} \right] \cdots$$

$$= \; \int_{-\infty}^{\infty} \frac{dy_0}{\sqrt{2\pi\hbar^2\beta/M}} \prod_{n=1}^{\infty} \left[\int_{-\infty}^{\infty} \int_{-\infty}^{\infty} \frac{d\mathrm{Re}\, y_n \, d\mathrm{Im}\, y_n}{\pi/(M\beta\nu_n^2)} \right] \cdots \,. \tag{4.228}$$

The measure (4.228) in quantum statistical path integrals is generally valid for Gaussian fluctuation modes in systems with standard Lagrangians. Modifications are necessary for modes with zero or quasi-zero eigenvalues (cf. Chapters 16 and 17).

4.3.5   Partition function of the damped harmonic oscillator revisited

In Subsection 4.3.3 we have calculated the reduced partition function (4.225) via determination of the eigenvalues of the fluctuation modes. Complementary understanding of structure and specific form of the reduced partition function may be achieved by study of the reduction of the partition function Z_{tot} of the full system-plus-reservoir complex [198]. We rigorously have

$$Z^{(0)} = Z_{\text{tot}}^{(0)}/Z_{\text{R}}, \tag{4.229}$$

where Z_{R} is the partition function of the unperturbed reservoir given in Eq. (4.29).

To this purpose we start out from the Hamiltonian(3.11) with $V(q) = \frac{1}{2}m\omega_0^2 q^2$. Upon introducing mass weighted coordinates,

$$u_0 = M^{1/2}q, \qquad u_\alpha = m_\alpha^{1/2}x_\alpha \qquad (\alpha = 1, 2, \cdots, N), \tag{4.230}$$

the Hamiltonian of $N+1$ coupled stable oscillators takes the form

$$H = \frac{1}{2}\sum_{\alpha=0}^{N}\sum_{\alpha'=0}^{N}\left[\delta_{\alpha\alpha'}\,\dot{u}_\alpha\dot{u}_{\alpha'} + U_{\alpha\alpha'}^{(0)}\,u_\alpha u_{\alpha'}\right]. \tag{4.231}$$

The force constant matrix $\boldsymbol{U}^{(0)}$ with spring coefficients $U_{\alpha\alpha'}^{(0)}$ is given by

$$\boldsymbol{U}^{(0)} = \begin{pmatrix} \tilde{C}_0 & \tilde{C}_1 & \tilde{C}_2 & \cdots & \tilde{C}_N \\ \tilde{C}_1 & \omega_1^2 & 0 & \cdots & 0 \\ \tilde{C}_2 & 0 & \omega_2^2 & \cdots & 0 \\ \vdots & \vdots & \vdots & \ddots & \vdots \\ \tilde{C}_N & 0 & 0 & \cdots & \omega_N^2 \end{pmatrix}. \tag{4.232}$$

Here we have put

$$\tilde{C}_0 \equiv \omega_0^2 + \sum_{\alpha'=1}^{N}\frac{\tilde{C}_{\alpha'}^2}{\omega_{\alpha'}^2}, \qquad \tilde{C}_\alpha \equiv -\frac{c_\alpha}{(m_\alpha M)^{1/2}}, \qquad (\alpha = 1, 2, \cdots, N). \tag{4.233}$$

Denoting the eigenfrequencies of the stable harmonic system by μ_α $(\alpha = 0, 1, \cdots, N)$, the Hamiltonian (4.231) in normal mode representation reads

$$H = \frac{1}{2}\sum_{\alpha=0}^{N}\left(\dot{\rho}_\alpha^2 + \mu_\alpha^2\rho_\alpha^2\right), \tag{4.234}$$

and the partition function of the global harmonic system may be written in the form

$$Z_{\text{tot}}^{(0)} = \prod_{\alpha=0}^{N}\frac{1}{2\sinh(\beta\hbar\mu_\alpha/2)} = \prod_{\alpha=0}^{N}\left(\frac{1}{\beta\hbar\mu_\alpha}\prod_{n=1}^{\infty}\frac{\nu_n^2}{\mu_\alpha^2+\nu_n^2}\right). \tag{4.235}$$

In the second equality we have used again the infinite product representation of the hyperbolic sine function.

Consider next the dynamical matrix $\boldsymbol{A}^{(0)}$ for $\hbar\beta$-periodic functions in imaginary time associated with the harmonic Hamiltonian (4.231),

$$\boldsymbol{A}^{(0)}(\nu_n) \equiv \boldsymbol{U}^{(0)} + \nu_n^2\boldsymbol{1} = \begin{pmatrix} \overline{C}_0 & \tilde{C}_1 & \tilde{C}_2 & \cdots & \tilde{C}_N \\ \tilde{C}_1 & \Omega_1^2 & 0 & \cdots & 0 \\ \tilde{C}_2 & 0 & \Omega_2^2 & \cdots & 0 \\ \vdots & \vdots & \vdots & \ddots & \vdots \\ \tilde{C}_N & 0 & 0 & \cdots & \Omega_N^2 \end{pmatrix}. \tag{4.236}$$

We have
$$\overline{C}_0 = \tilde{C}_0 + \nu_n^2, \qquad \text{and} \qquad \Omega_\alpha^2 = \omega_\alpha^2 + \nu_n^2. \tag{4.237}$$
In diagonal basis, the determinant of $\boldsymbol{A}^{(0)}(\nu_n)$ has the form
$$\det \boldsymbol{A}^{(0)}(\nu_n) = \prod_{\alpha=0}^{N} (\mu_\alpha^2 + \nu_n^2). \tag{4.238}$$

On the other hand, we may calculate $\det \boldsymbol{A}^{(0)}(\nu_n)$ by employing the matrix representation (4.236). We then find
$$\det \boldsymbol{A}^{(0)}(\nu_n) = \left(\overline{C}_0 - \sum_{\alpha'=1}^{N} \frac{\tilde{C}_{\alpha'}^2}{\Omega_{\alpha'}^2} \right) \prod_{\alpha=1}^{N} \Omega_\alpha^2. \tag{4.239}$$

By means of the definitions (4.233) and (4.237), this expression is transformed into
$$\begin{aligned}
\det \boldsymbol{A}^{(0)}(\nu_n) &= \left(\omega_0^2 + \nu_n^2 + \frac{\nu_n^2}{M} \sum_{\alpha'=1}^{N} \frac{c_{\alpha'}^2}{m_{\alpha'}\omega_{\alpha'}^2} \frac{1}{(\omega_{\alpha'}^2 + \nu_n^2)} \right) \prod_{\alpha=1}^{N} \left(\omega_\alpha^2 + \nu_n^2 \right) \\
&= \left(\omega_0^2 + \nu_n^2 + |\nu_n|\hat{\gamma}(|\nu_n|) \right) \prod_{\alpha=1}^{N} \left(\omega_\alpha^2 + \nu_n^2 \right).
\end{aligned} \tag{4.240}$$

In the second form, we have used the representation (3.28) for the Laplace transform $\hat{\gamma}(\nu_n)$ of the damping kernel $\gamma(t)$. Upon equating (4.238) with (4.240), we get
$$\prod_{\alpha=0}^{N} (\mu_\alpha^2 + \nu_n^2) = \left(\omega_0^2 + \nu_n^2 + |\nu_n|\hat{\gamma}(|\nu_n|) \right) \prod_{\alpha=1}^{N} \left(\omega_\alpha^2 + \nu_n^2 \right). \tag{4.241}$$
Putting $n = 0$ in Eq. (4.241) and observing that $\nu_0 = 0$, we obtain a simple relation between the set of eigenfrequencies $\{\mu_\alpha\}$, the system frequency ω_0, and the set of reservoir frequencies $\{\omega_\alpha\}$,
$$\det \boldsymbol{U}^{(0)} = \prod_{\alpha=0}^{N} \mu_\alpha^2 = \omega_0^2 \prod_{\alpha=1}^{N} \omega_\alpha^2. \tag{4.242}$$

With use of the relations (4.241) and (4.242), the expression (4.235) takes the form
$$Z_{\text{tot}}^{(0)} = \prod_{\alpha=1}^{N} \left\{ \frac{1}{\beta\hbar\omega_\alpha} \prod_{n=1}^{\infty} \frac{\nu_n^2}{\omega_\alpha^2 + \nu_n^2} \right\} \frac{1}{\beta\hbar\omega_0} \prod_{n=1}^{\infty} \frac{\nu_n^2}{\omega_0^2 + \nu_n^2 + \nu_n\hat{\gamma}(\nu_n)}. \tag{4.243}$$

The upshot of this analysis now is that the N-fold product of the curly bracket term is the partition function Z_{R} of the reservoir, Eq. (4.29), and the residual set of factors is the product representation (4.225) of the reduced partition function $Z^{(0)}$ of the damped quantum oscillator. Thus indeed, the partition function of the global system emerges just in the factorized form
$$Z_{\text{tot}}^{(0)} = Z_{\text{R}} \, Z^{(0)}. \tag{4.244}$$

In conclusion, we have shown by two different lines of reasoning that the partition function of the damped system is given by the expression (4.225).

4.4 Quantum statistical expectation values in phase space

We may ask ourselves if there is a phase space representation of quantum statistical expectation values[7] for general operator functions $\hat{A}(\hat{p}, \hat{q})$. A classical particle is described by a probability density function in phase space $f^{(\mathrm{cl})}(p, q)$. The average of a function of momentum and position can then be expressed as

$$\langle A \rangle_{\mathrm{cl}} = \int\!\!\int \frac{dp\,dq}{2\pi\hbar}\, A(p, q) f^{(\mathrm{cl})}(p, q) \,. \tag{4.245}$$

Because of the uncertainty principle, one cannot define a true phase space probability distribution for a quantum mechanical particle. Nevertheless, "quasi-probability distribution functions" which bear resemblance to classical phase space distribution functions are very useful as a calculational tool. In addition, they elucidate the nature of quantum mechanics. Within the framework of a quasi-probability distribution $f^{(\mathrm{qm})}(p, q)$, the quantum statistical average takes the form

$$\langle A \rangle_{\mathrm{qm}} \equiv \mathrm{tr}\,(\hat{A}\hat{\rho}) = \int\!\!\int \frac{dp\,dq}{2\pi\hbar}\, A(p, q) f^{(\mathrm{qm})}(p, q) \,. \tag{4.246}$$

First of all, the question arises whether an operator function $\hat{A}(\hat{p}, \hat{q})$ which is defined in terms of a certain ordering prescription can unambiguously be represented as an ordinary function in phase space. One positive answer is the generalized Weyl correspondence. This is discussed in the following subsection.

Since the pioneering work by Weyl [200], the association of distribution functions in phase space with operator ordering rules for operators $\hat{p}$ and $\hat{q}$ has been the subject of many studies (cf. the review article [201]). The quasi-probability distribution which is appropriate to the Weyl correspondence between operator-valued and ordinary functions is the Wigner distribution $f^{(\mathrm{W})}(p, q)$ [202].

4.4.1 Generalized Weyl correspondence

We now aim at establishing a correspondence between quantum mechanical operators $\hat{A}(\hat{p}, \hat{q})$ and ordinary functions $A(p, q)$ in phase space. As a specific example, we choose the case of a point-like particle in one space dimension. The generalization to a coordinate space with d dimensions and curvature is discussed in Ref. [203]. We start with the expansion of the operator $\hat{A}(\hat{p}, \hat{q})$ in position and momentum eigenstates,

$$\hat{A}(\hat{p}, \hat{q}) = \int dp'\,dp''\,dq'\,dq''\, |p''\!><\!p''|q''\!><\!q''|\hat{A}(\hat{p}, \hat{q})|q'\!><\!q'|p'\!><\!p'| \,. \tag{4.247}$$

Next, we consider a one-parameter family of transformations of the phase space variables with Jacobian unity,

$$\begin{aligned}
p'' &= p + \tfrac{1}{2}(1 - \gamma)\eta \,; & p' &= p - \tfrac{1}{2}(1 + \gamma)\eta \,, \\
q'' &= q + \tfrac{1}{2}(1 + \gamma)y \,; & q' &= q - \tfrac{1}{2}(1 - \gamma)y \,.
\end{aligned} \tag{4.248}$$

---

[7]There is a neat monograph by W. P. Schleich on quantum physics in phase space [199].

The range of the parameter γ is $-1 \leq \gamma \leq 1$. With use of the relations

$$< q|q' > = \delta(q-q') ; \quad < p|p' > = \delta(p-p') ; \quad < q|p > = (2\pi\hbar)^{-1/2} e^{ipq/\hbar} , \quad (4.249)$$

we then obtain

$$\hat{A}(\hat{p}, \hat{q}) = \iint \frac{dp\,dq}{2\pi\hbar}\, A_\gamma(p, q)\hat{\Delta}_\gamma(p, q) , \quad (4.250)$$

where

$$A_\gamma(p, q) = \int dy\, e^{-ipy/\hbar} < q + \tfrac{1}{2}(1+\gamma)y \,|\, \hat{A}(\hat{p}, \hat{q}) \,|\, q - \tfrac{1}{2}(1-\gamma)y > , \quad (4.251)$$

$$\hat{\Delta}_\gamma(p, q) = \int d\eta\, e^{-iq\eta/\hbar} |\, p + \tfrac{1}{2}(1-\gamma)\eta >< p - \tfrac{1}{2}(1+\gamma)\eta \,| . \quad (4.252)$$

The function $A_\gamma(p, q)$ is the generalized Weyl transform of the operator-valued function $\hat{A}(\hat{p}, \hat{q})$. The correspondence is one-to-one and is denoted by the symbol $\longleftrightarrow$. The functional form of $A_\gamma(p, q)$ depends on the ordering of the operators $\hat{p}$ and $\hat{q}$ in $\hat{A}(\hat{p}, \hat{q})$, and on the parameter γ. By calculation one finds, e.g., the correspondence

$$\hat{p}^m \hat{F}(\hat{q})\hat{p}^n \quad \longleftrightarrow \quad \left(p - i\hbar\,\frac{1-\gamma}{2}\,\frac{\partial}{\partial q}\right)^m \left(p + i\hbar\,\frac{1+\gamma}{2}\,\frac{\partial}{\partial q}\right)^n F(q) . \quad (4.253)$$

Hence the ordinary function in phase space usually differs from the corresponding operator function. It is useful to define a one-parameter family of operator ordering schemes in such a way that the functional form of the operator function coincides exactly with the ordinary function in phase space. Denoting this family symbolically by $\mathcal{Z}_\gamma\{\hat{A}(\hat{p}, \hat{q})\}$, we thus have the direct correspondence

$$\mathcal{Z}_\gamma\{\hat{A}(\hat{p}, \hat{q})\} \quad \longleftrightarrow \quad A(p, q) . \quad (4.254)$$

By definition, $\mathcal{Z}_\gamma$ is an operator which orders the operators $\hat{p}$ and $\hat{q}$ in an operator function $\hat{A}(\hat{p}, \hat{q})$ regardless of the commutation relations such that the correspondence (4.254) holds. There follows with Eq. (4.253) that the operator corresponding to the transform $p^m F(q)$ is given by

$$\mathcal{Z}_\gamma\{\hat{p}^m \hat{F}(\hat{q})\} =: \sum_{l=0}^{m} \binom{m}{l} \left(\frac{1+\gamma}{2}\hat{p}\right)^{m-l} \hat{F}(\hat{q}) \left(\frac{1-\gamma}{2}\hat{p}\right)^{l} . \quad (4.255)$$

With this form, we can find $\mathcal{Z}_\gamma\{\hat{A}(\hat{p}, \hat{q})\}$ for any operator function $\hat{A}(\hat{p}, \hat{q})$ of which the Weyl transform can be represented in the form of a power series in the momentum, $A(p, q) = \sum_j F_j(q)\, p^j$,

$$\mathcal{Z}_\gamma\{\hat{A}(\hat{p}, \hat{q})\} =: \sum_j \sum_{l=0}^{j} \binom{j}{l} \left(\frac{1+\gamma}{2}\hat{p}\right)^{j-l} \hat{F}_j(\hat{q}) \left(\frac{1-\gamma}{2}\hat{p}\right)^{l} . \quad (4.256)$$

Equation (4.255) or (4.256) defines the ordering prescription for a whole family of functions parametrized by a continuous parameter γ in the range $-1 \leq \gamma \leq 1$.

The special value $\gamma = 0$ represents the case of Weyl ordering. For example, if we put $m = 2$ in Eq. (4.255), we have

$$\mathcal{Z}_0\{\hat{p}^2\hat{F}(\hat{q})\} \ =: \ \frac{1}{4}\left(\hat{p}^2\hat{F}(\hat{q}) + 2\hat{p}\hat{F}(\hat{q})\hat{p} + \hat{F}(\hat{q})\hat{p}^2\right) . \tag{4.257}$$

The cases $\gamma = 1$ and $\gamma = -1$ represent the antistandard ($\hat{p}\hat{q}$) and standard ($\hat{q}\hat{p}$) ordering prescription, respectively. With use of (4.252) and (4.249), we get

$$< q''|\hat{\Delta}_\gamma(p, x)|q' > \ = \ \delta(x - q)\,e^{ip(q''-q')/\hbar} , \tag{4.258}$$

where

$$q \ = \ \tfrac{1}{2}\left(q'' + q'\right) - \tfrac{1}{2}\gamma(q'' - q') . \tag{4.259}$$

As γ is tuned from $+1$ to -1, the position q runs from the initial point q' to the end point q''. Insertion of (4.258) in the coordinate representation of (4.250) and observance of the correspondence (4.254) results in the expression

$$< q''|\mathcal{Z}_\gamma\{\hat{A}(\hat{p}, \hat{q})\}|q' > \ = \ (2\pi\hbar)^{-1}\int dp\, A(p, q)\,e^{ip(q''-q')/\hbar} . \tag{4.260}$$

Inversion of this relation yields

$$A(p, q) \ = \ \int dy\, e^{-ipy/\hbar} < q + \tfrac{1}{2}(1 + \gamma)y\,|\mathcal{Z}_\gamma\{\hat{A}(\hat{p}, \hat{q})\}|\, q - \tfrac{1}{2}(1 - \gamma)y > . \tag{4.261}$$

Thus, the generalized Weyl transform $A(p, q)$ is calculated by regarding the matrix element $< q''|\mathcal{Z}_\gamma\{\hat{A}(\hat{p}, \hat{q})\}|q' >$ as a function of q and y, and then taking the Fourier transform with respect to y. For symmetric (Weyl) ordering, $\gamma = 0$, we have

$$A(p, q) \ = \ \int dy\, e^{-ipy/\hbar} < q + \tfrac{1}{2}y\,|\mathcal{Z}_0\{\hat{A}(\hat{p}, \hat{q})\}|q - \tfrac{1}{2}y > . \tag{4.262}$$

In summary of the results obtained so far, we have found for the operator ordering prescription (4.255) a unique representation in phase space.

4.4.2   Generalized Wigner function and expectation values

Wigner's function is a phase space representation of the reduced density matrix $\rho_\beta(q'', q')$ in thermal equilibrium. Assume that $\rho_\beta(q'', q')$ is expressed in terms of the variables $y = q'' - q'$ and $q = \tfrac{1}{2}(q'' + q')$. Then, Wigner's function $f^{(W)}(p, q)$ is the Fourier transform with respect to the relative coordinate y,

$$f^{(W)}(p, q) \ = \ \int dy\, e^{-ipy/\hbar}\rho_\beta(q + \tfrac{1}{2}y,\ q - \tfrac{1}{2}y) . \tag{4.263}$$

The quasi-probability distribution $f^{(W)}(p, q)$ is the appropriate density function in phase space for operator functions with Weyl ordering.

If we aim at the phase space representation of quantum statistical expectation values for general operator functions $\mathcal{Z}_\gamma\{\hat{A}(\hat{p}, \hat{q})\}$, we have to make a subtle generalization of Wigner's function (4.263),

$$f^{(\gamma)}(p, \bar{q}) \equiv \int dy \, e^{-ipy/\hbar} \, \rho_\beta\left(\bar{q} + \tfrac{1}{2}(1 - \gamma)y, \, \bar{q} - \tfrac{1}{2}(1 + \gamma)y\right). \tag{4.264}$$

Here we have again $y = q'' - q'$, but we use as second variable

$$\bar{q} = \tfrac{1}{2}(q'' + q') + \tfrac{1}{2}\gamma(q'' - q'), \tag{4.265}$$

which differs from the expression (4.259) by the replacement $\gamma \to -\gamma$. The reason for this subtle change will become clear shortly. The generalized Wigner function at $\gamma = 0$, $f^{(0)}(p, \bar{q})$, coincides with $f^{(W)}(p, q)$. The expression (4.264) is inverted to yield

$$\rho_\beta(q'', q') = (2\pi\hbar)^{-1} \int dp \, f^{(\gamma)}(p, \bar{q}) \, e^{ip(q''-q')/\hbar}. \tag{4.266}$$

The quasi-probability distributions $f^{(W)}(p, q)$ and $f^{(\gamma)}(p, \bar{q})$ have properties similar to the classical density function $f(p, q)$ in phase space,

$$\frac{1}{2\pi\hbar} \int dp \, f^{(W)}(p, q) = \frac{1}{2\pi\hbar} \int dp \, f^{(\gamma)}(p, q) = \, <q|\hat{\rho}_\beta|q> \, \equiv P(q),$$

$$\int dq \, f^{(W)}(p, q) = \int d\bar{q} \, f^{(\gamma)}(p, \bar{q}) = \, <p|\hat{\rho}_\beta|p> \, \equiv \tilde{P}(p). \tag{4.267}$$

The diagonal elements $P(q)$ and $\tilde{P}(p)$ represent the probability for finding the damped particle in thermal equilibrium at position q and with momentum p, respectively. Although the quasi-probability distribution $f^{(W)}(p, q)$ $[f^{(\gamma)}(p, \bar{q})]$ satisfies (4.267), it does not describe the probability for finding the particle at position q $[\bar{q}]$ with momentum p, because the function can become negative for some values of p and q $[\bar{q}]$.

We are now ready to consider the phase space representation of the quantum statistical average of the operator $\mathcal{Z}_\gamma\{\hat{A}(\hat{p}, \hat{q})\}$,

$$\langle A_\gamma \rangle =: \left\langle \mathcal{Z}_\gamma\{\hat{A}(\hat{p}, \hat{q})\} \right\rangle = \int dq'' \int dq' \, <q''|\mathcal{Z}_\gamma\{\hat{A}(\hat{p}, \hat{q})\}|q'> \rho_\beta(q', q''). \tag{4.268}$$

Because $\rho_\beta(q', q'')$, instead of $\rho_\beta(q'', q')$, appears in Eq. (4.268), we used Eq. (4.265) in Eq. (4.266), instead of Eq. (4.259). As a result, the Wigner function related to $\rho_\beta(q', q'')$ is a function of q instead of $\bar{q}$. Substituting (4.260) and (4.266) we find

$$\langle A_\gamma \rangle = \int\int \frac{dp' \, dq}{2\pi\hbar} \int\int \frac{dp \, dy}{2\pi\hbar} \, A(p, q) f^{(\gamma)}(p', q) \, e^{iy(p-p')/\hbar}, \tag{4.269}$$

where we have interchanged q'' with q' in Eq. (4.266) and observed Eqs. (4.265) and (4.259). Since the y integral gives $\delta(p - p')$, we find in the end

$$\langle A_\gamma \rangle = \int\int \frac{dp \, dq}{2\pi\hbar} \, A(p, q) f^{(\gamma)}(p, q). \tag{4.270}$$

Thus, when the quasi-probability distribution in Eq. (4.246) is chosen to be $f^{(\gamma)}(p, q)$, the correspondence between $A(p, q)$ and $\hat{A}(\hat{p}, \hat{q})$ is that given by Eq. (4.254). Notice that the expression (4.270) applies to the family of operator ordering prescriptions

$\mathcal{Z}_\gamma\{\hat{A}(\hat{p},\hat{q})\}$ among which Weyl ordering, standard ordering, and anti-standard ordering are special cases. The main criterion for the choice of the parameter γ of the quasi-probability distribution function for a particular problem is convenience.

The phase space representation of quantum statistical expectation values is especially convenient if the density matrix in coordinate representation is already known. Then the remaining problem is limited to integrations of ordinary functions. We will make use of the result (4.270) in Section 8.2.

5. Real-time path integrals and nonequilibrium dynamics

5.1   Statement of the problem and general concepts

In the preceding chapter, I have formulated quantum statistical mechanics for a number of systems in contact with a thermal reservoir using the imaginary-time or Euclidean path integral representation. I have also shown some techniques useful in the approximate evaluation of path integrals. In this chapter we are concerned with the description of non-equilibrium time-dependent phenomena. I shall outline the general formalism with the simple phenomenological model introduced in Section 3.1. The proceeding will also illuminate the appropriate modifications in the passage from imaginary to real time for the various microscopic models introduced previously. Thus I take them up only briefly in the summary at the end of the chapter. Since we are interested in the evolution of mixed states, the proper vehicle is the density matrix representation.

Generic nonequilibrium problems are as follows: (i) The system is prepared in a nonequilibrium inital state and one is interested in the relaxation dynamics of a physical observable O at later time. (ii) The system is initially in the canonical equilibrium state and then time-dependent perturbations are switched on. The quantity of interest in both cases is the average value $\langle O(t)\rangle = \mathrm{tr}\{\hat{\rho}(t)\hat{O}\}$ as function of time.

In the next section, I study the evolution of the damped system under the assumption that it was prepared initially in a product state of the system-plus-reservoir complex. In the third section, I squeeze in a brief discussion on the physical significance of the influence functional obtained in the second section. The fourth section deals with other initial preparations which are appropriate in specific experiments. The general concepts will be applied in the fifth section to describe the evolution of a damped system for any form of initial preparation. This leads, in general, to a complex-time triple path integral. In the next section, I show that for ergodic systems we can regain a real time path integral, but for paths which evolve from the infinite past. In the seventh section it is shown that all these cases can be compactly presented as single-path integrals along closed time contours. In the remaining sections I discuss the semiclassical regime, in which we end up with a Langevin-type description of the quantum stochastic process, the stochastic unraveling of the influence functional, and the combination of semiclassics with stochastic unraveling. This will turn out as a promising route to handle non-Markovian dynamics.

Generally, the analysis of this chapter will display that the Hamiltonian dynamics of the global system induces a non-unitary dynamics for the reduced density matrix.

Let us examine the possibility of obtaining closed formal expressions for the dynamics of a dissipative quantum mechanical system. We assume that the underlying global system is governed by the Hamiltonian (3.11). Like in the static problem discussed in the preceding section, it is convenient to carry out the reduction within the functional integral description, a technique introduced by Feynman and Vernon [204] already in 1963. Our starting point in performing the reduction is the operator relation for the density operator of the global system

$$\hat{W}(t) = e^{-i\hat{H}t/\hbar}\,\hat{W}(0)\,e^{i\hat{H}t/\hbar}\,. \tag{5.1}$$

In coordinate representation we may write

$$< q_{\mathrm{f}}, \boldsymbol{x}_{\mathrm{f}}|\,\hat{W}(t)\,|q_{\mathrm{f}}', \boldsymbol{x}_{\mathrm{f}}'> \; = \; \int dq_{\mathrm{i}}\,dq_{\mathrm{i}}'\,d\boldsymbol{x}_{\mathrm{i}}\,d\boldsymbol{x}_{\mathrm{i}}'\,K(q_{\mathrm{f}}, \boldsymbol{x}_{\mathrm{f}}, t; q_{\mathrm{i}}, \boldsymbol{x}_{\mathrm{i}}, 0) \tag{5.2}$$

$$\times \; < q_{\mathrm{i}}, \boldsymbol{x}_{\mathrm{i}}|\,\hat{W}(0)\,|q_{\mathrm{i}}', \boldsymbol{x}_{\mathrm{i}}'> K^{*}(q_{\mathrm{f}}', \boldsymbol{x}_{\mathrm{f}}', t; q_{\mathrm{i}}', \boldsymbol{x}_{\mathrm{i}}', 0)\,.$$

Here, the N-component vector $\boldsymbol{x}_{\mathrm{i/f}}$ stands again for $(x_{\mathrm{i/f},1}, \ldots, x_{\mathrm{i/f},N})$, and K is the coordinate representation of the time evolution operator

$$K(q_{\mathrm{f}}, \boldsymbol{x}_{\mathrm{f}}, t; q_{\mathrm{i}}, \boldsymbol{x}_{\mathrm{i}}, 0) \; = \; < q_{\mathrm{f}}, \boldsymbol{x}_{\mathrm{f}}|\,e^{-i\hat{H}t/\hbar}|q_{\mathrm{i}}, \boldsymbol{x}_{\mathrm{i}}> \,, \tag{5.3}$$

which may be represented as a path integral,

$$K(q_{\mathrm{f}}, \boldsymbol{x}_{\mathrm{f}}, t; q_{\mathrm{i}}, \boldsymbol{x}_{\mathrm{i}}, 0) \; = \; \int \mathcal{D}q(\cdot)\,\mathcal{D}\boldsymbol{x}(\cdot)\;e^{iS[q(\cdot),\boldsymbol{x}(\cdot)]/\hbar}\,,$$

$$\tag{5.4}$$

$$K^{*}(q_{\mathrm{f}}', \boldsymbol{x}_{\mathrm{f}}', t; q_{\mathrm{i}}', \boldsymbol{x}_{\mathrm{i}}', 0) \; = \; \int \mathcal{D}q'(\cdot)\,\mathcal{D}\boldsymbol{x}'(\cdot)\;e^{-iS[q'(\cdot),\boldsymbol{x}'(\cdot)]/\hbar}\,.$$

The functional integrations in Eq. (5.4) extend over all paths with endpoints

$$q(0) = q_{\mathrm{i}}\,, \qquad q(t) = q_{\mathrm{f}}\,, \qquad q'(0) = q_{\mathrm{i}}'\,, \qquad q'(t) = q_{\mathrm{f}}'\,, \tag{5.5}$$

$$\boldsymbol{x}(0) = \boldsymbol{x}_{\mathrm{i}}\,, \qquad \boldsymbol{x}(t) = \boldsymbol{x}_{\mathrm{f}}\,, \qquad \boldsymbol{x}'(0) = \boldsymbol{x}_{\mathrm{i}}'\,, \qquad \boldsymbol{x}'(t) = \boldsymbol{x}_{\mathrm{f}}'\,. \tag{5.6}$$

The action in Eq. (5.4) is given by system, reservoir, and interaction part,

$$S \; = \; S_{\mathrm{S}} + S_{\mathrm{R}} + S_{\mathrm{I}} \; = \; \int_{0}^{t} dt'\,\Big(\mathcal{L}_{\mathrm{S}}(t') + \mathcal{L}_{\mathrm{R}}(t') + \mathcal{L}_{\mathrm{I}}(t')\Big)\,, \tag{5.7}$$

where [we add a source term in $\mathcal{L}_{\mathrm{S}}(t)$, as in Eq. (4.20)]

$$\mathcal{L}_{\mathrm{S}}(t) \; = \; \frac{1}{2}M\dot{q}^{2}(t) - V[q(t)] - \mathcal{J}(t)q(t)\,,$$

$$\mathcal{L}_{\mathrm{R}}(t) \; = \; \frac{1}{2}\sum_{\alpha=1}^{N} m_{\alpha}\,[\,\dot{x}_{\alpha}^{2}(t) - \omega_{\alpha}^{2}x_{\alpha}^{2}(t)\,]\,, \tag{5.8}$$

$$\mathcal{L}_{\mathrm{I}}(t) \; = \; \sum_{\alpha=1}^{N} \Big(c_{\alpha}x_{\alpha}(t)q(t) - \frac{1}{2}\frac{c_{\alpha}^{2}}{m_{\alpha}\omega_{\alpha}^{2}}\,q^{2}(t)\Big)\,.$$

The expression (5.2) for the density matrix describes the dynamics of the system-plus-environment complex as a whole. However, in most cases of interest the only information we wish to have is the system's dynamics under the reservoir's influence. Then the quantity we are really interested in is the reduced density matrix [11, 24, 23]

$$
\begin{aligned}
\rho(q_f, q_f'; t) &\equiv \int d\boldsymbol{x}_f < q_f, \boldsymbol{x}_f | \hat{W}(t) | q_f', \boldsymbol{x}_f > \\
&= \int dq_i \, dq_i' \, d\boldsymbol{x}_i \, d\boldsymbol{x}_i' \, d\boldsymbol{x}_f \, K(q_f, \boldsymbol{x}_f, t; q_i, \boldsymbol{x}_i, 0) \\
&\quad \times < q_i, \boldsymbol{x}_i | \hat{W}(0) | q_i', \boldsymbol{x}_i' > K^*(q_f', \boldsymbol{x}_f, t; q_i', \boldsymbol{x}_i', 0) \ .
\end{aligned}
\tag{5.9}
$$

We are now well prepared to work out the formal path integral expression of the reduced density matrix.

5.2   Feynman-Vernon method for a product initial state

Suppose that at time $t = 0$ the system and the bath are uncoupled, and the bath to be in its thermal equilibrium state. Then the density operator of the global system at $t = 0$ is in a product initial state (pis)

$$
\hat{W}_{\mathrm{pis}}(0) = \hat{\rho}(0) \otimes \hat{W}_{\mathrm{R}}(0) = \hat{\rho}(0) \otimes Z_{\mathrm{R}}^{-1} e^{-\beta \hat{H}_{\mathrm{R}}} \ .
\tag{5.10}
$$

where $\hat{\rho}$ is the reduced density operator. The initial state (5.10) is free of correlations between the system and the bath. Since the system-bath coupling is not at the disposal of the experimenter, the product initial state is somewhat artificial in many applications. However, the product form is very convenient for computations. It may be experimentally prepared in the past with the aid of suitable biasing forces, which are then switched off at time zero.

Assume that the system-bath coupling is suddenly switched on at time $t = 0^+$ and consider the dynamics of $\rho(t)$ for $t \geq 0$. Substituting Eq. (5.10) in Eq. (5.9) and using Eqs. (5.4) – (5.8), we obtain for the reduced density matrix the expression

$$
\rho(q_f, q_f'; t) = \int dq_i \, dq_i' \, J_{\mathrm{FV}}(q_f, q_f', t; q_i, q_i', 0) \, \rho(q_i, q_i'; 0) \ ,
\tag{5.11}
$$

where J_{FV} is a *propagating function* describing the time evolution of the reduced density matrix. The path integral for the propagating function is

$$
J_{\mathrm{FV}}(q_f, q_f', t; q_i, q_i', 0) = \int \mathcal{D}q \, \mathcal{D}q' \, \exp\left\{ \frac{i}{\hbar} \left(S_{\mathrm{S}}[q] - S_{\mathrm{S}}[q'] \right) \right\} \mathcal{F}_{\mathrm{FV}}[q, q'] \ .
\tag{5.12}
$$

The still formal expression

$$
\begin{aligned}
\mathcal{F}_{\mathrm{FV}}[q(\cdot), q'(\cdot)] &\equiv \int d\boldsymbol{x}_f \, d\boldsymbol{x}_i \, d\boldsymbol{x}_i' \, W_{\mathrm{R}}(\boldsymbol{x}_i, \boldsymbol{x}_i') \int_{\boldsymbol{x}(0)=\boldsymbol{x}_i}^{\boldsymbol{x}(t)=\boldsymbol{x}_f} \mathcal{D}\boldsymbol{x}(\cdot) \int_{\boldsymbol{x}'(0)=\boldsymbol{x}_i'}^{\boldsymbol{x}'(t)=\boldsymbol{x}_f} \mathcal{D}\boldsymbol{x}'(\cdot) \\
&\quad \times \exp\left\{ \frac{i}{\hbar} \left(S_{\mathrm{R}}[\boldsymbol{x}(\cdot)] + S_{\mathrm{I}}[\boldsymbol{x}(\cdot), q(\cdot)] - S_{\mathrm{R}}[\boldsymbol{x}'(\cdot)] - S_{\mathrm{I}}[\boldsymbol{x}'(\cdot), q'(\cdot)] \right) \right\}
\end{aligned}
\tag{5.13}
$$

is the so-called Feynman-Vernon influence functional [204]. It can be written as

$$\mathcal{F}_{\mathrm{FV}}[q(\cdot), q'(\cdot)] = \int d\boldsymbol{x}_{\mathrm{f}} \, d\boldsymbol{x}_{\mathrm{i}} \, d\boldsymbol{x}'_{\mathrm{i}} \, W_{\mathrm{R}}(\boldsymbol{x}_{\mathrm{i}}, \boldsymbol{x}'_{\mathrm{i}}) \, F(q; \boldsymbol{x}_{\mathrm{f}}, \boldsymbol{x}_{\mathrm{i}}) \, F^*(q'; \boldsymbol{x}_{\mathrm{f}}, \boldsymbol{x}'_{\mathrm{i}}) \; . \qquad (5.14)$$

The amplitude F is a real-time path integral over the path $\boldsymbol{x}(t')$,

$$F[q(\cdot); \boldsymbol{x}_f, \boldsymbol{x}_{\mathrm{i}}] = \int_{\boldsymbol{x}(0)=\boldsymbol{x}_{\mathrm{i}}}^{\boldsymbol{x}(t)=\boldsymbol{x}_{\mathrm{f}}} \mathcal{D}\boldsymbol{x}(\cdot) \, \exp\left\{ \frac{i}{\hbar}\left(S_{\mathrm{R}}[\boldsymbol{x}(\cdot)] + S_{\mathrm{I}}[\boldsymbol{x}(\cdot), q(\cdot)] \right) \right\} \; . \qquad (5.15)$$

For the harmonic reservoir model described by $\mathcal{L}_{\mathrm{R}}$ in Eq. (5.8), both the canonical density matrix W_{R} and the functional F are in product form. Each factor represents the contribution of an individual bath oscillator,

$$\begin{aligned} W_{\mathrm{R}}(\boldsymbol{x}_{\mathrm{i}}, \boldsymbol{x}'_{\mathrm{i}}) &= \prod_{\alpha=1}^{N} W_{\mathrm{R}}^{(\alpha)}(x_{\mathrm{i},\alpha}, x'_{\mathrm{i},\alpha}) \; , \\ F[q(\cdot); \boldsymbol{x}_{\mathrm{f}}, \boldsymbol{x}_{\mathrm{i}}] &= \prod_{\alpha=1}^{N} F_\alpha[q(\cdot); x_{\mathrm{f},\alpha}, x_{\mathrm{i},\alpha}] \; . \end{aligned} \qquad (5.16)$$

Here, $W_{\mathrm{R}}^{(\alpha)}$ is the canonical density matrix for the αth oscillator, and F_α is the propagator of this oscillator under the influence of an external force $c_\alpha q(s)$. The respective Gaussian functional integrals in imaginary time τ and real time s read

$$W_{\mathrm{R}}^{(\alpha)}(x_{\mathrm{i},\alpha}, x'_{\mathrm{i},\alpha}) = \frac{1}{Z_{\mathrm{R}}^{(\alpha)}} \int_{\bar{x}_\alpha(0)=x'_{\mathrm{i},\alpha}}^{\bar{x}_\alpha(\hbar\beta)=x_{\mathrm{i},\alpha}} \mathcal{D}\bar{x}_\alpha(\cdot) \, \exp\left\{ -\frac{1}{\hbar} \int_0^{\hbar\beta} d\tau \, \frac{m_\alpha}{2} \left(\dot{\bar{x}}_\alpha^2 + \omega_\alpha^2 \bar{x}^2 \right) \right\}, \quad (5.17)$$

$$\begin{aligned} F_\alpha[q(\cdot); x_{\mathrm{f},\alpha}, x_{\mathrm{i},\alpha}] = &\int_{x_\alpha(0)=x_{\mathrm{i},\alpha}}^{x_\alpha(t)=x_{\mathrm{f},\alpha}} \mathcal{D}x_\alpha(\cdot) \\ &\times \, \exp\left\{ \frac{i}{\hbar} \int_0^t ds \left[\frac{m_\alpha}{2} \left(\dot{x}_\alpha^2 - \omega_\alpha^2 x_\alpha^2 \right) + c_\alpha q \, x_\alpha - \frac{c_\alpha^2 q^2}{2 m_\alpha \omega_\alpha^2} \right] \right\} \; . \end{aligned} \qquad (5.18)$$

Working off the Gaussian integrals one finds the exact analytic expressions [204, 87]

$$W_{\mathrm{R}}^{(\alpha)}(x_{\mathrm{i},\alpha}, x'_{\mathrm{i},\alpha}) = \frac{1}{Z_{\mathrm{R}}^{(\alpha)}} \sqrt{\frac{m_\alpha \omega_\alpha}{2\pi\hbar \sinh(\beta\hbar\omega_\alpha)}} \qquad (5.19)$$

$$\times \, \exp\left\{ -\frac{m_\alpha \omega_\alpha}{2\hbar \sinh(\beta\hbar\omega_\alpha)} \left[(x_{\mathrm{i},\alpha}^2 + x_{\mathrm{i},\alpha}'^2) \cosh(\beta\hbar\omega_\alpha) - 2x_{\mathrm{i},\alpha} x'_{\mathrm{i},\alpha} \right] \right\} \; ,$$

$$F_\alpha[q(\cdot); x_{\mathrm{f},\alpha}, x_{\mathrm{i},\alpha}] = \sqrt{\frac{m_\alpha \omega_\alpha}{2\pi i\hbar \sin(\omega_\alpha t)}} \exp\left(\frac{i}{\hbar} \phi_\alpha[q(\cdot); x_{\mathrm{f},\alpha}, x_{\mathrm{i},\alpha}] \right) \; , \qquad (5.20)$$

The action in presence of the external force takes the form

$$\phi_\alpha[q(\cdot); x_{\mathrm{f},\alpha}, x_{\mathrm{i},\alpha}] \;=\; \frac{m_\alpha \omega_\alpha}{2\sin(\omega_\alpha t)} \left[(x_{\mathrm{i},\alpha}^2 + x_{\mathrm{f},\alpha}^2)\cos(\omega_\alpha t) - 2 x_{\mathrm{i},\alpha} x_{\mathrm{f},\alpha} \right] \tag{5.21}$$

$$+ \; \frac{x_{\mathrm{i},\alpha} c_\alpha}{\sin(\omega_\alpha t)} \int_0^t dt' \, \sin\left[\omega_\alpha(t - t')\right] q(t')$$

$$+ \; \frac{x_{\mathrm{f},\alpha} c_\alpha}{\sin(\omega_\alpha t)} \int_0^t dt' \, \sin(\omega_\alpha t')\, q(t') \;-\; \frac{c_\alpha^2}{2 m_\alpha \omega_\alpha^2} \int_0^t dt' \, q^2(t')$$

$$- \; \frac{c_\alpha^2}{m_\alpha \omega_\alpha \sin(\omega_\alpha t)} \int_0^t dt' \int_0^{t'} dt'' \, \sin\left[\omega_\alpha(t - t')\right] \sin(\omega_\alpha t'')\, q(t')\, q(t'') \,.$$

The partition function of the unperturbed reservoir is in product form,

$$Z_{\mathrm{R}} \;=\; \prod_{\alpha=1}^{N} Z_{\mathrm{R}}^{(\alpha)} \;=\; \prod_{\alpha=1}^{N} \left\{ 2\sinh(\beta\hbar\omega_\alpha/2) \right\}^{-1} . \tag{5.22}$$

Substitution of Eqs. (5.16) – (5.21) in Eq. (5.14) yields Gaussian integrals over the endpoints that are easily evaluated. In the end, we find that the pre-exponential factors complement each other to a factor unity. The resulting expression for the Feynman-Vernon influence functional is conveniently written in the form

$$\mathcal{F}_{\mathrm{FV}}[q(\cdot), q'(\cdot)] \;=\; \exp\left\{ -\mathcal{S}_{\mathrm{FV}}[q(\cdot), q'(\cdot)]/\hbar \right\}, \tag{5.23}$$

The influence action is

$$\mathcal{S}_{\mathrm{FV}}[q, q'] \;=\; \int_0^t dt' \int_0^{t'} dt'' \left\{ q(t') - q'(t') \right\} \left\{ L(t' - t'')\, q(t'') - L^*(t' - t'')\, q'(t'') \right\}$$

$$+ \, i\,\frac{\mu}{2} \int_0^t dt' \left\{ q^2(t') - q'^2(t') \right\}. \tag{5.24}$$

The kernel L has for complex time $z = t - i\tau$ and real time t, respectively, the spectral representations

$$L(z) \;=\; \sum_{\alpha=1}^{N} \frac{c_\alpha^2}{2 m_\alpha \omega_\alpha} \frac{\cosh[\omega_\alpha(\frac{1}{2}\hbar\beta - iz)]}{\sinh(\frac{1}{2}\omega_\alpha \hbar\beta)} \;=\; \int_0^\infty d\omega \, \frac{J(\omega)}{\pi} \frac{\cosh[\omega(\frac{1}{2}\hbar\beta - iz)]}{\sinh(\frac{1}{2}\omega\hbar\beta)} , \tag{5.25}$$

$$L(t) \;\equiv\; L'(t) + iL''(t) = \frac{1}{\pi} \int_0^\infty d\omega\, J(\omega) \left(\coth(\beta\hbar\omega/2)\cos(\omega t) - i\,\sin(\omega t) \right). \tag{5.26}$$

Inspection of the expression (4.54) with (3.55) reveals that the kernel $L(z)$ is the analytic continuation of the imaginary-time kernel $K(\tau)$ in the influence action (4.50),

$$L(z) \;=\; K(\tau = iz) \,. \tag{5.27}$$

The imaginary part $L''(t)$ of the Feynman-Vernon kernel is related to the damping kernel $\gamma(t)$ in the classical equation of motion [cf. Eq. (3.31)] by

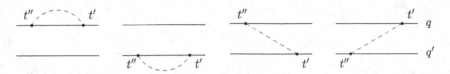

Figure 5.1: Graphical representation of the four contributions to the influence function $\mathcal{S}_{\mathrm{FV}}[q, q']$. The dashed line is the propagator $L(t' - t'')$ in the first and third graph, and $L^*(t' - t'')$ in the second and fourth graph.

$$\Theta(t)\, L''(t) = (M/2)\, \dot\gamma(t) . \tag{5.28}$$

The last term in Eq. (5.24) is due to the potential counter term in $\mathcal{L}_{\mathrm{I}}(t)$ given in Eq. (5.8). According to Eq. (4.51) and with use of Eq. (5.28) we have

$$\mu \equiv \sum_\alpha \frac{c_\alpha^2}{m_\alpha \omega_\alpha^2} = M\gamma(t = 0^+) = -2 \int_0^\infty dt\, L''(t) . \tag{5.29}$$

The Feynman-Vernon influence action (5.24) can be split up into three contributions

$$\mathcal{S}_{\mathrm{FV}}[q, q'] = \mathcal{S}_{\mathrm{FV, self}}[q] + \mathcal{S}^*_{\mathrm{FV, self}}[q'] + \mathcal{S}_{\mathrm{FV, cross}}[q, q'] , \tag{5.30}$$

where

$$\mathcal{S}_{\mathrm{FV, self}}[q] = \int_0^t dt' \int_0^{t'} dt''\, q(t')\, L(t' - t'')\, q(t'') + i\frac{\mu}{2} \int_0^t dt'\, q^2(t') ,$$

$$\mathcal{S}_{\mathrm{FV, cross}}[q, q'] = -\int_0^t dt' \int_0^{t'} dt'' \left\{ q'(t')L(t' - t'')\, q(t'') + q(t')\, L^*(t' - t'')\, q'(t'') \right\} .$$

The four diagrams contributing to $\mathcal{S}_{\mathrm{FV}}[q, q']$ are shown in Fig. 5.1. The first two graphs are the self-interactions of the paths q and q', respectively, and the other two describe cross interactions between the paths. The exponentiation of $\mathcal{S}_{\mathrm{FV}}[q, q']$ in $\mathcal{F}_{\mathrm{FV}}[q, q']$ gives all kinds of irreducible and reducible diagrams with any number of exchanged lines made up of the four fundamental contributions.

With use of (5.28) in (5.24) and integration by parts, the resulting boundary term cancels the potential counter term. The imaginary part of $\mathcal{S}_{\mathrm{FV}}[q, q']$ takes the form

$$\mathrm{Im}\, \mathcal{S}_{\mathrm{FV}}[q, q'] = \frac{M}{2} \int_0^t dt' \int_0^{t'} dt'' \left(q(t') - q'(t') \right)\gamma(t' - t'')\left(\dot q(t'') + \dot q'(t'') \right)$$

$$+ \frac{M}{2} \left(q(0) + q'(0) \right) \int_0^t dt'\, \gamma(t') \left(q(t') - q'(t') \right) . \tag{5.31}$$

At this point, it is useful to introduce symmetric and antisymmetric coordinates,

$$r(t) = \tfrac{1}{2}\left[q(t) + q'(t) \right] , \qquad \text{and} \qquad y(t) = q(t) - q'(t) . \tag{5.32}$$

Figure 5.2: Graphical representation of the two contributions to $S_{\mathrm{FV}}[r, y]$. The left diagram is the self-interaction of the off-diagonal path $y(t)$ and represents $\mathcal{S}^{(\mathrm{N})}[y]$. The exchanged line is the propagator $L'(t'-t'')$. The right diagram illustrates the correlations between $y(t)$ and $\dot{r}(t)$ in $\mathcal{S}^{(\mathrm{F})}[r, y]$. The exchanged line represents the friction kernel $\gamma(t'-t'')$.

The path $r(t)$ measures propagation of the system along the diagonal of the density matrix and is therefore termed *quasiclassical* path. The path $y(t)$ is book-keeping the system's off-diagonal excursions while propagating. This path describes quantum fluctuations. With use of (5.31), the influence action (5.24) may be written as

$$\mathcal{S}_{\mathrm{FV}}[q, q'] = \mathcal{S}^{(\mathrm{N})}[y] + i\mathcal{S}^{(\mathrm{F})}[r, y] ,$$

$$\mathcal{S}^{(\mathrm{N})}[y; t, t_{\mathrm{i}}] = \int_{t_{\mathrm{i}}}^{t} dt' \int_{t_{\mathrm{i}}}^{t'} dt'' \, y(t') \, L'(t'-t'') \, y(t'') , \qquad (5.33)$$

$$\mathcal{S}^{(\mathrm{F})}[r, y; t, t_{\mathrm{i}}] = M \int_{t_{\mathrm{i}}}^{t} dt' \int_{t_{\mathrm{i}}}^{t'} dt'' \, y(t')\gamma(t'-t'')\dot{r}(t'') + Mr(t_{\mathrm{i}}) \int_{t_{\mathrm{i}}}^{t} dt' \, y(t')\gamma(t'-t_{\mathrm{i}}) .$$

The expression (5.33) is the influence action for a product state at initial time t_{i}. The real part $\mathcal{S}^{(\mathrm{N})}[y]$ is dubbed *noise action*, and the imaginary part $\mathcal{S}^{(\mathrm{F})}[r, y]$ is termed *friction action*. The form (5.33) has been given first in Ref. [87].[1]

Next we write the system's action in terms of the variables r and y

$$\Sigma_{\mathrm{S}}[r, y; t, t_{\mathrm{i}}] \equiv S_{\mathrm{S}}[q] - S_{\mathrm{S}}[q'] = \int_{t_{\mathrm{i}}}^{t} dt' \left(M\dot{r}\dot{y} - V(r + \tfrac{1}{2}y) + V(r - \tfrac{1}{2}y) \right) . \quad (5.34)$$

Then the $\{r, y\}$-representation of the propagating function takes the form

$$J(r_{\mathrm{f}}, y_{\mathrm{f}}, t; r_{\mathrm{i}}, y_{\mathrm{i}}, 0) = \int \mathcal{D}r \, \mathcal{D}y \; e^{i\Sigma_{\mathrm{S}}[r, y; t, 0]/\hbar - \mathcal{S}^{(\mathrm{N})}[y; t, 0]/\hbar - i\mathcal{S}^{(\mathrm{F})}[r, y; t, 0]/\hbar} . \quad (5.35)$$

The actions $\mathcal{S}^{(\mathrm{N})}[y]$ and $\mathcal{S}^{(\mathrm{F})}[r, y]$ are sketched diagrammatically in Fig. 5.2. Exponentiation of the action takes into account graphs with any number of the two fundamental diagrams both stringed together and in nested form. Before turning to more general initial conditions, let us briefly reflect the significance of the influence functional and the effects caused by it.

In summary, the time evolution of the reduced density matrix for a product initial state may be calculated from Eq. (5.11) either with use of the $\{q, q'\}$-representation (5.12) with (5.23) and (5.30) for the propagating function, or with use of the $\{r, y\}$-representation (5.35) with (5.34) and (5.33). We see that the need to calculate the

---

[1]The last term in the friction action (5.33) is missing in the work by Caldeira and Leggett [205].

time-evolution with a double-path integral complicates the nonequilibrium theory in comparison with the equilibrium dynamics in imaginary time, Eq. (4.60) with (4.61). In actual calculations, the difficulties may be substantially reduced by choosing the $\{r, y\}$-representation which separates friction and quantum noise.

5.3 Decoherence and friction

The coupling of the system to the reservoir manifests itself in two different ways. On the one hand, the influence action $S^{(F)}[r, y]$ brings friction into the system. It provides the friction force $M \int_0^t dt' \, \gamma(t - t') \dot{r}(t')$ in the equation of motion for the quasiclassical path $r(t)$, as we shall see below in Subsection 5.8.1 within a stationary phase analysis. Pictorially, friction is conveyed by the right diagram in Fig. 5.2. On the other hand, the noise action $S^{(N)}[y]$ provides a Gaussian stochastic force which reflects random pumping of energy back and forth between system and reservoir, and thus bears loss of coherence in the system.

Quantum interference is a phenomenon directly related to the fact that transition amplitudes representing two different paths contributing to the propagation of a system may accumulate a phase difference φ. The interference terms of the squared amplitude come with phase factors $e^{\pm i\varphi}$. In the presence of the bath coupling, the phase φ is fluctuating. Then the relevant object is the statistical average of it, the "decoherence" factor $\langle e^{\pm i\varphi} \rangle$ [206, 207]. The extinction of quantum coherence for the pair of paths $q(t)$ and $q'(t)$ is determined by the noise action $S^{(N)}[y]$,

$$\langle e^{\pm i\varphi} \rangle = e^{-S^{(N)}[y]/\hbar}, \tag{5.36}$$

which is independent of the quasiclassical path $r(t)$. The noise functional acts as a Gaussian filter which quenches off-diagonal quantum fluctuations. Physically, the environment is continuously measuring the position of the system. It therefore weakens quantum interference between different position eigenstates and pushes the system towards classical behavior. Diagrammatically, $S^{(N)}[y]$ is represented by the left diagram in Fig. 5.2. To elucidate in some more detail the way $S^{(N)}[y]$ works, consider the Ohmic case $J(\omega) = \eta \omega$ at high temperature. In this white noise limit, we have $L'(t) = (2\eta/\hbar\beta) \, \delta(t)$, as follows from the expression (5.26). Thus we find for the case of two different localized states with spatial separation q_0 the decoherence factor

$$\langle e^{-i\varphi} \rangle = \exp(-\gamma_{\text{decoh}} t), \qquad \text{where} \qquad \gamma_{\text{decoh}} = \eta q_0^2 k_B T / \hbar^2. \tag{5.37}$$

The rate γ_{decoh} is the inverse time scale for decoherence. Pure dephasing is studied in some detail in Section 22.4. On the other hand, the strength of Ohmic friction is characterized by the damping rate

$$\gamma_{\text{damping}} = \eta/M. \tag{5.38}$$

The ratio of the two rates is

$$\gamma_{\text{decoh}}/\gamma_{\text{damping}} = q_0^2 M k_B T / \hbar^2 = q_0^2/\lambda_{\text{th}}^2, \tag{5.39}$$

where $\lambda_{\text{th}} = \hbar/p_{\text{th}} = \hbar/\sqrt{Mk_{\text{B}}T}$ is a thermal wave length. The ratio (5.39) relates the range of spatial coherence with the thermal wave length of the particle. For a typical macroscopic situation, $M = 1\,\text{g}$, $T = 300\,\text{K}$, $q_0 = 1\,\text{mm}$, a value of 10^{38} for the ratio $\gamma_{\text{decoh}}/\gamma_{\text{damping}}$ results. Thus for practically all macroscopic systems, friction is fully negligible on time scales where quantum coherence occurs. The ratio between the inverse time scales for decoherence and friction, Eq. (5.39), is generally valid for all systems which are in contact with an Ohmic heat bath with thermal energy large compared to the relevant internal energy scales of the system.[2] Suppression of quantum coherence for interfering paths is discussed in Chapter 9.

Comparison of the expression (2.33) with (5.26) shows that the kernel $L(t)$ in the influence action (5.24) is directly related to the correlation $\mathcal{X}(t) = \langle \hat{\xi}(t)\hat{\xi}(0)\rangle_\beta$ of the random force $\hat{\xi}(t)$ and to the correlation $\langle \hat{\mathfrak{E}}(z)\hat{\mathfrak{E}}(0)\rangle_\beta$ of the polarization energy or collective bath mode $\mathfrak{E}(t) = q_0\xi(t)$ defined in Eq. (3.136),

$$L(t) = \mathcal{X}(t)/\hbar = (1/q_0^2\,\hbar)\,\langle \hat{\mathfrak{E}}(t)\hat{\mathfrak{E}}(0)\rangle_\beta\,. \tag{5.40}$$

We shall refer to $L(t)$ as the *bath correlation function*. Here we have tacitly assumed that the fluctuation $\xi(t)$ is the deviation from the mean value for the shifted canonical distribution (3.45).

For a linearly responding bath, the random force obeys Gaussian statistics. Therefore, all influences of the reservoir exerted on the relevant system are fully captured by the autocorrelation function $\langle \xi(t)\xi(0)\rangle_\beta$. Hence a conventional molecular dynamics simulation may be used to determine this correlation function in the classical limit, which then defines $J(\omega)$ via the relations (2.6) and (3.26). The quantum mechanical response of the bath is then obtained by calculation of the expression (2.33) or (5.26).

With the determination of the Feynman-Vernon influence functional (5.33) within the model (5.8) we have all pieces together: the dynamics of the reduced density matrix is given by the relation (5.11) with (5.12), and the double functional integral in (5.12) is over paths $q(t')$ and $q'(t')$ with the endpoint constraints given in (5.5).

Finally, we briefly consider an ergodic open system which forgets about its initial state in the course of time [cf. Section 3.2]. The system relaxes to the thermal equilibrium state $\rho_\beta(q_{\text{f}}, q_{\text{f}}')$ independently of the particular initial state $\rho(q_{\text{i}}, q_{\text{i}}'; 0)$ chosen. Thus we obtain with use of the relation (5.11)

$$\rho_\beta(q_{\text{f}}, q_{\text{f}}') = \lim_{t\to\infty} \rho(q_{\text{f}}, q_{\text{f}}'; t) = \lim_{t\to\infty} \int dq_{\text{i}}\, dq_{\text{i}}'\, J_{\text{FV}}(q_{\text{f}}, q_{\text{f}}', t; q_{\text{i}}, q_{\text{i}}', 0)\, \rho(q_{\text{i}}, q_{\text{i}}'; 0)\,. \tag{5.41}$$

This is a remarkably useful relation: instead of calculating the thermal equilibrium state of the damped system with the imaginary-time path integral (4.60), we may find the equilibrium density matrix $\rho_\beta(q_{\text{f}}, q_{\text{f}}')$ from the real-time propagation of the density matrix at asymptotic time.

The dynamical method for the calculation of the equilibrium density matrix has considerable advantage over the imaginary-time method in many specific physical situations. The detailed discussion is deferred until Section 22.9.

---

[2]See Sections 21.3 and 22.3 for the discussion of decoherence in the dissipative two-state system.

5.4 General initial states and preparation function

Often in experiments, the "small" system is prepared, say at $t = 0$, by first letting it equilibrate with the bath and then measuring certain dynamical variables of it. The measurement leads to a reduction of the equilibrium density matrix of the system-plus-reservoir complex. Initial states of this type are generated by a linear transformation of the equilibrium density matrix of the global system. We write

$$< q_i, \boldsymbol{x}_i| \hat{W}(0) |q_i', \boldsymbol{x}_i' > \; = \; \int d\bar{q}\, d\bar{q}' \; \lambda(q_i, q_i'; \bar{q}, \bar{q}') \; <\bar{q}, \boldsymbol{x}_i| \hat{W}_\beta |\bar{q}', \boldsymbol{x}_i' > , \qquad (5.42)$$

where $\hat{W}_\beta = e^{-\beta \hat{H}}/Z$. The linking function $\lambda(q, q'; \bar{q}, \bar{q}')$ is a *preparation function* which depends on the specific measurement apparatus at $t = 0$ [87].

Initial states of the form (5.42) are not in the product form (5.10). Thus we cannot apply the procedure described in Section 5.2. Before we give the corresponding generalization for correlated initial states, we briefly discuss specific initial preparations which may have certain experimental relevance.

Let the effect of the measuring device be described by operators $\hat{O}_j, \hat{O}_j{}'$ which act on the system only and leave the reservoir unaffected. Then the preparation function is in the form

$$\lambda(q_i, q_i'; \bar{q}, \bar{q}') \; = \; \sum_j < q_i|\hat{O}_j|\bar{q} > < \bar{q}'|\hat{O}_j'|q_i' > . \qquad (5.43)$$

Consider briefly three important cases. First, if we are interested in the response of a system in thermal equilibrium to an external force switched on at time $t = 0$, we have $\hat{W}(0) = \hat{W}_\beta$. Then the preparation function $\lambda(q_i, q_i'; \bar{q}, \bar{q}')$ takes the simple form

$$\lambda_\beta(q_i, q_i'; \bar{q}, \bar{q}') \; = \; \delta(q_i - \bar{q})\, \delta(q_i' - \bar{q}') . \qquad (5.44)$$

The second important example is that we perform initially a measurement of a dynamical variable of the system. In accordance with the principles of quantum measurement theory (see, for instance, Refs. [208, 57, 209]), the effect of an "observation" is to project the density matrix that described the system just prior to observation onto the manifold corresponding to the measured interval of eigenvalues. For an ideal measurement at time $t = 0^-$, the initial density matrix is then given by

$$\hat{W}(0) \; = \; \hat{P}\hat{W}_\beta\hat{P} , \qquad (5.45)$$

where $\hat{P}$ is the projection operator corresponding to the measurement. For example, the measurement of the position of a particle with the outcome q_0 with uncertainty Δq is described by the projection operator

$$\hat{P}_q \; = \; \int_{q_0 - \Delta q/2}^{q_0 + \Delta q/2} dq\, |q> <q| . \qquad (5.46)$$

Thirdly, the structure factor investigated in scattering experiments is closely related to the Fourier transform of the equilibrium coordinate autocorrelation function of the scatterer [210]. Now, an equilibrium correlation function, say $\langle A(t) B(0) \rangle$, may be formally looked upon as the expectation value of $\hat{A}$ at time t for a system starting out from the initial state $\hat{W}(0) = \hat{B} \hat{W}_\beta$. In this case, the specific form of the preparation function follows directly from Eq. (5.42). For the thermal coordinate autocorrelation function $C_{qq}(t) = \text{tr} \{q(t)q(0)W_\beta\}$ the preparation function reads

$$\lambda_q(q_i, q_i'; \bar{q}, \bar{q}') = q_i \lambda_\beta(q_i, q_i'; \bar{q}, \bar{q}') , \qquad (5.47)$$

while for the coordinate-momentum correlation function in thermal equilibrium $C_{qp}(t) = \text{tr} \{q(t) p(0) W_\beta\}$ the preparation function is

$$\lambda_p(q_i, q_i'; \bar{q}, \bar{q}') = \frac{\hbar}{i} \frac{\partial}{\partial q_i} \lambda_\beta(q_i, q_i'; \bar{q}, \bar{q}') . \qquad (5.48)$$

The two cases correspond to $\hat{W}(0) = \hat{q} \hat{W}_\beta$ and $\hat{W}(0) = \hat{p} \hat{W}_\beta$, respectively. These expressions are not proper density matrices since they are not positive definite. However, they belong to the class of initial conditions specified by the relation (5.43).

5.5   Complex-time path integral for the propagating function

For the general class of initial states discussed in the preceding section, the quantum mechanical stochastic process is characterized by a *preparation function* $\lambda(q_i, q_i'; \bar{q}, \bar{q}')$ and by a *propagating function* $J(q_f, q_f', t; q_i, q_i'; \bar{q}, \bar{q}')$. The propagation function describes the time evolution of the reduced system starting from an initial state which is procured by the preparation function. The time evolution of the prepared initial state is described by the relation

$$\rho(q_f, q_f'; t) = \int dq_i \, dq_i' \, d\bar{q} \, d\bar{q}' \, J(q_f, q_f', t; q_i, q_i'; \bar{q}, \bar{q}') \, \lambda(q_i, q_i'; \bar{q}, \bar{q}') . \qquad (5.49)$$

Insertion of (5.42) into (5.9) and comparison with (5.49) yields for J the expression

$$
\begin{aligned}
J(q_f, q_f'; t; q_i, q_i'; \bar{q}, \bar{q}') &= \int d\boldsymbol{x}_i \, d\boldsymbol{x}_i' \, d\boldsymbol{x}_f \, K(q_f, \boldsymbol{x}_f, t; q_i, \boldsymbol{x}_i, 0) \\
&\quad \times <\bar{q}, \boldsymbol{x}_i | W_\beta | \bar{q}', \boldsymbol{x}_i' > K^*(q_f', \boldsymbol{x}_f, t; q_i', \boldsymbol{x}_i', 0) .
\end{aligned}
$$

With use of the functional integral expressions (4.25) for the thermal density matrix and (5.4) for the propagators we may write the propagating function (5.50) as

$$J(q_f, q_f'; t; q_i, q_i'; \bar{q}, \bar{q}') = Z^{-1} \int \mathcal{D}q \, \mathcal{D}q' \, \mathcal{D}\bar{q} \qquad (5.50)$$

$$\times \exp \left\{ \frac{i}{\hbar} \left(S_S[q] - S_S[q'] \right) - \frac{1}{\hbar} S_S^{(E)}[\bar{q}] \right\} \mathcal{F}_I[q, q', \bar{q}] ,$$

where Z is the reduced partition function, and where

$$\mathcal{F}_{\mathrm{I}}[q,q',\bar{q}] \;=\; \int d\boldsymbol{x}_{\mathrm{f}}\, d\boldsymbol{x}_{\mathrm{i}}\, d\boldsymbol{x}'_{\mathrm{i}}\, \frac{1}{Z_{\mathrm{R}}} \int \mathcal{D}\boldsymbol{x}\,\mathcal{D}\boldsymbol{x}'\,\mathcal{D}\bar{\boldsymbol{x}} \tag{5.51}$$

$$\times\, \exp\left\{ i\big(S_{\mathrm{R}}[\boldsymbol{x}] + S_{\mathrm{I}}[\boldsymbol{x},q] - S_{\mathrm{R}}[\boldsymbol{x}'] - S_{\mathrm{I}}[\boldsymbol{x}',q']\big)/\hbar - \big(S_{\mathrm{R}}^{(\mathrm{E})}[\bar{\boldsymbol{x}}] + S_{\mathrm{I}}^{(\mathrm{E})}[\bar{\boldsymbol{x}},\bar{q}]\big)/\hbar \right\}$$

represents the influence functional. The functional integrations in Eq. (5.51) are evaluated for paths $\boldsymbol{x}(s)$, $\boldsymbol{x}'(s')$, $\bar{\boldsymbol{x}}(\tau)$ which form a closed path on a contour in the complex-time plane with the endpoint conditions

$$\boldsymbol{x}(t) = \boldsymbol{x}'(t) = \boldsymbol{x}_{\mathrm{f}}\,, \qquad \boldsymbol{x}(0) = \bar{\boldsymbol{x}}(\hbar\beta) = \boldsymbol{x}_{\mathrm{i}}\,, \qquad \bar{\boldsymbol{x}}(0) = \boldsymbol{x}'(0) = \boldsymbol{x}'_{\mathrm{i}}\,. \tag{5.52}$$

We can now proceed as in Section 5.2 by writing the expression (5.51) in the form

$$\mathcal{F}_{\mathrm{I}}[q(\cdot), q'(\cdot), \bar{q}(\cdot)] \;=\; \prod_{\alpha=1}^{N} \frac{1}{Z_R^{(\alpha)}} \int dx_{\mathrm{f},\alpha}\, dx_{\mathrm{i},\alpha}\, dx'_{\mathrm{i},\alpha} \tag{5.53}$$

$$\times\; F_\alpha[q(\cdot), x_{\mathrm{f},\alpha}, x_{\mathrm{i},\alpha}]\, F_\alpha^*[q'(\cdot), x_{\mathrm{f},\alpha}, x'_{\mathrm{i},\alpha}]\, F_\alpha^{(\mathrm{E})}[\bar{q}(\cdot), x_{\mathrm{i},\alpha}, x'_{\mathrm{i},\alpha}]\,.$$

Here the amplitude F_α is the path sum expression given in Eq. (5.18), and $F_\alpha^{(\mathrm{E})}$ is the respective path sum in the imaginary-time interval $0 \leq \tau < \hbar\beta$. The resulting analytic expression for F_α is put down in Eq. (5.20) with (5.21). The analytic expression for $F_\alpha^{(\mathrm{E})}$ is found from that for F_α by the substitution $t \to -i\,\hbar\beta$.

With these forms, the integrand of the expression (5.53) is a Gaussian in the variables $x_{\mathrm{f},\alpha}, x_{\mathrm{i},\alpha}$ and $x'_{\mathrm{i},\alpha}$. The evaluation of the Gaussian integrals is straightforward but laborious. In the end one finds [87]

$$\mathcal{F}_{\mathrm{I}}[q,q',\bar{q}] \;=\; \exp\left\{ -\,\mathcal{S}_{\mathrm{I}}[q,q',\bar{q}]/\hbar \right\}\,, \tag{5.54}$$

where

$$\mathcal{S}_{\mathrm{I}}[q,q',\bar{q}] \;=\; -\int_0^{\hbar\beta} d\tau \int_0^{\tau} d\tau'\, K(\tau-\tau')\,\bar{q}(\tau)\,\bar{q}(\tau') + \frac{M\gamma(0^+)}{2} \int_0^{\hbar\beta} d\tau\, \bar{q}^2(\tau)$$

$$-\; i\int_0^{\hbar\beta} d\tau \int_0^{t} dt'\, L^*(t'-i\tau)\,\bar{q}(\tau)\left\{q(t') - q'(t')\right\}$$

$$+\; \int_0^{t} dt' \int_0^{t'} dt''\, \left\{q(t')-q'(t')\right\}\left\{L(t'-t'')q(t'') - L^*(t'-t'')\,q'(t'')\right\}$$

$$+\; i\frac{\mu}{2} \int_0^{t} dt'\, \left\{q^2(t') - q'^2(t')\right\}\,, \tag{5.55}$$

and where $K(\tau)$ and $L(t)$ are defined in Eqs. (4.54) and (5.26), respectively. The triple path integral (5.50) with Eqs. (5.54) and (5.55) is the exact formal solution for the propagating function $J(q_{\mathrm{f}}, q'_{\mathrm{f}}; t; q_{\mathrm{i}}, q'_{\mathrm{i}}; \bar{q}, \bar{q}')$. The paths $q(t')$, $q'(t')$ and $\bar{q}(\tau)$ obey the constraints

$$q(0) \;=\; q_{\mathrm{i}} \,, \quad q'(0) \;=\; q'_{\mathrm{i}} \,, \quad \bar{q}(0) \;=\; \bar{q}' \,,$$
$$q(t) \;=\; q_{\mathrm{f}} \,, \quad q'(t) \;=\; q'_{\mathrm{f}} \,, \quad \bar{q}(\hbar\beta) \;=\; \bar{q} \,. \tag{5.56}$$

The results obtained in this section shed light on the evolution of a general initial state. The dynamics is specified by a preparation function $\lambda(q_{\mathrm{i}}, q'_{\mathrm{i}}; \bar{q}, \bar{q}')$ and by a propagating function $J(q_{\mathrm{f}}, q'_{\mathrm{f}}, t; q_{\mathrm{i}}, q'_{\mathrm{i}}; \bar{q}, \bar{q}')$. The preparation function fixes the particular initial state $\rho_0 = \rho(0)$ within the preparation class through the relation

$$\rho_0(q_{\mathrm{i}}, q'_{\mathrm{i}}) \;=\; \int d\bar{q}\, d\bar{q}'\, \lambda(q_{\mathrm{i}}, q'_{\mathrm{i}}; \bar{q}, \bar{q}')\, \rho_\beta(\bar{q}, \bar{q}') \,, \tag{5.57}$$

where $\rho_\beta(\bar{q}, \bar{q}')$ is the thermal reduced density matrix given in Eq. (4.60). The relation (5.57) is obtained by tracing out the reservoir coordinates in Eq. (5.42). It also follows from Eq. (5.49) in the limit $t \to 0$ with the initial condition [cf. Eq. (5.50)]

$$\lim_{t \to 0} J(q_{\mathrm{f}}, q'_{\mathrm{f}}, t; q_{\mathrm{i}}, q'_{\mathrm{i}}; \bar{q}, \bar{q}') \;=\; \delta(q_{\mathrm{f}} - q_{\mathrm{i}})\, \delta(q'_{\mathrm{f}} - q'_{\mathrm{i}}) \rho_\beta(\bar{q}, \bar{q}') \,. \tag{5.58}$$

The preparation function $\lambda(q_{\mathrm{i}}, q'_{\mathrm{i}}; \bar{q}, \bar{q}')$ represents the quantum analogue of the initial classical phase space distribution. Note that it is not uniquely specified through the relation (5.57). It is rather determined by the initial density matrix of the total system through relation (5.42). In quantum mechanics, the propagating function $J(q_{\mathrm{f}}, q'_{\mathrm{f}}, t; q_{\mathrm{i}}, q'_{\mathrm{i}}; \bar{q}, \bar{q}')$ has taken over the role of the classical conditional probability. From the considerations above it is clear that the propagating function is determined by the equilibrium properties of the reduced system. This is in correspondence with the classical stationary conditional probability.

5.6   Real-time path integral for the propagating function

In the formal solution (5.50) with (5.54) and (5.55) of the propagating function, one faces difficulties which arise from the correlations of the forward and backward paths $q(t')$ and $q'(t')$ with the imaginary-time path $\bar{q}(\tau)$. The complications caused by this coupling can be simplified to some extent in actual calculations for *ergodic* open systems, as we explain in the sequel.

We start with the observation that the propagating function (5.50) has the initial condition (5.58). With use of the relation (5.49) we then obtain

$$\lim_{t \to 0} \rho(q_{\mathrm{f}}, q'_{\mathrm{f}}; t) \;=\; \int d\bar{q}\, d\bar{q}'\, \lambda(q_{\mathrm{f}}, q'_{\mathrm{f}}; \bar{q}, \bar{q}')\, \rho_\beta(\bar{q}, \bar{q}') \,. \tag{5.59}$$

Suppose now that the system has been prepared at some large negative time $-|t_{\mathrm{p}}|$ in a particular initial state and recall that the ergodic open system relaxes to thermal equilibrium independently of the chosen initial preparation [cf. Section 3.2 and the discussion given at the end of Section 5.3]. Then, the system will have equilibrated with the reservoir at time $t = 0$, if it started out in the infinite past from some non-equilibrium initial state. Therefore, we may choose in the infinite past without loss of generality a product initial state of the form

$$\lim_{t_p \to -\infty} \hat{W}(t_p) = \hat{\rho}_p \otimes e^{-\beta \hat{H}_R}/Z_R . \tag{5.60}$$

We then have at time $t = 0^-$ in correspondence with the relation (5.41)

$$\rho(\bar{q}, \bar{q}'; 0^-) = \rho_\beta(\bar{q}, \bar{q}') = \lim_{t_p \to -\infty} \int dq_p \, dq_p' \, J_{FV}(\bar{q}, \bar{q}', 0^-; q_p, q_p'; t_p) \, \rho_p(q_p, q_p') , \tag{5.61}$$

where the propagating function J_{FV} is given by Eq. (5.12) with (5.24) and (5.5).

After having determined the correlated equilibrium state of the damped particle at time $t = 0^-$, it is straightforward to describe the evolution of the reduced density matrix. We have

$$\rho(q_f, q_f'; t) = \int dq_i \, dq_i' \, d\bar{q} \, d\bar{q}' \, \lambda(q_i, q_i'; \bar{q}, \bar{q}') \int dq_p \, dq_p' \, \rho_p(q_p, q_p')$$

$$\times \lim_{t_p \to -\infty} J(q_f, q_f', t; q_i, q_i', 0^+; \bar{q}, \bar{q}', 0^-; q_p, q_p', t_p) . \tag{5.62}$$

The propagating function J is given by the real-time double path integral

$$J(q_f, q_f', t; q_i, q_i', 0^+; \bar{q}, \bar{q}', 0^-; q_p, q_p', t_p) = \int \mathcal{D}q \, \mathcal{D}q' \, e^{i(S_S[q] - S_S[q'])/\hbar} \, e^{-S_{II}[q,q']/\hbar} \tag{5.63}$$

with the influence action

$$S_{II}[q, q'] = \int_{t_p}^t dt' \int_{t_p}^{t'} dt'' \left\{ q(t') - q'(t') \right\} \left\{ L(t' - t'') q(t'') - L^*(t' - t'') q'(t'') \right\}$$

$$+ i \frac{\mu}{2} \int_{t_p}^t dt' \left\{ q^2(t') - q'^2(t') \right\} . \tag{5.64}$$

The functional integrations in (5.63) are over paths $q(t')$ and $q'(t')$ which are subject to the constraints

$$q(t_p) = q_p , \qquad q(0^-) = \bar{q} , \qquad q(0^+) = q_i , \qquad q(t) = q_f ,$$
$$q'(t_p) = q_p' , \qquad q'(0^-) = \bar{q}' , \qquad q'(0^+) = q_i' , \qquad q(t) = q_f' . \tag{5.65}$$

Observe that the paths $q(t')$ and $q'(t')$ are discontinuous at time $t = 0$.

It is again convenient to introduce the symmetric and antisymmetric variables $r(t)$ and $y(t)$ defined in Eq. (5.32). The expression (5.49) is then written as

$$\rho(r_f, y_f; t) = \int dr_i \, dy_i \, d\bar{r} \, d\bar{y} \, \lambda(r_i, y_i; \bar{r}, \bar{y}) \int dr_p \, dy_p \, \rho_p(r_p, y_p)$$

$$\times \lim_{t_p \to -\infty} J(r_f, y_f, t; r_i, y_i, 0^+; \bar{r}, \bar{y}, 0^-; r_p, y_p, t_p) , \tag{5.66}$$

and the propagating function (5.63) is expressed by the path integral

$$J(r_f, y_f, t; r_i, y_i, 0^+; \bar{r}, \bar{y}, 0^-; r_p, y_p, t_p) = \int \mathcal{D}r \, \mathcal{D}y \, e^{i \Sigma_{tot}(r, y; t, t_p)/\hbar} \tag{5.67}$$

with the total action

$$\Sigma_{\text{tot}}(r, y; t, t_{\text{p}}) = \Sigma_{\text{S}}[r, y; t, t_{\text{p}}] + i\,\mathcal{S}^{(\text{N})}[y; t, t_{\text{p}}] - \mathcal{S}^{(\text{F})}[r, y; t, t_{\text{p}}] - \mathcal{S}^{(\text{ini})}[y; t, 0] . \quad (5.68)$$

The actions Σ_{S}, $\mathcal{S}^{(\text{N})}$ and $\mathcal{S}^{(\text{F})}$ are defined in Eqs. (5.34) and (5.33). The boundary conditions of the paths $y(t')$ and $r(t')$ are

$$r(t_{\text{p}}) = r_{\text{p}}, \quad r(0^-) = \bar{r}, \quad r(0^+) = r_{\text{i}}, \quad r(t) = r_{\text{f}},$$
$$y(t_{\text{p}}) = y_{\text{p}}, \quad y(0^-) = \bar{y}, \quad y(0^+) = y_{\text{i}}, \quad y(t) = y_{\text{f}}. \quad (5.69)$$

The additional contribution

$$\mathcal{S}^{(\text{ini})}[y; t, 0] = \Theta(t)\,M(r_{\text{i}} - \bar{r}) \int_0^t dt'\, y(t')\gamma(t') \quad (5.70)$$

is due to partial integration and accounts for the discontinuity of $r(t)$ at $t = 0$.

The limit $t_{\text{p}} \to -\infty$ in Eq. (5.66) is quite subtle. As $t_{\text{p}} \to -\infty$, the double path integral expression in Eq. (5.67) becomes a δ–distribution in the variable y_{p} which is centered at $y_{\text{p}} = 0$, and at the same time it becomes independent of the variable r_{p}. We then have

$$\int dr_{\text{p}}\, \rho_{\text{p}}(r_{\text{p}}, y_{\text{p}} = 0) = \text{tr}\,\rho_{\text{p}} = 1. \quad (5.71)$$

Thus the effects of the particular initial distribution $\rho_{\text{p}}(r_{\text{p}}, y_{\text{p}})$ die out completely in the expression (5.66) as $t_{\text{p}} \to -\infty$. As a result, $\rho(r_{\text{f}}, y_{\text{f}}; t)$ is in fact independent of the particular initial state chosen in the infinite past.

It is worth closing this section with the remark that we ended up with a pure real-time path integral expression for the propagating function for the class of preparations discussed in Section 5.2. In this formulation, the effects of the system-reservoir correlations of the equilibrium state at time $t = 0^+$ are described by interactions between the positive and negative time branches of the functional integral expression (5.63). The interactions are contained in the influence action (5.64). The real-time approach has been applied successfully, e. g., to the calculation of equilibrium correlation functions for dissipative two-state systems [219] and to the study of universal low temperature properties [220] (see Chapter 21).

5.7   Closed time contour representation

The propagating functions discussed in Sections 5.2, 5.5 and 5.6 may be written in compact form in terms of a path $\tilde{q}(z)$ defined on a closed time contour. The system's Minkowskian action for the path $\tilde{q}(z)$ reads

$$\mathcal{S}_{\text{S}}[\tilde{q}(\cdot)] = \int dz \left[\frac{M}{2} \left(\frac{\partial \tilde{q}}{\partial z} \right)^2 - V(\tilde{q}) \right] . \quad (5.72)$$

The influence action for this path with $L(z)$ given in Eq. (5.25) is

$$\mathcal{S}_{\text{infl}}[\tilde{q}(\cdot)] = \int_{z > z'} dz \int dz'\, L(z - z')\, \tilde{q}(z)\, \tilde{q}(z') + i\,\frac{\mu}{2} \int dz\, \tilde{q}^2(z) . \quad (5.73)$$

The constraint $z > z'$ means time ordering along the path $\tilde{q}(z)$.

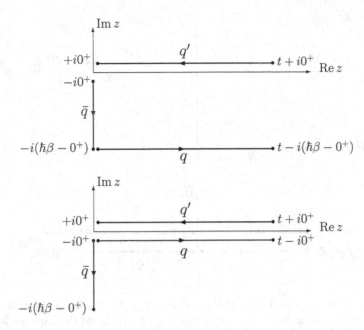

Figure 5.3: Schematic depiction of the contours $\mathcal{C}_\mathrm{I}$ (top) and $\mathcal{C}_\mathrm{II}$ (bottom). The complex time runs in the direction indicated by the arrows and the sequence is q', $\bar{q}$, q for $\mathcal{C}_\mathrm{I}$, and q, q', $\bar{q}$ for $\mathcal{C}_\mathrm{II}$. The filled circles indicate the endpoint conditions (5.56). The infinitesimal shifts of the path segments q and q' away from the real axis serve to have proper convergence behavior of the factors $e^{iS_\mathrm{S}[q]/\hbar}$ and $e^{-iS_\mathrm{S}[q']/\hbar}$. The Feynman-Vernon contour $\mathcal{C}_\mathrm{FV}$ for a product initial state differs from the contour $\mathcal{C}_\mathrm{I\,(II)}$ by omission of the path segment $\bar{q}(\tau)$.

5.7.1 Complex-time path

Consider a contour $\mathcal{C}_\mathrm{I}$ and a contour $\mathcal{C}_\mathrm{II}$ as sketched in Fig. 5.3 top and bottom, respectively. The path $\tilde{q}(z)$ with z on the contour $\mathcal{C}_\mathrm{I}$ is

$$
\tilde{q}(z) = \begin{cases} q'(s) & \text{for} \quad z = s + i\,0^+ , & 0 \leq s \leq t , \\ \bar{q}(\tau) & \text{for} \quad z = -i\tau , & 0 \leq \tau \leq \hbar\beta , \\ q(s) & \text{for} \quad z = -i\,(\hbar\beta - 0^+) + s , & 0 \leq s \leq t , \end{cases} \tag{5.74}
$$

and on the contour $\mathcal{C}_\mathrm{II}$

$$
\tilde{q}(z) = \begin{cases} q(s) & \text{for} \quad z = s - i\,0^+ , & 0 \leq s \leq t , \\ q'(s) & \text{for} \quad z = s + i\,0^+ , & 0 \leq s \leq t , \\ \bar{q}(\tau) & \text{for} \quad z = -i\tau , & 0 \leq \tau \leq \hbar\beta . \end{cases} \tag{5.75}
$$

The propagating function for the path $\mathcal{C}_\mathrm{I}$ ($\mathcal{C}_\mathrm{II}$) now reads

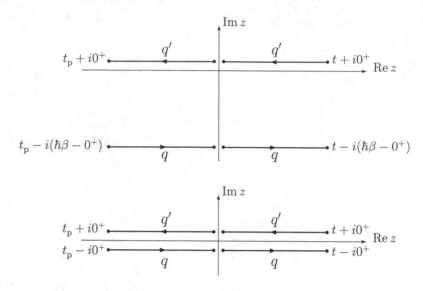

Figure 5.4: Schematic representation of the contours $\mathcal{C}_{\mathrm{III}}$ (top) and $\mathcal{C}_{\mathrm{IV}}$ (bottom). The complex time runs again in the direction indicated by the arrows and the sequence is q, q' both for $\mathcal{C}_{\mathrm{III}}$ and $\mathcal{C}_{\mathrm{IV}}$. The filled circles symbolize the endpoint conditions (5.65).

$$J(q_f, q'_f; t; q_i, q'_i; \bar{q}, \bar{q}') \;=\; Z^{-1} \int_{\mathcal{C}_{\mathrm{I}}\,(\mathcal{C}_{\mathrm{II}})} \mathcal{D}\tilde{q}(\cdot)\; \mathrm{e}^{i S_{\mathrm{s}}[\tilde{q}(\cdot)]/\hbar - S_{\mathrm{infl}}[\tilde{q}(\cdot)]/\hbar} \,. \tag{5.76}$$

The equivalence of the contours $\mathcal{C}_{\mathrm{I}}$ and $\mathcal{C}_{\mathrm{II}}$ follows with use of symmetry relations resulting from the definition (5.25),

$$L\left(t + i\tau - i\hbar\beta\right) = L^*\left(t - i\tau\right) = L\left(-t - i\tau\right), \qquad L\left(-i\tau\right) = K(\tau)\,. \tag{5.77}$$

Further, with these relations it is straightforward to see that the concise expression (5.76) is equivalent to the former result (5.50) with (5.54) and (5.55). The propagation of a product initial state, Eq. (5.12), is defined on a contour $\mathcal{C}_{\mathrm{FV}}$ which differs from $\mathcal{C}_{\mathrm{I}}$ or $\mathcal{C}_{\mathrm{II}}$ by omission of the path segment $\bar{q}(\tau)$ and omission of the factor Z^{-1}. Schwinger was the first who observed that nonequilibrium quantum field dynamics proceeds along a closed time contour [211]. The contour $\mathcal{C}_{\mathrm{I}}$ has been discussed and applied to quantum Brownian motion in Ref. [87], while $\mathcal{C}_{\mathrm{II}}$ is the standard Kadanoff-Baym contour [212]. The latter formulation is closely related to the Green function technique introduced by Keldysh [213]. The change of variables from (q, q') to (r, y) corresponds to the Keldysh rotation [214]. The closed-time-path Green function approach has been developed further and has been applied to various equilibrium and non-equilibrium problems. For reviews we refer to the work by Chou *et al.* [215], and by Rammer and Smith [216], and to the book by Rammer [217].

5.7.2 Real-time path

Consider next a contour $\mathcal{C}_{\text{III}}$ and a contour $\mathcal{C}_{\text{IV}}$ as illustrated in Fig. 5.4 top and bottom. The path $\tilde{q}(z)$ with z along the contour $\mathcal{C}_{\text{III}}$ is given by

$$\tilde{q}(z) = \begin{cases} q'(s) & \text{for} \quad z = s + i\,0^+\,, & t_{\text{p}} \le s \le t\,, \\ q(s) & \text{for} \quad z = -i\,(\hbar\beta - 0^+) + s\,, & t_{\text{p}} \le s \le t\,, \end{cases} \tag{5.78}$$

and the path with z along the contour $\mathcal{C}_{\text{IV}}$ by

$$\tilde{q}(z) = \begin{cases} q'(s) & \text{for} \quad z = s + i\,0^+\,, & t_{\text{p}} \le s \le t\,, \\ q(s) & \text{for} \quad z = s - i\,0^+\,, & t_{\text{p}} \le s \le t\,. \end{cases} \tag{5.79}$$

For these closed time contours the propagating function J takes the compact form

$$J(q_{\text{f}}, q'_{\text{f}}; t; q_{\text{i}}, q'_{\text{i}}, 0^+; \bar{q}, \bar{q}', 0^-; q_{\text{p}}, q'_{\text{p}}, t_{\text{p}}) = \int_{\mathcal{C}_{\text{III}}(\mathcal{C}_{\text{IV}})} \mathcal{D}\tilde{q}(\cdot)\, e^{iS_{\text{S}}[\tilde{q}(\cdot)]/\hbar - S_{\text{infl}}[\tilde{q}(\cdot)]/\hbar}\,, \tag{5.80}$$

where $S_{\text{S}}[\tilde{q}(\cdot)]$ is defined in Eq. (5.72) and $S_{\text{infl}}[\tilde{q}(\cdot)]$ in Eq. (5.73). With the relations (5.77) the compact expression (5.80) both for the contour $\mathcal{C}_{\text{III}}$ and $\mathcal{C}_{\text{IV}}$ is in correspondence with the expression (5.63) with (5.64).

For an ergodic system, the expression (5.80) for the propagating function is exactly equivalent to the former expression (5.76). The crucial point behind the flexibility to employ either path integration along the contours $\mathcal{C}_{\text{I}}$ or $\mathcal{C}_{\text{II}}$ depicted in Fig. 5.3, or along the contours $\mathcal{C}_{\text{III}}$ or $\mathcal{C}_{\text{IV}}$ shown in Fig. 5.4 is that for ergodic open systems the bath correlation function $L(t)$ drops sufficiently fast at asymptotic time so that differing terms vanish as $t_{\text{p}} \to -\infty$. An explicit demonstration of the equivalence has been given for specific models such as the damped harmonic oscillator [218] and the spin-boson system [219].

5.8 Semiclassical regime

5.8.1 Extremal paths

In the semiclassical limit, we may discard the imaginary-time branch and the negative-time branch, respectively. The functional integral describing the time evolution of the reduced density matrix is dominated by paths for which the action is stationary. Stationarity of the total action (5.68) under variation of the paths $y(s)$ and $r(s)$ with fixed endpoints yields in the time regime $0 < s < t$ the equations of motion

$$M\ddot{r}(s) + M\int_0^s dt'\,\gamma(s - t')\dot{r}(t') + \frac{d}{dy}\Big(V(r + \tfrac{1}{2}y) - V(r - \tfrac{1}{2}y)\Big)$$
$$= i\int_0^t dt'\,L'(s - t')y(t')\,, \tag{5.81}$$

$$M\ddot{y}(s) - M \int_s^t dt'\, \dot{y}(t')\gamma(t'-s) + \frac{1}{M}\frac{\partial}{\partial r}\left(V(r + \tfrac{1}{2}y) - V(r - \tfrac{1}{2}y)\right) = 0 \,. \quad (5.82)$$

If the system is initially and finally in a diagonal state of the reduced density matrix, the boundary conditions are

$$y(0) = y(t) = 0 \,. \quad (5.83)$$

For these constraints, the solution of Eq. (5.82) is the trivial fluctuation path $y(s) = 0$. Hence the off-diagonal elements of the reduced density matrix are negligible small in the semiclassical limit. With this fact the equation of motion for the stationary quasiclassical path takes the familiar form

$$M\ddot{r}(s) + M \int_0^s dt'\, \gamma(s-t')\dot{r}(t') + \frac{dV(r)}{dr} = 0 \,, \quad (5.84)$$

which is the deterministic equation of motion for a damped particle in the external potential $V(r)$.

5.8.2   Quasiclassical Langevin equation

We now go beyond the deterministic description by taking into account the Gaussian fluctuations $y(s)$ about the minimal action path. Again, we are interested only in the probability distribution $w(r_f; t)$ evolving from the initial distribution $w(r_i; 0)$. We devise that the evolution of $w(r_f; t)$ is represented by a double path integral of the form

$$w(r_f; t) = \int dr_i\, w(r_i; t) \int \mathcal{D}r\, \mathcal{D}y \exp\left(\frac{i}{\hbar}\Sigma^{(\mathrm{sc})}[r, y]\right) \,, \quad (5.85)$$

in which the yet to be determined action $\Sigma^{(\mathrm{sc})}$ is a quadratic form in y. The functional integral extends over all paths $y(t')$ and $r(t')$ which obey $y(0) = y(t) = 0$, $r(0) = r_i$ and $r(t) = r_f$. We proceed with the observation that the noise action $\mathcal{S}^{(\mathrm{N})}[y]$ in Eq. (5.33) allocates in the functional integral (5.67) the weight function

$$p[y(\cdot)] = \exp\left\{-\frac{1}{2\hbar}\int_0^t dt' \int_0^t dt''\, y(t')L'(t'-t'')y(t'')\right\} \,. \quad (5.86)$$

The function $p[y(\cdot)]$ represents a Gaussian filter function which suppresses large deviations from the classical path in the semiclassical limit. Since the fluctuations $y(s)$ are small, it is reasonable to expand the potential term in the action about $y = 0$ and to keep the lowest nontrivial order only,

$$-V(r + \tfrac{1}{2}y) + V(r - \tfrac{1}{2}y) = -V'(r)\, y + \mathcal{O}(y^3) \,. \quad (5.87)$$

With this approximation in the action $\Sigma_S[r, y]$ and with partial integration of the kinetic term in $\Sigma_S[r, y]$ the action $\Sigma_{\mathrm{tot}}(r, y; t, 0)$ given in Eq. (5.68) reduces to the semiclassical action

$$\Sigma^{(\mathrm{sc})}[r,y] = -\int_0^t ds\, y(s) \left[M\ddot{r}(s) + M \int_0^s du\, \gamma(s-u)\,\dot{r}(u) + V'(r) \right]$$
$$+ \frac{i}{2} \int_0^t ds \int_0^t du\, y(s) L'(s-u) y(u) \,. \tag{5.88}$$

The path probability in the region from $r(t)$ to $r(t) + \mathcal{D}r(t)$,

$$W[r]\,\mathcal{D}r \equiv \left\{ \int \mathcal{D}y(\cdot)\, \exp\left(\frac{i}{\hbar} \Sigma^{(\mathrm{sc})}[r, y(\cdot)] \right) \right\} \mathcal{D}r \tag{5.89}$$

is a Gaussian functional integral in the fluctuation $y(\cdot)$ which can be evaluated by "completing the square". Omitting an irrelevant normalization constant, we obtain

$$W[r]\,\mathcal{D}r = \exp\left(-\frac{1}{2\hbar} \int_0^t ds \int_0^t du\, \xi[r(s)]\, L'^{-1}(s-u)\, \xi[r(u)] \right) \mathcal{D}r \,. \tag{5.90}$$

Here, $L'^{-1}(s-u)$ is the inverse of $L'(s-u)$ and the functional $\xi[r(s)]$ is defined by

$$\xi[r(s)] \equiv M\ddot{r}(s) + M \int_0^s du\, \gamma(s-u)\,\dot{r}(u) + V'[r(s)] \,. \tag{5.91}$$

The Gaussian weight function (5.90) suggests to introduce a measure of path integration $\mathcal{D}\xi(t)$ associated with the $\xi(t)$ fluctuations. The functional Jacobian of the transformation $\mathcal{D}r = \mathcal{J}^{-1}\mathcal{D}\xi$ is given by

$$\mathcal{J} = \det\left[\left(M\frac{\partial^2}{\partial t^2} + V''(r(t)) \right) \delta(t-t') + M\int_0^t dt'\, \gamma(t-t')\frac{\partial}{\partial t'} \right] \,. \tag{5.92}$$

Schmid [54] has argued by discretization of the differential operator on the sliced time axis that the Jacobian is a constant independent of the choice of the potential $V(r)$.[3] Thus, we may consider $\xi(t)$ as an independent stochastic variable. As $\xi(t)$ in fact does not depend on the path $r(t)$, we may regard Eq. (5.91) as a standard Langevin equation with memory-friction and colored-noise force,

$$M\ddot{r}(t) + M\int_0^t dt'\, \gamma(t-t')\dot{r}(t') + V'(r(t)) = \xi(t) \,. \tag{5.93}$$

The fluctuating force $\xi(t)$ undergoes the Gaussian stochastic process

$$W[\xi(\cdot)]\,\mathcal{D}\xi(\cdot) = \exp\left(-\frac{1}{2\hbar} \int_0^t ds \int_0^t du\, \xi(s) L'^{-1}(s-u)\xi(u) \right) \mathcal{D}\xi(\cdot) \,. \tag{5.94}$$

Average with this Gaussian weight function yields the statistical properties

$$\langle \xi(t) \rangle = 0 \,, \tag{5.95}$$
$$\mathcal{X}(t) \equiv \mathrm{Re}\,\langle \xi(t)\xi(0) \rangle = \hbar L'(t) = \frac{\hbar}{\pi} \int_0^\infty d\omega\, J(\omega) \coth(\beta\hbar\omega/2) \cos(\omega t) \,.$$

---

[3]The criticism stated in Ref. [6], p.1300, on the slicing in Ref. [54] does not spoil our reasoning.

The power spectrum of the correlation function $\mathcal{X}(t)$ is found to read

$$\tilde{\mathcal{X}}(\omega) \equiv \int_{-\infty}^{\infty} dt\, \mathcal{X}(t)\, \cos(\omega t) \;=\; M\hbar\omega \coth(\beta\hbar\omega/2)\, \tilde{\gamma}'(\omega)\,, \qquad (5.96)$$

where we have employed the relation (3.26). The noise characteristics of the Langevin force derived here agrees with that derived above in Subsection 3.1.4.

The relation (5.96) is the quantum mechanical version of Kubo's "Second Fluctuation-Dissipation Theorem". As $\hbar \to 0$, it becomes equivalent to the classical FDT of the second kind, Eq. (2.8). Further, since the quadratic form in y of the action $\Sigma^{(\text{sc})}[r, y]$ becomes exact in the classical limit, the Langevin equation (5.91) becomes exact as well.

In conclusion, in the semiclassical limit, the Brownian particle can be described by a Langevin equation in which the quantum mechanical stochastic nature of the Langevin force is given by the expressions (5.95). From the above derivation it is clear that the quasiclassical Langevin equation is exact for a Brownian particle with Gaussian statistics when the external force is linear in the coordinate. The results of the quasiclassical Langevin equation are also reliable when the anharmonic part of the potential can be treated as a perturbation. On this level of approximation, squeezing of thermal and quantum fluctuations by time-dependent external forces has been studied in Ref. [221]. For a discussion of the validity of the quasiclassical Langevin equation for anharmonic potentials we refer to Refs. [54] and [55], and to the remarks given in Subsection 2.3.4.

General quantum kinetic equations were derived in Ref.[222]. For weak dissipation, they can be reduced, upon employing a generalized Born approximation, to relaxation equations for the diagonal and off-diagonal components of the reduced density matrix, as familiar from nuclear magnetic relaxation theory [32]. In the classical limit, one is led to the energy-diffusion version of the Fokker-Planck equation studied below in Section 11.4. On the other hand, when coarse grained, the kinetic equation gives the full Fokker-Planck equation capturing the relevant physics in the classical limit.

The quasiclassical Langevin equation is easily generalized to state-dependent damping [150, b]. Particular cases are described by the action given in Eq. (4.66) or Eq. (4.187), and the relevant quasiclassical Langevin equation is of the form (3.19). Important examples in mesoscopic physics are quasiparticle tunneling between superconductors and strong electron tunneling through small junctions or metallic grains. When the fluctuations of the electronic tunneling current are strong, fluctuations of the conjugate phase variable φ are weak, and therefore the phase dynamics can well be described in terms of a Langevin equation for the phase variable φ with a state-dependent stochastic force $\xi_1(t) \cos\varphi + \xi_2(t) \sin\varphi$, where $\xi_1(t)$ and $\xi_2(t)$ are two independent stochastic variables obeying Gaussian statistics with a correlator of the form (5.95) with Eq. (5.96). For a recent application of a quasiclassical Langevin equation with state-dependent noise to "strong electron tunneling", see Ref. [223].

5.9 Stochastic unraveling of influence functionals

Present exact numerical techniques such as quantum Monte Carlo (QMC) methods keep struggling with the so-called dynamical sign problem. The numerical load is due to the fact that complex-valued probability amplitudes with varying phases rather than probabilities have to be summed up in a quantum mechanical computation.

Path integral representations of the reduced density matrix are intimately connected to stochastic processes for pure quantum states. It has been shown that there is a close connection between influence functionals and non-Markovian generalizations of quantum state diffusion [224] as well as Langevin-like stochastic dynamics of the reduced density matrix [225].

Quantum state diffusion (QSD) has been established as an alternative approach to quantum dissipation in the perturbative weak-coupling regime [65, 66, 60, 61]. In this method, *pure* quantum states are propagated stochastically. This yields an appealing picture of individual quantum trajectories for open systems and facilitates effective numerical calculations (see also Section 2.4). Extension of this method to the case of non-perturbative, non-Markovian dynamics paves the way to overcome many of the above-mentioned numerical limitations. A corresponding generalization of quantum state diffusion has been proposed by Strunz, Diósi, and Gisin [224]. In their approach, the cross term $\exp\{-\mathcal{S}_{\mathrm{FV,\,cross}}[q, q']/\hbar\}$ of the Feynman-Vernon influence functional (5.23) with (5.30) is unraveled into a stochastic influence functional. However, the resulting propagation amplitude $G[q]$ still experiences quantum memory effects arising from the self-interaction term $\exp\{-\mathcal{S}_{\mathrm{FV,\,self}}[q]/\hbar\}$. As a result of that, the respective stochastic Schrödinger equation is equipped with a time-nonlocal functional derivative, for which a general solution strategy, even numerically, is not known, except for restrictive limits. Stockburger and Grabert overcame these difficulties by unraveling the full influence functional into a stochastic one [226]. Following these lines, they derived exact c-number representations of non-Markovian quantum dissipation in terms of stochastic Schrödinger equations or equivalent stochastic Liouville-von Neumann equations, which are free of quantum memory effects.

To catch the principle of stochastic unraveling of the influence functional we take the contour path integral (5.76) for the propagating function J as starting point, and we choose the contour $\mathcal{C}_{\mathrm{I}}$ specified in Eq. (5.74). To proceed, we combine the system's action $S_{\mathrm{S}}[\tilde{q}]$ with the action due to the potential counter term in the controlled action $S_{\mathrm{contr}}[\tilde{q}]$ and subjoin a noise action originating from a stochastic force $\xi(z)$,

$$S_{\mathrm{contr}}[\tilde{q}(\cdot)] \;=\; S_{\mathrm{S}}[\tilde{q}(\cdot)] - \tfrac{1}{2}\mu \int_{\mathcal{C}_{\mathrm{I}}} dz \, \tilde{q}^2(z) \,,$$

$$S_{\xi}[\tilde{q}(\cdot)] \;=\; S_{\mathrm{contr}}[\tilde{q}(\cdot)] + \int_{\mathcal{C}_{\mathrm{I}}} dz \, \xi(z)\tilde{q}(z) \,, \tag{5.97}$$

where the path $\tilde{q}(z)$ is specified in Eq. (5.74). For given force $\xi(z)$, the time evolution of the propagating function J_{ξ} and ultimately of the density matrix ρ_{ξ} is given by

$$J_{\xi}(q_{\mathrm{f}}, q_{\mathrm{f}}'; t; q_{\mathrm{i}}, q_{\mathrm{i}}'; \bar{q}, \bar{q}') \;=\; \frac{1}{Z} \int_{\mathcal{C}_{\mathrm{I}}} \mathcal{D}\tilde{q}(\cdot) \; e^{iS_{\xi}[\tilde{q}(\cdot)]/\hbar} \,. \tag{5.98}$$

The propagating function $J(q_f, q_f'; t; q_i, q_i'; \bar{q}, \bar{q}')$, as well as the reduced density matrix $\rho(q_f, q_i; t)$, is the thermal average of the individual noise realizations,

$$J(q_f, q_f'; t; q_i, q_i'; \bar{q}, \bar{q}') = \left\langle J_\xi(q_f, q_f'; t; q_i, q_i'; \bar{q}, \bar{q}') \right\rangle_\beta . \qquad (5.99)$$

For Gaussian statistics, the average of the noise action factor is an exponential expressed by the autocorrelation function of the stochastic force $\xi(t)$,

$$\left\langle \exp\left(\frac{i}{\hbar} \int_{C_1} dz\, \xi(z) \tilde{q}(z) \right) \right\rangle_\beta = \exp\left(-\frac{1}{\hbar^2} \int_{C_1} dz \int_{z > z'} dz'\, \tilde{q}(z) \left\langle \xi(z)\xi(z') \right\rangle_\beta \tilde{q}(z') \right) . \quad (5.100)$$

From this we see that the noise-averaged propagating function (5.99) coincides with the former expression (5.76), if we identify [cf. Eq. (5.40)]

$$\left\langle \xi(z)\xi(z') \right\rangle_\beta = \hbar L(z - z') , \qquad \text{for} \qquad z \geq z' . \qquad (5.101)$$

In case of a factorizing initial state the Feynman-Vernon contour C_{FV} sketched in Fig. 5.3 applies. Then we can replace the noise force $\xi(z)$ with complex-time argument z by two noise forces with real-time argument, one along the forward path, $\xi_1(s) = \xi(s - i\hbar\beta)$, and the other along the backward path, $\xi_2(s) = \xi^*(z = s)$. If we assume that the random forces have correlations (we recall that s is the running real time)

$$\left\langle \xi_{1/2}(s)\xi_{1/2}(0) \right\rangle_\beta = \hbar L(s) , \qquad \text{and} \qquad \left\langle \xi_1(s)\xi_2(0) \right\rangle_\beta = \hbar L^*(s) , \qquad (5.102)$$

then the stochastic influence functional (5.98) for the contour C_{FV} is unraveling the reduced dynamics according to

$$\rho(q_f, q_f'; t) = \int dq_i \int dq_i' \left\langle G_{\xi_1}(q_f, q_i; t)\, G_{\xi_2}^*(q_f', q_i'; t) \right\rangle_\beta \rho(q_i, q_i'; 0) . \qquad (5.103)$$

Here $G_{\xi_j}(q_f, q_i; t)$ is the stochastic propagator

$$G_{\xi_j}(q_f, q_i; t) = \int_{q(0)=q_i}^{q(t)=q_f} \mathcal{D}q(\cdot)\, e^{iS_\xi[q(\cdot)]/\hbar} , \qquad (5.104)$$

which is free of memory load. Next, it is expedient to introduce the linear combinations $\zeta(s) = \frac{1}{2}[\xi_1(s) + \xi_2^*(s)]$ and $\nu(s) = [\xi_1(s) - \xi_2^*(s)]/\hbar$. The correlations of the new random forces are

$$\left\langle \zeta(s)\zeta(0) \right\rangle_\beta = \hbar L'(s) , \qquad \left\langle \nu(s)\nu(0) \right\rangle_\beta = 0 ,$$

$$\left\langle \zeta(s)\nu(0) \right\rangle_\beta = 2i\,\Theta(s)\, L''(s) = i\, M\, \dot{\gamma}(s) , \qquad\qquad (5.105)$$

where $L(s) = L'(s) + i\, L''(s)$ is given in Eq. (5.26). In the second line we have employed the relation (5.28).

Since the propagator of the forward path differs from that of the backward path, the propagation of a stochastic sample of the density matrix with initial state $\rho_{ini} = |\psi_1(0) > < \psi_2(0)|$ is determined by two different stochastic Schrödinger equations,

$$i\hbar|\dot{\psi}_1(t)> = \left\{H_S - [\zeta(t) + \tfrac{1}{2}\hbar\nu(t)]\,q + \tfrac{1}{2}\mu q^2\right\}|\psi_1(t)>,$$

$$i\hbar|\dot{\psi}_2(t)> = \left\{H_S - [\zeta^*(t) - \tfrac{1}{2}\hbar\nu^*(t)]\,q + \tfrac{1}{2}\mu q^2\right\}|\psi_2(t)>.$$

(5.106)

These equations are local in time and their solution renders the propagation amplitudes (5.104). All effects of the quantum mechanical memory are contained in the averaging of the product of stochastic propagators in Eq. (5.103).

The other correlations $\langle\zeta(s)\zeta^*(0)\rangle_\beta$, $\langle\zeta(s)\nu^*(0)\rangle_\beta$ and $\langle\nu(s)\nu^*(0)\rangle_\beta$ are left undetermined by the model. They can be chosen according to computational convenience. The noise force $\zeta(s)$ can be chosen real, at which the noise force $\nu(s)$ is imaginary.

The dynamics governed by the stochastic Schrödinger equations (5.106) can be rewritten as a stochastic Liouville-von Neumann (SLN) equation in which the noise terms are in the form of a commutator $[\cdot,\cdot]$ and an anti-commutator $\{\cdot,\cdot\}$,

$$i\hbar\dot{\rho}_\xi(t) = [H_S + \tfrac{1}{2}\mu q^2, \rho_\xi(t)] - \zeta(t)[q, \rho_\xi(t)] - \tfrac{1}{2}\hbar\nu(t)\{q, \rho_\xi(t)\}. \qquad (5.107)$$

The linear Schrödinger equations (5.106) and the SLN equation (5.107) are formally exact. Monte Carlo sampling of $\rho_\xi(t)$ for the realizations of $\zeta(s)$ and $\nu(s)$ with statistical weight (5.105) will eventually converge to the exact time evolution of the reduced density matrix $\rho(t)$.

The stochastic linear Schrödinger equations (5.106), as well as other stochastic linear Schrödinger equations, do not preserve the norm of the states, since the noise force $\nu(t)$ is complex. The norm $\operatorname{tr}\rho(t)$ is preserved, however, after stochastic sampling. The diffusive spread of the sample trace slows down numerical convergence and limits practical applications. This flaw can be softened or even eliminated by subtraction of a suitably chosen guide trajectory $\bar{q}(t)$ in the anti-commutator term of Eq. (5.107), $q \to q - \bar{q}(t)$. The function $\bar{q}(t)$ may be any causal function of $\zeta(s)$ and of $\nu(s)$. Turning to the normalized density matrix $\hat{\rho}_\xi(t) = \rho_\xi(t)/\operatorname{tr}\rho_\xi(t)$, the factor $\operatorname{tr}\rho_\xi(t) = \exp[i\int_0^t ds\,\nu(s)\bar{q}(s)]$ must be included in the Gaussian integration measure of the stochastic forces. The linear shift leads to noise forces with the same variance but displaced mean values. Upon completing the square in the exponent one obtains the original Gaussian probability measure for shifted variables. This shift can be compensated by a corresponding shift of the stochastic force $\zeta(t)$ in the SLN equation, $\zeta(t) \to \tilde{\zeta}(t) = \zeta(t) - M\int_0^t ds\,\dot{\gamma}(t-s)\bar{q}(s)$. Upon performing integration by parts in the second term, the SLN equation for the normalized density $\hat{\rho}_\xi(t)$ takes the form

$$i\hbar\dot{\hat{\rho}}_\xi(t) = [H_S + \tfrac{1}{2}\mu(q - \bar{q}(t))^2, \hat{\rho}_\xi(t)] - \tfrac{1}{2}\hbar\nu(t)\{q - \bar{q}(t), \hat{\rho}_\xi(t)\}$$

$$- \left(\zeta(t) - M\int_0^t ds\,\gamma(t-s)\dot{\bar{q}}(s) - M\gamma(t)\bar{q}(0)\right)[q, \hat{\rho}_\xi(t)].$$

(5.108)

The random force $\zeta(t)$ has vanishing mean, and noise correlations are as in Eq. (5.105). The SLN equation (5.108) conserves the normalization of $\hat{\rho}_\xi(t)$ for each sample. A natural choice for the guide trajectory is $\bar{q}(t) = \operatorname{tr}\{q\hat{\rho}(t)\} = <\psi_2(t)|q|\psi_1(t)>$.

In the classical limit, terms $\propto q - \bar{q}$ become negligibly small. Then the dynamics of $q(t)$ resulting from Eq. (5.108) is described by the Langevin equation (3.19).

The implementation of a numerical simulation scheme based on the LVN equation (5.108) is discussed in Ref. [227]. The transition from linear QSD approaches to stochastic processes with normalized quantum states proved to be a decisive step towards robust numerical simulations [226, 227].

5.10   Non-Markovian dissipative dynamics in the semiclassical limit

5.10.1   Van Vleck and Herman-Kluk propagator

Shortly after the discovery of Schrödinger's equation in 1927, Van Vleck wrote down in 1928 a complex-valued function which nowadays is called the semi-classical or quasi-classical propagator of quantum mechanics [228]. In n space dimensions it reads

$$K_{sc}(\mathbf{q}_f, t; \mathbf{q}_i, 0) = (2\pi i\hbar)^{-n/2} \sqrt{C(\mathbf{q}_f, t; \mathbf{q}_i, 0)} \, \exp[i\, S_{cl}(\mathbf{q}_f, t; \mathbf{q}_i, 0)/\hbar - i\,\phi]. \quad (5.109)$$

Here $S_{cl}(\mathbf{q}_f, t; \mathbf{q}_i, 0)$ is the classical action, and the density $C(\mathbf{q}'', t; \mathbf{q}', 0)$ is the so-called Van Vleck determinant

$$C(\mathbf{q}'', t; \mathbf{q}', 0) = \left| \frac{-\partial^2 S_{cl}(\mathbf{q}'', t; \mathbf{q}', 0)}{\partial q_i'' \, \partial q_j'} \right|. \quad (5.110)$$

The root in Eq. (5.109) is taken on the absolute value of the Van Vleck determinant, and the phase ϕ, which was absent in Van Vleck's original work, is given as $\pi/2$ times the number of conjugate points along the classical trajectory from $\mathbf{q}_i$ to $\mathbf{q}_f$ [229].

The path integral for the quantum mechanical propagator

$$K(\mathbf{q}_f, t; \mathbf{q}_i, 0) \equiv <\mathbf{q}_f|e^{-iHt/\hbar}|\mathbf{q}_i> = \int_{\mathbf{q}(0)=\mathbf{q}_i}^{\mathbf{q}(t)=\mathbf{q}_f} \mathcal{D}\mathbf{q}(\cdot)\, \exp\{i\, S[\mathbf{q}(\cdot)]/\hbar\}, \quad (5.111)$$

or its discretized version analogous to Eq. (4.16) with (4.15), can be worked out in the neighborhood of the classical path, if we take only the second variation of the action into acount. The resulting Gaussian integral over the quadratic fluctuations yields exactly the Van Vleck propagator [230]. The semiclassical propagator satisfies a fixed-point property on stationary phase approximation (spa) of the integral,

$$K_{sc}(\mathbf{q}'', t''; \mathbf{q}', t') = \int_{spa} d^n\mathbf{q}\, K_{sc}(\mathbf{q}'', t''; \mathbf{q}, t) K_{sc}(\mathbf{q}, t; \mathbf{q}', t'). \quad (5.112)$$

Above all, it fulfills the time-dependent Schrödinger equation except for a potential correction $\mathcal{V}_{qu}$, $(i\hbar\partial/\partial t - H - \mathcal{V}_{qu})K_{sc}(\mathbf{q}, t; \mathbf{q}_i, 0) = 0$. The correction term is of order $\hbar^2$, $\mathcal{V}_{qu} = \hbar^2(\Delta\sqrt{C})/(2M\sqrt{C})$ [230]. Thus it vanishes in the classical limit.

The semiclassical propagator (5.109) requires to determine classical trajectories $\mathbf{q}_{cl}(s)$ that obey specific boundary conditions at time zero and at time t, $\mathbf{q}_{cl}(0) = \mathbf{q}_i$ and $\mathbf{q}_{cl}(t) = \mathbf{q}_f$. This fact numerically costs root searches. This drawback is avoided in the time-dependent semiclassical initial value formalism introduced by Heller [231] and justified later by Herman and Kluk [232]. The (nowadays called) Herman-Kluk propagator propagates Gaussian non-spreading or frozen wave packets according to the laws of classical mechanics. In one space dimension it takes the form

$$K_{\mathrm{HK}}(q_{\mathrm{f}}, t; q_{\mathrm{i}}, 0) = \int \frac{dp\, dx}{2\pi\hbar} <q_{\mathrm{f}}|g_\sigma(p_t, x_t)> R(p_t, x_t)\, e^{i\, S_{\mathrm{cl}}(x, p; t)/\hbar} <g_\sigma(p, x)|q_{\mathrm{i}}> \,.$$

The classical action is given by

$$S_{\mathrm{cl}}(p, q; t) = \int_0^t dt' \left\{ p(t')\dot{q}(t') - H[p(t'), q(t'); t'] \right\},\tag{5.113}$$

and $<q|g_\sigma>$ is a complex-valued Gaussian wave packet of fixed real width σ,

$$< q|g_\sigma(p, x) > = (\pi\sigma^2)^{-1/4}\, e^{-(q-x)^2/2\sigma^2}\, e^{i\, p(q-x)/\hbar}\,.\tag{5.114}$$

The width σ may, in principle, be chosen arbitrarily. The momentum p_t and position x_t evolve according to Hamilton's equations of motion. The prefactor in the HK integrand is expressed in terms of the stability matrix elements as

$$R(p_t, x_t) = \sqrt{\frac{1}{2}\left(\frac{\partial x_t}{\partial x} + \frac{\partial p_t}{\partial p} - \frac{i\hbar}{\sigma^2}\frac{\partial x_t}{\partial p} - \frac{\sigma^2}{i\hbar}\frac{\partial p_t}{\partial x} \right)}\,.\tag{5.115}$$

In view of doubts about its validity, the HK propagator was proven to be the leading term in a uniform semiclassical expansion [233]. The crucial advantage of the HK propagator compared to the Van Vleck propagator (5.109) is that the former constitutes a classical initial value problem which is free of root searches for trajectories.

5.10.2 Semiclassical dissipative dynamics

Recently, it has been proposed to combine the stochastic quantum dynamics presented in the preceding section with semiclassical propagation based on the frozen Gaussian approximation [234]. In this approach the density matrix $\rho(q_{\mathrm{f}}, q_{\mathrm{i}}; t)$ involves three Monte Carlo integrations, two over the forward and backward phase spaces of the HK propagators and the residual one over the noise distribution. The classical path in the HK propagator is found from the quasiclassical dynamics of the frozen Gaussians under the transformed versions of Eqs. (5.106). The non-dispersive Gaussians are, up to an irrelevant phase factor, coherent states $|\alpha>_j = e^{-|\alpha_j^2|/2}\, e^{\alpha_j \hat{a}^\dagger}|0>$ with

$$\alpha_j = \sqrt{1/2\sigma^2}\,(q_j + i\, p_j\sigma^2/\hbar)\,, \qquad j = 1, 2\,.\tag{5.116}$$

The classical equations of motion derived from Eqs. (5.106) read

$$\dot{\alpha}_1 = \frac{1}{\sqrt{2}\,\sigma}\left\{ \frac{p_1}{M} + \frac{\sigma^2}{i\hbar}\left[V'(q_1) + \mu q_1 - \zeta + M\int_0^t dt'\, \dot{\gamma}(t-t')\bar{q}(t') - \tfrac{1}{2}\hbar\nu \right] \right\},$$

$$\dot{\alpha}_2 = \frac{1}{\sqrt{2}\,\sigma}\left\{ \frac{p_2}{M} + \frac{\sigma^2}{i\hbar}\left[V'(q_2) + \mu q_2 - \zeta^* + M\int_0^t dt'\, \dot{\gamma}(t-t')\bar{q}(t') + \tfrac{1}{2}\hbar\nu^* \right] \right\}.$$

These equations have to be solved for real q_j and p_j. If we disregard the ν-fluctuations we find upon elimination of p exactly the Langevin equation (3.19) with the force correlation $\langle \zeta(s)\zeta(0)\rangle_\beta = \hbar L'(s)$.

The guide trajectory $\bar{q}(t)$ is again found by the requirement that the ν-dependent terms leave the trace of the sample density matrix $\rho_\xi(t)$ invariant. This leads to a relation which includes the pair of trajectories α_1 and α_2 as

$$\bar{q}(t) \;=\; \sqrt{\sigma^2/2}\,[\,\alpha_1(t) + \alpha_2^*(t)\,].\tag{5.117}$$

In the numerical algorithm, it is therefore expedient to combine the integrations over the two HK phase spaces and the space of noise trajectories $\zeta(t), \nu(t)$ into a joint Monte Carlo sampling scheme, where new initial phase space points and a new noise trajectory are chosen for each sample. Wave packet dynamics in a Morse potential has been studied with this method over many oscillation periods in Ref. [234]. The analysis showed that dissipative dynamics is accurately described even when non-Markovian effects are significant. As a perspective, systematic quantum corrections to the HK propagator [233, 235] may be implemented in the numerical scheme.

5.11   Brief summary and outlook

So far we have presented the general real-time formalism for the density matrix with the phenomenological model described by the Hamiltonian (3.11). In the first section we have familiarized ourselves with the general formalism by studying the evolution of a product initial state. Then we have addressed the important question of the preparation of initial states. We have introduced a preparation function which describes the reduction of the equilibrium density matrix of the global system due to a measurement of certain variables of the system carried out at time $t = 0$. The subsequent evolution is determined by a propagating function.

In the third section we have studied the propagating function and found the exact formal solution in the compact form of a path integral where the progression in time is along a certain contour in the complex-time plane. We then have discussed a real-time formulation of the propagating function in which the system is evolving since the infinite past and then is subject to a measurement of its position at time zero, and again at time t. The appropriate modifications required for the various microscopic models introduced in Section 3.4 are easy to understand, and the implementation is left to the reader. At this, the key point is that the action $-iS_S[\tilde{q}] + \mathcal{S}_{\text{infl}}[\tilde{q}]$ in Eq. (5.76) or (5.80) for the various microscopic models results from the corresponding imaginary-time action $S_{\text{eff}}^{(E)}[q]$ by the substitution $\tau = iz$ and $q(\tau) = \tilde{q}(z)$.

PART II
MISCELLANEOUS APPLICATIONS

So far I have employed functional integral methods in order to set up exact formal expressions which describe the thermodynamics and dynamics of damped quantum systems. Now we move to new grounds and consider the physical properties of exactly solvable damped linear systems, i.e. the damped quantum oscillator and quantum Brownian particle. Subsequently, I discuss a variational method which makes it possible for us to treat nonlinear damped quantum systems in much the same way as linear systems. I conclude this part with a brief general discussion of quantum decoherence.

6. Damped linear quantum mechanical oscillator

The interplay of friction and quantum-statistical fluctuations with the implications for thermodynamics and dynamics can be studied quantitatively on a microscopic basis for *linear* systems. Of particular interest is a harmonic oscillator subjected to linear (state-independent) friction of arbitrary frequency-dependence.

The understanding of the quantum-mechanical damped oscillator is very important because of its universal relevance. This model applies for every macroscopic quantum system which is slightly displaced from a stable local potential minimum. Applications are, e.g., current oscillations in circuits with impedances (cf. Sect. 3.5).

The problem of a damped harmonic quantum mechanical oscillator has been studied intensively since the early works in the 1960s [20, 21, 75]. A satisfactory understanding in the classical regime and in the weak-coupling limit has been reached in the 1970s [23]–[27]. Substantial progress beyond the limitations of the weak-coupling approach was made in the 1980s, in particular with phenomenological methods [236], with study of the microscopic dynamics [237], and with Feynman-Vernon path integral techniques [238]. These and related works [54, 56, 205], [239]–[242] disclosed unexpected behaviors at low temperature, such as, e.g., nonexponential decay of correlation functions. Haake and Reibold derived from the microscopic dynamics for general memory friction and factorizing initial state an exact second order partial differential equation for the Wigner function, and they transformed it into the corresponding quantum master equation for the reduced density operator [237]. Later, the master equation was rederived by Hu, Paz, and Zhang with the Feynman-Vernon path integral method [243]. Grabert, Schramm, and Ingold determined with the Feynman-Vernon method along the lines given in Sections 4.2 and 5.2 the exact time evolution of the reduced density matrix for general memory friction and general initial preparation [87]. Subsequently, Karrlein and Grabert showed that the time-dependent Liouville operator of the respective quantum master equation depends on the initial preparation, and they determined it for thermal and factorizing initial conditions, and for different types of damping [48].

In the presence of a time dependent external force $f_{\text{ext}}(t)$ coupled linearly to the coordinate of the particle, the Hamiltonian of the global system reads

$$H(t) = \frac{p^2}{2M} + \frac{M\omega_0^2 q^2}{2} - q f_{\text{ext}}(t) + \sum_{\alpha=1}^{N} \left[\frac{p_\alpha^2}{2m_\alpha} + \frac{m_\alpha \omega_\alpha^2}{2} \left(x_\alpha - \frac{c_\alpha}{m_\alpha \omega_\alpha^2} q \right)^2 \right]. \quad (6.1)$$

It describes a central harmonic oscillator with eigenfrequency ω_0 in a spatially constant force field and bi-linearly coupled to a bath of harmonic oscillators. Clearly, the Hamiltonian of the global system can be diagonalized directly, as we have already discussed in Section 4.3.5. Since the action of the global system is quadratic, path integrals for thermodynamical and dynamical quantities can be calculated exactly. Above all, the Hamiltonian (6.1) gives *linear* equations of motion, which can be solved in analytic form. Because of the linearity, a damped linear oscillator is a system simple enough that it allows for an exact study of the quantum mechanical stochastic process in the entire range of parameters by using only phenomenological considerations. This is possible since the quantum mechanical equation for $q(t)$ is in the form of a classical Langevin equation with quantum mechanical colored noise.

Before embarking on the specific problem, we consider an important theorem which is generally valid for any particular linear and nonlinear quantum system.

6.1   Fluctuation-dissipation theorem

The fluctuation-dissipation theorem (FDT) is a cornerstone of linear response theory. The theorem relates relaxation of a weakly perturbed system to the spontaneous fluctuations in thermal equilibrium. Well-known special cases of the FDT in the classical regime are the Einstein relation [244], which connects the diffusion constant with the viscosity of a Brownian particle, and the Johnson-Nyquist formula [245, 246], which relates thermal current fluctuations in an electrical circuit to the impedance.

To illustrate how the FDT ensues from principles of quantum statistical mechanics, consider a system described by a Hamiltonian H, say a particle in a potential $V(q)$, which has reached the canonical equilibrium state. Imagine that at a particular time a weak time-dependent external force $\delta f_{\text{ext}}(t)$ is switched on. By the perturbation, the particle is thrown off the equilibrium state. With the additional force, the Hamiltonian is time-dependent,

$$\mathcal{H}(t) = H + \delta H(t) = H - q \, \delta f_{\text{ext}}(t). \quad (6.2)$$

The time evolution of the average position $\langle q(t) \rangle$ for a mixed initial state is carried by the density matrix $\rho(t)$ according to

$$\langle q(t)\rangle \; = \; \mathrm{tr}\,\{\rho(t)\,q\}\;, \tag{6.3}$$

where $\mathrm{tr}\,\{\rho(t)\} \;=\; 1$. Here, $\mathrm{tr}\,\{\cdots\}$ denotes the usual quantum mechanical trace operation. The density matrix $\rho(t)$ obeys the equation of motion

$$i\hbar\,\dot{\rho}(t) \;=\; [\,\mathcal{H}(t),\,\rho(t)\,] \;=\; [\,H,\,\rho(t)\,] + [\,\delta H(t),\,\rho(t)\,]\;, \tag{6.4}$$

where $[\,A,B\,] = AB - BA$. We assume that the system is prepared in the thermal equilibrium state in the infinite past and then the perturbation $\delta H(t)$ is turned on,

$$\lim_{t\to-\infty}\rho(t) \;=\; \rho_{\mathrm{eq}} \;=\; \mathrm{e}^{-\beta H}/Z\;,$$

$$\lim_{t\to-\infty}\langle q(t)\rangle \;=\; \langle q\rangle_\beta \;=\; \mathrm{tr}\,\{\mathrm{e}^{-\beta H}q\}/Z\;, \tag{6.5}$$

$$Z \;=\; \mathrm{tr}\,\{\mathrm{e}^{-\beta H}\}\;.$$

For weak external perturbation $\delta H(t)$, the average displacement of the particle from the equilibrium position is a linear functional of the history of the time-dependent external force $\delta f_{\mathrm{ext}}(t)$,

$$\delta\langle q(t)\rangle \;\equiv\; \langle q(t)\rangle - \langle q\rangle_\beta \;=\; \int_{-\infty}^{+\infty} dt'\,\chi(t-t')\,\delta f_{\mathrm{ext}}(t')\;. \tag{6.6}$$

The response function $\chi(t)$ is a retarded temperature Green function called generalized susceptibility. To determine $\chi(t)$, we put

$$\rho(t) \;=\; \rho_{\mathrm{eq}} + \delta\rho(t)\;. \tag{6.7}$$

The change of the density matrix $\delta\rho(t)$ caused by the perturbation obeys the equation

$$i\hbar\,\delta\dot{\rho}(t) \;=\; [\,\delta H(t),\rho_{\mathrm{eq}}\,] + [\,H,\delta\rho(t)\,]\;. \tag{6.8}$$

This equation is easily solved and gives

$$\delta\rho(t) \;=\; \frac{1}{i\hbar}\int_{-\infty}^{t} dt'\, \mathrm{e}^{-iH(t-t')/\hbar}\,[\,\delta H(t'),\,\rho_{\mathrm{eq}}\,]\,\mathrm{e}^{iH(t-t')/\hbar}\;. \tag{6.9}$$

Upon inserting the expression (6.9) into $\delta\langle q(t)\rangle = \mathrm{tr}\,\{\delta\rho(t)\,q\}$, we find that $\chi(t)$ is connected with the antisymmetric part of the position auto-correlation function in thermal equilibrium by the relation

$$\chi(t) \;=\; \frac{i}{\hbar}\Theta(t)\,\langle\,[\,q(t)q(0) - q(0)q(t)\,]\,\rangle_\beta\;. \tag{6.10}$$

The function $\chi(t)$ describes the linear response of the equilibrated system to an external force. The step function implicit in Eq. (6.10) saves *causality*. Since we did not use specific properties of the system, the expression (6.10) is a general result for any quantum system within linear response.

It is convenient to define the position autocorrelation function $C^\pm(t)$ with a suitable subtraction in order that the Fourier transform is well-defined,

$$C^+(t) \;\equiv\; C_{qq}(t) \;\;\equiv\;\; \langle q(t)q(0)\rangle_\beta - \langle q(t)\rangle_\beta \langle q(0)\rangle_\beta \,,$$

$$C^-(t) \;\equiv\; C_{qq}(-t) \;\equiv\; \langle q(0)q(t)\rangle_\beta - \langle q(0)\rangle_\beta \langle q(t)\rangle_\beta \,.$$

(6.11)

The thermal expectation value is independent of time, $\langle q(t)\rangle_\beta = \langle q(0)\rangle_\beta = \langle q\rangle_\beta$. With the subtraction in Eq. (6.11) , we ensure that $C^\pm(t) \to 0$ as $t \to \infty$. Since $C^{+*}(t) = C^-(t)$, we may write

$$C^\pm(t) \;=\; S(t) \pm i\,A(t)\,.$$

(6.12)

The real part $S(t)$ is the symmetrized correlation function

$$S(t) \;=\; S(-t) \;=\; \frac{1}{2}\langle\,[\,q(t)q(0) + q(0)q(t)\,]\,\rangle_\beta - \langle q\rangle_\beta^2\,,$$

(6.13)

and the imaginary part is expressed in terms of the expectation value of the anti-commutator as

$$A(t) \;=\; -A(-t) \;=\; \frac{1}{2i}\langle\,[\,q(t)q(0) - q(0)q(t)\,]\,\rangle_\beta\,.$$

(6.14)

We see from the expressions (6.10) and (6.14) that the response function $\chi(t)$ and the function $A(t)$ are related by

$$\chi(t) \;=\; -\,(2/\hbar)\Theta(t)A(t)\,.$$

(6.15)

It is illustrative to specify this relation in Fourier space. By dint of the transforms

$$\tilde{C}^\pm(\omega) \;\equiv\; \int_{-\infty}^{\infty} dt\, C^\pm(t)\,e^{i\omega t} \;=\; \tilde{S}(\omega) \pm i\,\tilde{A}(\omega)\,,$$

$$\tilde{\chi}(\omega) \;\equiv\; \int_{0}^{\infty} dt\,\chi(t)\,e^{i\omega t}\,,$$

(6.16)

the relation (6.15) is rephrased as

$$\tilde{\chi}''(\omega) \;=\; \frac{i}{\hbar}\tilde{A}(\omega) \;=\; \frac{1}{2\hbar}\left(\tilde{C}^+(\omega) - \tilde{C}^-(\omega)\right)\,,$$

(6.17)

where $\tilde{\chi}''(\omega)$ is the absorptive part of the dynamical susceptibility.

We now prepare the ground to proof the fluctuation-dissipation theorem. This can be done in two different ways. In the first, we utilize that the canonical operator $e^{-\beta H}$ is an imaginary-time translational operator, $e^{-\beta H}q(0)\,e^{\beta H} = q(i\hbar\beta)$, and then employ invariance of the trace under cyclic permutation. This yields

$$\mathrm{tr}\,\{\,e^{-\beta H}q(0)\,q(t)\,\} \;=\; \mathrm{tr}\,\{\,q(i\hbar\beta)\,e^{-\beta H}q(t)\,\} \;=\; \mathrm{tr}\,\{\,e^{-\beta H}q(t)\,q(i\hbar\beta)\,\}\,.$$

(6.18)

Since the correlations in thermal equilibrium are time translation invariant,

$$\langle q(t)q(t')\rangle_\beta \;=\; \langle q(t-t')q(0)\rangle_\beta\,,$$

(6.19)

we find from Eq. (6.18) the relation

$$C^-(t) = C^+(t - i\hbar\beta). \tag{6.20}$$

With use of the relation (6.20) we then immediately obtain in Fourier space as a central result

$$\tilde{C}^-(\omega) = e^{-\beta\hbar\omega}\,\tilde{C}^+(\omega). \tag{6.21}$$

In the alternative second way, we write the correlation functions (6.11) in energy representation and then switch to the respective Fourier transforms. This yields

$$\tilde{C}^+(\omega) = \frac{2\pi\hbar}{Z}\sum_{n,m}e^{-\beta E_m} < n|q|m >< m|q|n > \delta(E_m - E_n + \hbar\omega),$$

$$\tilde{C}^-(\omega) = \frac{2\pi\hbar}{Z}\sum_{n,m}e^{-\beta E_n} < n|q|m >< m|q|n > \delta(E_m - E_n + \hbar\omega). \tag{6.22}$$

Since we have the constraint $E_n = E_m + \hbar\omega$, the expressions (6.22) are indeed in agreement with relation (6.21).

Insertion of Eq. (6.21) into Eq. (6.17) yields the fluctuation-dissipation theorem discovered first by Callen and Welton in 1951 [247],

$$\tilde{\chi}''(\omega) = \frac{1}{2\hbar}\left(1 - e^{-\beta\hbar\omega}\right)\tilde{C}^+(\omega). \tag{6.23}$$

We can also relate the Fourier transform of the symmetrized autocorrelation function to $\tilde{\chi}''(\omega)$. Upon using the relation $\tilde{S}(\omega) = \frac{1}{2}[\tilde{C}^+(\omega) + \tilde{C}^-(\omega)]$ and Eq. (6.21), we find

$$\tilde{S}(\omega) = \hbar\coth(\beta\hbar\omega/2)\,\tilde{\chi}''(\omega). \tag{6.24}$$

There are two important limits of the fluctuation-dissipation theorem (6.24):
(i) in the extreme quantum limit $k_{\mathrm{B}}T \ll \hbar\omega$, there are pure quantum fluctuations,

$$\tilde{S}(\omega) = \hbar\,\tilde{\chi}''(|\omega|). \tag{6.25}$$

According to the relation (6.25), the function $\tilde{S}(\omega)$ is non-analytic at the origin. Hence the position autocorrelation function decays algebraically with time at zero temperature.
(ii) for low frequency or high temperature, $\hbar\omega \ll k_{\mathrm{B}}T$, the spectral function $\tilde{S}(\omega)$ takes the classical ($\hbar$-independent) form ,

$$\tilde{S}(\omega) = 2k_{\mathrm{B}}T\frac{\tilde{\chi}''(\omega)}{\omega}. \tag{6.26}$$

Special cases of this formula are the Einstein relation, which relates the diffusion coefficient D to the mobility μ, $D = k_{\mathrm{B}}T\mu$ (see Section 7.3 for linear systems and Subsection 25.4.2 for a nonlinear system), and the Johnson-Nyquist formula, which relates the power spectrum of the current fluctuations $\tilde{S}_{\mathrm{II}}(\omega) = \int_{-\infty}^{\infty} dt\,\langle I(t)I(0)\rangle_\beta\,e^{i\omega t}$ in an electrical circuit to the admittance $\tilde{Y}(\omega)$, $\tilde{S}_{\mathrm{II}}(\omega) = 2k_{\mathrm{B}}T\,\mathrm{Re}\,\tilde{Y}(\omega)$. The quantum version of the Johnson-Nyquist formula is

$$\tilde{S}_{\mathrm{II}}(\omega) \;=\; \hbar\omega \coth\left(\frac{\hbar\omega}{2k_{\mathrm{B}}T}\right) \operatorname{Re}\tilde{Y}(\omega) \;=\; 2\left(\frac{\hbar\omega}{2} + \frac{\hbar\omega}{e^{\beta\hbar\omega} - 1}\right)\operatorname{Re}\tilde{Y}(\omega)\,. \qquad (6.27)$$

The quantum version was already anticipated by Nyquist in his 1928 paper [246], but devoid of the vacuum energy term in the second equality.

It is appropriate to emphasize that at no step of the derivation we used particular properties of a linear system. Thus, the FDT relations (6.23) and (6.24) are generally valid for any open quantum system in thermal equilibrium. Therefore, they have broad implications far beyond the particular linear model systems discussed here.

6.2   Stochastic modeling

Straight knowledge about many important properties of linear dissipative quantum systems is easily gathered from a simple stochastic modeling [236] which is based upon the following three principles:

  i. The mean values obey the classical equations of motion (*Ehrenfest's theorem*).

 ii. The response function and the equilibrium autocorrelation function are related by the *fluctuation-dissipation theorem*.

iii. The stochastic process is a *stationary Gaussian process*.

While the first two principles are generally valid, the assumption of a strict Gaussian process is limited to linear systems. For *linear* systems, the response to an external perturbation is *linear* for arbitrary strength of the perturbation. Thus, all we said in the previous subsection for an infinitesimal perturbation $\delta f_{\mathrm{ext}}(t)$ is generally valid for an external force $f_{\mathrm{ext}}(t)$ of any strength in Eq. (6.1).

By virtue of Ehrenfest's theorem, the average position $\langle q(t)\rangle$ of a damped quantum oscillator obeys the equation of motion

$$\langle \ddot{q}(t)\rangle + \int_{-\infty}^{t} dt'\,\gamma(t - t')\langle \dot{q}(t')\rangle + \omega_0^2\langle q(t)\rangle \;=\; \frac{1}{M}\,f_{\mathrm{ext}}(t)\,. \qquad (6.28)$$

To be general, we have permitted memory-friction. The response of $\langle q(t)\rangle$ to a general force $f_{\mathrm{ext}}(t)$ is given by [cf. Eq. (6.6)]

$$\langle q(t)\rangle \;=\; \int_{-\infty}^{+\infty} dt'\,\chi(t - t')f_{\mathrm{ext}}(t')\,, \qquad (6.29)$$

where $\chi(t)$ is the causal response function, $\chi(t) = 0$ for $t \leq 0$. Causality requires that the dynamical susceptibility

$$\tilde{\chi}(\omega) \;=\; \int_{-\infty}^{\infty} dt\,\chi(t)\,e^{i\omega t} \qquad (6.30)$$

is an analytic function of ω in the upper complex ω–half-plane $(\operatorname{Im}\omega > 0)$. This is the origin of the significant Kramers-Kronig dispersion relations

$$\tilde{\chi}'(\omega) = \mathcal{P}\int_{-\infty}^{\infty} \frac{d\nu}{\pi} \frac{\tilde{\chi}''(\nu)}{\nu - \omega} , \qquad \tilde{\chi}''(\omega) = -\mathcal{P}\int_{-\infty}^{\infty} \frac{d\nu}{\pi} \frac{\tilde{\chi}'(\nu)}{\nu - \omega} . \tag{6.31}$$

These relations are employed in nearly all areas of physics as they express dispersion in terms of absorption and vice versa.

For a *linear* quantum system, the Ehrenfest equation (6.28) agrees exactly with the corresponding classical equation of motion. This implicates that the quantum mechanical response function and the classical one are identical, $\chi(t) = \chi_{cl}(t)$.

For the linear equation of motion (6.28), the dynamical susceptibility reads

$$\tilde{\chi}(\omega) \equiv \tilde{\chi}'(\omega) + i\,\tilde{\chi}''(\omega) = \frac{1}{M}\frac{1}{\omega_0^2 - \omega^2 - i\,\omega\tilde{\gamma}(\omega)} , \tag{6.32}$$

where $\tilde{\gamma}(\omega) = \int_0^\infty dt\, \gamma(t)\, e^{i\omega t}$ is the frequency-dependent damping coefficient. To complete the picture, we also give the response function in Laplace representation,

$$\chi(t) = \frac{1}{2\pi i}\int_{-i\infty+c}^{i\infty+c} dz\, \hat{\chi}(z)\, e^{zt} , \tag{6.33}$$

$$\hat{\chi}(z) = \frac{1}{M}\frac{1}{\omega_0^2 + z^2 + z\hat{\gamma}(z)} . \tag{6.34}$$

The constant c in the integral expression (6.33) must be chosen such that all singularities of $\hat{\chi}(z)$ are lying to the left of the integration path.

If we wish to calculate the thermal position autocorrelation function[1]

$$C_{qq}(t) \equiv \langle q(t)q(0)\rangle_\beta \tag{6.35}$$

with the help of the fluctuation-dissipation theorem (6.24), knowledge of the absorptive part of the dynamical susceptibility is required. The expression (6.32) yields

$$\tilde{\chi}''(\omega) = M\omega\tilde{\gamma}'(\omega)\tilde{\chi}(\omega)\tilde{\chi}^*(\omega) = \frac{1}{M}\frac{\omega\tilde{\gamma}'(\omega)}{[\omega^2 - \omega_0^2 - \omega\tilde{\gamma}''(\omega)]^2 + \omega^2\tilde{\gamma}'^2(\omega)} . \tag{6.36}$$

For frequency-independent damping, $\tilde{\gamma}(\omega) = \gamma$, the expression (6.36) is essentially a superposition of two Lorentzians of width $\gamma/2$ centered at $\omega = \pm\sqrt{\omega_0^2 - \gamma^2/4}$. In the limit $\gamma \to 0$, the Lorentzians reduce to δ-functions located at $\omega = \pm\omega_0$.

Looking back on these findings, we draw an important conclusion for *linear* systems: since the dynamical susceptibility $\tilde{\chi}(\omega)$ is purely classical, quantum mechanics enters solely by the fluctuation-dissipation theorem (6.24).

In a stationary stochastic process, the time correlations for dynamical variables do *not* depend on absolute time. Time translation invariance of correlations is reflected by the property (6.19). Thus, once we have gained knowledge about the position correlation function $C_{qq}(t)$, some other pair correlation functions can be found by differentiation. Upon relating the momentum to the time derivative of the position and using Eq. (6.19), we obtain the simple relations

---

[1]Note that there is no subtraction in $C_{qq}(t)$ since the equilibrium average $\langle q\rangle_\beta$ is zero.

$$C_{pq}(t) \equiv \langle p(t)q(0)\rangle_\beta = M\frac{d}{dt}\langle q(t)q(0)\rangle_\beta \,,$$

$$C_{qp}(t) \equiv \langle q(t)p(0)\rangle_\beta = -M\frac{d}{dt}\langle q(t)q(0)\rangle_\beta \,, \qquad (6.37)$$

$$C_{pp}(t) \equiv \langle p(t)p(0)\rangle_\beta = -M^2\frac{d^2}{dt^2}\langle q(t)q(0)\rangle_\beta \,.$$

It follows from property (iii) that correlation functions with an odd number of variables (position or momentum) vanish whereas correlation functions with an even number of variables can be written as a sum of combinations of factorized pair correlation functions. Within each pair, the original order of the variables is preserved [248]. For illustration, we give an example,

$$\langle q(t)p(t')p(t'')\,q(0)\rangle_\beta = \langle q(t)p(t')\rangle_\beta\langle p(t'')q(0)\rangle_\beta \qquad (6.38)$$

$$+ \langle q(t)p(t'')\rangle_\beta\langle p(t')q(0)\rangle_\beta + \langle q(t)q(0)\rangle_\beta\langle p(t')p(t'')\rangle_\beta \,.$$

Thus, every correlation function with an even number of dynamical variables can be expressed in terms of the position autocorrelation function and of its time derivatives. In this way we can determine the complete quantum mechanical stochastic process of the damped harmonic system.

6.3 Susceptibility

6.3.1 Ohmic friction

For *Ohmic* or frequency-independent damping, the dynamical susceptibility (6.32) has two poles in the lower ω half-plane at $\omega = -i\lambda_1^{(0)}$ and at $\omega = -i\lambda_2^{(0)}$, where

$$\lambda_{1,2}^{(0)} = \tfrac{1}{2}\gamma \pm i\omega_\gamma \,, \qquad \omega_\gamma = \sqrt{\omega_0^2 - \tfrac{1}{4}\gamma^2} = \sqrt{1-\alpha^2}\,\omega_0 \,. \qquad (6.39)$$

In the second form we have introduced the dimensionless damping parameter

$$\alpha = \frac{\gamma}{2\omega_0} \,. \qquad (6.40)$$

In Laplace space we have

$$\hat{\chi}(z) = \frac{1}{M}\frac{1}{\omega_0^2 + z^2 + \gamma z} = \frac{1}{M}\frac{1}{(z+\lambda_1^{(0)})(z+\lambda_2^{(0)})} \,. \qquad (6.41)$$

Use of Cauchy's integral theorem in the expression (6.33) then yields

$$\chi(t) = -(2/\hbar)\,\Theta(t)\,A(t) \,,$$

$$A(t) = -(\hbar/2M\omega_\gamma)\sin(\omega_\gamma t)\,e^{-\gamma|t|/2} \,. \qquad (6.42)$$

The response function $\chi(t)$ performs damped oscillations in the underdamped regime $0 < \gamma < 2\omega_0$. In the overdamped regime $\gamma > 2\omega_0$, the quantities $\lambda_{1/2}$ are real and hence $A(t)$ and $\chi(t)$ decay exponentially with relaxation rates $\tfrac{1}{2}\gamma \pm \sqrt{\tfrac{1}{4}\gamma^2 - \omega_0^2}$.

6.3.2 Ohmic friction with Drude cutoff

When $\hat{\gamma}(z)$ is rational in z, the characteristic equation determining the poles of $\hat{\chi}(z)$ is algebraic. For the familiar Drude damping (3.67), the susceptibility (6.34) takes the form

$$\hat{\chi}(z) = \frac{1}{M} \frac{z + \omega_D}{(\omega_0^2 + z^2)(\omega_D + z) + \gamma \omega_D z} . \tag{6.43}$$

This expression is conveniently written in the factorized form

$$\hat{\chi}(z) = \frac{1}{M} \frac{z + \omega_D}{(z^2 + \Gamma z + \Omega_0^2)(z + \Omega_D)} . \tag{6.44}$$

Equating the expressions (6.43) and (6.44) we obtain the relations

$$\omega_0^2 = \frac{\Omega_D}{\Omega_D + \Gamma} \Omega_0^2 , \qquad \gamma = \frac{\Omega_D^2 + \Gamma \Omega_D + \Omega_0^2}{(\Omega_D + \Gamma)^2} \Gamma , \qquad \omega_D = \Omega_D + \Gamma . \tag{6.45}$$

One may consider these relations as expressions for the parameters ω_0, γ, and ω_D in terms of the parameters Ω_0, Γ and Ω_D. Conversely, one may solve these relations for the parameters Ω_0, Γ and Ω_D. In the most interesting regime $\omega_0 \ll \omega_D$ and $\gamma \ll \omega_D$ we find

$$\Omega_0 = \left(1 + \frac{\gamma}{2\omega_D} + +\frac{7\gamma^2}{8\omega_D^2} + \mathcal{O}(1/\omega_D^3)\right) \omega_0 ,$$

$$\Gamma = \left(1 + \frac{\gamma}{\omega_D} + \frac{2\gamma^2 - \omega_0^2}{\omega_D^2} + \mathcal{O}(1/\omega_D^3)\right) \gamma , \tag{6.46}$$

$$\Omega_D = \left(1 - \frac{\gamma}{\omega_D} - \frac{\gamma^2}{\omega_D^2} + \mathcal{O}(1/\omega_D^3)\right) \omega_D .$$

The poles of $\hat{\chi}(z)$ are all in the half-plane $\mathrm{Re}\,(z) < 0$ at $z = -\lambda_1$, $z = -\lambda_2$, and at $z = -\lambda_3$, where

$$\lambda_3 = \Omega_D , \qquad \text{and} \qquad \lambda_{1,2} = \tfrac{1}{2}\Gamma \pm i\Omega \qquad \text{with} \qquad \Omega = \sqrt{\Omega_0^2 - \tfrac{1}{4}\Gamma^2} . \tag{6.47}$$

The root λ_3 is positive throughout and gives rise to a pure relaxation contribution in the time regime. For $\Gamma < 2\Omega_0$, the roots λ_1 and λ_2 are complex conjugate and yield the damped oscillatory part of the response function. In the regime $\Gamma > 2\Omega_0$ the oscillation is overdamped. Altogether we find for the response function in the underdamped regime the analytic expression

$$\chi(t) = \frac{\Theta(t)}{M} \frac{1}{\Omega^2 + (\Omega_D - \tfrac{1}{2}\Gamma)^2} \Big\{ (\Omega_D - \omega_D) \left[\cos(\Omega t)\, e^{-\Gamma t/2} - e^{-\Omega_D t}\right]$$

$$+ \left[\Omega + (\Omega_D - \tfrac{1}{2}\Gamma)(\omega_D - \tfrac{1}{2}\Gamma)/\Omega\right] \sin(\Omega t)\, e^{-\Gamma t/2} \Big\} . \tag{6.48}$$

6.3.3 Radiation damping

We have shown in Subsection 3.1.10 that the coupling of a charged particle to the radiation field in dipole approximation with form factor $f(\omega) = 1/\sqrt{1 + \omega^2/\omega_D^2}$ is described by the spectral density $J(\omega) = (2e^2/3c^3)\,\omega^3/(1 + \omega^2/\omega_D^2)$. With this form we obtain from Eq. (3.28) the spectral damping function as

$$\hat{\gamma}(z) = \gamma_3 \frac{z}{z + \omega_D} \qquad \text{with} \qquad \gamma_3 = \frac{2e^2\omega_D^2}{3c^3 M}. \tag{6.49}$$

With this the dynamical susceptibility reads

$$\hat{\chi}(z) = \frac{1}{M} \frac{z + \omega_D}{(\omega_0^2 + z^2)(\omega_D + z) + \gamma_3 z^2}. \tag{6.50}$$

Again we may write this expression in factorized form as in Eq. (6.44),

$$\hat{\chi}(z) = \frac{1}{M} \frac{z + \omega_D}{(z^2 + \Gamma z + \Omega_0^2)(z + \Omega_D)}. \tag{6.51}$$

The expression (6.51) is in correspondence with Eq. (6.50) under the relations

$$\omega_0^2 = \Omega_0^2 + \Gamma\Omega_D, \qquad \gamma_3 = \frac{\Omega_D^2 + \Gamma\Omega_D + \Omega_0^2}{\Omega_0^2 + \Gamma\Omega_D}\Gamma, \qquad \omega_D = \frac{\Omega_0^2}{\Omega_0^2 + \Gamma\Omega_D}\Omega_D. \tag{6.52}$$

Solving these for the parameters Ω_0, Γ and Ω_D, we find in the limit of large cutoff

$$\Omega_0 = \left(1 - \frac{\gamma_3}{2\omega_D} + \frac{3\gamma_3^2}{8\omega_D^2} + \mathcal{O}(1/\omega_D^3)\right)\omega_0,$$

$$\Gamma = \left(1 - \frac{2\gamma_3}{\omega_D} + \mathcal{O}(1/\omega_D^2)\right)\frac{\omega_0^2}{\omega_D^2}\gamma_3, \tag{6.53}$$

$$\Omega_D = \left(1 + \frac{\gamma_3}{\omega_D} + \mathcal{O}(1/\omega_D^3)\right)\omega_D.$$

The expression (6.49) yields the mass contribution $\Delta M = (\gamma_3/\omega_D)M$, as we infer from Eq. (3.72). The renormalization of the mass results in a negative frequency shift $\Delta\omega = -(\Delta M/2M)\omega_0$. This is the leading contribution to $\Omega_0 - \omega_0$ in Eq. (6.53).

The poles of $\hat{\chi}(z)$ are again all in the half-plane $\mathrm{Re}\,(z) < 0$ at $z = -\lambda_1$, $z = -\lambda_2$, and at $z = -\lambda_3$, where

$$\lambda_3 = \Omega_D, \qquad \text{and} \qquad \lambda_{1,2} = \tfrac{1}{2}\Gamma \pm i\Omega \qquad \text{with} \qquad \Omega = \sqrt{\Omega_0^2 - \tfrac{1}{4}\Gamma^2}. \tag{6.54}$$

Since the susceptibilities (6.51) and (6.44) match each other, the response function for radiation damping coincides with that for Drude-regularized Ohmic friction, Eq. (6.48). The distinction is in the relations to the original parameters. The important difference to the Ohmic case is that for radiation damping the dynamics of $\chi(t)$ is always underdamped, $\Gamma/2 < \Omega_0$.

Beyond that, all dynamical and thermodynamical results presented in this chapter for Drude-regularized Ohmic friction can directly be translated with the relations (6.52) – (6.54) into corresponding results for radiation damping.

6.4 The position autocorrelation function

With the aid of the FDT (6.23), the position autocorrelation function $C^+(t)$ defined in Eq. (6.11) may be written as the Fourier integral

$$C^+(t) = \frac{\hbar}{\pi} \int_{-\infty}^{\infty} d\omega\, \tilde{\chi}''(\omega) \frac{e^{-i\omega t}}{1 - e^{-\beta\hbar\omega}}. \tag{6.55}$$

Using the expression (6.36), we obtain for the damped harmonic oscillator

$$C^+(t) = \frac{\hbar}{\pi M} \int_{-\infty}^{\infty} d\omega \frac{\omega \tilde{\gamma}'(\omega)}{[\omega^2 - \omega_0^2 - \omega\tilde{\gamma}''(\omega)]^2 + \omega^2 \tilde{\gamma}'^2(\omega)} \frac{e^{-i\omega t}}{1 - e^{-\beta\hbar\omega}}. \tag{6.56}$$

Interestingly, this function may be analytically continued to complex time $z = t - i\tau$ in the strip $0 < \tau < \hbar\beta$. The resulting Euclidean or imaginary-time correlation function

$$C^{(\mathrm{E})}(\tau) \equiv C^+(z = -i\tau) \tag{6.57}$$

is real. Further, since $\tilde{\chi}''(\omega)$ is an odd function of ω, we have $C^{(\mathrm{E})}(0) = C^{(\mathrm{E})}(\hbar\beta)$. If we periodically continue $C^{(\mathrm{E})}(\tau)$ outside of the principal interval $0 \leq \tau < \hbar\beta$, we may write it as Fourier series

$$C^{(\mathrm{E})}(\tau) = \frac{1}{\hbar\beta} \sum_{n=-\infty}^{+\infty} c_n\, e^{i\nu_n \tau}, \tag{6.58}$$

where the frequencies ν_n are the bosonic Matsubara frequencies

$$\nu_n = \frac{2\pi}{\hbar\beta} n, \qquad n = \pm 1, \pm 2, \cdots. \tag{6.59}$$

The Fourier coefficient c_n in the series (6.58) may be related to the dynamical susceptibility $\tilde{\chi}(\omega)$. Use of Eqs. (6.57) and (6.55) yields

$$
\begin{aligned}
c_n &= \int_0^{\hbar\beta} d\tau\, C^{(\mathrm{E})}(\tau)\, e^{-i\nu_n \tau} \\
&= \frac{\hbar}{\pi} \int_{-\infty}^{\infty} d\omega \frac{\tilde{\chi}''(\omega)}{\omega + i\nu_n} = \frac{\hbar}{\pi} \mathrm{Im} \int_{-\infty}^{\infty} d\omega \frac{\omega\, \tilde{\chi}(\omega)}{\omega^2 + \nu_n^2}.
\end{aligned} \tag{6.60}
$$

Observing that $\tilde{\chi}(\omega)$ is analytic in the upper half-plane, we obtain by contour integration [2]

$$c_n = c_{-n} = \hbar\tilde{\chi}(\omega = i\,|\nu_n|) = \hbar\hat{\chi}(z = |\nu_n|), \tag{6.61}$$

and finally with the explicit form (6.32),

$$c_n = \frac{\hbar}{M} \frac{1}{\omega_0^2 + \nu_n^2 + |\nu_n|\,\hat{\gamma}(|\nu_n|)}. \tag{6.62}$$

---

[2] We recall that the Fourier transform is denoted by a tilde and the Laplace transform by a hat.

Separating Eq. (6.55) into real and imaginary parts, $C^+(t) = S(t) + i\,A(t)$, the symmetrized and anti-symmetrized correlation functions (6.13) and (6.14) are found as

$$S(t) \;=\; \frac{\hbar}{2\pi} \int_{-\infty}^{\infty} d\omega\, \tilde{\chi}''(\omega)\, \coth(\beta\hbar\omega/2)\, e^{-i\omega t} \;, \tag{6.63}$$

$$A(t) \;=\; -\frac{\hbar}{2\pi} \int_{-\infty}^{\infty} d\omega\, \tilde{\chi}''(\omega)\, \sin(\omega t) \;. \tag{6.64}$$

With use of the relation (6.15), the response function may thus be written as

$$\chi(t) \;=\; \Theta(t)\, \frac{1}{\pi} \int_{-\infty}^{\infty} d\omega\, \tilde{\chi}''(\omega)\, \sin(\omega t) \;. \tag{6.65}$$

Consider next the initial value $S(0)$. First, we observe that $\tilde{\chi}(\omega)$ is analytic in the upper ω half-plane and that the coth-function can be written as a series of fractions,

$$\coth\!\left(\frac{\beta\hbar\omega}{2}\right) \;=\; \frac{2\omega}{\hbar\beta}\left(\frac{1}{(\omega + i\,0^+)(\omega - i\,0^+)} + 2\sum_{n=1}^{\infty} \frac{1}{(\omega + i\,\nu_n)(\omega - i\,\nu_n)}\right). \tag{6.66}$$

Observing that $\tilde{\chi}(-\omega) = \tilde{\chi}^*(\omega)$, and closing the integration contour such that all poles of the series (6.66) with positive imaginary part are included, we obtain

$$S(0) \;=\; \frac{\hbar}{2\pi i} \int_{-\infty}^{\infty} d\omega\, \coth(\beta\hbar\omega/2)\, \tilde{\chi}(\omega) \;=\; \frac{1}{\beta} \sum_{n=-\infty}^{\infty} \hat{\chi}(|\nu_n|) \;. \tag{6.67}$$

In the classical limit, the coordinate autocorrelation function takes the form

$$C_{\mathrm{cl}}^+(t) \;=\; S_{\mathrm{cl}}(t) \;=\; \frac{2}{\beta\pi} \int_0^{\infty} d\omega\, \frac{\tilde{\chi}''(\omega)}{\omega}\, \cos(\omega t) \;. \tag{6.68}$$

Analytic expressions for the functions $S(t)$, $A(t)$ and $\chi(t)$ may be derived for special forms of the damping function $\tilde{\gamma}(\omega)$.

6.4.1   Ohmic friction

For Ohmic friction, $\hat{\gamma}(|\nu_n|) = \gamma$, explicit expressions for $\chi(t)$ and $A(t)$ are given above in Eq. (6.42). Consider now the symmetrized position autocorrelation function $S(t)$. For $t > 0$ the integration contour in Eq. (6.63) must be closed in the lower ω half-plane. The contour integral picks up two contributions,

$$S(t) \;=\; S_1(t) + S_2(t) \;. \tag{6.69}$$

The one, say $S_1(t)$, is the contribution of the poles of $\tilde{\chi}''(\omega)$ located at $\omega = -i\,\lambda_{1,2}^{(0)}$

$$S_1(t) \;=\; \frac{\hbar}{2M}\, \frac{\coth(i\,\tfrac{1}{2}\beta\hbar\lambda_2^{(0)})\,e^{-\lambda_2^{(0)}|t|} - \coth(i\,\tfrac{1}{2}\beta\hbar\lambda_1^{(0)})\,e^{-\lambda_1^{(0)}|t|}}{i\,(\lambda_2^{(0)} - \lambda_1^{(0)})} \;. \tag{6.70}$$

The pole positions $\lambda_{1,2}^{(0)}$ are given in Eq. (6.39), Since $\lambda_1^{(0)}$ and $\lambda_2^{(0)}$ are either real or complex conjugate, the expression (6.70) is real. In the underdamped regime $\gamma < 2\omega_0$, the frequency $\omega_\gamma = \sqrt{\omega_0^2 - \gamma^2/4}$ is real, and $S_1(t)$ is conveniently written as

$$S_1(t) = \frac{\hbar}{2M\omega_\gamma} \frac{[\sinh(\beta\hbar\omega_\gamma)\cos(\omega_\gamma t) + \sin(\beta\hbar\gamma/2)\sin(\omega_\gamma|t|)]}{\cosh(\beta\hbar\omega_\gamma) - \cos(\beta\hbar\gamma/2)} e^{-\gamma|t|/2}. \tag{6.71}$$

The contribution $S_2(t)$ comes from the infinite sequence of simple poles of the function $\coth(\beta\hbar\omega/2)$ located at $\omega = -i\nu_n$ $(n = 1, 2, \ldots)$ [cf. Eq. (6.66)],

$$S_2(t) = -\frac{2\gamma}{M\beta} \sum_{n=1}^{\infty} \frac{\nu_n e^{-\nu_n|t|}}{(\omega_0^2 + \nu_n^2)^2 - \gamma^2\nu_n^2}. \tag{6.72}$$

By dint of partial fraction decomposition one finds that $S_2(t)$ can be expressed as a linear combination of four hypergeometric functions [87].

In the Ohmic case, the Matsubara sum (6.67) for the initial value takes the form

$$S(0) = \frac{1}{M\beta} \sum_{n=-\infty}^{\infty} \frac{1}{\omega_0^2 + \nu_n^2 + \gamma|\nu_n|}, \tag{6.73}$$

The interested reader may see for himself that $S_1(0)$ and $S_2(0)$ add up to the expression (6.73). At high temperature, $k_B T >> \hbar\omega_0$, we have

$$|S_2(0)|/S_1(0) = \zeta(3) \left(2\gamma/\omega_0\right)\left(\hbar\omega_0/2\pi k_B T\right)^3, \tag{6.74}$$

where $\zeta(3)$ is a Riemann number. Hence $S_2(t = 0)$ becomes very small compared to $S_1(t = 0)$ as we approach the classical regime. In addition, when $k_B T \gg \hbar\gamma/4\pi$, the function $S_2(t)$ drops to zero much faster than $S_1(t)$. For these two reasons, the contribution $S_2(t)$ is negligibly small at high temperature for all t. In the asymptotic high temperature limit $T \to \infty$, $S_1(t)$ becomes independent of Planck's constant while $S_2(t)$ drops to zero. The resulting classical correlation function reads

$$S(t) \xrightarrow{\hbar \to 0} S_{cl}(t) = \frac{1}{M\beta\omega_0^2} \left[\cos(\omega_\gamma t) + \left(\frac{\gamma}{2\omega_\gamma}\right)\sin(\omega_\gamma|t|)\right] e^{-\gamma|t|/2}. \tag{6.75}$$

Conversely, as temperature is lowered, quantum effects appear in $S_1(t)$ and above all the term $S_2(t)$ becomes increasingly important.

Historically, the leading quantum corrections were calculated with weak-coupling theories like the quantum master equation method [23, 25, 27, 30]. However, weak-coupling theories fail to describe the temperature regime $k_B T \lesssim \hbar\gamma/4\pi$, since the weak-coupling assumption is not valid anymore.

In the temperature regime $0 \leq k_B T < \hbar\gamma/4\pi$, we have $\nu_1 < \gamma/2$. Hence the decay of $S(t)$ at long time is determined by the contribution of the lowest Matsubara frequency in Eq. (6.72) which is $\propto e^{-\nu_1 t}$, and not anymore by the term $S_1(t) \propto e^{-\gamma t/2}$.

As temperature approaches absolute zero, the Matsubara sum of exponentials in Eq. (6.72) merges into the integral expression

$$S_2(t) = -\frac{\hbar\gamma}{M\pi} \int_0^\infty d\nu \, \frac{\nu \, e^{-\nu|t|}}{(\omega_0^2 + \nu^2)^2 - \gamma^2\nu^2} \, . \tag{6.76}$$

The integral is a linear combination of exponential-integral functions. Asymptotic evaluation of the integral (6.76) for time $t \gg 1/\omega_0$, γ/ω_0^2 yields algebraic decay. The leading term is

$$S_2(t) = -\frac{\hbar\gamma}{\pi M\omega_0^4} \frac{1}{t^2} = -\frac{M\hbar\gamma}{\pi} \frac{\chi_0^2}{t^2} \, . \tag{6.77}$$

In the last form, we have introduced the static susceptibility $\chi_0 = 1/M\omega_0^2$.

The algebraic long-time tail with power law t^{-2} of the correlation function $S(t)$ is characteristic for Ohmic friction. At low T, the algebraic decay $\propto 1/t^2$ occurs at intermediate times before the exponential decay $\propto e^{-\nu_1 t}$ sets in [238].

We refrain from writing down the corresponding expressions for the Drude model (3.67). These can easily be found with use of Eq. (6.43). Since a high-frequency cutoff does not change the low-frequency behavior of the dynamical susceptibility, the $1/t^2$ tail in Eq. (6.77) also holds for Ohmic friction with a Drude cutoff..

6.4.2   Non-Ohmic spectral density

We see from the expressions for $\hat\gamma(z)$ given in Eq. (3.83) that the Laplace transform $\hat\chi(z)$ in Eq. (6.34) has a branch point at $z = 0$ for $s \neq 1 + 2n$. The complex z-plane is cut along the negative real axis, and in the cut plane $\hat\chi(z)$ is single-valued. The principal branch of z^s is taken in the cut plane.

Coherent contribution

The two complex conjugate poles of $\hat\chi(z)$ for zero friction move away from the positions $z = \pm i\,\omega_0$ when friction is switched on. Depending on the particular values of the parameters chosen, they may or may not lie in the cut plane, depending on the parameters s, γ_s and ω_{ph}. The locations of the poles can not be given in analytic form for general s. If they lie in the cut plane, they lead to damped oscillatory contributions $A_{coh}(t)$, $\chi_{coh}(t) = -(2/\hbar)\Theta(t)A_{coh}(t)$ and $S_{coh}(t)$. Generally the amplitudes of $S_{coh}(t)$ are temperature-dependent.

Cut contribution

In addition to the coherent contribution of the simple poles in the cut plane there is an incoherent cut contribution

$$\chi_{cut}(t) = \frac{1}{\pi} \text{Im} \int_0^\infty d\nu \, \hat\chi(\nu \, e^{-i\pi}) \, e^{-\nu t} \, . \tag{6.78}$$

The leading asymptotic contribution from the cut expressed in terms of the static susceptibility χ_0 is

$$\chi_{cut}(t) = \frac{M\chi_0^2}{\pi} \text{Im} \int_0^\infty d\nu \, \nu \hat\gamma(\nu \, e^{-i\pi}) \, e^{-\nu t} \, . \tag{6.79}$$

Interestingly, this form holds for any *linear* and *nonlinear* system with a finite static susceptibility.

Next we observe that contributions to $z\hat{\gamma}(z)$ with positive integer powers of z, e.g. mass renormalization terms, do not contribute to the long-time tail of $\chi_{\rm cut}(t)$. We readily find with the relations (3.83) in the asymptotic regime $t \gg (\lambda_s\omega_{\rm ph}/\omega_0^2)^{1/s}/\omega_{\rm ph}$ for all s the universal behavior

$$\chi_{\rm cut}(t) \;=\; \frac{2\Gamma(1+s)\cos(\tfrac{1}{2}\pi s)}{\pi} M\gamma_s\omega_{\rm ph}^2 \frac{\chi_0^2}{(\omega_{\rm ph}t)^{1+s}} \;. \tag{6.80}$$

For odd s the prefactor vanishes and thus the algebraic tail is absent.

According to the relation (6.15), the anti-symmetrized position autocorrelation function at asymptotic time thus takes the form

$$A(t) \;=\; -{\rm sgn}(t)\,\frac{\Gamma(1+s)\cos(\tfrac{1}{2}\pi s)}{\pi} M\hbar\gamma_s\omega_{\rm ph}^2 \frac{\chi_0^2}{(\omega_{\rm ph}|t|)^{1+s}} \;. \tag{6.81}$$

Consider next the symmetrized position autocorrelation function at zero temperature. We obtain from Eq. (6.63)

$$S(t) \;=\; \frac{\hbar}{2\pi} \int_{-\infty}^{\infty} d\omega\, \chi''(\omega)\,{\rm sgn}(\omega)\,{\rm e}^{-i\omega t}\,, \tag{6.82}$$

from where we find with Eq. (6.36) at asymptotic time

$$S(t) \;=\; \frac{M\hbar\chi_0^2}{\pi} {\rm Re} \int_0^{\infty} d\omega\, \omega\tilde{\gamma}'(\omega)\,{\rm e}^{-i\omega t}\;. \tag{6.83}$$

With the power-law form $\tilde{\gamma}'(\omega) = \gamma_s(\omega/\omega_{\rm ph})^{s-1}$ this yields

$$S(t) \;=\; -\,\frac{\Gamma(1+s)\sin(\tfrac{1}{2}\pi s)}{\pi} M\hbar\gamma_s\omega_{\rm ph}^2 \frac{\chi_0^2}{(\omega_{\rm ph}|t|)^{1+s}} \;. \tag{6.84}$$

Again these expressions at asymptotic time hold for all s.

Now, the position correlation function $C^+(t)$ for zero temperature and asymptotic positive time may be assembled from the expressions (6.84) and (6.81),

$$C^+(t) \;=\; \frac{\Gamma(1+s)}{\pi} M\hbar\gamma_s\omega_{\rm ph}^2 \frac{\chi_0^2\,{\rm e}^{-i\pi(1+s)/2}}{(\omega_{\rm ph}t)^{1+s}} \;. \tag{6.85}$$

Clearly, we would have obtained the same expression if we had computed the imaginary-time correlation function $C^{({\rm E})}(\tau)$, Eq. (6.58) with (6.62) and (3.83) for $T = 0$ and τ very large, and had performed the analytical continuation $\tau \to it$.

Finally we remark that the result (6.85) also covers the case of radiation damping, $s = 3$, in which $\lim_{\omega \to 0} \tilde{\gamma}'(\omega)/\omega^2 = \gamma_3/\omega_{\rm D}^2$.

6.4.3 Shiba relation

The algebraic decay law of the position correlations at $T = 0$ and asymptotic time given in Eqs. (6.77), (6.80) and (6.84) can be written in a catchy way which will turn out to hold universally. To this aim, we measure length in units of the position spread of the undamped oscillator in the ground state, $\sqrt{\langle q^2 \rangle}|_{T=0,\gamma_s=0} \equiv \frac{1}{2}q_0 = \sqrt{\hbar/2M\omega_0}$, and we introduce a dimensionless friction coefficient $\delta_s \equiv M\gamma_s q_0^2/2\pi\hbar = \gamma/\pi\omega_0$ and normalized correlation functions $S_{\rm n}(t) = 4S(t)/q_0^2$ and $\chi_{\rm n}(t) = 4\chi(t)/q_0^2$. The normalized static susceptibility $\chi_{\rm n,0} = 4\chi_0/q_0^2$ has dimension inverse energy. We then have

$$\lim_{t\to\infty} S_{\rm n}(t)\, t^{1+s} = -2\delta_s\Gamma(1+s)\sin(\pi s/2)\omega_{\rm ph}^{1-s}(\hbar\chi_{\rm n,0}/2)^2 \,,$$

$$\lim_{t\to\infty} \chi_{\rm n}(t)\, t^{1+s} = 2\delta_s\Gamma(1+s)\cos(\pi s/2)\omega_{\rm ph}^{1-s}(\hbar\chi_{\rm n,0}/2)^2 \,. \tag{6.86}$$

The corresponding expressions in frequency space are the *generalized Shiba relations*

$$\lim_{\omega\to0^\pm} \hbar\,{\rm sgn}(\omega)\tilde{\chi}_{\rm n}''(\omega)/|\omega|^s = 2\pi\delta_s\omega_{\rm ph}^{1-s}(\hbar\chi_{\rm n,0}/2)^2 \,,$$

$$\lim_{\omega\to0^\pm} \tilde{S}_{\rm n}(\omega)/|\omega|^s = 2\pi\delta_s\omega_{\rm ph}^{1-s}(\hbar\chi_{\rm n,0}/2)^2 \,. \tag{6.87}$$

The standard Shiba relations pertain to the Ohmic case $s = 1$. They have been discovered first for the Anderson model [249].

We anticipate that the Shiba relations (6.87), and the analogues (6.86) in the time regime, are universally valid for any linear or nonlinear open system with spectral density of the form (3.72) which has a finite linear or nonlinear static susceptibility at zero temperature. The distinction of the particular system is in the static susceptibility $\chi_{n,0}$. For the linear oscillator model, e.g., we have $\chi_{n,0} = 2/\hbar\omega_0$.

The Shiba relations in the dissipative two-state system and in the fermionic Toulouse model are discussed in Subsections 22.6.5 and 22.7.2.

6.5 Partition function and implications

6.5.1 Partition function

The partition function $Z(\beta)$ of the damped linear quantum oscillator has been calculated with the imaginary-time path integral method in Section 4.3. There we obtained the partition function of the composite system in the factorized form $Z_{\rm tot} = Z_{\rm R}Z$, where $Z_{\rm R}$ and Z are the partition functions of the reservoir and of the damped oscillator, respectively. The reduced partition function was found to read [we use the Matsubara infinite-product representation (4.225) and drop the label (0) in $Z^{(0)}$]

$$Z(\beta) = \frac{1}{\beta\hbar\omega_0}\prod_{n=1}^{\infty}\frac{\nu_n^2}{\omega_0^2 + \nu_n^2 + \nu_n\hat{\gamma}(\nu_n)} \,. \tag{6.88}$$

This expression holds for arbitrary frequency-dependent damping. In order that the infinite product is convergent, we must have $\lim_{z\to\infty}\hat{\gamma}(z) \to 0$. Hence the strict Ohmic case is excluded. For the Drude-regularized damping function (3.67), the partition function of the damped oscillator takes the form

$$Z(\beta) = \frac{1}{\beta\hbar\omega_0} \prod_{n=1}^{\infty} \frac{\nu_n^2(\nu_n + \omega_D)}{(\nu_n + \lambda_1)(\nu_n + \lambda_2)(\nu_n + \lambda_3)}. \tag{6.89}$$

The λ_j ($j = 1, 2, 3$) are given in Eq. (6.47). Using the infinite product representation of the gamma function [88, 89] the expression (6.89) can be concisely written as

$$Z(\beta) = \frac{\beta\hbar\omega_0}{4\pi^2} \frac{\Gamma(\beta\hbar\lambda_1/2\pi)\Gamma(\beta\hbar\lambda_2/2\pi)\Gamma(\beta\hbar\lambda_3/2\pi)}{\Gamma(\beta\hbar\omega_D/2\pi)}. \tag{6.90}$$

The analytic expression (6.90) holds for arbitrary ω_0, γ, ω_D and T.

When ω_D is very large compared to ω_0 and γ, and to the thermal frequency ν_1, the partition function of the damped quantum oscillator is found from Eq. (6.90) as

$$Z(\beta) = \frac{1}{\beta\hbar\omega_0} \left(\frac{2\pi}{\beta\hbar\omega_D}\right)^{\beta\hbar(\omega_D - \Omega_D)/2\pi} \Gamma\left(1 + \frac{\beta\hbar\lambda_1^{(0)}}{2\pi}\right)\Gamma\left(1 + \frac{\beta\hbar\lambda_2^{(0)}}{2\pi}\right), \tag{6.91}$$

where $\lambda_{1,2}^{(0)} = \frac{1}{2}\gamma \pm i\left(\omega_0^2 - \frac{1}{4}\gamma^2\right)^{1/2}$.

With the expressions for λ_1, λ_2, and λ_3 given in Subsection 6.3.3, the expressions (6.90) and (6.91) also hold for the case of radiation damping.

6.5.2   Internal energy, free energy, and entropy

The analytic expressions (6.90) and (6.91) facilitate direct calculation of thermodynamic quantities, such as the free energy $F = -\ln Z/\beta$ and the internal energy $U = -\partial \ln Z/\partial\beta$, and derivatives of these quantities, for instance the entropy $S = -\partial F/\partial T$ and the specific heat $c_U = \partial U/\partial T = -k_B \beta^2 \partial U/\partial\beta$.

The free energy $f(\omega, \beta)$ and internal energy $u(\omega, \beta)$ of an individual oscillator with eigenfrequency ω at inverse temperature β are

$$f(\omega, \beta) = \ln[2\sinh(\tfrac{1}{2}\beta\hbar\omega)]/\beta, \qquad u(\omega, \beta) = \tfrac{1}{2}\hbar\omega\coth(\tfrac{1}{2}\beta\hbar\omega), \tag{6.92}$$

as follows from the partition function $z(\omega, \beta) = 1/[2\sinh(\tfrac{1}{2}\beta\hbar\omega)]$. In the eigenfrequency representation (4.235) of the system-plus-reservoir complex the total free energy and total internal energy thus take the forms

$$F_{\text{tot}} = \sum_{\alpha=0}^{N} f(\mu_\alpha, \beta), \qquad U_{\text{tot}} = \sum_{\alpha=0}^{N} u(\mu_\alpha, \beta), \tag{6.93}$$

To divide the internal energy into the reservoir's and the reduced system's part, we start out from the representation (4.243) of the product form $Z^{(\text{tot})} = Z_R Z$. Thus we may write $F_{\text{tot}} = F_R + F$ and $U_{\text{tot}} = U_R + U$, where F (U) is the free (internal) energy of the open system, and where F_R (U_R) is the free (internal) energy of the reservoir,

$$F_{\mathrm{R}} = \sum_{\alpha=1}^{N} f(\omega_\alpha, \beta), \qquad U_{\mathrm{R}} = \sum_{\alpha=1}^{N} u(\omega_\alpha, \beta). \qquad (6.94)$$

Consider now the internal and free energy of the open system. We obtain from the expression (6.88) for general frequency-dependent friction $\hat{\gamma}(z)$

$$U = \frac{1}{\beta} + \frac{1}{\beta} \sum_{n=1}^{\infty} \nu_n \frac{\partial}{\partial \nu_n} \left[\ln(\nu_n^2 \hat{\chi}(\nu_n)) \right] = \frac{1}{\beta} \sum_{n=-\infty}^{\infty} \left\{ 1 + \frac{\nu_n}{2} \frac{\partial}{\partial \nu_n} \ln[\hat{\chi}(|\nu_n|)] \right\}, \quad (6.95)$$

By analogy with Eq. (6.67) we may write the Matsubara sum as frequency integral,[3]

$$U = \int_0^\infty d\omega\, \xi(\omega) u(\omega, \beta). \qquad (6.96)$$

with the spectral density of the fictional oscillators of eigenfrequency ω

$$\xi(\omega) = \frac{1}{\pi} \mathrm{Im}\, \frac{\partial \ln[\tilde{\chi}(\omega)]}{\partial \omega}. \qquad (6.97)$$

Similarly, we obtain for the free energy upon integration by parts the representation

$$F = \int_0^\infty d\omega\, \xi(\omega) f(\omega, \beta). \qquad (6.98)$$

These remarkable expressions for the internal and free energy of the open system are exact. All the environmental influences are in the dynamical susceptibility.

Both for Drude regularized Ohmic friction and for radiation damping, Eq. (6.43), the internal energy U is found to be a linear combination of four digamma functions,

$$U = \frac{1}{\beta} + \frac{\hbar}{2\pi} \left\{ \omega_{\mathrm{D}}\, \psi(1 + \beta\hbar\omega_{\mathrm{D}}/2\pi) - \sum_{i=1}^{3} \lambda_i\, \psi(1 + \beta\hbar\lambda_i/2\pi) \right\}. \qquad (6.99)$$

In the remainder of this subsection we restrict ourselves to the Ohmic case. When the thermal energy $k_{\mathrm{B}}T$ is large compared to the energy scales of the damped oscillator, we may expand the curly in the expression (6.99) in a power series in β. The resulting high temperature series for the internal energy is

$$U = k_{\mathrm{B}}T \left\{ 1 + \frac{1}{12} \frac{\gamma\omega_{\mathrm{D}} + \omega_0^2}{\omega_0^2} \frac{1}{\vartheta^2} + \mathcal{O}\left(\frac{\gamma\omega_{\mathrm{D}}^2}{\omega_0^3} \frac{1}{\vartheta^3} \right) \right\}. \qquad (6.100)$$

Here we have used the scaled temperature

$$\vartheta = \frac{k_{\mathrm{B}}T}{\hbar\omega_0}. \qquad (6.101)$$

Evidently, the internal energy of the damped oscillator approaches in the classical limit the equipartition value $k_{\mathrm{B}}T$ as required.

For large cutoff frequency, $\omega_{\mathrm{D}} \gg \omega_0, \gamma, \nu_1$, we obtain from (6.91) or from (6.99)

---

[3]This exact result was derived by Ford et al. by two alternative methods [97, 250].

$$U = \frac{\hbar\gamma}{2\pi}\left\{1 + \ln\left(\frac{\beta\hbar\omega_D}{2\pi}\right)\right\}\left[1 + \mathcal{O}\left(\frac{\gamma}{\omega_D}\right)\right]$$

$$+ \frac{1}{\beta} - \frac{\hbar}{2\pi}\sum_{i=1}^{2}\lambda_i\,\psi\left(1 + \frac{\beta\hbar\lambda_i}{2\pi}\right). \tag{6.102}$$

Consider next the asymptotic low-temperature regime. As $T \to 0$, the sum in Eq. (6.95) turns into an integral. The ground-state energy $E_0 = \lim_{\beta\to\infty} U(\beta)$ is

$$E_0 \equiv \hbar\varepsilon_0 = \frac{\hbar}{2\pi}\int_0^\infty d\nu\,\ln\left(\frac{\omega_0^2 + \nu^2 + \nu\hat{\gamma}(\nu))}{\nu^2}\right). \tag{6.103}$$

This form holds for arbitrary frequency-dependent damping. For Drude regularized Ohmic friction and for radiation damping, we find either from Eq. (6.103) or from the asymptotic limit of the expression (6.99)

$$\hbar\varepsilon_0 = \frac{\hbar}{2\pi}\left[\lambda_1\ln(\omega_D/\lambda_1) + \lambda_2\ln(\omega_D/\lambda_2) + \lambda_3\ln(\omega_D/\lambda_3)\right]. \tag{6.104}$$

In the regime $\omega_D \gg \omega_0, \gamma$, the ground state energy takes the form

$$\hbar\varepsilon_0 = \frac{\hbar\omega_0}{2}\left\{(1 - \alpha^2)\,g(\alpha) + \frac{2\alpha}{\pi}\left[1 + \ln(\omega_D/\omega_0)\right] + \mathcal{O}\left(\frac{\gamma}{\omega_D}\right)\right\}, \tag{6.105}$$

where

$$g(\alpha) = \frac{2}{\pi}\frac{\omega_0}{\lambda_1^{(0)} - \lambda_2^{(0)}}\left[\ln(\lambda_1^{(0)}/\omega_0) - \ln(\lambda_2^{(0)}/\omega_0)\right]. \tag{6.106}$$

With the explicit forms (6.39) for $\lambda_{1,2}^{(0)}$, and with $\alpha = \gamma/2\omega_0$, we obtain

$$g(\alpha) = \begin{cases} \dfrac{1}{\sqrt{1 - \alpha^2}}\left(1 - \dfrac{2}{\pi}\arcsin\alpha\right), & \text{for} \quad \alpha < 1, \\[2ex] \dfrac{2}{\pi}\dfrac{\ln\left(\alpha + \sqrt{\alpha^2 - 1}\,\right)}{\sqrt{\alpha^2 - 1}}, & \text{for} \quad \alpha > 1. \end{cases} \tag{6.107}$$

We see from Eqs. (6.105) - (6.107) that the vacuum energy increases monotonously with α and with ω_D. For strong damping $\alpha \gg 1$, we have $\varepsilon_0 = (\gamma/2\pi)\ln(\omega_D/\gamma)$.

With use of the asymptotic series of the digamma function [89], the low temperature expansion of the internal energy is found to read

$$U = \hbar\varepsilon_0 + a_1\hbar\omega_0\,\vartheta^2 + a_2\hbar\omega_0\,\vartheta^4 + \mathcal{O}(\vartheta^6), \tag{6.108}$$

where

$$a_1 = \frac{\pi}{6}\frac{\gamma}{\omega_0}, \qquad a_2 = \frac{\pi^3}{15}\frac{\gamma}{\omega_0}\left(3 - \frac{\gamma^2}{\omega_0^2}\right) + \mathcal{O}\left(\frac{\gamma^2}{\omega_0\omega_D}\right). \tag{6.109}$$

The corresponding expansion for the free energy $F = -k_B T \ln Z$ is

$$F = \hbar\varepsilon_0 - a_1\hbar\omega_0\,\vartheta^2 - \tfrac{1}{3}a_2\hbar\omega_0\,\vartheta^4 + \mathcal{O}(\vartheta^6). \tag{6.110}$$

The leading thermal variation of U and $F \propto T^2$ at low T is a characteristic feature of Ohmic dissipation. This contribution can be rewritten in terms of the static susceptibility $\chi_0 = 1/M\omega_0^2$ as

$$\Delta U = -\Delta F = (\pi/6\hbar)\gamma M\chi_0(k_{\mathrm{B}}T)^2 \ . \qquad (6.111)$$

This form of the T^2-term universally holds for any linear and nonlinear Ohmic system which has a non-vanishing static susceptibility at $T = 0$ (see Section 22.9).

The series (6.110) yields for the entropy $S = -\partial F/\partial \mathrm{T}$ the low temperature series

$$S/k_{\mathrm{B}} = 2a_1\vartheta + \tfrac{4}{3}a_2\vartheta^3 + \mathcal{O}(\vartheta^5) \ . \qquad (6.112)$$

Hence the entropy vanishes at zero temperature in accordance with the Third Law of thermodynamics.

6.5.3 Specific heat and Wilson ratio

With the various results found for the internal energy U for Drude memory friction it is only a small move towards the specific heat $c_U \equiv \partial U/\partial T$. First, we find from the expression (6.96) with (6.97) the remarkable exact integral representation

$$c_U/k_{\mathrm{B}} = \int_0^\infty d\omega\, \xi(\omega) \left(\frac{\beta\hbar\omega/2}{\sinh(\beta\hbar\omega/2)} \right)^2 \ . \qquad (6.113)$$

At very high T, $\nu_1 \gg \omega_0, \gamma, \omega_{\mathrm{D}}$, we obtain either from Eq. (6.113) or from Eq. (6.100)

$$c_U/k_{\mathrm{B}} = 1 - \frac{1}{12}\frac{\gamma\omega_{\mathrm{D}} + \omega_0^2}{\omega_0^2}\frac{1}{\vartheta^2} + \mathcal{O}\left(\frac{\gamma\omega_{\mathrm{D}}^2}{\omega_0^3}\frac{1}{\vartheta^3}\right) \ . \qquad (6.114)$$

When the Drude cutoff ω_{D} is very large compared to ω_0, γ, and ν_1, we find from the expression (6.102)

$$c_U/k_{\mathrm{B}} = 1 - \frac{\beta\hbar\gamma}{2\pi} + \sum_{i=1}^2 \left(\frac{\beta\hbar\lambda_i^{(0)}}{2\pi}\right)^2 \psi'\left(1 + \frac{\beta\hbar\lambda_i^{(0)}}{2\pi}\right) \ . \qquad (6.115)$$

This form yields in the regime $\omega_{\mathrm{D}} \gg \nu_1 \gg \omega_0, \gamma$ the series

$$c_U/k_{\mathrm{B}} = 1 - \frac{\gamma}{2\pi\omega_0}\frac{1}{\vartheta} - \frac{1}{24}\frac{2\omega_0^2 - \gamma^2}{\omega_0^2}\frac{1}{\vartheta^2} + \mathcal{O}\left(\frac{\gamma}{\omega_0}\frac{1}{\vartheta^3}\right) \ . \qquad (6.116)$$

The main difference in the high-temperature expansions (6.114) and (6.116) is the appearance of the $1/\vartheta$-term as the leading quantum contribution in the latter series.

The low-temperature series for the specific heat is found from Eq. (6.108) as

$$c_U/k_{\mathrm{B}} = 2a_1\vartheta + 4a_2\vartheta^3 + \mathcal{O}(\vartheta^5) \ . \qquad (6.117)$$

Finally, we consider the case of non-Ohmic spectral density $J_{\mathrm{lf}}(\omega) = M\gamma_s\omega_{\mathrm{ph}}^{1-s}\omega^s$. By inserting the low-frequency series of the spectral density found from Eq. (6.97) with Eqs. (6.32) and (3.26),

$$\xi(\omega) \;=\; \frac{Ms\gamma_s\chi_0}{\pi}(\omega/\omega_{\rm ph})^{s-1}\left[\,1+\mathcal{O}(\omega^s,\,\omega^2)\,\right]\,, \tag{6.118}$$

into the integral (6.113), we obtain for the specific heat the low temperature series

$$c_{\rm U}/k_{\rm B} \;=\; \frac{s\Gamma(2+s)\zeta(1+s)}{\pi}\,M\gamma_s\omega_{\rm ph}\chi_0\left(\frac{k_{\rm B}T}{\hbar\omega_{\rm ph}}\right)^s\left[\,1+\mathcal{O}(T^s,\,T^2)\,\right]\,. \tag{6.119}$$

Hence the asymptotic low temperature dependence is T^s for $c_{\rm U}$, and T^{1+s} both for the internal energy and the free energy.

The leading temperature dependence of the specific heat may be written in a form which is universally valid. To this end we introduce, as in Subsection 6.4.3, a scaled static susceptibility $\chi_{\rm n,0}$ according to $\chi_0 = \tfrac{1}{4}q_0^2\chi_{\rm n,0}$, and a dimensionless friction coefficient $\delta_s = M\gamma_s q_0^2/2\pi\hbar$. Further, we define the so-called generalized Wilson ratio

$$R_s \;\equiv\; \lim_{T\to0}\frac{4c(T)/k_{\rm B}}{\chi_{\rm n,0}(\hbar\omega_{\rm ph})^{1-s}(k_{\rm B}T)^s}\,. \tag{6.120}$$

Thus the specific heat at low T, Eq. (6.119), is characterized by the Wilson ratio

$$R_s \;=\; 2s\Gamma(2+s)\zeta(1+s)\delta_s\,. \tag{6.121}$$

This reduces in the Ohmic case to

$$R_1 \;=\; 2\delta_1\pi^2/3\,. \tag{6.122}$$

Interestingly, the Wilson ratio (6.121) holds for any dissipative system which has finite linear or nonlinear static susceptibility at zero temperature [cf. Section 22.9].

The power laws in T of the internal and free energy, and of the specific heat indicate that there is now energy gap above the ground state. This conjecture is confirmed in the following subsection by a direct analysis of the energy spectrum.

6.5.4 Spectral density of states

Generally, the partition function provides access to the spectrum of the quantum mechanical system. In analogy to the case of a closed system, the reduced partition function $Z(\beta)$ may be regarded as the Laplace transform of a sort of density of states $\rho(\varepsilon)$ of the open system

$$Z(\beta) \;=\; \int_0^\infty d\varepsilon\,\rho(\varepsilon)\,{\rm e}^{-\beta\hbar\varepsilon}\,. \tag{6.123}$$

The inversion is accomplished by the inversion integral

$$\rho(\varepsilon) \;=\; \frac{\hbar}{2\pi i}\int_{c-i\infty}^{c+i\infty}d\beta\,Z(\beta)\,{\rm e}^{\beta\hbar\varepsilon}\,. \tag{6.124}$$

Here c is a real constant that exceeds the real part of all singularities of $Z(\beta)$. To be consistent with the conception of a density, $\rho(\varepsilon)$ must be a positive function of ε.

Equipped with the integral expression (6.124) with (6.90), we can now study how the sharp lines of the discrete spectrum of the undamped oscillator,

$$\rho_0(\varepsilon) = \sum_{n=0}^{\infty} \delta\left(\varepsilon - \left(n + \tfrac{1}{2}\right)\omega_0\right), \qquad (6.125)$$

spread out and melt into another when damping is turned on and raised.

Direct numerical computation of the integral in Eq. (6.124) is difficult to perform since the integrand is rapidly oscillating along the integration contour. Fortunately, some valuable insights can be obtained by analytic methods.

First, it is immediately obvious from the low temperature expansion of the free energy, Eq. (6.110), that the ground state remains as a separate δ-function at the frequency ε_0,

$$\rho(\varepsilon) = \delta(\varepsilon - \varepsilon_0) + \Theta(\varepsilon - \varepsilon_0)\,\Phi(\varepsilon). \qquad (6.126)$$

The δ-peak is shifted with increasing damping strength towards higher frequency as specified by the expression (6.104). The function $\Phi(\varepsilon)$ represents the density of states above the ground state. It is zero for $\varepsilon < \varepsilon_0$. The series expansion of $\Phi(\varepsilon)$ about ε_0 for $\varepsilon > \varepsilon_0$ is found from the asymptotic expansion of F, Eq. (6.110), as

$$\Phi(\varepsilon) = \frac{1}{\omega_0}\left\{a_1 + \frac{a_1^2}{2}\frac{\varepsilon - \varepsilon_0}{\omega_0} + \left(\frac{a_1^3}{12} + \frac{a_2}{6}\right)\left(\frac{\varepsilon - \varepsilon_0}{\omega_0}\right)^2 + \mathcal{O}\left[\left(\frac{\varepsilon - \varepsilon_0}{\omega_0}\right)^3\right]\right\}. \qquad (6.127)$$

From this we draw the following conclusions. First, we see that the gap above the ground state in the energy spectrum is washed out by the dissipative coupling. The function $\Phi(\varepsilon)$ makes a jump at $\varepsilon = \varepsilon_0$ of height $\pi\alpha/3\omega_0$. The leading thermal enhancement $\propto T^2$ of the internal energy is a direct implication of the nonzero density of states above ϵ_0. Secondly, the derivative of $\Phi(\varepsilon)$ at $\varepsilon = \varepsilon_0$ is positive for all α, whereas the curvature of $\Phi(\varepsilon)$ at $\varepsilon = \varepsilon_0$ changes sign at $\alpha = 0.88$ (see Fig. 6.1).

The density $\Phi(\varepsilon)$ well above ε_0 is given by the sum of contributions collected from the infinite series of simple poles in the expression (6.90) [251],

$$\Phi(\varepsilon) = \frac{1}{\omega_0} + \sum_{k=1}^{3}\sum_{n=1}^{\infty} R_{n,k}\, e^{-2\pi n\varepsilon/\lambda_k}. \qquad (6.128)$$

The residues are

$$R_{n,1} = \frac{(-1)^{n-1}}{(n-1)!}\frac{\Gamma(-n\lambda_2/\lambda_1)\Gamma(-n\lambda_3/\lambda_1)}{\Gamma(-n\omega_{\mathrm{D}}/\lambda_1)}\frac{\omega_0}{\lambda_1^2}, \qquad (6.129)$$

and the expressions for $R_{n,2}$ and $R_{n,3}$ result by cyclic permutation of the indices.

The analysis yields that the series (6.127) and (6.128) have an overlapping region of convergence. Further, for $\alpha > 0$, the contribution $\Phi(\varepsilon)$ is an absolutely continuous function of ε in the interval $\varepsilon_0 < \varepsilon < \infty$. Therefore, apart from the ground state, there are no other discrete states embedded in the continuum. In the underdamped regime $\Gamma < 2\Omega_0$ in Eq. (6.47), in which two of the λ_k are complex conjugate to each other, the density of states shows roughly equidistant resonances with exponentially reduced amplitudes towards higher frequencies. The transition from the bumpy to the smooth behavior occurs roughly at the critical damping $\Gamma = 2\Omega_0$. For $\Gamma > 2\Omega_0$, the spectral density increases initially from the threshold value $\Phi(\epsilon_0) = \pi\alpha/3\omega_0$ to a

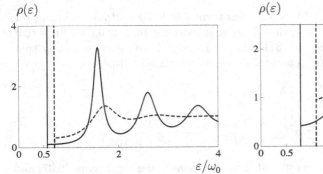
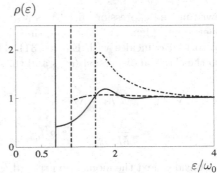

Figure 6.1: The density of states $\rho(\varepsilon)$ is plotted versus ε/ω_0 for $\omega_D = 10\,\omega_0$ for weak and strong damping. The left plot shows the cases $\alpha = 0.1$ (full curve) and $\alpha = 0.3$ (dashed curve), and the right plot the cases $\alpha = 0.4$ (full curve), $\alpha = 0.9$ (dashed curve) and $\alpha = 1.8$ (dashed-dotted curve). Both the ground-state energy and the height of the jump increase monotonously with α. The respective ground-state values are $\varepsilon/\omega_0 = 0.576, 0.721, 0.790, 1.108$, and 1.607.

maximum and then decreases monotonously to the classical value $\Phi_{\rm cl} = 1/\omega_0$ reached at high frequency $\epsilon \gg \omega_0$. The characteristic features of the density of states in the underdamped and overdamped regime are shown in Fig. 6.1.

Experimentally, the density of states can be found by measuring the absorbed microwave power of the externally driven oscillator.

Finally, consider the spectral density of states for non-Ohmic spectral coupling $J(\omega) \propto \omega^s$. In this general case the ground state contributes again a δ-peak $\delta(\epsilon - \epsilon_0)$. However, now the spectral density of states slightly above the ground state energy $\hbar\epsilon_0$ has power-law form $\Phi(\varepsilon) \propto (\varepsilon - \varepsilon_0)^{s-1}$. Hence the density of states above the ground state grows smoothly with increasing frequency from zero in the super-Ohmic case, whereas it is singular at the threshold for sub-Ohmic environmental coupling.

6.6 Mean square of position and momentum

6.6.1 General expressions for colored noise

The dispersion of the position in the equilibrium state is given by

$$\langle q^2 \rangle = S(0) = \frac{\hbar}{\pi} \int_0^\infty d\omega\, \chi''(\omega) \coth(\beta\hbar\omega/2) , \qquad (6.130)$$

as follows from Eq. (6.63). A representation of $\langle q^2 \rangle$ in terms of a Matsubara sum is found from the imaginary-time correlation function (6.58) with Eq. (6.62),

$$\langle q^2 \rangle = C^{\rm E}(0) = \frac{1}{M\beta} \sum_{n=-\infty}^{\infty} \frac{1}{\omega_0^2 + \nu_n^2 + |\nu_n|\hat{\gamma}(|\nu_n|)} . \qquad (6.131)$$

Rewriting the expression (6.130) as an integral with a closed contour encircling the upper ω half-plane and observing that $\chi(\omega)$ is analytic for $\text{Im}\,\omega > 0$, one finds that Eq. (6.130) coincides with Eq. (6.131). The expression (6.131) is conveniently split into the classical dispersion $\langle q^2 \rangle_{\text{cl}}$ and the quantum mechanical dispersion $\langle q^2 \rangle_{\text{qm}}$. We have

$$\langle q^2 \rangle = \langle q^2 \rangle_{\text{cl}} + \langle q^2 \rangle_{\text{qm}} \,,$$

$$\langle q^2 \rangle_{\text{cl}} = \frac{1}{M\beta\omega_0^2} \,, \qquad \langle q^2 \rangle_{\text{qm}} = \frac{2}{M\beta} \sum_{n=1}^{\infty} \frac{1}{\omega_0^2 + \nu_n^2 + \nu_n \hat{\gamma}(\nu_n)} \,. \qquad (6.132)$$

Consider next the momentum spread $\langle p^2 \rangle$. By means of the expressions (6.37) and (6.55) the mean value of p^2 is found as

$$\langle p^2 \rangle = \frac{\hbar M^2}{\pi} \int_0^{\infty} d\omega \, \omega^2 \tilde{\chi}''(\omega) \coth(\beta\hbar\omega/2) \,. \qquad (6.133)$$

Alternatively, we may start from the Fourier series (6.58) with (6.62) of $\langle q(\tau)q(0) \rangle_\beta$, which carries cusps at $\tau = m\hbar\beta \quad (m = 0, \pm 1, \cdots)$. Hence the periodically continued momentum correlation function has δ-function singularities at $\tau = m\hbar\beta$. We find

$$\langle p(\tau)p(0) \rangle = -\hbar M : \delta(\tau) : + \langle p(\tau)p(0) \rangle_{\text{reg}} \,,$$

$$\langle p(\tau)p(0) \rangle_{\text{reg}} = \frac{M}{\beta} \sum_{n=-\infty}^{\infty} \frac{\omega_0^2 + |\nu_n| \hat{\gamma}(|\nu_n|)}{\omega_0^2 + \nu_n^2 + |\nu_n| \hat{\gamma}(|\nu_n|)} \, e^{i\nu_n\tau} \,. \qquad (6.134)$$

Here $:\delta(\tau):$ is the periodically continued δ-function (3.58). Now, only the regular part of $\langle p(\tau)p(0) \rangle$ is an analytic continuation of the real-time momentum autocorrelation function in the strip $0 < \text{Re}\,\tau < \hbar\beta$. Thus, the Matsubara representation of the momentum dispersion is found to read

$$\langle p^2 \rangle = \lim_{\tau \to 0^+} \langle p(\tau)p(0) \rangle_{\text{reg}} = \frac{M}{\beta} \sum_{n=-\infty}^{\infty} \frac{\omega_0^2 + |\nu_n| \hat{\gamma}(|\nu_n|)}{\omega_0^2 + \nu_n^2 + |\nu_n| \hat{\gamma}(|\nu_n|)} \,. \qquad (6.135)$$

This expression can also be derived from Eq. (6.133) by contour integration in the upper ω half-plane bearing in mind that $\tilde{\chi}(\omega)$ is analytic in this region.

The dispersions of position and momentum can also be related to the partition function $Z(\beta)$ of the damped linear oscillator. The product representation (6.88) reveals that the dispersions (6.131) and (6.135) may be expressed in the form

$$\langle q^2 \rangle = -\frac{1}{M\beta\omega_0} \frac{d}{d\omega_0} \ln Z(\beta) \,, \qquad (6.136)$$

$$\langle p^2 \rangle = -\frac{M}{\beta} \left(\omega_0 \frac{d}{d\omega_0} + 2\gamma \frac{d}{d\gamma} \right) \ln Z(\beta) \,. \qquad (6.137)$$

In the second relation we have employed the representation $\hat{\gamma}(z) = \gamma g(z)$. The expressions of this subsection are generally valid for any form of linear memory-friction.

6.6.2 Ohmic friction

For strict Ohmic friction $\hat{\gamma}(|\nu_n|) = \gamma$, the expression (6.131) may be written as

$$\langle q^2 \rangle = \frac{1}{M\beta\omega_0^2} + \frac{2}{M\beta(\lambda_2^{(0)} - \lambda_1^{(0)})} \sum_{n=1}^{\infty} \left(\frac{1}{\nu_n + \lambda_1^{(0)}} - \frac{1}{\nu_n + \lambda_2^{(0)}} \right) , \qquad (6.138)$$

where $\lambda_1^{(0)}$ and $\lambda_2^{(0)}$ are the characteristic frequencies defined in Eq. (6.39). The sum in Eq. (6.138) is a linear combination of digamma functions,

$$\langle q^2 \rangle = \frac{1}{M\beta\omega_0^2} + \frac{\hbar}{M\pi(\lambda_2^{(0)} - \lambda_1^{(0)})} \left(\psi(1 + \lambda_2^{(0)}/\nu) - \psi(1 + \lambda_1^{(0)}/\nu) \right) , \qquad (6.139)$$

where again $\nu \equiv \nu_1 = 2\pi/\hbar\beta$. In the absence of damping, this reduces to the familiar expression $\langle q^2 \rangle = (\hbar/2m\omega_0)\coth(\beta\hbar\omega_0/2)$.

In the high-temperature regime $k_{\mathrm{B}}T \gg \hbar|\lambda_{1,2}^{(0)}|/2\pi$, the quantum mechanical contribution $\langle q^2 \rangle_{\mathrm{qm}}$ is found from the expression (6.139) as

$$\langle q^2 \rangle_{\mathrm{qm}} = \frac{1}{12} \frac{\hbar^2}{Mk_{\mathrm{B}}T} \left[1 + \mathcal{O}\left(\frac{\hbar\gamma}{k_{\mathrm{B}}T} \right) \right] . \qquad (6.140)$$

This shows again that damping becomes irrelevant at high temperature.

On the other hand, in the low temperature limit the asymptotic expansion of the digamma functions applies. Then the term $\langle q^2 \rangle_{\mathrm{cl}}$ is cancelled by a counter term stemming from $\langle q^2 \rangle_{\mathrm{qm}}$. The resulting series is

$$\langle q^2 \rangle = \frac{\hbar}{2M\omega_0} \left\{ g(\alpha) + \frac{2\pi}{3} \frac{\gamma}{\omega_0} \left(\frac{k_{\mathrm{B}}T}{\hbar\omega_0} \right)^2 + \mathcal{O}\left[\left(\frac{k_{\mathrm{B}}T}{\hbar\omega_0} \right)^4 \right] \right\} . \qquad (6.141)$$

The function $g(\alpha)$ is given in Eq. (6.107).

The expression (6.141) reproduces the correct form $\langle q^2 \rangle_{T=0} = \hbar/2M\omega_0$ for the dispersion of an undamped oscillator in the ground state.

At finite T, the dispersion of the position is enhanced compared with the zero temperature value. The thermal contributions grow algebraically with T for a damped system. The leading T^2 law at low T is again a signature of Ohmic friction. For a spectral density $J(\omega \to 0) \propto \omega^s$, the leading thermal enhancement grows as T^{1+s}.

For large friction $\gamma \gg \omega_0$ and temperature in the range

$$k_{\mathrm{B}}T \gg \hbar\omega_0^2/2\pi\gamma , \qquad (6.142)$$

the leading quantum mechanical contribution to the coordinate dispersion is

$$\langle q^2 \rangle_{\mathrm{qm}} = \frac{\hbar}{\pi M\gamma} \left(\psi(1 + \beta\hbar\gamma/2\pi) - \psi(1) \right) . \qquad (6.143)$$

Interestingly enough, the quantum mechanical part of the coordinate dispersion does not depend in leading order on properties of the potential for strong friction and temperature in the regime (6.142). At temperature $T \gg \hbar\gamma/2\pi k_{\mathrm{B}}$, the expression (6.143) reduces to the form (6.140).

In the temperature regime $\hbar\omega_0^2/2\pi\gamma \ll k_{\mathrm{B}}T \ll \hbar\gamma/2\pi$, the leading quantum contribution to the coordinate dispersion is found from the expression (6.143) as

$$\langle q^2 \rangle_{\mathrm{qm}} = \frac{\hbar}{\pi M\gamma} \ln\left(\frac{\beta\hbar\gamma}{2\pi}\right). \tag{6.144}$$

In classical physics, the evolution of a Brownian particle in phase space effectively reduces for very large friction to a time evolution in position space. The corresponding dynamics is well described by the Smoluchowski diffusion equation, which is discussed below in Section 11.3. In the temperature range (6.142), quantum effects are important. However it is still possible to describe the time-evolution of the position distribution in terms of a diffusion equation. The corresponding quantum Smoluchowski equation (QSE) is discussed in Subsection 15.3. We shall see that the size of quantum effects in the QSE is controlled by the strength of $\langle q^2 \rangle_{\mathrm{qm}}$.

6.6.3   Ohmic friction with Drude cutoff

For strict Ohmic damping, the integral in Eq. (6.133) as well as the sum in Eq. (6.135) have a logarithmic ultra-violet divergence. The divergence of the momentum dispersion shows that care has to be exercised with the Markov assumption. Here the assumption of a memoryless reservoir is unphysical. There is always a "microscopic" time scale below which the inertia of the environment becomes relevant. In its simplest form, this is represented by a high-frequency cutoff in $\tilde{\gamma}(\omega)$. By this, the thermal average of $\langle p^2 \rangle$ is regularized. For the Drude model (3.67) we find

$$
\begin{aligned}
\langle q^2 \rangle &= \frac{1}{M\beta} \sum_{n=-\infty}^{+\infty} \frac{1}{\omega_0^2 + \nu_n^2 + \gamma\omega_{\mathrm{D}}|\nu_n|/(\omega_{\mathrm{D}} + |\nu_n|)}, \\
\langle p^2 \rangle &= \frac{M}{\beta} \sum_{n=-\infty}^{+\infty} \frac{\omega_0^2 + \gamma\omega_{\mathrm{D}}|\nu_n|/(\omega_{\mathrm{D}} + |\nu_n|)}{\omega_0^2 + \nu_n^2 + \gamma\omega_{\mathrm{D}}|\nu_n|/(\omega_{\mathrm{D}} + |\nu_n|)}.
\end{aligned}
\tag{6.145}
$$

The partial fraction decomposition yields

$$\langle q^2 \rangle = \frac{1}{\beta M\omega_0^2} + \frac{\hbar}{M\pi} \sum_{j=1}^{3} a_j \psi(1 + \lambda_j/\nu), \tag{6.146}$$

$$\langle p^2 \rangle = M^2\omega_0^2\langle q^2 \rangle + \Pi^2, \tag{6.147}$$

$$\Pi^2 = \frac{M\gamma\hbar\omega_{\mathrm{D}}}{\pi} \sum_{j=1}^{3} b_j \psi(1 + \lambda_j/\nu), \tag{6.148}$$

where

$$a_1 = \frac{\lambda_1 - \omega_{\mathrm{D}}}{(\lambda_1 - \lambda_2)(\lambda_1 - \lambda_3)}, \qquad b_1 = \frac{\lambda_1}{(\lambda_1 - \lambda_2)(\lambda_1 - \lambda_3)}, \tag{6.149}$$

and where a_2, a_3 and b_2, b_3 are defined by cyclic permutation of the indices.

We see with use of Eq. (6.47) with Eq. (6.46) that the expression (6.146) for $\langle q^2 \rangle$ differs from the previous result (6.139) by terms of order ω_0/ω_D and γ/ω_D.

For extremely high temperature $T \gg \hbar\omega_D/k_B$, we may expand the digamma functions in Eq. (6.148) about unity. We then obtain

$$\Pi^2 = \frac{M\hbar\gamma}{12}\frac{\hbar\omega_D}{k_B T}\left\{1 + \mathcal{O}\left(\frac{\hbar\omega_D}{k_B T}\right)\right\}. \tag{6.150}$$

From this we see that the term Π^2 is negligibly small at high temperature. Hence the expression (6.147) has the proper classical limit, $\langle p^2 \rangle \to M^2\omega_0^2\langle q^2 \rangle \to Mk_B T$.

In the temperature range $\hbar\omega_D/k_B \gg T \gg \hbar\lambda_{1,2}/k_B$, the leading contribution to Π^2 comes from the $j = 3$ term in Eq. (6.148). This yields

$$\Pi^2 = (M\hbar\gamma/\pi)\ln(\beta\hbar\omega_D). \tag{6.151}$$

For very low temperature $T \ll \hbar\lambda_{1,2}/k_B$, we may use for the three digamma functions in Eq. (6.148) the respective asymptotic expansions. We then find

$$\Pi^2 = \frac{M\hbar\gamma}{\pi}\ln\left(\frac{\omega_D}{\omega_0}\right) - M\hbar\omega_0\,\alpha^2 g(\alpha) - \frac{\pi M\hbar\gamma}{3}\left(\frac{k_B T}{\hbar\omega_0}\right)^2, \tag{6.152}$$

where terms of order ω_0/ω_D, γ/ω_D, and $(k_B T/\hbar\omega_0)^4$ have been disregarded. Now, Eq. (6.152) combines with Eq. (6.141) to give for $\langle p^2 \rangle$ at zero temperature

$$\langle p^2 \rangle_{T=0} = M^2\omega_0^2(1 - 2\alpha^2)\langle q^2 \rangle_{T=0} + (2\alpha M\hbar\omega_0/\pi)\ln(\omega_D/\omega_0). \tag{6.153}$$

Since, in a sense, the reservoir is continuously measuring the position of the particle, the position spread becomes smaller whereas momentum spread and mean kinetic energy become larger as damping is increased.[4] For large damping $\gamma \gg \omega_0$ we have

$$\langle q^2 \rangle_{T=0} = \frac{2\hbar}{M\pi\gamma}\ln(\gamma/\omega_0), \qquad \langle p^2 \rangle_{T=0} = \frac{M\hbar\gamma}{\pi}\ln(\omega_D/\gamma). \tag{6.154}$$

The expression (6.153) reduces at $\alpha = 0$ to the momentum dispersion in the ground state of the undamped system, $\langle p^2 \rangle_{T=0} = M\hbar\omega_0/2$. In Fig. 6.3 (left diagram) the normalized dispersions $2M\omega_0\langle q^2 \rangle/\hbar$ (dashed-dotted line) and $2\langle p^2 \rangle/M\hbar\omega_0$ (dashed line), and the normalized uncertainty $2\sqrt{\langle q^2 \rangle\langle p^2 \rangle}/\hbar$ (full line) in the ground state are plotted as functions of the damping parameter α. The uncertainty reaches a plateau fairly above $\alpha = 2$. There, it has only weak logarithmic dependence on α.

The low temperature series for $\langle q^2 \rangle$ and $\langle p^2 \rangle$ may be calculated from the integral expressions (6.130) and (6.133) by writing $\coth(\beta\hbar\omega/2) = 1+2/(e^{\beta\hbar\omega}-1)$, and taking the low-frequency series for $\tilde{\chi}''(\omega)$. Since $\tilde{\chi}''(\omega)$ is odd in ω, the low temperature series is a power series in T^2. The leading term of $\langle q^2 \rangle$ is T^2, while that of $\langle p^2 \rangle$ is T^4.

---

[4]If the reservoir would couple to the momentum of the oscillator, the features of coordinate and momentum dispersion would perform role reversal.

The dispersions $\langle q^2 \rangle$ and $\langle p^2 \rangle$ of the harmonic oscillator with Drude-regularized Ohmic damping are depicted as functions of temperature in Fig. 6.2. The findings are as one would expect intuitively. For fixed temperature, the dispersion of the position decreases as the damping is increased while the dispersion of the momentum gets larger. At high temperature, the variances are independent of the damping strength, and they vary linearly with T. This is in agreement with the equipartition law.

A final consideration of this section concerns the mean energy of the damped oscillator. With use of the expressions (6.131) and (6.135) we have

$$\langle E \rangle \equiv \frac{\langle p^2 \rangle}{2M} + \frac{1}{2} M\omega_0^2 \langle q^2 \rangle = \frac{1}{\beta} + \frac{1}{\beta} \sum_{n=1}^{\infty} \frac{2\omega_0^2 + \nu_n \hat{\gamma}(\nu_n)}{\nu_n^2 + \nu_n \hat{\gamma}(\nu_n) + \omega_0^2} . \tag{6.155}$$

At first glimpse one might have guessed that the mean energy $\langle E \rangle$ coincides with the internal energy $U = -\partial \ln Z / \partial \beta$. However they differ in reality. With use of the expressions (6.95), (6.131) and (6.135) we obtain

$$
\begin{aligned}
U - \langle E \rangle &= \frac{1}{\beta} \sum_{n=1}^{\infty} \frac{\nu_n^2}{\omega_0^2 + \nu_n^2 + \nu_n \hat{\gamma}(\nu_n)} \left(-\frac{\partial \hat{\gamma}(\nu_n)}{\partial \nu_n} \right) \\
&= \int_0^{\infty} d\omega \, u(\omega, \beta) \, \mathrm{Re} \left\{ \frac{M\omega \tilde{\chi}(\omega)}{\pi} \frac{\partial \tilde{\gamma}(\omega)}{\partial \omega} \right\} .
\end{aligned}
\tag{6.156}
$$

As a result, the specific heats $c_U = \partial U / \partial T$ and $c_E = \partial \langle E \rangle / \partial T$ differ, barring the Ohmic case $\hat{\gamma}(z) = \gamma$. Since $Z = Z_{\text{tot}} / Z_R$ [cf. Eq. (4.244)], we see that the internal energy U and the specific heat c_U include the system-reservoir interaction as a whole. The discrepancy shows that the system-reservoir interaction is only partially taken into account in $\langle E \rangle$ and in c_E. The difference originates from non-Ohmic parts of the friction kernel. For the Drude-regularized Ohmic damping kernel (3.67), it is now a simple undertaking to derive with use of the results (6.99) and (6.146) - (6.148) explicit expressions for $U - \langle E \rangle$ and the difference in the specific heat in terms of polygamma functions. We shall conduct this study for the free Brownian particle in Section 7.5.

6.7 Equilibrium density matrix

6.7.1 Derivation of the action

For a Gaussian process, the normalized equilibrium density matrix is completely specified by the position dispersion $\langle q^2 \rangle$ and the momentum dispersion $\langle p^2 \rangle$. Therefore, the correct form for any linear dissipative mechanism can be inferred from the density matrix of the undamped oscillator

$$< q | \hat{\rho}_\beta | q' > = \frac{1}{\sqrt{2\pi \langle q^2 \rangle}} \exp\left(-\frac{(q+q')^2}{8\langle q^2 \rangle} - \frac{\langle p^2 \rangle}{2\hbar^2} (q-q')^2 \right) , \tag{6.157}$$

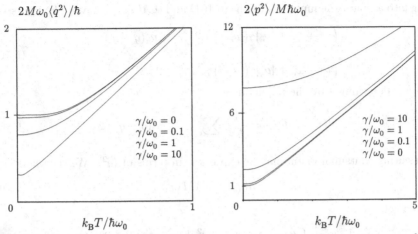

Figure 6.2: The normalized equilibrium variances of position and momentum are plotted versus temperature for the Drude model. The cutoff frequency is $\omega_D = 10\,\omega_0$. The damping strength γ/ω_0 varies between 0 and 10.

where we have used the normalization $\mathrm{tr}\,\hat{\rho}_\beta = 1$. Let us now see how the exponent in Eq. (6.157) emerges from the action (4.61). For a harmonic system we have

$$S_{\mathrm{eff}}^{(\mathrm{E})}[q] = \int_0^{\hbar\beta} d\tau \left(\frac{M}{2}\dot{q}^2 + \frac{1}{2}M\omega_0^2 q^2 \right) + \int_0^{\hbar\beta} d\tau \int_0^\tau d\tau'\, k(\tau - \tau') q(\tau) q(\tau') . \qquad (6.158)$$

The Fourier expansion method is somewhat complicated by the fact that the endpoints of the path $q(\tau)$ are different and therefore the periodically continued path

$$q(\tau) = \frac{1}{\hbar\beta} \sum_{n=-\infty}^{\infty} q_n e^{i\nu_n \tau} \qquad (6.159)$$

involves periodically repeated jumps from q back to q' and periodically repeated cusps, both of them at times $\tau = m\hbar\beta$ $(m = 0, \pm1, \cdots)$. It is convenient to write

$$q(\tau) = q^{(1)}(\tau) + q^{(2)}(\tau) , \qquad (6.160)$$

where $q^{(1)}(\tau)$ makes jumps, and $q^{(2)}(\tau)$ has cusps which result in jumps of the velocity,

$$q^{(1)}(0^- + m\hbar\beta) - q^{(1)}(0^+ + m\hbar\beta) = q - q' , \qquad (6.161)$$

$$\dot{q}^{(2)}(0^- + m\hbar\beta) = -\dot{q}^{(2)}(0^+ + m\hbar\beta) = v . \qquad (6.162)$$

Away from the discontinuities, the equation of motion for the extremal path reads

$$-M\ddot{q}(\tau) + M\omega_0^2 q(\tau) + \int_0^\tau d\tau'\, k(\tau - \tau') q(\tau') = 0 . \qquad (6.163)$$

Taking into account the jump conditions (6.161) and (6.162), we have in Fourier space

$$
\begin{aligned}
\left(\nu_n^2 + \omega_0^2 + |\nu_n|\hat{\gamma}(|\nu_n|)\right)q_n^{(1)} &= i\nu_n(q - q') \,, \\
\left(\nu_n^2 + \omega_0^2 + |\nu_n|\hat{\gamma}(|\nu_n|)\right)q_n^{(2)} &= 2v \,,
\end{aligned}
\tag{6.164}
$$

where v is determined by the requirement

$$
q^{(2)}(\tau = m\hbar\beta) = \frac{1}{\hbar\beta}\sum_{n=-\infty}^{\infty}q_n^{(2)} = (q + q')/2 \,.
\tag{6.165}
$$

The resulting Matsubara sum may be expressed in terms of $\langle q^2 \rangle$. We then find

$$
v = \hbar(q + q')/4M\langle q^2 \rangle \,.
\tag{6.166}
$$

Thus we get

$$
\begin{aligned}
q_n = \frac{1}{\hbar\beta}\sum_{n=-\infty}^{\infty}\Bigg\{&\left(-\frac{1}{i\nu_n} + \frac{1}{i\nu_n}\frac{\omega_0^2 + |\nu_n|\hat{\gamma}(|\nu_n|)}{\omega_0^2 + \nu_n^2 + |\nu_n|\hat{\gamma}(|\nu_n|)}\right)(q - q') \\
&+ \frac{1}{\nu_n^2 + \omega_0^2 + |\nu_n|\hat{\gamma}(|\nu_n|)}\frac{\hbar}{2M\langle q^2 \rangle}(q + q')\Bigg\} \,.
\end{aligned}
$$

With this we now have

$$
\begin{aligned}
q(0^+) &= q' \,, \qquad\qquad q(0^-) = q \,, \\
\dot{q}(0^\pm) &= \frac{\langle p^2 \rangle}{M\hbar}(q - q') \pm \frac{\hbar}{4M\langle q^2 \rangle}(q + q') \,.
\end{aligned}
\tag{6.167}
$$

Since the extremal path satisfies the equation of motion (6.163), the harmonic action (6.158) along this path may be put into the forms

$$
S_{\text{eff}}^{(E)}[q] = \int_0^{\hbar\beta}d\tau\,\frac{M}{2}\frac{d}{d\tau}\Big(q(\tau)\dot{q}(\tau)\Big) = \frac{M}{2}\Big(q(0^-)\dot{q}(0^-) - q(0^+)\dot{q}(0^+)\Big) \,,
\tag{6.168}
$$

where we have used $q(\hbar\beta - 0^+) = q(0^-)$. This combines with Eq. (6.167) to yield

$$
S_{\text{eff}}^{(E)}[q]/\hbar = \frac{1}{8\langle q^2 \rangle}(q + q')^2 + \frac{\langle p^2 \rangle}{2\hbar^2}(q - q')^2 \,,
\tag{6.169}
$$

which is (with a minus sign) the desired exponent in the expression (6.157).

6.7.2 Purity

After having substantiated the explicit form of the equilibrium density matrix of the damped oscillator, we can now check the purity,

$$
\text{pur}\,(\alpha) \equiv \text{tr}\,\hat{\rho}^2 = \int dx\,dx'\,\rho(x, x')\,\rho(x', x) = \frac{\hbar}{2\sqrt{\langle p^2 \rangle\,\langle q^2 \rangle}} \,.
\tag{6.170}
$$

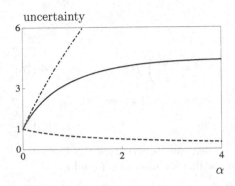
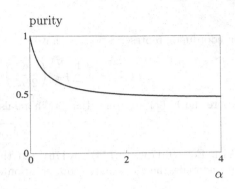

Figure 6.3: In the left figure the normalized variances $2M\omega_0\langle q^2\rangle/\hbar$ (dashed curve) and $2\langle p^2\rangle/M\hbar\omega_0$ (dashed-dotted curve), and the normalized uncertainty $2\sqrt{\langle q^2\rangle\langle p^2\rangle}/\hbar$ (full curve) of the ground state are plotted versus the damping parameter α. In the right figure, the purity pur(α) at zero temperature is plotted versus α. The Drude cutoff is $\omega_D = 100\,\omega_0$.

For a pure quantum state, we have pur $= 1$. With use of the expressions for $\langle q^2\rangle_{T=0}$ and $\langle p^2\rangle_{T=0}$ derived in previous subsections, we see that already weak damping makes the ground state rather impure. Hence, pure quantum states never exist in real life. On the other hand, for γ somewhat above ω_0 the purity reaches a plateau which is still of the order of one even for large α. There, the dependence of the purity on α is only logarithmic [cf. Fig 6.3]. This indicates that the damped oscillator stays quantum mechanical even for very large damping.

Let us now study what the effective mass and effective frequency of a fictional linear quantum oscillator with a discrete equidistant spectrum would be, if it had the same coordinate and momentum dispersions as specified in the preceding chapter. The respective statistical operator with normalization tr$\hat{\rho}_\beta = 1$ is

$$\hat{\rho}_\beta = \tilde{Z}^{-1}(\beta)\,\mathrm{e}^{-\beta\hat{H}_{\mathrm{eff}}}\,, \tag{6.171}$$

$$\hat{H}_{\mathrm{eff}} = \frac{1}{2M_{\mathrm{eff}}}\hat{p}^2 + \frac{1}{2}M_{\mathrm{eff}}\omega_{\mathrm{eff}}^2\hat{q}^2\,. \tag{6.172}$$

Here the parameters ω_{eff} and M_{eff} are fixed by the requirement that Eq. (6.171) has the coordinate representation (6.157). The density matrix can be represented as

$$\hat{\rho}_\beta = \frac{1}{\tilde{Z}(\beta)}\sum_{n=0}^{\infty}\mathrm{e}^{-\beta E_n}\,|n><n|\,, \tag{6.173}$$

where $|n>$ is an eigenstate of the effective Hamiltonian

$$\hat{H}_{\mathrm{eff}}\,|n> = \hbar\omega_{\mathrm{eff}}(n+\tfrac{1}{2})\,|n>\,. \tag{6.174}$$

Now the partition function has the familiar form

$$\tilde{Z}(\beta) = \left[\,2\sinh(\beta\hbar\omega_{\mathrm{eff}}/2)\,\right]^{-1}\,. \tag{6.175}$$

In coordinate representation, we have

$$< q|n > = \left(\frac{1}{\pi}\right)^{1/4} \left(\frac{c}{2^n n!}\right)^{1/2} \exp\left(-\frac{c^2 q^2}{2}\right) H_n(cq) , \qquad (6.176)$$

where the $H_n(x)$ are Hermitian polynomials of degree n, and where

$$c = \sqrt{M_{\text{eff}}\omega_{\text{eff}}/\hbar} . \qquad (6.177)$$

With Eqs. (6.173), (6.174), (6.176) and the generating function of the Hermitian polynomials, the coordinate representation of the density matrix is found as [4]

$$< q|\hat{\rho}_\beta|q' > = \frac{c}{\tilde{Z}\sqrt{2\pi \sinh\Omega}} \exp\left\{-\frac{c^2 \left[(q^2 + q'^2)\cosh\Omega - 2qq'\right]}{2\sinh\Omega}\right\} , \qquad (6.178)$$

where $\Omega = \omega_{\text{eff}}\hbar\beta$. The quantities ω_{eff} and M_{eff} can be determined upon comparing Eq. (6.157) with Eq. (6.178). This gives for the effective parameters the expressions

$$\omega_{\text{eff}} = \frac{1}{\hbar\beta} \ln\frac{\sqrt{\langle p^2\rangle\langle q^2\rangle} + \hbar/2}{\sqrt{\langle p^2\rangle\langle q^2\rangle} - \hbar/2} , \qquad M_{\text{eff}} = \sqrt{\langle p^2\rangle/\langle q^2\rangle}/\omega_{\text{eff}} . \qquad (6.179)$$

At high temperatures, the effective quantities ω_{eff} and M_{eff} approach their bare values ω_0 and M. At low temperatures, ω_{eff} and M_{eff} strongly differ from their bare values. Substituting the zero temperature values for $\langle q^2\rangle$ and $\langle p^2\rangle$, we find

$$\hbar\omega_{\text{eff}} = 2 k_B T \operatorname{arcoth} \sqrt{g(\alpha)\left[(1 - 2\alpha^2)g(\alpha) + (4\alpha/\pi)\ln(\omega_D/\omega_0)\right]} , \qquad (6.180)$$

where the function $g(\alpha)$ is given in Eq. (6.107). Observe that ω_{eff} vanishes linearly with T as $T \to 0$. Hence the discrete energy levels in Eq. (6.174) become very narrowly spaced near zero temperature. In fact, the occupation probabilities $p_n = \tilde{Z}^{-1} \exp(-\beta E_n)$ remain finite in the limit $T \to 0$. From this we infer that the ground state of the fictional oscillator is not a pure state but a mixture.

We conclude with the remark that the partition function of the damped oscillator Z and the partition function of the fictional oscillator $\tilde{Z}$ drastically differ at low temperatures for $\alpha > 0$. They approach each other as temperature is increased and coincide in the classical limit.

6.8   Quantum master equations for the reduced density matrix

So far we have considered selected thermal averages, e.g., the mean squares of displacement and momentum. For a full characterization of the statistics of the central oscillator, knowledge of the reduced density matrix (RDM) or the Wigner function is essential. Of particular interest are the dynamical equations of these quantities. The path integral expression (5.50) with (5.51) for the propagating function of the damped oscillator has been calculated in analytic form in Ref. [87]. For a given preparation function, the reduced density matrix is then given by the integral expression (5.49).

Starting out from these expressions, Karrlein and Grabert have studied various quantum master equations for the RDM for different types of the initial preparation. They found that the RDM is described in terms of the "quasiclassical" coordinate $r = \frac{1}{2}(q + q')$ and the "fluctuation" coordinate $y = q - q'$ by the master equation [48]

$$\frac{\partial}{\partial t}\rho(y, r; t) = \mathcal{L}\left(y, r, \frac{\partial}{\partial y}, \frac{\partial}{\partial r}; t\right)\rho(y, r; t) \qquad (6.181)$$

with the time-dependent Liouville operator

$$\mathcal{L} = \frac{i\hbar}{M}\frac{\partial^2}{\partial y \partial r} - \frac{iM}{\hbar}\gamma_q(t)ry - \gamma_p(t)y\frac{\partial}{\partial y} - \frac{iM}{\hbar}D_q(t)y\frac{\partial}{\partial r} - \frac{M^2}{\hbar^2}D_p(t)x^2 . \qquad (6.182)$$

The functions $\gamma_q(t)$ and $\gamma_p(t)$ are time-dependent drift coefficients and the functions $D_q(t)$ and $D_p(t)$ are time-dependent diffusion coefficients. The analysis in Ref. [48] showed that these functions depend on the preparation function. Hence there is no unique exact Liouville operator.

6.8.1 Thermal initial condition

Consider now initial conditions specified by the preparation function (cf. Section 5.4)

$$\lambda(y_i, r_i, \bar{y}, \bar{r}) = f(y_i, r_i)\lambda_\beta(y_i, r_i, \bar{y}, \bar{r}) - f(y_i, r_i)\delta(\bar{y} \quad y_i)\delta(\bar{r} - r_i) \qquad (6.183)$$

In this case the diffusion and drift coefficients are expressed in terms of the symmetrized part $S(t)$ and the antisymmetrized part $A(t) = -(\hbar/2M)G(t)$ of the equilibrium position autocorrelation function $C^+(t) \equiv \langle q(t)q(0)\rangle_\beta = S(t) - i(\hbar/2M)G(t)$ discussed in Sections 6.3 and 6.4. One finds [48]

$$\gamma_q(t) = \frac{\ddot{G}(t)\dot{S}(t) - \dot{G}(t)\ddot{S}(t)}{\dot{G}(t)S(t) - G(t)\dot{S}(t)}, \qquad \gamma_p(t) = \frac{G(t)\dddot{S}(t) - \ddot{G}(t)S(t)}{\dot{G}(t)S(t) - G(t)\dot{S}(t)}, \qquad (6.184)$$

$$D_q(t) = \gamma_q(t)\langle q^2\rangle - \langle p^2\rangle/M^2 , \qquad D_p(t) = \gamma_p(t)\langle p^2\rangle/M^2 .$$

The dispersions $\langle q^2\rangle$ and $\langle p^2\rangle$ are given in Eqs. (6.131) and (6.135). By virtue of the assignments

$$y \to [q, \cdot] , \quad r \to \{q, \cdot\}/2 , \quad \partial/\partial y \to (i/2\hbar)\{p, \cdot\} , \quad \partial/\partial r \to (i/\hbar)[p, \cdot] , \qquad (6.185)$$

the master equation (6.181) with (6.184) can be written as

$$\dot{\rho}(t) = -(iM/\hbar)\gamma_q(t)\left[q, \{q, \rho(t)\}/2 + (i/\hbar)\langle q^2\rangle[p, \rho(t)]\right]$$

$$- (i/\hbar)\gamma_p(t)\left[q, \{p, \rho(t)\}/2 - (i/\hbar)\langle p^2\rangle[q, \rho(t)]\right] \qquad (6.186)$$

$$- (i/M\hbar)\left[p, \{p, \rho(t)\}/2 - (i/\hbar)\langle p^2\rangle[q, \rho(t)]\right] ,$$

where $[u, v] = uv - vu$ is the commutator and $\{u, v\} = uv + vu$ is the anti-commutator.

The master equation (6.186) can be transformed into a generalized Fokker-Planck equation for the Wigner function $w(p, q; t) = \int dy \rho(y, q; t)e^{-iyp/\hbar}$. With use of the correspondences

$$[q, \cdot] \to -\frac{\hbar}{i} \frac{\partial}{\partial p} \,, \quad \{q, \cdot\} \to 2q \,, \quad [p, \cdot] \to \frac{\hbar}{i} \frac{\partial}{\partial q} \,, \quad \{p, \cdot\} \to 2p \,, \tag{6.187}$$

we then obtain

$$\dot{w}(p, q; t) \;=\; \left\{ \frac{\partial}{\partial p} M\gamma_q(t) \left[q + \frac{\partial}{\partial q} \langle q^2 \rangle \right] \right.$$
$$\left. + \left[\frac{\partial}{\partial p} \gamma_p(t) - \frac{\partial}{\partial q} \frac{1}{M} \right] \left[p + \frac{\partial}{\partial p} \langle p^2 \rangle \right] \right\} w(p, q; t) \,. \tag{6.188}$$

To see what we have obtained so far, consider now the classical limit. The classical position autocorrelation function $S_{cl}(t)$ is given by the integral expression (6.68). Now, the derivative $\dot{S}_{cl}(t)$ is directly related to the function $G(t)$ as $G(t) = M\beta \dot{S}_{cl}(t)$. Further we have $\langle q^2 \rangle_{cl} = 1/M\beta\omega_0^2$ and $\langle p^2 \rangle_{cl} = M^2\omega_0^2 \langle q^2 \rangle_{cl}$. With these expressions the generalized Fokker-Planck equation (6.188) takes exactly the form found by Adelman [252]. He derived this equation from the generalized classical Langevin equation (2.4) with (2.8). The dynamical equation (6.188) with (6.184) represents the exact quantum mechanical generalization of Adelman's Fokker-Planck equation.

6.8.2   Product initial state

In many works it has been assumed for simplicity that the initial density matrix is in factorized form [204, 205, 237, 243]. The time evolution is then given by the expression (5.11) with (5.12). The resulting master equation is again in the form (6.181) with (6.182), but now the drift and diffusion functions are given by

$$\gamma_q(t) \;=\; \frac{\ddot{G}^2(t) - \dot{G}(t)\dddot{G}(t)}{\dot{G}^2(t) - G(t)\ddot{G}(t)} \,, \qquad \gamma_p(t) \;=\; \frac{G(t)\dddot{G}(t) - \dot{G}(t)\ddot{G}(t)}{\dot{G}^2(t) - G(t)\ddot{G}(t)}] \,,$$

$$D_q(t) \;=\; \frac{\hbar}{M} \left[\frac{1}{2}\dot{K}_q(t) - K_p(t) + \gamma_q(t)K_q(t) + \frac{1}{2}\gamma_p(t)\dot{K}_q(t) \right] \,, \tag{6.189}$$

$$D_p(t) \;=\; \frac{\hbar}{M} \left[\frac{1}{2}\dot{K}_p(t) + \gamma_p(t)K_p(t) + \frac{1}{2}\gamma_q(t)\dot{K}_q(t) \right] \,.$$

$$K_q(t) \;=\; \frac{1}{M} \int_0^t dt' \int_0^t dt'' \, G(t')L'(t' - t'')G(t'') \,,$$
$$K_p(t) \;=\; \frac{1}{M} \int_0^t dt' \int_0^t dt'' \, \dot{G}(t')L'(t' - t'')\dot{G}(t'') \,, \tag{6.190}$$

where $L'(t)$ is the real part of the real-time influence kernel $L(t)$ given in Eq. (5.26). The master equation (6.188) with (6.189) and (6.190) corresponds to that by Haake and Reibold [237] who derived it from the microscopic dynamics. Hu, Paz and Zhang found this equation later with the Feynman-Vernon approach [243]. The sudden switch-on of the interaction with the bath at time zero in the product initial state as compared to the thermal initial state leads to artifacts which persist until long time.

As a result of this, the master equation for a product initial state does not reduce in the classical limit to a master equation that is equivalent to Adelman's Fokker-Planck equation.

6.8.3 Approximate time-independent Liouville operators

The drift coefficients $\gamma_p(t)$ and $\gamma_q(t)$ in Eq. (6.184) and in Eq. (6.189) become time independent when $G(t)$ and $S(t)$ are of the form

$$G(t) = a_1 e^{-\lambda_1 t} + a_2 e^{-\lambda_2 t}, \qquad S(t) = b_1 e^{-\lambda_1 t} + b_2 e^{-\lambda_2 t}, \qquad (6.191)$$

where λ_1 and λ_2 are complex conjugate or real frequencies, and where the coefficients a_1, a_2 and b_1, b_2 are complex conjugate pairs or real, respectively. We then obtain both from Eq. (6.184) and from Eq. (6.189)

$$\gamma_q = \lambda_1 \lambda_2, \qquad \text{and} \qquad \gamma_p = \lambda_1 + \lambda_2. \qquad (6.192)$$

We see from the expressions in Subsection 6.3.1 that $G(t)$ for strictly Ohmic friction is exactly of this form. However, the strict Ohmic case is excluded since it would yield a divergent variance of momentum, as we have seen in Subsection 6.6.3. In addition, the function $S(t)$ has additional contributions from the poles of the coth-function [cf. Eq. (6.63) with Eq. (6.66)]. Hence time-independent Liouville operators are only approximately valid.

Evidently, for Ohmic friction with a Drude cutoff, the third pole of $\hat{\chi}(z)$ in Eq. (6.43) located at $z = -\lambda_3$, and, in addition, the contribution $S_2(t)$ of $S(t)$ given in Eq. (6.72) must be disregarded in order that the forms (6.191) hold. Hence for large Drude cutoff $\omega_D \gg \omega_0, \gamma$ and high temperature $k_B T \gg \hbar\omega_0, \hbar\gamma$, the dynamics can be described by a time-independent Liouville operator of the form (6.182) with the coefficients (6.192), where the frequencies λ_1 and λ_2 are given in Eq. (6.47) with Eq. (6.46). Explicit expressions for the variances $\langle q^2 \rangle$ and $\langle p^2 \rangle$ are discussed in Subsection 6.6.3. The resulting approximate master equation has the form [237]

$$
\begin{aligned}
\dot{\rho}(t) = & -\frac{iM}{\hbar}\lambda_1\lambda_2 \left[q, \frac{1}{2}\{q, \rho(t)\} + \frac{i}{\hbar}\langle q^2 \rangle [p, \rho(t)] \right] \\
& -\frac{i}{\hbar}(\lambda_1 + \lambda_2) \left[q, \frac{1}{2}\{p, \rho(t)\} - \frac{i}{\hbar}\langle p^2 \rangle [q, \rho(t)] \right] \\
& -\frac{i}{M\hbar} \left[p, \frac{1}{2}\{p, \rho(t)\} - \frac{i}{\hbar}\langle p^2 \rangle [q, \rho(t)] \right].
\end{aligned}
\qquad (6.193)
$$

The classical regime with strictly Ohmic friction is reached by first taking the high temperature limit and subsequently the limit $\omega_D \to \infty$. This yields

$$\gamma_q = \omega_0^2, \qquad \gamma_p = \gamma, \qquad D_q = 0, \qquad D_p = \gamma k_B T/M. \qquad (6.194)$$

The corresponding master equation can be transformed into a Fokker-Planck equation for the Wigner function $w(p, q; t)$,

$$\dot{w}(p,q;t) \;=\; \left\{ -\frac{1}{M}\frac{\partial}{\partial q}\,p \;+\; \frac{\partial}{\partial p}\!\left[\frac{\partial V(q)}{\partial q}+\gamma\,p\right] \;+\; \gamma k_{\mathrm{B}}TM\frac{\partial^2}{\partial p^2} \right\} w(p,q;t)\,. \quad (6.195)$$

Here we have put $M\omega_0^2 q = \partial V(q)/\partial q$. Writing Eq. (6.195) in position and velocity space, it becomes the Klein-Kramers equation derived by Kramers already in 1940 [253]. We shall study Kramers' flux solution of this equation in Section 11.2.

For weak damping and arbitrary frequency dependence, the contribution $S_2(t)$ resulting from the poles of the expression (6.66) is negligibly small. The relevant poles of $\hat{G}(z)$ and $\hat{S}_1(z)$ are at $z = -\lambda_{1,2}$, where $\lambda_{1,2} = \frac{1}{2}\gamma' \pm i\,(\omega_0 + \frac{1}{2}\gamma'')$, and where $\gamma'+i\,\gamma'' = \hat{\gamma}(z=-i\,\omega_0)$. Thus the detailed frequency-dependence of $\tilde{\gamma}(\omega)$ is irrelevant. With these poles, the functions $G(t)$ and $S(t)$ read as in Eq. (6.191). As a result one ends up again with a quantum master equation of the form (6.193).

6.8.4 Connection with Lindblad theory

We see from the derivation that the time-independent quantum master equation (6.193) holds only for times $t > t_0$ where $t_0 > 1/\omega_{\mathrm{D}}$, $\hbar\beta/2\pi$. During the initial interval $0 < t < t_0$ the fast components decay and after this the density matrix is reduced to a subspace in which the fast components are absent. The Markovian master equation (6.193) holds only for density matrices within this subspace.

On the other hand, Lindblad theory requires validity for any reduced density matrix. The Lindblad master equation (2.20) preserves the positivity of density operators for all times. Now, the time-independent Liouville operators derived in the previous subsection are not of Lindblad form, as can be seen with use of findings by Sandulescu and Scutaru [254]. This is not disquieting, as we have seen in Subsection 2.3.2: a slippage of fast components must occur before the dynamics effectively is Markovian. Therefore, complete positivity postulated by Lindblad theory is not necessary, as was emphasized by Pechukas [47].

Further coarse-graining was studied in Ref. [48] for weak damping $\gamma \ll \omega_0$. It was found that in the time regime $\omega_0^{-1} \ll t \ll \gamma^{-1}$ the above time-independent Liouville operators can be cast indeed into Lindblad form.

7. Quantum Brownian free motion

Historically, the Langevin equation (2.3) was used to analyze the irregular motion (*Zitterbewegung*) of a heavy particle moving in a thermally equilibrated molecular medium. This phenomenon is known as *Brownian motion*, named after Robert Brown, who observed the random motion of pollen grains immersed in a fluid. In many cases, the systematic external force can be disregarded. Then we are left with the problem of free Brownian motion. Often, the memory times of the fluctuating forces are negligibly short compared to the time scale over which one observes the Brownian particle. The assumption of memoryless friction is usually referred to as Markov limit.

In this chapter I consider quantum Brownian free motion for general frequency-dependent damping $\tilde{\gamma}(\omega)$. I begin with an important remark: since the equation of motion is linear, the particle is classical, i.e., the susceptibility is devoid of $\hbar$. Planck's constant enters solely via the fluctuation-dissipation theorem, as already emphasized after Eq. (6.36). A brief discussion of genuine quantum effects resulting from state-dependent dissipation is postponed to Subsection 9.3.1. Intrinsic quantum effects are pertinent to nonlinear systems, and they are decisive for linear and nonlinear transport properties, as we shall see in Part V.

The simplest microscopic model for a free Brownian particle with bare mass M_0 is described by the Hamiltonian (3.11) with $V(q) = 0$ and $M = M_0$. The classical equation of motion reads

$$M_0 \ddot{q}(t) + M_0 \int_{-\infty}^{t} dt' \, \gamma(t - t') \, \dot{q}(t') \; = \; \xi(t) \,. \tag{7.1}$$

Here $\xi(t)$ is the fluctuating force with properties as discussed in Section 3.1.2. As emphasized previously, the dynamical equation (7.1) holds exactly also in the quantum regime, but with the addition that the fluctuating force complies with the quantum mechanical force auto-correlation function (2.33). Additional light is shed on the underlying physics by introducing new coordinates y_α and masses μ_α for the bath oscillators [255],

$$y_\alpha = \frac{m_\alpha \omega_\alpha^2}{c_\alpha} x_\alpha \,; \qquad \mu_\alpha = \frac{c_\alpha^2}{m_\alpha \omega_\alpha^4} \,. \tag{7.2}$$

With these definitions the Hamiltonian (3.11) with $V(q) = 0$ becomes

$$H \; = \; \frac{p^2}{2\,M_0} + \frac{1}{2} \sum_{\alpha=1}^{N} \mu_\alpha \left(\dot{y}_\alpha^2 + \omega_\alpha^2 \, (y_\alpha - q)^2 \right) \,. \tag{7.3}$$

We see that the model describes a particle of mass M_0 with all of the reservoir's effective masses μ_α attached with springs to its position coordinate q. This form elucidates the translational invariance of the model.

7.1 Spectral density, damping function and mass renormalization

In this section, I briefly discuss renormalization of the mass of the Brownian particle induced by the environmental coupling.

With the substitution (7.2) the spectral density of the coupling takes the form

$$J(\omega) \; = \; \frac{\pi}{2} \sum_{\alpha=1}^{N} \mu_\alpha \omega_\alpha^3 \, \delta(\omega - \omega_\alpha) \,. \tag{7.4}$$

We can model any desired frequency dependence of $J(\omega)$ by an appropriate choice of the spectral distribution of the oscillators.

If $\sum_\alpha \mu_\alpha$ is finite, the total energy associated with the Hamiltonian (7.3) is the sum of an internal energy, which is independent of the total momentum, and the kinetic energy of the center of mass. The latter is the square of the total momentum divided by twice the mass of the total system. The total mass is

$$M_{\text{tot}} = M_0 + \sum_{\alpha=1}^{N} \mu_\alpha = M_0 + \frac{2}{\pi} \int_0^\infty d\omega \, \frac{J(\omega)}{\omega^3} \, . \tag{7.5}$$

In the second form, we have taken the continuum limit. Clearly, the notion of a total mass M_{tot} is only meaningful when the sum or the integral in Eq. (7.5) is finite.

Consider now the case when the spectral density $J(\omega)$ is split in low and high frequency parts with crossover frequency ω_c, as given in Eq. (3.69) with (3.70). For the purpose of this section, details of the high-frequency part $J_{\text{hf}}(\omega)$ are irrelevant. For times $t \gg \omega_c^{-1}$, the only effect of the high frequency modes is mass renormalization,

$$M = M_0 + \Delta M_{\text{hf}} = M_0 + \frac{2}{\pi} \int_0^\infty d\omega \, \frac{J_{\text{hf}}(\omega)}{\omega^3} \, . \tag{7.6}$$

Now we are left with the problem of a Brownian particle of mass M coupled to a heat bath environment described by the spectral density $J_{\text{lf}}(\omega)$ given in Eq. (3.70).

If the exponent s in Eq. (3.70) exceeds 2, the integral in Eq. (7.5) is infrared-convergent. Then the environment in its entirety equips the particle with a polaron cloud which enhances inertia of the particle. The increment of the inertial mass due to the low-frequency spectral density (3.72) is

$$\Delta M_s \equiv \frac{2}{\pi} \int_0^\infty d\omega \, \frac{J_{\text{lf}}(\omega)}{\omega^3} = \Gamma(s-2) \frac{2}{\pi} \frac{\gamma_s}{\omega_{\text{ph}}} \left(\frac{\omega_c}{\omega_{\text{ph}}} \right)^{s-2} M \, . \tag{7.7}$$

With Eq. (7.7), the spectral function $\tilde{\gamma}(\omega)$ behaves analogous to Eq. (3.77) as

$$\lim_{\omega \to 0} \tilde{\gamma}(\omega)/\omega = -i \, \Delta M_s/M \, . \tag{7.8}$$

The dynamical susceptibility of the Brownian particle

$$\tilde{\chi}(\omega) = \frac{1}{M} \frac{1}{-\omega^2 - i\,\omega\tilde{\gamma}(\omega)]} \, , \tag{7.9}$$

which results from Eq. (6.32) by setting $\omega_0 = 0$, has then the low-frequency behavior

$$\lim_{\omega \to 0} \tilde{\chi}(\omega)\,\omega^2 = -\frac{1}{(M + \Delta M_s)} \, . \tag{7.10}$$

From this we infer that friction effectively vanishes at asymptotic times when $s > 2$. Then the Brownian particle behaves as a free particle with the renormalized mass

$$M_{\text{r}} = M + \Delta M_s \, . \tag{7.11}$$

On the other hand, the integral in Eq. (7.7) is infrared-divergent if $s \leq 2$. This indicates that the low-energy excitations of the reservoir produce real dynamical effects beyond simple mass renormalization even at asymptotic time. The leading low-frequency behavior of the spectral damping function is [cf. Eq. (3.83)]

$$\tilde{\gamma}(\omega) = [\gamma_s/\sin(\pi s/2)]\,(-i\omega/\omega_{\text{ph}})^{s-1} \, . \tag{7.12}$$

For $s < 2$, $\tilde{\gamma}(\omega) \propto \omega^{s-1}$ outstrips linear dependence on ω as $\omega \to 0$. Hence for $s < 2$ damping is effective at all times. In the sequel, we shall regard M as the mass which is already dressed by the reservoir's high frequency modes and consider the effects induced by the spectral coupling function $J_{\text{lf}}(\omega)$.

7.2 Displacement correlation and response function

If we take the limit $\omega_0 \to 0$ in the expression (6.131), the mean square of the position diverges. This singular behavior is natural since the position of a free particle is not bounded. In the displacement correlation function the divergent initial value is subtracted,

$$
\begin{aligned}
D^+(t) &\equiv C^+(0) - C^+(t) = \langle [q(0) - q(t)]\, q(0) \rangle_\beta , \\
D^-(t) &\equiv C^-(0) - C^-(t) = \langle q(0)\, [q(0) - q(t)] \rangle_\beta ,
\end{aligned}
\tag{7.13}
$$

The function $D^\pm(t)$ describes correlations between the displacement $q(0) - q(t)$ of the particle at time t and the initial position $q(0)$. In the split-up

$$
D^\pm(t) = \tfrac{1}{2} D_{\text{th}}(t) \mp i\, A(t) ,
\tag{7.14}
$$

the real function $D_{\text{th}}(t)$ is the mean square displacement. It is related to the symmetrized position correlation function $S(t)$, which is defined in Eq. (6.13), as

$$
D_{\text{th}}(t) \equiv \langle [q(t) - q(0)]^2 \rangle_\beta = 2[S(0) - S(t)] .
\tag{7.15}
$$

The imaginary part $A(t)$ is connected with the thermal expectation value of the commutator in the usual way,

$$
A(t) = \frac{1}{2i} \langle [\, q(t), q(0)\,] \rangle_\beta .
\tag{7.16}
$$

The response function is [cf. Eq. (6.15)]

$$
\chi(t) = -(2/\hbar)\, \Theta(t)\, A(t) .
\tag{7.17}
$$

Since $\langle q(t) \rangle_\beta = 0$, the variance $\sigma^2(t)$ coincides with the mean square displacement,

$$
\sigma^2(t) = D_{\text{th}}(t) \qquad \text{for} \qquad t > 0 .
\tag{7.18}
$$

The Fourier integral representations of $D_{\text{th}}(t)$ and of $\chi(t)$ are

$$
D_{\text{th}}(t) = \frac{\hbar}{\pi} \int_{-\infty}^{+\infty} d\omega\, \tilde{\chi}''(\omega) \coth(\beta\hbar\omega/2)\, [1 - \cos(\omega t)] ,
\tag{7.19}
$$

$$
\chi(t) = \Theta(t) \frac{1}{\pi} \int_{-\infty}^{+\infty} d\omega\, \tilde{\chi}''(\omega) \sin(\omega t) .
\tag{7.20}
$$

Comparison of Eq. (7.19) with Eq. (7.20) shows that at asymptotic time the function $\chi(t)$ is connected with the change of $D_{\text{th}}(t)$ per unit time by the simple relation

$$
\lim_{t \to \infty} \frac{1}{\chi(t)} \frac{d}{dt} D_{\text{th}}(t) = 2k_{\text{B}}T .
\tag{7.21}
$$

The relation (7.21) is a version of the *Einstein relation*, as we shall see shortly. Note that in the derivation of Eq. (7.21) we made no specific assumptions on $\tilde{\gamma}(\omega)$, and not even on $\tilde{\chi}(\omega)$. Therefore, the relation (7.21) holds for nonlinear systems alike.

The significance of the response function may be illustrated by the effect of a constant force F applied to the Brownian particle for $t \geq 0$. There follows from Eq. (6.6) with the substitution $\delta f_{\text{ext}}(t) \to \Theta(t)\,F$, where F need not be infinitesimal, that the drift velocity of the particle is connected with $\chi(t)$ by

$$\langle \dot{q}(t) \rangle \;=\; \chi(t)\,F\,. \tag{7.22}$$

We shall use this relation when we discuss the Brownian dynamics at long time.

For convenience in Subsection 25.7.3, we also consider the Laplace transform of $D_{\text{th},+}(t) = \Theta(t)\,D_{\text{th}}(t)$. With use of the expression (7.19) we obtain

$$\hat{D}_{\text{th},+}(\lambda) \;=\; \frac{\hbar}{\pi\lambda} \int_{-\infty}^{+\infty} d\omega'\, \frac{\omega'^2}{\lambda^2 + \omega'^2}\, \coth\!\left(\frac{\beta\hbar\omega'}{2}\right) \operatorname{Im}\tilde{\chi}(\omega')\,. \tag{7.23}$$

This relation is a variant of the fluctuation-dissipation theorem and therefore holds for any linear or nonlinear quantum transport system.

7.3   Ohmic friction

We now study Brownian motion in the case of Ohmic friction in some detail.

7.3.1   Response function

It is practical knowledge from daily experience that for a constant force F and Ohmic friction, the drift velocity becomes asymptotically constant,

$$\lim_{t \to \infty} \langle \dot{q}(t) \rangle \;=\; \mu\,F\,. \tag{7.24}$$

The quantity μ is the dc mobility of the Brownian particle. The mobility μ is related to the response function $\chi(t)$ as

$$\mu \;=\; \lim_{t \to \infty} \chi(t) \;=\; \lim_{\omega \to 0} \omega\,\tilde{\chi}''(\omega = 0)\,. \tag{7.25}$$

The first form follows from Eq. (7.22). The second form is found with use of Eq. (7.20) and the representation $\delta(\omega) = \lim_{t \to \infty} \sin(\omega t)/(\pi\omega)$. Putting $\tilde{\gamma}(\omega) = \gamma$ in Eq. (7.9), the integral (7.20) takes the familiar form

$$\chi(t) \;=\; \Theta(t)\,\frac{1}{M\gamma}\Big(1 - e^{-\gamma t}\Big)\,. \tag{7.26}$$

Thus we have

$$\mu \;=\; \frac{1}{M\gamma} \;=\; \frac{1}{\eta}\,. \tag{7.27}$$

Hence the mobility of an Ohmic Brownian particle neither depends on temperature nor captures quantum effects. Finally, the velocity response to an impulsive force is

$$\mathcal{R}(t) \;=\; M\dot{\chi}(t) \;=\; \Theta(t)\,e^{-\gamma t}\,. \tag{7.28}$$

7.3.2 Mean square displacement

The time-dependence of the mean square displacement can be found either by direct contour integration of the expression (7.19), or by taking the limit $\omega_0 \to 0$ in the result for $S(t)$ given in Subsection 6.4.1. Either way yields

$$\sigma^2(t) = \frac{2}{M\beta\gamma} t - \frac{\hbar}{M\gamma} \cot(\beta\hbar\gamma/2) \left[1 - e^{-\gamma t} \right] + \frac{4\gamma}{M\beta} \sum_{n=1}^{\infty} \frac{1 - e^{-\nu_n t}}{\nu_n(\gamma^2 - \nu_n^2)} . \qquad (7.29)$$

In the high-temperature limit, this expression reduces to the classical mean square displacement

$$\sigma_{\mathrm{cl}}^2(t) = \frac{2}{M\beta\gamma^2} \left(\gamma t - 1 + e^{-\gamma t} \right) . \qquad (7.30)$$

At short time, $\sigma_{\mathrm{cl}}^2(t)$ increases $\propto t^2$, while at asymptotic time it grows $\propto t$. We see from Eq. (7.29) that quantum signatures in $\sigma^2(t)$ are relevant only for low temperature, $\beta\hbar\gamma/2\pi > 1$, and for short to intermediate time. With decreasing temperature, the regime where quantum fluctuations are relevant extends to longer time. However at any *finite* temperature, the dynamics at asymptotic time is dominated by the first term in Eq. (7.29), which is purely classical. We find for the diffusion coefficient

$$D \equiv \frac{1}{2} \lim_{t\to\infty} \frac{d}{dt} \sigma^2(t) = \frac{1}{M\beta\gamma} = \frac{k_{\mathrm{B}}T}{M\gamma} . \qquad (7.31)$$

Thus, the diffusion coefficient vanishes at zero temperature. This indicates that the growth of the mean square displacement is slowing down when T approaches zero.

The remaining task is to link the (linear) mobility μ to the diffusion coefficient. Upon inserting Eqs. (7.25) with (7.27) and (7.31) into (7.21), or by comparison of Eq. (7.27) with Eq. (7.31), we obtain the Einstein relation in the familiar form

$$D = k_{\mathrm{B}}T\mu . \qquad (7.32)$$

As $T \to 0$, only the last term in Eq. (7.29) remains, and the Matsubara sum turns into an integral,

$$\lim_{T\to 0} \frac{1}{\beta} \sum_{n=1}^{\infty} f(\nu_n) = \frac{\hbar}{2\pi} \int_0^{\infty} d\nu \, f(\nu) . \qquad (7.33)$$

With this, the time derivative of $\sigma_{T=0}^2(t)$ takes the form

$$\frac{d}{dt} \sigma_{T=0}^2(t) = \frac{2\hbar\gamma}{\pi M} \int_0^{\infty} d\nu \, \frac{e^{-\nu t}}{\gamma^2 - \nu^2} . \qquad (7.34)$$

The integral in Eq. (7.34) can be expressed in terms of the exponential integral function $\mathrm{Ei}(x)$. For x not on the positive real axis, $\mathrm{Ei}(x)$ is defined by the integral

$$\mathrm{Ei}(x) = \int_{-\infty}^{x} dx' \, \frac{e^{x'}}{x'} . \qquad (7.35)$$

The function $\mathrm{Ei}(x)$ is analytically continued to x on the positive real axis as

$$\overline{\mathrm{Ei}}(x) = \lim_{\epsilon \to 0} \frac{1}{2} \Big[\mathrm{Ei}(x + i\epsilon) + \mathrm{Ei}(x - i\epsilon) \Big] . \tag{7.36}$$

Integration of Eq. (7.34) with initial condition $\sigma^2_{T=0}(0) = 0$ then yields the integral representation [87]

$$\sigma^2_{T=0}(t) = \frac{\hbar}{\pi M} \int_0^t ds \left(\overline{\mathrm{Ei}}(\gamma s)\, e^{-\gamma s} - \mathrm{Ei}(-\gamma s)\, e^{\gamma s} \right) . \tag{7.37}$$

In the limit $t \to \infty$, the integral can be evaluated asymptotically. We then find that the mean square displacement at $T = 0$ shows sluggish logarithmic growth in the regime $t \gg 1/\gamma$,

$$\sigma^2_{T=0}(t) = \left[2\hbar/\pi M\gamma \right] \ln(\gamma t) . \tag{7.38}$$

The logarithmic growth of $\sigma^2_{T=0}(t)$ at asymptotic time was obtained first by Hakim and Ambegaokar [255].

We shall see in Subsection 25.7.2 that the sluggish logarithmic growth of the position spread holds for all quantum transport systems which have a finite linear mobility at zero temperature,

$$\mu_0 = \mu(T = 0) = \lim_{\omega \to 0} \omega\, \tilde{\chi}''(\omega, T = 0) . \tag{7.39}$$

The mean square displacement at zero temperature asymptotically behaves as

$$\sigma^2_{T=0}(t \to \infty) = (2\hbar\mu_0/\pi) \ln(t/t_0) , \tag{7.40}$$

where μ_0 is the limiting value (7.39), and where t_0 is the system's longest characteristic time scale. It is obvious from the derivation that the logarithmic growth of the mean square displacement at zero temperature has its origin in the *fluctuation-dissipation theorem* (6.25). In addition, the $\hbar$-dependence in Eq. (7.40) has the same root. Hence it does not reflect an intrinsic quantum mechanical feature of the particle.

7.3.3 Momentum spread

With the dynamical susceptibility (7.9), the integral expression (6.133) for the mean value of p^2 takes the form

$$\langle p^2 \rangle = \frac{\hbar M}{\pi} \int_0^\infty d\omega \, \frac{\omega\, \tilde{\gamma}'(\omega)}{[\omega - \tilde{\gamma}''(\omega)]^2 + \tilde{\gamma}'^2(\omega)} \coth(\beta\hbar\omega/2) . \tag{7.41}$$

Alternatively, we may put $\omega_0 = 0$ in Eq. (6.135). The resulting series is

$$\langle p^2 \rangle = \frac{M}{\beta} \left(1 + 2 \sum_{n=1}^{\infty} \frac{\hat{\gamma}(\nu_n)}{\nu_n + \hat{\gamma}(\nu_n)} \right) . \tag{7.42}$$

In the strict Ohmic case, $\hat{\gamma}(z) = \gamma$, both the integral and the series expression are ultraviolet divergent. Hence the required cutoff in the spectral coupling is actually a measurable quantity. Choosing the Drude regularization (3.67) we obtain from (7.41) or from (7.42) the exact analytic expression expression

$$\langle p^2 \rangle = \frac{M}{\beta} + \frac{M\hbar}{\pi} \frac{\lambda_1 \lambda_2}{\lambda_1 - \lambda_2} \left[\psi\left(1 + \frac{\beta\hbar\lambda_1}{2\pi}\right) - \psi\left(1 + \frac{\beta\hbar\lambda_2}{2\pi}\right) \right] , \tag{7.43}$$

where

$$\lambda_{1,2} = \tfrac{1}{2}\omega_{\mathrm{D}} \pm \tfrac{1}{2}\sqrt{\omega_{\mathrm{D}}^2 - 4\gamma\omega_{\mathrm{D}}} . \tag{7.44}$$

The first term in Eq. (7.43) is the leading one in the high-temperature regime. The residual terms become increasingly important as temperature is decreased. For zero temperature, we find in the limit $\omega_{\mathrm{D}} \gg 4\gamma$ logarithmic cutoff dependence

$$\langle p^2 \rangle_{T=0} = (M\hbar\gamma/\pi) \ln(\omega_{\mathrm{D}}/\gamma) . \tag{7.45}$$

7.4 Frequency-dependent friction

For general power-law forms of frequency-dependent damping, the response function can still be expressed in terms of known functions. To our knowledge, this is not possible for the mean square displacement, except for long times where the dominant contribution to the Fourier integral of $\sigma^2(t)$ comes from frequencies near $\omega = 0$.

7.4.1 Response function and mobility

For memory-friction it is convenient to base the discussion of response functions on their Laplace integral representations. Putting $\omega_0 = 0$ in Eq. (6.33), we obtain for the velocity response $\mathcal{R}(t)$ and position response $\chi(t)$ the Bromwich contour integrals

$$\mathcal{R}(t) = \frac{1}{2\pi i} \int_{\mathcal{C}} dz \, \frac{e^{zt}}{z + \hat{\gamma}(z)} , \quad \text{and} \quad \chi(t) = \frac{1}{2\pi i} \int_{\mathcal{C}} dz \, \frac{e^{zt}}{Mz[z + \hat{\gamma}(z)]} , \tag{7.46}$$

where the $\mathcal{C}$ is a vertical contour in the complex z-plane chosen such that all singularities of the integrand lie to the left of it. Consider now the various cases given for $\hat{\gamma}(z)$ in Eq. (3.83). In the regime $s < 2$, the integral (7.46) for $\chi(t)$ takes the form

$$\chi(t) = \frac{1}{2\pi i} \int_{\mathcal{C}} dz \, \frac{e^{zt}}{Mz[z + \omega_s(z/\omega_s)^{s-1}]} , \tag{7.47}$$

$$= \Theta(t) \frac{t}{M} (\omega_s t)^{s-2} \frac{1}{2\pi i} \int_{\mathcal{C}} dy \, \frac{e^y}{y^s} \frac{1}{1 + (y/\omega_s t)^{2-s}} , \tag{7.48}$$

where

$$\omega_s = (\lambda_s/\omega_{\mathrm{ph}})^{1/(2-s)} \omega_{\mathrm{ph}} . \tag{7.49}$$

The integral is a generalized Mittag-Leffler function [256]

$$E_{\alpha,\beta}(x) = \frac{1}{2\pi i} \int_{\mathcal{C}} dy \, \frac{y^{\alpha-\beta} e^y}{y^\alpha - x} = \sum_{k=0}^{\infty} \frac{x^k}{\Gamma(\alpha k + \beta)} . \tag{7.50}$$

Specifically, we have $\alpha = 2 - s$, $\beta = 2$, and $x = -(\omega_s t)^{2-s}$, and hence

$$\chi(t) = \Theta(t) \frac{t}{M} E_{2-s,2}\left[-(\omega_s t)^{2-s} \right] . \tag{7.51}$$

For $s < 1$, the integrand in Eq. (7.47) has three singularities:

(1) A complex conjugate pair of simple poles on the principal sheet at $z = -\Gamma_s \pm i\Omega_s$,

$$\Omega_s = \omega_s \cos\varphi_s \,, \qquad \Gamma_s = \omega_s \sin\varphi_s \,, \qquad \varphi_s = \frac{\pi}{2}\frac{s}{(2-s)} \,. \qquad (7.52)$$

(2) A branch point at $z = 0$ with a cut in the complex plane along the negative
real axis. The branch cut contributes to $\chi(t)$ an algebraic long-time tail which
dominates the behavior of $\chi(t)$ at long time.

The resulting expression for $\chi(t)$ at asymptotic time is

$$\chi(t) = \frac{2}{(2-s)}\frac{\sin(\Omega_s t - \varphi_s)\,e^{-\Gamma_s t}}{M\omega_s} + \frac{t}{M}\sum_{n=1}^{\infty}\frac{(-1)^{n-1}}{\Gamma[2-(2-s)n]}\frac{1}{(\omega_s t)^{(2-s)n}} \,. \qquad (7.53)$$

For $s = 1$, the branch cut in the integrand of Eq. (7.47) degrades into a pole at
$z = 0$ and the other simple pole is at $z = -\gamma$. The contributions of these poles yield
the familiar expression (7.26).

In the regime $1 < s < 2$, the poles of the integrand in Eq. (7.47) are not anymore
on the principal sheet. The residual cut contribution is a completely monotonic function of t. At long time, it is given by the asymptotic series in Eq. (7.53). The leading
asymptotic term is

$$\chi(t) = \frac{\sin(\pi s/2)}{M\gamma_s\Gamma(s)}(\omega_{\mathrm{ph}} t)^{s-1}\left(1 + \mathcal{O}\left[(\omega_{\mathrm{ph}}/\gamma_s)(\omega_{\mathrm{ph}} t)^{s-2}\right]\right) \,. \qquad (7.54)$$

This term includes the Ohmic result $\chi(t\to\infty) = 1/M\gamma$ as special case.

The asymptotic form (7.54) suggests to define a generalized linear dc mobility as

$$\mu_\ell^{(s)} \equiv \lim_{t\to\infty}\chi(t)/(\omega_{\mathrm{ph}} t)^{s-1} \,. \qquad (7.55)$$

Use of the expression (7.54) yields

$$\mu_\ell^{(s)} = \frac{\sin(\pi s/2)}{M\gamma_s\Gamma(s)} \qquad \text{for} \qquad 0 < s < 2 \,. \qquad (7.56)$$

Alternatively, we may consider the linear ac mobility

$$\tilde{\mu}(\omega) = -i\omega\tilde{\chi}(\omega) \,. \qquad (7.57)$$

For $s < 2$, we find with use of Eq. (3.83) the low-frequency behavior

$$\tilde{\mu}(\omega) = \frac{1}{M\tilde{\gamma}(\omega)} = \frac{\sin(\pi s/2)}{M\gamma_s}\left(\frac{\omega_{\mathrm{ph}}}{-i\omega}\right)^{s-1} \,. \qquad (7.58)$$

For $s = 2$, the ac mobility takes the form

$$\tilde{\mu}(\omega) = \frac{\pi}{2M\gamma_2}\frac{i\omega_{\mathrm{ph}}}{\omega\ln(\omega_{\mathrm{c}}/\omega)} \qquad \text{for} \qquad \omega \ll \omega_{\mathrm{c}} \,. \qquad (7.59)$$

Accordingly, the response function at time $t \gg 1/\omega_{\mathrm{c}}$ reads

$$\chi(t) \ = \ \frac{\pi}{2M\gamma_2}\,\frac{\omega_{\mathrm{ph}}t}{\ln(\omega_c t)}\,, \tag{7.60}$$

which shows logarithmic modification of the free particle behavior.

We have discussed already in Section 7.1 that for $s > 2$ the main effect of the environmental coupling on the asymptotic dynamics is renormalization of the particle's mass M, $M_{\mathrm{r}} = M + \Delta M_s$, where ΔM_s is defined in Eq. (7.7). With the expression (3.83) for $\hat{\gamma}(z)$ in the regime $2 < s < 4$, we find that the transient behavior of $\chi(t)$ is described again in terms of a generalized Mittag-Leffler function $E_{\alpha,\beta}(x)$, but now with $\alpha = s - 2$, $\beta = 2$, and $x = -(M_{\mathrm{r}}/M)(\omega_s t)^{s-2}$. The response function reads

$$\chi(t) \ = \ \Theta(t)\,t\left(1 - E_{s-2,2}[-(M_{\mathrm{r}}/M)(\omega_s t)^{s-2}]\right)\Big/M_{\mathrm{r}}\,, \tag{7.61}$$

where ω_s is given in Eq. (7.49). It is straightforward to write down the asymptotic expansion of the expression (7.61) analogous to Eq. (7.53). The leading contribution at asymptotic time is

$$\chi(t) \ = \ \frac{t}{M_{\mathrm{r}}}\left(1 - \frac{M}{M_{\mathrm{r}}}\,\frac{1}{\Gamma(4 - s)}\,\frac{1}{(\omega_s t)^{s-2}} + \mathcal{O}\left[(\omega_s t)^{4-2s}\right]\right). \tag{7.62}$$

The corresponding analysis in the regime $2n < s < 2n+2$, where $n = 2, 3, \cdots$, is left to the interested reader.

We may also also specify the velocity response function $\mathcal{R}(t) = M\dot{\chi}(t)$. In the most relevant regime $0 < s < 4$ we find the analytic expressions

$$\begin{aligned}
\mathcal{R}(t) \ &= \ \Theta(t)\,E_{2-s,1}[-(\omega_s t)^{2-s}]\,, && 0 < s < 2\,, \\[6pt]
\mathcal{R}(t) \ &= \ \Theta(t)\,\frac{M}{M_{\mathrm{r}}}\left(1 - E_{s-2,1}[-(M_{\mathrm{r}}/M)(\omega_s t)^{s-2}]\right), && 2 < s < 4\,.
\end{aligned} \tag{7.63}$$

7.4.2 Mean square displacement

The findings of the preceding subsection may be profitably deployed to obtain the asymptotic time-dependence of the mean square displacement. At finite temperature, we can infer $\sigma^2(t)$ at asymptotic time from the first integral of the relation (7.21),

$$\sigma^2(t) \ = \ \frac{2}{\beta}\int_0^t dt'\,\chi(t')\,. \tag{7.64}$$

Above all, this relation holds for all times in the classical regime, in which the thermal frequency $k_{\mathrm{B}}T/\hbar$ exceeds the characteristic frequencies of the open system. With the expressions (7.51) and (7.61), we then obtain

$$\begin{aligned}
\sigma_{\mathrm{cl}}^2(t) \ &= \ \frac{2}{M\beta}\,t^2\,E_{2-s,3}[-(\omega_s t)^{2-s}]\,, && 0 < s < 2\,, \\[6pt]
\sigma_{\mathrm{cl}}^2(t) \ &= \ \frac{1}{M_{\mathrm{r}}\beta}\,t^2\left\{1 - 2E_{s-2,3}[-(M_{\mathrm{r}}/M)(\omega_s t)^{s-2}]\right\}, && 2 < s < 4\,.
\end{aligned} \tag{7.65}$$

These expressions together with Eq. (7.60) give the asymptotic time-dependence as

$$
\sigma^2(t) = \begin{cases}
\dfrac{2\sin(\pi s/2)}{M\beta\gamma_s\omega_{\text{ph}}\Gamma(s+1)}(\omega_{\text{ph}}t)^s\,, & s < 2\,, \\[2ex]
\dfrac{\pi}{2M\beta\gamma_2\omega_{\text{ph}}}\dfrac{(\omega_{\text{ph}}t)^2}{\ln(\omega_c t)}\,, & s = 2\,, \\[2ex]
\dfrac{1}{M_{\text{r}}\beta}t^2\,, & s > 2\,.
\end{cases}
\tag{7.66}
$$

We may summarize the results as follows. For sub-Ohmic damping ($s < 1$), the ac mobility, the response function, and the mean square displacement in thermal equilibrium show sub-diffusive behavior: the growth at asymptotic time is slower than in the diffusive regime. In contrast, in the super-Ohmic regime $1 < s < 2$, the growth of these quantities is faster than in the diffusive regime. For instance, the ac mobility $\tilde{\mu}(\omega)$ vanishes for $s < 1$ and diverges for $s > 1$ in the zero frequency limit. In the case $s = 2$ we find logarithmic modification of the free particle behavior. For $s > 2$ the asymptotic dynamics is that of a free particle with a renormalized mass. For a related discussion concerning ergodicity we refer to Section 3.2.

Finally, we discuss the position dispersion at zero temperature. In the limit $T \to 0$, the expression (7.19) takes the form

$$
\sigma_{T=0}^2(t) = \frac{2\hbar}{\pi}\int_0^\infty d\omega\,\tilde{\chi}''(\omega)\,[1 - \cos(\omega t)]\,.
\tag{7.67}
$$

We may relate this function to the response function $\chi(t)$ by use of the integral representation [88]

$$
1 - \cos a = \frac{2}{\pi}\int_0^\infty dx\,\frac{\sin(ax)}{x(1-x^2)}\,, \qquad a > 0\,,
\tag{7.68}
$$

and use of the representation (7.20). We then obtain

$$
\sigma_{T=0}^2(t) = \frac{2\hbar}{\pi}\int_0^\infty dx\,\frac{1}{x(1-x^2)}\chi(xt)\,.
\tag{7.69}
$$

Given the dynamics of $\chi(t)$ at long times, the asymptotic behavior of $\sigma_{T=0}^2(t)$ can now be derived from the relation (7.69). With the forms for $\chi(t)$ given in Eqs. (7.54), (7.60) and (7.62), one obtains in the various damping regimes [87]

$$
\sigma_{T=0}^2(t \to \infty) = \begin{cases}
\sigma_\infty^2\,, & 0 < s < 1\,, \\[1ex]
a_1\,\ln(\gamma_1 t)\,, & s = 1\,, \\[1ex]
a_s\,(\omega_{\text{ph}}t)^{s-1}\,, & 1 < s < 2\,, \\[1ex]
a_2\,\omega_{\text{ph}}t/\ln^2(\omega_c t)\,, & s = 2\,, \\[1ex]
a_s\,(\omega_{\text{ph}}t)^{3-s}\,, & 2 < s < 3\,, \\[1ex]
a_3\,\ln(\omega_c t)\,, & s = 3\,, \\[1ex]
\bar{\sigma}_\infty^2\,, & s > 3\,.
\end{cases}
\tag{7.70}
$$

The reasons are as follows. In the regime $s < 1$, the response function goes to zero as $t \to \infty$. Thus, with the substitution $u = xt$ in Eq. (7.69) we find that $\sigma_{T=0}^2(t)$ approaches asymptotically the constant

$$\sigma_\infty^2 = \frac{2\hbar}{\pi} \int_0^\infty du \, \frac{\chi(u)}{u} = \frac{2\hbar}{\pi} \int_0^\infty dz \, \hat{\chi}(z) \, . \tag{7.71}$$

Upon inserting $\hat{\chi}(z) = M^{-1}[z^2 + z\hat{\gamma}(z)]^{-1}$ and the expression (3.83) for $\hat{\gamma}(z)$, the integral can be evaluated exactly. We then find

$$\sigma_\infty^2 = \frac{2}{(2-s)\sin[\pi/(2-s)]} \left(\frac{\omega_{\mathrm{ph}} \sin(\pi s/2)}{\gamma_s} \right)^{1/(2-s)} \frac{\hbar}{M\omega_{\mathrm{ph}}} \, . \tag{7.72}$$

In the regime $1 \leq s \leq 3$, we can calculate the x-integral in Eq. (7.69) with the respective asymptotic forms for $\chi(xt)$. We find the behavior (7.70) with coefficients

$$a_s = \begin{cases} \dfrac{2}{\pi} \dfrac{\hbar}{M\gamma_1}, & \text{for} \quad s = 1 \, , \\[2ex] \dfrac{\sin^2(\pi s/2)}{|\cos(\pi s/2)|\Gamma(s)} \dfrac{\hbar}{M\gamma_s}, & \text{for} \quad 1 < s < 2 \, , \\[2ex] \dfrac{\pi^2}{4} \dfrac{\hbar}{M\gamma_2}, & \text{for} \quad s = 2 \, , \\[2ex] \dfrac{1}{|\cos(\pi s/2)|\Gamma(s)} \dfrac{\hbar M\gamma_s}{M_{\mathrm{r}}^2\omega_{\mathrm{ph}}^2}, & \text{for} \quad 2 < s < 3 \, , \\[2ex] \dfrac{2}{\pi} \dfrac{\hbar M\gamma_3}{M_{\mathrm{r}}^2\omega_{\mathrm{ph}}^2}, & \text{for} \quad s = 3 \, . \end{cases} \tag{7.73}$$

For $s = 1$, we recover the previous result (7.38). In the region $1 < s < 2$, the mean square displacement for $T = 0$ grows as t^{s-1}, whereas for $2 < s < 3$ we have the behavior $\propto t^{3-s}$. In the cases $s = 2$ and $s = 3$, $\sigma_{T=0}^2(t \to \infty)$ is not described by a simple power-law. For $s > 3$, the leading contribution to $\chi(t)$ grows linearly with t. This term does not contribute to $\sigma_{T=0}^2(t \to \infty)$. Observing that the sub-leading term $\propto t^{3-s}$ in $\chi(t)$ vanishes at long time for $s > 3$, the case is similar to the sub-Ohmic one, apart from subtraction of the free particle term. As a result, we find that $\sigma_{T=0}^2(t)$ approaches a constant $\overline{\sigma}_\infty^2$ as $t \to \infty$,

$$\overline{\sigma}_\infty^2 = \frac{2\hbar}{\pi} \int_0^\infty dz \left(\hat{\chi}(z) - \frac{1}{M_{\mathrm{r}}z^2} \right) \, . \tag{7.74}$$

In difference to the sub-Ohmic case, the constant $\overline{\sigma}_\infty^2$ is not determined by the low-frequency behavior of $\hat{\gamma}(z)$. For $s > 3$, the behavior of the mean square displacement at asymptotic time depends on the high-frequency properties of the memory-friction.

7.5   Partition function and thermodynamic properties

7.5.1   Partition function

Consider a free spinless particle with dispersion relation $E = (\hbar k)^2/2M$ confined to a one-dimensional box of length L. Mode counting in k-space yields that the number of quantum states in the interval between E and $E + dE$ is $g(E) = \sqrt{M/2E}\, L/\pi\hbar$. The respective partition function then is

$$Z_0(\beta) = \int_0^\infty dE\, g(E)\, e^{-\beta E} = \frac{L}{\hbar} \sqrt{\frac{M}{2\pi\beta}} \,. \tag{7.75}$$

For a damped particle, $Z_0(\beta)$ must be upgraded by the contribution of the quantum fluctuations. This part can be derived from the infinite product in Eq. (6.88) by putting $\omega_0 = 0$. Accordingly, the reduced partition function of the free Brownian particle reads

$$Z(\beta) = Z_0(\beta) \prod_{n=1}^\infty \frac{\nu_n}{\nu_n + \hat{\gamma}(\nu_n)} \,. \tag{7.76}$$

As in Eq. (6.88), the strict Ohmic case is excluded. In the Drude case (3.67) we obtain

$$Z(\beta) = Z_0(\beta) \prod_{n=1}^\infty \frac{\nu_n(\nu_n + \omega_D)}{(\nu_n + \lambda_1)(\nu_n + \lambda_2)} = Z_0(\beta) \frac{\Gamma(1 + \frac{\beta\hbar\lambda_1}{2\pi})\Gamma(1 + \frac{\beta\hbar\lambda_2}{2\pi})}{\Gamma(1 + \frac{\beta\hbar\omega_D}{2\pi})} \,, \tag{7.77}$$

where λ_1, λ_2 are given in Eq. (7.44). Of prevailing interest is the regime $\omega_D \gg \gamma, \nu_1$, in which we have

$$Z(\beta) = Z_0(\beta) \left(\frac{2\pi}{\beta\hbar\omega_D}\right)^{\beta\hbar\gamma/2\pi} \Gamma\left(1 + \frac{\beta\hbar\gamma}{2\pi}\right) \,. \tag{7.78}$$

The expression (7.77) for the reduced partition function holds for arbitrary ω_D, while Eq. (7.78) applies when ω_D is by far the largest frequency of the reduced system.

7.5.2   Internal and free energy

The internal energy $U = -\partial \ln Z/\partial\beta$ is found from the expression (7.76) as

$$U = \frac{1}{2\beta} + \frac{1}{\beta} \sum_{n=1}^\infty \frac{\hat{\gamma}(\nu_n) - \nu_n \partial\hat{\gamma}(\nu_n)/\partial\nu_n}{\nu_n + \hat{\gamma}(\nu_n)} \,. \tag{7.79}$$

As an alternative form, we may write U as the spectral representation (6.96) in which the spectral density $\xi(\omega)$ is expressed in terms of the spectral velocity response function $\tilde{R}(\omega) = 1/[-i\omega + \tilde{\gamma}(\omega)]$ [cf. Eq. (3.115)] as

$$\xi(\omega) = \frac{1}{\pi} \text{Im} \frac{\partial \ln[\tilde{R}(\omega)]}{\partial\omega} \,, \tag{7.80}$$

For Ohmic friction with algebraic Drude cutoff, the density takes the form

$$\xi(\omega) = \frac{1}{\pi}\left[\frac{\lambda_1}{\omega^2 + \lambda_1^2} + \frac{\lambda_2}{\omega^2 + \lambda_2^2} - \frac{\omega_D}{\omega^2 + \omega_D^2}\right]. \tag{7.81}$$

With this form employed in Eq. (6.96), or by differentiation of the second form in Eq. (7.77), we obtain the analytic expression

$$U = \frac{1}{2\beta} + \frac{\hbar\omega_D}{2\pi}\psi\left(1 + \frac{\beta\hbar\omega_D}{2\pi}\right) - \frac{\hbar\lambda_1}{2\pi}\psi\left(1 + \frac{\beta\hbar\lambda_1}{2\pi}\right) - \frac{\hbar\lambda_2}{2\pi}\psi\left(1 + \frac{\beta\hbar\lambda_2}{2\pi}\right). \tag{7.82}$$

When ω_D is very large compared to γ and to ν_1, this form reduces to

$$U = \frac{1}{2\beta} + \frac{\hbar\gamma}{2\pi}\left[1 + \ln\left(\frac{\beta\hbar\omega_D}{2\pi}\right) - \psi\left(\frac{\beta\hbar\gamma}{2\pi}\right)\right]. \tag{7.83}$$

Like for the damped oscillator, the mean energy $\langle E\rangle = \langle p^2\rangle/2M$ of the Brownian particle differs from the internal energy. We obtain with use of Eqs. (7.79) and (7.42)

$$\langle E\rangle = \frac{1}{2\beta} + \frac{1}{\beta}\sum_{n=1}^{\infty}\frac{\hat{\gamma}(\nu_n)}{\nu_n + \hat{\gamma}(\nu_n)} = \int_0^{\infty}d\omega\,\xi_E(\omega)u(\omega,\beta)\,,$$

$$\xi_E(\omega) = \frac{1}{\pi}\frac{\lambda_1\lambda_2}{\lambda_1 - \lambda_2}\left(\frac{1}{\omega^2 + \lambda_2^2} - \frac{1}{\omega^2 + \lambda_1^2}\right). \tag{7.84}$$

Again, $\langle E\rangle$ does not fully take into account the system-reservoir interaction.

Next, consider the low-temperature series for the internal energy U and the free energy F. Writing in the latter case $Z = \kappa\,e^{-\beta F}$, where $\kappa = L\sqrt{M\gamma/2\pi\hbar}$, we obtain from the expression (7.77) the asymptotic series[1]

$$U = \hbar\epsilon_0 + b_1\frac{(k_BT)^2}{\hbar\gamma} - b_2\frac{(k_BT)^4}{(\hbar\gamma)^3} + \mathcal{O}(T^6)\,, \tag{7.85}$$

$$F = \hbar\epsilon_0 - b_1\frac{(k_BT)^2}{\hbar\gamma} + \frac{b_2}{3}\frac{(k_BT)^4}{(\hbar\gamma)^3} + \mathcal{O}(T^6)\,. \tag{7.86}$$

Here $\hbar\epsilon_0 = \hbar[\lambda_1\ln(\omega_D/\lambda_1) + \lambda_2\ln(\omega_D/\lambda_2)]/2\pi$ is the ground-state energy, and

$$b_1 = \frac{\pi}{6}\left(1 - \frac{\gamma}{\omega_D}\right)\,, \quad\text{and}\quad b_2 = \frac{\pi^3}{15}\left(1 - 3\frac{\gamma}{\omega_D} - \frac{\gamma^3}{\omega_D^3}\right). \tag{7.87}$$

The disturbing point with the series (7.85) and (7.86) is that the coefficient b_1 is negative in the regime $\omega_D < \gamma$ [257]. Prima facie, one might be worried with the thermodynamic stability of the open system in this regime. However, with the global system represented by the partition function $Z_{tot} = Z\,Z_R$ in mind, it is evident that the internal energies of the open system and the reservoir add up, $U_{tot} = U + U_R$, and equally the free energies, $F_{tot} = F + F_R$. Now, since the total system is perfectly stable, the free energy U_{tot} is a monotonically increasing function of T. Apparently, this property is not required by physical principles for a subsystem. This baffling

---

[1]To stay within the continuum limit of the density of states, we must have $k_BTL^2 \gg \hbar^2\pi^2/2M$.

phenomenon is not restricted to the Drude cutoff. With the low frequency expansion $\hat{\gamma}(z) = \hat{\gamma}(0) + \hat{\gamma}'(0)z + \mathcal{O}(z^2)$ one finds from the integral representation (6.96)

$$b_1 = \pi[1 + \hat{\gamma}'(0)]/6 . \qquad (7.88)$$

Thus the internal energy of the open system associated with the partition function (7.77) decreases with increasing temperature at very low T when $1 + \hat{\gamma}'(0) < 0$.

We see from the expression (6.96) with (7.81) that the annoying contribution to the internal energy comes from the last term in Eq. (7.81). Since this term does not depend on the coupling parameter γ, it can be eliminated by adding a counter term taken from the reservoir. If we augment the energies U and F with the reservoir contributions ΔU_R and ΔF_R,

$$
\begin{aligned}
\tilde{U} &= U + \Delta U_R , \\
\tilde{F} &= F + \Delta F_R ,
\end{aligned}
\qquad (7.89)
$$

where

$$
\begin{aligned}
\Delta U_R &= \int_0^\infty d\omega \, \frac{1}{\pi} \left(\frac{\omega_D}{\omega^2 + \omega_D^2} - \frac{\bar{\omega}_D}{\omega^2 + \bar{\omega}_D^2} \right) \left(u(\omega, \beta) - \tfrac{1}{2}\hbar\omega \right) , \\
\Delta F_R &= \int_0^\infty d\omega \, \frac{1}{\pi} \left(\frac{\omega_D}{\omega^2 + \omega_D^2} - \frac{\bar{\omega}_D}{\omega^2 + \bar{\omega}_D^2} \right) \left(f(\omega, \beta) - \tfrac{1}{2}\hbar\omega \right) ,
\end{aligned}
\qquad (7.90)
$$

the improper terms are cancelled and $\tilde{U}$ and $\tilde{F}$ are well behaved if we choose $\bar{\omega}_D > \gamma$. In particular, the coefficient of the T^2 contributions in Eqs. (7.85) and (7.86) becomes $b_1 = \pi(1 - \gamma/\bar{\omega}_D)/6 > 0$. The additional term ΔU_R is zero for $T = 0$ as well as in the classical or high temperature limit, $\tilde{U}_{cl} = U_{cl} = 1/2\beta$.

The reservoir contributions in Eq. (7.89) are taken into account if we modify the partition function (7.77) as

$$Z(\beta) \quad \to \quad \tilde{Z}(\beta) = Z(\beta)\, e^{-\beta \Delta F_R} . \qquad (7.91)$$

7.5.3 Specific heat

There are different ways to identify the specific heat [257]. In the first, the specific heat is deduced from the internal energy of the open system, $c_U/k_B = -\beta^2 \partial U / \partial \beta$, while in the second it is derived from the mean energy, $c_E/k_B = -\beta^2 \partial \langle E \rangle / \partial \beta$. Quite generally, whether c_U or c_E is relevant depends on the measurement prescription.

Consider now first the case of Drude friction $\hat{\gamma}(z) = \gamma/(1 + z/\omega_D)$. Use of the expressions (7.82) and (7.43) yields

$$c_U/k_B = \frac{1}{2} + \frac{\hbar^2\beta^2}{4\pi^2} \left\{ \sum_{i=1}^2 \lambda_i^2 \psi'\left(1 + \frac{\beta\hbar\lambda_i}{2\pi}\right) - \omega_D^2 \psi'\left(1 + \frac{\beta\hbar\omega_D}{2\pi}\right) \right\} , \qquad (7.92)$$

$$c_E/k_B = \frac{1}{2} + \frac{\hbar^2\beta^2}{4\pi^2} \frac{\lambda_1\lambda_2}{\lambda_1 - \lambda_2} \left\{ \lambda_2 \, \psi'\left(1 + \frac{\beta\hbar\lambda_2}{2\pi}\right) - \lambda_1 \, \psi'\left(1 + \frac{\beta\hbar\lambda_1}{2\pi}\right) \right\} . \qquad (7.93)$$

If we send ω_D to infinity, the energies U and $\langle E \rangle$ diverge logarithmically with ω_D, whereas the specific heats c_U and c_E become independent of ω_D and coincide in this limit,

$$c_U/k_B = c_E/k_B = \frac{1}{2} - \frac{\beta\hbar\gamma}{2\pi} + \frac{(\beta\hbar\gamma)^2}{4\pi^2}\psi'\left(1 + \frac{\beta\hbar\gamma}{2\pi}\right). \tag{7.94}$$

From this we see that the energy scale which separates the low and high temperature behavior of the specific heat is set by $\hbar\gamma$.

Consider next the specific heat in the high-temperature regime $\beta\hbar\gamma \ll 1$ and $\beta\hbar\omega_D \ll 1$. We then find from Eqs. (7.92) and (7.93) agreement in the classical limit, but already difference by a factor 2 in the leading quantum correction,

$$c_U/k_B = \frac{1}{2} - \frac{1}{12}\frac{\hbar^2\gamma\omega_D}{(k_B T)^2} + \mathcal{O}(T^{-3}),$$

$$c_E/k_B = \frac{1}{2} - \frac{1}{24}\frac{\hbar^2\gamma\omega_D}{(k_B T)^2} + \mathcal{O}(T^{-3}). \tag{7.95}$$

As $T \to 0$, the specific heat vanishes linearly with T, but the prefactors in c_U and c_E are different. We have

$$c_U/k_B = \frac{\pi}{3}\frac{k_B T}{\hbar\gamma}\left(1 - \frac{\gamma}{\omega_D}\right) - \frac{4\pi^3}{15}\left(\frac{k_B T}{\hbar\gamma}\right)^3\left[1 - 3\frac{\gamma}{\omega_D} - \frac{\gamma^3}{\omega_D^3}\right] + \mathcal{O}(T^5), \tag{7.96}$$

$$c_E/k_B = \frac{\pi}{3}\frac{k_B T}{\hbar\gamma} - \frac{4\pi^3}{15}\left(\frac{k_B T}{\hbar\gamma}\right)^3\left[1 - 2\frac{\gamma}{\omega_D}\right] + \mathcal{O}(T^5). \tag{7.97}$$

Already the leading terms $\propto T$ show an important difference. While the term in c_E is positive definite and independent of the cutoff ω_D, the corresponding term in c_U changes sign at $\gamma = \omega_D$ [257]. Apparently, the annoying term $-(\pi/3)k_B T/\hbar\omega_D$ is replaced by $-(\pi/3)k_B T/\hbar\bar{\omega}_D$ if we add to c_U/k_B the term $-\beta^2\partial\Delta U_R/\partial\beta$. With this bath contribution, the specific heat is completely positive if we choose $\bar{\omega} > \gamma$.

At last we study the specific heat at low T for sub- and super-Ohmic friction, $J(\omega) \propto \omega^s$, $s \gtrless 1$. It is convenient to base the analysis on the integral representation

$$c_{U,E}/k_B = \int_0^\infty d\omega\, \xi_{U,E}(\omega)\left(\frac{\beta\hbar\omega/2}{\sinh(\beta\hbar\omega/2)}\right)^2 \tag{7.98}$$

with the spectral densities

$$\xi_E(\omega) \equiv \frac{1}{\pi}\mathrm{Re}\,\tilde{\mathcal{R}}(\omega) = \frac{\sin^2(\frac{1}{2}\pi s)}{\gamma_s}\left(\frac{\omega}{\omega_{ph}}\right)^{1-s}\left[1 + \mathcal{O}(\omega^{2-s})\right],$$

$$\xi_U(\omega) \equiv \frac{1}{\pi}\mathrm{Im}\,\frac{\partial\ln\tilde{\mathcal{R}}(\omega)}{\partial\omega} = (2-s)\frac{\sin^2(\frac{1}{2}\pi s)}{\gamma_s}\left(\frac{\omega}{\omega_{ph}}\right)^{1-s}\left[1 + \mathcal{O}(\omega^{2-s})\right], \tag{7.99}$$

where $\tilde{\mathcal{R}}(\omega) \equiv -i\,\omega M\tilde{\chi}(\omega) = 1/[-i\,\omega + \tilde{\gamma}(\omega)]$ is the spectral velocity response function. With the power law form $\hat{\gamma}(z) = \lambda_s(z/\omega_{ph})^{s-1}$ holding in the regime $0 < s < 2$

[cf. Eq. (3.83)], we obtain the second expressions as the leading terms at low frequency. With these expressions we find the leading low-temperature behavior of the specific heat in the regime $0 < s < 2$ as

$$
c_{\mathrm{E}}/k_{\mathrm{B}} \;=\; \frac{\Gamma(4-s)\zeta(3-s)}{\pi}\,\sin^2(\pi s/2)\frac{\omega_{\mathrm{ph}}}{\gamma_s}\Big(\frac{k_{\mathrm{B}}T}{\hbar\omega_{\mathrm{ph}}}\Big)^{2-s}\Big[1+\mathcal{O}(T^{2-s})\Big]\,,
$$

$$
c_{\mathrm{U}}/k_{\mathrm{B}} \;=\; (2-s)\,c_{\mathrm{E}}/k_{\mathrm{B}}\,,
$$

(7.100)

where $\zeta(z)$ is the Riemann function. These results coincide in the Ohmic case $s = 1$, $\omega_{\mathrm{D}} \to \infty$, with the expressions (7.96) and (7.97).

In summary, the specific heats c_{U} and c_{E} differ in the Ohmic case $s = 1$ by terms depending on the cutoff ω_{D}, and they coincide in the limit $\omega_{\mathrm{D}} \to \infty$. In contrast, in the regime $0 < s < 2$, $s \ne 1$, the specific heats c_{U} and c_{E} already differ in the contributions which are independent of the cutoff.

Finally we compare the specific heat of the free Brownian particle with that of the damped oscillator in the regime $0 < s < 2$. In the former case, the specific heat depends inversely on the friction coefficient γ_s and varies as T^{2-s} at low T. In the latter case, it depends linearly on the friction coefficient and varies as T^s. The qualitative difference is because the static susceptibility at zero temperature is finite for the oscillator, while it is infinite for the free Brownian particle.

7.5.4   Spectral density of states

The anomaly of the internal energy in Eq. (7.85) and of the specific heat c_{U} in Eq. (7.96) appears also in the spectral density of states. Following the lines presented in Subsection 6.5.4 and employing the low-temperature series (7.86) for the free energy, we obtain

$$
\rho(\varepsilon) \;=\; \delta(\varepsilon - \varepsilon_0) + \Theta(\varepsilon - \varepsilon_0)\,\Phi(\varepsilon)\,,
\tag{7.101}
$$

where $\Phi(\varepsilon)$ represents the density of states above the ground state. The resulting low-temperature series is

$$
\Phi(\varepsilon) \;=\; \frac{1}{\gamma}\Big\{b_1 + \frac{b_1^2}{2}\frac{\varepsilon-\varepsilon_0}{\gamma} + \Big(\frac{b_1^2}{12}+\frac{b_2}{6}\Big)\Big(\frac{\varepsilon-\varepsilon_0}{\gamma}\Big)^2 + \mathcal{O}\Big[\Big(\frac{\varepsilon-\varepsilon_0}{\gamma}\Big)^3\Big]\Big\}\,.
\tag{7.102}
$$

From this we see that the spectral density of states $\rho(\varepsilon)$ is negative close above the ground state when $\omega_{\mathrm{D}} < \gamma$, since the coefficient b_1 given in Eq. (7.87) is negative in this regime. Hence the meaning of $\rho(\varepsilon)$ as density of states breaks down in this parameter range. In Subsection 7.5.2, a loophole was shown how the defect can be healed: with the modified partition function $\tilde{Z}(\beta)$ given in Eq. (7.91) as the Laplace transform of $\rho(\varepsilon)$, the spectral density of states is consistently positive.

8. The thermodynamic variational approach

It is appealing to study quantum statistical expectation values with concepts familiar from classical statistical mechanics. The phase space representation of the reduced density matrix in terms of Wigner's function discussed in Subsection 4.4.2 is just one example. Another important concept which goes back to Feynman [258] is the *effective classical potential.*

In the effective classical potential method, all physical effects related to nonlinearity are exactly taken into account in the classical regime. In the quantum regime, the effective potential is optimized by a variational principle with a trial action which is quadratic in the dynamical variables. Feynman originally used a "free particle" trial action. Considerable improvement has been achieved by Giachetti and Tognetti [259], and independently by Feynman and Kleinert [260], by adding a harmonic potential term in the trial action, thereby introducing a second variational parameter. Further generalization is obvious: since linear dissipation is rendered by a "quadratic" action functional, it can be included in the variational approach with minor ancillary efforts.

In the following I present the main ideas. To be general, the discussion will be given for open systems with arbitrary linear dissipation.

8.1 Centroid and the effective classical potential

8.1.1 Centroid

A useful path decomposing concept has been developed by Feynman [3, 4]. The basic idea is to divide up the paths into equivalence classes and to rewrite the original path sum as a sum over paths belonging to the same equivalence class and an additional sum over the equivalence classes. In connection with the *effective classical potential*, it is convenient to consider as equivalence class all those paths which have the same mean position $q_c[q(\cdot)]$, usually referred to as the *centroid*,[1]

$$q_c[q(\cdot)] = \frac{1}{\hbar\beta} \int_0^{\hbar\beta} d\tau \, q(\tau) . \tag{8.1}$$

Within the centroid method, the path sum (4.60) for the reduced density matrix in thermal equilibrium is rearranged as a sum over all paths which are constrained to the same centroid and an additional ordinary integral over the centroid coordinate. We then have (in this chapter we use the normalization $\mathrm{tr}\,\rho = Z$)

$$\rho(q'', q') = \int_{-\infty}^{\infty} dq_c \, \rho(q'', q'; q_c) ,$$

$$\rho(q'', q'; q_c) \equiv \int_{q(0)=q'}^{q(\hbar\beta)=q''} \mathcal{D}q(\cdot) \, \delta\Big(q_c - q_c[q(\cdot)]\Big) \exp\Big(- S_{\mathrm{eff}}^{\mathrm{E}}[q(\cdot)]/\hbar \Big) . \tag{8.2}$$

---

[1]The meaning of the central dot in $q(\cdot)$ is explained in the footnote to Eq. (4.22).

Correspondingly, we may also write the partition function $Z = \int dq'\, \rho(q', q')$ in the form of an integral over the centroid density,

$$Z = \int_{-\infty}^{\infty} dq_c\, \rho_c(q_c) \,, \tag{8.3}$$

$$\rho_c(q_c) \equiv \oint \mathcal{D}q(\cdot)\, \delta\Big(q_c - q_c[q(\cdot)]\Big) \exp\Big(-S_{\mathrm{eff}}^{\mathrm{E}}[q(\cdot)]/\hbar\Big) \,. \tag{8.4}$$

To proceed, we Fourier-expand the periodic path $q(\tau)$ with period $\hbar\beta$,

$$q(\tau) = q_c + \tilde{q}(\tau) = Q_0 + \sum_{\substack{n=-\infty \\ n\neq 0}}^{\infty} Q_n\, e^{i\nu_n \tau} \,. \tag{8.5}$$

The Fourier coefficient Q_0 represents the centroid q_c, as follows from definition (8.1). Next, we use the functional integral measure (4.228) and write the path sum in (8.2) as an infinite product of ordinary integrals. We then get the centroid density as

$$\rho_c(q_c) = \Big(M/2\pi\hbar^2\beta\Big)^{1/2} \exp\Big(-\beta\mathcal{V}(q_c)\Big) \,, \tag{8.6}$$

in which $\mathcal{V}(q_c)$ is the *effective classical potential*,

$$
\begin{aligned}
e^{-\beta\mathcal{V}(q_c)} = \oint \mathcal{D}\tilde{q} \,\exp\bigg(&- M\beta \sum_{n=1}^{\infty} \big[\nu_n^2 + \nu_n \hat{\gamma}(\nu_n)\big]\,|Q_n|^2 \\
&- \frac{1}{\hbar}\int_0^{\hbar\beta} d\tau\, V\,[\,q_c + \tilde{q}(\tau)\,]\bigg) \,.
\end{aligned}
\tag{8.7}
$$

The functional integral measure is limited to pure quantum fluctuations. It reads

$$\oint \mathcal{D}\tilde{q} \times \cdots \equiv \prod_{n=1}^{\infty} \int_{-\infty}^{\infty}\int_{-\infty}^{\infty} \frac{d\mathrm{Re}\,Q_n\, d\mathrm{Im}\,Q_n}{\pi/M\beta\nu_n^2} \times \cdots \,. \tag{8.8}$$

The expression (8.6) with (8.7) and (8.8) is still exact. In this decomposition, the quantum nature of the system is contained in the effective classical potential $\mathcal{V}(q_c)$.

8.1.2   The effective classical potential

Due to the kinetic action term $M\beta\nu_n^2|Q_n|^2$ in the exponent of (8.7), the Q_n are effectively of order $1/(\nu_n\sqrt{M\beta}) \propto 1/\sqrt{T}$. Therefore, the quantum fluctuations $\tilde{q}(\tau)$ in the potential term in Eq. (8.7) decrease when temperature is increased. In the classical limit, $V[q(\tau)]$ becomes independent of $\tilde{q}(\tau)$. Hence in this limit all the integrals in Eq. (8.7) are Gaussian. Since, in addition, $\hat{\gamma}(\nu_n)/\nu_n \to 0$ in the classical limit, the fluctuation integrals simply give a factor unity. Thus we have

$$\lim_{T\to\infty} \mathcal{V}(q_c) = V(q_c) \,, \tag{8.9}$$

and hence

$$\lim_{T\to\infty} Z \to Z^{(\mathrm{cl})} = \Big(M/2\pi\hbar^2\beta\Big)^{1/2} \int_{-\infty}^{\infty} dq_c\, \exp\Big(-\beta V(q_c)\Big) \,. \tag{8.10}$$

This consideration shows that the quantum statistical partition function of any particular smooth potential reduces to the classical partition function in the high temperature limit. Because of the similarity of Eq. (8.6) with (8.10), the function $\mathcal{V}(q_c)$ has been dubbed by Feynman and Kleinert [260] the *effective classical potential*.

To obtain first insights, consider the effective classical potential $\mathcal{V}(q_c)$ for the damped harmonic oscillator. Now, all the integrals contained in Eq. (8.7) with (8.8) are in strict Gaussian form at any temperature. We then get

$$\mathcal{V}_{\mathrm{osc}}(q_c) = V_{\mathrm{osc}}(q_c) - (1/\beta) \ln \mu_{\mathrm{osc}}(\omega_0) \,, \tag{8.11}$$

where μ_{osc} is the pure quantum part of the partition function (6.88),

$$\mu_{\mathrm{osc}}(\omega_0) = \prod_{n=1}^{\infty} \frac{\nu_n^2}{\nu_n^2 + \nu_n \hat{\gamma}(\nu_n) + \omega_0^2} = \beta \hbar \omega_0 Z_{\mathrm{osc}}(\beta) \,. \tag{8.12}$$

For the Drude model (3.67), μ_{osc} can be expressed in terms of gamma functions,

$$\mu_{\mathrm{osc}}(\omega_0) = \left(\frac{\beta \hbar \omega_0}{2\pi} \right)^2 \frac{\Gamma(\lambda_1/\nu_1)\,\Gamma(\lambda_2/\nu_1)\,\Gamma(\lambda_3/\nu_1)}{\Gamma(\omega_{\mathrm{D}}/\nu_1)} \,, \tag{8.13}$$

where λ_1, λ_2, and λ_3 are given in Eq. (6.47) with Eqs. (6.45) and (6.46). In the absence of damping, we have

$$\mu_{\mathrm{osc}}(\omega_0) = \frac{\hbar \omega_0 / 2 k_{\mathrm{B}} T}{\sinh(\hbar \omega_0 / 2 k_{\mathrm{B}} T)} \,. \tag{8.14}$$

Alternatively, we may express $\mathcal{V}_{\mathrm{osc}}(q_c)$ in terms of the free energy $F_{\mathrm{osc}} = -\ln Z_{\mathrm{osc}}/\beta$,

$$\mathcal{V}_{\mathrm{osc}}(q_c) = V_{\mathrm{osc}}(q_c) - (1/\beta) \ln(\beta \hbar \omega_0) + F_{\mathrm{osc}} \,. \tag{8.15}$$

Thus for the damped harmonic oscillator, $\mathcal{V}_{\mathrm{osc}}(q_c)$ differs from $V_{\mathrm{osc}}(q_c)$ only by a temperature-dependent energy shift. At high temperatures, the shift vanishes as $1/T$, $\mathcal{V}_{\mathrm{osc}}(q_c) - V_{\mathrm{osc}}(q_c) = \hbar^2 [\omega_0^2 + \gamma \omega_{\mathrm{D}}]/12 k_{\mathrm{B}} T + \mathcal{O}(1/T^3)$. In the opposite limit $T \to 0$, we have

$$\mathcal{V}_{\mathrm{osc}}(q_c) = V_{\mathrm{osc}}(q_c) + \hbar \varepsilon_0 \,, \tag{8.16}$$

where $\hbar \varepsilon_0$ is the ground state energy of the oscillator given in Eq. (6.103) or (6.104).

Utilization of the effective classical potential $\mathcal{V}(q_c)$ is physically very appealing. Use of $\mathcal{V}(q_c)$ for anharmonic potentials may seem of little aid, however, since it is impossible to carry out the integrals over all Q_n in Eq. (8.7) in analytic form. Fortunately, there is a variational method in which the path integral in question is compared with the path integral of a simpler problem solvable in analytic form. There is a minimum principle which leads to an upper bound for the effective classical potential. Upon choosing a suitable trial action, the partition function can be calculated quite accurately in the whole temperature range for many anharmonic systems.

8.2 Variational method

I now discuss a method which combines the path integral approach with a variational principle. The minimum principle is based on *Jensen's inequality* for convex functions $f(x)$, e.g. $f(x) = e^{-x}$, in probability theory.[2] For x being random, the average value

---

[2]For a discussion of bounds in quantum partition functions, cf. Ref. [261].

of $f(x)$ always exceeds or equals the function of the average value of x, as long as x is real and the distribution of the underlying stochastic process is positive,

$$\langle f(x) \rangle \geq f(\langle x \rangle) . \tag{8.17}$$

This inequality for convex functions can be generalized to an exponential functional. Therefore it is very powerful in quantum statistical mechanics.

8.2.1 Variational method for the free energy

Suppose we have a trial action S_{tr} that is easier to work with. Then we may write

$$Z \equiv e^{-\beta F} = \oint \mathcal{D}q \, e^{-(S-S_{\text{tr}})/\hbar} \, e^{-S_{\text{tr}}/\hbar} = e^{-\beta F_{\text{tr}}} \left\langle e^{-(S-S_{\text{tr}})/\hbar} \right\rangle_{\text{tr}} , \tag{8.18}$$

where $Z_{\text{tr}} \equiv \exp(-\beta F_{\text{tr}}) = \oint \mathcal{D}q \exp(-S_{\text{tr}}/\hbar)$ is the partition function connected with S_{tr}, and $\langle O \rangle_{\text{tr}}$ is the average of the observable $O[q]$ with weight $\exp(-S_{\text{tr}}[q]/\hbar)$,

$$\langle O \rangle_{\text{tr}} \equiv \frac{\oint \mathcal{D}q(\cdot) \, O[q(\cdot)] \exp\left(-S_{\text{tr}}[q(\cdot)]/\hbar\right)}{\oint \mathcal{D}q(\cdot) \exp\left(-S_{\text{tr}}[q(\cdot)]/\hbar\right)} . \tag{8.19}$$

Using (8.17), we obtain from (8.18) the lower bound for the partition function

$$\exp(-\beta F) \geq \exp(-\beta F_{\text{tr}}) \exp\left(-\langle S - S_{\text{tr}} \rangle_{\text{tr}}/\hbar\right) . \tag{8.20}$$

This implies Feynman's upper bound for the free energy [3, 4],

$$F \leq F_{\text{tr}} + \langle S - S_{\text{tr}} \rangle_{\text{tr}}/\hbar\beta . \tag{8.21}$$

The minimum principle (8.21) has been used to determine the "best" free energy from a trial action $S_{\text{tr}}[q(\cdot)]$ for which the path integral can be worked out.

8.2.2 Variational method for the effective classical potential

Similar to the derivation of a global bound for the free energy, we can also deduce a local bound for the effective classical potential $\mathcal{V}(q_c)$, which is somewhat simpler to study. Since we wish to perform the variational principle in the quantum fluctuation part, we introduce a trial potential $V_{\text{tr}}(\tilde{q}; q_c)$ depending on $\tilde{q} = q - q_c$, and also the corresponding trial action $S_{\text{tr}}[\tilde{q}(\cdot); q_c]$. Here the extra variable q_c is to indicate parametric dependence of the trial potential and of the trial action on the centroid q_c. By this ansatz, both the trial potential and the trial action become *nonlocal*.

For notational distinction from the full average $\langle O \rangle$, we introduce the double bracket symbol $\langle\!\langle O[q(\cdot)] \rangle\!\rangle_{q_c}^{(\text{tr})}$ in order to denote the average of the observable $O[q(\cdot)]$ over the pure quantum fluctuations $\tilde{q}(\cdot)$ with weight $\exp\{-S_{\text{tr}}[\tilde{q}(\cdot); q_c]/\hbar\}$,

$$\langle\!\langle O[q(\cdot)] \rangle\!\rangle_{q_c}^{(\mathrm{tr})} \equiv \frac{\oint \mathcal{D}\tilde{q}(\cdot)\, O[q_c + \tilde{q}(\cdot)] \exp\left\{ - S_{\mathrm{tr}}[\tilde{q}(\cdot); q_c]/\hbar \right\}}{\oint \mathcal{D}\tilde{q}(\cdot)\, \exp\left\{ - S_{\mathrm{tr}}[\tilde{q}(\cdot); q_c]/\hbar \right\}}. \tag{8.22}$$

With the definition (8.22), the effective classical potential may be written as

$$e^{-\beta \mathcal{V}(q_c)} = \mu_{\mathrm{tr}}(q_c) \left\langle\!\left\langle \exp\left\{ - \int_0^{\hbar\beta} d\tau \left(V[q(\tau)] - V_{\mathrm{tr}}[\tilde{q}(\tau); q_c] \right)/\hbar \right\} \right\rangle\!\right\rangle_{q_c}^{(\mathrm{tr})}, \tag{8.23}$$

where $\mu_{\mathrm{tr}}(q_c)$ is the pure quantum contribution to the trial partition function,

$$\mu_{\mathrm{tr}}(q_c) \equiv e^{-\beta F_{\mathrm{tr}}(q_c)} = \oint \mathcal{D}\tilde{q}(\cdot)\, \exp\left\{ - S_{\mathrm{tr}}[\tilde{q}(\cdot); q_c]/\hbar \right\}. \tag{8.24}$$

Jensen's inequality (8.17) yields $\exp[-\beta \mathcal{V}(q_c)] \geq \exp[-\beta \mathcal{V}_{\mathrm{tr}}(q_c)]$, which implies for the effective classical potential the upper bound

$$\mathcal{V}(q_c) \leq \mathcal{V}_{\mathrm{tr}}(q_c). \tag{8.25}$$

The trial effective potential is given by

$$\mathcal{V}_{\mathrm{tr}}(q_c) = \left\langle\!\left\langle \int_0^{\hbar\beta} d\tau \left(V[q(\tau)] - V_{\mathrm{tr}}[\tilde{q}(\tau); q_c] \right)/\hbar\beta \right\rangle\!\right\rangle_{q_c}^{(\mathrm{tr})} - \frac{1}{\beta} \ln \mu_{\mathrm{tr}}(q_c). \tag{8.26}$$

Consider first an undamped system and use the trial action of a free particle,

$$S_{\mathrm{tr}}[\tilde{q}; q_c] \;\rightarrow\; S_0[q] = S_0[\tilde{q}] = M\hbar\beta \sum_{n=1}^{\infty} \nu_n^2 |Q_n|^2. \tag{8.27}$$

The simple trial action (8.27) does not depend on the centroid. Using the functional integration measure (8.8), we find from Eq. (8.24)

$$\mu_0 = 1, \tag{8.28}$$

showing that μ is purely classical for a free particle trial action. The computation of the average in Eq. (8.26) is straightforward by writing $V(q)$ as Fourier integral and by completing the square of the Q_n in the exponent. The resulting upper bound for the effective potential is a fuzzy copy of the original potential $V(q)$ [4],

$$\mathcal{V}_0(q_c) = \langle\!\langle V[q(\cdot)] \rangle\!\rangle_{q_c}^{(0)}. \tag{8.29}$$

Substituting the form (8.27) into Eq. (8.22), we obtain

$$\langle\!\langle V[q(\cdot)] \rangle\!\rangle_{q_c}^{(0)} \equiv \frac{1}{\sqrt{2\pi \langle \xi^2 \rangle_0}} \int_{-\infty}^{\infty} dq\, V(q) \exp\left(- \frac{(q - q_c)^2}{2\langle \xi^2 \rangle_0} \right). \tag{8.30}$$

The convolution integral smears the original potential $V(q)$ out over a length scale $\sqrt{\langle \xi^2 \rangle_0}$. The mean square spread of the Gaussian weight function in Eq. (8.30) is

$$\langle \xi^2 \rangle_0 \equiv \langle \tilde{q}^2 \rangle_0 = \frac{2}{M\beta} \sum_{n=1}^{\infty} \frac{1}{\nu_n^2} = \frac{\hbar^2 \beta}{12M} . \tag{8.31}$$

The Gaussian smearing of the original potential in the trial effective classical potential $V_0(q_c)$ on the scale (8.31) is determined by the quantum fluctuations of a free particle.

A considerable improvement on Feynman's original variational method has been put forward by Giachetti and Tognetti [259], and independently by Feynman and Kleinert [260]. The essence of the refined variational ansatz is to include a harmonic potential in the trial action,

$$V_h[q(\tau); q_c] = \frac{M}{2} \Omega^2(q_c) \Big(q(\tau) - q_c \Big)^2 = \frac{M}{2} \Omega^2(q_c) \, \tilde{q}^2(\tau) , \tag{8.32}$$

where $\Omega(q_c)$ is a local trial frequency to be optimally chosen. The trial potential $V_h[q; q_c]$ is nonlocal because of the parametric dependence on the centroid q_c.

At this stage, we find it appropriate to generalize the discussion to include dissipation. For damped systems, it is natural to incorporate the dissipative action into the trial action. Thus we choose as trial action the general harmonic form

$$S_{tr}[\tilde{q}; q_c]/\hbar \quad \rightarrow \quad S_h[\tilde{q}; q_c]/\hbar = M\beta \sum_{n=1}^{\infty} \Big(\nu_n^2 + \nu_n \hat{\gamma}(\nu_n) + \Omega^2(q_c) \Big) |Q_n|^2 . \tag{8.33}$$

Substituting this form into Eq. (8.24), we find

$$\mu_h[\Omega(q_c)] \equiv e^{-\beta F_h[\Omega(q_c)]} = \prod_{n=1}^{\infty} \frac{\nu_n^2}{\nu_n^2 + \nu_n \hat{\gamma}(\nu_n) + \Omega^2(q_c)} = \beta\hbar\Omega(q_c) Z_h(\beta) . \tag{8.34}$$

The quantity $F_h[\Omega(q_c)]$ is the quantum part of the free energy, and $Z_h(\beta)$ is the full partition function of the trial damped harmonic oscillator [cf. Eq.(6.88)]. Performing the average in Eq. (8.26) with the form (8.33) in the exponent of the weight function, we find the variational classical effective potential

$$V_h(q_c) = \langle\!\langle V[q(\cdot)] \rangle\!\rangle_{q_c}^{(h)} + F_h[\Omega(q_c)] - \frac{M}{2} \Omega^2(q_c) \langle \xi^2 \rangle_h . \tag{8.35}$$

Here $\langle\!\langle V[q(\cdot)] \rangle\!\rangle_{q_c}^{(h)}$ is the original potential $V(q)$ smeared with a Gaussian of half-width $\langle \xi^2 \rangle_h$ analogous to Eq. (8.30). The mean spread is given by

$$\langle \xi^2 \rangle_h \equiv \langle \tilde{q}^2 \rangle_h = \frac{2}{M\beta} \sum_{n=1}^{\infty} \frac{1}{\nu_n^2 + \nu_n \hat{\gamma}(\nu_n) + \Omega^2(q_c)} . \tag{8.36}$$

Comparing the expression (8.36) with Eq. (6.131) we see that $\langle \xi^2 \rangle_h$ differs from the full coordinate dispersion by the missing $n = 0$ term, which is the classical dispersion $\langle q^2 \rangle_{\Omega}^{(cl)} = 1/M\beta\Omega^2(q_c)$. While the missing term grows linearly with T, the fluctuation width $\langle \xi^2 \rangle_h$ shrinks with $1/T$ at large T, as can be seen from Eq. (6.140). The width $\langle \xi^2 \rangle_h$ is the pure quantum coordinate dispersion of the damped harmonic oscillator which is finite at all T,

$$\langle \xi^2 \rangle_{\mathrm{h}} = \langle q^2 \rangle_{\Omega} - \langle q^2 \rangle_{\Omega}^{(\mathrm{cl})} . \tag{8.37}$$

The optimal trial function $\Omega^2(q_c)$ is determined by minimization of $\mathcal{V}_{\mathrm{h}}(q_c)$,

$$\frac{d\mathcal{V}_{\mathrm{h}}(q_c)}{d\Omega^2(q_c)} = \frac{\partial \mathcal{V}_{\mathrm{h}}(q_c)}{\partial \Omega^2(q_c)} + \frac{\partial \mathcal{V}_{\mathrm{h}}(q_c)}{\partial \langle \xi^2 \rangle_{\mathrm{h}}} \frac{\partial \langle \xi^2 \rangle_{\mathrm{h}}}{\partial \Omega^2(q_c)} = 0 . \tag{8.38}$$

Using the expressions (8.34) and (8.36), we obtain $\partial F_{\mathrm{h}}/\partial \Omega^2 = \frac{1}{2}M\langle \xi^2 \rangle_{\mathrm{h}}$. With this we find from the defining expression (8.35) that $\partial \mathcal{V}_{\mathrm{h}}(q_c)/\partial \Omega^2(q_c)$ vanishes. Thus Eq. (8.38) reduces to $\partial \mathcal{V}_{\mathrm{h}}(q_c)/\partial \langle \xi^2 \rangle_{\mathrm{h}} = 0$. From this we see that the best choice is

$$\Omega^2(q_c) = \frac{2}{M} \frac{\partial \langle\!\langle V[q(\cdot)] \rangle\!\rangle_{q_c}^{(\mathrm{h})}}{\partial \langle \xi^2 \rangle_{\mathrm{h}}} . \tag{8.39}$$

Combining Eq. (8.39) with (8.36), we get in general a transcendental equation for the function $\Omega^2(q_c)$, which has to be solved numerically. The expressions (8.36) and (8.39) represent the optimal choice for the functions $\langle \xi^2 \rangle_{\mathrm{h}}$ and $\Omega^2(q_c)$ within the trial ansatz (8.33). With the optimal choice for $\langle \xi^2 \rangle_{\mathrm{h}}$ and $\Omega(q_c)$, the function $\mathcal{V}_{\mathrm{h}}(q_c)$ gives the best upper local bound for the true effective classical potential $\mathcal{V}(q_c)$, $\mathcal{V}(q_c) \leq \mathcal{V}_{\mathrm{h}}(q_c)$. Accordingly, we find for the partition function the best lower bound

$$Z \geq \left(\frac{M}{2\pi\hbar^2\beta} \right)^{1/2} \int_{-\infty}^{\infty} dq_c \, \exp[-\beta \mathcal{V}_{\mathrm{h}}(q_c)] . \tag{8.40}$$

For anharmonic systems at high temperature, the centroid integral has to be treated in most cases numerically. The second derivative of $\mathcal{V}_{\mathrm{h}}(q_c)$ with respect to $\Omega^2(q_c)$ is non-negative, as can be easily demonstrated. This implies that the extremal effective potential $\mathcal{V}_{\mathrm{h}}(q_c)$ is a local minimum. Clearly, for the particular case of the damped harmonic oscillator, $V(q) \propto q^2$, the variational method reproduces the exact partition function. On the other hand, in the limit $\Omega \to 0$, Feynman's original (non-optimal) choice is recovered. By construction, the variational result $\mathcal{V}_{\mathrm{h}}(q_c)$ for $\mathcal{V}(q_c)$ becomes more accurate as the temperature is increased. Janke has shown [262] that the high temperature expansion of $\mathcal{V}_{\mathrm{h}}(q_c)$ agrees with the Wigner expansion of $\mathcal{V}(q_c)$ up to terms of order β^3. In the classical limit, the mean spread $\langle \xi^2 \rangle_{\mathrm{h}}$ becomes zero and $\mu_{\mathrm{h}}(q_c) \to 1$, so that $\langle\!\langle V[q(\cdot)] \rangle\!\rangle_{q_c}^{(\mathrm{h})}$ is reduced to the original potential $V(q_c)$ and (8.40) maps on the classical partition function (8.10). In the opposite limit $T \to 0$, the variational method corresponds to a Gaussian trial ansatz for the ground state [260]. The Gaussian function is known to give extremely good ground-state energies for smooth single-well potentials. Thus, the effective classical potential turns out to be a good concept both in the high- and in the low-temperature limit, and in many cases at any temperature. Numerically, the approach gives quite reliable results for many smooth potentials in the whole temperature range as emphasized in Refs. [6], [262] – [264]. The variational method works surprisingly well for hard-core potentials as well as for singular potentials except at very low temperatures where the cusp-like shape of the true wave functions becomes more important. The method has also been applied successfully to nonlinear field theories [265]. A friction term has been included in the variational method in Ref. [266].

In the path integral approach, imaginary-time position correlation functions are treated by adding the source term $\int_0^{\hbar\beta} d\tau'\, j(\tau') q(\tau')$ to the Euclidean action and performing functional differentiations with respect to the source $j(\tau)$. Using this method, we can also study position correlation functions within the variational method. The corresponding treatment is straightforward [6]. Analytic continuation to real time may be performed in the end.

A different view on the effective classical potential method has been put forward recently. It has been shown in Ref. [267] that, for standard Hamiltonians, the variational method is equivalent to a scheme in which the pure quantum fluctuations are treated self-consistently in a harmonic approximation (PQSCHA method). For nonstandard Hamiltonians like those describing magnetic systems, the Feynman-Jensen inequality is generally not applicable. The PQSCHA in phase space representation (see Subsection 8.2.4) is still meaningful for nonstandard systems and allows to construct an effective classical potential. For a review and survey of applications of the PQSCHA, see Ref. [267, b].

8.2.3 Variational perturbation theory

A systematic improvement of the variational approach consists in performing a variational perturbation expansion [268, 6]. In the first step, the anharmonic action is expanded in powers of the fluctuations about the centroid, $\tilde{q}(\tau) = q(\tau) - q_c$,

$$S[q(\cdot)] = \hbar\beta V(q_c) + S_{\mathrm{h}}[\tilde{q}(\cdot); q_c] + S_{\mathrm{int}}[\tilde{q}(\cdot); q_c]\,, \tag{8.41}$$

$$S_{\mathrm{int}}[\tilde{q}(\cdot); q_c] = \int_0^{\hbar\beta} d\tau\, V_{\mathrm{int}}[\tilde{q}(\tau); q_c]\,, \tag{8.42}$$

where $V_{\mathrm{int}}(\tilde{q}; q_c)$ is the deviation of the full potential from the trial harmonic potential,

$$V_{\mathrm{int}}(\tilde{q}(\tau); q_c) \equiv \frac{1}{2!}\Big(V''(q_c) - M\Omega^2(q_c)\Big)\tilde{q}^2(\tau) + \frac{1}{3!}V'''(q_c)\,\tilde{q}^3(\tau) + \cdots\,, \tag{8.43}$$

and where the prime denotes differentiation with respect to the coordinate. Substituting the expansion (8.43) into Eq. (8.23), we obtain the variational perturbation expansion for the effective classical potential in the form of a cumulant expansion,

$$\mathcal{V}(q_c) = \mathcal{V}_{\mathrm{h}}(q_c) - \frac{1}{2!\,\hbar^2\beta}\left[\langle\!\langle S_{\mathrm{int}}^2\rangle\!\rangle_{q_c}^{(\mathrm{h})} - \big(\langle\!\langle S_{\mathrm{int}}\rangle\!\rangle_{q_c}^{(\mathrm{h})}\big)^2 \right] \tag{8.44}$$

$$+ \frac{1}{3!\,\hbar^3\beta}\left[\langle\!\langle S_{\mathrm{int}}^3\rangle\!\rangle_{q_c}^{(\mathrm{h})} - 3\,\langle\!\langle S_{\mathrm{int}}^2\rangle\!\rangle_{q_c}^{(\mathrm{h})}\langle\!\langle S_{\mathrm{int}}\rangle\!\rangle_{q_c}^{(\mathrm{h})} + 2\,\big(\langle\!\langle S_{\mathrm{int}}\rangle\!\rangle_{q_c}^{(\mathrm{h})}\big)^3 \right] \pm\cdots\,.$$

The expression $\langle\!\langle S_{\mathrm{int}}^m\rangle\!\rangle_{q_c}^{(\mathrm{h})}$ involves the average

$$\langle\!\langle\, V_{\mathrm{int}}[\tilde{q}(\tau_1); q_c]\, V_{\mathrm{int}}[\tilde{q}(\tau_2); q_c]\cdots V_{\mathrm{int}}[\tilde{q}(\tau_m); q_c]\,\rangle\!\rangle_{q_c}^{(\mathrm{h})} \tag{8.45}$$

over the quantum fluctuations $\tilde{q}(\cdot)$ with weight function $\exp\{-S_{\mathrm{h}}[\tilde{q}(\cdot); q_c]/\hbar\}$, where $S_{\mathrm{h}}[\tilde{q}(\cdot); q_c]$ is given in Eq. (8.33). The smearing formula can be found upon spatial Fourier expansion of $V_{\mathrm{int}}[\tilde{q}(\tau); q_c]$. The average of the Fourier factors yields

$$\left\langle\!\!\left\langle \prod_{\ell=1}^{m} \exp[\, i k_\ell \tilde{q}(\tau_\ell)\,] \right\rangle\!\!\right\rangle_{q_c}^{(h)} = \exp\left[-\frac{1}{2} \sum_{i=1}^{m} \sum_{j=1}^{m} k_i \xi_{\tau_i,\tau_j}^2 k_j \right], \qquad (8.46)$$

where we have introduced the pure quantum coordinate autocorrelation function

$$\xi_{\tau,\tau'}^2 \equiv \langle \tilde{q}(\tau)\tilde{q}(\tau') \rangle_{\mathrm{h}} = \frac{2}{M\beta} \sum_{n=1}^{\infty} \frac{\cos[\,\nu_n(\tau-\tau')\,]}{\nu_n^2 + \nu_n \hat{\gamma}(\nu_n) + \Omega^2(q_c)} . \qquad (8.47)$$

The Fourier inversion of Eq. (8.46) gives again a Gaussian form, but with the inverse matrix. Thus we find in generalization of Eq. (8.30) the smearing formula [269]

$$\langle\!\langle F_1[q(\tau_1)]\, F_2[q(\tau_2)]\cdots F_m[q(\tau_m)] \rangle\!\rangle_{q_c}^{(h)} = \left(\prod_{k=1}^{m} \int_{-\infty}^{\infty} dq_k\, F_k(q_k) \right) \qquad (8.48)$$

$$\times \frac{1}{\sqrt{(2\pi)^m \det\left[\,\xi_{\tau,\tau'}^2\,\right]}} \exp\left[-\frac{1}{2} \sum_{i=1}^{m} \sum_{j=1}^{m} (q_i - q_c)\xi_{\tau_i,\tau_j}^{-2}(q_j - q_c) \right],$$

where $\xi_{\tau,\tau'}^{-2}$ is the inverse of the $m \times m$-matrix $\xi_{\tau,\tau'}^2$.

Truncation of the series (8.44) after the nth order defines the effective classical potential in order n, $\mathcal{V}^{(n)}(q_c)$. The first order corresponds to the variational effective classical potential, $\mathcal{V}^{(1)}(q_c) = \mathcal{V}_{\mathrm{h}}(q_c)$. The infinite sum (8.44) is independent of the variational frequency $\Omega(q_c)$ by construction. For any finite order n, the optimal choice for $\Omega(q_c)$ is determined by the relation

$$\partial \mathcal{V}^{(n)}(q_c)/\partial \Omega(q_c) = 0 . \qquad (8.49)$$

The optimal frequency $\Omega^{(n)}(q_c)$ for a given order n has been dubbed *frequency of least dependence*. The variational perturbation expansion (8.44) is an alternating series, and the last cumulant in each order is positive for odd orders. This property leads to the following behavior. For odd n, $n = 2m + 1$, $\mathcal{V}^{(2m+1)}(q_c)$ is a minimum in Ω at the optimal frequency, whereas the second derivative for $n = 2m$ even, $\partial^2 \mathcal{V}^{(2m)}(q_c)/(\partial\Omega)^2$, may change sign at the optimal frequency. With increasing odd or even n, the potential $\mathcal{V}^{(n)}(q_c)$ develops as a function of Ω a flat plateau at the optimal frequency. It has been demonstrated for the anharmonic oscillator and quartic double well that the third order effective classical potential $\mathcal{V}^{(3)}(q_c)$ is by far more accurate than the first order effective classical potential $\mathcal{V}^{(1)}(q_c)$.

A discussion of the smearing formula (8.48) and application of the variational perturbation method to the Coulomb potential for zero friction is given in Ref. [269].

For the anharmonic oscillator $V(q) = M\omega^2 q^2/2 + g\, q^4/4$, the energy eigenvalues $\{E_\nu(g)\}$ possess an essential singularity at $g = 0$. Thus, ordinary perturbation expansion in g has zero radius of convergence. The physical origin of the singular behavior is the possibility for tunneling when g is negative. Therefore, the form of the singularity may be analyzed using semiclassical methods. The important point now is that the variational perturbation expansion involves convergent infinite-order summations. This changes the ordinary expansion into a systematic expansion which has uniform convergence for arbitrary coupling strength, including strong-coupling [6].

8.2.4 Expectation values in coordinate and phase space

The results of Subsection 8.2.1 can be used directly to formulate quantum statistical averages of operators being functions of the position operator, $\hat{A}(\hat{q})$, or more generally of the position and momentum operator, $\hat{A}(\hat{p}, \hat{q})$. In the remainder of this subsection, we restrict the attention to Weyl ordering $\mathcal{Z}_0\{\hat{A}(\hat{p}, \hat{q})\}$, which is the case $\gamma = 0$ in Subsection 4.4.1. For the harmonic trial ansatz (8.33) or the PQSCHA, the optimally chosen restricted equilibrium density matrix $\rho(q'', q'; q_c)$ defined in Eq. (8.2) takes the particular form [270]

$$\rho(q'', q'; q_c) = \frac{\sqrt{M}\, e^{-\beta \mathcal{V}_{\rm h}(q_c)}}{2\pi\hbar\sqrt{\beta \langle \xi^2 \rangle_{\rm h}}} \exp\left(-\frac{(q'' + q' - 2q_c)^2}{8\langle \xi^2 \rangle_{\rm h}} - \frac{\langle p^2 \rangle_{\rm h}}{2\hbar^2}(q'' - q')^2 \right) . \quad (8.50)$$

The effective classical potential $\mathcal{V}_{\rm h}(q_c)$ and the mean square variation $\langle \xi^2 \rangle_{\rm h}$ are given in Eqs. (8.35) and (8.36). The momentum dispersion of the trial damped oscillator is

$$\langle p^2 \rangle_{\rm h} = \frac{M}{\beta} \sum_{n=-\infty}^{\infty} \frac{\Omega^2(q_c) + |\nu_n| \hat{\gamma}(|\nu_n|)}{\Omega^2(q_c) + \nu_n^2 + |\nu_n| \hat{\gamma}(|\nu_n|)} , \quad (8.51)$$

which is discussed in some detail in Section 6.6. The form (8.50) immediately follows from undoing the integration over the centroid in the expression (6.157) [cf. Eq. (8.2)].

The restricted Wigner function is a phase space density distribution which is connected with $\rho(q'', q'; q_c)$ as given by the correspondence (4.263). The restricted Wigner function resulting from Eq. (8.50) reads

$$f^{(\rm W)}(p, q; q_c) = \frac{\exp\left(-\beta \mathcal{V}_{\rm h}(q_c) - (q - q_c)^2 / 2\langle \xi^2 \rangle_{\rm h} - p^2 / 2\langle p^2 \rangle_{\rm h} \right)}{\sqrt{2\pi\hbar^2\beta/M}\,\sqrt{2\pi\langle \xi^2 \rangle_{\rm h}}\,\sqrt{2\pi\langle p^2 \rangle_{\rm h}}} . \quad (8.52)$$

In the self-consistent harmonic approximation (8.50) or (8.52), the quantum statistical average of an observable $A(q)$ takes the form of an integral over the centroid q_c,

$$\langle A \rangle = \frac{(M/2\pi\hbar^2\beta)^{1/2}}{Z_{\rm h}} \int_{-\infty}^{\infty} dq_c \exp\left(-\beta \mathcal{V}_{\rm h}(q_c) \right) \langle\!\langle A[q(\cdot)] \rangle\!\rangle_{q_c}^{(\rm h)} , \quad (8.53)$$

in which $\langle\!\langle A[q(\cdot)] \rangle\!\rangle_{q_c}^{(\rm h)}$ captures the average over the quantum fluctuations. The function $\langle\!\langle A[q(\cdot)] \rangle\!\rangle_{q_c}^{(\rm h)}$ is a smeared portrayal of the original function $A(q)$ like in Eq. (8.30), but now the smearing width is $\sqrt{\langle \xi^2 \rangle_{\rm h}}$,

$$\langle\!\langle A[q(\cdot)] \rangle\!\rangle_{q_c}^{(\rm h)} = \frac{1}{\sqrt{2\pi\langle \xi^2 \rangle_{\rm h}}} \int_{-\infty}^{\infty} dq\, A(q) \exp\left(-\frac{(q - q_c)^2}{2\langle \xi^2 \rangle_{\rm h}} \right) . \quad (8.54)$$

Consider next the quantum statistical average of an observable $A(p, q)$ depending on position and momentum. The average of the operator in Weyl form $\mathcal{Z}_0\{\hat{A}(\hat{p}, \hat{q})\}$ over the pure quantum fluctuations in harmonic approximation turns out with the use of Eq. (8.52) as a double Gaussian average with half-widths $\langle \xi^2 \rangle_{\rm h}$ and $\langle p^2 \rangle_{\rm h}$,

$$\langle\!\langle A[q(\cdot), p(\cdot)]\rangle\!\rangle^{(\mathrm{h})}_{q_c} \equiv \frac{1}{2\pi} \iint_{-\infty}^{\infty} \frac{dq\,dp}{\sqrt{\langle\xi^2\rangle_{\mathrm{h}}\langle p^2\rangle_{\mathrm{h}}}} A(p, q) \exp\left(-\frac{(q - q_c)^2}{2\langle\xi^2\rangle_{\mathrm{h}}} - \frac{p^2}{2\langle p^2\rangle_{\mathrm{h}}}\right) .$$

The final integral over the centroid is as given in Eq. (8.53). It should be remarked that in the case of Ohmic dissipation, both the quantum part μ_{h} of the partition function and the momentum dispersion $\langle p^2\rangle_{\mathrm{h}}$ must be Drude-regularized since otherwise they diverge logarithmically (cf. Section 6.5 and 6.6). A different possibility for regularization of the effective classical potential would be to subtract the contribution of a free Brownian particle $(1/\beta)\ln(1/\mu_{\mathrm{free}})$ with $1/\mu_{\mathrm{free}} = \prod_{n=1}^{\infty}[1 + \hat\gamma(\nu_n)/\nu_n]$.

One remark is appropriate. Since the local density matrix $\rho(q'', q'; q_c)$ in Eq. (8.50) has a factor $\exp[-(q'' + q' - 2q_c)^2/8\langle\xi^2\rangle_{\mathrm{h}}]$, the fluctuation width of the centroid is generally small at low and intermediate temperatures, and it becomes zero in the classical limit. Thus the special role of the centroid is less important for the density matrix than for the partition function. In numerical computations of the variational perturbation expansion, it turns out that one may treat the centroid in the same way as the quantum fluctuations. By this, the numerical efforts are reduced enormously.[3]

Historically, the variational method has been introduced by Feynman when he studied the polaron problem. For the action (4.124) with (4.126) there are two major reasons that a potential term in the trial action is not the appropriate representative to master the physical situation. First, the polaron is free to wander around in a crystal without giving preference to any particular place. Second, the polaron at position $q(\tau)$ interacts with itself at positions taken at any previous imaginary time τ'.

A trial action for the polaron which has these properties (except for a different spatial dependence) and which is analytically tractable is [3, 4]

$$S_0 = \frac{M}{2} \int_0^{\hbar\beta} d\tau\, \dot{\boldsymbol{q}}^2(\tau) + \frac{C}{2} \int_0^{\hbar\beta} d\tau \int_0^{\tau} d\tau'\, D_\omega(\tau - \tau')|\boldsymbol{q}(\tau) - \boldsymbol{q}(\tau')|^2 , \qquad (8.55)$$

where $D_\omega(\tau)$ is given in (3.56), and where C and ω are taken as adjustable parameters. According to the variational principle (8.21), we have for the ground-state energy

$$E \leq E_0 + \lim_{\beta\to\infty} \frac{1}{\hbar\beta}\langle S - S_0\rangle_0 , \qquad (8.56)$$

where E_0 is the ground-state energy of the model (8.55), and $\langle\cdots\rangle_0$ means average with weight $\exp(-S_0/\hbar)$. The lowest upper bound found by variation of the parameters ω and C gives for E values that are better than those obtained by of all other methods, and in fact for all values of the polaron coupling constant α [4]. Unfortunately, Jensen's inequality does *not* hold in general for a polaron in a magnetic field (magneto-polaron) since the action is complex [271]. A slightly modified variational method tailored to the magneto-polaron case has been suggested in Ref. [272].

---

[3]I am indebted to A. Pelster for this communication.

9. Suppression of quantum coherence

There are two main lines of investigations on decoherence in quantum mechanics. On the one hand, one is interested in fundamental problems in the interpretation of quantum theory, in particular in questions connected with measurement and the quantum to classical transition (cf., e.g., Ref. [57]). The importance of decoherence in the appearance of a classical world in quantum theory is well known [273]. In cosmology, the decoherence which must have been occurred during the inflationary era of the Universe has been attributed to quantum fluctuations. On the other hand, environment-induced decoherence is omnipresent in the microscopic world. It is found, e.g., for an atom confined in a quantum optical trap, or for electron propagation in a mesoscopic device. Striking examples for coherence effects are the anomalous magnetoresistance in disordered systems due to weak localization [274], and Aharonov-Bohm type oscillations in the resistance of mesoscopic rings [275].

A system prepared in a non-equilibrium state and in contact with the environment relaxes to equilibrium. The decay of the off-diagonal states of the reduced density matrix (coherences) is denoted as decoherence or loosely speaking as dephasing [cf. the remark at the end of Section 9.1]. Evidently, the origin of decoherence is entanglement and interaction with a fluctuating environment. Therefore, decoherence depends on the spectral density of the environmental coupling, on temperature and on specific properties of the system. There follows from the formal structure of the influence functional that response and equilibrium correlation functions decay roughly on the same time scale as the coherences. We shall discuss this in some detail in Chapter 21.

There are experimental indications that dephasing of electrons in mesoscopic conductors persists down to zero temperature [276, 277, 278]. On the theoretical side, Altshuler and coworkers advocated that there is no dephasing at $T = 0$ [279], while Golubev and Zaikin found in their studies on diffusive conductors that dephasing of electrons levels off at low temperatures [280, 281]. There is still an ongoing debate about dephasing of electrons in diffusive conductors at zero temperature [282] - [285].

In a multitude of model systems, decoherence persists down to zero temperature, as we shall see in Section 9.3. This has been found also by others [167, 282, 286].

To define decoherence quantitatively, it is natural to introduce a time scale τ_φ over which interference effects are suppressed. Basically, phase randomization of the system's wave function occurs through energy exchange processes with environmental modes, e.g., electron-electron or electron-phonon scattering.

Two complementary views about dephasing are possible [206]. Either one may treat the problem by studying the changes induced by the interfering particle in the states of the environment, or one may study the accumulation of the phase uncertainty due to quantum fluctuations of the reservoir coupling. First I discuss the important differences between a non-dynamical and a dynamical environment with regard to dephasing. Then I explain that there is no universal decoherence in models with delocalized bath modes. In the third section, I introduce a model with spatially

localized reservoir modes in which the actual geometry is irrelevant for the decoherence process and I calculate semiclassically the time scale on which interference
effects are suppressed. Finally I dwell on electron dephasing in a diffusive conductor,
and on the delicate question whether dephasing persists down to $T = 0$.

9.1 Nondynamical versus dynamical environment

An environment with no significant dynamics of its own acts upon the system as if it
were a classically fluctuating potential. To study the loss of phase memory, consider
a particle whose dynamics is governed by the Hamiltonian $H = H_0 + V_0(q) + \delta V(t)$.
We assume that $\delta V(t)$ is a stochastic potential term obeying Gaussian statistics with
covariance $\langle \delta V(t) \delta V(0) \rangle_{\mathrm{av}} = \langle \delta V^2 \rangle_{\mathrm{av}} \rho(t)$ where $\rho(0) = 1$. The stochastic contribution
$\delta V(t)$ gives rise to a fluctuating phase of the propagator,

$$ K(q_2, q_1; t) = e^{-i\varphi(t)} K_0(q_2, q_1; t) \quad \text{with} \quad \varphi(t) = \frac{1}{\hbar} \int_0^t d\tau \, \delta V(\tau) \, . \tag{9.1} $$

Taking the statistical average, we have $\langle K(q_2, q_1; t) \rangle_{\mathrm{av}} = \langle e^{-i\varphi(t)} \rangle_{\mathrm{av}} K_0(q_2, q_1; t)$ with

$$ \langle e^{-i\varphi(t)} \rangle_{\mathrm{av}} = \exp \left[-\tfrac{1}{2} \langle \varphi^2(t) \rangle_{\mathrm{av}} \right] , \tag{9.2} $$

$$ \langle \varphi^2(t) \rangle_{\mathrm{av}} = \frac{2}{\hbar^2} \langle \delta V^2 \rangle_{\mathrm{av}} \int_0^t dt' \int_0^{t'} dt'' \, \rho(t' - t'') \, . \tag{9.3} $$

We now assume that the observation time t is large compared to the width of the
correlation function $\rho(t)$. Then the phase uncertainty increases linearly with t. Writing
$\frac{1}{2} \langle \varphi^2(t) \rangle_{\mathrm{av}} = \gamma_{\mathrm{deph}} t$, we then have

$$ \langle e^{-i\varphi(t)} \rangle_{\mathrm{av}} = e^{-\gamma_\varphi t} \, . \tag{9.4} $$

It is natural to interpret γ_φ as the inverse time scale for dephasing. We now have

$$ \gamma_\varphi = \frac{\langle \delta V^2 \rangle_{\mathrm{av}}}{\hbar^2} \int_0^\infty d\tau \, \rho(\tau) \, . \tag{9.5} $$

In a fully quantized model with *dynamical* fluctuations, the situation is more
intricate. Assume that for the global model (3.1) the initial state is factorized into a
system state $|\psi_i\rangle$ and a bath state $|\chi\rangle$. The transition probability from state $|\psi_i\rangle$ to
final state $|\psi_f\rangle$ is

$$ P_{fi} = \sum_{\chi'} |\langle \psi_f | \langle \chi' | \exp[-i(H_{\mathrm{R}} + H_{\mathrm{S}} + H_{\mathrm{I}}) t/\hbar] |\chi\rangle |\psi_i\rangle|^2 \, . \tag{9.6} $$

When the coupling is weak enough to leave the bath unchanged, we can substitute
the expectation values for the bath operators. Assigning the expectation value $\langle H_{\mathrm{R}} \rangle_\chi$
to the phase of the bath state, we find for the transition amplitude of the system

$$ A_{fi} = \langle \psi_f | \exp[-i(H_{\mathrm{S}} + \langle H_{\mathrm{I}} \rangle_\chi) t/\hbar] |\psi_i\rangle \, . \tag{9.7} $$

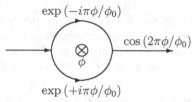

Figure 9.1: Schematic sketch of an Aharonov-Bohm device threaded by a magnetic flux ϕ.

The amplitude is conveniently written as modulus times phase factor. Next assume that the initial bath state is given in the form of a mixture with weight c_χ for the state $|\chi\rangle$. Then, in analogy with Eq. (9.2), we take the average of the phase factor as

$$\langle e^{-i\varphi} \rangle = \sum_\chi c_\chi |A_{fi}(\chi)| \, e^{-i\varphi(\chi)} \Big/ \sum_\chi c_\chi |A_{fi}(\chi)| \, . \tag{9.8}$$

Evidently, this proceeding breaks down for stronger coupling since the environmental states are not any more dwelling on their initial distribution. For a dynamically fluctuating potential it is therefore not possible to split the total phase into a system and a reservoir part. This demonstrates that in general it is impossible to assign to the subsystem a definite phase. From this we may draw the important conclusion that the whole concept of phase memory generally fails for an open system. Therefore, one should speak of *decoherence* rather than of *dephasing*.

9.2 Suppression of transversal and longitudinal interferences

We recall to the reader's attention the Feynman-Vernon method for the reduced density matrix in which the motion of the particle in contact with the reservoir is described by the propagating function [cf. Eq. (5.12)]

$$J_{\mathrm{FV}}(q_f, q_f', t; q_i, q_i', 0) = \int \mathcal{D}q \, \mathcal{D}q' \exp\left\{\frac{i}{\hbar}\Big(S_{\mathrm{S}}[q] - S_{\mathrm{S}}[q']\Big)\right\} \mathcal{F}_{\mathrm{FV}}[q, q'] \, . \tag{9.9}$$

Let us now study whether decoherence depends on the properties of the reservoir coupling alone, or also on the particular geometry of the quantum interference system. Consider first an Aharonov-Bohm device in which a metallic ring is threaded by a magnetic flux ϕ as sketched in Fig. 9.1. The traversing paths are divided into two groups that pass the ring clockwise and anti-clockwise, thereby picking up a phase difference $\varphi = 2\pi\phi/\phi_0$, where ϕ_0 is the flux quantum. Since the paths are spatially separated in transversal direction, we shall refer to this type of interferences as *transversal*. For the global model (3.11), the decoherence factor is given by the noise action, Eq. (5.36) with (5.33). Assuming that the paths $q_1(t)$ and $q_2(t)$ are coupled to statistically independent environments, the cross terms in the noise action $\Phi^{(\mathrm{N})}$ are absent. Then we have

$$\langle e^{-i\varphi(t)} \rangle = \exp\left(-\sum_{j=1}^{2} \frac{1}{\hbar} \int_0^t dt' \int_0^{t'} dt'' \, q_j(t') L'(t' - t'') q_j(t'')\right) \, . \tag{9.10}$$

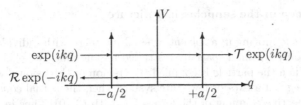

Figure 9.2: A resonant tunneling structure made up of two δ-potentials.

The same expression holds for a two-slit device. In the white noise limit, we recover for a slit distance q_0 the previous result (5.37).

The other device is a resonant tunneling structure with reflection and transmission coefficients $\mathcal{R}$ and $\mathcal{T}$, sketched in its simplest form in Fig. 9.2. In this device, the directly through-going path is interfering with a path which has made an additional roundtrip between the barriers. For δ-potentials, the two paths evolve in the scattering region as sketched in Fig. 9.3. Because they travel through the same spatial region and are only separated with respect to time, this interference type has been termed *longitudinal* [166]. Obviously, the interfering paths are influenced by the same environment. Putting $y(t) = q_1(t) - q_2(t)$, the resulting decoherence factor is

$$\langle e^{-i\varphi(t)} \rangle = \exp\left(-\frac{1}{\hbar} \int_0^t dt' \int_0^{t'} dt'' \, y(t') L'(t' - t'') y(t'') \right). \tag{9.11}$$

The expressions (9.10) and (9.11) have been analyzed in Ref. [166]. One finds that they behave quite differently. For a ballistic path, e.g., the former exponent grows with the third power of t, whereas the latter increases quadratically in t. Observe also that Eq. (9.11) is space translational invariant while Eq. (9.10) is not. In models with delocalized bath modes, the suppression of interferences depends also strongly on spatial dimension, as emphasized in Refs. [206, 287]. In conclusion, in models with delocalized bath modes, the decoherence process depends on the geometry of the interference device. Hence there is *no universal* decoherence behavior in these models.

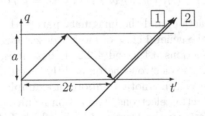

Figure 9.3: Sketch of two interfering path in the resonant tunneling device in Fig. 9.2.

9.3   Decoherence in the semiclassical picture

Physically, a particle moving in a solvent or solid interacts with individual modes of the environment only within a spatially restricted region which we characterize by a length scale λ. When the particle is confined to a region of extension d where $d \ll \lambda$, as it may occur, e.g., in a two- or few-state system, then the actual coupling may be replaced by an effective coupling of the form (3.4) with (3.10). Thus in this case the simplified model with *delocalized* bath modes, Eq. (3.11), is appropriate. However, when the particle travels over distances large compared to the interaction range λ, we have to take into account the *localized* nature of the bath modes.

9.3.1   A model with localized bath modes

In this subsection we limit ourselves to the case of a quasi one-dimensional system. We assume that the interaction potential is linear in the bath coordinates. In accordance with Eqs. (3.4) and (3.7), we take the interaction in the form

$$H_I = -\left(\sum_\alpha c_\alpha x_\alpha \right) \lambda \sum_{q_0} u\left(\frac{q - q_0}{\lambda} \right) . \tag{9.12}$$

The form factor $u(x)$ is a dimensionless interaction potential with range of order unity. The form (9.12) describes interaction of the particle with individual bath modes within a range of length λ about the center q_0. To restore translational invariance of the coupling, the scatterers in Eq. (9.12) are assumed to be uniformly distributed in space. Then the interaction is characterized by the spatial autocorrelation function

$$\lambda^2 \int_{-\infty}^{\infty} dx'\, u(x')u(q/\lambda - x') = \lambda^2 U(q) . \tag{9.13}$$

The dimensionless interaction $U(q)$ is an even function of the variable q and has a range λ. The elimination of the bath modes may be effected analogously to the case of delocalized modes discussed in Section 5.2. We then get as a generalization of Eq. (5.33) for the present model the influence action

$$\Phi_{\text{infl}}[q_1, q_2] = \lambda^2 \sum_{j=1}^{2} (-1)^{j-1} \int_0^t dt' \int_0^{t'} dt'' \tag{9.14}$$

$$\times \left\{ L(t' - t'')\, U\left(|q_j(t') - q_1(t'')| \right) - L^*(t' - t'')\, U\left(|q_j(t') - q_2(t'')| \right) \right\} .$$

The real part is the noise action, and the imaginary part is the friction action.

The model belongs to the general class of models describing state-dependent dissipation discussed in Subsections 3.1.2 and 4.2.2. It describes, e.g., the interaction with extended field modes. The electron-photon and the electron-phonon interaction, Eq. (3.168), are of the form (9.12). Another important example is the interaction of a charged particle with conduction electrons. The action resulting from the imaginary-time action (4.146) is exactly in the form (9.14), in which $L(t' - t'')$ is the Ohmic heat bath kernel, and $\lambda^2 U(\Delta q) = (3/2k_{\text{F}}^2) \sin^2(k_{\text{F}} \Delta q)/(k_{\text{F}} \Delta q)^2$.

The Zwanzig–Caldeira-Leggett model with delocalized modes, Eq. (3.11), is contained in Eq. (9.14) as the special case $U(|q_j - q_i|) = -\frac{1}{2}(q_j - q_i)^2/\lambda^2$. If we substitute this form, the action (9.14) reduces to the familiar expression (5.24).

The model with localized bath modes has been introduced in Ref. [166] in order to study suppression of interferences without dependence on geometry (see below). The same model has been used in Ref. [167] to study state-dependent quantum effects in Brownian motion. A number of new quantum effects have been found. For instance, the logarithmic law (7.40) for the spatial spreading in the Ohmic case at $T = 0$ holds for this model only on a scale $\sigma^2(t) \ll \lambda^2$. On a larger scale, there is a smooth crossover from a Gaussian to a frozen exponential profile. This genuine quantum effect can be understood from the interplay of the negative time correlations of the quantum noise with the spatial correlations described by the function $U(q)$.

In the remainder of this section, we base the discussion of the suppression of quantum interference on the action (9.14). First of all, we observe that the self-terms are added up in the noise action $\Phi^{(N)} = \operatorname{Re} \Phi_{\mathrm{infl}}$, whereas they appear with opposite sign in the friction action $\Phi^{(F)} = \operatorname{Im} \Phi_{\mathrm{infl}}$. Suppose for simplicity that the self-terms of the two paths give the same contribution. Now assume that the distance between the two interfering paths is large compared to the interaction range λ for most of the time, $\lambda \gg |q_1(t') - q_2(t'')|$. Then the self-terms in Eq. (9.14) are very large compared to the cross terms. Since the self-terms cancel out in the friction action, the action depends only on the cross terms. Therefore we have

$$| \Phi^{(F)}[q_1, q_2] | \ll \Phi^{(N)}[q_1, q_2] . \tag{9.15}$$

Thus in the present model friction plays only a secondary role in the decoherence process. In view of Eq. (9.15), we disregard the friction action in the remainder of this section.[1] There is a controversy whether Eq. (9.15) holds when fermionic many-body effects are relevant (see subsequent section). The noise action now reads

$$\Phi^{(N)}[q] = 2\lambda^2 \int_0^t dt' \int_0^{t'} dt'' \, L'(t' - t'') \, U\left(|q(t') - q(t'')|\right) . \tag{9.16}$$

The actual geometry is irrelevant for this noise action. Decoherence by scattering events is as effective for small separation as for large separation of the interfering paths. While the action in Eq. (9.10) is proportional to the square of the slit distance q_0, the action (9.16) is proportional to λ^2. Since λ is a microscopic length, we usually have $q_0 \gg \lambda$. Hence the suppression of interferences is universal in this regime.

9.3.2 Dephasing rate formula

In a mesoscopic conductor, the electron moves under the concerted influences of static disorder and dynamical fluctuations induced by the surroundings. A stochastic

---

[1]In the model with delocalized bath modes, Eq. (3.11), the situation is different. In this model, also the friction action $\Phi^{(F)}$ may become relevant for dephasing, as noticed in Ref. [207].

static potential for the electron is generated by randomly distributed impurities. The dynamical fluctuations entail a dephasing factor of the form

$$\left\langle e^{-i\varphi(t)} \right\rangle = \exp\left(-\Phi^{(N)}[\boldsymbol{q}]/\hbar\right) . \tag{9.17}$$

Following Chakravarty and Schmid [274], the average of paths in the random impurity potential is performed in a quasiclassical approach. To this we define for the pair of trajectories $\boldsymbol{q}(t')$ and $\boldsymbol{q}(t'')$ in n space dimensions the probability distribution

$$P(\boldsymbol{r}, t' - t'') = a^{n-3}\left\langle \delta^n\left[\boldsymbol{r} - \boldsymbol{q}(t') + \boldsymbol{q}(t'')\right]\right\rangle_{\text{imp}} . \tag{9.18}$$

For $n = 2$, the parameter a is the film thickness, and for $n = 1$ the quantity a^2 is the cross-section of the wire. Taking now the impurity average of Eq. (9.17) we have

$$\left\langle\!\left\langle e^{-i\varphi(t)}\right\rangle\!\right\rangle_{\text{imp}} = \left\langle\exp\left(-\Phi^{(N)}[\boldsymbol{q}]/\hbar\right)\right\rangle_{\text{imp}} \approx \exp\left(-\left\langle\Phi^{(N)}[\boldsymbol{q}]/\hbar\right\rangle_{\text{imp}}\right) . \tag{9.19}$$

On condition that the observation time t is large compared to the width of the function $L'(\tau)$, we find, analogous to the proceeding in Eqs. (9.3) - (9.5), that $\Phi^{(N)}[\boldsymbol{q}]$ increases linearly with time. Observing the relation (9.4) we obtain

$$\gamma_\varphi = \lim_{t\to\infty} \left\langle \Phi^{(N)}[\boldsymbol{q}]/\hbar t\right\rangle_{\text{imp}} = \frac{\lambda^2}{\hbar}\int_{-\infty}^{\infty} d\tau\, L'(\tau)\langle U(|\boldsymbol{q}(\tau) - \boldsymbol{q}(0)|)\rangle_{\text{imp}} . \tag{9.20}$$

Generally, the dephasing rate can be written in position and $\boldsymbol{k}$ space as[2]

$$\gamma_\varphi = \int_{-\infty}^{\infty} d\tau \int d^n r\, R(\boldsymbol{r}, \tau) P(\boldsymbol{r}, \tau) = \int_{-\infty}^{\infty} d\tau \int \frac{d^n k}{(2\pi)^n}\, \tilde{R}(\boldsymbol{k}, \tau)\tilde{P}(-\boldsymbol{k}, \tau) . \tag{9.21}$$

Here $P(\boldsymbol{r}, \tau)$ characterizes the motion of the particle and $R(\boldsymbol{r}, \tau)$ the spatial and temporal correlations of the environment. In the present model we have

$$\tilde{P}(\boldsymbol{k}, \tau) = a^{n-3}\left\langle e^{-i\boldsymbol{k}\cdot[\boldsymbol{r}(\tau)-\boldsymbol{r}(0)]}\right\rangle_{\text{imp}} , \tag{9.22}$$

$$\tilde{R}(\boldsymbol{k}, \tau) = \lambda^2 L'(\tau)\tilde{U}(|\boldsymbol{k}|)/\hbar . \tag{9.23}$$

Here we have put $\tilde{U}(|\boldsymbol{k}|) = a^{3-n}\int d^n r\, U(|\boldsymbol{r}|)\, e^{-i\boldsymbol{k}\cdot\boldsymbol{r}}$. Writing $R(\boldsymbol{r}, \tau)$ and $P(\boldsymbol{r}, \tau)$ as space-time Fourier integrals with transforms $\tilde{R}(\boldsymbol{k}, \omega)$ and $\tilde{P}(\boldsymbol{k}, \omega)$, we obtain

$$\gamma_\varphi = \frac{1}{(2\pi)^{n+1}}\int_{-\infty}^{\infty} d\omega \int d^n k\, \tilde{R}(\boldsymbol{k}, \omega)\tilde{P}(-\boldsymbol{k}, -\omega) , \tag{9.24}$$

There follow with Eq. (5.26)

$$\tilde{R}(\boldsymbol{k}, \omega) = \lambda^2 \tilde{\mathcal{X}}(\omega)\tilde{U}(\boldsymbol{k})/\hbar^2 = \lambda^2 M\tilde{\gamma}'(\omega)\omega\coth(\tfrac{1}{2}\beta\hbar\omega)\tilde{U}(|\boldsymbol{k}|)/\hbar . \tag{9.25}$$

The function $\tilde{\mathcal{X}}(\omega)$ is the power spectrum of the fluctuating force given in Eq. (2.34). The interaction $\tilde{U}(|\boldsymbol{k}|)$ restricts the k-integegral essentially to the volume of radius

$$|\boldsymbol{k}| \lesssim 1/\lambda . \tag{9.26}$$

For either ballistic or diffusive motion, most of the contribution comes from environmental modes with wave numbers near to $1/\lambda$.

---

[2]Corresponding dephasing rate formulas in the fermionic many-body context are discussed in Ref. [288]. The crucial physics of Fermi blocking is being controversially discussed (see Section 9.4).

9.3.3 Statistical average of paths

The motion of a particle with velocity v is ballistic on the scale of the interaction range λ when the mean free path ℓ is large compared to λ. With the diffusion coefficient in 3 dimensions $D = v\ell/3$ we then have as condition for ballistic motion

$$\kappa \equiv \lambda/\ell = \lambda v/3D \ll 1. \tag{9.27}$$

In the ballistic regime (9.27) we have $\boldsymbol{r}(\tau) = \boldsymbol{r}(0) + \boldsymbol{v}\tau$. Since scattering processes are disregarded, the probability function (9.22) takes the simple form

$$\tilde{P}_{\text{bal}}(\boldsymbol{k}, \tau) = a^{n-3} e^{-i\boldsymbol{k}\cdot\boldsymbol{v}\tau}. \tag{9.28}$$

Hence the spectral probabilitiy function in the ballistic regime for a single particle is

$$\tilde{P}_{\text{bal}}(\boldsymbol{k}, \omega) = a^{n-3} 2\pi \delta(\omega - \boldsymbol{k}\cdot\boldsymbol{v}). \tag{9.29}$$

In contrast, the motion of a particle is diffusive on the interaction range λ when

$$\kappa \gg 1. \tag{9.30}$$

In the regime (9.30), the average over many collisions in Eq. (9.22) leads to diffusive spread of the probability distribution,

$$P_{\text{diff}}(\boldsymbol{r}, \tau) = a^{n-3} (4\pi D|\tau|)^{-n/2} e^{-|\boldsymbol{r}|^2/4D|\tau|} \tag{9.31}$$

$$\tilde{P}_{\text{diff}}(\boldsymbol{k}, \tau) = a^{n-3} e^{-D|\boldsymbol{k}|^2|\tau|}. \tag{9.32}$$

Hence the power spectrum of a diffusive particle has the familiar Lorentzian form

$$\tilde{P}_{\text{diff}}(\boldsymbol{k}, \omega) = a^{n-3} \frac{2D|\boldsymbol{k}|^2}{(D|\boldsymbol{k}|^2)^2 + \omega^2}. \tag{9.33}$$

We are now primed to study the dephasing rate (9.24) in various limits. We assume in the remainder of this chapter that the cutoff in the spectral bath coupling is the largest frequency of the problem and therefore plays no particular role.

9.3.4 Ballistic motion

In the ballistic case, the characteristic frequency scale ω_{b} and temperature scale T_{b} are determined by the inverse time scale for traversing the interaction region,

$$\omega_{\text{b}} \equiv v/\lambda, \qquad \text{and} \qquad T_{\text{b}} \equiv \hbar\omega_{\text{b}}/k_{\text{B}}. \tag{9.34}$$

In one dimension, the decoherence rate (9.24) with Eq. (9.29) takes the form [166]

$$\gamma_\varphi = \frac{\lambda^2}{\pi\hbar v} \int_0^\infty d\omega\, J(\omega) \coth(\tfrac{1}{2}\beta\hbar\omega) \tilde{U}(\omega/v). \tag{9.35}$$

The frequency integral is effectively cut off at $\omega \approx \omega_b$, and the cutoff in the spectral density $J(\omega)$ is irrelevant. In case the environmental spectral density has a low-frequency cutoff ω_{min}, the decoherence rate drops to zero when the particle's velocity v falls below $\lambda \omega_{min}$.

The decoherence rate γ_φ can be calculated in analytic form for an algebraic spectral density $J(\omega) \propto \omega^s$ and a Lorentzian form of the spatial correlation function,

$$U(r) = 1/[1 + r^2/\lambda^2] \qquad\qquad \tilde{U}(k) = \pi \lambda e^{-\lambda |k|} . \qquad (9.36)$$

With the form (3.72) for $J(\omega)$ the integral can be expressed in terms of Riemann's zeta function $\zeta(q, z)$. We find for the decoherence rate the analytic expression

$$\gamma_\varphi = \left\{ 1 + 2(T/T_b)^{s+1} \zeta(s+1, 1+T/T_b) \right\} \gamma_{\varphi,0} , \qquad (9.37)$$

where $\gamma_{\varphi,0}$ is the decoherence rate at zero temperature,

$$\gamma_{\varphi,0} = 2\pi \Gamma(1+s)(\eta_s \lambda^2/2\pi\hbar)(\omega_b/\omega_{ph})^{s-1} \omega_b . \qquad (9.38)$$

The quantity $\eta_s \lambda^2/2\pi\hbar$ is a dimensionless damping parameter analogous to the Kondo parameter K introduced in Eq. (4.155). Since the spectral density of the environmental coupling is gapless, there is decoherence even at zero temperature. The rate is roughly independent of T in the temperature range $0 \le T \ll T_b$, and it varies with the velocity v of the particle as v^s. Well above T_b, the decoherence rate is determined by classical noise and thus depends linearly on T,

$$\gamma_\varphi = \frac{2T}{sT_b} \gamma_{\varphi,0} = 4\pi\Gamma(s) \frac{\eta_s \lambda^2}{2\pi\hbar} \left(\frac{\omega_b}{\omega_{ph}} \right)^{s-1} \frac{k_B T}{\hbar} . \qquad (9.39)$$

The form (9.37) describes the crossover between the limiting cases (9.38) and (9.39).

In the Ohmic case, $s = 1$, the decoherence rate (9.37) takes the form

$$\gamma_\varphi = \left\{ 1 + 2(T/T_b)^2 \psi'(1+T/T_b) \right\} \eta \lambda^2 \omega_b/\hbar , \qquad (9.40)$$

where $\psi(z)$ is the trigamma function. This reduces in the low and high temperature regimes to the forms

$$\gamma_\varphi = \eta\lambda v/\hbar = (\eta/M)\lambda/\lambda_B , \qquad\qquad 0 \le T \ll T_b , \qquad (9.41)$$

$$\gamma_\varphi = 2\eta\lambda^2 k_B T/\hbar^2 = 2(\eta/M)\lambda^2/\lambda_{th}^2 , \qquad\qquad T \gg T_b . \qquad (9.42)$$

The second forms are written in terms of the de Broglie wave length $\lambda_B = \hbar/Mv$ and thermal wave length $\lambda_{th} = \hbar/(Mk_B T)^{1/2}$, respectively.

At zero temperature, the dephasing rate is proportional to the velocity of the particle. In the white-noise limit, Eq. (9.42), the rate is independent of the velocity and formally agrees with the result for the model with delocalized oscillators, Eq. (5.37). However, the respective length parameters q_0 and λ may differ drastically.

9.3.5 Diffusive motion

For diffusive motion of the particle in n space dimensions with isotropic spatial correlation function $U(|\boldsymbol{r}|)$, the decoherence rate (9.24) with (9.25) takes the form

$$
\gamma_\varphi = \frac{\lambda^2}{\pi\hbar} \int_0^{1/\tau_c} d\omega \, J(\omega) \coth(\tfrac{1}{2}\hbar\beta\omega) \frac{\Omega_n a^{n-3}}{(2\pi)^n} \int_0^\infty dk \, k^{n-1} \frac{2Dk^2}{(Dk^2)^2 + \omega^2} \tilde{U}(k) \,, \quad (9.43)
$$

where Ω_n is the total solid angle and $k = |\boldsymbol{k}|$. The characteristic frequency and temperature scales for diffusive motion are

$$
\omega_d \equiv D/\lambda^2 = \omega_b/3\kappa \,, \quad \text{and} \quad T_d \equiv \hbar\omega_d/k_B = T_b/3\kappa \,. \quad (9.44)
$$

The ω-integral should be cut at the inverse collision time $1/\tau_c = v/\ell = v^2/3D$, since the assumption of diffusive transport is not valid anymore in the frequency range $\omega > 1/\tau_c$. We have $1/\tau_c = 3\kappa^2\omega_d$ and hence in the diffusive regime $\omega_d\tau_c \ll 1$.

Consider now first the case of one space dimension, $n = 1$. In the classical regime, we may put $\coth(\hbar\beta\omega/2) \to 2/\hbar\beta\omega$. With this, and with the spectral density $J(\omega) \propto \omega^s$ the frequency integral in Eq. (9.43) is ultraviolet-convergent in the regime $s < 2$ without cutoff. On the other hand, the high-frequency cutoff at $1/\tau_c$ is essential when $s \geq 2$. This shows that the classical regime corresponds the temperature regime $T \gg T_d$ for $s < 2$ and to the regime $T \gg \kappa^2 T_d$ for $s > 2$. With the space correlation function (9.36) we readily find

$$
\gamma_\varphi =
\begin{cases}
\dfrac{\eta_s \lambda^2}{\pi\hbar} \dfrac{2\pi\Gamma(2s-1)}{\sin(\pi s/2)} \left(\dfrac{\omega_d}{\omega_{ph}}\right)^{s-1} \dfrac{k_B T}{\hbar} \,, & \text{for} \quad \tfrac{1}{2} < s < 2 \,, \\[3ex]
\dfrac{\eta_2 \lambda^2}{\pi\hbar} 16 \ln(\kappa) \left(\dfrac{\omega_d}{\omega_{ph}}\right) \dfrac{k_B T}{\hbar} \,, & \text{for} \quad s = 2 \,, \\[3ex]
\dfrac{\eta_s \lambda^2}{\pi\hbar} 8\Gamma(s-2) \left(\dfrac{\kappa^2 \omega_d}{\omega_{ph}}\right)^{s-1} \dfrac{k_B T}{\hbar} \,, & \text{for} \quad s > 2 \,.
\end{cases}
\quad (9.45)
$$

In the Ohmic case, $s = 1$, we recover the white noise limit, Eq. (9.42).

In the limit $T \to 0$, the cutoff ω_c is relevant in the regime $s \geq 1$ and irrelevant when $s < 1$. We find from Eq. (9.43) the expressions

$$
\gamma_\varphi =
\begin{cases}
\dfrac{\eta_s \lambda^2}{\pi\hbar} \dfrac{\pi\Gamma(1+2s)}{\cos(\pi s/2)} \left(\dfrac{\omega_{ph}}{\omega_d}\right)^{1-s} \omega_d \,, & \text{for} \quad s < 1 \,, \\[3ex]
\dfrac{\eta_1 \lambda^2}{\pi\hbar} 8 \ln(\kappa) \, \omega_d \,, & \text{for} \quad s = 1 \,, \\[3ex]
\dfrac{\eta_s \lambda^2}{\pi\hbar} 4\Gamma(s-1) \left(\dfrac{\kappa^2 \omega_d}{\omega_{ph}}\right)^{s-1} \omega_d \,, & \text{for} \quad s > 1 \,.
\end{cases}
\quad (9.46)
$$

These expressions practically hold in the temperature regime $T \ll T_d$ for $s < 1$ and $T \ll \kappa^2 T_d$ for $s \geq 1$. Thus, this model with localized bath modes yields a finite decoherence rate in the diffusive regime at zero temperature.

The analysis can easily be generalized to n space dimensions. Assume that the correlation function in k-space behaves as $\tilde{U}(|k| \to 0) \propto |k|^{\sigma-n}$. For spatial short-range interaction of Gaussian form, we have $\sigma = n$. For spatial long-range power-law form, σ is less than n. In the classical limit, $\coth(\beta\hbar\omega/2) \to 2/\beta\hbar\omega$, the ω-integral resulting from the k-integration in Eq. (9.43) is IR-convergent for $\sigma > 2 - 2s$ [289]. If this condition is not met, the loophole is to place a time-dependent IR cutoff, $\omega_{ir} = 1/t$, as explained below after Eq. (9.60).

An expression for γ_φ similar to Eq. (9.43) is discussed in Refs. [167, 286]. Saturation of τ_φ at $T = 0$ is also found in models which are free of form factors in the system-reservoir interaction [290]. Loss and Mullen presented a model for dephasing in which quantum fluctuations of the particle's path are coupled with a spatially localized environment [207], and they compared their findings with those of Ref. [206].

In the above semiclassical study of the model with localized bath modes, decoherence persists down to $T = 0$ when $J(\omega)$ is nonzero at frequencies below ω_b for ballistic motion, and below ω_d for diffusive motion. Decoherence endures because the reservoir can randomly absorb arbitrarily small amounts of energy down to $T = 0$ when the spectral density is gapless. Since $\omega_{ph} \gg \omega_m \gg \omega_d$, the dephasing rate at zero temperature is usually very small, and it becomes even smaller as the parameter s is raised from the sub-Ohmic sector $s < 1$ to the super-Ohmic regime $s > 1$. Decoherence decreases at low T with increasing s because of the much weaker super-Ohmic coupling at low frequencies compared to the Ohmic and sub-Ohmic case.

In Aharonov-Bohm ring devices with a quantum dot (QD) built in one arm of the ring, the phase shift of the electron passing through the quantum dot has been analyzed [291]. The experiments demonstrated that the electron can propagate coherently through a quantum dot. Such device offers the possibility to control dephasing rates by modifying the electromagnetic environment of the QD. An additional wire with a quantum point contact located close to the QD may act as sensitive measurement device to detect when the electron passes through the QD. This opens the possibility to build a "Which Path?" interferometer [292]. The influence of the measurement apparatus on dephasing is twofold. First, creation of real electron-hole pairs in the wire measures which path the electron took around the ring. Secondly, Ohmic damping resulting from the creation of virtual electron-hole pairs (cf. Subsection 4.2.8) leads to power-law suppression of the Aharonov-Bohm oscillations.

9.4 Decoherence of electrons

The weak localization of electrons by coherent backscattering in a disordered conductor is a significant interference effect and manifests itself in a characteristic contribution to the magnetoconductivity. The study of decoherence in weak-localization systems is most easily described in a path integral description with an influence functional accounting for the interacting electrons. The influence functional method was originally devised for single-particle decoherence. The difficulty now is that the constituents of the reservoir and the dephasing electron are indistinguishable fermions and hence Fermi statistics has to be properly taken into account. For electrons in metals the dominant inherent interaction at low temperature is electron-electron scattering. As temperature is decreased, inelastic scattering processes get more and more rare, and the scattering time τ_{in} diverges as $1/T^2$. Because of this feature, it is largely thought that the electron decoherence time τ_φ should diverge as T approaches zero.

About ten years ago, Golubev and Zaikin (GZ) developed an influence functional approach for interacting fermions in a disordered conductor and studied the magnetoconductance in the weak-localization regime. They showed that, when focussing on the single-particle response, the interaction with all the other fermions can be viewed as an effective environment. The Fermi statistics reveals itself in a complicated form of the resulting effective action which depends on the Fermi distribution function. With this they calculated the dephasing rate, and found it to be finite at zero temperature, $\gamma_\varphi^{GZ}(T \to 0) = \gamma_{\varphi,0}^{GZ}$ [280, 281]. However, this result is in contradiction to the widely accepted findings of Altshuler, Aronov, and Khmelnitskii (AAK) that $\gamma_\varphi^{AAK}(T \to 0) \to 0$ [287]. There arose a fulminant controversy with impetuous critique from Altshuler and coworkers [279] and equally resolute rejection by GZ. An informative overview from GZ's point of view with most relevant references is given in Ref. [282]. Recently, von Delft rederived the controversial influence functional of GZ for interacting electrons in disordered metals and confirmed that it is exact and takes proper account of the Pauli principle [283] (see also Ref. [293]). Since then the dispute is relocated to the question how one correctly evaluates the resulting path integral. Approximations utilized by GZ have been criticized [279, 283], in particular those concerning terms connected with Pauli blocking. While GZ elaborated the crucial terms in the time regime, von Delft analyzed them in the frequency regime. He then argued that this would allow for a somewhat more accurate treatment of recoil effects. By this, GZ's approach would readily reproduce the results of AAK for the dephasing rate. Heuristic arguments to reach the same conclusions as Ref. [283] were given in Ref. [284]. Subsequently, GZ objected that uncontrolled manipulations with diagrams were performed in Ref. [283], and that the resulting influence functional violates fundamental principles of quantum-statistical mechanics [285]. Let us now briefly embark on the subtle problem of decoherence in weak localization.

The density matrix of a quantum particle evolves according to the expression (5.11). For a single electron of charge $-e$, which moves in an impurity potential $U(\boldsymbol{r})$ and is subject to an external electric potential $V_x(\boldsymbol{r}, t)$, the propagating function is

$$
J(\boldsymbol{r}_{1,f}, \boldsymbol{r}_{2,f}, t_f; \boldsymbol{r}_{1,i}, \boldsymbol{r}_{2,i}, t_i) = \int_{\boldsymbol{r}_1(t_i)=\boldsymbol{r}_{1,i}}^{\boldsymbol{r}_1(t_f)=\boldsymbol{r}_{1,f}} \mathcal{D}r_1(\cdot) \int_{\boldsymbol{r}_2(t_i)=\boldsymbol{r}_{2,i}}^{\boldsymbol{r}_2(t_f)=\boldsymbol{r}_{2,f}} \mathcal{D}r_2(\cdot)
$$

$$
\times \exp\left\{\frac{i}{\hbar}\left[S_0[\boldsymbol{r}_1(\cdot)] - S_0[\boldsymbol{r}_2(\cdot)] + \int_{t_i}^{t_f} dt' \left[eV_x(\boldsymbol{r}_1(t')) - eV_x(\boldsymbol{r}_2(t'))\right]\right]\right\},
$$

(9.47)

where $S_0[\boldsymbol{r}] = \int_{t_i}^{t_f} dt' \left[\frac{1}{2}m\dot{\boldsymbol{r}}^2(t') - U(\boldsymbol{r}(t'))\right]$. The current density is found from the density matrix as $\boldsymbol{j}(\boldsymbol{r},t) = i\,(e/m)[\boldsymbol{\nabla}_{\boldsymbol{r}_1}\rho(\boldsymbol{r}_1,\boldsymbol{r}_2,t) - \boldsymbol{\nabla}_{\boldsymbol{r}_2}\rho(\boldsymbol{r}_1,\boldsymbol{r}_2,,t)]|_{\boldsymbol{r}_1=\boldsymbol{r}_2=\boldsymbol{r}}$. Upon assuming a constant electrical field, $V_{\mathrm{x}} = -\boldsymbol{E}\cdot\boldsymbol{r}$, the linear conductivity is found as

$$\sigma = \frac{e^2}{3m}\int_{-\infty}^{t} dt' \int d\boldsymbol{r}_1' \, d\boldsymbol{r}_2' \, [\boldsymbol{\nabla}_{\boldsymbol{r}_{1,\mathrm{f}}} - \boldsymbol{\nabla}_{\boldsymbol{r}_{2,\mathrm{f}}}] J_0(\boldsymbol{r}_{1,\mathrm{f}},\boldsymbol{r}_{2,\mathrm{f}},t;\boldsymbol{r}_1',\boldsymbol{r}_2',t')$$
$$\times [\boldsymbol{r}_1' - \boldsymbol{r}_2']\,\rho_\beta(\boldsymbol{r}_1',\boldsymbol{r}_2')\Big|_{\boldsymbol{r}_{2,\mathrm{f}}=\boldsymbol{r}_{1,\mathrm{f}}}, \tag{9.48}$$

where J_0 is the propagating function for $V_{\mathrm{x}} = 0$, and ρ_β is the thermal density matrix. For a random potential $U(\boldsymbol{r})$ the contributions of different paths $\boldsymbol{r}_1(t') \neq \boldsymbol{r}_2(t')$ average each other out in the double path sum (9.47). Hence the dominant contribution to the propagating function J_0 comes from pairs of identical classical paths $\boldsymbol{r}_1^{\mathrm{cl}}(t') = \boldsymbol{r}_2^{\mathrm{cl}}(t')$, and from Gaussian fluctuations around them. In the corresponding quasiclassical approximation for the propagating function (see Subsection 5.8.2), the expression (9.48) is equivalent to the standard linear response Kubo formula

$$\sigma = \frac{2e^2 N_0}{3}\int_0^\infty d\tau \, \langle \boldsymbol{v}(\tau)\boldsymbol{v}(0)\rangle_\beta, \tag{9.49}$$

where $N_0 = p_{\mathrm{F}}^2/(2\pi^2\hbar^3 v_{\mathrm{F}})$ is the density of states per volume and per energy interval (and per spin polarization) at the Fermi surface. In classical transport theory we have $\langle \boldsymbol{v}(\tau)\boldsymbol{v}(0)\rangle_\beta = v_{\mathrm{F}}^2\,\mathrm{e}^{-\tau/\tau_{\mathrm{c}}}$, where $\tau_{\mathrm{c}} = \ell/|\boldsymbol{v}_{\mathrm{F}}|$ is the collision time. Hence we obtain the Drude conductivity $\sigma = 2e^2 N_0 D$, where $D = \frac{1}{3}v_{\mathrm{F}}^2\tau_{\mathrm{c}} = \frac{1}{3}v_{\mathrm{F}}\ell$ is the diffusion constant.

The leading quantum correction to the conductivity due to "weak localization" arises from quantum interference between self-crossing paths (Cooperon), in which an electron propagates in the clock-wise and counter-clockwise direction around a loop. Since the quantum phases of these paths cancel each other, these quantum interference terms, which otherwise would be random in sign, outlast disorder averaging [287, 274]. It is much more likely to find self-crossing paths in low dimensions. Therefore weak-localization effects are most significant in low dimensions (films and wires). The Cooperon yield the conductivity correction

$$\delta\sigma_{\mathrm{WL}} = -\frac{2e^2 D}{\pi}\int_{\tau_{\mathrm{c}}}^\infty dt \, P_{\mathrm{diff}}(0,t)\,\mathrm{e}^{-f_{\mathrm{d}}(t)}. \tag{9.50}$$

Here $P_{\mathrm{diff}}(0,t)$, which is given in Eq. (9.31), is the classical probability for a diffusive particle to be found after time t at the initial point. The exponential factor $\mathrm{e}^{-f_{\mathrm{d}}(t)}$ is due to electron interactions and is expressed in terms of the noise and friction action [see Eq. (5.33)] as $\mathrm{e}^{-f_{\mathrm{d}}(t)} = \big\langle \mathrm{e}^{-[\mathcal{S}^{(\mathrm{N})}+i\,\mathcal{S}^{(\mathrm{F})}]/\hbar}\big\rangle_{\mathrm{imp}} = \mathrm{e}^{-\langle[\mathcal{S}^{(\mathrm{N})}+i\,\mathcal{S}^{(\mathrm{F})}]/\hbar\rangle_{\mathrm{imp}}}$. The effective action $\mathcal{S}^{(\mathrm{N})} + i\,\mathcal{S}^{(\mathrm{F})}$, which properly takes into account the Pauli principle, has been calculated by GZ [280] and has been rederived by von Delft [283]. It was argued by GZ in Ref. [280] that the friction action $\mathcal{S}^{(\mathrm{F})}$ for time-reversed paths vanishes when the impurity average is performed so that $f_{\mathrm{d}}(t) = \langle \mathcal{S}^{(\mathrm{N})}(t)/\hbar\rangle_{\mathrm{imp}}$. The noise action $\mathcal{S}^{(\mathrm{N})}(t)$ can be analyzed in a simplified approach in which a colored noise source emulates the electron interaction [282]. The noise correlator of the fluctuating voltage $\delta V(\boldsymbol{r},t)$ obeying Gaussian statistics is

$$\langle \delta V(r,t)\delta V(0,0)\rangle/\hbar \equiv I(r,t) = \int \frac{d\omega\, d^3k}{(2\pi)^4} \frac{\tilde{\chi}_{ee}(\omega)}{|k|^2} e^{ik\cdot r - i\omega t}. \tag{9.51}$$

Here $\tilde{\chi}_{ee}(\omega)$ is the power spectrum of the stochastic force analogous to Eq. (2.34),

$$\tilde{\chi}_{ee}(\omega) = \mathrm{Im}\Big(\frac{-4\pi}{\epsilon_{ee}(k,\omega)}\Big)\coth(\tfrac{1}{2}\beta\hbar\omega), \tag{9.52}$$

where $\epsilon_{ee}(k,\omega)$ is the electronic contribution to the dielectric susceptibility.

Expressed in terms of the noise correlator (9.51), the noise action takes the form

$$S^{(N)}(t) = \frac{e^2}{2}\int_0^t dt' \int_0^t dt'' \Big\{ I[r_1(t')-r_1(t''), t'-t''] + I[r_2(t')-r_2(t''), t'-t''] $$
$$- I[r_1(t')-r_2(t''), t'-t''] - I[r_2(t')-r_1(t''), t'-t'']\Big\}. \tag{9.53}$$

Upon performing the impurity average, the function $f_d(t)$ is found to read

$$f_d(t) = \frac{e^2}{\hbar}\int_0^t dt' \int_0^t dt'' \int d^n r\, P_{\mathrm{diff}}(r,|t'-t''|)[I(r,t'-t'')-I(r,t'+t''-t)]. \tag{9.54}$$

The analysis now gives that the first term in the squared bracket is the leading one as $t \to \infty$. With the condition $f_d(\tau_\varphi) \approx 1$, the function $f_d(t)$ is found as

$$f_d(t) \equiv \frac{t}{\tau_\varphi^{\mathrm{GZ}}(\tau_{\mathrm{ir}})} = t\,\frac{2e^2}{\hbar}\int_{1/\tau_{\mathrm{ir}}}^{1/\tau_c} \frac{d\omega}{2\pi} \int \frac{d^n k}{(2\pi)^n}\, \tilde{\chi}_{\mathrm{GZ}}(\omega)\frac{P_{\mathrm{diff}}(k,\omega)}{|k|^2}, \tag{9.55}$$

where $\tilde{\chi}_{\mathrm{GZ}}(\omega)$ is the symmetrized power spectrum connected with the noise action. To avoid possible IR and UV divergences, we have placed cutoffs in Eq. (9.55). The high frequency cutoff $1/\tau_c = v_F/\ell$ is put for the reason stated after Eq. (9.44). Appropriate choices for τ_{ir} which fix τ_φ selfconsistently are $\tau_{\mathrm{ir}} = \tau_\varphi$ [280] and $\tau_{\mathrm{ir}} = t$ [283] (see below). For a diffusive conductor with conductivity σ, the electronic contribution to the dielectric function is $\epsilon_{ee}(k,\omega) = 1 + 4\pi\sigma/(-i\omega + D|k|^2)$. Since the first term can be disregarded for typical metals, the power spectrum (9.52) takes the form

$$\tilde{\chi}_{\mathrm{GZ}}(\omega) = \frac{\omega}{\sigma}\coth(\tfrac{1}{2}\beta\hbar\omega) = \frac{\omega}{\sigma}\Big(1 + \frac{e^{-\beta\hbar\omega/2}}{\sinh(\tfrac{1}{2}\beta\hbar\omega)}\Big). \tag{9.56}$$

The first term in the round bracket yields a finite decoherence time at zero temperature. This is the result which sparked the controversy. It has been argued that this term is erroneous and originates from the uncontrollable procedure of the semiclassical averages [279]. Recently it was stated that the standard AAK result of divergent decoherence time at $T = 0$ can be reproduced using GZ's method by accounting for recoil effects when the electron emits or absorbs a photon [279, 283, 284]. This would result in the substitution $\coth(\tfrac{1}{2}\beta\hbar\omega) \to \coth(\tfrac{1}{2}\beta\hbar\omega) + \tanh[\tfrac{1}{2}\beta\hbar(\epsilon - \omega)]$, where $\hbar\omega$ is the energy exchange of the electron with energy $\hbar\epsilon$ at an interaction vertex. The recoil term ensures that processes with $\omega > \epsilon$, $\omega \gg k_B T/\hbar$ are suppressed, and thus the frequency integral in Eq. (9.55) would be UV-convergent. This would entail absence of decoherence at $T = 0$. In the influence functional approach, the recoil terms are supposed to arise from the friction action [284, 283].

Taking the thermal average of the electron's energy, Aleiner, Altshuler and Gershenson (AAG) found the modified power spectrum [279, 282]

$$\tilde{\chi}_{\text{AAG}}(\omega) = \frac{\omega}{\sigma}\frac{\beta\hbar}{4}\int_{-\infty}^{\infty} d\epsilon \frac{\coth(\frac{1}{2}\beta\hbar\omega) + \tanh[\frac{1}{2}\beta\hbar(\epsilon - \omega)]}{\cosh^2(\frac{1}{2}\beta\hbar\epsilon)} = \frac{\omega}{\sigma}\frac{\frac{1}{2}\beta\hbar\omega}{\sinh^2(\frac{1}{2}\beta\hbar\omega)} . \quad (9.57)$$

Interestingly, the power spectra $\tilde{\chi}_{\text{GZ}}(\omega)$ and $\tilde{\chi}_{\text{AAG}}(\omega)$ have the same high temperature limit $2/\beta\hbar\sigma$ while subleading terms are different. Let us finally study the decoherence rate for the power spectrum with limiting low and high temperature behaviors

$$\tilde{\chi}(\omega) = \omega/\sigma + 2/\beta\hbar\sigma . \quad (9.58)$$

The first term is the controversial GZ contribution leading to decoherence at $T = 0$.

Consider now first a quasi-onedimensional wire. After performing the k integration, the dephasing time for IR cutoff $1/\tau_\varphi$ is self-consistently determined by

$$\begin{aligned}
\frac{1}{\tau_\varphi} &= \frac{e^2\sqrt{2D}}{\hbar\sigma_1}\int_{1/\tau_\varphi}^{1/\tau_c}\frac{d\omega}{2\pi}\frac{1}{\sqrt{\omega}}\left(1 + \frac{2}{\beta\hbar\omega}\right) \\
&= \frac{e^2}{\pi\hbar\sigma_1}\sqrt{2D/\tau_c}\left[1 + 2\sqrt{\tau_c\tau_\varphi}\frac{k_{\text{B}}T}{\hbar}\right],
\end{aligned} \quad (9.59)$$

where $\sigma_n = a^{3-n}\sigma$. At temperature $T > T_{\text{c},1} = (ev_{\text{F}}/2)\sqrt{\hbar/\pi\sigma_1\ell}/k_{\text{B}}$, the second term dominates. This yields for the decoherence rate the classical AAK result [287]

$$\frac{1}{\tau_\varphi} = \left(\frac{2\sqrt{2D}\,e^2}{\pi\hbar\sigma_1}\frac{k_{\text{B}}T}{\hbar}\right)^{2/3}, \quad (9.60)$$

which describes Cooperon decay according to e^{-t/τ_φ}. However, the Cooperon is known to decay as $e^{-(t/\bar{\tau}_\varphi)^{3/2}}$, i.e., with power $\frac{3}{2}$ in the exponent [287]. This time-dependence is exactly found if we assume the decay law $e^{-t/\tau_\varphi(t)}$ in which the time dependence of $\tau_\varphi(t)$ enters via the time-dependent IR cutoff $1/\tau_{\text{ir}} = 1/t$ in Eq. (9.55).

For temperature $T < T_{\text{c},1}$, the additive term $+1$ in the square bracket in Eq. (9.59) is the leading one. This term yields the temperature-independent rate contribution

$$\tau_{\varphi,0}^{\text{GZ}} = \pi\hbar\sigma_1/(\sqrt{2}\,e^2 v_{\text{F}}) . \quad (9.61)$$

This is the result by GZ which sparked the dispute.

Corresponding analysis of the expression (9.55) with (9.58) in two and three dimensions yield the limiting low- and high-temperature forms

$$\frac{1}{\tau_\varphi} = \frac{e^2}{4\pi\sigma_2\tau_c}\left[1 + 2\ln(k_{\text{B}}T\tau_c/\hbar)\,k_{\text{B}}T\tau_c/\hbar\right], \qquad n = 2 , \quad (9.62)$$

$$\frac{1}{\tau_\varphi} = \frac{e^2}{3\pi^2\hbar\sigma_3\sqrt{2D}\,\tau_c^{3/2}}\left[1 + 6(k_{\text{B}}T\tau_c/\hbar)^{3/2}\right], \qquad n = 3 . \quad (9.63)$$

These expressions describe again dephasing at zero temperature.

PART III
QUANTUM STATISTICAL DECAY

In the second part, I considered the exactly solvable case of linear dissipative quantum systems, the thermodynamic variational approach useful for nonlinear quantum systems, and issues of quantum decoherence. Now I turn to a study of the frequent situation in which a metastable state is separated from the outside region by a free energy barrier. The decay of a metastable state plays a central role in many scientific areas including low-temperature physics, nuclear physics, chemical kinetics and transport in biomolecules. At high temperatures, the system escapes from the metastable well predominantly over the barrier by thermal activation. At zero temperature on the other hand, the system is in the localized ground state of the metastable well and can escape only by quantum mechanical tunneling through the barrier. I shall focus the discussion on the influence of friction on the decay process. The treatment will be based on an effective method which allows to study this problem in a unified manner for temperatures ranging from $T = 0$ up to the classical regime.

10. Introduction

There are many processes in physics, chemistry, and biology in which a system makes transitions between different states by traversing a barrier. The theory of rate coefficients for barrier crossing has a long history since the days of Arrhenius [294]. H. A. Kramers' article of 1940 [253] represents a cornerstone in the quantitative analysis of thermally activated rate processes. This work provided a thorough theoretical description in the classical regime both for very *weak* and for *moderate-to-strong damping*. It includes important limiting cases such as the transition state theory or the Smoluchowski model of diffusion controlled processes.

The investigation of quantum mechanics for macroscopic variables has been stimulated considerably by Leggett's discussion concerning the validity of quantum mechanics at the macroscopic level [295, 296]. Strong impulse to the field was given further by the work of Caldeira and Leggett [77], who studied *quantum tunneling in the presence of dissipation* at zero temperature. From both the analytical and numerical point of view, quantum reaction theory has also been one of the most active and challenging areas in chemical physics and theoretical chemistry. In the recent past, the functional integral method has provided a unified description of the thermal quantum decay in the entire temperature range [297]. A comprehensive review of many

$V(q)$

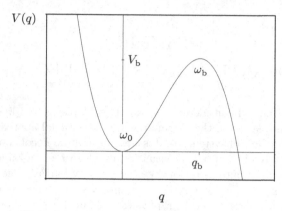

q

Figure 10.1: A metastable "quadratic-plus-cubic" potential well.

of the theoretical concepts and ideas in reaction rate theory extending from classical rate theory to more recent quantum versions has been given by Hänggi, Talkner
and Borkovec [298]. I also refer to a recent account of the semiclassical approach to
quantum tunneling in complex systems by Ankerhold [299].

Assume that the metastable system in question can be characterized by a generalized coordinate q. It may be visualized as a particle of mass M moving in an external
potential $V(q)$ which has a single metastable minimum at a point which I choose at
the origin of q. The bottom of the metastable potential is chosen to lie at zero, i.e.,
$V(0) = 0$. I assume that the potential $V(q)$ is fairly smooth and has the general form
depicted in Fig. 10.1. In particular, $V(q)$ is taken to be negative in the region $q > q_{\text{ex}}$,
where q_{ex} is the nonzero value of q for which $V(q) = 0$. The point q_{ex} is called the
"exit point" from the barrier region. Once the particle has left the metastable well,
it will not return in finite time. With this assumption, we can disregard quantum
coherence effects. In physical systems, the coordinate q is the tunneling degree of
freedom, and the thermal initial state is characterized by the partition function Z_0
of the well region. In chemical reactions, one thinks of q as the reaction coordinate
and Z_0 denotes the partition function for reactants. Among the chemical applications
are dissociation and recombination reactions and transfer processes of atoms and
electrons.

The concept of metastability is useful when the barrier is large enough that the
decay time of the metastable state is very long compared with all other characteristic
time scales of the system dynamics. There exist many time scales such as the cor-
relation time τ_{c} of the noise, the time of relaxation in the locally stable well τ_{r}, the
thermal time $\hbar\beta$, and the time scales $\tau_0 = \omega_0^{-1}$ and $\tau_{\text{b}} = \omega_{\text{b}}^{-1}$ which are related to the
curvature of the potential at the metastable minimum and at the barrier top. Here,
ω_0 is the angular frequency of small oscillations around the metastable minimum

$$\omega_0 = [V''(0)/M]^{1/2} \, , \tag{10.1}$$

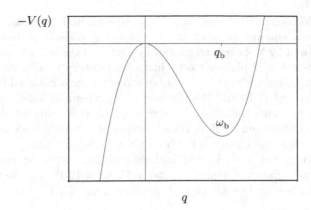

Figure 10.2: The upside-down potential $-V(q)$.

and the unstable barrier frequency

$$\omega_{\mathrm{b}} = [-V''(q_{\mathrm{b}})/M]^{1/2} \qquad (10.2)$$

characterizes the width of the parabolic top of the barrier hindering the decay process. It represents the angular frequency of small oscillations around the minimum of the upside-down potential $-V(q)$ (cf. Fig. 10.2). All these various time scales will become important in a precise description of the escape rate. Weak metastability implies that the activation energy V_{b} is by far the largest energy of the problem, in particular

$$V_{\mathrm{b}} \gg k_{\mathrm{B}}T , \qquad \text{and} \qquad V_{\mathrm{b}} \gg \hbar\omega_0 . \qquad (10.3)$$

In the sequel, I shall not put restrictions on the numerical value of the ratio $\hbar\omega_0/k_{\mathrm{B}}T$. It is convenient to parameterize the rate of escape from the metastable well in the form

$$k = A\,\mathrm{e}^{-B} . \qquad (10.4)$$

The quantity B is a dimensionless measure of the immenseness of the barrier the particle has to overcome. The prefactor A is a kind of an attempt frequency of the particle in the well towards the barrier. We shall be concerned with the question on which scale the parameters A and B are influenced by dissipation. The findings are briefly summarized as follows. In the classical regime, the exponent B is independent of damping and only the attempt frequency A is modified. In the quantum regime, both the exponent and the prefactor of the rate expression crucially depend on the strength and on the spectral properties of the dissipation. As the damping is increased, the classical regime extends to lower temperatures.

In the last decade, the problem of quantum tunneling of macroscopic variables has attracted a lot of interest. The research was considerably stimulated by experiments on macroscopic quantum tunneling in Josephson systems (cf., e.g., Ref. [300], and references therein), and by theoretical work put forward by Caldeira and Leggett [77]. As often in science, the problem of tunneling in the presence of coupling to (infinitely) many degrees of freedom, e.g., to phonons, quasiparticles, and electrons, has several precursors. Among them are the work on deep-inelastic collisions of heavy ions by Brink *et al.* [301], and by Möhring and Smilansky [302]. On the other hand, quantum tunneling in the presence of phonon modes has a long history in solid state physics. The study of multi-phonon effects in the polaron problem was initiated by Holstein [129] and by Emin and Holstein [303]. Further development includes the study of polaron effects on quantum tunneling which received enormous impact through the work by Flynn and Stoneham [132]. A critical review of the early work which relied mainly on the *Condon approximation* has been given by Sethna [169].

Before embarking on a closer look at multidimensional thermal quantum decay, I roam the classical regime in Chapter 11, and in Chapter 12 I cover the theoretical methods in the various regimes of thermal quantum decay for the one-dimensional case.

11. Classical rate theory: a brief overview

11.1 Classical transition state theory

Let us begin the discussion with the simplified description of thermally activated decay by transition state theory (TST). In its simplest form, which is the case of a single degree of freedom in the absence of a reservoir, one makes two *ad hoc* assumptions. First, thermal equilibrium prevails in the well, e.g., through the action of Maxwell's demon, so that the metastable state is represented by a canonical equilibrium distribution. Secondly, the particle will never return once it has crossed the barrier top even infinitesimally. Upon identifying the rate with the probability current across the barrier, and taking the Boltzmann weight of the normalized current density at the barrier top, we obtain

$$k_{\text{cl}}^{(\text{TST})} = \frac{\dfrac{1}{2\pi\hbar} \displaystyle\int dp\, dq\, \exp\left(-\beta\left[p^2/2M + V(q)\right]\right)\delta(q - q_{\text{b}})\,\Theta(p)\,p/M}{\dfrac{1}{2\pi\hbar}\displaystyle\int_{q<q_{\text{b}}} dp\, dq\, \exp\left(-\beta\left[p^2/2M + V(q)\right]\right)}. \tag{11.1}$$

Here I have chosen the barrier to the right of the well with the maximum $V_{\text{b}} = V(q_{\text{b}})$ at $q = q_{\text{b}}$. The expression (11.1) yields the well-known transition state formula

$$k_{\text{cl}}^{(\text{TST})} = f_{\text{cl}}^{(\text{TST})}\, e^{-V_{\text{b}}/k_{\text{B}}T} \qquad \text{with} \qquad f_{\text{cl}}^{(\text{TST})} = \frac{1}{2\pi\hbar}\frac{k_{\text{B}}T}{Z_0} = \frac{\omega_0}{2\pi}, \tag{11.2}$$

where $Z_0 = k_{\text{B}}T/\hbar\omega_0$ is the classical partition function of the well region. The Arrhenius law for the escape rate $k \propto e^{-V_{\text{b}}/k_{\text{B}}T}$ reflects the exponentially small tail of the canonical initial state at the threshold energy $E = V_{\text{b}}$. Note that the classical transition state formula (11.2) does not depend on the width of the barrier. The attempt frequency $f_{\text{cl}}^{(\text{TST})} = \omega_0/2\pi$ is just the frequency of small oscillations in the well. The classical rate vanishes as temperature is lowered to absolute zero.

11.2 Moderate-to-strong-damping regime

In a famous paper published in 1940, Kramers [253] reported his careful study of the classical escape of a particle from a metastable well in various limits. In his treatment of the reaction rate, the starting point is nonlinear Brownian motion in phase space, which is analogous to the Markovian Langevin equation in coordinate space, Eq. (2.3) with (2.1) and (2.2). Following Kramers, the stochastic dynamics for the reaction coordinate q and the velocity $v = \dot{q}$ (or momentum) is conveniently described in terms of a probability density $p(q, v; t)$. The dynamical equation found for $p(q, v; t)$ is of the Fokker-Planck type, nowadays termed Klein-Kramers equation,

$$\frac{\partial p(q, v; t)}{\partial t} = \left(-\frac{\partial}{\partial q} v + \frac{\partial}{\partial v} \left(\frac{1}{M} \frac{\partial V(q)}{\partial q} + \gamma v \right) + \gamma \frac{k_B T}{M} \frac{\partial^2}{\partial v^2} \right) p(q, v; t) . \quad (11.3)$$

A comprehensive review of the Fokker-Planck equation and of the common techniques for tracing solutions is given in the book by Risken [304]. In Subsection 6.8.2 I have sketched the derivation of the Kramers equation (11.3) for the damped linear oscillator by taking the classical limit of the corresponding quantum master equation. Equation (11.3) gives a complete description of the classical stochastic process described by the Langevin equation (2.1) – (2.3).

The Klein-Kramers equation (11.3) has two different time-independent solutions. One solution satisfying $\partial p(q, v; t)/\partial t = 0$ is the canonical equilibrium state

$$p_{eq}(q, v) = N \exp \left\{ -\left[\tfrac{1}{2} M v^2 + V(q) \right] \middle/ k_B T \right\} , \quad (11.4)$$

which describes the thermal distribution of q and v in the well (N is a normalization constant). The population of the well is found in Gaussian approximation as

$$n \equiv \int_{-\infty}^{q_b} dq \int_{-\infty}^{\infty} dv \, p_{eq}(q, v) = \frac{2\pi k_B T}{M \omega_0} N . \quad (11.5)$$

The other stationary solution of Eq. (11.3) represents a steady state with nonzero probability flow over the barrier. The steady state describes the following situation. Particles are continuously injected at the bottom of the well. Afterwards they stay sufficiently long in the well region so that they thermalize. Finally, those particles which have gained the barrier energy leave the barrier region. Outside the barrier region, they are absorbed by a particle sink. To obtain the steady-flow state for a potential of the form sketched in Fig. 10.1, I follow Kramers[253] and make the ansatz

$$p_{flow}(q, v) = G[u(q, v)] \, p_{eq}(q, v) , \quad (11.6)$$

where $G[u(q = 0, v)]$ captures the flow property. I impose the boundary conditions

$$G[u(q = 0, v)] = 1 , \quad \text{and} \quad G[u(q = q_+, v)] \approx 0 . \quad (11.7)$$

Here, q_+ is a point far beyond the barrier region ($q_+ \gg q_b$). The first condition determines that the steady state (11.6) matches on the equilibrium state (11.4) in the well region. The second condition arranges that the particles are removed on the other side of the barrier. The constraints (11.7) imply that the function $G[u(q, v)]$ depends on q and v only through a linear combination of q and v. With the choice

$$u(q, v) = q - q_b - \varrho v/\omega_b \qquad (11.8)$$

in the ansatz (11.6), the Klein-Kramers equation (11.3) is transformed into an ordinary differential equation for $G(u)$, if we choose the parameter ϱ as a solution of the quadratic equation

$$\varrho^2 + (\gamma/\omega_b)\varrho - 1 = 0 . \qquad (11.9)$$

The resulting differential equation for $G(u)$ reads

$$\kappa u G'(u) + G''(u) = 0 \qquad (11.10)$$

with

$$\kappa = M\omega_b^3/(\varrho\gamma k_B T) . \qquad (11.11)$$

The solution of Eq. (11.10) with the boundary conditions (11.7) is

$$G(u) = \sqrt{\frac{\kappa}{2\pi}} \int_u^\infty du' \exp\left(-\frac{\kappa u'^2}{2}\right) . \qquad (11.12)$$

The quantity κ must be positive in order that $G(u)$ vanishes for q near to q_+. This entails that we must choose the positive root of Eq. (11.9) which is

$$\varrho = \sqrt{1 + \left(\frac{\gamma}{2\omega_b}\right)^2} - \frac{\gamma}{2\omega_b} . \qquad (11.13)$$

Consider now the outgoing flow $\mathcal{F}$ of the steady state $p_{\text{flow}}(q, v)$ at the barrier top,

$$\mathcal{F}(q_b) \equiv \int_{-\infty}^\infty dv \, v p_{\text{flow}}(q_b, v) . \qquad (11.14)$$

Insertion of the form (11.6) with (11.12) yields an integral expression which can be transformed upon integration by parts into a standard Gauss integral. We readily get

$$\mathcal{F}(q_b) = \varrho \frac{k_B T}{M} N \exp\left(-V_b/k_B T\right) . \qquad (11.15)$$

The rate of escape over the barrier equals outgoing flux at the barrier top divided by the population of the well. Upon using the expressions (11.5) and (11.14), we obtain

$$k_{cl} \equiv \frac{\mathcal{F}(q_b)}{n} = f_{cl} \exp\left(-V_b/k_B T\right) , \qquad (11.16)$$

where the classical attempt frequency f_{cl} is given by

$$f_{cl} = \varrho f_{cl}^{(\text{TST})} = \varrho \frac{\omega_0}{2\pi} . \qquad (11.17)$$

The form (11.16) with (11.17) is the celebrated expression for the classical escape rate found by Kramers for moderate-to-strong damping. Strictly speaking, the form

(11.17) holds when damping is strong enough that the particle relaxes to the thermal equilibrium state in the well before it escapes over the barrier. The attempt frequency (11.17) differs from the transition state value $f_{\mathrm{cl}}^{(\mathrm{TST})}$ by the transmission factor ϱ, which captures all effects of damping in the classical regime.

In the light of the derivation given here in the framework of the Klein-Kramers equation, the transmission factor is reduced, $\rho < 1$, because of multiple re-crossing runs at the barrier which the particle's noisy trajectory makes.

Alternative derivations of the Kramers rate formula (11.16) with (11.17) are given below. In Section 15.1 I employ the Im F method, and in Section 15.4 I consider multi-dimensional transition state theory of the system-plus-environment complex. This will yield a physical interpretation of the transmission factor $\rho < 1$ which is different from the conception of recrossing processes at the barrier top [cf. concluding remarks in Section 15.4].

11.3   Strong damping regime

For strong damping or large viscosity, the inertia term in Eq. (2.3) becomes negligibly small. It is then expected that, starting from an arbitrary initial distribution $p(q, v; t)$, a Maxwell velocity distribution will arise within short time for every value of q. After that, a slow diffusion of the density distribution in coordinate space will take place. The time evolution of the reduced probability density

$$p(q, t) \equiv \int_{-\infty}^{\infty} dv \, p(q, v; t) \tag{11.18}$$

is described in the classical regime by the Smoluchowski diffusion equation [298, 304]

$$\frac{\partial p(q, t)}{\partial t} = \frac{1}{M\gamma} \frac{\partial}{\partial q} \hat{L}_{\mathrm{cl}}(q) \, p(q, t) \tag{11.19}$$

with the Smoluchowski operator

$$\hat{L}_{\mathrm{cl}}(q) \equiv \frac{\partial V(q)}{\partial q} + k_{\mathrm{B}}T \frac{\partial}{\partial q} . \tag{11.20}$$

For state-dependent diffusion, the Smoluchowski operator takes the form

$$\hat{L}_{\mathrm{cl}}(q) \equiv \frac{\partial V(q)}{\partial q} + \frac{\partial}{\partial q} D(q) . \tag{11.21}$$

The particle current or flow connected with the distribution $p(q, t)$ reads

$$\mathcal{F}(q, t) = -\frac{1}{M\gamma} \left(V'(q) + \frac{\partial}{\partial q} D(q) \right) p(q, t) . \tag{11.22}$$

The Langevin equation which corresponds to the diffusion equation (11.19) with the Smoluchowski operator (11.21) reads in the Ito representation [304]

$$M\gamma\dot{q}(t) = -V'(q) + \sqrt{2M\gamma D(q)}\,\xi(t) \,, \tag{11.23}$$

where $\xi(t)$ is Gaussian white noise with zero mean and correlation $\langle\xi(t)\xi(0)\rangle = \delta(t)$.

One stationary solution of the Smoluchowski equation (11.19) with (11.21) is the equilibrium state in the potential well

$$p_{\text{eq}}(q) = N\,e^{-\phi(q)} \,, \tag{11.24}$$

where N is a normalization constant, and the "effective potential"

$$\phi(q) = \ln D(q) + \psi(q) \qquad \text{with} \qquad \psi(q) = \int_0^q dq'\, \frac{V'(q')}{D(q')} \,. \tag{11.25}$$

The other stationary solution of the Smoluchowski equation is a steady-flow state with constant current or flow $\mathcal{F}$,

$$p_{\text{flow}}(q) = M\gamma\,\mathcal{F}\,e^{-\phi(q)} \int_q^{q_+} dq'\, e^{\psi(q')} \,. \tag{11.26}$$

The steady state $p_{\text{flow}}(q)$ is exponentially small at the other side of the potential barrier. In the steady-flow state, the population of the well is

$$n \equiv \int_{-\infty}^{q_b} dq\, p_{\text{flow}}(q) \,. \tag{11.27}$$

Next, we use that the integrations over the well and the barrier regions decouple and that they can be done in Gaussian approximation. Observing further that the escape rate k_{cl} is the current divided by the population of the well we obtain the expression

$$k_{\text{cl}} = \frac{\mathcal{F}}{n} = \frac{D(0)}{2\pi M\gamma}\sqrt{\phi''(0)|\psi''(q_b)|}\,e^{\psi(0)-\psi(q_b)} \,, \tag{11.28}$$

where we have assumed $\phi'(0) = \psi'(q_b) = 0$.

In the case of state-independent diffusion $D = k_{\text{B}}T$, this reduces to the expression

$$k_{\text{cl}} = \frac{\omega_0\omega_b}{2\pi\gamma}\exp\left(-V_b/k_{\text{B}}T\right) \,. \tag{11.29}$$

Thus we find that the escape rate in the Smoluchowski limit is again in the form (11.16) with (11.17), in which the transmission factor ϱ is inversely proportional to the friction coefficient,

$$\varrho = \omega_b/\gamma \,. \tag{11.30}$$

This form coincides with the strong-friction limit of Kramers' expression (11.13).

Finally I remark that we take up the expression (11.29) in Section 15.3. There we discuss quantum corrections of the escape rate in the Smoluchowski limit.

11.4 Weak-damping regime

For very weak damping, the treatment given in the previous two subsections fails since damping is not strong enough to maintain thermal equilibrium in the well region. In the steady-flux state, the escape over the barrier is limited by the short supply of energy which is required to raise the particle to the barrier top. The escape over the barrier results in a depletion compared with the canonical population in an energy band of width $k_B T$ just below the barrier top. For weak friction, the motion of the particle in the well is oscillatory, and the effect of damping is a gradual change of the distribution in energy space. The mean energy loss ΔE during one round trip in the well is found from the deterministic limit of Eq. (2.3) as

$$\Delta E = \gamma \mathcal{I}(E), \tag{11.31}$$

where

$$\mathcal{I}(E) = \oint dq\, p = M \oint dt\, \dot{q}^2(t) = 2 \int_{q_1}^{q_2} dq\, \sqrt{2M[E - V(q)]} \tag{11.32}$$

is the abbreviated (Minkowskian) action[1] for one round trip at total energy E in the well with turning points q_1 and q_2.

The energy E is a slowly varying variable undergoing a diffusion process. The diffusion equation for the probability density $p(E, t)$ may be written as [298]

$$\dot{p}(E, t) = \gamma \frac{\partial}{\partial E} \mathcal{I}(E) \left(1 + k_B T \frac{\partial}{\partial E} \right) \frac{\omega(E)}{2\pi} p(E, t), \tag{11.33}$$

where $\omega[\mathcal{I}(E)] \equiv \omega(E)$ is the angular frequency at abbreviated action $\mathcal{I}(E)$.

When the particle has built up energy as large as the threshold energy V_b, it escapes from the well with probability 1. The rate is therefore given by the particle flow in energy space through the energy point V_b, divided by the population n of the well. The current $\mathcal{F}$ associated with the steady-flow state $p_{\text{flow}}(E)$ of Eq. (11.33) is

$$\mathcal{F} = -\gamma \mathcal{I}(E) \left(1 + k_B T \frac{\partial}{\partial E} \right) \frac{\omega(E)}{2\pi} p_{\text{flow}}(E). \tag{11.34}$$

With use of the relation

$$p(E) = (\partial \mathcal{I} / \partial E)\, \tilde{p}(\mathcal{I}) = [2\pi / \omega(E)]\, \tilde{p}(\mathcal{I}), \tag{11.35}$$

the current $\mathcal{F}$ may be expressed in terms of the steady-flow state $\tilde{p}_{\text{flow}}(\mathcal{I})$,

$$\mathcal{F} = -\gamma \mathcal{I} \left(1 + \frac{2\pi k_B T}{\omega(E)} \frac{\partial}{\partial \mathcal{I}} \right) \tilde{p}_{\text{flow}}(\mathcal{I}). \tag{11.36}$$

Removing the particle at $\mathcal{I} = \mathcal{I}(E = V_b) \equiv \mathcal{I}_b$ from the well region means putting $\tilde{p}_{\text{flow}}(\mathcal{I} = \mathcal{I}_b) = 0$. With this boundary condition, the solution of Eq. (11.36) reads

---

[1] The Minkowskian [Euclidean] abbreviated action is denoted by $\mathcal{I}(E)$ [$W(E)$].

$$\tilde{p}_{\text{flow}}(\mathcal{I}) \;=\; \frac{\mathcal{F}}{2\pi\gamma k_{\mathrm{B}}T}\, e^{-E(\mathcal{I})/k_{\mathrm{B}}T} \int_{\mathcal{I}}^{\mathcal{I}_{\mathrm{b}}} d\mathcal{I}'\, \frac{\omega(\mathcal{I}')}{\mathcal{I}'}\, e^{E(\mathcal{I}')/k_{\mathrm{B}}T}\,. \tag{11.37}$$

The escape rate is again flow over population, this time in abbreviated-action space,

$$k_{\mathrm{cl}} \;=\; \mathcal{F} \Big/ \int_{0}^{\mathcal{I}_{\mathrm{b}}} d\mathcal{I}\, \tilde{p}_{\text{flow}}(\mathcal{I})\,. \tag{11.38}$$

Next we insert the expression (11.37) and calculate the integrals in Gaussian approximation. We then obtain the rate k_{cl} again in the form (11.16) with Eq. (11.17), in which the transmission factor ϱ is given by

$$\varrho = \gamma\, \mathcal{I}(V_{\mathrm{b}})/k_{\mathrm{B}}T\,. \tag{11.39}$$

Here $\mathcal{I}(V_{\mathrm{b}})$ is the abbreviated action for one round trip in the well at energy $E = V_{\mathrm{b}}$.

In summary, the transmission is reduced compared to the transition state value $\varrho = 1$ by two physically different effects. For moderate-to-strong damping, ϱ is diminished because the trajectory of the particle is stochastic and may cross the barrier point q_{b} several times before the particle eventually escapes. The corresponding expression is given in Eq. (11.13). In the opposite weak-damping limit, $\gamma < k_{\mathrm{B}}T/\mathcal{I}(V_{\mathrm{b}})$, the stationary distribution near the barrier top, $p(E \approx V_{\mathrm{b}})$, is depleted compared with the canonical distribution. As a result, the transmission factor decreases linearly with γ as in Eq. (11.39).

As already noted by Kramers, the two limiting behaviors (11.13) and (11.39) imply a maximum of the rate at a certain damping value intermediate in strength between the above two limits. The actual transmission factor ϱ undergoes a *turnover* near $\gamma = k_{\mathrm{B}}T/\mathcal{I}(V_{\mathrm{b}})$ in the form of a bell-shaped curve as sketched in Fig. 11.1. The turnover theory bridging between the spatial-diffusion-controlled and the energy-diffusion-limited formulas for the activated escape will be discussed in some detail in Section 14.2. There, light is shed also on the multi-dimensional aspects of the classical decay process.

The simplest analytic form of a metastable potential is the "quadratic-plus-cubic" potential sketched in Fig. 10.1,

$$V(q) \;=\; \tfrac{1}{2} M\omega_0^2 q^2 \left(1 - \frac{2q}{3q_{\mathrm{b}}}\right) \;=\; \tfrac{1}{2} M\omega_0^2 q^2 \left(1 - \frac{q}{q_{\mathrm{ex}}}\right)\,. \tag{11.40}$$

Here, q_{b} is the position of the barrier top. The exit point or tunneling distance for $E = 0$ is $q_{\mathrm{ex}} = 3q_{\mathrm{b}}/2$. The barrier height and barrier frequency are

$$V_{\mathrm{b}} \;=\; M\omega_0^2 q_{\mathrm{b}}^2/6\,, \qquad \text{and} \qquad \omega_{\mathrm{b}} = \omega_0\,. \tag{11.41}$$

The potential has the symmetry

$$V(q_{\mathrm{b}} - q) \;=\; V_{\mathrm{b}} - V(q)\,. \tag{11.42}$$

The round trip path in the potential $V(q)$ at energy $E = V_{\mathrm{b}}$, which leaves the point $q = q_{\mathrm{b}}$ at time $-\infty$, bounces back at $q = -q_{\mathrm{b}}/2$ at time zero, and then returns to the starting point at time $+\infty$, reads

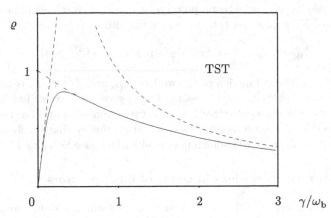

Figure 11.1: Turnover of the transmission factor ϱ from energy-diffusion-limited behavior [Eq. (11.39), straight dashed line] to spatial-diffusion-controlled behavior [Eq. (11.13), dashed-dotted curve]. The Smoluchowski limit $\varrho = \omega_b/\gamma$ is approached for large damping.

$$q_{E=V_b}(t) \;=\; q_b - (3q_b/2)\operatorname{sech}^2(\omega_0 t/2)\,. \tag{11.43}$$

The abbreviated action of this path is

$$\mathcal{I}(V_b) = 36V_b/5\omega_0\,. \tag{11.44}$$

We find with use of Eq. (11.44) that Kramers' formula (11.16) with (11.17) and (11.13) is correct in the classical regime when

$$\gamma/\omega_b \gtrsim 5k_BT/36V_b\,. \tag{11.45}$$

We see from the conditions (10.3) and (11.45) that the moderate-to-strong damping transmission factor (11.13) is valid down to considerably weak damping.

As the temperature is lowered, the classical escape rate (11.16) decreases exponentially fast so that at very low temperature the metastable state can only decay via quantum tunneling.

12. Quantum rate theory: basic methods

With the advent of quantum mechanics, the first who introduced quantum tunneling was Friedrich Hund in 1927 when he described the intramolecular rearrangement in ammonia molecules [305]. Shortly later, the tunneling effect was popularized by Oppenheimer [306], who used it to explain the ionization of atoms in strong electric fields, and by Gamow [307], and Gurney and Condon [308] when they explained the radioactive decay of nuclei. Perhaps the oldest guess of a quantum transition state theory was made by Wigner who proposed as a quantum mechanical generalization of the classical TST expression (11.1) the formula [309]

$$k_W = \frac{1}{Z_0} \frac{1}{2\pi\hbar} \int dp \, dq \, \frac{p}{M} \Theta(p) \delta(q - q_b) f^{(W)}(p, q) , \qquad (12.1)$$

where Z_0 is the partition function of the well region, and $f^{(W)}(p, q)$ is the quantum Wigner distribution (4.263). The expression (12.1) gives the correct first order quantum correction to the classical transition state rate expression and has been used by Miller [310] to derive a semiclassical transition state theory that involves a periodic classical trajectory in the upside-down potential surface (see Sections 12.3 and 15.4).

12.1   Formal rate expressions in terms of flux operators

A concise quantum mechanical rate expression in the form of a Boltzmann average of the reactive cross sections can be found from an exact quantum scattering calculation [311, 312]. It reads

$$k = \mathrm{Re} \left\{ \mathrm{tr} \left(e^{-\beta \hat{H}} \hat{F} \hat{\mathcal{P}} \right) \right\} / Z_0 = \mathrm{tr} \left(e^{-\beta \hat{H}/2} \hat{F}_s \, e^{-\beta \hat{H}/2} \hat{\mathcal{P}} \right) / Z_0 . \qquad (12.2)$$

Here, $\hat{H}$ is the Hamiltonian, Z_0 is the quantum partition function of the reactant, i.e. of the well region, and tr denotes a quantum mechanical trace. The operator $\hat{F}_s$ is the symmetrized flux through dividing surface operator. For the metastable potential of the form sketched in Fig. 10.1, the flux operator $\hat{F}_s$ is given by

$$\hat{F}_s = \tfrac{1}{2}\left(\hat{F} + \hat{F}^\dagger\right), \qquad \hat{F} = \delta\left(\hat{q} - q_b\right) \hat{p}/M , \qquad (12.3)$$

where p is the momentum conjugate to q. Finally, $\hat{\mathcal{P}}$ is a projection operator projecting onto outgoing states, which are states with positive momentum in the infinite future,

$$\hat{\mathcal{P}} = \lim_{t\to\infty} e^{i\hat{H}t/\hbar} \Theta(\hat{p}) \, e^{-i\hat{H}t/\hbar} = \lim_{t\to\infty} e^{i\hat{H}t/\hbar} \Theta(\hat{q} - q_b) \, e^{-i\hat{H}t/\hbar} . \qquad (12.4)$$

The equivalence of both forms was shown in Ref. [311]. A major advantage of the expression (12.2) is that the exact thermal quantum rate expression can be directly given without necessity of first calculating energy-dependent rates and then perform a thermal averaging. The expression (12.2) for the quantum crossing rate involves the Hamiltonian in two ways: first, in the Boltzmann operator $e^{-\beta\hat{H}}$, and secondly through the projection operator $\hat{\mathcal{P}}$ in Eq. (12.4). The infinite time limit of this projection can only be handled correctly by a quantum scattering calculation.

The expression for the crossing rate given in Eq. (12.2) is dynamically exact, with the only assumption that the total system is initially in thermal equilibrium. Thus, the formula covers both the spatial-diffusion-limited regime for moderate-to-strong damping and the energy-diffusion-limited regime for weak damping.

Other formally exact expressions for thermal rate constants were derived from Eq. (12.2) in Ref. [311]. An equivalent form of the rate which is similar to a result obtained by Yamamoto using Kubo's linear response formalism is [313]

$$k = \frac{1}{Z_0} \int_0^\infty dt\, C(t) , \qquad (12.5)$$

where $C(t)$ is the flux-flux autocorrelation function

$$C(t) = \mathrm{tr}\left[e^{i\hat{H}t/\hbar}\, e^{-\beta\hat{H}/2}\, \hat{F}_s\, e^{-\beta\hat{H}/2}\, e^{-i\hat{H}t/\hbar}\, \hat{F}_s \right] . \qquad (12.6)$$

To determine the rate expression (12.5) with (12.6), it is necessary to deal with the quantum scattering problem. This amounts to the evaluation of the matrix element of the complex-time evolution operator $< q'|\, e^{-i\hat{H}(t-i\hbar\beta/2)/\hbar}\, |q'' >$. The propagation of this quantity is the source of difficulty in the numerically exact calculation of reaction rates. It has been suggested to replace the time propagation with the full Hamiltonian by the propagation with that of a parabolic barrier [314]. This leads to the simpler projection operator

$$\hat{\mathcal{P}} = \Theta\left(\hat{q} - q_b + \hat{p}/M\omega_b\right) . \qquad (12.7)$$

With the replacement of the expression (12.4) by the approximation (12.7), the thermal rate may be written as

$$k = \mathrm{tr}\left(e^{-\beta\hat{H}/2}\hat{F}_s\, e^{-\beta\hat{H}/2}\Theta(\hat{q} - q_b + \hat{p}/M\omega_b)\right)\Big/ Z_0 , \qquad (12.8)$$

which depends only on matrix elements of the thermal density operator $e^{-\beta\hat{H}}$.

The flux operator $\hat{F}_s$ in Eq. (12.6) is of low rank. In one dimension, there are only two nonzero eigenvalues, one positive and one negative, corresponding to flux in forward and backward direction. The eigenvalues of the thermal flux operator $e^{-\beta\hat{H}/2}\,\hat{F}_s\, e^{-\beta\hat{H}/2}$ can be calculated efficiently using the iterative Lanczos scheme [315]. This allows to compute only those eigenvalues which are nonzero and contribute to the rate. The calculated basis functions are then propagated without numerical difficulties. The low rank of $\hat{F}_s$ implies a similar low rank for the full operator in Eq. (12.6). The trace can then be computed in this much smaller basis. Numerical stability of the real-time propagation can be considerably improved by introducing a complex absorbing potential. Following these lines, the numerically exact solution of Eq. (12.5) with (12.6) is feasible for systems with several degrees of freedom [316].

12.2 Quantum transition state theory

The quantum transition state approximation invokes the basic transition state idea in the rate expression (11.1): there are only outgoing particles at the barrier top. This is provided by the projection operator [312, 317]

$$\hat{\mathcal{P}} = \Theta(\hat{p}) . \qquad (12.9)$$

With this we have in Eq. (12.2) the substitution

$$\hat{F}\hat{\mathcal{P}} \xrightarrow{\mathrm{TST}} \delta\left(\hat{q} - q_b\right)\frac{\hat{p}}{M}\Theta(\hat{p}) = \frac{1}{2M}\delta\left(\hat{q} - q_b\right)\left(|\hat{p}| + \hat{p}\right) . \qquad (12.10)$$

The expression (12.9) is easier than the form (12.7) since it involves only the momentum. The second term in the bracket of Eq. (12.10) does not contribute in thermal equilibrium because of detailed balance. The physical meaning of the approximation is as in the classical case: any trajectory with positive momentum at the dividing surface (which is the position of the barrier top) does not turn around to make recrossings. The replacement prescription (12.9) eliminates the troublesome projection operator (12.4). As a result, the modified rate expression involves only matrix elements of the canonical operator $e^{-\beta \hat{H}}$.

The transition state approximation (12.9) circumvents the need to know the full scattering dynamics. The full Hamiltonian is retained however in the canonical operator $e^{-\beta \hat{H}}$. Upon employing Eq. (12.10) and the identity

$$ e^{-\beta \hat{H}} = \frac{1}{\pi \hbar} \lim_{\epsilon \to 0^+} \mathrm{Im} \int_0^\infty dE \, e^{-\beta E} \int_0^\infty d\tau \, e^{(E+i\epsilon-\hat{H})\tau/\hbar} \, , \tag{12.11} $$

we may rewrite the formal expression (12.2) in the form

$$ k^{\mathrm{(TST)}} = \frac{1}{Z_0} \frac{1}{2\pi\hbar} \int_0^\infty dE \, p(E) \, e^{-\beta E} \, , \tag{12.12} $$

where $p(E)$ is the dimensionless distribution function

$$ p(E) = \lim_{\epsilon \to 0^+} \mathrm{Im} \int_0^\infty d\tau \, e^{(E+i\epsilon)\tau/\hbar} $$
$$ \times \int dq \, \delta(q-q_{\mathrm{b}}) \, |\dot{q}|_{q=q_{\mathrm{b}}} < q \, | \, e^{-\hat{H}\tau/\hbar} | q > \, . \tag{12.13} $$

The expression (12.12) with (12.13) represents the quantum mechanical rate expression for a one-dimendional system in the transition state approximation. It is free of subsidiary approximations. The function $p(E)$ describes the transmission probability at energy E. From now on we write k for $k^{\mathrm{(TST)}}$.

12.3 Semiclassical limit

In the semiclassical quantum transition state theory [310], the *semiclassical* expression for the imaginary-time propagator $< q \, | \, e^{-\hat{H}\tau/\hbar} | q' >$ is employed in Eq. (12.13).[1] This expression is found from the real-time Van Vleck propagator (5.109) with (5.110) by analytic continuation from real time t to imaginary time $\tau = it$. In addition, it is pertinent to use the relation[2]

$$ -\frac{\partial^2 S_{\mathrm{cl}}(q'',\tau;q',0)}{\partial q'' \, \partial q'} = \frac{1}{\dot{q}(\tau)\,\dot{q}(0)} \frac{\partial^2 S_{\mathrm{cl}}(q'',\tau;q',0)}{\partial \tau^2} \, . \tag{12.14} $$

---

[1] For diverse applications of semiclassical propagators and periodic orbit theory to quantum chaos we refer to the monographs by M. Gutzwiller [229] and by F. Haake [318].

[2] Throughout Part III, the Euclidean action is denoted by S [without the superscript (E)]. The Euclidean and Minkowskian abbreviated actions are termed $W(E)$ and by $\mathcal{I}(E)$ [Eq. (11.32)].

We then have in the semiclassical limit

$$< q'' | e^{-\hat{H}\tau/\hbar} | q' > e^{S_{cl}(\tau)/\hbar} = \frac{e^{-i\phi}}{\sqrt{2\pi\hbar}} \sqrt{-\frac{\partial^2 S_{cl}}{\partial q'' \partial q'}} = \frac{e^{-i\phi}}{\sqrt{2\pi\hbar}} \sqrt{\frac{\partial^2 S_{cl}/d\tau^2}{\dot{q}(\tau)\,\dot{q}(0)}} , \quad (12.15)$$

where

$$S_{cl}(\tau) = \int_0^\tau d\tau' \left(\tfrac{1}{2}M\dot{q}^2(\tau') + V[q(\tau')] \right) \quad (12.16)$$

is the Euclidean action for the path obeying the equation of motion $M\ddot{q}(\tau) = \partial V/\partial q$ with constraints $q(0) = q'$ and $q(\tau) = q''$. The phase ϕ is $\pi/2$ times the number of conjugate points along the path from q' to q''. Equation (12.13) with (12.15) yields

$$p(E) = \frac{i\,\mathcal{J}_1}{\sqrt{2\pi\hbar}} \int_C d\tau \, \sqrt{|\partial^2 S_{cl}/\partial\tau^2|} \, e^{-[S_{cl}(\tau)-E\tau]/\hbar} . \quad (12.17)$$

The second step of the semiclassical approximation consists in the calculation of $p(E)$ the time integral in Eq. (12.13) with the method of steepest descent. The relevant exponential factor $\exp\{-[S_{cl}(\tau) - E\tau]/\hbar\}$ is stationary for Euclidean time $\tau = \bar{\tau}$,

$$\frac{\partial S_{cl}}{\partial\tau}\bigg|_{\tau=\bar{\tau}} = E . \quad (12.18)$$

The stationary points are periodic orbits with energy E in the upside-down potential $-V(q)$. Upon expanding the exponent about the stationary points, we obtain

$$S_{cl}(\tau) - E\tau = W(E) + \frac{1}{2}\frac{\partial^2 S_{cl}}{\partial\tau^2}\bigg|_{\tau=\bar{\tau}} (\tau - \bar{\tau})^2 + \mathcal{O}\big((\tau - \bar{\tau})^3\big) , \quad (12.19)$$

where $W(E)$ is the abbreviated *Euclidean* action

$$W(E) = M \oint d\tau \, \dot{q}^2(\tau) = 2 \int_{q_1}^{q_2} dq \, \sqrt{2M[V(q) - E]} . \quad (12.20)$$

The limits q_1 and q_2 are the zeros of the integrand. They form the turning points of the classical motion in the upside-down potential $-V(q)$ at conserved energy

$$E = V[q(\tau)] - \tfrac{1}{2}M\dot{q}^2(\tau) . \quad (12.21)$$

Since $\partial^2 S_{cl}/\partial\tau^2 = 1/(\partial\tau/\partial E) < 0$ at $\tau = \bar{\tau}$, the periodic orbits are unstable against small perturbations. The saddle points are located on the positive-real axis of the complex-τ plane at $\bar{\tau} \equiv \tau(E) = -\partial W(E)/\partial E$, and the direction of steepest descent is perpendicular to the imaginary-time axis in the complex τ-plane. The contribution of the periodic orbit with one cycle at energy E is

$$p_1(E) = -\mathcal{J}_1 \, e^{-W(E)/\hbar} . \quad (12.22)$$

The motion of the particle in the well of the upside-down potential is bounded and periodic for $E \geq 0$. For $E \ll V_b$, the particle spends most of the time in the vicinity of the inner turning point q_1 and it bounces back from the outer turning point q_2. For this reason, the periodic path satisfying Eq. (12.21) is called *bounce* [107]. We shall denote this path by $q_{\rm B}(\tau)$. It is convenient to choose the phase of the bounce such that $q_{\rm B}(\tau) = q_{\rm B}(-\tau)$, and $q_{\rm B}(\tau = 0) = q_2$. For a bounce with total energy $E = 0$ we have $q_2 = q_{\rm ex}$ and $q_1 = 0$. The additional factor $\mathcal{J}_1$ originates from the conjugate points. Each turning point along the orbit contributes a phase shift $\pi/2$ to the phase ϕ in Eq. (12.15). A periodic path with n bounces at energy E and total time $\tau_n = n\bar{\tau}$ has abbreviated action $nW(E)$, and the phase factor is $\mathcal{J}_n = {\rm e}^{-i\pi n} = (-1)^n$. The sum of contributions from the infinite sequence of stationary points is

$$p(E) = \sum_{n=1}^{\infty} (-1)^{n-1}\,{\rm e}^{-nW(E)/\hbar} = \frac{1}{1 + {\rm e}^{W(E)/\hbar}}\,. \tag{12.23}$$

In the semiclassical limit $\hbar \to 0$ with $T/T_0 < 1$ held fixed, where

$$k_{\rm B}T_0 = \frac{1}{\beta_0} \equiv \frac{\hbar\omega_{\rm b}}{2\pi}\,, \tag{12.24}$$

the Boltzmann average in Eq. (12.12) is dominated by the periodic orbit with one cycle of period $\tau \equiv -\partial W/\partial E = \hbar\beta$,

$$k = \frac{1}{Z_0}\frac{1}{2\pi\hbar}\int_0^{V_{\rm b}} dE\, p_1(E)\,{\rm e}^{-\beta E} = \frac{1}{Z_0}\frac{1}{2\pi\hbar}\int_0^{V_{\rm b}} dE\, {\rm e}^{-\beta E - W(E)/\hbar}\,. \tag{12.25}$$

Calculation of the integral by steepest descent yields

$$k = \frac{1}{Z_0}\frac{1}{\sqrt{2\pi\hbar|\tau_{\rm B}'|}}\,{\rm e}^{-S_{\rm B}/\hbar}\,, \qquad T < T_0\,, \tag{12.26}$$

$$\tau_{\rm B}' \equiv \frac{\partial\tau}{\partial E}\Big|_{E=E_{\hbar\beta}} = -\frac{\partial^2 W(E)}{\partial E^2}\Big|_{E=E_{\hbar\beta}} = \frac{1}{\partial^2 S/\partial\tau^2}\Big|_{\tau=\hbar\beta}\,. \tag{12.27}$$

The quantity $S_{\rm B}$ is the action (12.16) of the bounce trajectory with period $\hbar\beta$ and associated total energy $E = E_{\hbar\beta}$. Using Eqs. (12.16) and (12.21), we have

$$S_{\rm B} \equiv S_{\rm cl}(\hbar\beta) = W(E_{\hbar\beta}) + E_{\hbar\beta}\hbar\beta\,. \tag{12.28}$$

The expression (12.26) with Eqs. (12.27) and (12.28) represents the semiclassical quantum rate formula in transition state theory [319]. This form is generally valid in the temperature regime $0 \le T < T_0$. The semiclassical result (12.26) is rounded off with the semiclassical partition function of the well region

$$Z_0 = \frac{1}{2\sinh(\beta\hbar\omega_0/2)}\,. \tag{12.29}$$

In the temperature regime $T > T_0$, the integral in Eq. (12.12) is dominated by energies $E \gtrsim V_{\rm b}$. In this regime, the expression (12.23) applies in which $W(E)$ is the abbreviated action for the motion in the parabolic potential matched with the actual barrier,

$$p(E) = \frac{1}{1 + e^{2\pi(V_b - E)/\hbar\omega_b}} . \tag{12.30}$$

With use of the forms (12.29) and (12.30) in Eq. (12.12), and with extension of the lower integration limit to $E = -\infty$, we readily obtain

$$k = \frac{\omega_b}{2\pi} \frac{\sinh(\frac{1}{2}\beta\hbar\omega_0)}{\sin(\frac{1}{2}\beta\hbar\omega_b)} e^{-\beta V_b} , \qquad T > T_0 . \tag{12.31}$$

In the limit $T \gg \hbar\omega_0/k_B$, $\hbar\omega_b/k_B$ this expression reduces to the classical transition state formula (11.2). As T approaches T_0 from above, the expression (12.31) diverges $\propto 1/(T - T_0)$ or $\propto 1/(\beta_0 - \beta)$, respectively. We have at T slightly above T_0

$$k = \frac{1}{Z_0} \frac{1}{2\pi\hbar(\beta_0 - \beta)} e^{-\beta V_b} , \qquad T \gtrsim T_0 . \tag{12.32}$$

The singularity is removed by taking into account that the actual barrier is wider than the parabolic one. This yields for E close to V_b the modified abbreviated action

$$W(E) = \frac{2\pi}{\omega_b}\left(V_b - E\right) + \frac{|\tau_b'|}{2}\left(V_b - E\right)^2 + \mathcal{O}\left((V_b - E)^3\right) . \tag{12.33}$$

where $\tau_b' = \partial\tau/\partial E|_{E=V_b}$. With use of this form in the integral expression

$$k = \frac{1}{Z_0} \frac{1}{2\pi\hbar} \int_{-\infty}^{V_b} dE\, e^{-\beta E - W(E)/\hbar} , \tag{12.34}$$

and by completion of the square in the exponent, we get

$$k = \frac{1}{2Z_0} \frac{1}{\sqrt{2\pi\hbar|\tau_b'|}} \operatorname{erfc}\left(\frac{\hbar(\beta_0 - \beta)}{\sqrt{2\hbar|\tau_b'|}}\right) \exp\left(-\beta V_b + \frac{\hbar(\beta - \beta_0)^2}{2|\tau_b'|}\right) , \tag{12.35}$$

where $\operatorname{erfc}(z)$ is the complementary error function [89]

$$\operatorname{erfc}(z) \equiv \frac{2}{\sqrt{\pi}} \int_z^\infty dt\, e^{-t^2} . \tag{12.36}$$

The expression (12.35) smoothly interpolates between the rate expression (12.26) valid for temperatures below T_0 and the expression (12.32) which holds for temperatures slightly above T_0. The bridging expression (12.35) is required in a narrow temperature regime about T_0 which is $|T - T_0|/T_0 \lesssim k_B T_0 \sqrt{2|\tau_b'|}/\hbar$. We shall consider this regime in some more detail in Section 16.3.

12.4 Quantum tunneling regime

Consider now the decay rate in the low-temperature regime $T \ll T_0$ more closely. In the regime $E \ll V_b$, the abbreviated Euclidean action (12.20) can be expanded as

$$W(E) = W_h(E) + W(0) - W_h(0) + E\left[\partial W/\partial E - \partial W_h/\partial E\right]_{|E=0} + \cdots , \tag{12.37}$$

where $W_h(E)$ is the abbreviated action for the harmonic potential $V_h(q) = \frac{1}{2}M\omega_0^2 q^2$. The action of the bounce with period $\hbar\beta = \infty$ and associated energy $E = 0$ is

$$S_0 \equiv S_{cl}(\infty) = W(0) = 2\int_0^{q_{ex}} dq\,\sqrt{2MV(q)}\,, \tag{12.38}$$

where the bounds $q = 0$ and $q = q_{ex}$ are the zeros of the potential $V(q)$. The bounce path $q_B(\tau)$ at $E = 0$ is the solution of $\dot{q} = \sqrt{2V(q)/M}$. With the symmetric choice $q_B(\tau) = q_B(-\tau)$, it hits the bounce point q_{ex} at $\tau = 0$ and approaches the origin $q = 0$ with exponential slow-down

$$\lim_{\tau\to\pm\infty} q_B(\tau) = C_0\sqrt{\frac{S_0}{2M\omega_0}}\,e^{-\omega_0|\tau|}\,. \tag{12.39}$$

Here we have parametrized the pre-exponential factor conveniently. The constant C_0 is a numerical factor which depends on the shape of the barrier. With Eqs. (12.38) and (12.39) we obtain from Eq. (12.37) the low-energy expansion

$$W(E) = S_0 - \frac{E}{\omega_0} - \frac{E}{\omega_0}\ln\left(\frac{C_0^2 S_0\omega_0}{E}\right) + \mathcal{O}\left[(E/V_b)^2\right]\,. \tag{12.40}$$

We shall use this expression shortly.

Suppose now that the particle is initially prepared in the ground state of the metastable well. The probability per unit time that it escapes from this state by quantum tunneling through the barrier is given by the standard WKB formula [107]

$$\gamma_0 = \omega_0 C_0\sqrt{S_0/2\pi\hbar}\,e^{-S_0/\hbar}\,. \tag{12.41}$$

If the particle is initially placed in the nth excited state of the well with energy $E_n = \hbar\omega_0(n + \frac{1}{2})$, the rate of escape from the well in the semiclassical limit is [108]

$$\gamma_n = \frac{1}{n!}\left(C_0^2\frac{S_0}{\hbar}\right)^n \gamma_0\,. \tag{12.42}$$

For the potential (11.40), the bounce with $E = 0$ takes the bell-shaped form

$$q_B(\tau) = \frac{3}{2}q_b\,\text{sech}^2(\tfrac{1}{2}\omega_0\tau)\,. \tag{12.43}$$

Because of the symmetry relation (11.42), the Euclidean action S_0 coincides with the Minkowskian action (11.32) $I(E = V_b)$. The Euclidean action S_0 and the numerical factor C_0 are given by

$$S_0 = \mathcal{I}(V_b) = \frac{36}{5}\frac{V_b}{\omega_0} = \frac{6}{5}M\omega_0 q_b^2\,, \qquad \text{and} \qquad C_0 = \sqrt{60}\,. \tag{12.44}$$

We now define a thermal rate by averaging the decay rates for the individual energy levels in the well with the canonical distribution,

$$k \equiv \sum_n \gamma_n\,e^{-\beta E_n}\Big/\sum_m e^{-\beta E_m}\,. \tag{12.45}$$

Upon inserting the semiclassical expression (12.42), the sum in Eq. (12.45) can be carried out explicitly. We obtain

$$k = \frac{e^{-\hbar\omega_0/2k_B T}}{Z_0} \exp\left(C_0^2 \frac{S_0}{\hbar} e^{-\hbar\omega_0/k_B T}\right) \gamma_0 . \tag{12.46}$$

The expression (12.46) represents the leading thermal enhancement of the quantum statistical rate due to thermal occupation of excited states in the well. It applies when the thermal energy is very small compared to $\hbar\omega_0$. At higher temperature, the sum of partial rates in Eq. (12.45) is replaced by the integral expression

$$k = \frac{1}{Z_0} \int_0^\infty dE\, \rho(E)\gamma(E)\, e^{-\beta E} . \tag{12.47}$$

Here the density of states is constant, $\rho(E) = 1/\hbar\omega_0$, and $\gamma(E)$ is the WKB expression of the transmission probability

$$\gamma(E) = \frac{\omega_0}{2\pi} e^{-W(E)/\hbar} , \tag{12.48}$$

which applies for elevated energy $E \gg \hbar\omega_0$.

The rate expression (12.47) with Eq. (12.48) coincides with the expression (12.25). Hence the steepest descent approximation for the energy integral yields the semiclassical formula (12.26), which we rewrite for convenience,

$$k = \frac{1}{Z_0} \frac{1}{\sqrt{2\pi\hbar|\tau_B'|}} e^{-S_B/\hbar} , \qquad T < T_0 . \tag{12.49}$$

We now show that the expression (12.49) is valid down to zero temperature. To this we first calculate the energy of the bounce with period $\hbar\beta$. Inversion of the relation $|\partial W/\partial E|_{E=E_{\hbar\beta}} = \hbar\beta$ with $W(E)$ given in Eq. (12.40) yields

$$E_{\hbar\beta} = C_0^2 S_0 \omega_0 e^{-\beta\hbar\omega_0} . \tag{12.50}$$

With this expression and with Eq. (12.40) the action (12.28) of the bounce with period $\hbar\beta \gg 1/\omega_0$ is found as

$$S_B = \left(1 - C_0^2 e^{-\beta\hbar\omega_0}\right) S_0 . \tag{12.51}$$

With use of the last relation in Eq. (12.27) we readily get

$$|\tau_B'|^{-1} = |\partial^2 S/\partial\tau^2|_{\tau=\hbar\beta} = C_0^2 S_0 \omega_0^2 e^{-\beta\hbar\omega_0} . \tag{12.52}$$

Now all constituents of the expression (12.49) are on hand. With the expressions (12.51) and (12.52) we directly obtain the previous result (12.46). This verifies that the expression (12.49) is valid down to zero temperature [108, 319].

The agreement of Eq. (12.49) with the expression (12.46) at low T is surprising for two reasons. First, in the initial equation (12.47) the discreteness of the low-lying states is disregarded. Secondly, the steepest-descent transmission formula (12.48) is not correct for low-lying states [108].[3] Apparently, the defects cancel out each other when we proceed via steepest descent approximation to the formula (12.49).

---

[3]The formula (12.48) for $\gamma(E_n)$ follows from the expression (12.42) by substituting Stirling's asymptotic formula for the factorial $n!$. This approximation is only justified when n is large.

12.5   Free energy method

An independent and powerful line of reasoning is to calculate the quantum rate with a pure thermodynamic equilibrium method. The thermodynamic method was pioneered by Langer [320] in 1967 and reviewed by himself in 1980 in Ref. [321]. In the original treatment, Langer calculated the classical nucleation rate governing the early stage of a first-order phase transition. In this approach the quantity of interest is the free energy of the metastable system. Because of states of lower energy on the other side of the barrier, the partition function is defined by means of an analytical continuation from a stable potential to the metastable one depicted in Fig. 10.1. The corresponding deformation of the integration contour leads to a unique imaginary part of the free energy associated with the metastable state. This quantity is then related to the decay probability of the system, analogous to the interpretation of imaginary energies of resonances in quantum field theory. Interestingly and importantly enough, this method is not restricted to the classical regime. As we shall see, all the results of Section 12.3 are reproduced by the free energy ($\mathrm{Im}\,F$) method. In a sense, the $\mathrm{Im}\,F$ method is easier than the method presented in Section 12.3 since the energy and time integrals in Eqs. (12.12) and (12.13) are obviated.

In this section, we briefly sketch the $\mathrm{Im}\,F$ method for the onedimensional case. The $\mathrm{Im}\,F$ method turns out to be particularly appropriate in the case of a system-plus-environment complex where the number N of bath degrees of freedom is very large or even infinity. The respective treatment is given in Chapters 15 – 17.

The general argument is as follows [108, 319]. For a metastable system, the partition function may be written as

$$Z = \sum_n e^{-\beta z_n} = \sum_n e^{-\beta(E_n - i\hbar\gamma_n/2)} , \qquad (12.53)$$

where the $z_n = E_n - i\hbar\gamma_n/2$ are the complex energies of the individual states, and $\sum_n$ denotes summation over all states. For weak metastability, $V_\mathrm{b}/\hbar\omega_0 \gg 1$, we have $\hbar\gamma_n \ll E_n$ for all relevant n. Thus we may write

$$Z = Z' + iZ'' \approx \sum_n e^{-\beta E_n} + i\frac{\hbar\beta}{2}\sum_n \gamma_n e^{-\beta E_n} . \qquad (12.54)$$

The imaginary part Z'' of the partition function is generally small compared to the real part Z'. The real part Z' is (apart from exponentially small corrections) determined by properties of the well, $Z' = Z_0$, and the imaginary part Z'' is determined by properties of the barrier, $Z'' = \mathrm{Im}\,Z_\mathrm{b}$. Using the defining relation of the free energy

$$F = -\ln Z/\beta , \qquad (12.55)$$

we obtain with the expression (12.54)

$$\mathrm{Im}\,F = -\frac{1}{\beta}\frac{\mathrm{Im}\,Z_\mathrm{b}}{Z_0} = -\frac{\hbar}{2}\frac{\sum_n \gamma_n\, e^{-\beta E_n}}{\sum_m e^{-\beta E_m}} . \qquad (12.56)$$

Thus the quantum-statistical rate (12.45) is found in the Im F-method as

$$k = -\frac{2}{\hbar} \operatorname{Im} F = \frac{2}{\hbar\beta} \frac{\operatorname{Im} Z_b}{Z_0} . \qquad (12.57)$$

The formula (12.57) is the generalization of the ground state relation $\gamma_0 = -2 \operatorname{Im} E_0/\hbar$ to finite temperature. Actually, the relation (12.57) is only valid in the regime $T < T_0$, as we shall see shortly.

Let us study Im F in the semiclassical limit. The dominant stationary point in function space for the barrier contribution Z_b to the partition function in the regime $T < T_0$ is the periodic bounce path with period $\hbar\beta$ and associated action S_B. To second order in the fluctuations about the extremal path, $q(\tau) = q_B(\tau) + \xi(\tau)$, the action takes the form

$$S[q] = S_B + \frac{M}{2} \int_0^{\hbar\beta} d\tau\, \xi(\tau)\, \mathbf{\Lambda}[q_B(\tau)]\, \xi(\tau) \qquad (12.58)$$

with the fluctuation operator $\mathbf{\Lambda}[q_B(\tau)] = -\partial^2/\partial\tau^2 + V''[q_B(\tau)]/M$ acting in the space of $\hbar\beta$-periodic functions. Expanding $\xi(\tau)$ into the eigenmodes $\{\chi_n(\tau)\}$ of $\mathbf{\Lambda}[q_B(\tau)]$ with orthonormality relation $\int_0^{\hbar\beta} d\tau\, \chi_n(\tau)\chi_m(\tau)/\hbar\beta = \delta_{nm}$,

$$\xi(\tau) = \sum_n c_n \chi_n(\tau) , \qquad (12.59)$$

the action becomes

$$S[q(\cdot)] = S_B + \frac{M\hbar\beta}{2} \sum_n \lambda_n[q_B]\, c_n^2 . \qquad (12.60)$$

The $\lambda_n[q_B]$ are the eigenvalues of $\mathbf{\Lambda}[q_B(\tau)]$ for periodic boundary conditions. According to the analysis in Subsection 4.3.2, the ratio $Z^{(b)}/Z_0$ would then be

$$\frac{Z^{(b)}}{Z_0} = \sqrt{\frac{\det\left[-\partial^2/\partial\tau^2 + \omega_0^2\right]}{\det\left[-\partial^2/\partial\tau^2 + V''[q_B(\tau)]\right]}}\, e^{-S_B/\hbar} = \prod_n \sqrt{\frac{\lambda_n^{(0)}}{\lambda_n[q_B]}}\, e^{-S_B/\hbar} . \qquad (12.61)$$

However, this expression is not quite correct for two reasons.

(1) Differentiation of the equation of motion $M\ddot{q}_B(\tau) - V'[q_B(\tau)]/M = 0$ with respect to time τ shows that $\dot{q}_B(\tau)$ is an eigenmode of the fluctuation operator $\mathbf{\Lambda}[q_B(\tau)]$ with eigenvalue zero. The normalized zero mode is

$$\chi_1(\tau) = \sqrt{M\hbar\beta/W_B}\, \dot{q}_B(\tau) , \qquad (12.62)$$

where

$$W_B \equiv M \int_0^{\hbar\beta} d\tau\, \dot{q}_B^2(\tau) . \qquad (12.63)$$

The zero mode $\chi_1(\tau)$ induces an infinitesimal time translation of the bounce,

$$q_B(\tau + \delta\tau_1) = q_B(\tau) + \sqrt{W_B/M\hbar\beta}\, \chi_1(\tau)\, \delta\tau_1 . \qquad (12.64)$$

The eigenvalue of the mode $\chi_1(\tau)$ is zero because the bounce action is invariant under changes of the phase. As a result, the integral over the zero mode amplitude c_1 diverges for the second order action (12.60). With Eq. (12.64) however, the flawed fluctuation integral can be transcribed into a good-natured integral over the range of the arbitrary phase time τ_1 of the bounce $q_B(\tau + \tau_1)$ [310, 322, 323]. Upon disassembling the divergent zero eigenvalue term and using the Jacobian $dc_1/d\tau_1 = \sqrt{W_B/M\hbar\beta}$, we readily obtain the substitution rule

$$\frac{1}{\sqrt{\lambda_1[q_B]}} \;\to\; \sqrt{\frac{M\hbar\beta}{2\pi\hbar}} \int_{-\infty}^{\infty} dc_1 \exp\left(-M\hbar\beta\lambda_1[q_B]c_1^2/2\hbar\right)$$

$$\to \; \sqrt{\frac{W_B}{2\pi\hbar}} \int_0^{\hbar\beta} d\tau_1 \;=\; \sqrt{\frac{W_B}{2\pi\hbar}}\; \hbar\beta \;. \tag{12.65}$$

(2) While the bounce $q_B(\tau)$ is free of nodes, the zero mode (12.62) has one node. Thus, by the node-counting theorem, there exists a nodeless fluctuation mode $c_0\chi_0(\tau)$ with a negative eigenvalue $-|\lambda_0[q_B]|$. The negative eigenvalue manifests that the bounce is a saddle point of the action, and not a minimum This feature is owed to the fact that we are dealing with the partition function of a metastable well. The mode χ_0 must be dealt with by distortion of the c_0-contour into the half-plane $\mathrm{Im}\,c_0 > 0$. Thereby the contribution of this mode to the barrier partition function $Z^{(b)}$ gathers up an imaginary part. This term is given by

$$\sqrt{\frac{M\hbar\beta}{2\pi\hbar}} \int_0^{i\infty} dc_0 \exp\left(M\hbar\beta|\lambda_0[q_B]|c_0^2/2\hbar\right) \;=\; i\,\frac{1}{2}\,\frac{1}{\sqrt{|\lambda_0[q_B]|}} \;. \tag{12.66}$$

The factor i originates from the analytic continuation, which is required if one deforms an absolute stable potential to a metastable one [324]. The additional factor $\frac{1}{2}$ appears because the integration in Eq. (12.66) covers only one wing of the Gaussian function.

With items (1) and (2) taken into account, the flawed expression (12.61) is converted into the correct semiclassical result

$$\frac{\mathrm{Im}\,Z_b}{Z_0} \;=\; \frac{\hbar\beta}{2}\sqrt{\frac{W_B}{2\pi\hbar}} \left|\frac{\det\{-\partial^2/\partial\tau^2 + \omega_0^2\}}{\det'\{-\partial^2/\partial\tau^2 + V''[q_B(\tau)]\}}\right|^{1/2} e^{-S_B/\hbar} \;. \tag{12.67}$$

The determinants are calculated for eigenfunctions obeying periodic boundary conditions as discussed in Subsection 4.3.2. The prime in $\det'[\cdots]$ indicates that the zero eigenvalue is omitted.

Clearly, it would be inconsistent to keep a real contribution to $Z^{(b)}/Z^{(0)}$ resulting from $Z^{(b)}$, since it would be exponentially small by the factor $e^{-\beta V_b}$, while non-Gaussian corrections to $Z^{(0)}$ are disregarded anyhow. On the other hand, it is consistent to keep the exponentially small imaginary part (12.67) as it is the leading one.

Alternatively, we may calculate the bounce contribution to the partition function by starting from the semiclassical imaginary-time propagator (12.15). Putting $\tau =$

$\hbar\beta$, and taking into account that the distortion of the integration contour at the saddle point into the direction of steepest descent yields an factor $i/2$ compared with Eq. (12.15), we obtain

$$\text{Im} < q|e^{-\hat{H}\beta}|q >_{|\text{barrier}} = \frac{1}{2} \frac{1}{\sqrt{2\pi\hbar}} \frac{1}{|\dot{q}_\text{B}(q)|} \sqrt{\left|\frac{\partial^2 S_\text{cl}(q,\tau;q,0)}{\partial\tau^2}\right|_{\tau=\hbar\beta}} \; e^{-S_\text{B}/\hbar}. \quad (12.68)$$

Writing the trace integral $\oint dq \cdots$ as $\int_0^{\hbar\beta} d\tau \, \dot{q}(\tau) \cdots$, and observing the relations (12.27), we obtain

$$\text{Im}\, Z_\text{b} = \frac{1}{2} \frac{\hbar\beta}{\sqrt{2\pi\hbar|\tau_\text{B}'|}} \; e^{-S_\text{B}/\hbar}. \quad (12.69)$$

Upon equating Eq. (12.67) with Eq. (12.69), we can express the ratio of the determinants in Eq. (12.67) directly in terms of the classical mechanics of the bounce,

$$Z_0^2 \left|\frac{\det\{-\partial^2/\partial\tau^2 + \omega_0^2\}}{\det'\{-\partial^2/\partial\tau^2 + V''[q_\text{B}(\tau)]\}}\right| = \frac{1}{W_\text{B}|\tau_\text{B}'|}. \quad (12.70)$$

This relation agrees with the result of a direct computation [325].

By inserting Eq. (12.69) into Eq. (12.57) we see that the resulting rate expression coincides with the form (12.26). This consistency affirms validity of the relation (12.57) in the temperature regime $T \leq T_0$.

For $T > T_0$, the saddle point pertaining to the constant path $q = q_\text{b}$ determines the barrier partition function. Now, the zero mode is missing, but there is again a mode with negative eigenvalue. The obligatory deformation of the respective integration contour in turn leads to an imaginary contribution from the barrier region to the partition function,

$$\text{Im}\, F = -\frac{1}{\beta} \frac{\text{Im}\, Z_\text{b}}{Z_0} \quad \text{with} \quad \text{Im}\, Z_\text{b} = \frac{1}{4\sin(\beta\hbar\omega_\text{b}/2)} \; e^{-\beta V_\text{b}}. \quad (12.71)$$

The expressions (12.71) and (12.31) are in agreement if there holds the relation

$$k = -\frac{2}{\hbar} \frac{\beta}{\beta_0} \text{Im}\, F = \frac{2}{\hbar\beta_0} \frac{\text{Im}\, Z_\text{b}}{Z_0} \quad \text{for} \quad T > T_0. \quad (12.72)$$

These findings, which go back to Affleck [319], are summarized as follows. The relation (12.57) is appropriate in the temperature regime $T \leq T_0$, in which $\text{Im}\, Z_\text{b}$ is determined by the periodic bounce trajectory $q_\text{B}(\tau)$. Because the phase of the bounce $q_\text{B}(\tau)$ is arbitrary, there is a zero mode in the deviations about $q_\text{B}(\tau)$. For this reason, $\text{Im}\, Z_\text{b}$ is proportional to $\hbar\beta$. The factor β cancels out in the expression (12.69) for $\text{Im}\, F$. Above T_0, there are no zero mode fluctuations since the saddle point trajectory is the constant path $q(\tau) = q_\text{b}$. Because of the absence of the zero mode above T_0, the rate for $T > T_0$ is given by the modified formula (12.72). Clearly, this is merely a plausible argument and certainly not a proof. The expressions (12.57) and (12.72) can be written in the combined form

$$k = -2\sigma\, \text{Im}\, F/\hbar, \quad (12.73)$$

where σ is a piecewise constant numerical factor given by

$$
\begin{aligned}
\sigma &= 1 && \text{for} && \beta\hbar\omega_{\text{b}} \geq 2\pi\,, \\
\sigma &= \beta\hbar\omega_{\text{b}}/2\pi && \text{for} && \beta\hbar\omega_{\text{b}} < 2\pi\,.
\end{aligned}
\tag{12.74}
$$

We note that in multidimensional tunneling systems an effective barrier frequency ω_{R} takes the place of the bare barrier frequency ω_{b} (see Section 14.3).

A general remark is in order. We have seen that quantum tunneling opens new channels for barrier crossing as the temperature is lowered below the classical regime. Therefore, the escape rate is enhanced compared with the classical rate expression. At intermediate temperature, a rough estimate of the rate is obtained naively by adding to the classical rate a quantum mechanical counterpart, $k = k_{\text{cl}} + k_{\text{qm}}$ [326]. In a first guess, k_{qm} may be identified with the ground-state tunneling rate (12.41). With similar reasoning, a plausible criterion for the crossover temperature T_0, at which roughly the transition from thermally activated to quantum mechanical decay occurs, was given by Goldanskii already in 1959 [327]. By equating the Arrhenius exponent with the Gamow factor for the transmission of a parabolic barrier with height V_{b} and barrier frequency ω_{b},

$$
\beta_0 V_{\text{b}} = 2\pi V_{\text{b}}/\hbar\omega_{\text{b}}\,,
\tag{12.75}
$$

the crossover temperature of an undamped system is found just in the form (12.24).

The above considerations shed light on the significance of the crossover temperature T_0. For $T < T_0$, there is a periodic bounce trajectory with a period $\hbar/k_{\text{B}}T$ in the upside-down potential $-V(q)$. At the crossover temperature, $T = T_0$, the bounce takes the shortest possible period $2\pi/\omega_{\text{b}}$. At this temperature, the bounce path represents a harmonic oscillation about q_{b} in the inverted potential $-V(q)$, and the respective actions are the same. Interestingly, these actions coincide also with the action of the constant path $q = q_{\text{b}}$ at $T = T_0$, as we see from Eq. (12.75).

The crossover temperature T_0 can be quite large for very light particles, e.g., above room temperature for electrons. On the other hand, for macroscopic quantum tunneling phenomena in Josephson systems, the effective mass is typically very high so that the crossover temperature can well be in the milli-Kelvin region [300].

Since the $\text{Im}\,F$ method is basically a thermodynamic method, it is employable when thermal equilibrium prevails in the well region. This implies that the time scale for relaxation in the well is short compared to the average time for escape from the well. The $\text{Im}\,F$ method can also be applied in certain limits, when the barrier is time-dependent. First, it is applicable in the adiabatic limit of slow-frequency driving [328]. In this case, the escaping particle is subject to the current barrier potential, and the total rate is found from the average of the relevant potential configurations. Secondly, the $\text{Im}\,F$ method is also applicable in the case of fast driving, in which the escape occurs in an effective potential resulting from an average over one period of driving. Clearly, the $\text{Im}\,F$ method is not applicable when the driving frequency is of the order of the bare or dressed well frequency.

Let us conclude this section with a general remark on a recent controversy. Tunneling has been investigated by considering classical real-time trajectories with energies higher than the barrier top [329]. In this approach, the smallness of the tunneling rate results from destructive interference of the various path contributions. Subsequently, it has been shown that the numerical real-time Fourier integral method which employs input only from the energy region above the barrier top is inadequate in the deep tunneling regime [330]. In the semiclassical quantum transition state theory discussed above and in subsequent chapters, tunneling is regarded as an imaginary-time process which can be quantitatively described in terms of a classical trajectory on the upside-down potential surface and of the fluctuations about this path.

12.6 Centroid method

Dynamical simulations of barrier crossing processes are aggravated by the fact that the tunneling paths are rare events. The efficiency of the computation is enormously enhanced by filtering out irrelevant paths that do not cross the surface dividing the reactant from the reaction product and therefore do not contribute to the crossing rate. An efficient numerical strategy consists in a direct sampling of the centroid density $\rho_c(q_c)$ discussed in Section 8.1 by employing imaginary-time path integral Monte Carlo techniques for the constrained path integral in Eq. (8.4). Upon tuning the centroid variable q_c, the effective potential $\mathcal{V}(q_c)$ can be computed. Generally, the effective potential $\mathcal{V}(q_c)$ may not necessarily have the same shape as the original potential $V(q)$. As noticed by Gillan [331] and subsequently treated more rigorously by Voth et al. [332], the reaction rate is directly connected with the change of the effective classical potential $\mathcal{V}(q_c)$ when the centroid is tuned from the reactant state to the barrier top (dividing surface).

Assume that the effective potential has a local minimum at $q_c = q_{min}$ with well frequency ω_{min}, and a maximum at $q_c = q_{max}$ with barrier frequency ω_{max}. Unless $V(q)$ is symmetric, q_{max} is different from the position q_b of the barrier top of $V(q)$. Following the analysis given in Section 12.5, the point $q_c = q_{min}$ is a stable stationary point, and integration over q_c in the well region [cf. Eq. (8.3) with Eq. (8.6)] gives the partition function of reactants in the Gaussian approximation, $Z_0 = (1/\hbar\beta\omega_{min})\exp[-\beta\mathcal{V}(q_{min})]$. In contrast, the point q_{max} is a saddle point which requires distortion of the integration contour into the complex q_c-plane. We then find in steepest descent the barrier contribution

$$\mathrm{Im}\, Z_b = \frac{1}{2\beta\hbar\omega_{max}}\, e^{-\beta\mathcal{V}(q_{max})} . \tag{12.76}$$

Using Eq. (12.73), we obtain for the rate the expression

$$k = \frac{\sigma}{\hbar\beta}\frac{\omega_{min}}{\omega_{max}}\, e^{-\beta[\mathcal{V}(q_{max})-\mathcal{V}(q_{min})]} , \tag{12.77}$$

which for $\sigma = 1$ is Gillan's centroid formula [331]. For temperatures $k_B T \geq \hbar\omega_{max}/2\pi$, Affleck's factor σ in Eq. (12.77) is $\hbar\beta\omega_{max}/2\pi$ [cf. Eq. (12.74)], which is the centroid result obtained in Refs. [332, 333].

Instead of computing the effective classical potential numerically using quantum Monte Carlo techniques, one may estimate the effective potential by employing the variational methods discussed in Subsection 8.2.2 or one may apply semiclassical methods [334]. In the semiclassical limit, we have [cf. Eq. (8.11) with Eq. (8.14), and corresponding expressions for the parabolic barrier]

$$
\begin{aligned}
\mathcal{V}(q_{min}) &= V(q_{min}) + \ln[\sinh(\beta\hbar\omega_{min}/2)/(\beta\hbar\omega_{min}/2)]/\beta\,, \\
\mathcal{V}(q_{max}) &= V(q_{max}) + \ln[\sin(\beta\hbar\omega_{max}/2)/(\beta\hbar\omega_{max}/2)]/\beta\,.
\end{aligned}
\tag{12.78}
$$

It is straightforward to see that Eq. (12.77) with Eq. (12.78) agrees for $T > T_0$ with the previous result in Eq. (12.31).

We remark again that the effective potential includes all quantum effects of the partition function. The result (12.77) involves only two Gaussian approximations and therefore seems "less approximate" than the semiclassical expression in which all fluctuation modes are treated in the Gaussian approximation. Generally one finds from numerical studies that the centroid approximation gives reasonable results for a symmetric or almost symmetric barrier and not too low temperatures [334]. It is seemingly superior to semiclassical methods for low barriers. Being nearly exact about T_0, the accuracy of the centroid method deteriorates as temperature is decreased. For very low T, the semiclassical bounce or periodic orbit method appears to be more accurate. With increasing asymmetry, the centroid approximation becomes less reliable. For a metastable potential that is not bounded from below, the centroid density diverges because of the unlimited potential drop beyond the barrier leading to infinite values for the rate. The generalization of the centroid method to the multidimensional case along the lines given in Section 8.1 and Chapters 14 – 16 is straightforward.

13. Multidimensional quantum rate theory

13.1   The global metastable potential

We now generalize the discussion of quantum-statistical decay to a system with $N+1$ degrees of freedom and eventually take the limit $N \to \infty$. The multidimensional system is assumed to be described by the system-plus-reservoir model given in Eq. (3.11). We shall base the discussion on the thermodynamic Im F method. Some of the results will be confirmed by the multidimensional generalization of the semiclassical method discussed in Section 12.3.

The thermodynamic rate calculation starts out from the functional integral representation of the partition function of the damped system. The exact formal path integral expression for the reduced partition function Z is written in Eq. (4.210), and the effective action for the motion in the upside-down potential is given in Eq. (4.61). The multi-dimensional upside-down potential landscape $-V(q, \boldsymbol{x})$, where $V(q, \boldsymbol{x})$ is defined in Eq. (3.13) and replicated in Eq. (13.1), is sketched in Fig. 13.1.

For $N+1$ degrees of freedom, the potential $-V(q, \boldsymbol{x})$ is concave up in one direction, giving rise to periodic motion, but concave down in N directions and thus unstable with respect to small perturbations in these directions. The total energy

$$E = V(q, \boldsymbol{x}) - \frac{M}{2}\dot{q}^2 - \frac{1}{2}\sum_{\alpha=1}^{N} m_\alpha \dot{x}_\alpha^2 ,$$

$$V(q, \boldsymbol{x}) = V(q) + \frac{1}{2}\sum_{\alpha=1}^{N} m_\alpha \omega_\alpha^2 \left(x_\alpha - \frac{c_\alpha}{m_\alpha \omega_\alpha^2}q \right)^2 . \tag{13.1}$$

is usually the only constant of motion. The periodic orbit is then characterized by N stability angular frequencies [335].

The tunneling contribution to the partition function is dominated by the periodic orbit in the $(q, \boldsymbol{x})$-plane. At zero temperature, the total energy is zero and the period is infinity. The condition $V(q, \boldsymbol{x}) = 0$ defines the point P_{m} situated at the maximum of the parabolic hill of the upside-down potential ($q_{\mathrm{m}} = 0$, $\boldsymbol{x}_{\mathrm{m}} = 0$), and also the outer N-dimensional hyperbolic surface $\Sigma_{\mathrm{ex}}(q, \boldsymbol{x})$. The point P_{m} corresponds to the turning point q_1 in Eq. (12.20) for $E = 0$. The surface Σ_{ex} is the generalization of the turning point q_2 for N bath degrees of freedom. The surface separates the well and barrier region from the classically accessible outside region. On the periodic orbit, the particle stays infinitely long in the infinite past at the point P_{m} and then gradually starts moving towards the well of the upside-down potential. Eventually, it rushes through the well (which actually is a saddle point in the presence of the bath coordinates) and then bounces back from the surface Σ_{ex} at time zero. The particle hits the surface Σ_{ex} perpendicularly at a particular point denoted by P_{ex}. On the way back it rushes again through the well on the same path, and finally comes to rest at time infinity at the starting point P_{m}. The orbit of the particle near the surface Σ_{ex} is parabolic and therefore is very sensitive to a small perturbation. This reflects the fact that the periodic orbit is unstable. The envelope surface of the parabolic paths is the caustic which touches the surface Σ_{ex} at the point P_{ex}. If the particle would not approach the surface Σ_{ex} perpendicularly, it could not return on the same path, but would run down for ever the slope of the potential $-V(q, \boldsymbol{x})$ [cf. Fig. 13.1, left]. The particular role of the unstable periodic orbit in multi-dimensional quantum decay has been stressed by Banks et al. already in 1973 [336]. The periodic orbit is often referred to as the *most probable escape path* (MPEP).

13.2   Periodic orbit and bounce

At finite temperature, the period of the periodic orbit is $\hbar\beta$, and the energy range is $0 < E < V_{\mathrm{b}}$ in Eq. (13.1). Hence the point P_{m} is dissolved into a N-dimensional

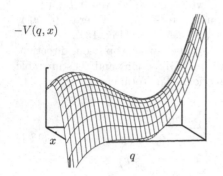
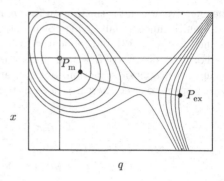

Figure 13.1: Potential landscape of the upside-down potential $-V(q, \boldsymbol{x})$ for a single bath degree of freedom (left). The contour lines of equal potential height are shown for the potential $-V(q, \boldsymbol{x})$ (right). The contour lines around the top (symbol $\circ$) are elliptic, while those separating the barrier region from the outside region are hyperbolic. The curved periodic orbit for a particular period with turning points P_m on Σ_m and P_ex on Σ_ex is also sketched.

elliptic surface Σ_m around $q = 0$, $\boldsymbol{x} = 0$. The inner surface Σ_m as well as the outer surface Σ_ex are defined by the condition $E - V(q, \boldsymbol{x}) = 0$. On the periodic orbit or MPEP, the particle leaves the surface Σ_m perpendicularly at a particular point (denoted again by P_m) at time $-\hbar\beta/2$, rushes through the well of the upside-down potential and hits the outer surface Σ_ex perpendicularly at time zero at a particular point P_ex. After this it returns to the starting point on Σ_m at time $\hbar\beta/2$. The periodic orbit meets the same points in configuration space on the ways forth and back.

In Fig. 13.1, we have sketched the potential landscape (left) and the contour lines of equal potential height (right). We have also drawn the periodic orbit for a particular value of E with turning points on the surfaces Σ_m and Σ_ex, respectively. The periodic orbit or MPEP is a curved path on the $\{q, \mathbf{x}\}$-plane. The projection of this path onto the q-axis is the bounce path $q_\mathrm{B}(\tau)$. This path is a stationary point of the action $S_\mathrm{eff}^\mathrm{E}[q]$ defined in Eq. (4.61) and satisfies the equation of motion

$$- M\,\ddot{q}_\mathrm{B}(\tau) + \frac{\partial V[q_\mathrm{B}(\tau)]}{\partial q_\mathrm{B}(\tau)} + \int_0^{\hbar\beta} d\tau'\, k(\tau - \tau') q_\mathrm{B}(\tau') = 0\,. \qquad (13.2)$$

Because of the projection onto the q-axis, which is tantamount to the elimination of the reservoir coordinates $\boldsymbol{x}$, the equation of motion for the bounce $q_\mathrm{B}(\tau)$ is time-nonlocal, and there is *no* conserved energy for this path.

In the presence of dissipation there is again a crossover temperature at which the bounce action coincides with the action of the trivial constant path $q = q_\mathrm{b}$, $x_\alpha = (c_\alpha/m_\alpha\omega_\alpha^2)q_\mathrm{b}$ ($\alpha = 1, \cdots, N$). The quantum fluctuations about the bounce $q_\mathrm{B}(\tau)$, apart from the zero mode, will be considered in Gaussian approximation. Only a narrow temperature window about the crossover temperature requires special treatment since there another non-Gaussian fluctuation mode is relevant. The crossover region will be discussed in Chapter 16.

For a globally stable potential, the stationary points are minima of the action. In contrast, the bounce path occurring in a metastable potential is a saddle point, i.e., there is a direction in function space along which the action gets smaller. This phenomenon is because the potential has another local minimum beyond the barrier. As a result, there appears a fluctuation mode about the bounce which has a negative eigenvalue: the nodeless amplitude fluctuation mode. In order to get a meaningful expression for the bounce contribution to the barrier partition function, the mode with the negative eigenvalue must be treated by analytical continuation. By deforming the respective integration contour from a stable to the actual unstable situation, the barrier partition function acquires an exponentially small imaginary part which can be assigned to the free energy as in Eq. (12.56).

Since the Im F method relies on equilibrium thermodynamics, it cannot describe effects which are related to nonthermal occupation of the states in the well. Such deviations may occur for extremely weak friction where the condition (11.45) is violated and the internal relaxation times for restoring equilibrium are of the order of the escape times or larger. The Im F method gives the proper classical limit of the escape in the damping regime (11.45). In the quantum regime, nonequilibrium effects are weaker. Hence the Im F method is valid down to considerably weaker damping than specified by the constraint (11.45). As we shall see, the rate formula for a damped system obtained by the Im F method approaches at high temperatures Kramers' expression (11.16) with Eq. (11.17) and Eq. (11.13). Moreover, the effects of frequency-dependent damping [337] as well as quantum corrections to thermally activated decay calculated by dynamical methods [338] are reproduced *quantitatively* by the thermodynamic method.

Among the first researchers who studied the effects of dissipation on the quantum mechanical decay of a metastable ground state were Caldeira and Leggett in 1983 [77]. They used the "bounce"-technique within the functional integral formulation [107]. Shortly later, the bounce method was found to be also an effective scheme for the calculation of dissipative quantum tunneling at finite temperature [339] – [342]. A unified description of quantum statistical decay of a metastable state in the entire temperature range was given in Ref. [297].

It is worth noting that some confusion in the literature about whether dissipation increases or decreases the tunneling probability is related to the absence or presence of the potential counter term $\Delta V(q)$, Eq. (3.6) with (3.10), in the system-reservoir coupling term defined in Eq. (3.4). As long as one wishes to study solely the effects of friction and decoherence, and not the addition of potential renormalization effects induced by the coupling, one has to choose $\Delta V(q)$ according to Eq. (3.6). Then one finds, as we shall see, that damping *always* decreases the decay probability. This can be simply understood as follows. While the environmental coupling in the Hamiltonian (3.11) does not change the barrier height, the tunneling distance along the curved periodic orbit gets larger with damping (see Fig. 13.1). On the other hand, if we choose $\Delta V(q) = 0$, the barrier is effectively reduced by the coupling. The effect of weakening the barrier overcompensates the opposite curvature effect, so that after all for $\Delta V(q) = 0$ the decay rate is increased by the coupling.

The functional integral method which we shall present in the following four chapters provides a unified approach to weak metastability of a dissipative system in large regions of the parameter space.

14. Crossover from thermal to quantum decay

Under the condition of weak metastability, Eq. (10.3), the functional integral (4.210) for the partition function is dominated by the stationary points of the action. Besides the periodic bounce $q_{\mathrm{B}}(\tau)$, there are two trivial constant solutions of Eq. (4.211). On the one hand, we have $\bar{q}(\tau) = q_{\mathrm{b}}$, in which the particle sits in the minimum of the upside-down potential $-V(q)$, and accordingly on the barrier top of the original potential. On the other hand, there is the constant path $\bar{q}(\tau) = 0$, in which the particle sits in the minimum of the original potential, and accordingly at the local maximum of the upside-down potential (see Figs. 10.1 and 10.2).

For temperatures below the crossover temperature T_0, the bounce with period $\hbar/k_{\mathrm{B}}T$ exists. In the absence of damping, the shortest period is $\hbar\beta_0 = 2\pi/\omega_{\mathrm{b}}$, at which the bounce is a harmonic oscillation about q_{b}. The respective temperature $T_0 = \hbar\omega_{\mathrm{b}}/2\pi k_{\mathrm{B}}$ is in agreement with Goldanskii's conjecture, Eq. (12.75). At temperatures below T_0, the action S_{B} of the bounce $q_{\mathrm{B}}(\tau)$ is smaller than the action $\hbar\beta V_{\mathrm{b}}$ of the constant path $\bar{q}(\tau) = q_{\mathrm{b}}$. For this reason the barrier contribution to the partition function below T_0 is dominated by the bounce $q_{\mathrm{B}}(\tau)$, instead by the trivial constant path $\bar{q}(\tau) = q_{\mathrm{b}}$. From this we infer that T_0 is the temperature where roughly the transition between thermally activated decay and quantum tunneling decay occurs.

We now turn to the discussion of a damped particle. Again, the crossover temperature is defined as the temperature at which the action of the bounce trajectory and the action of the constant path $q = q_{\mathrm{b}}$ merge. For a study of this regime it is expedient to perform a normal mode analysis of the global system at the barrier top.

14.1 Normal mode analysis at the barrier top

The dynamics near the barrier top can be solved in analytic form with a normal mode analysis of the system-plus-reservoir Hamiltonian for a parabolic barrier [343]. The mixed quadratic form of the Hamiltonian is

$$H^{(b)} = \frac{1}{2}M\dot{q}^2 + \frac{1}{2}\sum_{\alpha=1}^{N} m_\alpha \dot{x}_\alpha^2 + V^{(b)}(q, \boldsymbol{x}) \,,$$

$$V^{(b)}(q, \boldsymbol{x}) = V_b - \frac{1}{2}M\omega_b^2(q - q_b)^2 + \frac{1}{2}\sum_{\alpha=1}^{N} m_\alpha \omega_\alpha^2 \left(x_\alpha - \frac{c_\alpha}{m_\alpha \omega_\alpha^2}q \right)^2 \,.$$ (14.1)

The diagonal form in the mass weighted coordinates (4.230) reads

$$H^{(b)} = V_b + \frac{1}{2}\sum_{\alpha=0}^{N} \left(\dot{y}_\alpha^2 + \mu_\alpha^{(b)\,2} y_\alpha^2 \right) \,.$$ (14.2)

Here, $y_0(\tau)$ is the unstable mode with squared frequency $\mu_0^{(b)\,2} \equiv -\omega_R^2$, and the normal mode $y_\alpha(\tau)$, $\alpha \neq 0$, is a stable bath modes with frequency $\mu_\alpha^{(b)}$. The normal mode frequencies are related to the original frequencies by the relations

$$\omega_R^2 = \omega_b^2 \Big/ \left(1 + \frac{1}{M}\sum_{\alpha'=1}^{N} \frac{c_{\alpha'}^2}{[m_{\alpha'}\omega_{\alpha'}^2(\omega_{\alpha'}^2 + \omega_R^2)]} \right) \,,$$ (14.3)

$$\mu_\alpha^{(b)\,2} = \omega_\alpha^2 \Big/ \left(1 + \frac{1}{M}\sum_{\alpha'=1}^{N} \frac{c_{\alpha'}^2}{[m_{\alpha'}\omega_{\alpha'}^2(\omega_{\alpha'}^2 - \mu_\alpha^{(b)\,2})]} \right) \,, \qquad \alpha \neq 0 \,.$$ (14.4)

The force constant matrix at the barrier top $\boldsymbol{U}^{(b)}$ is related to the corresponding matrix in the well regime $\boldsymbol{U}^{(0)}$, which is given in Eq. (4.232), by

$$U_{ij}^{(b)} = U_{ij}^{(0)} - \delta_{i0}\delta_{j0}(\omega_0^2 + \omega_b^2) \,.$$ (14.5)

Similar to the relation (4.242) we may express the determinant of the force constant matrix $\boldsymbol{U}^{(b)}$ in terms of the original frequencies as

$$\det \boldsymbol{U}^{(b)} = \prod_{\alpha=0}^{N} \mu_\alpha^{(b)\,2} = -\omega_b^2 \prod_{\alpha=1}^{N} \omega_\alpha^2 \,.$$ (14.6)

The original system coordinate q is expressed in terms of the mass weighted normal coordinates $\{y_\alpha\}$ by the orthogonal transformation

$$q = q_b + \frac{1}{\sqrt{M}}u_{00}y_0 + \frac{1}{\sqrt{M}}\sum_{\alpha=1}^{N} u_{\alpha0}y_\alpha \,,$$ (14.7)

with the matrix elements

$$u_{00} = \left(1 + \frac{1}{M}\sum_{\alpha'=1}^{N} \frac{c_{\alpha'}^2}{m_{\alpha'}(\omega_{\alpha'}^2 + \omega_R^2)^2} \right)^{-1/2} \,,$$ (14.8)

$$u_{\alpha0} = \left(1 + \frac{1}{M}\sum_{\alpha'=1}^{N} \frac{c_{\alpha'}^2}{m_{\alpha'}(\omega_{\alpha'}^2 - \mu_\alpha^{(b)\,2})^2} \right)^{-1/2} \,.$$ (14.9)

The frequency ω_R of the unstable mode $y_0(\tau)$ and the matrix element u_{00} may be equivalently expressed in terms of the spectral density $J(\omega)$ or in terms of the friction kernel $\hat{\gamma}(z)$. We obtain from (14.3) upon using the definitions (3.24) and (3.28)

$$\omega_R^2 = \omega_b^2 \left(1 + \frac{2}{\pi M} \int_0^\infty d\omega \, \frac{J(\omega)}{\omega} \frac{1}{(\omega^2 + \omega_R^2)} \right)^{-1} = \frac{\omega_b^2}{1 + \hat{\gamma}(\omega_R)/\omega_R} . \qquad (14.10)$$

Since $\hat{\gamma}(\omega_R) > 0$, the effective barrier frequency ω_R is smaller than the bare barrier frequency ω_b. Similarly, the coefficient u_{00} is given by

$$u_{00}^2 = \left(1 + \frac{2}{\pi M} \int_0^\infty d\omega \, \frac{\omega J(\omega)}{(\omega^2 + \omega_R^2)^2} \right)^{-1} . \qquad (14.11)$$

The relation (14.10) may be rewritten as

$$\omega_R^2 + \omega_R \hat{\gamma}(\omega_R) = \omega_b^2 . \qquad (14.12)$$

Equation (14.12) has a unique real positive solution for ω_R. The renormalized barrier frequency ω_R is the familiar Kramers-Grote-Hynes frequency appearing in the theory of non-Markovian rate processes [337].

14.2   Turnover theory for activated rate processes

The development of a unified theory for the classical attempt frequency bridging between the spatial-diffusion limited transmission factor $\varrho = \omega_R/\omega_b$ [cf. Eq. (11.13) in the Ohmic case] and the energy-diffusion limited expression (11.39) is known as the *Kramers turnover problem*. Near to the barrier top, the total energy E of the global system becomes the sum of the normal mode energies E_α, $E = \sum_{\alpha=0}^N E_\alpha$. Since the normal modes are not coupled, the probability to cross the barrier depends entirely on the energy E_0 left in the unstable mode y_0. When $E_0 < V_b$, the particle moving along the escape coordinate y_0 encounters a turning point and thereafter it returns to the well. When $E_0 > V_b$, the particle leaves the well in each attempt with probability one. The particular role of the unstable mode has been utilized in Refs. [344, 345]. By calculating the energy left in the unstable mode y_0, rather than in the physical coordinate q, inconsistencies in an earlier theory [346] could be resolved. Following Kramers [253], imagine that particles are constantly injected near to the bottom of the well. Then the system will approach a steady flux state. When the probability in the well is normalized as to describe a single particle, the flux equals the classical escape rate k_{cl}. Let $p(E_0)dE_0$ be the probability per unit time to find the particle in the unstable mode y_0 near the barrier top in the energy interval between E_0 and $E_0 + dE_0$. Since particles with $E_0 > V_b$ escape, the rate is then given by the expression

$$k_{cl} = \int_{V_b}^\infty dE_0 \, p(E_0) . \qquad (14.13)$$

In the general case, $p(E_0)$ is a nonequilibrium probability distribution.

Now assume as a first crude guess a thermal distribution in the well, in fact also in the vicinity of the barrier top. Upon writing the canonical distribution in the normal mode representation (14.2) at the barrier top, and the normalization factor in the normal mode representation (4.234) of the well region, the equilibrium distribution of the unstable mode takes the form

$$
W_{\text{eq}}(\dot{y}_0, y_0) = \frac{\prod_{\alpha=1}^{N} \int d\dot{y}_\alpha \, dy_\alpha \exp\left(-\tfrac{1}{2}\beta\left[\dot{y}_\alpha^2 + \mu_\alpha^{(b)\,2} y_\alpha^2\right]\right)}{\prod_{\alpha=0}^{N} \int d\dot{\rho}_\alpha \, d\rho_\alpha \exp\left(-\tfrac{1}{2}\beta\left[\dot{\rho}_\alpha^2 + \mu_\alpha^{(0)\,2} \rho_\alpha^2\right]\right)}
\tag{14.14}
$$
$$
\times \ \exp\left(-\beta\left[\tfrac{1}{2}\dot{y}_0^2 + V_{\text{b}} - \tfrac{1}{2}\omega_{\text{R}}^2 y_0^2\right]\right).
$$

With the relations (4.242) and (14.6), and with use of the energy E_0 accumulated in the unstable mode, $E_0 = \tfrac{1}{2}\dot{y}_0^2 + V_{\text{b}} - \tfrac{1}{2}\omega_{\text{R}}^2 y_0^2$, we readily obtain

$$
p_{\text{eq}}(E_0) \equiv W_{\text{eq}}(\dot{y}_0, y_0) = \frac{\beta\omega_0}{2\pi}\frac{\omega_{\text{R}}}{\omega_{\text{b}}}\, e^{-\beta E_0}.
\tag{14.15}
$$

Insertion of the equilibrium probability (14.15) in Eq. (14.13) yields for the escape rate the form (11.16),

$$
k_{\text{cl}} = f_{\text{cl}}\, e^{-\beta V_{\text{b}}},
\tag{14.16}
$$

with the classical attempt frequency

$$
f_{\text{cl}} = \varrho\,\frac{\omega_0}{2\pi} \qquad \text{with} \qquad \varrho = \frac{\omega_{\text{R}}}{\omega_{\text{b}}}.
\tag{14.17}
$$

This expression is the generalization of the Ohmic form to general frequency-dependent friction. We see that memory-friction manifests itself through the Grote-Hynes frequency ω_{R}, which is implicitly given by the relation (14.12) [337].

Consider now the probability distribution $p(E_0)$ more closely. Because of the outgoing flux, the actual distribution $p(E_0)$ may deviate from the thermal distribution $p_{\text{eq}}(E_0)$ (14.15). In general, $p(E_0)$ obeys the steady-state condition

$$
p(E_0) = \int_0^{V_{\text{b}}} dE_0'\, P(E_0|E_0')\, p(E_0').
\tag{14.18}
$$

Here $P(E_0|E_0')$ is the conditional probability that the particle leaving the barrier region with energy E_0' in the unstable mode y_0 returns to the barrier with an energy E_0. The conditional probability satisfies detailed balance,

$$
P(E_0|E_0')\, e^{-\beta E_0'} = P(E_0'|E_0)\, e^{-\beta E_0}.
\tag{14.19}
$$

As a result of the relation (14.19), the distribution $p(E_0)$ matches with $p_{\text{eq}}(E_0)$ at energy E_0 considerably below V_{b}.

The key quantity now is the energy transferred from the unstable mode to the bath as the particle moves from the barrier to the well and then back to the barrier. The equation of motion for the stable mode α governed by the full Hamiltonian is

$$\ddot{y}_\alpha(t) + \mu_\alpha^{(b)\,2} y_\alpha(t) = g_\alpha F(t), \qquad \alpha \neq 0, \tag{14.20}$$

where

$$g_\alpha = u_{\alpha 0}/u_{00}. \tag{14.21}$$

The force pulse $F(t)$ comes from the anharmonic part of the potential

$$V^{(\mathrm{nl})}(q) = V(q) - V_{\mathrm{b}} + \tfrac{1}{2} M \omega_{\mathrm{b}}^2 (q - q_{\mathrm{b}})^2. \tag{14.22}$$

The coupling of the unstable mode to the reservoir is weak on the condition

$$\varepsilon \equiv \sum_{\alpha=1}^{N} g_\alpha^2 = \frac{1}{u_{00}^2} - 1 \ll 1. \tag{14.23}$$

In the regime $\varepsilon \ll 1$ we may disregard the back-action of the stable modes on the force pulse in Eq. (14.20). Then the nonlinear force $F(t)$ depends on the time dependence of the unstable mode according to the relation

$$F(t) = -u_{00} \frac{1}{M} \frac{\partial V^{(\mathrm{nl})}(q)}{\partial q}\bigg|_{q = q_{\mathrm{b}} + u_{00} y_0(t)/\sqrt{M}}. \tag{14.24}$$

Equation (14.20) describes a forced harmonic oscillator. It is easily solved for arbitrary initial values of y_α and $\dot{y}_\alpha$ at time $t = 0$. The energy transferred on average per cycle from the unstable mode y_0 to the stable mode y_α is

$$\begin{aligned}
\Delta E_{\mathrm{cycle}} &= \frac{1}{2} g_\alpha^2 \int_{-\infty}^{\infty} dt \int_{-\infty}^{\infty} dt' \, \cos\left[\mu_\alpha^{(b)}(t - t')\right] F(t)F(t') \\
&+ g_\alpha \int_{-\infty}^{\infty} dt \left[\dot{y}_\alpha(0) \cos\left(\mu_\alpha^{(b)} t\right) - y_\alpha \mu_\alpha^{(b)} \sin\left(\mu_\alpha^{(b)} t\right)\right] F(t).
\end{aligned} \tag{14.25}$$

The time integrations cover the history of the unstable trajectory $y_0(t)$ which starts at time $t = -\infty$ at the barrier, then moves through the well, and finally returns to the barrier at time $t = +\infty$. To leading order, the equation of motion of the unstable mode is decoupled from the bath modes and takes the form

$$\ddot{y}_0(t) - \omega_{\mathrm{R}}^2 y_0 + u_{00} \frac{1}{M} \frac{\partial V^{(\mathrm{nl})}(q)}{\partial q}\bigg|_{q = q_{\mathrm{b}} + u_{00} y_0(t)/\sqrt{M}} = 0. \tag{14.26}$$

Next, we choose thermal initial conditions according to

$$\begin{aligned}
\mu_\alpha^{(b)\,2} \langle y_\alpha^2(0) \rangle &= \langle \dot{y}_\alpha^2(0) \rangle = k_{\mathrm{B}} T, \\
\langle y_\alpha(0) \rangle &= \langle \dot{y}_\alpha(0) \rangle = \langle y_\alpha(0) \dot{y}_\alpha(0) \rangle = 0.
\end{aligned} \tag{14.27}$$

Then the energy transferred on average from the particle to the reservoir per cycle is

$$\Delta E \equiv \langle \Delta E_{\text{cycle}} \rangle = \frac{1}{2} \int_{-\infty}^{\infty} dt \int_{-\infty}^{\infty} dt' \, \Phi(t - t') F(t) F(t') \tag{14.28}$$

with the dissipation kernel

$$\Phi(t - t') = \sum_{\alpha=1}^{N} g_\alpha^2 \cos \left[\mu_\alpha^{(b)}(t - t') \right]. \tag{14.29}$$

The Laplace transform of the kernel $\Phi(t - t')$ may be written as

$$\hat{\Phi}(z) = \sum_{\alpha=1}^{N} g_\alpha^2 \frac{z}{z^2 + \mu_\alpha^{(b)\,2}} = \frac{1}{u_{00}} \frac{z}{z^2 + z\hat{\gamma}(z) - \omega_b^2} - \frac{z}{z^2 - \omega_R^2}. \tag{14.30}$$

The first expression is the Laplace transform of (14.29). The second form expresses $\hat{\Phi}(z)$ in terms of the spectral damping function $\hat{\gamma}(z)$. The derivation of this form is less direct and involves the inversion of the force constant matrix $\boldsymbol{U}^{(b)}$ [345]. Finally, the uncertainty of the mean square energy loss for the initial state is

$$\delta E^2 \equiv \langle \Delta E_{\text{cycle}}^2 \rangle - \langle \Delta E_{\text{cycle}} \rangle^2 = 2k_B T \, \Delta E. \tag{14.31}$$

Observing the relations (14.28) and (14.31), the conditional probability $P(E_0|E_0')$ is well described by the Gaussian distribution

$$P(E_0|E_0') = \left(4\pi k_B T \Delta E \right)^{-1/2} \exp \left[-\left(E_0 - E_0' + \Delta E \right)^2 / 4k_B T \Delta E \right]. \tag{14.32}$$

Considering that the solution of Eq. (14.18) matches the thermal distribution $p_{\text{eq}}(E_0)$ in the regime $V_b - E_0 \gg k_B T$, it is convenient to put

$$p(E_0) = p_{\text{eq}}(E_0) \, e^{\beta(E_0 - V_b)/2} \, \phi[\beta(E_0 - V_b)]. \tag{14.33}$$

The ansatz (14.33) transforms Eq. (14.18) into a Wiener-Hopf equation with a symmetric kernel that can be solved by standard techniques [346]. With the resulting expression for $p(E)$ we obtain the rate expression (14.13) in the form (11.16) with the attempt frequency (11.17), in which the transmission factor ϱ is given by

$$\varrho = \frac{\omega_R}{\omega_b} \exp \left\{ \frac{1}{\pi} \int_{-\infty}^{\infty} \frac{dx}{1 + x^2} \ln \left[1 - \exp \left(-\frac{1 + x^2}{4} \frac{\Delta E}{k_B T} \right) \right] \right\}. \tag{14.34}$$

For $\Delta E \gg k_B T$, the transmission factor approaches the Grote-Hynes value $\varrho = \omega_R/\omega_b$ exponentially fast. The corrections are of order $\exp(-\Delta E/4k_B T)$. Nonequilibrium effects in $p(E_0)$ are only important for ΔE of the order of $k_B T$ or smaller. In the opposite limit $\Delta E \ll k_B T$, we find from Eq. (14.34) linear dependence on ΔE,

$$\varrho = \frac{\omega_R}{\omega_b} \frac{\Delta E}{k_B T}. \tag{14.35}$$

For weak damping, ΔE coincides with the energy loss given in Eq. (11.31) and the transmission factor (14.35) agrees with Kramers' result given in Eq. (11.39).

A deeper analytical and numerical study of the energy loss formula (14.28) with (14.29) is given in Ref. [345]. A multidimensional generalization of the turnover theory for activated rate processes is reported in Ref. [347]. This concludes our discussion of the Kramers turnover problem.

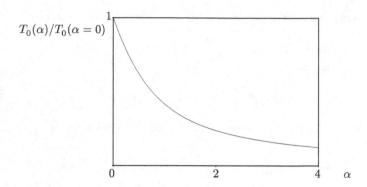

Figure 14.1: The normalized crossover temperature $T_0(\alpha)/T_0(\alpha = 0)$ is shown for Ohmic damping as a function of the parameter $\alpha = \gamma/2\omega_{\rm b}$.

14.3   The crossover temperature

At zero temperature, the period of the bounce is infinity, and the excursion it makes under the barrier is maximal. With increasing temperature, both the period and the amplitude of the bounce decrease. Eventually, the bounce has contracted to a small oscillation around the barrier point $q_{\rm b}$. In the $N + 1$-dimensional configuration space, this is a periodic harmonic oscillation in imaginary time with frequency $\omega_{\rm R}$ of the mode y_0. There is no periodic solution in the well of the upside-down potential $-V(q, \boldsymbol{x})$ with a frequency smaller than $\omega_{\rm R}$. The crossover temperature [348, 340]

$$T_0 \equiv \hbar\omega_{\rm R}/2\pi k_{\rm B} , \qquad (14.36)$$

where $\omega_{\rm R}$ is the positive root of Eq. (14.12), is the temperature at which the action of the bounce $q_{\rm B}(\tau)$ coincides with the action of the harmonic oscillation about $q_{\rm b}$, and with the action of the trivial path $\bar{q}(\tau) = q_{\rm b}$. Roughly speaking, at temperature $T = T_0$ there is the crossover between thermal hopping and quantum tunneling. The narrow temperature window about T_0 will be considered more closely in Chapter 16.

The relation (14.36) with (14.12) is valid for any linear dissipation mechanism. Since $\hat{\gamma}(\omega_{\rm R})$ in Eq. (14.12) is positive, the crossover temperature T_0 is always lowered with onset of damping. Further, T_0 increases monotonously towards the value $\hbar\omega_{\rm b}/2\pi k_{\rm B}$ of the undamped case as the memory-friction relaxation time is increased while $\hat{\gamma}(\omega = 0)$ is held fixed. In the particular case of Ohmic damping, $\hat{\gamma}(z) = \gamma$, and a cubic potential of the form (11.40), for which $\omega_{\rm b} = \omega_0$, we obtain from Eq. (14.12)

$$\omega_{\rm R} = \omega_{\rm b} \left(\sqrt{1 + \alpha^2} - \alpha\right) , \qquad (14.37)$$

where $\alpha = \gamma/2\omega_0 = \gamma/2\omega_{\rm b}$ is the usual dimensionless damping parameter. From this we find the monotonous behavior

$$T_0 = (\hbar\omega_{\rm b}/2\pi k_{\rm B}) \left(\sqrt{1 + \alpha^2} - \alpha\right) . \qquad (14.38)$$

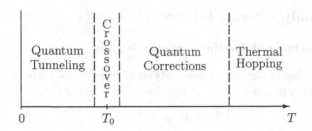

Figure 14.2: Sketch of the dominant escape mechanism as a function of temperature.

In the strong damping limit, Eq. (14.38) simplifies to the form

$$T_0 = \frac{\hbar}{2\pi k_B} \frac{\omega_b^2}{\gamma} . \tag{14.39}$$

The monotonous decrease of the crossover temperature with increasing damping strength is sketched in Fig. 14.1.

With the effective barrier frequency ω_R, the general relation (12.73) between the rate and the imaginary part of the free energy takes the form

$$k = -2\sigma \operatorname{Im} F /\hbar , \tag{14.40}$$

where
$$\sigma = 1 \qquad \text{for} \qquad \beta\hbar\omega_R \geq 2\pi ,$$
$$\sigma = \beta\hbar\omega_R/2\pi \qquad \text{for} \qquad \beta\hbar\omega_R < 2\pi . \tag{14.41}$$

Additional reasons for the validity of the expressions (14.40) with (14.41) are presented in Section 15.4. The unified formula (14.40) with (14.41) has also been derived in a study of the reactive flux through a parabolic barrier [333].

The various temperature regions in quantum statistical decay are shown in Fig. 14.2. Below T_0, the functional integral is dominated by the bounce trajectory $q_B(\tau)$, and the predominant escape mechanism is quantum mechanical tunneling. Above T_0, the functional integral is dominated by the constant trajectory $\bar{q}(\tau) = q_b$, which leads to the Arrhenius factor $e^{-\beta V_b}$, and the relevant decay process is thermally activated escape over the barrier. The fluctuation modes about $\bar{q}(\tau) = q_b$ yield quantum corrections in the prefactor of the rate formula thereby enhancing the rate above the classical expression. They become increasingly important as the temperature is lowered. Special care is required in the evaluation of the functional integral in a narrow crossover region about T_0 (cf. Chapter 16). We have now prepared the ground to work out the rate formulas in the various temperature regimes.

15. Thermally activated decay

15.1   Rate formula above the crossover regime

Consider the decay of the metastable state at temperatures fairly above T_0 so that thermal activation prevails. The rate may be written in the form

$$k = f_{cl} c_{qm} \, e^{-V_b/k_B T} \,, \tag{15.1}$$

where f_{cl} is the attempt frequency in the classical limit and c_{qm} is a (dimensionless) factor describing enhancement of the classical rate by quantum fluctuations. We now compute f_{cl} and c_{qm} for arbitrary frequency-dependent damping.

We proceed by splitting the partition function (4.210) of the damped system into the well contribution $Z^{(0)}$ and the barrier contribution $Z^{(b)}$, which result from the Gaussian fluctuations about the constant paths $\bar{q}(\tau) = 0$ and $\bar{q}(\tau) = q_b$, respectively,

$$Z = Z^{(0)} + Z^{(b)} = Z^{(0)}\left(1 + Z^{(b)}/Z^{(0)}\right) \,. \tag{15.2}$$

Now, we recall the analysis of the fluctuation modes for the partition function of a damped harmonic system given in Section 4.3. A path with $\hbar\beta$-periodic fluctuations about the constant path $\bar{q}(\tau) = 0$ is written as

$$q(\tau) \longrightarrow q^{(0)}(\tau) = \sum_{n=-\infty}^{\infty} Y_n \, e^{i\nu_n \tau} \,, \tag{15.3}$$

where $\nu_n = 2\pi n/\hbar\beta$. Inserting Eq. (15.3) into the expression (4.61) for the "dissipative" action and disregarding terms of third and higher order in Y_n, we find[1]

$$S[q^{(0)}(\cdot)] = \frac{1}{2}M\hbar\beta \sum_{n=-\infty}^{\infty} \Lambda_n^{(0)} \, Y_n Y_{-n} \,. \tag{15.4}$$

The $\{\Lambda_n^{(0)}\}$ are the eigenvalues of the fluctuation operator $\mathbf{\Lambda}^{(0)}$ defined in Eq. (4.219). With the findings in Subsection 4.3.3 [see Eq. (4.220)] we have

$$\Lambda_n^{(0)} = \nu_n^2 + \omega_0^2 + |\nu_n|\hat{\gamma}(|\nu_n|) \,. \tag{15.5}$$

Similarly, the path $q^{(b)}(\tau)$ with $\hbar\beta$-periodic fluctuations about q_b

$$q(\tau) \longrightarrow q^{(b)}(\tau) = q_b + \sum_{n=-\infty}^{\infty} X_n \, e^{i\nu_n \tau} \tag{15.6}$$

has the second order action

$$S[q^{(b)}(\cdot)] = \hbar\beta V_b + \frac{1}{2}M\hbar\beta \sum_{n=-\infty}^{\infty} \Lambda_n^{(b)} \, X_n X_{-n} \tag{15.7}$$

---

[1]Throughout the following we shall use the simpler notation $S[q]$ for $S_{eff}^{(E)}[q]$.

with

$$\Lambda_n^{(b)} = \nu_n^2 - \omega_b^2 + |\nu_n|\, \hat{\gamma}(|\nu_n|) \,. \tag{15.8}$$

The modes $\{Y_n\}$ and $\{X_n\}$, apart from the mode X_0, have positive eigenvalues. The contributions of these modes to the partition function is found in the usual way by carrying out the Gaussian integrals over the amplitude sets $\{Y_n\}$ and $\{X_n\}$ with use of the functional measure (4.228). The mode X_0 needs special treatment. Since the corresponding eigenvalue $\Lambda_0^{(b)} = -\omega_b^2$ is negative, the integral over X_0 is divergent. The divergence is related to the fact that the action for the constant path $\bar{q}(\tau) = q_b$ is a saddle point in function space with the *unstable* direction along X_0. In fact, this should not be a surprise since we are trying to evaluate the free energy of an unstable system. Langer [320] was the first who showed that the functional integral can still be defined by deforming the integration contour of the variable X_0 into the upper half of the complex plane along the direction of steepest descent. This leads to a positive imaginary part of the barrier contribution $Z^{(b)}$ with the familiar factor $\frac{1}{2}$, as we have explained in item (2) in Section 12.5. The resulting imaginary part of the free energy

$$\mathrm{Im}\, F = -\mathrm{Im}\, Z^{(b)}/(\beta\, Z^{(0)}) \tag{15.9}$$

is expressed again in terms of a ratio of determinants,

$$\mathrm{Im}\, F = -\frac{1}{2\beta}\sqrt{D^{(0)}/|D^{(b)}|}\; \mathrm{e}^{-\beta V_b} \,. \tag{15.10}$$

Here, $D^{(0)}$ and $D^{(b)}$ are the determinants related to the second order actions (15.4) and (15.7) for the well and barrier regions (see Subsections 4.3.2 and 4.3.3),

$$D^{(0)} = \omega_0^2 \prod_{n=1}^{\infty} \left(\Lambda_n^{(0)}\right)^2 \,, \qquad D^{(b)} = -\omega_b^2 \prod_{n=1}^{\infty} \left(\Lambda_n^{(b)}\right)^2 \,. \tag{15.11}$$

With use of Eqs. (12.72), (15.10) and (14.36) the rate above T_0 is found to read

$$
\begin{aligned}
k &= \frac{1}{\hbar\beta_0}\sqrt{\frac{D^{(0)}}{|D^{(b)}|}}\; \mathrm{e}^{-\beta V_b} = \frac{1}{\hbar\beta_0}\frac{\omega_0}{\omega_b}\prod_{n=1}^{\infty}\frac{\Lambda_n^{(0)}}{\Lambda_n^{(b)}}\; \mathrm{e}^{-\beta V_b}\\[2mm]
&= \frac{\omega_0}{2\pi}\frac{\omega_R}{\omega_b}\prod_{n=1}^{\infty}\frac{\nu_n^2 + \omega_0^2 + \nu_n\hat{\gamma}(\nu_n)}{\nu_n^2 - \omega_b^2 + \nu_n\hat{\gamma}(\nu_n)}\; \mathrm{e}^{-\beta V_b} \,.
\end{aligned}
\tag{15.12}
$$

Upon comparing Eq. (15.12) with Eq. (15.1), we find for the classical attempt frequency the previous form (14.17), and for the quantum mechanical enhancement factor the infinite product form [340]

$$c_{\mathrm{qm}} = \prod_{n=1}^{\infty}\frac{\Lambda_n^{(0)}}{\Lambda_n^{(b)}} = \prod_{n=1}^{\infty}\frac{\nu_n^2 + \omega_0^2 + \nu_n\hat{\gamma}(\nu_n)}{\nu_n^2 - \omega_b^2 + \nu_n\hat{\gamma}(\nu_n)} \,. \tag{15.13}$$

The result (15.13) is corroborated by an independent dynamical approach [338]. In the classical limit $T/T_0 \to \infty$, the factor c_{qm} approaches unity. Hence the rate expression (15.12) reduces to the proper classical form [cf. Eq. (14.16) with (14.17)].

$$k_{\text{cl}} = \frac{\omega_0}{2\pi} \frac{\omega_{\text{R}}}{\omega_{\text{b}}} \, e^{-\beta V_{\text{b}}} \,. \tag{15.14}$$

Compared with the TST formula (11.2), the classical rate is reduced by the transmission factor $\varrho = \omega_{\text{R}}/\omega_{\text{b}}$, which describes the effect of *recrossings* of the barrier top by the particle in the moderate to large damping region. The result agrees with the rate expression obtained from the extension of Kramers' approach to the case of memory friction [337].[2] The expression (15.14) also agrees with the classical rate expression obtained below in Section 15.4 from multidimensional transition state theory. There, however, the interpretation of the reduction factor ϱ will be different.

The effect of memory friction on the rate is reflected by the renormalization of the barrier frequency. Interestingly, the renormalized frequency ω_{R} is the same which enters the definition of the crossover temperature T_0. For frequency-independent damping, $\hat{\gamma}(z) = \gamma$, the attempt frequency (14.17) simplifies to Kramers' celebrated result, Eq. (11.17) with Eq. (11.13).

In summary, the pivotal assumptions underlying the generalized Kramers formula (15.14) are as follows [298]:

(1) The coupling to the environment is so strong that the rate is controlled by the diffusive dynamics near to the barrier top.

(2) The effective potential is harmonic near the bottom of the well and has a parabolic shape around the barrier top.

(3) The stochastic dynamics is represented by the *linear* Langevin equation (2.4) with a Gaussian random force with classical power spectrum (2.8).

15.2 Quantum corrections in the pre-exponential factor

At temperatures well below the classical regime but still above the crossover temperature, the rate is enhanced by quantum corrections,

$$k = c_{\text{qm}} k_{\text{cl}} \,, \tag{15.15}$$

where c_{qm} is given in Eq. (15.13).

The leading quantum correction at high T is found upon rewriting the expression (15.13) as an exponential of a sum of logarithms and expanding each logarithm in powers of $1/\hbar\beta$. One then finds in leading order in $1/\hbar\beta$ a surprisingly simple expression which is independent of damping for arbitrary spectral coupling [348],

$$c_{\text{qm}} = \exp\left((\omega_0^2 + \omega_{\text{b}}^2) \sum_{n=1}^{\infty} \frac{1}{\nu_n^2} \right) = \exp\left(\frac{\hbar^2 \, (\omega_0^2 + \omega_{\text{b}}^2)}{24 \, (k_{\text{B}}T)^2} \right) \,. \tag{15.16}$$

---

[2]In the chemical literature, the factor $\varrho = \omega_{\text{R}}/\omega_{\text{b}}$, where ω_{R} is determined by Eq. (14.12), is known as the Grote-Hynes correction to classical TST.

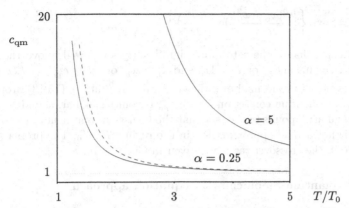

Figure 15.1: The quantum correction factor c_{qm} is shown as a function of the scaled temperature T/T_0 for a system with $\omega_0 = \omega_b$ and frequency-independent damping for $\alpha = 0.25$ and $\alpha = 5$. The dashed curve is a plot of the formula (15.16) for $\alpha = 0.25$.

We see that the escape is enhanced by two different quantum effects. First, quantum fluctuations increase the mean energy in the well. Second, when a particle is thermally excited almost to the barrier top, quantum fluctuations allow for tunneling through the remaining small barrier. Both effects lead to an effective reduction of the barrier.

For frequency-independent damping, the infinite product in Eq. (15.13) can be expressed in terms of gamma functions [338]

$$c_{qm} = \frac{\Gamma(1 + \lambda_b^+/\nu)\,\Gamma(1 + \lambda_b^-/\nu)}{\Gamma(1 + \lambda_0^+/\nu)\,\Gamma(1 + \lambda_0^-/\nu)} , \qquad (15.17)$$

where $\nu = \nu_1 = 2\pi k_B T/\hbar$ is the smallest Matsubara frequency, and where

$$\lambda_b^\pm = \frac{\gamma}{2} \pm \left(\frac{\gamma^2}{4} + \omega_b^2\right)^{1/2} , \qquad \lambda_0^\pm = \frac{\gamma}{2} \pm \left(\frac{\gamma^2}{4} - \omega_0^2\right)^{1/2} . \qquad (15.18)$$

Plots of the enhancement factor c_{qm} versus temperature are shown in Fig. 15.1.

At strong friction $\gamma \gg \omega_0,\, \omega_b$ and $T \gg T_0$, we get from the expression (15.17)

$$c_{qm} = \exp\left\{\frac{\hbar\beta(\omega_b^2 + \omega_0^2)}{2\pi\gamma}\left[\psi\left(1 + \frac{\hbar\beta\gamma}{2\pi}\right) - \psi(1)\right]\right\} , \qquad (15.19)$$

where $\psi(z)$ is the digamma function. This form reduces in the high temperature regime $T \gg \hbar\gamma/2\pi k_B$ to the leading quantum correction given above in Eq. (15.16).

In the temperature regime

$$T_0 \ll T \ll \hbar\gamma/2\pi k_B , \qquad (15.20)$$

where T_0 is given in Eq. (14.39), the expression (15.17) takes the form

$$c_{\mathrm{qm}} = \exp\left\{\frac{\hbar\beta(\omega_{\mathrm{b}}^2 + \omega_0^2)}{2\pi\gamma}\ln\left(\frac{\hbar\beta\gamma}{2\pi}\right)\right\} = \left(\frac{\gamma^2 T_0}{\omega_{\mathrm{b}}^2 T}\right)^{(1+\omega_0^2/\omega_{\mathrm{b}}^2)T_0/T}. \tag{15.21}$$

This factor can enhance the rate quite significantly, even well above the crossover temperature. For instance, for $T = 4T_0$ and $\omega_0 = \omega_{\mathrm{b}}$, one gets $c_{\mathrm{qm}} = \gamma/2\omega_{\mathrm{b}} \gg 1$.

As T gets close to T_0 from above, the eigenvalue $\Lambda_1^{(\mathrm{b})}$ in Eq. (15.13) drops to zero. As a result, the quantum correction factor c_{qm} becomes singular, as indicated in Fig. 15.1. The singularity points to the fact that the Gaussian treatment of the fluctuation modes $Y_{\pm1}$ is not a valid approximation in the regime $T \gtrsim T_0$. The proper treatment of this mode in the crossover regime is given in Chapter 16.

15.3   The quantum Smoluchowski equation approach

At strong friction $\gamma \gg \omega_0$, the evolution of the probability density in coordinate space $p(q,t)$ of a classical Brownian particle is described by the Smoluchowski diffusion equation (11.19). In the temperature regime (15.20), the dynamics of the Brownian particle can still be described by a diffusion equation of the form (11.19), but the diffusion coefficient D is above the classical value $D_{\mathrm{cl}} = k_{\mathrm{B}}T$ because of quantum fluctuations. When the drift potential is harmonic, $V(q) = \frac{1}{2}M\omega_0^2 q^2$, it is quite natural to assume that the diffusion coefficient is increased by quantum fluctuations in accordance with the enhancement of the position spread,

$$D_{\mathrm{cl}} = k_{\mathrm{B}}T = M\omega_0^2 \langle q^2 \rangle_{\mathrm{cl}} \longrightarrow \tilde{D}_{\mathrm{qm}} = M\omega_0^2[\langle q^2 \rangle_{\mathrm{cl}} + \lambda], \tag{15.22}$$

where $\lambda \equiv \langle q^2 \rangle_{\mathrm{qm}} = \langle q^2 \rangle - \langle q^2 \rangle_{\mathrm{cl}}$ is the quantum mechanical part of position spread at strong friction. Explicit expressions for λ are given in Eqs. (6.143) and (6.144).

For a nonlinear potential $V(q)$, it is now an obvious thing to generalize the expression (15.22) to a position-dependent diffusion coefficient $D_{\mathrm{qm}}(q) = \tilde{D}_{\mathrm{qm}}(q)$, where $\tilde{D}_{\mathrm{qm}}(q) = k_{\mathrm{B}}T + \lambda V''(q)$. We then arrive at the Smoluchowski diffusion equation

$$\frac{\partial p(q,t)}{\partial t} = \frac{1}{M\gamma}\frac{\partial}{\partial q}\hat{L}_{\mathrm{qm}}(q)p(q,t) \tag{15.23}$$

with the position-dependent quantum Smoluchowski flux operator

$$\hat{L}_{\mathrm{qm}}(q) = \frac{\partial V(q)}{\partial q} + \frac{\partial}{\partial q}D_{\mathrm{qm}}(q). \tag{15.24}$$

Here the quantum Smoluchowski equation (QSE) has been deduced in a heuristic manner. The quantum correction in the diffusion coefficient $D_{\mathrm{qm}}(q)$ has been substantiated by a path integral evaluation of the propagating function J introduced in Section 5.6 [349], and by different reasoning in Ref. [350].

Unfortunately, the QSE with the diffusion coefficient $D_{\mathrm{qm}}(q) = \tilde{D}_{\mathrm{qm}}(q)$ has a weakness which shows up in order λ^2 and in higher orders of λ. Namely, for a periodic potential $V(q) = V(q+nL)$ the diffusion coefficient $\tilde{D}_{\mathrm{qm}}(q)$ invalidates corresponding periodicity of the equilibrium potential $\psi(q) = \int_0^q dx\, V'(x)/\tilde{D}_{qm}(x)$ [cf. Eq. (11.25)] in order λ^2 and in higher orders of λ. As a result, for a periodic drift potential $V(q)$, there would be a nonzero stationary equilibrium current, which would violate the second law of thermodynamics in order λ^2. The appropriate modification of the diffusion coefficient, which coincides with $\tilde{D}_{qm}(q)$ in first order in λ, is [351]

$$D_{qm}(q) = k_B T/[1 - \lambda V''(q)/k_B T]. \qquad (15.25)$$

With this form, the QSE does not yield a stationary current in any orders of λ for a periodic drift potential. Hence the QSE is fully reconciled with the second law of thermodynamics. The resulting equilibrium position distribution is

$$p_{eq}(q) = \frac{N_0}{D_{qm}(q)}\, e^{-\psi(q)}, \qquad (15.26)$$

in which the equilibrium thermodynamic potential $\psi(q)$ is given by

$$\psi(q) = \beta\left\{V(q) - \tfrac{1}{2}\lambda\beta[V'(q)]^2\right\}. \qquad (15.27)$$

For a harmonic well, the potential $\psi(q)$ agrees in order λ with the exponent $\tfrac{1}{2}q^2/\langle q^2\rangle$ of the equilibrium position distribution $< q|\hat{\rho}_\beta|q >$ given in Eq. (6.157). Since $\langle q^2\rangle$ gets smaller as γ is increased, the equilibrium distribution for a particle in a well is squeezed by friction quite substantially.

We may now study the escape of the quantum Brownian particle from a metastable well within the QSE. With use of the expressions (15.27) and (15.25), we immediately obtain from Eq. (11.28) the expression

$$k = \frac{\sqrt{V''(0)|V''(q_b)|}}{2\pi M\gamma}\, e^{-\beta V_b}\, e^{\beta\lambda\{V''(0)+|V''(q_b)|\}/2}. \qquad (15.28)$$

This result differs from the expression (11.29) by the last exponential factor. This term describes enhancement of the classical rate in the overdamped Smoluchowski regime by quantum fluctuations. With the explicit forms (6.140), (6.143) or (6.144) applicable in the respective temperature regimes, the enhancement factor in the expression (15.28) coincides with the earlier expressions (15.16), (15.19) and (15.21), respectively, which have been calculated from the fluctuation determinant resulting from the path integral approach.

Finally, we mention that, in difference to the original work [349, 351], the QSE (15.23) with the flux operator (15.24) and the diffusion coefficient (15.25) has no quantum correction in the drift potential. If we would add the term $\Delta V_{qm} = \tfrac{1}{2}\lambda V''(q)$ to the drift potential, as done in Refs. [349, 351], the quantum mechanical enhancement in Eq. (15.28) would be doubled. The absence of quantum fluctuations in the QSE (15.23) with (15.24) has also been noticed in Refs. [299] and [350].

15.4   Multidimensional quantum transition state theory

With the above considerations, the escape problem for temperatures above T_0 is basically solved. However, the presence of the factor σ in the formula (12.73) or (14.40) has been questioned [352, 296]. In this regard, we find it useful to show that the results of the Im F method are confirmed by the periodic orbit approach.

An independent unique approach to quantum statistical decay is based on the multidimensional quantum transition state theory (MQTST) put forward by Miller [310]. In the semiclassical limit, the "thermal propagator" $< q \,|\, e^{-\hat{H}\tau/\hbar} \,|\, q' >$ is dominated by *periodic orbit* trajectories in the upside-down potential landscape $-V(q, \boldsymbol{x})$ of the $(N+1)$-dimensional configuration space. The periodic orbit is a solution of the equations of motion (4.34) and (4.35) analytically continued to imaginary time with total energy conserved, Eq. (13.1). The qualitative behavior of the periodic orbit has been discussed in Chapter 13.

Following Gutzwiller [335], we now introduce a coordinate z_0 which measures distance along the curved periodic trajectory and N coordinates $\{z_j\}$, $j = 1, \cdots, N$, being displacements locally orthogonal to z_0. We shall call z_0 the *escape coordinate*, and $z_0^{(PO)}(\tau)$ the periodic orbit trajectory, which progresses along the escape coordinate. It is convenient to choose the phase of the periodic orbit such that

$$z_0^{(PO)}(\pm\hbar\beta/2) = z_0^{(m)}, \qquad \text{and} \qquad z_0^{(PO)}(0) = z_0^{(ex)}, \qquad (15.29)$$

where $z_0^{(m)}$ and $z_0^{(ex)}$ are the turning points on the surfaces Σ_m and Σ_{ex} introduced in Chapter 13, respectively.

We choose a flat dividing surface Σ_b which delimits the domain of attraction about the metastable minimum from the exterior region. The escape path crosses the dividing plane Σ_b perpendicularly at $z_0^{(b)}$, where $z_0^{(b)}$ is the point on the escape coordinate at which the full potential $V(q, \boldsymbol{x})$ has a maximum. The transition state approximation implies that the escaping particle crosses the dividing surface Σ_b only once. With this choice of coordinates, we then have similar to Eq. (12.10)

$$\hat{F}\hat{\mathcal{P}} \xrightarrow{\text{TST}} \delta\left(\hat{z}_0 - z_0^{(b)}\right) \frac{\hat{p}_0}{M} \Theta(\hat{p}_0) = \frac{1}{2M}\delta\left(\hat{z}_0 - z_0^{(b)}\right) \left[\,|\hat{p}_0| + \hat{p}_0\,\right]. \qquad (15.30)$$

Again, the second term in the angular bracket does not contribute. The multidimensional generalization of Eq. (12.12) with (12.13) is

$$k = \frac{1}{Z_{tot}^{(0)}} \frac{1}{2\pi\hbar} \int_0^\infty dE\, p(E)\, e^{-\beta E}, \qquad (15.31)$$

where $Z_{tot}^{(0)}$ is the partition function of the global system, and

$$p(E) = \lim_{\epsilon \to 0^+} \text{Im}\left\{ \int_0^\infty d\tau\, e^{(E+i\epsilon)\tau/\hbar} \right.$$

$$\left. \times \int d\boldsymbol{z}\, \delta\left(z_0 - z_0^{(b)}\right) |\dot{z}_0|_{z_0 = z_0^{(b)}} < \boldsymbol{z} \,|\, e^{-\hat{H}\tau/\hbar} \,|\, \boldsymbol{z} > \right\}. \qquad (15.32)$$

The quantity $p(E)$ represents the inclusive transmission probability at the total energy E. As the quantum statistical decay rate (15.31) is the Boltzmann-averaged transmission probability, we should expect that $p(E)$ includes all possible partitions of the total energy E into the energy left in the escape coordinate and the individual energies in the transverse degrees of freedom. This is the case, indeed, as we shall see shortly [cf. Eq. (15.40) below].

According to Gutzwiller's semiclassical analysis [335], a single periodic orbit in the multi-dimensional upside-down potential $-V(q, \boldsymbol{x})$ [see Eq. (13.1)] yields in generalization of Eq. (12.17) the contribution

$$p_1(E) = \frac{i\,\mathcal{J}_1}{\sqrt{2\pi\hbar}} \int_C d\tau \left| \frac{\partial^2 S_{\text{cl}}}{\partial\tau^2} \right|^{1/2} \left(\prod_{\alpha=1}^{N} \frac{1}{2\sinh[\tau\mu_\alpha(z)/2]} \right) e^{-[S_{\text{cl}}(\tau) - E\tau]/\hbar} . \tag{15.33}$$

The Euclidean action for a single periodic orbit $z_0^{(\text{PO})}(\tau')$ with period $\bar{\tau}$ is

$$S_{\text{cl}}(\bar{\tau}) = W(E) + E\bar{\tau} , \tag{15.34}$$

where E is the conserved total energy (13.1) of this particular path, and

$$W(E) \equiv \int_0^{\bar{\tau}} d\tau' \left(M\dot{q}^2(\tau') + \sum_{\alpha=1}^{N} m_\alpha \dot{x}_\alpha^2(\tau') \right) \tag{15.35}$$

is the abbreviated action. The factor $\mathcal{J}_1 = -1$ keeps track of the phase factors $e^{-i\pi/2}$ picked up by the periodic orbit at each of the two conjugate points. Finally, the N-fold product in Eq. (15.33) results from integration of the Gaussian fluctuations perpendicular to the reaction coordinate. The set $\{\mu_\alpha(z)\}$ ($\alpha = 1, 2, \cdots, N$) represents the dynamical stability (angular) frequencies of the periodic orbit. Computation of the integral over τ in Eq. (15.33) by steepest descent, as in Section 12.3, yields

$$p_1(E) = -\mathcal{J}_1 \left(\prod_{\alpha=1}^{N} \frac{1}{2\sinh[\bar{\tau}(E)\mu_\alpha(E)/2]} \right) e^{-W(E)/\hbar} . \tag{15.36}$$

Consider now the contribution of n cycles at total energy E. Evidently, the abbreviated action is $nW(E)$, the total time spent is $\tau_n = n\bar{\tau}$, and the overall phase factor is $\mathcal{J}_n = (-1)^n$. Summation of the contributions from all numbers of cycles yields

$$p(E) = \sum_{n=1}^{\infty} (-1)^{n-1} e^{-nW(E)/\hbar} \prod_{\alpha=1}^{N} \frac{1}{2\sinh[n\bar{\tau}(E)\mu_\alpha(E)/2]} . \tag{15.37}$$

Next we expand each of the functions $\sinh^{-1}(\cdots)$-functions into a geometrical series of exponentials, and employ $\bar{\tau} = -W'(E)$. Then we can perform the sum over n in the expression (15.37). The resulting expression is

$$p(E) = \sum_{n_1=0}^{\infty} \sum_{n_2=0}^{\infty} \cdots \sum_{n_N=0}^{\infty}$$

$$\times \left\{ 1 + \exp\left[\left(W(E) - W'(E) \sum_{\alpha=1}^{N} \left(n_\alpha + \tfrac{1}{2} \right) \hbar\mu_\alpha(E) \right) / \hbar \right] \right\}^{-1} . \tag{15.38}$$

With the set of quantum numbers $\{n_\alpha\}$ for the transverse degrees of freedom, the energy left in the escape coordinate z_0 is

$$E_{\text{esc}} = E - E_\perp, \qquad \text{with} \qquad E_\perp = \sum_{\alpha=1}^{N} \left(n_\alpha + \tfrac{1}{2}\right)\hbar\mu_\alpha(E). \qquad (15.39)$$

With the observation that the argument of the exponential function in Eq. (15.38) represents the leading terms of a Taylor expansion of $W(E_{\text{esc}})$ around $W(E)$, the micro-canonical cumulative transmission probability can be written as [353, 298]

$$p(E) = \sum_{n_1=0}^{\infty} \sum_{n_2=0}^{\infty} \cdots \sum_{n_N=0}^{\infty} \frac{1}{e^{W(E_{\text{esc}})/\hbar} + 1}. \qquad (15.40)$$

For $T > T_0$, the main contribution to the integral in Eq. (15.31) comes from the energy range $E_{\text{esc}} \gtrsim V_b$. In this regime, the barrier is parabolic and the global system is conveniently given in the normal mode representation (14.2). Near the barrier top the escape coordinate z_0 coincides with the normal mode y_0, and the stability angular frequencies μ_α of the periodic orbit correspond to the eigenfrequencies $\mu_\alpha^{(b)}$, $(\alpha = 1, 2, \cdots, N)$. In the parabolic barrier approximation, the periodic orbit z_0 is oscillating along y_0 with period $2\pi/\omega_{\text{R}}$, and the abbreviated action simply is

$$W(E_{\text{esc}}) \approx W_{\text{harm}}(E_{\text{esc}}) = 2\pi(V_b - E_{\text{esc}})/\omega_{\text{R}}. \qquad (15.41)$$

Upon interchanging in Eq. (15.31) the integration over E with the summations in the expression (15.40) for $p(E)$, we obtain [353]

$$k = \frac{1}{Z_{\text{tot}}^{(0)}} \frac{1}{2\pi\hbar} \sum_{n_1=0}^{\infty} \sum_{n_2=0}^{\infty} \cdots \sum_{n_N=0}^{\infty} e^{-\beta(E_\perp + V_b)} \int_0^{\infty} dE \, \frac{e^{-\beta(E - E_\perp - V_b)}}{1 + e^{-\beta_0(E - E_\perp - V_b)}}, \qquad (15.42)$$

where $\beta_0 = 2\pi/\hbar\omega_{\text{R}}$, and where $E_\perp$ is the energy in the transverse modes.

The partition function $Z_{\text{tot}}^{(0)}$ of the well regime plus environment is conveniently expressed in terms of the eigenfrequencies $\mu_\alpha^{(0)}$, as discussed in Subsection 4.3.5 and specified in Eq. (4.235). Extending in Eq. (15.42) the lower integration limit to $E = -\infty$, the integral yields the function $\pi/[\beta_0 \sin(\pi\beta/\beta_0)]$. Corrections due to the actually finite lower bound of the integral in Eq. (15.42) are exponentially small in the range $T_0 < T \ll V_b/k_{\text{B}}$. It is now straightforward to perform the summations. We then find

$$k = \frac{\omega_{\text{R}}}{2\pi} \frac{\sinh(\beta\hbar\mu_0^{(0)}/2)}{\sin(\beta\hbar\omega_{\text{R}}/2)} \prod_{\alpha=1}^{N} \frac{\sinh(\beta\hbar\mu_\alpha^{(0)}/2)}{\sinh(\beta\hbar\mu_\alpha^{(b)}/2)} e^{-\beta V_b}. \qquad (15.43)$$

Next, the product of the sinh factors can be transformed to read

$$\prod_{\alpha=0}^{N} \frac{\sinh(\beta\hbar\mu_\alpha^{(0)}/2)}{|\sinh(\beta\hbar\mu_\alpha^{(b)}/2)|} = \prod_{\alpha=0}^{N} \left(\frac{\mu_\alpha^{(0)}}{|\mu_\alpha^{(b)}|} \prod_{n=1}^{\infty} \frac{\mu_\alpha^{(0)\,2} + \nu_n^2}{\mu_\alpha^{(b)\,2} + \nu_n^2} \right)$$

$$= \frac{\omega_0}{\omega_b} \prod_{n=1}^{\infty} \frac{\nu_n^2 + \omega_0^2 + \nu_n\hat{\gamma}(\nu_n)}{\nu_n^2 - \omega_b^2 + \nu_n\hat{\gamma}(\nu_n)}. \qquad (15.44)$$

In the first equality we have written the sinh function as infinite product representation (4.235). The second equality follows with Eqs. (4.238), (4.240) and (4.242), and with the corresponding relations in which $-\omega_{\mathrm{b}}^2$ is substituted for ω_0^2. Thus we find

$$k = \frac{\omega_0}{2\pi} \frac{\omega_{\mathrm{R}}}{\omega_{\mathrm{b}}} \prod_{n=1}^{\infty} \frac{\nu_n^2 + \omega_0^2 + \nu_n \hat{\gamma}(\nu_n)}{\nu_n^2 - \omega_{\mathrm{b}}^2 + \nu_n \hat{\gamma}(\nu_n)} \, \mathrm{e}^{-\beta V_{\mathrm{b}}} \, . \tag{15.45}$$

The expression (15.45) is in agreement with the result earlier result (15.12) found with the Im F method. Now, we have verified this formula by an independent method, and hence also the factor $\sigma = \beta \hbar \omega_{\mathrm{R}}/2\pi$ for $T > T_0$ in Eq. (14.40).

The periodic orbit approach applies also to the quantum tunneling regime $T < T_0$. In this case, the tunneling rate is again given by Eq. (15.31) with Eq. (15.40). However, the abbreviated action $W(E_{\mathrm{esc}})$ must take into account the actual barrier shape.

In the MQTST ansatz (15.30), recrossings of the dividing surface Σ_{b} are excluded. In this method, the reduction factor $\varrho = \omega_{\mathrm{R}}/\omega_{\mathrm{b}}$ appears because the barrier frequency ω_{R} of the potential $V(q, \boldsymbol{x})$ along the escape coordinate y_0 is smaller than the barrier frequency of the bare potential $V(q)$. In contrast to MQTST, in the standard treatment of the decay by Langevin or Fokker-Planck equation methods, the reduction factor $\varrho = \omega_{\mathrm{R}}/\omega_{\mathrm{b}}$ originates from diffusive recrossings of the barrier top along the particle's coordinate q [298]. Thus, although the two physical pictures are different, the results are in correspondence.

The form (15.43) displays the multi-dimensional character of the barrier crossing process. These features are somewhat hidden in the formula (15.45). The latter form is especially convenient for systems in which we eventually take the limit $N \to \infty$.

16. The crossover region

We have seen in Section 15.2 that the quantum correction factor c_{qm} increases with decreasing temperature, and that it diverges as $T \to T_0$ because the eigenvalue $\Lambda_1^{(b)} = \Lambda_{-1}^{(b)} = \nu_1^2 - \omega_{\mathrm{b}}^2 + |\nu_1| \hat{\gamma}(|\nu_1|)$ in Eq. (15.7) vanishes at this temperature. Near $T = T_0$, the eigenvalue $\Lambda_1^{(b)}$ may be expanded in powers of the parameter

$$\varepsilon \equiv (T_0 - T)/T_0 \, , \tag{16.1}$$

which is chosen negative above T_0 for later purposes. In leading order in ε we have from Eq. (15.8)

$$\Lambda_1^{(b)} = -\varepsilon \, \Omega^2 \, , \tag{16.2}$$

where

$$\Omega^2 = \omega_{\mathrm{b}}^2 + \omega_{\mathrm{R}}^2 [\, 1 + \partial \hat{\gamma}(\omega_{\mathrm{R}})/\partial \omega_{\mathrm{R}} \,] \, . \tag{16.3}$$

The rate expression (15.12) can be written in the form

$$k = \frac{\omega_{\mathrm{R}}}{2\pi} \frac{\Omega^2}{\Lambda_1^{(b)}} \, A \, \mathrm{e}^{-\beta V_{\mathrm{b}}} \, . \tag{16.4}$$

The smooth temperature dependence of the prefactor near $T = T_0$ is allocated to the dimensionless prefactor (we put again $\nu = \nu_1 = 2\pi/\hbar\beta$)

$$A \equiv \frac{\omega_0}{\omega_b} \frac{\nu^2 + \omega_0^2 + \nu\hat{\gamma}(\nu)}{\Omega^2} \prod_{n=2}^{\infty} \frac{\nu_n^2 + \omega_0^2 + \nu_n\hat{\gamma}(\nu_n)}{\nu_n^2 - \omega_b^2 + \nu_n\hat{\gamma}(\nu_n)} . \tag{16.5}$$

Substituting the expression (16.2) into Eq. (16.4), we obtain for T slightly above T_0

$$k = \frac{\omega_R}{2\pi} \frac{A}{-\varepsilon} e^{-\beta V_b} . \tag{16.6}$$

This expression is the multi-dimensional generalization of the former result (12.32). The prefactor A can equivalently be expressed in terms of the partition function of the global system $Z_{\text{tot}}^{(0)}$ and the eigenfrequencies $\mu_\alpha^{(b)}$ for $\alpha \geq 1$. Extracting the factor $1/\sin(\beta\hbar\omega_R/2) \approx -1/\pi\varepsilon$, and using the form (4.235), we find from Eq. (15.43)

$$A = \frac{1}{2\pi} \frac{1}{Z_{\text{tot}}^{(0)}} \prod_{\alpha=1}^{N} \frac{1}{2\sinh(\beta\hbar\mu_\alpha^{(b)}/2)} . \tag{16.7}$$

In the Ohmic case, the infinite product in Eq. (16.5) can be expressed again in terms of gamma functions. We find with Eqs. (15.17) and (15.18)

$$A = \frac{\omega_0}{\omega_b} \frac{\nu^2}{\Omega^2} \frac{\Gamma(2 + \lambda_b^+/\nu)\Gamma(2 + \lambda_b^-/\nu)}{\Gamma(1 + \lambda_0^+/\nu)\Gamma(1 + \lambda_0^-/\nu)} = \frac{\omega_0}{\omega_b} \frac{\nu^2 + \gamma\nu - \omega_b^2}{\Omega^2} c_{\text{qm}} . \tag{16.8}$$

The zero of the eigenvalue $\Lambda_1^{(b)}$ at $T = T_0$, Eq. (16.2), points to the fact that at this temperature the action of the constant path $q(\tau) = q_b$ is degenerate with the action of the periodic bounce path $q(\tau) = q_B(\tau)$.

It is obvious from the derivation that the singularity in Eq. (16.6) at $T = T_0$ is unphysical. Rather, it is an artifact of the Gaussian approximation for the fluctuation modes $X_{\pm 1}$ in Eq. (15.7). The singularity is absent if one evaluates these modes beyond the Gaussian approximation. The accurate calculation of the integral over the modes $X_{\pm 1}$ slightly above T_0 is different from the calculation slightly below T_0. The cases $T > T_0$ and $T < T_0$ were discussed in Refs. [342] and [340], respectively. The reader is also referred to the discussion in Ref. [297].

16.1   Beyond steepest descent above T_0

To regularize the annoying fluctuation integrals, the second-order action (15.7) must be supplemented by terms of higher order in the amplitudes X_1 and X_{-1}. Upon expanding the potential $V(q)$ about the barrier top beyond the parabolic barrier form,

$$V(q) = V_b - \frac{1}{2}M\omega_b^2(q - q_b)^2 + \sum_{k=3}^{\infty} \frac{1}{k}Mc_k(q - q_b)^k , \tag{16.9}$$

the extension of the second order action (15.7) is found as

$$S[q^{(b)}(\cdot)] = \hbar\beta V_b + \frac{1}{2}M\hbar\beta\left(\sum_{n=-\infty}^{\infty}\Lambda_n^{(b)}X_nX_{-n}\right.$$

$$\left. + 2c_3\left(X_{-2}X_1^2 + X_2X_{-1}^2 + 2X_0X_1X_{-1}\right) + 3c_4X_1^2X_{-1}^2\right),$$

where we have kept terms up to the fourth order in $X_{\pm 1}$. After integrating out the amplitudes X_0 and $X_{\pm n}$ ($n \geq 2$), the action of the residual modes $X_{\pm 1}$ becomes

$$\Delta S_1^{(b)} = \frac{1}{2}M\hbar\beta\left(2\Lambda_1^{(b)}X_1X_{-1} + B_4X_1^2X_{-1}^2\right) . \tag{16.10}$$

The coefficient B_4 measures the strength of the leading anharmonic contribution,

$$B_4 = 4c_3^2/\omega_b^2 - 2c_3^2/\Lambda_2^{(b)} + 3c_4 , \tag{16.11}$$

and is positive for the usual case of a barrier which is wider than a parabolic barrier.[1]

We now withdraw the Gaussian approximation for the fluctuation modes X_1 and X_{-1} which has lead us to the spurious form (16.1) near $T = T_0$. The removal amounts to the substitution [cf. the functional measure (4.228)]

$$\frac{1}{\Lambda_1^{(b)}} \longrightarrow \frac{1}{\tilde{\Lambda}_1^{(b)}} \equiv i\frac{M\beta}{2\pi}\int_{-\infty}^{\infty}dX_1\int_{-\infty}^{\infty}dX_{-1}\exp(-\Delta S_1^{(b)}/\hbar) . \tag{16.12}$$

Upon introducing polar coordinates (r, φ) defined by $X_{\pm 1} \equiv (r/\sqrt{2})\,e^{\pm i\varphi}$, we obtain

$$\frac{1}{\tilde{\Lambda}_1^{(b)}} = M\beta\int_0^{\infty}dr\,r\exp\left\{-\tfrac{1}{2}M\beta\Lambda_1^{(b)}r^2 - \tfrac{1}{8}M\beta B_4 r^4\right\} , \tag{16.13}$$

which may be written in terms of the function $\mathrm{erfc}(z)$ given in Eq. (12.36) as

$$\frac{1}{\tilde{\Lambda}_1^{(b)}} = \left(\frac{\pi M\beta}{2B_4}\right)^{1/2}\mathrm{erfc}\left(\Lambda_1^{(b)}\left(\frac{M\beta}{2B_4}\right)^{1/2}\right)\exp\left(\Lambda_1^{(b)2}\frac{M\beta}{2B_4}\right) . \tag{16.14}$$

With the form (16.2) for $\Lambda_1^{(b)}$, we then get the compact expression

$$\Omega^2/\tilde{\Lambda}_1^{(b)} = \sqrt{\pi}\,\kappa\,\mathrm{erfc}(-\kappa\varepsilon)\,e^{\kappa^2\varepsilon^2} , \tag{16.15}$$

in which the dimensionless parameter κ is given by

$$\kappa = \sqrt{\frac{M\Omega^4}{2B_4}}\beta . \tag{16.16}$$

Finally, upon substituting $\Omega^2/\tilde{\Lambda}_1^{(b)}$ for $\Omega^2/\Lambda_1^{(b)}$ in Eq. (16.4), the thermal rate is found in the form

$$k = \frac{\omega_R}{2\pi}A\sqrt{\pi}\kappa\,\mathrm{erfc}(-\kappa\varepsilon)\,e^{\kappa^2\varepsilon^2-\beta V_b} . \tag{16.17}$$

To summarize, the problematic term $1/\Lambda_1^{(b)}$ has been regularized by taking into account non-Gaussian fluctuations of the corresponding mode. The improved rate formula (16.17) is valid in the range extending from very high T, where pure thermal activation prevails, down to $T = T_0$. Before taking up discussion of the result achieved, let us consider the crossing rate for T slightly below T_0.

---

[1]A brief discussion of the role of the coefficient B_4 is given at the end of Section 16.3.

16.2   Beyond steepest descent below T_0

Below T_0, the bounce trajectory $q_B(\tau)$ exists as a third extremal action path besides the constant paths $q(\tau) = 0$ and $q(\tau) = q_b$. Since the bounce action $S_B \equiv S_{\text{eff}}^{(E)}[q_B]$ is smaller than the action $\hbar\beta V_b$ of the trivial saddle point $q(\tau) = q_b$, the bounce gives the leading contribution to $\operatorname{Im} Z^{(b)}$, and thus to $\operatorname{Im} F$. Because the bounce is a periodic path with period $\hbar\beta$, it may be written as Fourier series,

$$q_B(\tau) = q_b + \sum_{n=-\infty}^{\infty} Q_n \, e^{i\nu_n\tau} . \tag{16.18}$$

A change of the phase of the bounce $q_B(\tau) \to q_B(\tau + \tau_0)$ has no influence on the numerical value of the action. Because of the time translation invariance of the action, we encounter a *zero mode* in the quantum fluctuations about the bounce. It is convenient to choose $q_B(\tau) = q_B(-\tau)$ which entails $Q_n = Q_{-n}$. For T slightly below T_0, the amplitudes Q_n are small and they can be calculated from the equation of motion (13.2) as a power series in $\sqrt{\varepsilon}$, where again $\varepsilon \equiv (T_0 - T)/T_0$. The leading terms are

$$Q_1 = \left(\frac{\varepsilon\Omega^2}{B_4}\right)^{1/2} , \qquad Q_0 = -\frac{2c_3}{\omega_b^2} Q_1^2 , \qquad Q_2 = \frac{c_3}{\Lambda_2^{(b)}} Q_1^2 . \tag{16.19}$$

With these forms and with Eq. (16.16) the bounce action can be written as

$$S_B = \hbar\beta V_b - \hbar\kappa^2\varepsilon^2 + \mathcal{O}(\varepsilon^3) . \tag{16.20}$$

In the next step, we equip the bounce trajectory with quantum fluctuations,

$$q(\tau) = q_B(\tau) + \sum_{n=-\infty}^{\infty} F_n \, e^{i\nu_n\tau} , \tag{16.21}$$

and calculate the change of the action. We purposefully write

$$S[q] = S_B + S^{(1)} + S^{(2)} , \tag{16.22}$$

where $S^{(1)}$ is the second order action, and the relevant anharmonic contributions are in $S^{(2)}$. With use of the expressions (16.19) the action $S^{(1)}$ may be written as

$$S^{(1)} = \frac{M\hbar\beta}{2}\left(-\omega_b^2 G_0^2 + 2\Lambda_2^{(b)} G_2 G_{-2} + \varepsilon\Omega^2(F_1 + F_{-1})^2 + 2\sum_{n=3}^{\infty} \Lambda_n^{(b)} F_n F_{-n}\right) . \tag{16.23}$$

Here we have kept only the terms of leading order in ε, and we have introduced

$$G_0 \equiv F_0 - (2c_3/\omega_b^2)(\varepsilon\Omega^2/B_4)^{1/2}(F_1 + F_{-1}) ,$$

$$G_{\pm2} \equiv F_{\pm2} + (2c_3/\Lambda_2^{(b)})(\varepsilon\Omega^2/B_4)^{1/2}F_{\pm1} . \tag{16.24}$$

The fluctuation mode G_0 reduces the action. This verifies that the bounce is indeed a saddle point. The integration of the amplitude G_0 must again be carried out by deformation of the contour as described in Section 12.5. This leads to an imaginary contribution to the partition function from the barrier region with the already familiar extra factor $1/2$ [cf. Eq. (12.69)].

Next, observe that the eigenvalues $\Lambda_1^{(b)}$ and $\Lambda_{-1}^{(b)}$, which are degenerate above T_0 [see Eq. (15.8)], split into the eigenvalues $\Lambda_{1,-}^{(B)}$ and $\Lambda_{1,+}^{(B)}$ of the second order action (16.23) as T passes through T_0 from above. The degeneracy is removed because the constant path bifurcates into the bounce path at the crossover temperature. With substitution of $F_{\pm 1} \equiv (F_{1,+} \pm F_{1,-})/\sqrt{2}$ into Eq. (16.23), we find that the eigenvalue $\Lambda_{1,-}^{(B)}$ of the mode $F_{1,-}$ remains zero,[2] while the eigenvalue of the mode $F_{1,+}$ is $\Lambda_{1,+}^{(B)} = 2\varepsilon\Omega^2$. The eigenvalues with index $n \geq 2$ are independent of ε in leading order. In summary, the fluctuation determinant D_{B}' connected with the second-order action functional (16.23) is found as

$$|D_{\mathrm{B}}'| = 2\varepsilon\Omega^2 |D_{\mathrm{B}}''| \quad \text{with} \quad |D_{\mathrm{B}}''| = |D^{(b)''}| \equiv \omega_{\mathrm{b}}^2 \left(\prod_{n=2}^{\infty} \Lambda_n^{(b)} \right)^2 , \qquad (16.25)$$

where terms of order ε^2 are disregarded. The single prime in D_{B}' is to indicate that the zero eigenvalue is omitted.

Because of the smallness of the eigenvalues $\Lambda_{1,\pm}^{(B)}$, the Gaussian approximation for the modes $F_{1,\pm}$ breaks down. Hence the action (16.23) must be supplemented by terms of third and fourth order in the amplitudes F_1 and F_{-1}. These higher-order terms contain nonlinear couplings between the amplitudes F_1 and F_{-1} and with amplitudes F_0 and $F_{\pm 2}$. The relevant contribution has the form

$$\begin{aligned} S^{(2)} = &- M\hbar\beta \left[c_3 \left(F_2 F_{-1}^2 + F_{-2} F_1^2 + 2 F_0 F_1 F_{-1} \right) \right. \\ &\left. + 3 c_4 Q_1 \left(F_1^2 F_{-1} + F_1 F_{-1}^2 \right) + 3 c_4 F_1^2 F_{-1}^2/2 \right] . \end{aligned} \qquad (16.26)$$

The integrations over the amplitudes F_0 and $F_{\pm 2}$ are again Gaussian and can be performed by completing the square. Then we are left with the integral over the amplitudes F_1 and F_{-1} weighted with the action factor $\exp(-\Delta S_1^{(B)}/\hbar)$, where

$$\Delta S_1^{(B)} = \tfrac{1}{2} M B_4 \hbar\beta \left[Q_1^2 (F_1 + F_{-1})^2 + 2 Q_1 (F_1 + F_{-1}) F_1 F_{-1} + F_1^2 F_{-1}^2 \right] . \qquad (16.27)$$

Here the terms of third and fourth order are due to the action $S^{(2)}$. Upon introducing polar cordinates (ρ, φ) with origin at $-Q_1$,

$$F_{\pm 1} = (\rho/\sqrt{2}) \, e^{\pm i\varphi} - Q_1 , \qquad (16.28)$$

the action turns out to be independent of φ. Thus, the φ mode is the zero mode connected with changes of the phase of the bounce, while the ρ mode describes amplitude fluctuations. These fluctuations can be as large as the bounce amplitude Q_1. Therefore, we have to keep all terms up to the order ρ^4.

---

[2]The zero eigenvalue is due to the time-translation symmetry of the bounce mentioned before.

Abolishment of the Gaussian approximation amounts to the substitution

$$\frac{1}{\sqrt{\Lambda_{1,+}^{(B)}\Lambda_{1,-}^{(B)}}} \longrightarrow \frac{1}{\Lambda_1^{(B)}} \equiv i\frac{M\beta}{2\pi}\int_{-\infty}^{\infty}dF_1\int_{-\infty}^{\infty}dF_{-1}\exp\left(-\Delta S_1^{(B)}/\hbar\right)$$

$$= M\beta\int_0^{\infty}d\rho\,\rho\exp\left\{-\tfrac{1}{8}\beta MB_4\left(\rho^2-2Q_1^2\right)^2\right\}. \tag{16.29}$$

As in the case $T > T_0$, we thus obtain an expression with an error function,

$$1/\Lambda_1^{(B)} = (\sqrt{\pi}\kappa/\Omega^2)\operatorname{erfc}(-\kappa\varepsilon). \tag{16.30}$$

Altogether, the expression (15.10) is modified according to

$$\operatorname{Im}F = -\frac{1}{2\beta\Lambda_1^{(B)}}\left(\frac{D^{(0)}}{|D_B''|}\right)^{1/2}e^{-S_B/\hbar}, \tag{16.31}$$

where the determinant $|D_B''|$ is given in Eq. (16.25). With use of the relation (12.73) and the expressions (15.11), (16.5) and (16.20), the decay rate is readily found as

$$k = \frac{1}{\hbar\beta}A\sqrt{\pi}\kappa\operatorname{erfc}(-\kappa\varepsilon)\,e^{\kappa^2\varepsilon^2-\beta V_b}. \tag{16.32}$$

Now observe that at $T = T_0$, which is equivalent to $1/\hbar\beta = \omega_R/2\pi$, the formula (16.32) exactly coincides with the previous result (16.17) evaluated for temperatures slightly above T_0. From this we infer that the expression (16.32) is valid in the entire crossover region above and below T_0.

Below T_0, the coefficient B_4 defined in Eq. (16.11) can be related to the change of the bounce period per unit energy. This is shown as follows. First, we write down the formal expansion of the action about $\hbar\beta_0$,

$$S(\hbar\beta) = \hbar\beta V_b + \frac{1}{2}S''(\hbar\beta_0)\left(\hbar\beta - \hbar\beta_0\right)^2 + \mathcal{O}\left((\hbar\beta - \hbar\beta_0)^3\right). \tag{16.33}$$

In the second step, we identify Eq. (16.33) with Eq. (16.20) and use the relations[3]

$$S''(\tau) = -\frac{1}{W''(z)} = -\frac{1}{|\tau'(z)|}, \tag{16.34}$$

where $W(z)$ is the abbreviated action for one periodic orbit at total energy z and where $\tau(z)$ is the respective period [cf. Eq.(15.34)]. We then find

$$|\tau_b'| \equiv |\tau'(z = V_b)| = \frac{B_4}{M\Omega^4}\hbar\beta_0 = \frac{\hbar}{2}\frac{\beta_0^2}{\kappa^2}. \tag{16.35}$$

Solving Eq. (16.35) for κ and substituting this into Eq. (16.32), we obtain [353]

$$k = \frac{\pi A}{\sqrt{2\pi\hbar|\tau_b'|}}\operatorname{erfc}\left(\frac{\hbar(\beta_0-\beta)}{\sqrt{2\hbar|\tau_b'|}}\right)\exp\left(-\beta V_b + \frac{\hbar(\beta-\beta_0)^2}{2|\tau_b'|}\right). \tag{16.36}$$

We remark that this expression applies again for T around T_0. Substituting for A the form (16.7), we discover the multidimensional generalization of expression (12.35).

---

[3]The prime (double prime) denotes the first (second) derivative with respect to the argument.

16.3 The scaling region

As the energy $M\Omega^4/B_4$ is usually of the order of the barrier height, the parameter κ is of the order of $(V_b/\hbar\omega_R)^{1/2}$. Thus we have $\kappa \gg 1$ in the semiclassical limit. The sophisticated formula (16.17) [or (16.32)] is appropriate in the region $|\kappa\varepsilon| \lesssim 1$, or

$$|T - T_0| \lesssim T_0/\kappa, \qquad (16.37)$$

in which the argument of the erfc function is of order one or smaller. For $\kappa \gg 1$, the crossover region is narrow on the temperature scale T_0.

For T well above this regime, which is $\kappa\varepsilon < -1$, we may use the asymptotic form

$$\sqrt{\pi}\,\kappa\,e^{\kappa^2\varepsilon^2}\,\text{erfc}(-\kappa\varepsilon) \approx -1/\varepsilon. \qquad (16.38)$$

whereby Eq. (16.17) reduces to the expression (16.6).

Consider next the temperature region slightly below the crossover regime. We then have $\kappa\varepsilon > 1$ and $\text{erfc}(-\kappa\varepsilon) \approx 2$. The factor A may be written with use of Eq. (16.25) as $A = \sqrt{2\varepsilon D^{(0)}}/(|D'_B|\Omega)$. Next we relate the parameter ε to the normalization factor W_B of the zero mode [see Eq. (12.63)]

$$W_B = M\int_0^{\hbar\beta} d\tau\,\dot{q}_D^2(\tau). \qquad (16.39)$$

Observe that W_B differs from the (multidimensional) abbreviated action (15.35) by the missing kinetic contribution of the reservoir degrees of freedom. Using for the periodic bounce path the form (16.18) with (16.19), we obtain in leading order in ε

$$W_B = 2M\hbar\beta\nu_1^2 Q_1^2 = 8\pi^2\frac{M\Omega^2}{B_4\hbar\beta_0}\varepsilon \overset{(16.16)}{=} 16\pi^2\left(\frac{\kappa}{\hbar\beta_0\Omega}\right)^2\hbar\varepsilon. \qquad (16.40)$$

Solving this expression for ε, we readily may write the rate expression (16.32) as

$$k = \sqrt{\frac{W_B}{2\pi\hbar}\frac{D^{(0)}}{|D'_B|}}\,e^{-S_B/\hbar}, \qquad (16.41)$$

where $|D'_B|$ and S_B are given in Eqs. (16.25) and (16.20), respectively.

Another useful expression for the rate results from Eq. (16.36) with substitution of the form (16.7) for the quantity A. The result is

$$k = \frac{1}{Z_{\text{tot}}^{(0)}}\left(\prod_{\alpha=1}^{N}\frac{1}{2\sinh(\hbar\beta\mu_\alpha^{(b)}/2)}\right)\frac{1}{\sqrt{2\pi\hbar|\tau'_b|}}\,e^{-S_B/\hbar}. \qquad (16.42)$$

The expressions (16.41) and (16.42) hold in the regime $\kappa^{-1} < \varepsilon \ll 1$. They smoothly match well below T_0 on the semiclassical quantum tunneling rate formula which is discussed subsequently in Section 17.1. Finally, we note that the expression (16.42) represents the multidimensional generalization of the earlier result (12.26).

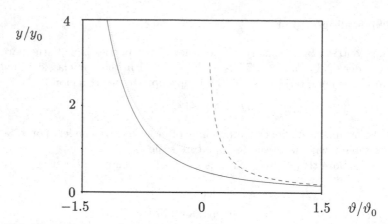

Figure 16.1: The scaled rate y/y_0 is shown as a function of the scaled temperature ϑ/ϑ_0. The high temperature formula (16.6) is represented by a dashed line and the low temperature formula (16.41) as a dotted line. The crossover function (16.46) smoothly matches onto these functions below and above the crossover region.

The quantum-statistical decay rate possesses in the crossover region a universal scaling behavior [297]. Defining a temperature scale ϑ_0 and a frequency scale y_0,

$$\vartheta_0 \equiv T_0/\kappa \,, \qquad y_0 = (\omega_R/2\pi)\sqrt{\pi}\,\kappa\,A \,, \qquad (16.43)$$

there follows from Eq. (16.32) that the quantity

$$y/y_0 \equiv (k/y_0)\,e^{\beta V_b} \qquad (16.44)$$

is represented by a *universal* function of $\vartheta/\vartheta_0 \equiv \kappa(T - T_0)/T_0$,

$$y/y_0 = U(\vartheta/\vartheta_0) \,. \qquad (16.45)$$

The scaling function $U(z)$ is defined by the integral representation

$$U(z) = \frac{2}{\sqrt{\pi}}\int_0^\infty dt\,e^{-t^2 - 2zt} = \operatorname{erfc}(z)\,e^{z^2} \,. \qquad (16.46)$$

Interestingly, the universal behavior (16.45) is independent of the specific form of the metastable potential and independent of the dissipative mechanism. Only the scale factors y_0 and ϑ_0 depend on the particular system under consideration. In Fig. 16.1, the scaled rate y/y_0 is shown as a function of the scaled temperature ϑ/ϑ_0 in comparison with the high and low temperature formulas (16.6) and (16.41).

Let us finally briefly discuss the significance of the anharmonic contribution in the action (16.10). When the coefficient B_4 is positive, the bounce amplitude and the bounce action increase monotonously with the bounce period. Hence the transition from the constant path to the periodic path at $T = T_0$ is a second order phase transition [341, 342]. In the limit $B_4 \to 0$, the parameter κ diverges, as we see from Eq. (16.16). Thus, the scaling function (16.46) becomes meaningless in this limit. To regularize the interpolating function in the case $B_4 = 0$, we have to include in the action $\Delta S_1^{(b)}$ in Eq. (16.10) the term of sixth order,

$$\Delta S_1^{(b)} = \frac{1}{2} M \hbar \beta \left(2 \Lambda_1^{(b)} X_1 X_{-1} + B_6 X_1^3 X_{-1}^3 \right) . \tag{16.47}$$

For the potential (16.9), the coefficient B_6 is

$$B_6 = 20 c_6/3 - 2 c_4^2 / \Lambda_3^{(b)} . \tag{16.48}$$

The significant eigenvalue $\tilde{\Lambda}_1^{(b)}$ is given by the expression (16.12), but now with the action (16.47). Eventually, we find again the universal scaling form (16.45) for the thermal-to-quantum crossover. However, the scaling function $U(z)$ and the scale factor κ are different. The dimensionless parameter κ is

$$\kappa = \Omega^2 \left(M^2 \beta^2 / 4 B_6 \right)^{1/3} , \tag{16.49}$$

and the scaling function $U(z)$ has the integral representation

$$U(z) = \frac{2}{\sqrt{\pi}} \int_0^\infty dt \, e^{-t^3 - 2zt} . \tag{16.50}$$

When the coefficient B_4 is negative, the barrier is narrower than the harmonic one. In this case, the bounce action is not any more a monotonous function of temperature. Rather, the classical-to-quantum transition has the characteristics of a phase transition of first order. Physically, this is due to the fact that tunneling near to the barrier top is less favorable than at a lower level. The case $B_4 < 0$ is exotic and of minor physical importance. Therefore, it will not be discussed further.

17. Dissipative quantum tunneling

We are now prepared for a study of the thermodynamic decay rate at temperature well below the crossover regime. In this temperature range, the escape from the well goes off by dissipative quantum tunneling through the barrier.

17.1 The quantum rate formula

At temperature T below T_0, the relevant stationary point accounting for the barrier region is the single-bounce trajectory $q_B(\tau)$. It obeys the equation of motion (13.2) with the dissipative kernel (4.45). To second order in the fluctuations about the extremal path, $q(\tau) = q_B(\tau) + \xi(\tau)$, the action takes the form

$$S[q] = S_B + \frac{M}{2} \int_0^{\hbar\beta} d\tau \, \xi(\tau) \, \mathbf{\Lambda}[q_B(\tau)] \, \xi(\tau) , \tag{17.1}$$

where the fluctuation operator $\Lambda[q_B(\tau)]$ is defined in Eq. (4.214). Expanding $\xi(\tau)$ into the normalized eigenmodes $\chi(\tau)$ according to

$$\xi(\tau) \;=\; \sum_n c_n \chi_n(\tau) \,, \tag{17.2}$$

where $\chi_n(\tau)$ is normalized on the interval $(0, \hbar\beta)$, the action becomes

$$S[q] \;=\; S_B + \frac{M}{2}\sum_n \Lambda_n[q_B]\, c_n^2 \,. \tag{17.3}$$

The $\Lambda_n[q_B]$ are the eigenvalues of $\Lambda[q_B]$ for periodic boundary conditions. The determinant of the operator $\Lambda[q_B]$ in diagonal representation reads

$$D[q_B] \;\equiv\; \det\left(\Lambda[q_B]\right) \;=\; \prod_n \Lambda_n[q_B] \,. \tag{17.4}$$

Since the bounce is time-periodic, there is again a zero mode due to the time-translational invariance of the action, and a mode with a negative eigenvalue accounting for metastability. These modes must be precisely dealt with as explained in items (1) and (2) in Section 12.5. For T well below T_0, even the smallest positive eigenvalue[1] is sufficiently large that the respective mode can be treated in Gaussian approximation.

Following the lines drawn in Section 12.5, the barrier contribution to the partition function becomes imaginary. Finally, with use of the relation $k = (2/\hbar\beta)\,\mathrm{Im}\,Z_b/Z_0$ the quantum decay rate at temperature T emerges concisely as

$$k \;=\; f_{qm}\, e^{-S_B/\hbar} \,. \tag{17.5}$$

The exponent is determined by the effective action (4.61) of a single bounce. We have $S_B = S_{\mathrm{eff}}^{(E)}[q_B(\cdot)]$. The bounce path $q_B(\tau)$ is a stationary point of this action. It is a periodic path with period $\hbar\beta$ in the upside-down potential obeying the equation of motion (13.2). The prefactor or quantum mechanical attempt frequency is

$$f_{qm} \;=\; \sqrt{\frac{W_B}{2\pi\hbar}\frac{D^{(0)}}{|D'[q_B]|}} \,. \tag{17.6}$$

Here, W_B is twice the kinetic part of the bounce action [see Eq. (16.39)], and $D'[q_B]$ is the determinant (17.4) connected with the Gaussian fluctuations about the bounce path, and the prime indicates that the zero eigenvalue is omitted.

The rate expression (17.5) with (17.6) applies down to zero temperature. It matches slightly below T_0 the previous expression (16.41) with the actions S_B and W_B given in Eq. (16.20) and Eq. (16.40), respectively. Further, slightly below T_0 we have $|D'[q_B]| = |D'_B|$, where $|D'_B|$ is given in Eq. (16.25).

---

[1]For T slightly below T_0, this is the quasi-zero eigenvalue $\Lambda_{1,+}^{(B)} = 2\varepsilon\Omega^2$ discussed in Section 16.2.

As in the case $T > T_0$ discussed in Section 15.4, it is possible to rewrite Eq. (17.5) with Eq. (17.6) in a form which directly displays the multi-dimensional character of the tunneling process. In the "many-dimensional WKB" approach the system-plus-reservoir complex is visualized as tunneling entity which moves along the (curved) most probable escape path (periodic orbit) $z_0^{(PO)}(\tau)$ in the $(N+1)$-dimensional configuration space. In this picture, the bounce trajectory $q_B(\tau)$ is the projection of the periodic orbit[2] onto the reaction coordinate q. In the multi-dimensional representation, the exponent in the tunneling formula is given by the expression [cf. Eqs. (15.34) and (15.35)]

$$S_B/\hbar = \frac{2}{\hbar} \int_{z_0^{(m)}}^{z_0^{(ex)}} dz_0 \sqrt{2M[V(q,\boldsymbol{x}) - E]} + E\beta , \qquad (17.7)$$

where dz_0 measures distance along the curved orbit with period $\hbar\beta$, and where E is the conserved energy of this path. The turning points are as in Eq. (15.29). At zero temperature we have $E = 0$ and $\hbar\beta = \infty$.

Consider next the pre-exponential factor of the tunneling rate. It is convenient to split the determinant of the fluctuations about the periodic orbit into the contribution from *longitudinal* fluctuations along the escape coordinate

$$\xi_{\|}(\tau) \equiv y_0(\tau) - y_0^{(PO)}(\tau) , \qquad (17.8)$$

and into the *transverse* determinant describing fluctuations in the coordinates $y_n(\tau)$ $(n = 1, 2, \cdots, N)$ which are locally perpendicular to the escape coordinate. The latter determinant is conveniently expressed in terms of the dynamical stability frequencies $\mu_\alpha^{(B)}$ $(\alpha = 1, 2, \cdots, N)$ of the periodic orbit. We then have [298, 353]

$$k = \frac{1}{Z_{tot}^{(0)}} \left(\prod_{\alpha=1}^{N} \frac{1}{2\sinh(\hbar\beta\mu_\alpha^{(B)}/2)} \right) \frac{1}{\sqrt{2\pi\hbar|\tau'(E_{\hbar\beta})|}} e^{-S_B/\hbar} , \qquad (17.9)$$

where $E_{\hbar\beta}$ is the total energy of the periodic orbit with period $\hbar\beta$. This form of the quantum statistical tunneling rate is valid down to zero temperature. It smoothly matches on the previous formula (16.42) at temperatures slightly below T_0.

An alternative derivation of the formula (17.5) with (17.6) has been given for zero temperature by Schmid [354] using a many-dimensional WKB approach for the quasi-stationary ground-state wave function. The generalization to finite temperatures is presented in Ref. [355].

For weak Ohmic dissipation, the bounce action and the attempt frequency may be calculated perturbatively in the Ohmic coupling $\alpha = \gamma/2\omega_0$. For the cubic metastable potential (11.41), one obtains as extension of the WKB result (12.41) with (12.44) at $T = 0$ the bounce action

$$S_B(T = 0) = \frac{36}{5} \frac{V_b}{\omega_0} \left(1 + \frac{45\zeta(3)}{\pi^3} \alpha + \mathcal{O}(\alpha^2) \right) , \qquad (17.10)$$

---

[2]As a result of the projection, the energy of the path $q_B(\tau)$ is not conserved, and the equation of motion becomes nonlocal in time.

where $\zeta(3)$ is a Riemann number, and the quantum mechanical prefactor [357]

$$f_{\text{qm}}(T = 0) \;=\; 12\sqrt{6\pi}\,\frac{\omega_0}{2\pi}\sqrt{\frac{V_{\text{b}}}{\hbar\omega_0}}\Big(1 + 2.86\,\alpha + \mathcal{O}(\alpha^2)\Big). \tag{17.11}$$

Dissipative quantum tunneling from the ground state in a metastable well was studied first by Caldeira and Leggett [77]. The qualitative conclusion of their study was that damping suppresses quantum tunneling by an exponential factor, which depends linearly on α for weak damping, as given in Eq. (17.10) .

17.2   Thermal enhancement of macroscopic quantum tunneling

When temperature is above absolute zero, there is a finite probability that the particle tunnels from an excited state in the well through the barrier. Hence the quantum-statistical decay at finite T is enhanced compared to the case $T = 0$. The leading thermal enhancement at low T arises from the temperature dependence of the bounce action. It is convenient to write

$$k(T) \;=\; k(0)\, e^{\mathcal{A}(T)} \qquad \text{with} \qquad \mathcal{A}(T) = [\, S_{\text{B}}(0) - S_{\text{B}}(T)\,]/\hbar\,. \tag{17.12}$$

The asymptotic expansion of the bounce action about zero temperature is discussed in some detail in Ref. [358] for a general metastable potential and arbitrary frequency-dependent damping. By use of the asymptotic Euler-Maclaurin expansion [89], the Fourier coefficients Q_n of the bounce at low T may be related to the Fourier representation $Q(\omega)$ of the zero temperature bounce path. With this it is found that the leading contribution to $\mathcal{A}(T)$ is due to the temperature dependence of the dissipative kernel $K(\tau)$ in the action $S_{\text{B}} = S_{\text{eff}}^{(\text{E})}[q_{\text{B}}(\cdot)]$ given in Eq. (4.61). We obtain

$$\begin{aligned}
\mathcal{A}(T) &= \frac{1}{2\hbar}\,[\,K_T(0) - K_{T=0}(0)\,]\int_{-\infty}^{\infty}\! d\tau\, q_{\text{B}}^{(0)}(\tau)\int_{-\infty}^{\infty}\! d\tau'\, q_{\text{B}}^{(0)}(\tau') \\[4pt]
&= \frac{q_{\text{b}}^2\tau_{\text{W}}^2}{2\pi\hbar}\int_0^{\infty}\! d\omega\, J(\omega)\Big(\coth(\tfrac{1}{2}\beta\hbar\omega) - 1\Big),
\end{aligned} \tag{17.13}$$

where $q_{\text{B}}^{(0)}(\tau)$ is the zero temperature bounce. In the second form, we have employed Eq. (4.54) with Eq. (3.55), and we have used the width or length of the zero temperature bounce,

$$\tau_{\text{W}} \equiv \frac{1}{q_{\text{b}}}\int_{-\infty}^{\infty}\! d\tau\, q_{\text{B}}^{(0)}(\tau)\,. \tag{17.14}$$

Since the metastable minimum is chosen at $q = 0$, we have $q_{\text{B}}^{(0)}(\tau \to \pm\infty) \to 0$. For a spectral density $J(\omega) \propto \omega^s$, the kernel $k(\tau)$ in the equation of motion decays asymptotically for $T = 0$ as $|\tau|^{-(1+s)}$. Therefore, we have $q_{\text{B}}(|\tau| \to \infty) \propto |\tau|^{-(1+s)}$. This verifies that the bounce width (17.14) is finite for $s > 0$.

For the spectral density $J(\omega) = M\gamma_s\omega_{\text{ph}}^{1-s}\omega^s$ [cf. Eqs (3.72) and (3.73)], the enhancement function (17.13) takes the form

$$\mathcal{A}(T) = 2\Gamma(1+s)\zeta(1+s)\frac{M\gamma_s q_b^2}{2\pi\hbar}\omega_{\rm ph}^2\tau_{\rm W}^2\left(\frac{T}{T_{\rm ph}}\right)^{1+s},\qquad(17.15)$$

where $\Gamma(z)$ is Euler's gamma function and $\zeta(z)$ is Riemann's zeta function. Interestingly, the thermal enhancement is sensitive to the spectral form of the dissipative mechanism, and it qualitatively differs from the undamped case. For an undamped system, the enhancement is exponentially weak, $\mathcal{A}(T) \propto e^{-\hbar\omega_0/k_BT}$, as we can see from the expression (12.46). In contrast, for a damped system, we have algebraic enhancement, $\mathcal{A}(T) \propto T^{1+s}$. The exponent of the algebraic law is independent of the particular form of the metastable potential and is therefore a distinctive feature of the current damping [339, 358]. Hence measurement of the power of the algebraic enhancement of the tunneling probability at low T gives direct information about the spectral features of the environmental coupling. The properties of the barrier and additional dependence on friction enter only through the square of the factor $\tau_{\rm W}q_{\rm b}$.

For the important case of Ohmic damping, $J(\omega) = M\gamma\omega$, we have

$$\mathcal{A}(T) = \frac{\pi^2}{3}\frac{M\gamma q_b^2}{2\pi\hbar}\left(\frac{\tau_{\rm W}k_BT}{\hbar}\right)^2,\qquad(17.16)$$

and the next-to-leading order term varies with temperature as T^4. Consider next the bounce width at $T = 0$. In the limit $\gamma \to 0$, the width for a cubic potential is obtained from Eq. (17.14) with Eq. (12.43) as $\tau_{\rm W} = 6/\omega_0$ For strong damping, $\alpha \equiv \gamma/2\omega_0 \gg 1$, we find $\tau_{\rm W} = 4\pi\alpha/\omega_0$, as follows from the analytic form of the bounce given below in Eq. (17.24). From these limiting cases we infer that the bounce width grows with increasing damping. This is again a signature of the fact that the tunneling distance in the $\{q, \boldsymbol{x}\}$ configuration space gets larger with increasing damping, as we have already discussed in Chapter 13.

The universal enhancement (17.16) is due to thermally excited states of the environment, and *not* to thermally excited states in the well. A discussion of the formula (17.16) from the view of thermal quantum noise-theory is given in Ref. [359]. Finally, we remark that a similar *universal* low temperature behavior $\propto T^{s+1}$ occurs in the free energy of damped quantum systems [220] (cf. Sections 6.5 and 22.9), and, e.g., in transport properties [360] (see Subsection 25.5.1).

17.3 Quantum decay in a cubic potential for Ohmic friction

In this section, we consider the quantum-statistical decay of the metastable in the cubic potential (11.40) for the case of Ohmic friction. There are two relevant dimensionless damping parameters,

$$\alpha = \frac{\gamma}{2\omega_0} = \frac{\eta}{2M\omega_0},\qquad\text{and}\qquad K = \frac{\eta q_{\rm ex}^2}{2\pi\hbar} = \frac{27}{2\pi}\frac{\alpha V_{\rm b}}{\hbar\omega_0},\qquad(17.17)$$

The parameter K is independent of the mass, whereas $\alpha \propto 1/\sqrt{M}$. Therefore, α and K are independent parameters.

In the strong-damping regime $\alpha \gg 1$, the inertia term in the equation of motion (13.2) and in the bounce action is small compared to the friction term and may therefore be disregarded. The semiclassical limit for strong damping $\alpha \gg 1$ corresponds to the regime $K \gg 1$.

For general α, the quantum mechanical rate expression (17.5) with (17.6) must be computed numerically. In the regime $\alpha \gg 1$, $K \gg 1$ and $T < T_0$, the tunneling rate can be calculated in analytic form for the cubic metastable potential and for the tilted cosine potential. We now turn to the discussion of the first system. The second system will be discussed in the subsequent section.

17.3.1 Bounce action and quantum mechanical prefactor

For the cubic metastable potential (11.40), the bounce obeys the equation of motion

$$-\ddot{q}_{\rm B}(\tau) + \omega_0^2 q_{\rm B}(\tau) - \frac{\omega_0^2}{q_{\rm b}} q_{\rm B}^2(\tau) + \frac{1}{M} \int_0^{\hbar\beta} d\tau' \, k(\tau - \tau') q_{\rm B}(\tau') \;=\; 0\,, \qquad (17.18)$$

or equivalently in Fourier space upon writing $q_{\rm B}(\tau) = \sum_\ell Q_\ell \, e^{i\nu_\ell \tau}$ with $\nu_\ell = 2\pi\ell/\hbar\beta$,

$$\left(\nu_\ell^2 + \gamma|\nu_\ell| + \omega_0^2\right) Q_\ell \;=\; \frac{\omega_0^2}{q_{\rm b}} \sum_m Q_{\ell+m} Q_m\,. \qquad (17.19)$$

With use of the equation of motion (17.19) the bounce action may be written as

$$S_{\rm B} \;=\; \frac{M\omega_0^2}{6q_{\rm b}} \, \hbar\beta \sum_{\ell,m} Q_\ell Q_m Q_{\ell+m}\,, \qquad (17.20)$$

and the zero mode normalization factor (16.39) is $W_{\rm B} = M\,\hbar\beta \sum_m \nu_m^2 Q_m^2$. The equation (17.19) may be solved numerically by successive iteration, starting for instance with the zero order ansatz $Q_m \propto q_{\rm b}\, e^{-|m|}$.

Consider next the fluctuations $y(\tau)$ about the bounce. Writing $y(\tau) = \sum_\ell Y_\ell \, e^{i\nu_\ell \tau}$, the eigenvalue equation for the fluctuation operator $\Lambda[q_{\rm B}(\tau)]$ reads

$$[\,\nu_\ell^2 + \gamma|\nu_\ell| + \omega_0^2\,] Y_\ell^{(n)} - \frac{2\omega_0^2}{q_{\rm b}} \sum_m Q_{\ell+m} Y_m^{(n)} \;=\; \Lambda_n \, Y_\ell^{(n)}\,. \qquad (17.21)$$

The lowest odd mode of this equation is the translational mode which has one node and eigenvalue zero, $y_{\rm zero}(\tau) = \dot{q}_{\rm B}(\tau) = i \sum_\ell \nu_\ell Q_\ell \, e^{i\nu_\ell \tau}$.

Since the Fourier coefficient Q_m is small for large m, the eigenvalue Λ_n for $n \gg 1$ can be calculated by iteration with the starting value $\Lambda_n = \Lambda_n^{(b)}$. For small n on the other hand, the eigenvalue Λ_n can be calculated numerically by diagonalization of the truncated dynamical matrix of Eq. (17.21). In the numerical implementation, low eigenvalues require large rank N of the dynamical matrix with N up to $N = 250$.

A numerical calculation of tunneling rates at $T = 0$ was given first by Chang and Chakravarty [364]. Accurate numerical calculations at finite temperatures in the range $0.1\, T_0 \leq T \leq T_0$ and $0.1 \leq \alpha \leq 1$ are presented in Ref. [297]. The work gives tables both for the exponent and the prefactor of the quantum rate formula (17.5) with (17.6) in this range of parameters. The numerical data are in remarkable agreement with the measured lifetime of the metastable zero-voltage state in the current-biased Josephson junction [300, 365].

The characteristic features of the quantum decay rate as a function of friction strength and temperature are summarized below in Section 17.5.

17.3.2 Analytic results for strong Ohmic dissipation

For strong Ohmic damping, $\alpha \gg 1$, the inertia term $\nu_\ell^2 Q_\ell$ in Eq. (17.19) can be disregarded. In the absence of the inertia term, the equation of motion for the Fourier coefficients Q_ℓ is solved in analytic form with the ansatz

$$Q_\ell = a\, e^{-b|\ell|} . \qquad (17.22)$$

The coefficients are found to read

$$a = q_{\rm b} T/T_0 , \qquad \text{and} \qquad b = \text{artanh}(T/T_0) , \qquad (17.23)$$

where $T_0 = \hbar\omega_0/(4\pi\alpha k_{\rm B})$ is the crossover temperature in the strong damping limit, Eq. (14.39). With the expressions (17.22) and (17.23), the bounce trajectory in the temperature regime $0 \leq T \leq T_0$ takes the analytic form [342]

$$q_{\rm B}(\tau) = q_{\rm b}\frac{T}{T_0}\sum_\ell e^{-b|\ell|}\, e^{i\nu_\ell\tau} = q_{\rm b}\,\frac{(T/T_0)^2}{1 - \sqrt{1-(T/T_0)^2}\cos(\nu\tau)} , \qquad (17.24)$$

where $\nu = \nu_1 = 2\pi/\hbar\beta$. At $T = T_0$, the bounce path (17.24) coincides with the constant path $q = q_{\rm b}$. For $T > T_0$, the solution (17.24) is not real anymore. The action (17.20) pertaining to the trajectory (17.24) is found in the analytic form

$$S_{\rm B}(T) = 6\pi\alpha\frac{V_{\rm b}}{\omega_0}\left[1 - \frac{1}{3}\left(\frac{T}{T_0}\right)^2\right], \qquad T \leq T_0 . \qquad (17.25)$$

In the limit $T \to 0$, the bounce (17.24) reduces to the Lorentzian form [341]

$$q_{\rm B}^{(0)}(\tau) = \frac{2\alpha}{\omega_0}q_{\rm b}\int_{-\infty}^{+\infty} d\nu\, e^{-2\alpha|\nu|/\omega_0}\, e^{i\nu\tau} = \frac{2q_{\rm b}}{1 + (\omega_0\tau/2\alpha)^2} . \qquad (17.26)$$

For this path, the bounce point or turning point is $q_{\rm turn} = 2q_{\rm b}$, while $q_{\rm turn} = 3q_{\rm b}/2$ for $\alpha = 0$. The qualitative conclusion is that the turning point $q_{\rm turn}$ is a monotonous function of α varying in the range $3q_{\rm b}/2 \leq q_{\rm turn} \leq 2q_{\rm b}$. We remark that $q_{\rm turn} > 3q_{\rm b}/2$ is possible because the bounce does not conserve energy when α is nonzero.[3]

The action at zero temperature is [297]

---

[3]The periodic orbit in the full $\{q, x\}$ space has conserved total energy. The point $q_{\rm turn}$ is the projection of the exit point $P_{\rm ex}$ of the periodic orbit onto the q-axis (cf. Fig. 13.1 in Chapter 13).

$$S_{\mathrm{B}}(T=0) \; = \; 6\pi\alpha\frac{V_{\mathrm{b}}}{\omega_0}\left[1+\frac{1}{4\alpha^2}+\mathcal{O}(\alpha^{-4})\right] , \qquad (17.27)$$

where we have also given the leading correction $\propto 1/\alpha^2$. The width (17.14) of the bounce path (17.26) is $\tau_{\mathrm{W}} = 4\pi\alpha/\omega_0$. With this result, the thermal enhancement expression (17.16) is in agreement with the T^2 contribution in the action (17.25). Interestingly enough, we see from the expression (17.25) that the T^2 law strictly holds in the range $0 \le T \le T_0$, and there is no other temperature dependence in the exponential factor of the rate for $\alpha \gg 1$.

Consider next the fluctuations $y(\tau)$ about the bounce. For the low-energy modes we may disregard the inertia term $\nu_\ell^2 Y_\ell$ in Eq. (17.21). The truncated eigenvalue equation for the even modes with zero and with two nodes can be solved in analytic form with the ansatz

$$Y_\ell^{(0)} \; = \; a\, e^{-b|\ell|} , \qquad\qquad Y_\ell^{(1,+)} \; = \; a(1+c|\ell|)\, e^{-b|\ell|} , \qquad (17.28)$$

respectively. The resulting eigenvalues Λ_0 and $\Lambda_{1,+}$ are

$$\Lambda_{0/1,+} \; = \; -\omega_0^2\left(1\pm\sqrt{1+4[1-(T/T_0)^2]}\right)/2 . \qquad (17.29)$$

Owing to metastability, the lowest eigenvalue Λ_0 is negative. Further, the odd mode with one node is the time-translational mode with eigenvalue $\Lambda_{1,-} = 0$.

Now we turn to the higher eigenvalues Λ_n with $|n| \ge 2$. First, we observe that at the crossover temperature $T = T_0$, at which the periodic bounce is degenerated into a point, $q_{\mathrm{B}}(\tau) = q_{\mathrm{b}}$, the eigenvalues Λ_n ($|n| \ge 2$) are given by [cf. Eq. (15.8)]

$$\Lambda_n \; = \; \nu_n^2 + \gamma|\nu_n| + \omega_0^2 - 2\omega_{\mathrm{b}}^2 \; = \; \nu_n^2 + \gamma|\nu_n| + \omega_0^2 - 4\pi\gamma/\hbar\beta_0 , \qquad (17.30)$$

where $\beta_0 = 1/k_{\mathrm{B}}T_0$. It is tempting to replace for $T < T_0$, the inverse crossover temperature β_0 in Eq. (17.30) by the actual inverse temperature β. This yields

$$\Lambda_n \; = \; \nu_n^2 + \gamma|\nu_n| + \omega_0^2 - 2\gamma\nu , \qquad (|n| \ge 2) , \qquad (17.31)$$

where $\nu = \nu_1 = 2\pi/\hbar\beta$. The result (17.31) found by this intuitive argument is indeed the right expression in the regime $0 < T \le T_0$, as was shown in Ref. [341]. We remark that in the expression (17.31), the kinetic term ν_n^2 is needed for $|n| \gg 1$, while it is irrelevant for small n. Readily, the action factor related to the zero mode normalization is calculated with the expression (17.24) as

$$W_{\mathrm{B}} \; = \; M\int_0^{\hbar\beta} d\tau\, \dot{q}_{\mathrm{B}}^2(\tau) \; = \; 6\pi\frac{V_{\mathrm{b}}}{\gamma}\left(1-\frac{T^2}{T_0^2}\right) . \qquad (17.32)$$

To obtain the pre-exponential factor (17.6), we gather up the expression (15.11) for the determinant D_0 and the expressions (17.29), (17.31) and (17.32). We then find the pre-exponential factor of the rate expression in the form

$$f_{\mathrm{qm}} \; = \; A_1 A_2\, \omega_0 , \qquad (17.33)$$

where A_1 includes the zero-mode normalization factor and the eigenvalues of the two lowest even modes, and A_2 covers all higher eigenvalues,

$$A_1 = \sqrt{\frac{W_B}{2\pi\hbar}} \frac{\omega_0^2}{\sqrt{|\Lambda_0|\Lambda_{1,+}}}, \qquad A_2 = \frac{\prod_{n=1}^{\infty}[\nu_n^2 + \gamma\nu_n + \omega_0^2]}{\omega_0^2 \prod_{n=2}^{\infty}[\nu_n^2 + \gamma\nu_n + \omega_0^2 - 2\gamma\nu]}. \tag{17.34}$$

Inserting Eqs. (17.29) and (17.32) into the expression for A_1, the temperature dependence cancels out and we obtain

$$A_1 = \sqrt{3V_b/\hbar\gamma}. \tag{17.35}$$

The infinite products in the expression for A_2 can be expressed again in terms of gamma functions as in Eq. (15.17),

$$A_2 = \frac{\omega_0^2 + \nu^2 - \gamma\nu}{\omega_0^2} \frac{\Gamma(1 + \lambda_B^+/\nu)\,\Gamma(1 + \lambda_B^-/\nu)}{\Gamma(1 + \lambda_0^+/\nu)\,\Gamma(1 + \lambda_0^-/\nu)}, \tag{17.36}$$

where
$$\lambda_0^{\pm} = \gamma/2 \pm \sqrt{\gamma^2/4 - \omega_0^2}, $$
$$\lambda_B^{\pm} = \gamma/2 \pm \sqrt{\gamma^2/4 - \omega_0^2 + 2\gamma\nu}. \tag{17.37}$$

For strong damping, $\alpha \gg 1$, we obtain from Eq. (17.37) the relations

$$\lambda_B^+ = \lambda_0^+ + 2\nu, \quad \text{and} \quad \lambda_B^- = \lambda_0^- - 2\nu, \tag{17.38}$$

with which the partial prefactor A_2 is readily evaluated as

$$A_2 = (\gamma/\omega_0)^4 = 16\,\alpha^4. \tag{17.39}$$

Thus, the pre-exponential factor of the rate turns out to be temperature-independent. Collecting the factors A_1 and A_2, and the bounce action (17.25), we readily obtain the thermal rate (17.5) in the regime $0 \leq T \leq T_0$ in analytic form,

$$k = 8\sqrt{6}\,\alpha^3\omega_0 \sqrt{\alpha\frac{V_b}{\hbar\omega_0}}\,\exp\left[-6\pi\frac{\alpha V_b}{\hbar\omega_0}\left(1 - \frac{T^2}{3T_0^2}\right)\right]. \tag{17.40}$$

Subleading contributions to the rate expression are given in Ref. [297]. This concludes our discussion of the quantum-statistical escape of an overdamped particle from the metastable well of a cubic potential in the semiclassical limit.

17.4 Quantum decay in a tilted cosine potential

The phenomenon of incoherent tunneling of a macroscopic variable has become clearly visible in superconducting quantum interference devices, in which the phase difference ψ of the Cooper pair wave function across the Josephson junction plays the role of the tunneling coordinate (cf. Subsections 3.3.2 and 4.2.10). The measured lifetimes of the zero voltage state in current-biased Josephson systems were found to be in excellent agreement [361] with the theoretical predictions.

The standard phenomenological model for a Josephson junction is the resistively shunted junction (RSJ) model [109] (cf. Subsection 3.3.2). Within the RSJ model, the deterministic equation of motion for the phase difference ψ across the junction, which has capacitance C and effective shunt resistance R, reads[4]

$$C\left(\frac{\phi_0}{2\pi}\right)^2 \frac{d^2\psi}{dt^2} + \frac{1}{R}\left(\frac{\phi_0}{2\pi}\right)^2 \frac{d\psi}{dt} + E_J \sin\psi - I_{\text{ext}}\frac{\phi_0}{2\pi} = 0, \qquad (17.41)$$

where $\phi_0 = h/2e$ is the flux quantum, I_{ext} is the externally applied bias current, and $E_J = I_c\phi_0/2\pi$ is the Josephson coupling energy. The current I_c is the maximum super-current which the junction can sustain. Equation (17.41) describes classical damped motion of the phase in the potential (4.196).

The dynamical equation (17.41) is formally equivalent to the equation of motion for a Brownian particle of mass M and position X in the absence of fluctuations,

$$M\ddot{X}(t) + M\gamma\dot{X}(t) + \frac{\partial V(X)}{\partial X} = 0. \qquad (17.42)$$

The potential $V(X)$ is of trigonometric form with a sloping background resulting in a potential drop $2\pi V_{\text{tilt}}$ per length X_0 as sketched in Fig. 17.1,

$$V(X) = -V_0 \cos\left(\frac{2\pi X}{X_0}\right) - V_{\text{tilt}}\frac{2\pi X}{X_0}. \qquad (17.43)$$

In the mapping of the two models, the equivalence relations are

$$\frac{X}{X_0} \;\hat{=}\; \frac{\psi}{2\pi}, \qquad MX_0^2 \;\hat{=}\; C\phi_0^2, \qquad \gamma \;\hat{=}\; \frac{1}{RC},$$
$$V_0 \;\hat{=}\; E_J, \qquad V_{\text{tilt}} \;\hat{=}\; I_{\text{ext}}\frac{\phi_0}{2\pi}. \qquad (17.44)$$

The model (17.42) with (17.43) is archetypal for many transport systems in condensed matter physics.

The two macroscopically distinguishable states in the current-biased Josephson junction are the zero voltage state and the voltage state. According to the relation (3.144), the first state corresponds to the particle being trapped in a well, and the second to the particle sliding down the cascade of wells. The particle in the well oscillates with the "plasma" frequency

$$\omega_0 = \frac{2\pi}{X_0}\frac{(V_0^2 - V_{\text{tilt}}^2)^{1/4}}{\sqrt{M}} \qquad \text{for} \qquad V_0 > V_{\text{tilt}}. \qquad (17.45)$$

For the potential $V(X)$ given in Eq. (17.43), the barrier frequency ω_b coincides again with the well frequency ω_0. The minima and maxima of the tilted washboard potential are located in the principal interval $0 \leq X < X_0$ at

---

[4]The RSJ model is Markovian and corresponds to the substitution $Y^*(\omega) = 1/R$ in Eq. (3.210).

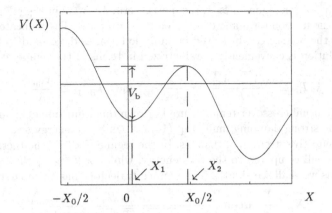

Figure 17.1: Sketch of the tilted washboard potential.

$$X_1 = (X_0/2\pi)\arcsin(V_{\text{tilt}}/V_0) , \quad \text{and} \quad X_2 = X_0/2 - X_1 , \tag{17.46}$$

and the barrier height is given by

$$V_{\text{b}} = V(X_2) - V(X_1) = 2\sqrt{V_0^2 - V_{\text{tilt}}^2} - 2V_{\text{tilt}}\arccos(V_{\text{tilt}}/V_0) . \tag{17.47}$$

For $V_{\text{tilt}} = V_0$, the cascade of barriers ceases to exist. In the regime $V_{\text{tilt}} > V_0$, the slope of the potential is negative everywhere, so that the particle slides all the way down and cannot be trapped even in the absence of inertia.

In the regime $V_{\text{tilt}} < V_0$, the usual dimensionless damping parameter is

$$\alpha \equiv \frac{\gamma}{2\omega_0} = \frac{\gamma\sqrt{M}X_0}{4\pi(V_0^2 - V_{\text{tilt}}^2)^{1/4}} . \tag{17.48}$$

For the potential (17.43), it is convenient to define the second dimensionless friction parameter K introduced in Eq. (17.17) as

$$K \equiv \frac{M\gamma X_0^2}{2\pi\hbar} = 4\pi\alpha\frac{\sqrt{V_0^2 - V_{\text{tilt}}^2}}{\hbar\omega_0} . \tag{17.49}$$

We note that the definition of K is in coincidence with the Kondo parameter K introduced in Eq. (4.155).

Consider now the crossing rate from the higher to the lower well which we denote by k^+ henceforth. We are interested in the regime $V_{\text{tilt}} < V_0$, $T < T_0$ in the semiclassical limit $K \gg 1$. In the tunneling regime, the crossing rate is dominated by the one-bounce contribution. The bounce trajectory in imaginary time obeys the equation of motion

$$-M\ddot{X}_{\text{B}}(\tau) + V_0\frac{2\pi}{X_0}\sin\left(\frac{2\pi}{X_0}X_{\text{B}}(\tau)\right) - \frac{2\pi V_{\text{tilt}}}{X_0} + \int_0^{\hbar\beta} d\tau' \, k(\tau-\tau')X_{\text{B}}(\tau') = 0 . \tag{17.50}$$

In the high-friction regime $\alpha \gg 1$, the inertia term $M\ddot{X}_{\mathrm{B}}(\tau)$ in Eq. (17.50) is small compared to the friction term and can therefore be dropped. The truncated equation of motion for the bounce can be solved in analytic form, as discovered by Korshunov [362]. The solution is conveniently parametrized in terms of the temperature scales

$$k_{\mathrm{B}}T_0 \equiv \frac{\sqrt{V_0^2 - V_{\mathrm{tilt}}^2}}{K} = \frac{\hbar\omega_{\mathrm{b}}}{4\pi\alpha}, \quad \text{and} \quad k_{\mathrm{B}}T_1 \equiv \frac{V_{\mathrm{tilt}}}{K}. \tag{17.51}$$

Here, T_0 is the usual crossover temperature between quantum tunneling and thermal hopping in the strong damping limit, Eq. (14.39), while the energy scale $k_{\mathrm{B}}T_1$ characterizes the effective strength of the bias. In the regime $T \ll T_1$, the thermal energy is negligibly small compared to the bias energy, while for $T \gg T_1$ the bias may be disregarded, as we shall see shortly. The bounce trajectory has the analytic form

$$X_{\mathrm{B}}(\tau) = X_1 + \frac{X_0}{\pi} \arctan\left(\frac{T^2/T_0 T_1}{1 - \sqrt{1 - (T/T_0)^2} \cos(2\pi\tau/\hbar\beta)} \right). \tag{17.52}$$

In the zero temperature limit, we obtain from Eq. (17.52) the bounce as

$$X_{\mathrm{B}}(\tau) = X_1 + \frac{X_0}{\pi} \arctan\left(\frac{2T_0/T_1}{1 + (\omega_0\tau/2\alpha)^2} \right). \tag{17.53}$$

In the opposite limit $T \to T_0$, the bounce (17.52) is constricted to the constant path $X_{\mathrm{B}}(\tau) = X_2$. The action of the bounce (17.52) is found in the range $0 \leq T \leq T_0$ as

$$S_{\mathrm{B}}/\hbar = K \ln\left(\frac{V_0^2}{V_{\mathrm{tilt}}^2 [1 + (T/T_1)^2]} \right) + 2K\left(1 - \frac{T_1}{T} \arctan\frac{T}{T_1}\right). \tag{17.54}$$

Consider next the determinant of the fluctuation operator. The eigenvalues of the symmetric modes with zero and with two nodes Λ_0 and $\Lambda_{1,+}$ are found as

$$\Lambda_{0/1,+} = \omega_0^2 \left\{ \frac{1}{2} - \frac{V_{\mathrm{tilt}}^2}{V_0^2}\left(1 + \frac{T^2}{T_1^2}\right) \mp \sqrt{\frac{1}{4} + \frac{V_{\mathrm{tilt}}^4}{V_0^4}\left(1 + \frac{T^2}{T_1^2}\right)\left(1 - \frac{T^2}{T_0^2}\right)} \right\}. \tag{17.55}$$

The lowest eigenvalue is again negative due to metastability. We now have

$$|\Lambda_0|\Lambda_{1,+} = \omega_0^4 \frac{V_{\mathrm{tilt}}^2}{V_0^2}\left(1 + \frac{T^2}{T_1^2}\right)\left(1 - \frac{T^2}{T_0^2}\right). \tag{17.56}$$

The eigenvalues above $\Lambda_{1,+}$ are in the form (17.31) with ω_0 given in Eq. (17.45). The action which determines the normalization of the zero mode is

$$W_{\mathrm{B}} \equiv M\int_0^{\hbar\beta} d\tau\, \dot{X}_{\mathrm{B}}^2(\tau) = 4\pi \frac{(V_0^2 - V_{\mathrm{tilt}}^2)^{3/2}}{\gamma V_0^2}\left(1 - \frac{T^2}{T_0^2}\right). \tag{17.57}$$

Writing the pre-exponential factor of the rate as in Eq. (17.33) with (17.34), we find

$$A_1 = \frac{1}{\sqrt{\alpha\hbar\omega_0}} \frac{M^{3/2}\omega_0^3}{V_{\mathrm{tilt}}}\left(\frac{X_0}{2\pi}\right)^3 \frac{1}{\sqrt{1 + (T/T_1)^2}}, \tag{17.58}$$

$$A_2 = \gamma^4/\omega_0^4 = 16\alpha^4.$$

After all, the quantum mechanical attempt frequency is found to read

$$f_{\text{qm}} = A_1 A_2 \omega_0 = \frac{16\alpha^{7/2} M^{3/2} \omega_0^4}{\sqrt{\hbar \omega_0} V_{\text{tilt}}} \left(\frac{X_0}{2\pi}\right)^3 \frac{1}{\sqrt{1 + (T/T_1)^2}} . \tag{17.59}$$

The expressions (17.54) and (17.59) are the required components which determine the semiclassical tunneling rate to the next lower well [cf. Ref. [362]],

$$k^+ = f_{\text{qm}} e^{-S_{\text{B}}/\hbar} . \tag{17.60}$$

In the limit $X_0, V_0, V_{\text{tilt}} \to \infty$ with the constraints

$$\lim_{X_0, V_0, V_{\text{tilt}} \to \infty} \left(\frac{2\pi}{X_0}\right)^3 (V_0 + V_{\text{tilt}}) = 24 \frac{V_{\text{b}}}{X_{\text{b}}^3} ,$$

$$\lim_{X_0, V_0, V_{\text{tilt}} \to \infty} \frac{2\pi}{X_0} (V_0 - V_{\text{tilt}}) = \frac{3}{2} \frac{V_{\text{b}}}{X_{\text{b}}} , \tag{17.61}$$

the tilted washboard potential mutates into the cubic potential given by Eq. (11.40) with Eq. (11.41) (up to a constant shift). It is straightforward to see that in this limit the results (17.52) – (17.59) coincide with the corresponding results for the cubic metastable potential given in Subsection 17.3.2. Accordingly, the rate expression (17.60) with (17.54) and (17.59) reduces in this limit to the expression (17.40).

In the remaining part of this section, we consider the opposite case of weak tilt in which the bounce consists of a weakly bound and widely spaced instanton–anti-instanton pair. The distance τ_s between the instanton and the anti-instanton is large compared to the width α/ω_{b} of the instanton. Hence the action of the bounce changes only weakly when the distance τ is moved away from the stationary point $\tau = \tau_s$. To study this case, we write the bounce action (17.54) as the action for two instantons at a distance τ,

$$S(\tau)/\hbar = 2 S_{\text{inst}}/\hbar + W(\tau) - \epsilon\tau . \tag{17.62}$$

The first term in Eq. (17.62) is twice the action of an instanton. The second term $W(\tau)$ represents the interaction between the instantons at distance τ. The last term is the bias action for potential drop $\hbar\epsilon$ per distance X_0. The bias frequency is

$$\epsilon = 2\pi V_{\text{tilt}}/\hbar . \tag{17.63}$$

The instanton interaction has the spectral representation (4.72). For Ohmic spectral density $G(\omega) = 2K\omega$ with cutoff at $\omega = \omega_{\text{c}}$ it takes the form [see also Eq. (18.56)]

$$W(\tau) = 2K \ln\left[(\hbar\beta\omega_{\text{c}}/\pi) \sin(\pi\tau/\hbar\beta) \right] . \tag{17.64}$$

The action (17.62) is extremal for $\tau = \tau_s$, where

$$\tau_s = (\hbar\beta/\pi) \operatorname{arccot} (\hbar\beta\epsilon/2\pi K) . \tag{17.65}$$

We now require that the action $S(\tau)$ at $\tau = \tau_s$ coincides with the bounce action S_{B} given in Eq. (17.54). With this postulation we obtain the instanton action as

$$S_{\text{inst}}/\hbar = K \left[1 + \ln(\pi V_0/K\hbar\omega_{\text{c}}) \right] . \tag{17.66}$$

As an interim result, the tunneling rate (17.60) with (17.59) may be written as

$$k^+ = \frac{K\gamma^2}{\pi} \left(\frac{\pi K}{\epsilon^2 + (2\pi K/\hbar\beta)^2} \right)^{1/2} e^{-S(\tau_s)/\hbar} . \tag{17.67}$$

17.4.1   The case of weak bias

When the instanton–anti-instanton pair is only weakly bound, the negative eigenvalue Λ_0 of the fluctuation operator is close to zero. We find from Eq. (17.55)

$$\Lambda_0 \approx -\omega_0^2 (V_{\text{tilt}}^2/V_0^2) \left(1 + T^2/T_1^2\right) . \tag{17.68}$$

Therefore, for weak bias the Gaussian approximation for the fluctuation mode with zero nodes, which is the breathing mode of the bounce, breaks down. This indicates that the formula (17.67) is not appropriate for weak bias, $V_{\text{tilt}} \ll V_0$, and low temperature, $T, T_1 \ll T_0$. To deal with this regime, we take one step back by withdrawing the execution of the breathing mode integral by steepest descent. At this earlier stage, the rate expression (17.67) takes the form

$$k^+ = \frac{K\gamma^2}{\pi} \operatorname{Im} \int_C d\tau \ e^{-S(\tau)/\hbar} , \tag{17.69}$$

where $S(\tau)$ is the bounce action (17.62) as function of the distance τ of the instanton pair, and C is a contour which may be deformed to pass through the stationary point of the action. The location of this point is given by Eq. (17.65). Since we have $\partial^2 W/\partial \tau^2|_{\tau=\tau_s} < 0$, the point τ_s is a saddle point, and the direction of steepest descent is perpendicular to the real axis of the complex τ-plane. Deforming the integration path at the saddle point along the direction of steepest descent, the expression (17.69) takes the form

$$k^+ = \frac{\Delta^2}{2} \operatorname{Im} \int_{\tau_s}^{\tau_s+i\infty} d\tau \ e^{\epsilon\tau - W(\tau)} . \tag{17.70}$$

Here, Δ is the usual tight-binding tunnel matrix element (cf. Sections 18.1 and 20.2) which is determined by the instanton trajectory,

$$\Delta/2 = f_{\text{inst}} \ e^{-S_{\text{inst}}/\hbar} . \tag{17.71}$$

Comparing Eq. (17.69) with Eqs. (17.70), (17.71), and substituing the instanton action (17.66) we obtain $f_{\text{inst}} = \sqrt{K/2\pi}\,\gamma$. The tunneling matrix element is found as

$$\Delta/2 = \sqrt{K/2\pi}\,\gamma e^{-K}\left(K\hbar\omega_c/\pi V_0\right)^K . \tag{17.72}$$

The relation (17.72) expresses the transfer matrix element Δ in terms of the parameters of the extended system. We shall derive this important result quite differently in Section 26.4 by use of a self-duality symmetry of the model.

If we had evaluated Eq. (17.70) by steepest descent, we would recover the previous result (17.67). Fortunately, it is possible to perform the integration exactly. Observing that the analytically continued function $Q(z) = W(\tau = iz)$ is an analytic function in the strip $0 > \operatorname{Im} z > -\hbar\beta$ of the complex z-plane, we may write

$$k^+ = \frac{\Delta^2}{4} \int_{-\infty-ic}^{\infty-ic} dz \ e^{i\epsilon z - Q(z)} , \tag{17.73}$$

where the constant c is chosen in the interval $0 < c < \hbar\beta$ such that the integration contour lies to the right of the singularities of the integrand. For $K < \frac{1}{2}$, we may put $c = 0^+$. We then have

$$k^+ = \frac{\Delta^2}{2} \left(\frac{\pi}{\hbar\beta\omega_c} \right)^{2K} \int_0^\infty dt \, \frac{\cos(\epsilon t - \pi K)}{\sinh^{2K}(\pi t/\hbar\beta)} . \tag{17.74}$$

The integration can be carried out exactly [89]. We then obtain the one-bounce contribution to the forward rate in the form

$$k^+(T, \epsilon) = \gamma \frac{K \, e^{-2K}}{4\pi^2} \beta\hbar\gamma \left(\frac{2K}{\beta V_0} \right)^{2K} \frac{|\Gamma(K + i\hbar\beta\epsilon/2\pi)|^2}{\Gamma(2K)} e^{\hbar\beta\epsilon/2} . \tag{17.75}$$

The expression (17.73) is a contour integral which encircles the singularity at $z = 0$. Thus, the final expression (17.75) is not limited to the regime $K < \frac{1}{2}$ but also holds for $K \geq \frac{1}{2}$.[5] Further discussion of this point is given below Eq. (20.79).

In the absence of the bias, $\epsilon = 0$, the crossing rate takes the form

$$k^+(T, 0) = \gamma \frac{K\Gamma(1 + K) \, e^{-2K}}{2\Gamma(\frac{1}{2})\Gamma(\frac{1}{2} + K)} \frac{\hbar\gamma}{\pi V_0} \left(\frac{K k_{\mathrm{B}} T}{V_0} \right)^{2K-1} . \tag{17.76}$$

In the opposite limit $T = 0$ and nonzero bias, we find

$$k^+(0, \epsilon) = \gamma \frac{K^2 \, e^{-2K}}{\Gamma(2K)} \frac{\hbar\gamma}{\pi V_0} \left(\frac{K\hbar\epsilon}{\pi V_0} \right)^{2K-1} . \tag{17.77}$$

The temperature $T_1 = \hbar|\epsilon|/2\pi K k_{\mathrm{B}}$, is a kind of a crossover temperature. In the low temperature regime $T \ll T_1$, the thermal energy is negibly small compared with the bias energy, so that the expression (17.77) is appropriate. For $T \gg T_1$, the bias energy may be disregarded, and the form (17.76) is adequate.

Up to now, we have considered the crossing rate from the higher to the lower well, $k = k^+$. The backward rate from the lower to the higher well obeys detailed balance,

$$k^-(T, \epsilon) = e^{-\hbar\epsilon/k_{\mathrm{B}}T} k^+(T, \epsilon) . \tag{17.78}$$

Within steepest descent for the breathing mode, the backward rate k^- is determined by a different branch of the arctan function in the action (17.54). A general discussion of the detailed balance property is given below in Subsection 20.2.1.

We remark that the leading thermal enhancement calculated from the expression (17.75) is in agreement with the formula (17.16).

These results can be applied directly to macroscopic quantum tunneling in a voltage-biased Josephson junction. The corresponding model is introduced in Subsection 3.5.3 [see also Eq. (17.41)]. We have the parameter identifications $K = 1/\rho$, $\hbar\epsilon = 2eV_{\mathrm{x}}/\rho$, and $\gamma = E_{\mathrm{c}}/\pi\hbar\rho$, where $\rho = R/R_{\mathrm{Q}}$ [$R_{\mathrm{Q}} = 2\pi\hbar/4e^2$ and $E_{\mathrm{c}} = 2e^2/C$

---

[5]The contour integral (17.73) with the singular integrand $e^{-W(iz)}$, where $W(\tau)$ is given in Eq. (17.64), is analogous to Hankel's contour integral of the reciprocal gamma function [363].

are the resistance quantum and charging energy for Cooper pairs]. The overdamped limit corresponds to $1/4\alpha^2 = 2\pi^2\rho^2 E_J/E_c \ll 1$, as follows with the correspondence relations (17.44). The dc current through the Ohmic impedance R in the circuit sketched in Fig. 3.3 (see Subsection 3.5.3) is $I = (V_x - V_a)/R$, where $V_a = \hbar\langle\dot\psi\rangle/2e$. For $\rho \ll 1$ and small V_x, the phase slip is determined by the single-bounce contribution (17.60), $\langle\dot\psi\rangle = 2\pi k^+$. At $T = 0$ and in the overdamped limit, the rate k^+ is given in Eq. (17.77). Thus we find

$$I(V_x) = \frac{V_x}{R}\left(1 - \frac{\pi}{2}(\pi\rho)^{-2/\rho-5/2}\left(\frac{eV_x}{E_J}\right)^{2/\rho}\left(\frac{E_c}{eV_x}\right)^2 + \mathcal{O}\left[\left(\frac{E_c}{eV_x}\right)^4\right]\right). \qquad (17.79)$$

When the junction is in the zero voltage state, the voltage drop occurs at the resistor, yielding Ohm's law. Macroscopic quantum tunneling into a finite voltage state contributes the *nonlinear* term in the $I(V_x)$ characteristics. Further discussion of the voltage-biased Josephson junction is given in Subsections 20.3.4, 26.3, and 26.5.1.

17.5   Concluding remarks

The imaginary-time functional integral approach has provided an almost complete description of the quantum statistical decay of a metastable state extending from thermally activated decay at high temperatures down to very low temperatures where the system tunnels out of the ground state. The main features of the decay are summarized in the Arrhenius plot shown in Fig. 17.2. In this diagram, the classical rate is represented by a falling straight line. The rate flattens out towards a finite value at $T = 0$ due to quantum tunneling. For an undamped system $(\alpha = 0)$, the transition between the classical and quantum regime is rather sharp. In the presence of damping, the classical rate is reduced only slightly because the attempt frequency gets smaller by the factor ω_R/ω_b. On the other hand, damping causes an exponentially strong suppression of the zero-temperature tunneling rate. Furthermore, the crossover temperature is lowered and the transition between thermally activated decay and tunneling becomes more gradual when damping is increased. For strong Ohmic dissipation, there is a large region in which thermal and quantum fluctuations interplay and thermal enhancement is governed by the power law given in Eq. (17.16).

    The qualitative features of the tunneling rate can be seen rather directly in the "many-dimensional WKB" approach. The minimum of the global potential $V(q, \boldsymbol{x})$ of the system-environment complex on the hyperplane of constant q is located away from the q-axis at $x_\alpha = (c_\alpha/m_\alpha\omega_\alpha^2)\,q$, and the value at the minimum is $V(q)$ and hence independent of the bath coupling, as we see from Eq. (3.13). The reduction of the tunneling rate by the environmental coupling can therefore be viewed as being due to the fact that the path under the barrier of the global potential $V(q, \boldsymbol{x})$, i.e. the path between the turning points in Eq. (17.7), is longer, both for zero and finite temperature, compared to the under-barrier-distance or tunneling length of the undamped system in the potential $V(q)$.

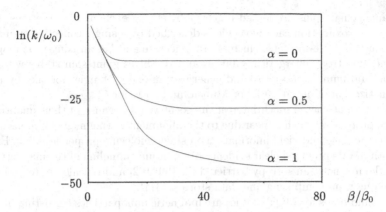

Figure 17.2: Arrhenius plot of the decay rate for a system with a cubic plus quadratic potential ($V_b = 5\hbar\omega_0$) and Ohmic damping ($\alpha = \gamma/2\omega_0$) for various values of α.

The phenomenon of macroscopic quantum tunneling has been observed experimentally in a large number of physical and chemical systems [298]. An especially attractive physical system is the current-biased Josephson junction or rf superconducting quantum interference device (SQUID) system. We have shown in Part I that for the case of a general linear impedance mechanism knowledge of the classical dissipative equation of motion of the flux uniquely determines its quantum-mechanical behavior. Since in Josephson devices all relevant parameters can be measured independently, these systems provided accurate tests of the MQT theory without adjustable parameters.

On a $\log k$ versus T^2 plot, the experimentally observed slope can be compared with the predicted behavior (17.16). Indeed, experiments have confirmed most of the theoretical predictions very accurately [361, 300]. Surveys of experimental results in MQT are given in Refs. [365, 366].

Evidence both for non-Ohmic and Ohmic dissipation has been discovered experimentally in diverse systems. Thermal enhancement with $s = 5$ in Eq. (17.15) was found in experiments on proton tunneling in hydrated protein powders [367]. The data analysis of tunneling rates for Li^+ impurities in diluted perovskite $K_{1-x}Li_xTaO_3$ was found to be consistent with $s = 3$ [368].

Dielectric relaxation of orientational defects in polycrystalline H_2O and D_2O ices in the crossover regime for different samples was shown to be in perfect agreement with the universal scaling law (16.45) with (16.46), and the thermal enhancement was found in the Ohmic form, Eq. (17.12) with (17.16) [369].

Single-molecule magnets (SMMs) opened about 20 years ago the new research field of nanomagnetism. SMMs are a novel class of materials in which nearly identical magnetic molecules form a regular assembly in large crystals. Every molecular cluster has a central complex consisting of magnetic metal ions. At low temperature the intramolecular spins are locked together by strong exchange interactions. In the giant spin approximation each molecule is described by a single collective spin S. For instance, in the single-molecule clusters Mn_{12} acetate and Fe_8 an Ising-type magneto-crystalline anisotropy energetically favors and stabilizes giant spin states with magnetic quantum number $m = \pm 10$ and generates an energy barrier for the reversal of the magnetization of about 70 K for $Mn_{12}ac$ and 25 K for Fe_8.

SMMs have attracted much interest in recent years because of their quantum interference properties which appear due to the alignment of the magnetic molecules on a macroscopic scale and yield information on single-molecule properties [370]. Experiments with SMMs provided clear evidence of quantum tunneling of the magnetization through the magnetic anisotropy barrier [157, 159, 162] and revealed potential use of SMMs in quantum computing and data storage [371].

Other candidates for MQT physics are magnetic nanoparticles consisting of 10^3 to 10^6 magnetic moments. Magnetic systems have the advantage that several parameters which can tune the tunneling probability can be varied in a controlled manner.

Towards a quantitative picture of tunneling rates in MQT and of decoherence rates in MQC it is an indispensable prerequisite to understand the interaction of the giant spin with the environment. In particular, interactions with nuclear spins and with the lattice vibrations of the crystal are sources of energy dissipation and decoherence and therefore are of particular interest. Microwave radiation induces transitions between different spin states and can therefore provide information on the spin dynamics of the molecules.

Given an effective Hamiltonian for the giant-spin-plus-reservoir complex, one may then derive, with the methods presented here, MQT rates for the diverse parameter regimes. Altogether, macroscopic quantum tunneling of magnetic particles is an effective field to check experimentally the theoretical predictions assembled in this part.

PART IV
THE DISSIPATIVE TWO-STATE SYSTEM

One of the most interesting aspects of quantum theory is the phenomenon of constructive and destructive interference. The phase coherence between different quantum mechanical states may lead to clockwise motion. The simplest model involving quantum coherence is a two-state system. The problem of a quantum system whose state is effectively confined to a two-dimensional Hilbert space is often encountered in physics and chemistry. For instance, imagine a quantum mechanical particle tunneling clockwise forth and back between two different localized states. In reality, such a system is strongly affected by the surroundings. We will see that the coupling can lead to qualitative changes in the behaviors: the environment-induced fluctuations can destroy quantum coherence and can even lead to a "phase transition" to a state in which quantum tunneling is quenched.

18. Introduction

Anderson et al. [119] and, independently, Phillips [120] postulated in 1972 the existence of two-level systems in glasses to explain low temperature anomalies of the specific heat in these amorphous materials. Although the true microscopic nature of the tunneling entities in glasses is unclear, they can be visualized by a particle tunneling in a double well along an (unknown) reaction coordinate. It turned out in ultrasonic experiments as an important difference between dielectric and metallic glasses that the lifetime of the tunneling eigenstates is drastically shorter in the case of metals. Golding et al. [372] explained this unexpected behavior by a nonadiabatic coupling of the tunneling entity to the conduction electrons in metals. Shortly later, Black and Fulde [373] extended this idea to describe relaxation processes in a superconducting environment by following the lines sketched above in Subsections 3.4.3 and 4.2.8. The theoretical predictions were confirmed by G. Weiss et al. [374] in ultrasonic experiments on superconducting amorphous metals. They showed that the lifetime is reduced when the environment is switched from superconducting to normalconducting, thus demonstrating the significance of the electronic coupling. A survey of the early developments and the perturbative treatment of this coupling is given in Ref. [121]. For a nonperturbative treatment see Ref. [375]. The nonlinear acoustic response of amorphous metals has been studied by Stockburger et al. [376], and the results for the dynamical susceptibility were found in good agreement with experiments [377]. A comprehensive review of the physics of tunneling systems in amorphous and crystalline solids with emphasis on the thermodynamic, acoustic, dielectric and optical properties has been given in Ref. [378].

The tunneling of light particles like small polarons, hydrogen isotopes or muons in solids has been thoroughly studied since several decades. While earlier work was mainly concerned with the significance of polaron effects [128] – [134], more recent attention has focused on the singular transient response of the fermionic environment in metals at low temperatures. The nonadiabatic influence of conduction electrons on the motion of interstitials in metals was proposed by Kondo [379] to explain the anomalous temperature dependence of muon diffusion in host metals at low temperature. This mechanism has been confirmed experimentally for incoherent tunneling of μ^+ and p in Cu, Al, and Sc [100], for defects in mesoscopic wires [104, 103], and for tunneling of H and D in metals and superconductors. Comprehensive reviews of experiment [380] and theory [381] are available. The particle tunnels either randomly or coherently between particular interstitial sites. Experimentally, one passes from the incoherent to the coherent dynamics by lowering temperature. The transition becomes apparent, e.g., in a change from quasi-elastic to inelastic neutron scattering.

Electron transfer reactions are ubiquitous in chemical and biological systems. In its simplest form, an electron localized at a donor site is tunneling to the acceptor site. Marcus' theory [105, 106] provides the appropriate scenario to describe such processes. Often the interaction between the charge and the polarization cloud of the environment is so strong that tunneling is possible only with the assistance of favorable equilibrium fluctuations in the environmental modes. Because of the small mass, electron transfer is strongly influenced by quantum effects even at room temperature.

Examples of quantum coherence in an effective two-state system are the inversion resonance of the NH_3 molecule, strangeness oscillations of a neutral K-meson beam, and coherent tunneling of light interstitials in tunneling centers. Another important, but apparently quite different example of a two-state system is a rf SQUID ring threaded by an external flux near half a flux quantum (cf. Subsection 3.3.2). Such a system might be the appropriate vehicle for the observation of "macroscopic quantum coherence" (MQC) on a true macroscopic scale [296, 382].

The behavior of the two-state system (TSS) is strongly influenced by the dissipative coupling to the heat bath's dynamical degrees of freedom. We shall consider linear couplings to the heat bath that are sensitive to the value of σ_z. For instance, a dipole-local-field coupling provides a simple physical model for this type of coupling. To be definite, we choose $H_I = -q \sum_\alpha c_\alpha x_\alpha$, where $q = \sigma_z q_0/2$ with q_0 being the spatial distance of the two localized states. It should be adequate in many cases of interest that the response of the environment to a perturbation can be considered as linear. Then, a bath which is represented by a set of harmonic oscillators with a coupling linear in the coordinates captures the essential physics we wish to describe. Since the TSS is like a spin, the corresponding model has become known in the literature as the "spin-boson" model. We have introduced this model already in

Section 3.3. The relevant Hamiltonian is given in Eq. (3.137) or Eq. (3.139).

Despite its apparent simplicity, the spin-boson model cannot be solved exactly by any known method (apart from some limited regimes of the parameter space). Not only is the spin-boson model nontrivial mathematically, it is also nontrivial physically.

The environment acts on the TSS by a fluctuating force $\xi(t) = \sum_\alpha c_\alpha x_\alpha(t)$. For a bath with linear response, the modes $x_\alpha(t)$ obey Gaussian statistics. Therefore the dynamics of the bath is fully characterized by the force autocorrelation function in thermal equilibrium $\langle \xi(t)\xi(0)\rangle_\beta$, which is simply a superposition of harmonic oscillator correlation functions [cf. Eq. (5.33)]. In the formal path integral expression for the reduced density matrix, the environment reveals itself through an influence functional $\mathcal{F}$. In Chapters 4 and 5 we have given several useful forms for $\mathcal{F}$ applicable to thermodynamics and dynamics, respectively.

18.1 Truncation of the double-well to the two-state system

18.1.1 Shifted oscillators and orthogonality catastrophe

In Section 3.3 we have already briefly addressed the reduction of a double well system to a two-state system. Before resuming the discussion of the reduction for the dissipative case let us first consider the adiabatic limit, in which the environmental modes instantaneously adapt themselves to the particle's position. The oscillator part of the Hamiltonian (3.137) for the two positions $\sigma_z = \pm 1$ of the particle reads

$$H_\pm = \frac{1}{2} \sum_\alpha \left(\frac{p_\alpha^2}{m_\alpha} + m_\alpha \omega_\alpha^2 x_\alpha^2 \mp q_0 c_\alpha x_\alpha \right) = \sum_\alpha \left[\hbar\omega_\alpha b_\alpha^\dagger b_\alpha \mp \tfrac{1}{2}\hbar\lambda_\alpha \left(b_\alpha + b_\alpha^\dagger \right) \right] \ , \quad (18.1)$$

where in the second line we have introduced the language of creation and annihilation operators, and where $c_\alpha = \sqrt{2\hbar m_\alpha \omega_\alpha}\, \lambda_\alpha/q_0$. Upon introducing the shifted operator

$$b_{\pm,\alpha} = b_\alpha \mp \tfrac{1}{2}\lambda_\alpha/\omega_\alpha \,, \quad (18.2)$$

the terms linear in b_α, $b_\alpha^\dagger$ cancel out, and we get the Hamiltonian $H_\pm$ in normal form

$$H_\pm = \sum_\alpha H_{\pm,\alpha} \quad \text{with} \quad H_{\pm,\alpha} = \hbar\left(\omega_\alpha b_{\pm,\alpha}^\dagger b_{\pm,\alpha} - \frac{\lambda_\alpha^2}{4\,\omega_\alpha} \right) . \quad (18.3)$$

Since the shifted operator obeys the same commutation relations as the original one, the two different vacuum states for the boson α are defined by $b_{\pm,\alpha}|0_{\pm,\alpha}\rangle = 0$. The normalized many-boson states $\{\, |n_{\pm,\alpha}\rangle \,\}$ can be created from the vacuum state $|0_{\pm,\alpha}\rangle$ in the usual way, $|n_{\pm,\alpha}\rangle = (b_{\pm,\alpha}^\dagger)^{n_{\pm,\alpha}}|0_{\pm,\alpha}\rangle/\sqrt{n_{\pm,\alpha}!}$. Application of the one annihilation operator on the vacuum state of the other yields

$$b_{\mp,\alpha}|0_{\pm,\alpha}\rangle = \pm \frac{\lambda_\alpha}{\omega_\alpha}\,|0_{\pm,\alpha}\rangle \,. \quad (18.4)$$

The Hamiltonians $H_{\pm,\alpha}$ describe two harmonic oscillators with eigenfrequency ω_α of which the centers have spatial distance

$$s_\alpha = \frac{c_\alpha}{m_\alpha \omega_\alpha^2} q_0 = \sqrt{\frac{2\hbar}{m_\alpha \omega_\alpha}} \frac{\lambda_\alpha}{\omega_\alpha} . \tag{18.5}$$

Hence the vacuum or ground states of $H_{-,\alpha}$ and $H_{+,\alpha}$ can mutually be transformed into each other with the displacement operation

$$|0_{\pm,\alpha}\rangle = e^{i\Omega_{\mp,\alpha}} |0_{\mp,\alpha}\rangle \quad \text{with} \quad \Omega_{\mp,\alpha} = \pm \frac{s_\alpha p_\alpha}{\hbar} = \pm i \frac{\lambda_\alpha}{\omega_\alpha} [b_{\mp,\alpha}^\dagger - b_{\mp,\alpha}] . \tag{18.6}$$

With use of the commutation relation of $b_{\pm,\alpha}$ with $b_{\pm,\alpha}^\dagger$ we find that the vacuum state of the one oscillator can be expressed in terms of a coherent state of the displaced other oscillator,

$$|0_{\pm,\alpha}\rangle = \exp\left(-\frac{\lambda_\alpha^2}{2\omega_\alpha^2}\right) \sum_{n=0}^{\infty} \frac{(\pm 1)^n}{\sqrt{n!}} \left(\frac{\lambda_\alpha}{\omega_\alpha}\right)^n |n_{\mp,\alpha}\rangle . \tag{18.7}$$

Thus, the probability to excite n bosons with energy $\hbar\omega_\alpha$ in a sudden transition from the ground state at $\sigma_z = -1$ to the state $\sigma_z = 1$ is the Poissonian distribution

$$p_{n,\alpha} = |\langle 0_{-,\alpha}|n_{+,\alpha}\rangle|^2 = \exp\left(-\frac{\lambda_\alpha^2}{\omega_\alpha^2}\right) \frac{1}{n!} \left(\frac{\lambda_\alpha}{\omega_\alpha}\right)^{2n} \tag{18.8}$$

with mean particle number $\bar{n}_\alpha = \sum_n n\, p_{n,\alpha} = \lambda_\alpha^2/\omega_\alpha^2$. The probability for transition without excitation of bosons with frequency ω_α is

$$p_{0,\alpha} = |\langle 0_{-,\alpha}|0_{+,\alpha}\rangle|^2 = \exp\left(-\bar{n}_\alpha\right) . \tag{18.9}$$

We can now use these findings to calculate the dressed amplitude at zero temperature for transitions between the states $\sigma_z = \pm 1$. Considering the boson drag in the adiabatic limit, the bare amplitude $\langle -|+\rangle = \Delta_0$ is renormalized by an exponential dressing factor, which is known as Franck-Condon factor,

$$\Delta = \Delta_0 \prod_\alpha \langle 0_{-,\alpha}|0_{+,\alpha}\rangle = \Delta_0 \exp\left(-\frac{1}{2}\sum_\alpha \frac{\lambda_\alpha^2}{\omega_\alpha^2}\right) . \tag{18.10}$$

The bare amplitude is reduced because of the displacement of the bosonic cloud in the transition. For a finite number of boson modes the dressed tunneling amplitude stays finite. In the continuum limit, the spectral coupling $G(\omega) = \sum_\alpha \lambda_\alpha^2 \delta(\omega - \omega_\alpha)$, introduced in Eq. (3.140), becomes a smooth function of ω. Then we obtain

$$\Delta = \Delta_0 \exp\left(-\frac{1}{2}\int_0^\infty d\omega \frac{G(\omega)}{\omega^2}\right) . \tag{18.11}$$

When the integral diverges, the two ground states $|0_{\pm,\alpha}\rangle$ with displaced bosons become orthogonal. In this case, they represent two different worlds which do not mutually interfere. This situation is dubbed *orthogonality catastrophe*. Ultraviolet-divergence of the integral is irrelevant, since there is always some physical cutoff in the spectral function $G(\omega)$, as we have discussed in Subsection 3.1.7.

In the super-Ohmic case $G(\omega) \propto \omega^s$ with $s > 1$, the integral is infrared-convergent and hence there is no orthogonality catastrophe. This is an indication for the possibility of elastic tunneling processes without dynamical involvement of the reservoir. The integral is infrared-divergent, however, when the spectral density is Ohmic or sub-Ohmic, $s \leq 1$. This case is more subtle since tunneling is quenched in the adiabatic limit discussed hitherto. This indicates that for $s \leq 1$ the low frequency modes must be treated nonadiabatically. The appropriate treatment is given in Section 20.2.

At finite T the adiabatic dressing of the bare tunneling matrix element is

$$\Delta = \Delta_0 \exp\left(-\tfrac{1}{2}[\rho_e + \rho_a]\right) , \qquad (18.12)$$

where ρ_e and ρ_a are the Huang-Rhys factors for emission and absorption of phonons [383] (see also the discussion in Subsection 20.2.2),

$$\rho_e = \int_0^\infty d\omega\, \frac{G(\omega)}{\omega^2} \frac{1}{e^{\beta\hbar\omega} - 1} , \qquad \rho_a = \int_0^\infty d\omega\, \frac{G(\omega)}{\omega^2} \frac{1}{1 - e^{-\beta\hbar\omega}} . \qquad (18.13)$$

The integrals are infrared-divergent in the parameter range $0 < s \leq 2$. This indicates that the corresponding low-frequency modes must be treated nonadiabatically.

18.1.2 Adiabatic renormalization

Consider a quantum particle moving in a double-well potential $V(q)$ as described in Sect. 3.3. Assume that the particle interacts with an environment as given in Eq. (3.11). We now wish to see under which conditions the continuous system can be reduced to a discrete two-state system. Roughly speaking, the reduction is possible if there is a wide separation of the relevant energy scales. In the first stage of the truncation procedure, the system is coupled to only high-frequency bosons, $\omega > \omega_c$, where ω_c is chosen to simultaneously satisfy the conditions

$$\hbar|\epsilon|, k_B T \ll \hbar\omega_c \ll \hbar\omega_R(\omega_c) , \quad \text{and} \quad \Delta(\omega_c) \ll \omega_c . \qquad (18.14)$$

Here, ω_R is the dressed well frequency [cf. Eq. (14.12) with $\omega_b \to \omega_0$], and $\hbar\Delta(\omega_c)$ is the dressed tunnel splitting. The frequencies ω_R and Δ are renormalized by the modes with $\omega > \omega_c$. Under conditions (18.14), the excited states in each well will not be significantly populated. Therefore, they can be ignored so that we can replace the Hamiltonian of the continuous double-well system by an effective two-state Hamiltonian with a renormalized tunnel splitting. In the second stage of the reduction scheme, the remaining low-frequency bosons with $\omega < \omega_c$ are coupled to the discrete system. As we shall see, these modes will severely influence transitions between the two low-lying states. Thus finally, we have reached the dissipative two-state or spin-boson Hamiltonian [cf. Eq. (3.139)]

$$H_{SB} = -\frac{\hbar\Delta}{2}\sigma_x - \frac{\hbar\epsilon}{2}\sigma_z - \frac{1}{2}\sigma_z \sum_{\alpha\,\varepsilon\,\mathrm{lf}} \hbar\lambda_\alpha \left(b_\alpha + b_\alpha^\dagger\right) + \sum_{\alpha\,\varepsilon\,\mathrm{lf}} \hbar\omega_\alpha b_\alpha^\dagger b_\alpha , \qquad (18.15)$$

in which the spectral density $G_{lf}(\omega)$ subsumes the coupling to modes with $\omega \lesssim \omega_c$,

$$G_{lf}(\omega) = \sum_{\alpha \in lf} \lambda_\alpha^2 \delta(\omega - \omega_\alpha) . \qquad (18.16)$$

The spectral density $G_{lf}(\omega)$ includes all bath modes that are slow on the time scale Δ^{-1} and hence cause long-ranged memory effects.

In contrast, fast modes are dynamically important only on short time scales. Since fast modes can quickly adjust themselves to the slow tunneling motion, they can be treated adiabatically as discussed in the preceding subsection. For $\Delta = 0$, the two lowest energy eigenfunctions of the *symmetric* global system at $T = 0$ are given by

$$\begin{aligned}|\psi_\pm\rangle &= \tfrac{1}{\sqrt{2}} \left(|+\rangle \prod_{\alpha \in hf} |0_{+,\alpha}\rangle \pm |-\rangle \prod_{\alpha \in hf} |0_{-,\alpha}\rangle \right) \\ &= \tfrac{1}{\sqrt{2}} \left(|+\rangle \prod_{\alpha \in hf} e^{i\Omega_\alpha/2} |0_\alpha\rangle \pm |-\rangle \prod_{\alpha \in hf} e^{-i\Omega_\alpha/2} |0_\alpha\rangle \right) ,\end{aligned} \qquad (18.17)$$

where $|\pm\rangle$ denotes the state $\sigma_z = \pm 1$ of the TSS, and $|0_{\pm,\alpha}\rangle$ are the shifted ground states of the α'th oscillator for these TSS states. The translation operator Ω_α is defined as in Eq. (18.6), but depends on the original creation and annihilation operators,

$$\Omega_\alpha = s_\alpha p_\alpha/\hbar = i(\lambda_\alpha/\omega_\alpha)[b_\alpha^\dagger - b_\alpha] . \qquad (18.18)$$

The level splitting dressed by the high frequency modes is given by

$$\begin{aligned}\Delta &= \Delta_0 \prod_{\alpha \in hf} \langle 0_\alpha | e^{i\Omega_\alpha} | 0_\alpha \rangle \\ &= \Delta_0 \exp\left(-\frac{1}{2} \sum_{\alpha \in hf} \frac{\lambda_\alpha^2}{\omega_\alpha^2} \right) = \Delta_0 \exp\left(-\frac{1}{2} \int_0^\infty d\omega \, \frac{G_{hf}(\omega)}{\omega^2} \right) .\end{aligned} \qquad (18.19)$$

We note that under condition $k_B T \ll \hbar\omega_c$ the terms with $n(\omega)$ occurring in Eq. (18.13) may safely be disregarded in Eq. (18.19). The Franck-Condon factor represents the polarization cloud of phonons with frequencies above ω_c in the adiabatic limit. In the approximation (18.19) for the high-frequency modes of the bath, it is not distinguished whether the Franck-Condon factor originates from a single bath mode, or whether there is in reality a continuum of bath states in the regime $\omega > \omega_c$ [384].

The reduction procedure relying on the expression (18.19) has been studied by Sethna [169] in the context of tunneling centers in solids which are described by a spectral density $G(\omega) \propto \omega^3$ at low frequencies (cf. Subsection 4.2.4). In this case, also low-frequency modes may be included in the Franck-Condon factor. The case of Ohmic dissipation is more subtle due to an inherent infrared divergence, as addressed already in the preceding subsection (see also Subsection 18.1.4).

In the sequel, we shall regard Δ as the tunneling matrix element which is already dressed by the high-frequency modes of the environment as given in Eq. (18.19).

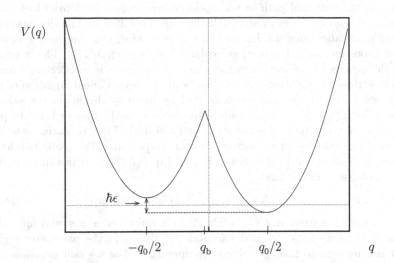

Figure 18.1: The asymmetric double well described in Eq. (18.22).

18.1.3 Instanton in a double parabolic well

Let us now study the instanton for Ohmic damping in a particular double well model potential, which is analytically solvable. The spectral density is [cf. Eq. (4.154)]

$$G_{\mathrm{lf}}(\omega) = 2K\omega \qquad (18.20)$$

with the Kondo coupling parameter [cf. Eqs. (17.17) and (17.49)]

$$K \equiv \frac{\eta q_0^2}{2\pi\hbar} = \frac{M\gamma q_0^2}{2\pi\hbar} = \frac{\alpha v}{\pi}, \qquad (18.21)$$

where η is the Ohmic viscosity. In the last form, we have introduced the usual dimensionless damping parameter $\alpha = \gamma/2\omega_0$ and the dimensionless "barrier height" $v \equiv M\omega_0^2 q_0^2/\hbar\omega_0$. In the semiclassical limit, we have $v \gg 1$.

For the slightly asymmetric double well potential sketched in Fig. 18.1,

$$V(q) = \begin{cases} \frac{1}{2}M\omega_0^2(q + \frac{1}{2}q_0)^2 + \frac{1}{2}\hbar\epsilon, & q < q_{\mathrm{b}}, \\ \frac{1}{2}M\omega_0^2(q - \frac{1}{2}q_0)^2 - \frac{1}{2}\hbar\epsilon, & q > q_{\mathrm{b}}, \end{cases} \qquad (18.22)$$

the renormalized tunnel matrix element [cf. Eq. (17.71)]

$$\Delta/2 = f_{\mathrm{inst}}\, e^{-S_{\mathrm{inst}}/\hbar} \qquad (18.23)$$

can be calculated in analytic form [385] for arbitrary damping strength α. The potential (18.22) has a cusp at the barrier top which is located at $q_{\mathrm{b}} = -\hbar\epsilon/(M\omega_0^2 q_0)$.

The bounce is the extremal path in the upside-down potential $-V(q)$ which starts near $-q_0/2$ and then moves forth and back through the valley, and finally returns to the starting point after time $\hbar\beta$. Because of the weak bias, the bounce is a weakly-bound instanton–anti-instanton pair, as explained in Subsection 17.4. The interval τ betweeen the centers of the instanton and the anti-instanton is a collective variable which parametrizes the breathing mode. The bounce obeys a linear equation of motion of the form (6.163) which is supplemented by jump conditions in the velocity and in the acceleration. The discontinuities are situated at the times where the path passes the cusp of the potential on the way forth and back. The calculation is similar to the proceeding explained in Subsection 6.7 and is reported in Ref. [385]. It is found that the bounce action can be decomposed as in Eq. (17.62). The instanton part of the action is obtained in the form

$$S_{\text{inst}}/\hbar = \left[v^2/2 - \pi^2 K^2 \right] \langle q^2 \rangle / q_0^2 + K \left[C_{\text{E}} + \ln(\omega_0/\omega_{\text{c}}) \right] , \qquad (18.24)$$

where ω_{c} is a reference frequency, $C_{\text{E}} = 0.5772\ldots$ is Euler's constant, and $\langle q^2 \rangle$ is the coordinate dispersion of the damped harmonic oscillator. In the parameter regime (18.14), it is consistent to take $\langle q^2 \rangle$ at zero temperature. The relevant expression for the dispersion $\langle q^2 \rangle$ is given in Eqs. (6.141) – (6.154). In the weak damping limit, $\alpha \ll 1$, we obtain from Eq. (18.24)

$$S_{\text{inst}}/\hbar = S_{\text{inst}}^{(0)}/\hbar + K \left[\ln(\omega_0/\omega_{\text{c}}) - c_0 \right] + \mathcal{O}(K^2) , \qquad (18.25)$$

where $S_{\text{inst}}^{(0)} = M\omega_0 q_0^2/4$ is the instanton action for zero damping. The numerical constant is $c_0 = 1/2 - C_{\text{E}} \approx -0.0772$.

In the opposite strong-damping limit, $\alpha \gg 1$, we find from Eq. (18.24)

$$S_{\text{inst}}/\hbar = K \left[\ln(\omega_0/\omega_{\text{c}}) + \ln v - \ln K - c_\infty \right] + \mathcal{O}\left[\ln(K/v) \right] , \qquad (18.26)$$

where $c_\infty = \ln(2\pi) - C_{\text{E}} \approx 1.2606$. For strong damping, the form (18.26) is generally valid. Only the numerical factor c_∞ depends on the particular shape of the potential. The terms $K \ln v$ and $K \ln K$ are independent of the shape of the double-well potential for fixed distance q_0 between the wells. This is because the instanton width in the limit $\alpha \gg 1$ is $\omega_{\text{R}}^{-1} = \alpha\omega_0^{-1} = (\pi K/v)\omega_0^{-1}$ [cf. Eq. (14.37) for $\alpha \gg 1$]. The inverse width ω_{R} provides the effective high-frequency cutoff of the problem.

The instanton action has been estimated for the quartic double-well potential, Eq. (3.123), in Refs. [386, 387]. The numerical constant for weak damping was obtained in Ref. [387], using a variational method, as $c_0 \approx -0.2392$, while in Ref. [386] it was found $c_0 \approx 0.2006$. In the strong damping limit, Dorsey et al. [387] found the form (18.26) with $c_\infty = 1.24$, while Chakravarty and Kivelson [386] gave $c_\infty = 1.594$.

For the cosine potential (17.43), the instanton action in the regime $\alpha \gg 1$ has been given already in Eq. (17.66). This expression is again in the form (18.26) with the constant $c_\infty = \ln(4\pi) \approx 2.5310$. For this particular potential, also the prefactor f_{inst} is known in analytic form for $\alpha \gg 1$ [cf. Eq. (17.72)],

$$f_{\text{inst}} = \sqrt{K/2\pi}\,\gamma = \sqrt{2\pi}K^{3/2}\omega_0/v . \qquad (18.27)$$

On the other hand, for the double well with the cusp at the barrier top, Eq. (18.22), one finds for $\alpha \gg 1$ [385]

$$f_{\text{inst}} = \sqrt{2.042 \ln(\pi K/v)/\pi K} \; v\omega_0/2\pi \; . \tag{18.28}$$

The quite different dependence of the prefactor f_{inst} on the parameters K and v in Eqs. (18.27) and (18.28) originates from the denominator of the determinantal factor [cf. Eq. (17.33)]

$$A_2 = \frac{\prod_{n=1}^{\infty}[\nu_n^2 + \gamma\nu_n + \omega_0^2]}{\omega_0^2 \prod_{n=2}^{\infty}\Lambda_n} \; . \tag{18.29}$$

The quantities Λ_n ($n \geq 2$) are the positive eigenvalues of the second-order fluctuation operator about the instanton–anti-instanton path.

For the cosine potential, we have $A_2 = 16\alpha^4 \propto (K/v)^4$ [cf. Eq. (17.58)] , whereas for the cusp potential $A_2 = 2.042 \ln(\pi K/v)$, as discussed in Ref. [385]. This explains why Eqs. (18.27) and (18.28) differ by a factor $\propto (K/v)^2/\sqrt{\ln(K/v)}$.

It is plausible to assume that the behavior $f_{\text{inst}} \propto K^{3/2}\omega_0/v$ for $\alpha \gg 1$ [cf. Eq. (18.27)] is generally found for any smooth double-well potential for which the barrier frequency ω_{b} is of the order of the well frequency ω_0. The prefactor (18.28) is different because the barrier frequency ω_{b} is singular for a cusp-shaped barrier.

It remains to demonstrate that the unphysical cutoff frequency ω_{c} cancels out in the calculation of physical quantities using the truncated model (18.15) with (18.16). In the Ohmic case, we have $\Delta \propto \exp[-K \ln(\omega_0/\omega_{\text{c}})]$, as follows from Eq. (18.23) with (18.24). Furthermore, in physical quantities, Δ and the instanton–anti-instanton interaction $W(\tau)$ are combined in the form $\Delta^2 \exp[-W(\tau)]$ [cf. the rate expression (17.70)]. Substituting the form (17.64) for $W(\tau)$, we see that ω_{c} cancels out. However, when we consider the pair interaction $W(\tau)$ in the core regime, $\omega_{\text{c}}\tau \lesssim 1$, the cancellation is only partial. From this we infer that the cancellation is only in leading order of $\epsilon/\omega_{\text{c}}$, $\omega_{\text{c}}/\omega_{\text{b}}$, $\omega_{\text{c}}/\omega_0$, and $1/\hbar\beta\omega_{\text{c}}$ [387].

18.1.4 Renormalized tunneling matrix element

We now resume the adiabatic renormalization scheme in which we have successively integrated out high frequencies. We intend to extend the adiabatic renormalization (18.19) to lower frequencies. To this, we define the dressed tunneling matrix element

$$\Delta_{\text{d}}(\omega_\ell) \equiv \Delta \exp\left(-\frac{1}{2}\int_{\omega_\ell}^{\infty} d\omega \, \frac{G_{\text{lf}}(\omega)}{\omega^2}\right) \; . \tag{18.30}$$

Since $\Delta_{\text{d}} < \Delta$, we may iteratively integrate out oscillators with frequencies in the range $\omega_\ell(\Delta) > \omega > \omega_\ell(\Delta_{\text{d}}) \equiv p\Delta_{\text{d}}$, where p is unspecified, but generally large, $p \gg 1$. The procedure converges to a dressed matrix element Δ_{r} which is self-consistently determined by the relation [85]

$$\Delta_{\text{r}} = \Delta \exp\left(-\frac{1}{2}\int_{p\Delta_{\text{r}}}^{\infty} d\omega \, \frac{G_{\text{lf}}(\omega)}{\omega^2}\right) \; . \tag{18.31}$$

The treatment of the slow modes can be improved by employing the flow equation formalism by Wegner [389] in which the Hamiltonian is diagonalized by continuous unitary transformations. This method leads to the self-consistent relation[1] [391, 392]

$$\Delta_{\mathrm{r}} = \Delta \exp \left(-\frac{1}{2} \int_0^\infty d\omega \, \frac{G_{\mathrm{lf}}(\omega)}{\omega^2 - \Delta_{\mathrm{r}}^2} \right), \qquad (18.32)$$

where the principal value of the improper integral has to be taken. The expression (18.31) disregards frequencies below $p\Delta_{\mathrm{r}}$. In contrast, in the form (18.32) slow modes $\omega < \Delta_{\mathrm{r}}$ lead to an increase of the renormalized tunneling matrix element. The difference of Eq. (18.32) from (18.31) is negligible in the Ohmic and super-Ohmic case.

Consider now a spectral density of the algebraic form[2] [cf. Eq. (3.72)]

$$G_{\mathrm{lf}}(\omega) = 2\delta_s \omega_{\mathrm{ph}}^{1-s} \omega^s \, f(\omega/\omega_{\mathrm{c}}) \qquad \text{with} \qquad f(\omega/\omega_{\mathrm{c}}) = e^{-\omega/\omega_{\mathrm{c}}}. \qquad (18.33)$$

In the super-Ohmic case, $s > 1$, the integral in Eq. (18.30) remains integrable in the limit $\omega_\ell \to 0$. It is then natural to introduce a Franck-Condon factor which includes *all* modes of the reservoir. Thus we define

$$\Delta_{\mathrm{eff}} \equiv \Delta_{\mathrm{d}}(\omega_\ell = 0) = \Delta e^{-B_s} \qquad \text{for} \qquad s > 1, \qquad (18.34)$$

$$B_s = \delta_s \Gamma(s-1)(\omega_{\mathrm{D}}/\omega_{\mathrm{ph}})^{s-1}, \qquad (18.35)$$

where we have used an exponential cutoff with the Debye frequency, $\omega_{\mathrm{c}} = \omega_{\mathrm{D}}$.

For Ohmic damping, we have $d\ln[\Delta_{\mathrm{d}}(\omega_\ell)]/d\ln(\omega_\ell) = K$. Therefore, the effects of the low-frequency modes at $T = 0$ crucially depend on the value of K. For $K > 1$, the renormalized tunneling matrix element Δ_{r} is iterated to zero, $\Delta_{\mathrm{r}} = \Delta_{\mathrm{d}}(\omega_\ell = 0) = 0$. This is the localization phenomenon obtained by Chakravarty [142] and by Bray and Moore [143] using renormalization group methods. For $K < 1$, we find from Eq. (18.31) or from Eq. (18.32), disregarding an unspecified numerical constant,

$$\Delta_{\mathrm{r}} = (\Delta/\omega_{\mathrm{c}})^{K/(1-K)} \Delta \qquad \text{for} \qquad K < 1. \qquad (18.36)$$

This form was also found with a variational treatment of the sluggish modes [388].

For subsequent convenience, we also introduce a slightly modified frequency scale in which Δ_r is multiplied by the factor $[\Gamma(1 - 2K)\cos(\pi K)]^{1/2(1-K)}$,

$$\Delta_{\mathrm{eff}} = [\Gamma(1 - 2K)\cos(\pi K)]^{1/2(1-K)}(\Delta/\omega_{\mathrm{c}})^{K/(1-K)} \Delta, \qquad K < 1. \qquad (18.37)$$

The additional factor is of order one for all $K \lesssim \frac{1}{2}$ and is singular at $K = 1$. The effective frequency Δ_{eff} is equal to Δ for $K = 0$ and equal to $\pi\Delta^2/2\omega_{\mathrm{c}}$ for $K = \frac{1}{2}$. We shall see in Subsection 22.1.1 that Δ_{eff} is the proper inverse time scale for $0 \le K < 1$.

In the sub-Ohmic case, $s < 1$, we may put $\omega_{\mathrm{c}} \to \infty$ since the adiabatic dressing integral is ultraviolet convergent. In the adiabatic scheme, Eq. (18.31), Δ_{r} is iterated

---

[1] A similar integral is also found for the finite-T modification, Eq. (19.42) with (19.43).

[2] Here we distinguish the frequency scale ω_{ph} from the cutoff ω_{c}. In the super-Ohmic case, we shall sometimes identify ω_{c} with the Debye frequency ω_{D}. For $s = 1$, the parameter δ_1 coincides with Kondo's parameter K and with the parameter α introduced by Leggett *et al.* [85].

to zero with an essential singularity. Thus we would expect that tunneling is quenched at zero temperature, i.e., the particle is effectively trapped in one of the two localized states. In contrast, the improved relation (18.32) predicts different behavior for very weak sub-Ohmic coupling [393] (c is a numerical constant of order unity)

$$\delta_s^{1/(1-s)} \omega_{\mathrm{ph}} < c\Delta . \tag{18.38}$$

In this regime, the relation (18.32) gives a nonzero renormalized tunneling matrix element. The findings are in accord with results from a study of the related Ising model [cf. Section 19.5] using a variational method [384].

At finite temperature the iteration procedure stops at $\omega_\ell = 1/\hbar\beta$. We then obtain $\Delta_{\mathrm{r}} = \Delta \exp[-\delta_s(\hbar\beta\omega_{\mathrm{ph}})^{1-s}/(1-s)]$, which is qualitatively correct [85], as we shall see in Subsection 20.2.7.

In connection with quantum state engineering the coherent sub-Ohmic regime (18.38) is of particular interest. The low-frequency modes $\omega \lesssim \Delta$ or $\omega \lesssim 1/\hbar\beta$ must be treated nonadiabatically in this regime, as we have seen above. One finds that these modes entail pure dephasing. Discussion of this issue is given in Section 22.4 and Subsection 22.5.3.

18.1.5 Polaron transformation

In particular cases it is useful to introduce a basis of dressed states in which the oscillator α is shifted by a displacement $-\frac{1}{2}s_\alpha\sigma_z$, where $s_\alpha = q_0 c_\alpha/m_\alpha\omega_\alpha^2$, as discussed in Subsection 18.1.1. The unitary operator which transforms to the basis of the displaced harmonic oscillator states is

$$U = \exp[-\tfrac{1}{2}i\sigma_z\Omega] \quad \text{with} \quad \Omega = \sum_\alpha \frac{s_\alpha p_\alpha}{\hbar} = i\sum_\alpha \frac{\lambda_\alpha}{\omega_\alpha}[b_\alpha^\dagger - b_\alpha] . \tag{18.39}$$

The polaron transformation diagonalizes the last three terms in the Hamiltonian (18.15). The transformed Hamiltonian $\tilde{H} = U^{-1}HU$ takes the exact form

$$\tilde{H} = -\frac{\hbar\Delta}{2}\Big(|R\rangle\langle L|\, e^{i\Omega} + |L\rangle\langle R|\, e^{-i\Omega}\Big) - \frac{\hbar\epsilon}{2}\sigma_z + \sum_\alpha \hbar\omega_\alpha b_\alpha^\dagger b_\alpha , \tag{18.40}$$

where we have dropped an irrelevant constant.

Alternatively, instead of transforming the Hamiltonian, we may transform the tunneling operator as $\tilde{\sigma}_x = U\sigma_x U^{-1}$. The polaron-transformed tunneling operator $\tilde{\sigma}_x$ acts in the full system-plus-reservoir space and takes the form

$$\tilde{\sigma}_x = |R\rangle\langle L|\, e^{-i\Omega} + \mathrm{h.c.} = |R\rangle\langle L|\, \prod_\alpha \int dx_\alpha\, |x_\alpha\rangle\langle x_\alpha - s_\alpha| + \mathrm{h.c.} . \tag{18.41}$$

The operation of $\tilde{\sigma}_x$ transfers the particle from one localized state to the other and simultaneously shifts the set of bath oscillators $\{\alpha\}$ by the set of displacements $\{s_\alpha\}$. In the dressed basis, pictorially, the particle drags behind it a polaronic cloud. The equilibrium autocorrelation function of the dressed tunneling operator evolving under the full untransformed Hamiltonian, $\tilde{\sigma}_x(t) = e^{iHt/\hbar}\,\tilde{\sigma}_x(0)\,e^{-iHt/\hbar}$, differs from the autocorrelation function of the bare σ_x. This will be discussed in Subsection 22.6.5.

18.2   Pair interaction in the charge picture

18.2.1   Analytic expression for spectral density with any power s

We have introduced a charge picture for the spin-boson model in Subsection 4.2.3. The charges are lined up with alternating sign, and the imaginary-time charge interaction $W(\tau)$ is the second integral of the kernel $\mathcal{K}(\tau)$ [cf. Eqs. (4.70) and (4.72)],

$$W(\tau) \;=\; \int_0^\infty d\omega \, \frac{G_{\text{lf}}(\omega)}{\omega^2} \, \frac{\cosh(\omega\hbar\beta/2) - \cosh[\omega(\hbar\beta/2 - \tau)]}{\sinh(\omega\hbar\beta/2)} \,, \qquad (18.42)$$

where the spectral density $G_{\text{lf}}(\omega)$ represents the coupling to modes with frequencies below ω_c. The function $W(\tau)$ is real for real τ and satisfies the reflection property

$$W(\tau) \;=\; W(\hbar\beta - \tau) \qquad \text{for} \qquad 0 \le \tau < \hbar\beta \,. \qquad (18.43)$$

On the other hand, dynamical expressions involve the function $W(\tau)$ analytically continued to real time, $Q(t) \equiv W(\tau = it)$. We have

$$Q(t) \;=\; \int_0^\infty d\omega \, \frac{G_{\text{lf}}(\omega)}{\omega^2} \left\{ \coth\left(\frac{\beta\hbar\omega}{2}\right)\left(1 - \cos(\omega t)\right) + i\sin(\omega t) \right\} . \qquad (18.44)$$

For complex time $z = t - i\tau$, the pair interaction $Q(z)$ has the symmetry

$$Q(-z - i\hbar\beta) \;=\; Q(z) \,. \qquad (18.45)$$

The function $Q(z)$ is analytic in the strip $0 \ge \text{Im}\, z > -\hbar\beta$.

Consider now the correlation function $Q(z)$ for the spectral density (3.141) with Eq. (3.72) [the coupling constant $\delta_s = M\gamma_s q_0^2/2\pi\hbar$ is dimensionless],

$$G_{\text{lf}}(\omega) \;=\; 2\delta_s \omega_{\text{ph}}^{1-s} \omega^s \, e^{-\omega/\omega_c} \,. \qquad (18.46)$$

With the form (18.46) the integration in Eq. (18.44) can be performed exactly. We find for the pair interaction for complex time z and general s the analytic form [394]

$$\begin{aligned} Q(z) \;=\;\; & 2\delta_s \Gamma(s-1)\left(\frac{\omega_c}{\omega_{\text{ph}}}\right)^{s-1}\left\{ \left(1 - (1 + i\,\omega_c z)^{1-s}\right) + 2(\beta\hbar\omega_c)^{1-s}\zeta(s-1,1+\kappa) \right. \\[4pt] & \left. - (\beta\hbar\omega_c)^{1-s}\left[\zeta\!\left(s-1,1+\kappa+i\,\frac{z}{\hbar\beta}\right) + \zeta\!\left(s-1,1+\kappa-i\,\frac{z}{\hbar\beta}\right)\right] \right\} . \quad (18.47) \end{aligned}$$

Here, $\zeta(z,q)$ is Riemann's generalized zeta function [88], $\Gamma(z)$ is Euler's gamma function, and

$$\kappa \;=\; 1/\beta\hbar\omega_c \,. \qquad (18.48)$$

The expression (18.47) satisfies the symmetry relation (18.45). For integer $s > 1$ ($s = 2, 3, \cdots$) and $\kappa \ll 1$, the temperature-dependent part of the pair interaction can be expressed in terms of polygamma functions. Explicit forms for $W(\tau)$ are easily obtained from Eq. (18.47) by analytic continuation. For later convenience, we also introduce the function $X(t) \equiv Q(t - i\hbar\beta/2)$,

$$X(t) \;\equiv\; X_1 - X_2(t) \;=\; \int_0^\infty d\omega \, \frac{G_{\text{lf}}(\omega)}{\omega^2}\left[\coth(\tfrac{1}{2}\beta\hbar\omega) - \frac{\cos(\omega t)}{\sinh(\tfrac{1}{2}\beta\hbar\omega)}\right] . \qquad (18.49)$$

The integral representation for $X(t)$ follows from Eq. (18.44). Using the functional relation $\zeta(x, q+1) = \zeta(x, q) - q^{-x}$, we find from Eq. (18.47) the explicit form

$$X(z) = 2\delta_s\Gamma(s-1)\left(\frac{\omega_c}{\omega_{ph}}\right)^{s-1}\left\{1 + 2(\beta\hbar\omega_c)^{1-s}\zeta(s-1, 1+\kappa)\right.$$

$$\left. - (\beta\hbar\omega_c)^{1-s}\left[\zeta\left(s-1, \frac{1}{2}+\kappa+i\frac{z}{\hbar\beta}\right) + \zeta\left(s-1, \frac{1}{2}+\kappa-i\frac{z}{\hbar\beta}\right)\right]\right\}. \quad (18.50)$$

The expressions (18.47) and (18.50) hold for general s and arbitrary temperature. The time-independent part is the Huang-Rhys factor for emission and absorption of bosons, which we have already introduced in Eq. (18.13),

$$X_1 = \rho_a + \rho_e = \int_0^\infty d\omega\, \frac{G_{lf}(\omega)}{\omega^2}\, \coth(\tfrac{1}{2}\beta\hbar\omega). \quad (18.51)$$

We see from the expression (18.44) that for a bosonic bath $Q'(t) = \operatorname{Re} Q(t)$ is temperature-dependent, whereas $Q''(t) = \operatorname{Im} Q(t)$ is temperature-independent. A spin-$\frac{1}{2}$ bath described by the spectral density Eq. (3.242) shows opposite behavior, i.e., $Q'(t)$ is temperature-independent, and the temperature dependence is captured by $Q''(t)$. The explicit calculation of the spin-bath correlation function for the various spectral densities and the study of the implications with respect to thermodynamics and dynamics is left to the reader.

18.2.2 Ohmic dissipation and universality limit

Upon taking in Eqs. (18.47) and (18.50) the limit $s \to 1$ we obtain the pair interaction for the Ohmic spectral density $G(\omega) = 2K\omega$ as [cf. footnote to Eq. (18.33)]

$$Q(z) = 2K\ln\left(\frac{\beta\hbar\omega_c\Gamma^2(1+\kappa)}{\Gamma(\kappa+iz/\hbar\beta)\Gamma(1+\kappa-iz/\hbar\beta)}\right), \quad (18.52)$$

$$X(z) = 2K\ln\left(\frac{\beta\hbar\omega_c\Gamma^2(1+\kappa)}{\Gamma(\frac{1}{2}+\kappa+iz/\hbar\beta)\Gamma(\frac{1}{2}+\kappa-iz/\hbar\beta)}\right). \quad (18.53)$$

We shall refer to the situation in which $\hbar\omega_c$ is the largest energy scale of the open system as the strict Ohmic case. Formally, this case corresponds to the limit $\kappa \to 0$ in Eqs. (18.52) and (18.53). We then arrive at the so-called scaling form[3]

$$Q(t) \equiv Q'(t) + iQ''(t) = 2K\ln\left(\frac{\beta\hbar\omega_c}{\pi}\sinh\frac{\pi|t|}{\hbar\beta}\right) + i\pi K\operatorname{sgn}(t), \quad (18.54)$$

$$X(t) = 2K\ln\left(\frac{\beta\hbar\omega_c}{\pi}\cosh\frac{\pi t}{\hbar\beta}\right). \quad (18.55)$$

Correspondingly, the charge interaction in imaginary time has the scaling form

$$W(\tau) = 2K\ln\left[(\beta\hbar\omega_c/\pi)\sin(\pi\tau/\hbar\beta)\right]. \quad (18.56)$$

---

[3]In the scaling limit, effects of the core regime of the pair interaction, $\tau \lesssim 1/\omega_c$, are disregarded.

We shall see in Subsection 19.1.1 that in the Coulomb gas representation for the partition function the charge interactions can be divided into dipole self-interactions and interdipole interactions. Each factor Δ^2 is associated with a dipole self-interaction, which for dipole length τ is $e^{-W(\tau)} \propto \omega_c^{-2K}$. Correspondingly, in the propagating function from a diagonal to a diagonal state of the RDM, each factor Δ^2 comes with a factor $e^{-Q'(t)}$, as can be seen from the exact formal expressions given below in Subsection 21.2.2. Thus, with the forms (18.54) and (18.56), Δ and ω_c occur only in the combination $\Delta^2/\omega_c^{2K} = \Delta_r^{2-2K}$, where Δ_r is defined in Eq. (18.36). We call an observable which is a function of Δ_r without other dependence on ω_c as universal. Any extra dependence on the cutoff ω_c is called non-universal. The universality or scaling limit is

$$\omega_c \to \infty \quad \text{with} \quad \Delta_r \quad \text{fixed} . \tag{18.57}$$

In the scaling limit, the Ohmic TSS is equivalent to the anisotropic Kondo model, and the energy scale $\hbar\Delta_r$ corresponds to the Kondo energy, up to a numerical factor which is of order 1 for $0 < K \lesssim \frac{1}{2}$. The correspondence is discussed in Section 19.4.

For $K \geq \frac{1}{2}$, the interaction factor $e^{-W(\tau)}$ with the form (18.56) must be regularized at short distance. It is convenient to employ a hard-sphere repulsion at a distance $\tau_c = 1/\omega_c$ between the charges [cf. Eqs. (19.101) and (19.109) given below]. As a result, the partition function explicitly depends on the cutoff for $K \geq \frac{1}{2}$. In contrast, the specific heat is found to be universal for $K < \frac{3}{2}$ and to depend on the cutoff for $K \geq \frac{3}{2}$ [cf. the discussion in Subsection 19.2.1].

19. Thermodynamics

In this chapter, we consider the effects of a dissipative environment on thermodynamic properties. First, we investigate the equilibrium properties of the open two-state system. Then we discuss, based on exact formal expressions for the partition function, the relationship of the Ohmic spin-boson model with variants of the Kondo model and with the $1/r^2$ Ising model. Finally, we calculate thermal tunneling rates by using the method of analytic continuation of the free energy. Emphasis is put on electron transfer in a solvent and on interstitial tunneling in solids.

19.1 Partition function and specific heat

We now study the partition function of the dissipative TSS in the form of a power series in Δ^2. We discuss various limits and introduce adequate approximations. Diverse applications will be given in subsequent sections.

19.1.1 Exact formal expression for the partition function

In the semiclassical limit, the partition function is determined by the classical paths in the upside-down potential. In an upside-down double well, the only classical paths available at low temperatures [apart from the trivial paths $q(\tau) = \pm q_0/2$] are sequences of well-separated instantons or kinks[1]. In the tight-binding two-state case, the path $\sigma(\tau) = (2/q_0)q(\tau)$ flips suddenly, say at times s_j, between the eigenvalues ± 1 of the Pauli matrix σ_z, as sketched in Fig. 4.1. Since each kink is followed by an anti-kink and vice versa, it is useful to group a given path consisting of $2m$ transitions into m kink–anti-kink pairs (bounces). We shall denote the lengths of the bounces by $\{\tau_j\}$ and the intervals between bounces by $\{\rho_j\}$,

$$\tau_j = s_{2j} - s_{2j-1} ; \qquad \rho_j = s_{2j+1} - s_{2j} , \qquad (j = 1, \ldots, m) , \qquad (19.1)$$

where $s_0 = 0$ and $s_{2m+1} = \hbar\beta$. In the equivalent charge picture, the bounces are dipoles, and the intervals τ_j and ρ_j are intra-dipole and inter-dipole lengths, respectively. Each kink–anti-kink pair or dipole contributes to the partition function a term $\Delta^2/4$ times a factor depending on the bias. The effects of the environment are in the interactions between the charges. It is now straightforward to write the partition function in the form of a power series in Δ^2. The contribution from the right/left well, which is the lower/higher well for $\epsilon > 0$, is [the label (E) refers to Euclidean quantities]

$$Z_{R/L} = \sum_{m=0}^{\infty} \left(\frac{\Delta}{2}\right)^{2m} \int_0^{\hbar\beta} ds_{2m} \int_0^{s_{2m}} ds_{2m-1} \cdots \int_0^{s_2} ds_1 \, B_{R/L,m}^{(E)}(\{s_j\}) \, \mathcal{F}_m^{(E)}(\{s_j\}) . \qquad (19.2)$$

The kink centers (charge positions) s_j are the collective coordinates of the problem. The function $B_{R/L,m}^{(E)}(\{s_j\})$ combines the dependence on the bias,

$$B_{R/L,m}^{(E)} = \exp\left(\pm \tfrac{1}{2}\hbar\beta\epsilon \mp \epsilon \sum_{j=1}^{2m} (-1)^j s_j\right) = \exp\left(\pm \tfrac{1}{2}\hbar\beta\epsilon \mp \epsilon \sum_{j=1}^{m} \tau_j\right) . \qquad (19.3)$$

The first form gives the bias factor in terms of the flip times $\{s_j\}$, whereas the second form is expressed in terms of the periods $\{\tau_j\}$ spent in the left/right well.

The factor $\mathcal{F}_m^{(E)}$ carries the environmental influences in the form of kink or charge interactions. The influence function $\mathcal{F}_m^{(E)}$ has been given in Eq. (4.72) and is

$$\mathcal{F}_m^{(E)} = \exp\left\{ \sum_{j=2}^{2m} \sum_{i=1}^{j-1} (-1)^{i+j} W(s_j - s_i) \right\} . \qquad (19.4)$$

The expression (19.2) with Eqs. (19.3) and (19.4) is the "Coulomb gas" representation of the partition function. It is in the form of a grand-canonical sum of unit charges ± 1 stringed with alternating sign. The influences of the environment are in the charge interaction $W(\tau)$ which is defined in Eq. (18.42).

---

[1]Instantons or kinks are paths that interpolate at times $-\infty$ and $+\infty$ between the two distinct ground states at $-q_0/2$ and $q_0/2$, respectively. The center of the instanton is the time where the transition occurs. In the tight-binding limit, the instanton has zero width and its shape is step-like.

Alternatively, the sequence of alternating charges may be viewed as a sequence of dipoles. Thus we may rewrite Eq. (19.4) as

$$
\mathcal{F}_m^{(E)} = \exp\left(-\sum_{j=1}^{m} W(\tau_j) \right) \exp\left(-\sum_{n=1}^{m-1}\sum_{j=1}^{m-n} \Lambda_{j+n,j}^{(E)} \right). \tag{19.5}
$$

Here the consecutively numbered dipoles are arranged in chronological order. The first exponential factor represents the self-interactions of the m dipoles, and the second one describes the interactions between these units. The terms $\Lambda_{j+1,j}^{(E)}$ are the nearest-neighbor terms, the terms $\Lambda_{j+2,j}^{(E)}$ are the next-nearest-neighbor terms, etc. The interaction of dipole j with a chronologically later dipole k is given by

$$
\Lambda_{k,j}^{(E)} = W(s_{2k} - s_{2j-1}) + W(s_{2k-1} - s_{2j}) - W(s_{2k} - s_{2j}) - W(s_{2k-1} - s_{2j-1}). \tag{19.6}
$$

We remark again that Δ is already adiabatically renormalized by the modes with $\omega > \omega_c$. Therefore, if the two-state system was actually obtained from an extended system as described in Section 18.1, the physically relevant quantity Δ_{eff} becomes independent of ω_c to leading order. Just as well, the spin-boson model may be considered as an independent model in which Δ is a free parameter which does not depend on ω_c at all. The latter view is taken in Sections 19.4 and 19.5 in which the spin-boson model is linked to the Kondo model and to the Ising model, respectively.

It is useful to write the time-ordered expression (19.2) as a Laplace integral,

$$
Z_{\text{R/L}} = \frac{1}{2\pi i} \int_{\mathcal{C}} d\lambda \, e^{\lambda \hbar \beta} \, z_{\text{R/L}}(\lambda), \tag{19.7}
$$

where $\mathcal{C}$ is the standard Bromwich contour, i. e., any contour from $-i\infty$ to $+i\infty$ lying entirely to the right of all singularities of $z_{\text{R/L}}(\lambda)$. We then obtain

$$
z_{\text{R/L}}(\lambda) = \frac{1}{\lambda \mp \epsilon/2} \sum_{m=0}^{\infty} \left(\frac{\Delta}{2} \right)^{2m}
$$

$$
\times \prod_{j=1}^{m} \left(\int_0^\infty d\tau_j \, e^{-(\lambda \pm \epsilon/2)\tau_j} \int_0^\infty d\rho_j \, e^{-(\lambda \mp \epsilon/2)\rho_j} \right) \mathcal{F}_m^{(E)}. \tag{19.8}
$$

The partition function of the two-state system is

$$
Z = Z_{\text{R}}(\epsilon) + Z_{\text{L}}(\epsilon) = Z_{\text{R}}(\epsilon) + Z_{\text{R}}(-\epsilon) = Z_{\text{L}}(\epsilon) + Z_{\text{L}}(-\epsilon). \tag{19.9}
$$

The expressions (19.7) – (19.9) represent the exact formal solution for the partition function of the spin-boson model. The result is valid for linear friction with arbitrary frequency dependence.

19.1.2 Static susceptibility and specific heat

The occupation probability of the right/left well in thermal equilibrium is given by

$$P_{\mathrm{R/L}}^{(\mathrm{eq})} = Z_{\mathrm{R/L}}/Z \,, \tag{19.10}$$

yielding for the thermal expectation value of σ_z

$$\langle \sigma_z \rangle^{(\mathrm{eq})} \equiv \frac{1}{Z} \operatorname{tr} \left\{ \sigma_z \mathrm{e}^{-\beta H} \right\} = P_{\mathrm{R}}^{(\mathrm{eq})} - P_{\mathrm{L}}^{(\mathrm{eq})} = \frac{Z_{\mathrm{R}} - Z_{\mathrm{L}}}{Z} \,. \tag{19.11}$$

Alternatively, we may express $\langle \sigma_z \rangle^{(\mathrm{eq})}$ directly in terms of the free energy F,

$$\langle \sigma_z \rangle^{(\mathrm{eq})} = \frac{2}{\hbar\beta} \frac{1}{Z} \frac{\partial Z}{\partial\epsilon} = \frac{2}{\hbar\beta} \frac{\partial\ln Z}{\partial\epsilon} = -\frac{2}{\hbar} \frac{\partial F}{\partial\epsilon} \,. \tag{19.12}$$

The response of the TSS to an external static bias is described by the static nonlinear susceptibility [2]

$$\overline{\chi}_z(T,\epsilon) = \frac{2}{\hbar} \frac{\partial\langle\sigma_z\rangle^{(\mathrm{eq})}}{\partial\epsilon} = \frac{4}{\beta\hbar^2} \frac{\partial^2\ln Z}{\partial\epsilon^2} = -\frac{4}{\hbar^2} \frac{\partial^2 F}{\partial\epsilon^2} \,. \tag{19.13}$$

In context with the Kondo problem, $\overline{\chi}_z$ describes the response of the impurity to a magnetic field perturbation $-g\mu_\mathrm{B}h\sigma_z/2$. In the two-state system, the static susceptibility $\overline{\chi}_z$ is the response to a static strain field or bias. We shall see in Subsection 21.2.4 that the static susceptibility can be also calculated in a dynamical approach.

Finally, the specific heat is readily obtained from the thermodynamic relation

$$c/k_\mathrm{B} = \beta^2 \, \partial^2 \ln Z/\partial\beta^2 \,. \tag{19.14}$$

The thermodynamic method presented in this chapter gives the partition function in the form of a power series in Δ^2. We see from the defining relations (19.11) – (19.14) that we have to perform a resummation of the resulting series in order to obtain the thermodynamic response functions in the form of power series representations in Δ^2. In Chapter 21, we shall consider the thermodynamic response functions by a different approach. There we derive within a *dynamical approach* directly exact formal series expressions in Δ^2 for $\langle\sigma_z\rangle^{(\mathrm{eq})}$ and for $\ln Z$. As an advantage over the series for Z, they give upon differentiation directly the respective series expansions for the thermodynamic response functions.

19.1.3   The self-energy method

In the absence of dissipation, we have $\mathcal{F}_m^{(\mathrm{E})} = 1$. Then the series (19.8) for the Laplace transform of the partition function of the right well is a geometrical series which can be summed to the form[3]

---

[2] We mark the static susceptibility by a bar, and the dynamical susceptibility by a tilde. Throughout part IV, we choose a normalization which differs by a factor of 4 compared to the 1993 edition in order to have agreement with the standard form (22.165) of the fluctuation-dissipation theorem for the σ_z autocorrelation function. There is also a factor 4 difference with respect to the literature on the s-d model, since there the corresponding susceptibility χ_{sd} is normalized according to the correlation function for the spin $S_z = \sigma_z/2$. The equivalence relation is $\overline{\chi}_z = 4\chi_{sd}/(g\mu_\mathrm{B})^2$.

[3] For later convenience, we write Δ instead of Δ_0.

$$z_{R/L}^{(0)}(\lambda) \;=\; \cfrac{1}{\left(\lambda \mp \epsilon/2\right)\left(1 - \cfrac{\Delta^2}{4}\cfrac{1}{(\lambda^2 - \epsilon^2/4)}\right)} \;=\; \frac{\lambda \pm \epsilon/2}{\lambda^2 - (\Delta^2 + \epsilon^2)/4} \,. \qquad (19.15)$$

The expression (19.15) has simple poles at $\lambda = \pm\Delta_{\rm b}/2$, where

$$\Delta_{\rm b} \;\equiv\; (\Delta^2 + \epsilon^2)^{1/2} \,. \qquad (19.16)$$

The transform $z_{R/L}^{(0)}(\lambda)$ is easily inverted to obtain $Z_{R/L}^{(0)}(\beta)$,

$$Z_{R/L}^{(0)}(\beta) \;=\; \cosh(\beta\hbar\Delta_{\rm b}/2) \pm (\epsilon/\Delta_{\rm b})\sinh(\beta\hbar\Delta_{\rm b}/2) \,. \qquad (19.17)$$

We remark that the same form for $Z_{R/L}^{(0)}(\beta)$ is found upon expanding in the defining expression $Z_{R/L}^{(0)}(\beta) = <R/L|\,{\rm e}^{-\beta H_{\rm TSS}}\,|R/L>$ the localized states $|R, L>$ in eigenstates of the Hamiltonian $H_{\rm TSS}$ as given in the transformation (3.130) with (3.131).

The specific heat resulting from $Z^{(0)}(\beta)$ shows the familiar Schottky anomaly

$$c/k_{\rm B} \;=\; (\beta\hbar\Delta_{\rm b}/2)^2 \,{\rm sech}^2(\beta\hbar\Delta_{\rm b}/2) \,. \qquad (19.18)$$

From this we find that the specific heat of the undamped system is exponentially small at low temperature $k_{\rm B}T \ll \hbar\Delta_{\rm b}$, $c/k_{\rm B} = (\hbar\Delta_{\rm b}/k_{\rm B}T)^2 \exp(-\hbar\Delta_{\rm b}/k_{\rm B}T)$, while it depends algebraically on temperature when $k_{\rm B}T \gg \hbar\Delta_{\rm b}$, $c/k_{\rm B} = (\hbar\Delta_{\rm b}/2k_{\rm B}T)^2$.

For the open system, we may write $z_{R/L}(\lambda)$ in the form

$$z_{R/L}(\lambda) \;=\; \frac{1}{\left[z_{R/L}^{(0)}(\lambda)\right]^{-1} - \Sigma_{R/L}(\lambda)} \,, \qquad (19.19)$$

where $\Sigma_{R/L}(\lambda)$ captures the effects of the interaction with the environment. Following the Green function terminology, we shall refer to the function $\hbar\Sigma_{R/L}(\lambda)$ as the self-energy. Formally, the analog of the self-energy is an isobaric ensemble of interacting kink–anti-kink pairs or dipoles. In graphical terms, the self-energy diagrams are *irreducible*, i.e., they cannot be divided into disconnected sub-diagrams without cutting inter-dipole interaction lines. To proceed, we define a modified influence functional which is irreducible by construction. Subtracting all reducible parts of $\mathcal{F}_n^{(E)}$, the irreducible influence functional reads

$$\tilde{\mathcal{F}}_n^{(E)} \;\equiv\; \mathcal{F}_n^{(E)} - \sum_{j=2}^{n}(-1)^j \sum_{m_1,\cdots,m_j \geq 1} \mathcal{F}_{m_1}^{(E)}\,\mathcal{F}_{m_2}^{(E)}\cdots\mathcal{F}_{m_j}^{(E)}\,\delta_{m_1+\cdots+m_j,n} \,. \qquad (19.20)$$

Here again we choose the convention that in each term the dipoles are consecutively numbered and arranged in chronological order. In the subtracted terms, the bath correlations are only inside of the individual factors $\mathcal{F}_{m_j}^{(E)}$. To illustrate the general expression (19.20), we explicitly give the case $n = 3$,

$$\tilde{\mathcal{F}}_3^{(E)} \;\equiv\; \mathcal{F}_3^{(E)} - \mathcal{F}_2^{(E)}\,\mathcal{F}_1^{(E)} - \mathcal{F}_1^{(E)}\,\mathcal{F}_2^{(E)} + \mathcal{F}_1^{(E)}\,\mathcal{F}_1^{(E)}\,\mathcal{F}_1^{(E)}$$

$$= \;\; {\rm e}^{-W(\tau_1)-W(\tau_2)-W(\tau_3)} \qquad (19.21)$$

$$\times \left\{ \left({\rm e}^{-\Lambda_{3,1}^{(E)}} - 1\right){\rm e}^{-\Lambda_{3,2}^{(E)} - \Lambda_{2,1}^{(E)}} + \left({\rm e}^{-\Lambda_{3,2}^{(E)}} - 1\right)\left({\rm e}^{-\Lambda_{2,1}^{(E)}} - 1\right) \right\} \,.$$

From this form, we can directly see the irreducibility of $\tilde{\mathcal{F}}_3^{(E)}$.

With the irreducible influence function (19.20), the series in Δ^2 for $\Sigma_{R/L}(\lambda)$ reads

$$\Sigma_{R/L}(\lambda) = \sum_{m=1}^{\infty} \Sigma_{R/L}^{(m)}(\lambda), \tag{19.22}$$

where

$$\Sigma_{R/L}^{(1)}(\lambda) = \frac{\Delta^2}{4} \int_0^{\infty} d\tau \, e^{-(\lambda \pm \epsilon/2)\tau} \left(e^{-W(\tau)} - 1 \right), \tag{19.23}$$

and for $m \geq 2$,

$$\Sigma_{R/L}^{(m)}(\lambda) = \left(\frac{\Delta^2}{4} \right)^m \prod_{j=1}^{m} \left(\int_0^{\infty} d\tau_j \, e^{-(\lambda \pm \epsilon/2)\tau_j} \right)$$

$$\times \prod_{k=1}^{m-1} \left(\int_0^{\infty} d\rho_k \, e^{-(\lambda \mp \epsilon/2)\rho_k} \right) \tilde{\mathcal{F}}_m^{(E)}. \tag{19.24}$$

Upon expanding the expression (19.19) in powers of $\Sigma_{R/L}(\lambda)$ and substituting the series expansion (19.22) with the definitions (19.23) and (19.24), the original series (19.8) is recovered. This verifies that the expression (19.19) with Eqs. (19.22) – (19.24) is a formally exact rearrangement of the original series (19.8).

In practical computations, one may truncate the series (19.22) for the self-energy at a given order, and then invert the Laplace transform (19.19) to obtain $Z_{R/L}(\beta)$. In cases where explicit inversion is difficult, much can nevertheless be learned from a study of the singularities of $z_{R/L}(\lambda)$.

19.1.4   The limit of high temperatures

For temperatures $T \gg \hbar\Delta_{\text{eff}}/k_B$, the leading contributions to the partition function are the terms with $m = 0$ and $m = 1$ in Eq. (19.2). We find from Eq. (19.2) or (19.8)

$$Z_{R/L} = e^{\pm \beta\hbar\epsilon/2} \left(1 + \frac{\Delta^2}{4} \int_0^{\hbar\beta} d\tau \, [\hbar\beta - \tau] \, e^{\mp \epsilon\tau - W(\tau)} + \mathcal{O}(\Delta^4) \right). \tag{19.25}$$

The symmetry property (18.43) suggests to map the interval $\hbar\beta/2 \leq \tau \leq \hbar\beta$ onto the interval $0 \leq \tau \leq \hbar\beta/2$. We then obtain for $Z = Z_R + Z_L$ the expression

$$Z(\beta) = 2\cosh(\tfrac{1}{2}\beta\hbar\epsilon) + \frac{\hbar\beta\Delta^2}{2} \int_0^{\hbar\beta/2} d\tau \, \cosh[\epsilon(\tfrac{1}{2}\hbar\beta - \tau)] \, e^{-W(\tau)}. \tag{19.26}$$

Thus, the leading high-temperature dependence of the partition function of the TSS is determined by the one-dipole contribution.

19.1.5 Noninteracting–kink-pair approximation

As temperature is decreased, subsequent terms of the series (19.2) become relevant. Let us now assume for the moment that the width of a typical kink–anti-kink pair is very much smaller than the typical interval between such pairs. This seems to be intuitively obvious since the average width $\langle \tau \rangle$ tends to be suppressed relative to the average interval $\langle \rho \rangle$ because of the self-interactions in Eq. (19.5). The picture we then have is that the kink pairs or dipoles form a dilute "gas" in the fixed "volume " $\hbar\beta$, and we may argue that the inter-dipole interactions $\Lambda_{j,k}^{(\mathrm{E})}$ may safely be ignored relative to the intra-dipole or self-interactions. Within the noninteracting–kink-pair approximation, we have $\mathcal{F}_{2m}^{(\mathrm{E})} = \exp[-\sum_{j=1}^{m} W(\tau_j)]$. With this form, the integrals in the series (19.2) are convolutions. The series (19.8) for the Laplace transform becomes a geometrical series which may be summed to the form

$$z_{\mathrm{R/L}}^{(1)}(\lambda) \;=\; \frac{1}{(\lambda \mp \epsilon/2)} \, \frac{1}{(1 - K_{\mathrm{R/L}}(\lambda)/[\lambda \mp \epsilon/2])} \,, \qquad (19.27)$$

$$K_{\mathrm{R/L}}(\lambda) \;=\; \frac{\Delta^2}{4} e^{i\varphi} \int_0^{\infty} du \, e^{-[\lambda \pm \epsilon/2]\tau(u)} \, e^{-W[\tau(u)]} \,, \qquad (19.28)$$

where $\tau(u) = u\, e^{i\varphi}$, and $0 \le \varphi \le \tfrac{1}{2}\pi$. Here we have used the analytic properties of the function $W(\tau)$. The original contour is $\varphi = 0$. In numerical computations, it is favorable to choose $\varphi = \pi/2$. The expression (19.27) can be written in the form (19.19). The resulting self-energy is the expression $\Sigma_{\mathrm{R/L}}^{(1)}(\lambda)$ given in Eq. (19.23). Thus, the noninteracting–kink-pair approximation is equivalent to consideration of the self-energy in order Δ^2. We have

$$Z_{\mathrm{R/L}}^{(1)}(\beta) \;=\; \frac{1}{2\pi i} \int_C d\lambda \, \frac{e^{\lambda\hbar\beta}}{\left[\, z_{\mathrm{R/L}}^{(0)}(\lambda) \,\right]^{-1} - \Sigma_{\mathrm{R/L}}^{(1)}(\lambda)} \,. \qquad (19.29)$$

Unfortunately, the kink–anti-kink pairs or dipoles do *not* actually form a dilute gas except for fairly strong damping. In this regime, the attractive self-interaction of the pair is strong enough at short distances that the breathing-mode integral $K_{\mathrm{R/L}}(\lambda)$ can be expanded about $\lambda = \pm\epsilon$,

$$K_{\mathrm{R/L}}(\lambda) \;=\; K_{\mathrm{R/L}}(\pm\epsilon) + (\lambda \mp \epsilon)\, K'_{\mathrm{R/L}}(\pm\epsilon) + \cdots . \qquad (19.30)$$

Now, the average dipole length $\langle \tau \rangle$, which can be estimated as $|K'_{\mathrm{R/L}}(\pm\epsilon)|/K_{\mathrm{R/L}}(\pm\epsilon)$, is small compared with the average interval $\langle \rho \rangle$ which is of the order of $1/K_{\mathrm{R/L}}(\pm\epsilon)$ provided that

$$|K'_{\mathrm{R/L}}(\pm\epsilon)| \;\equiv\; \frac{\Delta^2}{4} \int_0^{\infty} d\tau \, \tau \, e^{\mp \epsilon\tau - W(\tau)} \;\ll\; 1 \,. \qquad (19.31)$$

When this condition is met, the dipoles form a dilute gas. If we truncate the series (19.30) after the first term, then $z_{\mathrm{R/L}}(\lambda)$ has a simple pole at $\lambda = \pm\epsilon/2 + K_{\mathrm{R/L}}(\pm\epsilon)$. The transform $z_{\mathrm{R/L}}^{(1)}(\lambda)$ is easily inverted. The free energy associated with $Z_{\mathrm{R/L}}^{(1)}$ is

$$F_{R/L}^{(1)} = \mp \frac{\hbar\epsilon}{2} - \frac{\hbar\Delta^2}{4} \int_0^\infty d\tau\, e^{\mp\epsilon\tau - W(\tau)} \,. \tag{19.32}$$

If the condition of strong damping and/or high temperature is not met, the noninteracting–kink-pair approximation fails.[4] The approximation is inconsistent in particular regimes since it does not properly take into account the reflection property (18.43) of the interactions. Moreover, in the weak-coupling limit, the interactions $\Lambda_{j,k}^{(E)}$ render contributions to the first order in δ_s even for zero bias, as we can see from the expression (19.2) with (19.5). Thus, it is *not* consistent to make the noninteracting–kink-pair approximation in the regime of weak-to-moderate damping and/or low-to-moderate temperature.

19.1.6   Weak-damping limit

In the weak-damping limit, the coupling to the environment can be treated as a perturbation. It is clear that a perturbative expansion of the partition function itself is not very meaningful. Rather we aim at a perturbative calculation of the self-energy to first order in the coupling parameter δ_s. In the second stage, we then calculate the shift of the poles of $z_{R/L}(\lambda)$ to linear order in δ_s.

It is convenient to employ the renormalized frequency scale $\Delta_{\rm eff}$, as defined in Eqs. (18.34) and (18.37) for the super-Ohmic and Ohmic case, respectively. To proceed, we follow Ref. [395]. We expand $\Sigma_{R/L}(\lambda)$ as in Eq. (19.22) and consider each term of the series to linear order in δ_s. Thus we may write for weak coupling (wc)

$$\Sigma_{R/L}^{(wc)}(\lambda) = \Sigma_{R/L}'(\lambda) + \Sigma_{R/L}''(\lambda) \,, \tag{19.33}$$

where $\Sigma_{R/L}'(\lambda)$ is the contribution from the intra-dipole interaction,

$$\Sigma_{R/L}'(\lambda) = \frac{\Delta_{\rm eff}^2}{4} \int_0^\infty d\tau \left(2q - W_{T=0}(\tau)\right) e^{-(\lambda \pm \epsilon/2)\tau} \,, \tag{19.34}$$

which is of order $\Delta_{\rm eff}^2$. The term $2q$ is a counter term. It will be chosen such that the effective level spacing at $T = 0$ comes out correctly as $\hbar\Delta_b$, where

$$\Delta_b = \sqrt{\Delta_{\rm eff}^2 + \epsilon^2} \,. \tag{19.35}$$

The other term $\Sigma_{R/L}''(\lambda)$ in Eq. (19.33) captures the contributions from the inter-dipole interactions. We find from Eq. (19.24) the series

$$\Sigma_{R/L}''(\lambda) = -\sum_{m=2}^\infty \left(\frac{\Delta_{\rm eff}^2}{4}\right)^m \prod_{j=1}^m \left(\int_0^\infty d\tau_j\, e^{-(\lambda \pm \epsilon/2)\tau_j}\right)$$
$$\times \prod_{k=1}^{m-1} \left(\int_0^\infty d\rho_k\, e^{-(\lambda \mp \epsilon/2)\rho_k}\right) \Lambda_{m,1}^{(E)} \,, \tag{19.36}$$

---

[4]Compared to the noninteracting-blip approximation (NIBA) which is useful in the calculation of dynamical quantities [cf. Section 21.3], this approximation is valid in a smaller region of the parameter space. In the NIBA, there are partial cancellations among interblip correlations. Corresponding cancellations are absent in the present case.

where $\Lambda_{k,j}^{(E)}$ is defined in Eq. (19.6). The series (19.36) describes sequences of dipoles in which the first dipole interacts with the last one. The intermediate dipoles are noninteracting and form a grand-canonical ensemble in the "volume" ρ. For the term of order Δ_{eff}^{2m} we have

$$\rho = \sum_{j=2}^{m-1} \tau_j + \sum_{k=1}^{m-1} \rho_k \ . \tag{19.37}$$

The expression (19.36) is irreducible in the sense specified in Subsection 19.1.3. Note that the term $\Sigma_{\text{R/L}}''(\lambda)$ is disregarded in the noninteracting–kink-pair approximation. Physically, $\Sigma_{\text{R/L}}''(\lambda)$ describes one-phonon thermal processes in which the system may make any number of tunneling transitions in the interval between emission and absorption of the phonon. The grand-canonical ensemble of non-interacting kinks in the interval ρ can be summed in analytic form yielding the imaginary-time propagator with initial and final state $\sigma = +1$ (right well) [cf. Eq. (19.17)],

$$Z_{\text{R/L}}^{(0)}(\rho) = \cosh(\rho\Delta_{\text{b}}/2) \pm (\epsilon/\Delta_{\text{b}})\sinh(\rho\Delta_{\text{b}}/2) \ . \tag{19.38}$$

With this summation carried out, the quantity $\Sigma_{\text{R/L}}''(\lambda)$ takes the form

$$\Sigma_{\text{R/L}}''(\lambda) = -\frac{\Delta_{\text{eff}}^4}{16} \int_0^\infty d\tau_1 \, d\rho \, d\tau_2 \, e^{-(\lambda \pm \epsilon/2)(\tau_1+\tau_2)} \, e^{-(\lambda \mp \epsilon/2)\rho} \, Z_{\text{R/L}}^{(0)}(\rho) \, \Lambda_{2,1}^{(E)} \ , \tag{19.39}$$

where

$$\Lambda_{2,1}^{(E)} = W(\tau_1 + \rho + \tau_2) + W(\rho) - W(\tau_1 + \rho) - W(\rho + \tau_2) \ . \tag{19.40}$$

The self-energy correction $\hbar\Sigma_{\text{R/L}}^{(\text{wc})}(\lambda)$ shifts the poles of $z_{\text{R/L}}(\lambda)$ in Eq. (19.19) from $\lambda = \pm\Delta_{\text{b}}/2$ to $\lambda = \pm\Omega/2$. To linear order in δ_s, the frequency Ω is given by

$$\Omega^2 = \Delta_{\text{b}}^2 - 2(\Delta_{\text{b}} \pm \epsilon)\Sigma_{\text{R/L}}(\lambda = \Delta_{\text{b}}/2) \ . \tag{19.41}$$

The expressions (19.34) and (19.39) are conveniently evaluated by using the spectral representation (18.42) for the pair interaction $W(\tau)$ and by commuting the frequency integral with the time integrals. We readily obtain

$$\Omega^2 = \Delta_{\text{eff}}^2 + \epsilon^2 - 2\Delta_{\text{eff}}^2 \, \text{Re}\, v\left(\sqrt{\Delta_{\text{eff}}^2 + \epsilon^2}\right) \ , \tag{19.42}$$

where Δ_{eff} is the dressed tunneling matrix element at $T = 0$ introduced for the particular cases in Eqs. (18.34) and (18.37), and

$$v(z) \equiv \int_0^\infty d\omega \, \frac{G_{\text{lf}}(\omega)}{\omega^2 - z^2} \, \frac{1}{e^{\beta\hbar\omega} - 1} \ . \tag{19.43}$$

Thus, the frequency Ω turns out symmetric in ϵ. The last term in Eq. (19.42) vanishes at zero temperature. Picking up the two poles by the contour integral (19.7), we find $Z_{\text{R/L}}$ exactly in the form (19.17) in which Δ_{b} is replaced by Ω,

$$Z_{\text{R/L}}(\beta) = \cosh(\beta\hbar\Omega/2) \pm (\epsilon/\Omega)\sinh(\beta\hbar\Omega/2) \ . \tag{19.44}$$

To summarize, the functional form of the partition function is left unchanged for weak damping, but the level splitting becomes renormalized and depends on the spectral density of the coupling to the reservoir and on temperature.

19.1.7   The self-energy method revisited: partial resummation

In Subsection 19.1.3, we have explained the systematic series expansion in Δ^2 for the self-energy. In this section, we give an iterative scheme where in each step the self-energy is calculated with insertions of the imaginary-time propagator which has been obtained in the preceding step.

Assume that we have calculated the self-energy in the nth step of the iteration scheme up to terms with n irreducible dipoles. We then have the form

$$\Sigma_{\mathrm{R/L},n}(\lambda) = \Sigma_{\mathrm{R/L}}^{(1)}(\lambda) + \sum_{m=2}^{n} \Sigma_{\mathrm{R/L},n}^{(m)}(\lambda) . \tag{19.45}$$

The single-dipole contribution $\Sigma_{\mathrm{R/L}}^{(1)}(\lambda)$, which is defined in Eq. (19.23), is the same in each step. Further assume that in all terms with $m \geq 2$ we have made the insertions of the propagator, which has been calculated in the previous step, into the intervals the system spends in the right well. The self-energy (19.45) defines the imaginary-time propagator with boundary conditions $\sigma(0) = \sigma(\rho) = \pm 1$ (right/left state) according to the integral expression

$$Z_{\mathrm{R/L}}^{(n)}(\rho) = \frac{1}{2\pi i} \int_C d\lambda \, \frac{e^{\lambda \rho}}{\left[z_{\mathrm{R/L}}^{(0)}(\lambda) \right]^{-1} - \Sigma_{\mathrm{R/L},n}(\lambda)} . \tag{19.46}$$

In the next step of the iteration, we calculate the self-energy as

$$\Sigma_{\mathrm{R/L},n+1}(\lambda) = \Sigma_{\mathrm{R/L}}^{(1)}(\lambda) + \sum_{m=2}^{n+1} \Sigma_{\mathrm{R/L},n+1}^{(m)}(\lambda) , \tag{19.47}$$

in which $\Sigma_{\mathrm{R/L},n+1}^{(m)}(\lambda)$ for $m \geq 2$ is given by

$$
\begin{aligned}
\Sigma_{\mathrm{R/L},n+1}^{(m)}(\lambda) = {} & \left(\frac{\Delta^2}{4}\right)^m \prod_{j=1}^{m} \left(\int_0^\infty d\tau_j \, e^{-(\lambda \pm \epsilon/2)\tau_j} \right) \\
& \times \prod_{k=1}^{m-1} \left(\int_0^\infty d\rho_k \, e^{-(\lambda \mp \epsilon/2)\rho_k} Z_{\mathrm{R/L}}^{(n)}(\rho_k) \right) \\
& \times \exp\left(-\sum_{j=1}^{m} W(\tau_j) - \sum_{n=2}^{m-1}\sum_{j=1}^{m-n} \Lambda_{j+n,j}^{(\mathrm{E})} \right) \\
& \times \prod_{j=1}^{m-1} \left(\exp\left(-\Lambda_{j+1,j}^{(\mathrm{E})} \right) - 1 \right) .
\end{aligned}
\tag{19.48}
$$

Here, m is the number of dipoles in the "bare" self-energy term. The $\{\tau_j\}$ are the dipole lengths (periods spent in the left well), and the $\{\rho_k\}$ are the intervals between the dipoles. In Eq. (19.48), the bare self-energy term is dressed by inserting in each interval ρ_k the Euclidean propagator obtained in the previous step, Eq. (19.46). The first term in the second line contains the self-interactions of the dipoles and all inter-dipole interactions, except of those between nearest neighbours. The nearest-neighbour interactions are contained in the last term and we have subtracted in each interaction factor a one in order that the expression (19.48) is irreducible.

The procedure starts with the computation of the imaginary-time propagator $Z_{\mathrm{R/L}}^{(1)}(\rho)$ in the noninteracting–kink-pair approximation, Eq. (19.29). We then insert this form into the interdipole interval of two irreducible dipoles. Thus we obtain the improved self-energy $\Sigma_{\mathrm{R/L},2}(\lambda) = \Sigma_{\mathrm{R/L}}^{(1)}(\lambda) + \Sigma_{\mathrm{R/L},2}^{(2)}(\lambda)$ which gives the propagator $Z_{\mathrm{R/L}}^{(2)}(\rho)$ according to Eq. (19.46), and so on.

19.2 Ohmic dissipation

So far, the treatment has been quite general since we did not specify the spectral density of the coupling. Consider now the Ohmic case. In the most interesting regime $k_{\mathrm{B}}T \ll \hbar\omega_{\mathrm{c}}$, the pair interaction is given by the scaling form (18.56). We now give results in various limits. Then we turn to the exactly solvable case $K = 1/2$.

19.2.1 Specific heat and Wilson ratio

With the form (18.56) for $W(\tau)$, the integral in the high temperature formula (19.25) must be furnished with a hard-sphere cutoff at $\tau = 1/\omega_{\mathrm{c}}$ in the regime $K \geq \frac{1}{2}$. The cutoff is irrelevant for the specific heat in the regime $K < \frac{3}{2}$, whereas it is relevant for $K > \frac{3}{2}$ (see below). With use of Eqs. (19.9) and (19.14) we obtain for the specific heat in the regime $K < 1$ and zero bias ($\epsilon = 0$) the analytic expression

$$c/k_{\mathrm{B}} = \frac{(1-K)(1-2K)}{\cos(\pi K)} \left(\frac{(2\pi)^K}{2\Gamma(1-K)} \right)^2 \theta^{2K-2} + \mathcal{O}\left(\theta^{4K-4} \right) , \qquad (19.49)$$

where θ is a dimensionless temperature with the scale Δ_{eff} given in Eq. (18.37),

$$\theta = k_{\mathrm{B}}T/\hbar\Delta_{\mathrm{eff}} . \qquad (19.50)$$

The high temperature formula is valid for temperatures in the range $\theta \gg 1$.

For weak-damping $K \ll 1$ and general ϵ, the level splitting $\hbar\Omega$ has been calculated in Subsection 19.1.6. Substitution of the Ohmic form for the spectral density $G_{\mathrm{lf}}(\omega)$ [the case $s = 1$ in Eq. (18.46)] into Eq. (19.42) with Eq. (19.43) yields for all T

$$\Omega^2 = \Delta_{\mathrm{eff}}^2 + \epsilon^2 + 2K\Delta_{\mathrm{eff}}^2 \left\{ \mathrm{Re}\,\psi\left(\frac{i\hbar\sqrt{\Delta_{\mathrm{eff}}^2 + \epsilon^2}}{2\pi k_{\mathrm{B}}T} \right) - \ln\left(\frac{\hbar\sqrt{\Delta_{\mathrm{eff}}^2 + \epsilon^2}}{2\pi k_{\mathrm{B}}T} \right) \right\} , \quad (19.51)$$

where $\psi(z)$ is the digamma function. Inserting this form into Eq. (19.44) and using Eqs. (19.9) and (19.14), we obtain the specific heat in the regime $\theta \ll 1$ as the series

$$c/k_{\mathrm{B}} = K\varrho^2 \sum_{n=1}^{\infty} (2n-1)\pi^{2n} |B_{2n}| (2\varrho\theta)^{2n-1} \qquad \text{for} \qquad K \ll 1 , \qquad (19.52)$$

where B_m is a Bernoulli number and $\varrho = \Delta_{\text{eff}}/\sqrt{\Delta_{\text{eff}}^2 + \epsilon^2}$. The series (19.52) is the exact asymptotic expansion in the weak-damping limit. We remark that the noninteracting–kink-pair approximation discussed in Subsection 19.1.5 describes the low-temperature regime inadequately. For instance, it yields in the low-temperature expansion for the specific heat also even powers of θ. In reality, these terms are absent.

When the specific heat $c(T)$ varies linear with temperature and the static susceptibility $\overline{\chi}_z(T, \epsilon)$ defined in Eq. (19.13) approaches a constant as $T \to 0$, a useful quantity is the so-called Wilson ratio

$$R_1 \equiv \lim_{T \to 0} \frac{4\, c(T, \epsilon)/k_{\text{B}}}{\overline{\chi}_z(T, \epsilon) k_{\text{B}} T} \,, \tag{19.53}$$

where the factor of 4 is chosen for convenience [see footnote to Eq.(19.13)].

For weak damping and zero bias, Eq. (19.44) yields the form $\overline{\chi}_z(0, 0) = 2/(\hbar \Delta_{\text{eff}})$. Combining this expression with the result (19.52), we obtain the Wilson ratio for weak damping $K \ll 1$ as

$$R_1 = 2K\pi^2/3 \,. \tag{19.54}$$

We shall see in Section 22.9 that the form (19.54) holds for K in the regime $K < 1$.

In the language of the equivalent Kondo problem [see Section 19.4], the system shows Fermi liquid behavior at low temperature. Surprisingly, the weak-damping result (19.54) turns out to be exact for all K in the regime $0 < K < 1$ [220]. Moreover, it also holds for nonzero bias, and, in this case, it is even valid for $K > 1$. The corresponding proof is given in Section 22.9, following a different line of arguments.

The damped system behaves Schottky-like in the regime $0 < K < 1$ (cf. the discussion in Ref. [394]). The specific heat increases algebraically as the temperature is raised above zero, instead of the exponentially small enhancement in the zero-damping limit. It then approaches a maximum near $T = \hbar \Delta_{\text{eff}}/k_{\text{B}}$. Well above this value, c drops to zero again algebraically, and the exponents of the power series depend on the coupling constant K, as we have seen already in Eq. (19.49).

The regime $K > 1$ corresponds to the ferromagnetic sector of the Kondo problem. This case is mathematically easier since the perturbative series in Δ^2 is convergent at all temperatures. The formula (19.49) can be analytically continued to the regime $3/2 > K > 1$ using standard functional relations of the gamma functions. We find

$$c/k_{\text{B}} = (K-1)^2 \frac{\pi^{3/2}}{2} \frac{\Gamma(3/2 - K)}{\Gamma(2 - K)} \frac{\Delta^2}{\omega_c^2} \left(\frac{\pi k_{\text{B}} T}{\hbar \omega_c} \right)^{2K-2} \,. \tag{19.55}$$

This term is the leading one for all T in the regime $\hbar \beta \omega_c \gg 1$. Thus, the Schottky peak has faded away and the specific heat increases monotonously with T.

For strong damping $K > 3/2$, the leading contribution to the specific heat originates from the hard-sphere cutoff at $\tau = 1/\omega_c$ mentioned above and after Eq.(18.56). This term describes Fermi liquid behavior for all T below $\hbar \omega_c/k_{\text{B}}$ [cf. Ref. [394]],

$$c/k_{\text{B}} = \frac{\pi^2}{6} \frac{K}{(2K-3)} \frac{\Delta^2}{\omega_c^2} \frac{k_{\text{B}} T}{\hbar \omega_c} \qquad \text{for} \qquad K > 3/2 \,. \tag{19.56}$$

The two expressions (19.55) and (19.56) diverge in the limit $K \to \frac{3}{2}$. However, when taking the sum of the expressions (19.55) and (19.56), the singularities cancel each other. As a result, the specific heat is actually regular at the Kondo parameter $K = \frac{3}{2}$. We readily find the limiting form

$$c/k_{\mathrm{B}} = \frac{\pi^2}{12} \frac{\Delta^2}{\omega_{\mathrm{c}}^2} \frac{k_{\mathrm{B}}T}{\hbar\omega_{\mathrm{c}}} \qquad \text{at} \qquad K = \frac{3}{2} . \tag{19.57}$$

19.2.2   The special case $K = \frac{1}{2}$

For the special value $\frac{1}{2}$ of K, the influence interaction factor (19.4) or (19.5) with the pair interaction (18.56) can be decomposed into a sum of $m(m-1)$ terms which represent the combinatorical possibilities of dividing up m kinks and m anti-kinks into kink–anti-kink pairs. In each term, the given sequence of kinks and anti-kinks (cf. Fig. 4.1) is left unchanged and only the self-interactions of each pair or dipole are taken into account. For instance, the influence factor for two kink pairs or dipoles with dipole lengths τ_1 and τ_2 and distance ρ,

$$\mathcal{F}_2^{(\mathrm{E})} \equiv \mathrm{e}^{-W(\tau_1) - W(\tau_2) - W(\rho) - W(\tau_1 + \rho + \tau_2) + W(\tau_1 + \rho) + W(\rho + \tau_2)} , \tag{19.58}$$

can be decomposed exactly as

$$\mathcal{F}_2^{(\mathrm{E})} = \mathrm{e}^{-W(\tau_1) - W(\tau_2)} + \mathrm{e}^{-W(\rho) - W(\tau_1 + \rho + \tau_2)} . \tag{19.59}$$

The equivalence of Eq. (19.59) with Eq. (19.58) holds for the Ohmic scaling form (18.56) at $K = \frac{1}{2}$. It directly corresponds to the application of Wick's theorem in the related fermionic representation (19.67) with (19.68). By virtue of the respective decomposition of $\mathcal{F}_m^{(\mathrm{E})}$ in the series (19.2), the resulting integrals over the intervals $\{\tau_j\}$ and $\{\rho_i\}$ can be calculated in analytic form (cf. Ref. [394] for details). The resulting coefficients of the series expression for Z are expressed in terms of the integral

$$K_m = \int_{x_0}^{\infty} dx \, \frac{x^m}{\sinh x} , \tag{19.60}$$

where $x_0 = \pi/\hbar\beta\omega_{\mathrm{c}}$ is a hard-sphere cutoff at short distance. In reality, the cutoff is required only for K_0. The partition function for the symmetric system is found as

$$Z = 2\exp\left(\sum_{m=1}^{\infty} \frac{(-1)^{m-1}}{m!} \frac{K_{m-1}}{(2\pi\theta)^m} \right) = 2\exp\left(\int_{x_0}^{\infty} dx \, \frac{1 - \mathrm{e}^{-x/2\pi\theta}}{x \sinh x} \right) , \tag{19.61}$$

where θ is given in Eq. (19.50). For the particular case $K = \frac{1}{2}$ we have

$$\theta = \frac{k_{\mathrm{B}}T}{\hbar\gamma} , \qquad \text{where} \qquad \gamma \equiv \Delta_{\mathrm{eff}}(K = \tfrac{1}{2}) = \frac{\pi}{2} \frac{\Delta^2}{\omega_{\mathrm{c}}} . \tag{19.62}$$

Upon inserting the form (19.61) into Eq. (19.14), we obtain for the specific heat

$$c/k_{\mathrm{B}} = \frac{1}{2\pi\theta} - \frac{1}{(2\pi\theta)^2} \int_0^\infty dx\, \frac{x\,\mathrm{e}^{-x/2\pi\theta}}{\sinh x}\,, \tag{19.63}$$

which is free of a short-distance singularity. The integral can be expressed in terms of the trigamma function $\psi'(z)$ [88] as

$$c/k_{\mathrm{B}} = \frac{1}{2\pi\theta} - \frac{1}{2(2\pi\theta)^2}\,\psi'\left(\frac{1}{2} + \frac{1}{4\pi\theta}\right)\,. \tag{19.64}$$

At this point, we anticipate that the case $K = \frac{1}{2}$ corresponds to the "Toulouse limit" of the anisotropic Kondo model [141], which in turn is equivalent to the resonance level model [396, 398] with zero Coulomb interaction.[5] The respective model describes a d-level near the Fermi surface which is coupled to a bath of spinless fermions with a band width $2\hbar\omega_{\mathrm{c}}$ around the Fermi level. The coupling to the fermions is assumed constant within the band. We put $\hbar\omega_{\mathrm{c}} \ll E_{\mathrm{F}}$. Then the dispersion relation can be linearized about the Fermi wave vector. Measuring the momentum from the reference value $p_{\mathrm{F}} = mv_{\mathrm{F}}$, we have

$$E(k) - E_{\mathrm{F}} \equiv \hbar\epsilon_k = \hbar v_{\mathrm{F}} k\,. \tag{19.65}$$

The Toulouse Hamiltonian reads (L is a normalization length)

$$H_{\mathrm{T}} = \sum_k \hbar\epsilon_k c_k^\dagger c_k + \hbar\epsilon_d d^\dagger d + \frac{V}{\sqrt{L}}\sum_k \left(c_k^\dagger d + d^\dagger c_k\right)\,. \tag{19.66}$$

To show the equivalence, we first consider the partition function of this model. For simplicity, we place the d-level at the Fermi energy, $\epsilon_{\mathrm{d}} = 0$, for the moment. The perturbation series in the interaction picture reads

$$Z_{\mathrm{T}} = 2\sum_{m=0}^\infty \left(\frac{V^2}{L\hbar^2}\right)^m \int_0^{\hbar\beta - \tau_c} ds_{2m} \int_0^{s_{2m}-\tau_c} ds_{2m-1} \cdots \int_0^{s_2 - \tau_c} ds_1\, \mathcal{G}_m(\{s_j\})\,, \tag{19.67}$$

$$\mathcal{G}_m(\{s_j\}) \equiv \sum_{k_1,\cdots,k_{2m}} \exp\left(\sum_{j=1}^m \left(\epsilon_{k_{2j-1}} s_{2j-1} - \epsilon_{k_{2j}} s_{2j}\right)\right) \langle c_{k_{2m}} c_{k_{2m-1}}^\dagger \cdots c_{k_2} c_{k_1}^\dagger\rangle_\beta\,, \tag{19.68}$$

where $\langle \cdots \rangle_\beta$ means thermal average with respect to the fermionic reservoir, and where we have already extracted the time-dependent exponential factors. Next, we aim at writing the thermal average in terms of electron-hole excitations. These are expressed in terms of the Fermi function $f(\epsilon)$ as [cf. Eq. (4.131)]

$$\langle c_k^\dagger c_{k'}\rangle_\beta = \delta_{kk'} f(\epsilon_k)\,, \qquad \text{and} \qquad \langle c_k c_{k'}^\dagger\rangle_\beta = \delta_{kk'}[1 - f(\epsilon_k)]\,. \tag{19.69}$$

The thermal fermion propagator is given by

$$D(\tau) \equiv \sum_k \langle c_k^\dagger c_k\rangle_\beta\, \mathrm{e}^{\epsilon_k \tau} = \sum_k \langle c_k c_k^\dagger\rangle_\beta\, \mathrm{e}^{-\epsilon_k \tau}\,. \tag{19.70}$$

---

[5]The equivalence of the Ohmic two-state model to these models is discussed in Section 19.4.

Taking the continuum limit and observing that the density of states per unit length

$$\rho \equiv dk/2\pi d\epsilon_k = (2\pi v_{\rm F})^{-1} \tag{19.71}$$

is constant, we have

$$D(\tau) = \rho L \int_{-\omega_c}^{\omega_c} d\epsilon \, \frac{e^{-\epsilon\tau}}{1 + e^{-\hbar\beta\epsilon}} . \tag{19.72}$$

The integral in Eq. (19.72) is well-defined for infinite limits. Hence we may remove the ultraviolet cutoff in the integral, and instead of that regularize the propagator by introducing a hard-sphere repulsion at short distance $\tau = \tau_c \equiv 1/\omega_c$. Thus we find

$$D(\tau) = \Theta(\tau - \tau_c) \frac{\pi\rho L}{\hbar\beta} \frac{1}{\sin(\pi\tau/\hbar\beta)} = \Theta(\tau - \tau_c)\,\rho L \omega_c \, e^{-W_{K=1/2}(\tau)} . \tag{19.73}$$

In the second form, we have expressed the fermionic thermal propagator in terms of the kink pair interaction (18.56) for $K = \frac{1}{2}$.

The thermal average in Eq. (19.68) can be decomposed with Wick's theorem into a series which represents all possibilities of building products of electron-hole contractions. Using the correspondence relation (19.73), one finds that the Wick decomposition of $\mathcal{G}_m(\{s_j\})$ coincides with the before-mentioned decomposition of the influence function $\mathcal{F}_m^{(\rm E)}(\{s_j\})$. We have

$$\mathcal{G}_m(\{s_j\}) = (\rho L \omega_c)^m \mathcal{F}_m^{(\rm E)}(\{s_j\}) . \tag{19.74}$$

This is the key relation uncovering the correspondence of the resonance level model (19.66) with the Ohmic spin-boson model for the special damping parameter $K = \frac{1}{2}$. Substituting Eq. (19.74) into Eq. (19.67), the partition function takes the form

$$Z_{\rm T} = 2 \sum_{m=0}^{\infty} \left(\frac{\rho V^2 \omega_c}{\hbar^2} \right)^m \int_0^{\hbar\beta-\tau_c} ds_{2m} \int_0^{s_{2m}-\tau_c} ds_{2m-1} \cdots \int_0^{s_2-\tau_c} ds_1 \, \mathcal{F}_m^{(\rm E)}(\{s_j\}) . \tag{19.75}$$

Evidently, this series is identical in form with the corresponding series for the partition function of the Ohmic two-state system for $K = \frac{1}{2}$. The mapping is completed by the parameter identifications (for completeness we generalize to the case $\epsilon_{\rm d} \neq 0$)

$$\gamma \equiv \pi\Delta^2/2\omega_c = 2\pi\rho V^2/\hbar^2 , \qquad \text{and} \qquad \epsilon = \epsilon_{\rm d} . \tag{19.76}$$

The latter relation connects the bias of the TSS with the energy of the d-level.

The equations of motion for the single-particle imaginary-time Green functions may be solved in Fourier space, yielding the fermionic Matsubara sum representations

$$G^{(\rm E)}(\tau) \equiv \langle T_\tau d^\dagger(\tau) d(0) \rangle_\beta = \frac{1}{\hbar\beta} \sum_{n \, {\rm odd}} \frac{e^{-i\nu_n\tau}}{[-i\nu_n - \epsilon_{\rm d} - i\gamma\,{\rm sgn}(\nu_n)/2]} , \tag{19.77}$$

$$g_k^{(\rm E)}(\tau) \equiv \langle T_\tau c_k^\dagger(\tau) d(0) \rangle_\beta = \frac{1}{\hbar\beta} \sum_{n \, {\rm odd}} \frac{V/\hbar\sqrt{L}}{(-i\nu_n - \epsilon_k)} \frac{e^{-i\nu_n\tau}}{[-i\nu_n - \epsilon_{\rm d} - i\gamma\,{\rm sgn}(\nu_n)/2]} .$$

We are now ready to calculate the partition function. Putting $H(\lambda) = H_0 + \lambda H_1$, where H_1 is the interaction term in Eq. (19.66), the change of the free energy by the interaction is given by [168]

$$\Delta F = \frac{V}{\sqrt{L}} \int_0^1 d\lambda \sum_k \left\langle c_k^\dagger d + d^\dagger c_k \right\rangle_\lambda , \qquad (19.78)$$

where $\langle \cdots \rangle_\lambda$ denotes the *thermal average* $\langle H_1 \rangle_\lambda = \operatorname{tr}\{e^{-\beta H(\lambda)} H_1\}/\operatorname{tr} e^{-\beta H(\lambda)}$. Inserting the expression (19.77) for $g_k^{(E)}(0)$, the thermal expectation value and finally the free energy (19.78) is easily calculated. Thus we find for the partition function the expression [the hard core in Eq. (19.75) is replaced by a high-frequency cutoff]

$$Z_T = 2\cosh\left(\frac{\beta\hbar\epsilon_d}{2}\right) \exp\left\{ \frac{\beta\hbar}{2\pi} \int_{-\omega_c}^{\omega_c} d\omega \tanh\left(\frac{\beta\hbar\omega}{2}\right) \arctan\left(\frac{\gamma}{2(\omega - \epsilon_d)}\right) \right\} , \qquad (19.79)$$

where the first factor is due to the term $\hbar\epsilon_d d^\dagger d$ of the Hamiltonian, and the second factor results from the coupling term. To transform the "fermionic" expression (19.79) into the bosonic form (19.61), we employ the integral representations

$$\tanh\left(\frac{\beta\hbar\omega}{2}\right) = \frac{2}{\pi} \int_0^\infty dx \, \frac{\sin(\beta\hbar\omega x/\pi)}{\sinh x} ,$$

$$1 - e^{-x/2\pi\theta} = \frac{\hbar\beta}{\pi^2} \int_{-\infty}^\infty d\omega \, \sin\left(\frac{\beta\hbar\omega x}{\pi}\right) \arctan\left(\frac{\gamma}{2\omega}\right) , \qquad (19.80)$$

We then obtain from Eq. (19.79)

$$Z_T = 2\cosh\left(\frac{\beta\hbar\epsilon_d}{2}\right) \exp\left\{ \int_{x_0}^\infty dx \, \frac{[1 - e^{-x/2\pi\theta}]\cos(\mu x/2\pi\theta)}{x \sinh x} \right\} , \qquad (19.81)$$

where $\mu \equiv 2\epsilon_d/\gamma$. Finally, by virtue of the integral representation

$$\cosh\left(\frac{\beta\hbar\epsilon_d}{2}\right) = \lim_{x_0 \to 0} \exp\left\{ \int_{x_0}^\infty dx \, \frac{1 - \cos(\mu x/2\pi\theta)}{x \sinh x} \right\} , \qquad (19.82)$$

and the relation $\epsilon_d = \epsilon$, the expression (19.81) takes the form

$$Z_T = 2\exp\left\{ \int_{x_0}^\infty dx \, \frac{1 - e^{-x/2\pi\theta}\cos(\epsilon x/\pi\gamma\theta)}{x \sinh x} \right\} . \qquad (19.83)$$

The expression (19.83) for the partition function is exactly the generalization of the former result (19.61) to nonzero bias. It is straightforward to calculate from Eq. (19.83) the specific heat for nonzero bias. The resulting generalization of the expression (19.64) is

$$c/k_B = \frac{1}{2\pi\theta} - \frac{1}{8\pi^2\theta^2} \operatorname{Re}\left\{ \left(1 + 2i\epsilon/\gamma\right)^2 \psi'\left(\frac{1}{2} + \frac{1 + 2i\epsilon/\gamma}{4\pi\theta}\right) \right\} , \qquad (19.84)$$

where $\psi'(z) = d\psi(z)/dz$ is the trigamma function. From Eq. (19.84), we obtain for the specific heat the leading low-temperature behavior

$$c/k_{\rm B} \;=\; \frac{2\pi}{3}\,\frac{\gamma^2}{\gamma^2 + 4\epsilon^2}\,\frac{k_{\rm B}T}{\hbar\gamma}\;. \tag{19.85}$$

Inserting the expression (19.83) into Eq. (19.13) and taking the derivatives yields the static susceptibility $\overline{\chi}_z(T,\epsilon)$. We find that the linear susceptibility $\overline{\chi}_z^{(\ell)}(\theta) = \overline{\chi}_z(\theta, \epsilon = 0)$ is connected with the specific heat for zero bias by the relation $\overline{\chi}_z^{(\ell)}(\theta) = 8[1 - 2\pi\theta c(\theta)/k_{\rm B}]/\pi\hbar\gamma$. Thus the Wilson ratio for general T is found as

$$R_{1,\ell}(\theta) \;\equiv\; 4\frac{c(\theta)/k_{\rm B}}{\theta\gamma\overline{\chi}_z^{(\ell)}(\theta)} \;=\; \frac{\pi c(\theta)/k_{\rm B}}{2\theta[\,1 - 2\pi\theta c(\theta)/k_{\rm B}\,]}\;. \tag{19.86}$$

This expression has the limiting value $R_{1,\ell}(0) = \pi^2/3$, which is in agreement with the previous result (19.54).

Finally, with the expression (19.83) the static *nonlinear* susceptibility (19.13) is found in the analytic form

$$\overline{\chi}_z(T,\epsilon) \;=\; (2\beta/\pi^2)\,{\rm Re}\,\psi'[\,\tfrac{1}{2} + \beta\hbar(\gamma + 2i\epsilon)/4\pi\,]\;. \tag{19.87}$$

We shall take up again the case $K = \frac{1}{2}$ in Section 22.6.

19.3   Non-Ohmic spectral densities

In this section we study the thermodynamics of the open TSS for non-Ohmic spectral densities of the form (18.46). We restrict the discussion to the unbiased case $\epsilon = 0$.

19.3.1   The sub-Ohmic case

The imaginary-time pair correlation function at zero temperature for $s \neq 1$ can be found from the general expression (18.47) by analytic continuation. We obtain

$$W_{T=0}(\tau) \;=\; 2\delta_s\Gamma(s-1)\left(\frac{\omega_{\rm c}}{\omega_{\rm ph}}\right)^{s-1}\left(1 - (1+\omega_{\rm c}\tau)^{1-s}\right)\;. \tag{19.88}$$

An important feature that distinguishes the sub-Ohmic from the super-Ohmic case is that the pair correlation function at $T = 0$ increases monotonously with τ in the sub-Ohmic case $s < 1$, while it approaches a finite constant in the super-Ohmic case $s > 1$. As a result of this, sub-Ohmic damping is usually quite large down to zero temperature. We have in the regime $\omega_{\rm c}\tau \gg 1$

$$W_{T=0}(\tau) \;=\; 2\delta_s\frac{\Gamma(s)}{(1-s)}(\omega_{\rm ph}\tau)^{1-s} - 2B_s\;, \qquad B_s \;=\; \delta_s\frac{\Gamma(s)}{(1-s)}(\omega_{\rm ph}/\omega_{\rm c})^{1-s}\;.$$

Use of this form yields that the dilute dipole gas requirement (19.31) for zero bias corresponds to the condition

$$\rho_s^{1/(1-s)} \equiv \delta_s^{1/(1-s)} \omega_{\mathrm{ph}}/\Delta_{\mathrm{eff}} \gg 1 \,, \tag{19.89}$$

where $\Delta_{\mathrm{eff}} = \Delta\, e^{B_s}$. In the temperature regime (we put $T_{\mathrm{ph}} = \hbar\omega_{\mathrm{ph}}/k_{\mathrm{B}}$)

$$\delta_s^{(1+s)/(1-s)} \left(T_{\mathrm{ph}}/T \right)^{1+s} \gg 1 \,, \tag{19.90}$$

we may consider the temperature-dependent part of $W(\tau)$ in one-phonon approximation. In addition, we may expand this contribution about $\tau = 0$. We then have

$$e^{-W(\tau)} = e^{-W_{T=0}(\tau)} \left\{ 1 + 2\delta_s \Gamma(1+s)\zeta(1+s)(T/T_{\mathrm{ph}})^{s-1}\, (\tau/\hbar\beta)^2 \right\} \,, \tag{19.91}$$

where terms of order $(\tau/\hbar\beta)^4$ are disregarded. Use of the expression (19.91) in Eq. (19.26) then yields for the specific heat (19.14) in order Δ^2 the analytic form

$$c/k_{\mathrm{B}} = q_s\, e^{p_s} \Gamma\big[3/(1-s), \, p_s \big] (\Delta_{\mathrm{eff}}/2\omega_{\mathrm{ph}})^2 \, (T/T_{\mathrm{ph}})^s \,, \tag{19.92}$$

where $\Gamma(z, a)$ is the incomplete gamma function [88], and where

$$q_s = s^2(s+1)\zeta(s+1)\Big(2\delta_s \Gamma(s)/(1-s) \Big)^{(s+2)/(s-1)} \,, \tag{19.93}$$
$$p_s = 2\delta_s(\omega_{\mathrm{ph}}/\omega_{\mathrm{c}})^{1-s}\Gamma(s)/(1-s) \,.$$

In the regime (19.89), the condition (19.90) is satisfied not only for $k_{\mathrm{B}}T \ll \hbar\Delta$ but also for $k_{\mathrm{B}}T_{\mathrm{ph}} \gtrsim k_{\mathrm{B}}T \gg \hbar\Delta$. Thus, the Schottky peak is absent below T_{ph}, and the specific heat steadily increases like in the Ohmic strong-damping case. Eventually, we may perform the limit $\omega_{\mathrm{c}} \to \infty$ in Eq. (19.92). Then the specific heat takes the form

$$c/k_{\mathrm{B}} = q_s\Gamma[3/(1-s)]\, (\Delta_{\mathrm{eff}}/2\omega_{\mathrm{ph}})^2 \, (T/T_{\mathrm{ph}})^s \,. \tag{19.94}$$

Finally, consider the expression (19.92) in the limit $s \to 1^-$. Interestingly, the respective asymptotic series for the incomplete gamma function can be summed exactly in this limit in the damping regime $\delta_1 = K > 3/2$. In the end, the expression (19.92) turns into the Ohmic result (19.56).

19.3.2   The super-Ohmic case

Because of the low spectral density of low energy excitations in the super-Ohmic case, we are usually in the weak-coupling regime at low temperature. For weak damping, we may expand the self-energy in a power series in δ_s. With a dimensional analysis it is straightforward to see that the "one-phonon process", which is linear in δ_s, is the dominant one in the parameter regime $\delta_s(\omega_{\mathrm{c}}/\omega_{\mathrm{ph}})^{s-1} \ll 1$. Then we may use the expressions (19.42) – (19.44). From these we obtain for a symmetric system the asymptotic low temperature expansion

$$c/k_{\mathrm{B}} = \delta_s \left(\frac{\Delta_{\mathrm{eff}}}{\omega_{\mathrm{ph}}} \right)^{s-1} \sum_{n=0}^{\infty} (2n+s)\Gamma(2n+s+2)\zeta(2n+s+1)\, \theta^{2n+s} \,. \tag{19.95}$$

Here $\zeta(z)$ is Riemann's zeta function, and $\theta = k_B T/\hbar\Delta_{\text{eff}}$ with Δ_{eff} given in Eq. (18.34). Hence the specific heat behaves as T^s as $T \to 0$. In the Ohmic limit $s \to 1$, the expression simplifies to the series (19.52) with $\varrho = 1$. In the high-temperature limit $k_B T \gg \hbar\Delta_{\text{eff}}$ we find from Eqs. (19.42) – (19.44)

$$c/k_B = \frac{1}{4\theta^2}\left\{1 - 2\delta_s(3-s)(2-s)\Gamma(s-1)\zeta(s-1)\left(k_B T/\hbar\omega_{\text{ph}}\right)^{s-1}\right\}. \qquad (19.96)$$

For a discussion of other parameter regimes, we refer to Ref. [394].

19.4 Relation between the Ohmic TSS and the Kondo model

In this section we consider the relation between the spin-boson system with Ohmic dissipation and the Kondo problem. In its simplest form, the Kondo problem deals with a single magnetic impurity of spin 1/2 which interacts via an exchange scattering potential with a band of conduction electrons. The peculiarity of the model has its origin in the constant density of electron-hole excitations in the vicinity of the Fermi surface. In the adiabatic scheme, the constant density of states gives rise to Anderson's orthogonality catastrophe [399], and to logarithmic infrared divergences in a variety of physical quantities, e. g., in the static magnetic susceptibility, or in the soft X-ray absorption and emission of metals [186, 400, 141]. These phenomena have been popularized by Kondo as the Fermi surface effects [136].

The relationship between the spin-boson model and the Kondo model is due to the fact that the low-energy electron-hole excitations have bosonic character and can be interpreted in terms of density fluctuations [187]. In one dimension, there is an exact mapping between fermions and bosons [401]. Here we restrict our attention to the anisotropic Kondo model and to the resonance level model (see Ref. [141]). Generalizations to multi-channel Kondo impurity models which show non-Fermi liquid behavior, and the possible experimental relevance are reviewed in Ref. [402]. In Subsection 19.2.2 we have already discussed a particular case of the correspondence. There we have shown that the Ohmic spin-boson model with the Kondo parameter $K = \frac{1}{2}$ maps on the Toulouse model which is a special case of the Kondo model.

19.4.1 Anisotropic Kondo model

The Kondo model is, at first glance, a simple model in which one assumes that a magnetic impurity interacts with the conduction electrons via a pointlike exchange interaction, so that only s-wave scattering occurs. This reduces the problem to an essentially one-dimensional one. Further, it is assumed that, out of the many different bands which may exist in a solid, only a narrow band of half-width $\hbar\omega_c = \hbar/\tau_c$ which is chosen symmetric about the Fermi energy E_F interacts with the impurity. Linearizing the dispersion relation as in Eq. (19.65), the Kondo Hamiltonian is given by

$$H_K = \hbar v_F \sum_{k,\sigma} k c_{k,\sigma}^\dagger c_{k,\sigma} + J\boldsymbol{S} \cdot \boldsymbol{s}(0)/\hbar, \qquad (19.97)$$

which is also referred to as the s-d Hamiltonian. The operator $c^\dagger_{k,\sigma}$ creates a conduction electron with momentum $\hbar k$ and spin polarization $\sigma = \pm 1$. Further, $S_i = \frac{1}{2}\hbar\tau^i$ is the spin operator of the impurity, where τ^i ($i = 1,2,3$) are the Pauli matrices (to make a distinction from the TSS operators, we choose τ^i). The effective spin operator due to the conduction electrons at the impurity site $r = 0$ is conveniently expressed in terms of Wannier operators for electrons localized at the origin with spin polarization σ,

$$c^\dagger_\sigma(0) = L^{-1/2} \sum_k c^\dagger_{k,\sigma} , \tag{19.98}$$

where $k = 2\pi n/L$ ($n \in \{0, \pm 1, \ldots\}$), and L is a normalization length. We then have

$$s_i(0) = \frac{\hbar}{2} \sum_{\sigma,\sigma'} c^\dagger_\sigma(0)\tau^i_{\sigma,\sigma'} c_{\sigma'}(0) . \tag{19.99}$$

In the original Kondo problem, the exchange interaction satisfies rotational invariance, $J_\parallel = J_\perp$. However, in order to be able to relate the Kondo problem to the spin-boson problem, it is essential to generalize the isotropic coupling in the original model (19.97) to the case where the exchange constants $J_\parallel$ for the $S_z s_z$ term and $J_\perp$ for $S_x s_x + S_y s_y$ are independent parameters and arbitrarily large.

The anisotropic Kondo Hamiltonian takes the form

$$H_K = \hbar v_F \sum_{k,\sigma} k c^\dagger_{k,\sigma} c_{k,\sigma} + \frac{\hbar J_\parallel}{4} \tau_z \sum_\sigma \sigma c^\dagger_\sigma c_\sigma + \frac{\hbar J_\perp}{2} \left(\tau_+ c^\dagger_\downarrow c_\uparrow + \tau_- c^\dagger_\uparrow c_\downarrow \right) \tag{19.100}$$

with $\tau_\pm = (\tau_x \pm i\tau_y)/2$. The $J_\parallel$-term describes scattering of the fermions at the impurity in which the spin polarization is conserved while the $J_\perp$-term describes spin-flip scattering. The parameters $J_\parallel$ and $J_\perp$ have dimension velocity. The dimensionless coupling constants are $\rho J_\parallel$ and $\rho J_\perp$, and the constant density of states ρ is as in Eq. (19.71), $\rho = (2\pi v_F)^{-1}$.

To show the equivalence between the anisotropic Kondo model, Eq. (19.100), and the spin-boson model, we follow the steps of Yuval and Anderson [403], who have cast the partition function for the impurity into the "Coulomb gas" form

$$Z_K = \sum_{m=0}^\infty \left(\frac{\rho J_\perp \cos^2 \delta_K}{2\tau_c} \right)^{2m} \int_0^{\hbar\beta - \tau_c} ds_{2m} \int_0^{s_{2m} - \tau_c} ds_{2m-1} \cdots \int_0^{s_2 - \tau_c} ds_1 \tag{19.101}$$

$$\times \exp\left\{ 2\left(1 - \frac{2}{\pi}\delta_K\right)^2 \sum_{j>k=1}^{2m} (-1)^{j+k} \ln\left[\frac{\hbar\beta}{\pi\tau_c} \sin\left(\frac{\pi(s_j - s_k)}{\hbar\beta} \right) \right] \right\} .$$

Here, $\tau_c = 1/\omega_c$ is the hard-core length of the Coulomb charges, and δ_K is the scattering phase shift at the singular non-spin-flip potential $J_\parallel\delta(x)/4$. The series (19.101) corresponds to the Coulomb gas representation (19.2) with Eq. (19.4) and the Ohmic form (18.56) for $W(\tau)$, and $\epsilon = 0$. The correspondence of the parameters is

$$\Delta = \rho J_\perp \cos^2(\delta_K)/\tau_c ; \qquad K = (1 - 2\delta_K/\pi)^2 . \tag{19.102}$$

The scattering phase shift $\delta_K(J_\parallel)$ depends on the regularization prescription for the singular scattering potential [141]. If we choose a separable form, we obtain

$$\delta_K(J_\parallel) \; = \; \arctan(\pi\rho J_\parallel/4) \,, \tag{19.103}$$

while the smoothing prescription $\delta(x) \to (2\pi)^{-1} \int_{-k_c}^{k_c} dk\, e^{ikx}$, where k_c is large, gives $\delta_K = \pi\rho J_\parallel/4$. In Eq. (19.103), only the contribution which is linear in $J_\parallel$ is universal. This implies that the correspondence between the anisotropic Kondo model and the Ohmic TSS is universal in the regime $\rho J_\perp = \Delta/\omega_c \ll 1$ and $\rho|J_\parallel| = |1 - K| \ll 1$.

A local magnetic field h_m in z-direction which couples to the impurity spin, and not to the conduction electrons, gives to H_K the additional contribution $-\frac{1}{2}g\mu_B h_m\tau_z$. Thus, the bias energy $\hbar\epsilon$ in the TSS corresponds to $g\mu_B h_m$. Introducing the static susceptibility χ_{sd} with zero Landé factor for the electron band, the Wilson ratio is found as [cf. Eq. (57) in Ref. [396], and footnote to Eq. (19.13)]

$$R_1 \; = \; \lim_{T \to 0} \frac{(g\mu_B)^2\, c(T)/k_B}{T\chi_{sd}} \; = \; \frac{2\pi^2}{3}\left(1 - \frac{2\delta_K}{\pi}\right)^2, \qquad \frac{\chi_{sd}}{(g\mu_B)^2} = \frac{\bar{\chi}_z}{4}\,. \tag{19.104}$$

This result is in agreement with the Wilson ratio for the Ohmic TSS (see the discussion in Section 22.9). We see from Eq. (19.103) that the regime $-\infty < \rho J_\parallel < \infty$ maps on the regime $4 > K > 0$. The critical coupling $K = 1$ separates the ferromagnetic Kondo regime $\rho J_\parallel < 0$, which corresponds to the regime $K > 1$, from the more intriguing antiferromagnetic regime $\rho J_\parallel > 0$, i. e., $K < 1$. The special choice $\delta_K(J_\parallel) = (2 - \sqrt{2})\pi/4$, which corresponds to the case $K = \frac{1}{2}$, is known in the literature as the Toulouse limit of the anisotropic Kondo model. This limiting case is exactly solvable in closed form [141].

In the antiferromagnetic regime, the Kondo temperature $T_K \propto \Delta_r$ separates the perturbative regime $T > T_K$, in which physical quantities, e.g., the magnetic susceptibility, can be expanded in powers of $J_\perp$, from the nonperturbative regime. In the regime $\rho J_\perp \ll \rho J_\parallel \ll 1$, which is equivalent to $\Delta/\omega_c \ll 1 - K \ll 1$, the Kondo temperature is given by [141]

$$k_B T_K \; = \; \frac{2}{\pi}\left(\frac{J_\perp}{2J_\parallel}\right)^{1/\rho J_\parallel} \hbar\omega_c \; = \; \frac{2}{\pi}\left(\frac{1}{2(1-K)}\right)^{1/(1-K)} \hbar\Delta_r\,. \tag{19.105}$$

The second equality follows with the equivalence relations (19.102), and with (18.36). In the antiferromagnetic sector, the properties for $T \ll T_K$ can be described phenomenologically in the spirit of the Landau theory for a Fermi liquid [397].

19.4.2 Resonance level model

The other model of interest is the *resonance level model*. This model has been introduced as a generalization of the Toulouse model (19.66) by Schlottmann [398], and independenly by Vigmann and Finkels'teĭn [396] (for a review of the resonance level model, see Ref. [141]). The corresponding Hamiltonian describes a *d*-level near

the Fermi surface interacting with a band of spinless fermions. In generalization of
the model (19.66), the fermions interact with each other through a repulsive contact
potential which represents the (screened) Coulomb interaction. Linearizing again the
dispersion relation about the Fermi momentum, the Hamiltonian takes the form

$$H_{\text{RL}} = \hbar v_{\text{F}} \sum_k k c_k^\dagger c_k + \hbar \epsilon_d d^\dagger d + \frac{V}{\sqrt{L}} \sum_k \left(c_k^\dagger d + d^\dagger c_k \right)$$
$$+ \frac{U}{L} \sum_{k,k'} \left(c_k^\dagger c_{k'} - c_{k'} c_k^\dagger \right) \left(d^\dagger d - \tfrac{1}{2} \right) .$$

(19.106)

The partition function of this model is equivalent to Eq. (19.2) if we identify

$$\Delta^2 = 4 \rho \omega_{\text{c}} \frac{V^2}{\hbar^2} \cos^2(\delta_{\text{RL}}) ; \quad K = \frac{1}{2} \left(1 - \frac{2}{\pi} \delta_{\text{RL}} \right)^2 ; \quad \epsilon = \epsilon_{\text{d}} ,$$

(19.107)

where $\delta_{\text{RL}} = \pi \rho U / \hbar + \mathcal{O}(U^2)$ is the scattering phase for the contact potential $U\delta(x)$,
and where again $\rho = (2\pi v_{\text{F}})^{-1}$. In the absence of the repulsive interaction, $U = 0$,
we have $K = \frac{1}{2}$. This case corresponds to the Toulouse limit of the antiferromagnetic Kondo Hamiltonian. Perturbation expansion in ρU in the resonant level model
corresponds to perturbation expansion about $K = \frac{1}{2}$ in the two-state model.

Here we have discussed the similarity between the Ohmic two-state system and
the Kondo problem on the level of the partition function. The comparison of these
models on the basis of their Hamiltonians is set out in Ref. [85]. We should like to
remark that the mapping of the Ohmic two-state system neither to the anisotropic
Kondo model nor to the resonant level model is exact. However, since the low-energy
excitations of these models are very similar, the physical phenomena of these models
at low temperature are closely related.

19.5 Equivalence of the Ohmic TSS with the $1/r^2$ Ising model

It has been shown by Anderson and Yuval [404] that the Ising model with ferromagnetic $1/r^2$ pair interaction can be mapped on the Kondo model. The inverse square
ferromagnetic Ising model is another specific realization of the Coulomb gas model.
The Ising spins represent the imaginary time history of the single impurity spin of
the Kondo model. The short-time regularization $\tau_{\text{c}} = 1/\omega_{\text{c}}$ is provided by the lattice,
$N\tau_{\text{c}} = \hbar\beta$. With the correspondence of the Kondo model to the TSS explained in the
previous section, the inverse-square Ising model is also equivalent to the Ohmic TSS.
Direct correspondence of these models has been shown by Spohn and Dümcke with
functional integral techniques [384].

Consider the one-dimensional N-site Ising model for spins $\{S_j = \pm 1\}$ with ferromagnetic long-range interactions, a magnetic field term, and periodic boundary
conditions. In the state representation, the partition function reads

$$Z_{\text{Ising}} = \sum_{S_1, \cdots, S_N} \exp \left\{ \sum_{j > i} V(j - i) S_j S_i - \frac{h_{\text{m}}}{2} \sum_j S_j \right\} .$$

(19.108)

The first term in the exponent represents the spin interactions, and the second term is a magnetic field contribution. Next, we change over from the state representation (19.108) to the charge representation. Concerning the interaction term, this change is analogous to switching over in the influence functional from the form (4.71) to the form (4.72). The Coulomb gas representation of the partition function reads [405]

$$Z_{\text{Ising}} = \sum_{n=0}^{\infty} y^{2n} \int_0^{\hbar\beta - \tau_c} \frac{ds_{2n}}{\tau_c} \int_0^{s_{2n} - \tau_c} \frac{ds_{2n-1}}{\tau_c} \cdots \int_0^{s_2 - \tau_c} \frac{ds_1}{\tau_c} \qquad (19.109)$$

$$\times \exp \left\{ 4 \sum_{j=2}^{2n} \sum_{i=1}^{j-1} (-1)^{j+i} U[(s_j - s_i)/\tau_c] + h_{\text{m}} \sum_{j=1}^{2n} (-1)^j s_j/\tau_c \right\},$$

where $V(k)$ is the second derivative of $-U(k)$ in discretized form,

$$V(k) = 2U(k) - U(k+1) - U(k-1), \qquad (19.110)$$

and where $y = \exp[2U(0)]$ is the fugacity of the Coulomb gas. Upon requiring correspondence of Eq. (19.109) with the partition function of the Ohmic TSS, Eq. (19.2) with Eqs. (19.3), (19.4), and (18.56), we obtain

$$U(n) = \frac{K}{2} \ln \left[\left(\frac{N}{\pi} \right) \sin \left(\frac{\pi n}{N} \right) \right], \qquad n \geq 1, \qquad (19.111)$$

and

$$y = \Delta/2\omega_c, \qquad N = \hbar\beta/\tau_c, \qquad \epsilon = -h_{\text{m}}/\tau_c. \qquad (19.112)$$

Substituting the form (19.111) into Eq. (19.110), we find

$$V(n) = \frac{K}{2} \frac{(\pi/N)^2}{\sin^2(\pi n/N)} + \mathcal{O}[(\pi/N)^4], \qquad n \geq 2. \qquad (19.113)$$

The spin interaction $V(n)$ is (apart from a constant) the discretized version of the Ohmic kernel $K(\tau)$ in Eq. (4.57).

In the limit $N \to \infty$, we obtain $2V(1) = 2U(0) + KC$, where $C = \ln 2$.[6] In this limit, the interaction between the spin states depends inversely on the square of the distance. The Hamiltonian of the Ising model is found in the usual notation as

$$H_{\text{Ising}} = \frac{\hbar}{N\tau_c} \left(-\frac{J_{NN}}{2} \sum_j S_{j+1} S_j - \frac{J_{LR}}{2} \sum_{j>i} \frac{(\pi/N)^2 S_j S_i}{\sin^2[\pi(j-i)/N]} + \frac{h_{\text{m}}}{2} \sum_j S_j \right), \quad (19.114)$$

where $J_{LR} = K$, and $J_{NN} + J_{LR} = -2V(1)$. The correspondence relations with the Ohmic TSS are

$$\frac{\Delta}{2\omega_c} = \exp[-J_{NN} - (1+C)J_{LR}], \qquad K = J_{LR}, \qquad (19.115)$$

---

[6]The constant C depends on the particular hard-core regularization chosen.

and the relations (19.112). The equivalence of the Ohmic TSS with the $1/r^2$ Ising model has been utilized recently [406, 407]. In these studies, the imaginary-time correlation function is calculated by Monte Carlo simulations on the Ising system and then continued to real-time by employing a Padé approximation.

The equivalence of the TSS with a long-range ferromagnetic Ising model can also be established for non-Ohmic spectral coupling. For $G_{lf}(\omega) \propto \omega^s$, the spin interaction $V(j - i)$ in the Ising model at zero temperature falls off as $(j - i)^{-(1+s)}$. Using the correspondence, a number of properties proven rigorously for the Ising model [384, 408] can directly be transferred to the dissipative two-state system.

20. Electron transfer and incoherent tunneling

Before we turn in Chapter 21 to the real-time dynamics of the TSS , we first study quantum-statistical tunneling with the thermodynamic method discussed above in Part III. The results shall be confirmed in Chapter 21 by a dynamical approach.

20.1   Electron transfer

Electron transfer plays an important role in many processes in chemistry and biology. A striking example is the ultrafast primary electron transfer step found in photosynthetic reaction centers which is responsible for the high efficiency of the photosynthetic mechanism. Significant progress with the understanding of the essence of electron transfer (ET) processes in condensed matter has been made by Marcus [105, 106]. He was the first who discovered the importance of the solvent environment. To describe the effects of the solvent coupled to the electronic degree of freedom, he employed linear response theory. In the usual ET reaction, there is a free energy barrier separating reactants and products. The barrier is because the donor and acceptor states are strongly solvated, and the transfer of the electron then requires a reorganization of the environment. At normal temperature, the electron has to tunnel through the barrier from the reactant to the product state. The tunneling is only effective if suitable bath fluctuations bring reactant and product energy levels into resonance. In the classical limit, the transfer rate is determined by two factors. First, a Boltzmann factor with the activation energy required for bath fluctuations. Second, an attempt frequency prefactor describing the probability for tunneling once the levels are in resonance. The latter factor is mainly determined by the overlap between the electronic wave functions localized on different redox sites. For large electronic coupling the reaction is adiabatic, whereas the reaction is nonadiabatic for weak electronic coupling,

It is implied by the success of the classical Marcus theory that the electron can be described in terms of a discrete variable at all temperatures of interest. Here we restrict the attention to the important case of a two-state system or a spin-1/2 system with localized states at $q = \frac{1}{2}q_0\sigma$: a *donor* state D with eigenvalue $\sigma = -1$ of σ_z, and an *acceptor* state A with eigenvalue $\sigma = +1$, and we describe the solvent in terms of a linearly responding heat bath. Then the proper Hamiltonian is the spin-boson Hamiltonian, as given in Eq. (3.139) with (3.128) or in Eq. (18.15),

$$H = -\tfrac{1}{2}\hbar\Delta\,\sigma_x - \tfrac{1}{2}\hbar\epsilon\,\sigma_z + \tfrac{1}{2}\mu\mathcal{E}(t)\,\sigma_z + H_{\mathrm{R}}\,. \qquad (20.1)$$

Here, $H_{\mathrm{TSS}} = -\tfrac{1}{2}\hbar\Delta\,\sigma_x - \tfrac{1}{2}\hbar\epsilon\,\sigma_z$ describes the tunneling of the electron between the donor and the acceptor state, and H_{R} represents the solvent as a Gaussian reservoir. The collective bath mode $\mu\mathcal{E}(t) = -\tfrac{1}{2}q_0 \sum_\alpha c_\alpha x_\alpha(t)$ is coupled to the spin operator σ_z. It can be thought of as a fluctuating dynamical polarization energy. The polarization field $\mathcal{E}(t)$ vanishes on average, and μ is the difference of the dipole moments of the two electronic states.

All effects of the solvent on the electron transfer are contained in the properties of the bath correlation function [cf. Eqs. (4.70) and (18.44)]

$$\left(\frac{\mu}{\hbar}\right)^2 \big\langle\,\mathcal{E}(z)\mathcal{E}(0)\,\big\rangle_\beta = \frac{d^2 Q(z)}{dz^2} = \int_0^\infty d\omega\, G(\omega)\,\frac{\cosh[\omega(\hbar\beta/2 - iz)]}{\sinh[\omega\hbar\beta/2]}\,, \qquad (20.2)$$

where $z = t - i\tau$ is a complex time.

20.1.1 Adiabatic bath

An adiabatic bath is a sluggish bath. It is characterized by only zero frequency fluctuations and is therefore not dynamical. In practice, electron transfer (ET) systems are adiabatic when $\omega_c \ll \Delta$ and $\omega_c \ll k_{\mathrm{B}}T/\hbar$, where ω_c is a typical bath frequency, and when frequencies significantly higher than ω_c are absent in the bath. The case of an adiabatic bath can be solved in analytical form [409].

In the adiabatic regime, Eq. (20.2) reduces to

$$\mu^2\big\langle\mathcal{E}(z)\mathcal{E}(0)\big\rangle_\beta \;\longrightarrow\; \mu^2\big\langle|\mathcal{E}|^2\big\rangle_\beta = 2\Lambda_{\mathrm{cl}}/\beta\,, \qquad (20.3)$$

where

$$\Lambda_{\mathrm{cl}} = \hbar \int_0^\infty d\omega\, \frac{G(\omega)}{\omega} \qquad (20.4)$$

is the classical *reorganization energy*, or *solvation energy*, or *coincidence energy*.

For $G(\omega) = 2\delta_s\omega_{\mathrm{ph}}^{1-s}\omega^s\,e^{-\omega/\omega_c}$, the reorganization energy is given by

$$\Lambda_{\mathrm{cl}} = 2\delta_s\Gamma(s)\,(\omega_c/\omega_{\mathrm{ph}})^{s-1}\hbar\omega_c\,. \qquad (20.5)$$

In particular, the Ohmic case $s = 1$ gives with $\delta_1 = K$

$$\Lambda_{\mathrm{cl}} = 2K\hbar\omega_c\,. \qquad (20.6)$$

For a linearly responding bath, the polarization energy $\mu\mathcal{E}$ has the normalized Gaussian distribution with mean square (20.3),

$$\rho(\mathcal{E}) = \frac{\mu}{2}\sqrt{\frac{\beta}{\pi\Lambda_{\mathrm{cl}}}}\,\exp\left(-\beta\mu^2\mathcal{E}^2/4\Lambda_{\mathrm{cl}}\right)\,. \qquad (20.7)$$

In the adiabatic limit, the polarization is slow, and the Born-Oppenheimer method gives for the Hamiltonian $H_{TSS} + \frac{1}{2}\mu\mathcal{E}\sigma_z$ two electronic eigenstates with energies

$$E_\pm(\mathcal{E}) = \pm\tfrac{1}{2}\hbar\Omega(\mathcal{E}) , \qquad \Omega(\mathcal{E}) \equiv \sqrt{\Delta^2 + (\epsilon - \mu\mathcal{E}/\hbar)^2} . \qquad (20.8)$$

The adiabatic partition function is an average with the Gaussian probability distribution (20.7) of Boltzmann factors, where the activation energies are given by Eq. (20.8),

$$Z_{ad} = \int_{-\infty}^{\infty} d\mathcal{E}\, \rho(\mathcal{E}) \left(e^{-\beta E_+(\mathcal{E})} + e^{-\beta E_-(\mathcal{E})}\right) = 2\int_{-\infty}^{\infty} d\mathcal{E}\, \rho(\mathcal{E}) \cosh[\tfrac{1}{2}\beta\hbar\Omega(\mathcal{E})] . \quad (20.9)$$

The expression (20.9) is also the partition function of a 1D Ising model with *infinite range* interactions. To discuss the qualitative features, we rewrite Eq. (20.9) as

$$Z_{ad} = \frac{\mu}{2} \sqrt{\frac{\beta}{\pi\Lambda_{cl}}} \int_{-\infty}^{\infty} d\mathcal{E} \left(e^{-\beta F_+(\mathcal{E})} + e^{-\beta F_-(\mathcal{E})}\right) , \qquad (20.10)$$

where

$$F_\pm(\mathcal{E}) = \frac{\mu^2\mathcal{E}^2}{4\Lambda_{cl}} \pm \frac{\hbar}{2}\sqrt{\Delta^2 + (\epsilon - \mu\mathcal{E}/\hbar)^2} \qquad (20.11)$$

are the adiabatic potential energies for the polarization energy $\mu\mathcal{E}$.

Consider first the symmetric case $\epsilon - 0$, where the two redox states are solvated with the same energy. The adiabatic surfaces are qualitatively different depending on the size of the parameter

$$p = \Lambda_{cl}/\hbar\Delta . \qquad (20.12)$$

Measuring the polarization energy in units of $\hbar\Delta$, $\phi = \mu\mathcal{E}/\hbar\Delta$, we then have

$$F_\pm(\phi) = \frac{\hbar\Delta}{2} \left(\frac{\phi^2}{2p} \pm \sqrt{1+\phi^2}\right) . \qquad (20.13)$$

The potential $F_+(\phi)$ is monostable with its minimum at $\phi = 0$ for all p, while $F_-(\phi)$ is monostable with minimum at $\phi = 0$ for $p \le 1$ only. If $p > 1$, the potential $F_-(\phi)$ is bistable with minima $F_-^{(min)} = -[p + 1/p]\hbar\Delta/4$ at $\phi = \pm\phi_0$, where $\phi_0 = \sqrt{p^2 - 1}$. This behavior is illustrated in Fig. 20.1. Thus, if $p > 1$, the average spin $\langle\sigma_z\rangle$ is nonzero as $\beta \to \infty$. Hence the *adiabatic* spin-boson model exhibits broken symmetry in the ground state, which is a type of fluctuation-induced *self-trapping*. Whether trapping occurs for an adiabatic bath is decided by the competition between the reorganization energy Λ_{cl} and the energy for resonant tunneling $\hbar\Delta$. Slowly fluctuating fields may impede tunneling and may even prevent transitions. When spontaneous fluctuations are absent, the reorganization energy is zero, and then the particle tunnels clockwise.

The localization transition for an adiabatic bath, $\omega_c \ll k_B T/\hbar$, is not sensitive to whether the spectral density is sub-Ohmic, Ohmic or super-Ohmic, since only the reorganization energy, which is the value of the integral in Eq. (20.4), decisively matters. The situation is different for a *dynamical* bath, $\omega_c \gg k_B T/\hbar$ and $\omega_c \gg \Delta$. The localization phenomenon is conveniently studied either by considering the flow of the parameters in the renormalization group equations [85] or by employing a

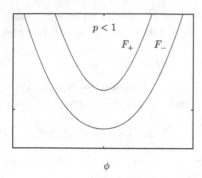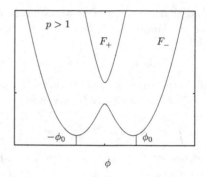

Figure 20.1: The adiabatic Born-Oppenheimer surfaces $F_+(\phi)$ and $F_-(\phi)$ for the symmetric spin-boson model, Eq.(20.13). The parameter p is chosen as $p = 1/2$ (left) and $p = 8$ (right). The distance between the surfaces at $\phi = 0$ is $\hbar\Delta$.

variational method [384]. One finds that the ground state exhibits symmetry breaking in the sub-Ohmic case when the inequality (18.38) is reversed [384, 393]. In the Ohmic case, the localization transition occurs at the Kondo parameter $K = 1$ [142, 143, 85]. Here again slow modes are responsible for the symmetry breaking. The phase diagram as a function of Δ/ω_c for $s \leq 1$ is discussed in Ref. [384]. It is shown by using an energy-entropy argument that there is *no* phase transition to localization when

$$\int_0^\infty d\omega\, G(\omega)/\omega^2 < \infty . \qquad (20.14)$$

Thus, the symmetry of the ground state is not spontaneously broken for super-Ohmic dissipation of any strength.

When the thermal bath is adiabatic, the dynamics can be solved in analytic form. The route of solution consists of two steps. Consider for instance the thermal equilibrium correlation function $C^+(t) \equiv \langle \sigma_z(t)\sigma_z(0) \rangle_\beta$. In the first step, the dynamics of the TSS is calculated for a constant polarization energy $\mu\mathcal{E}$. We obtain[1]

$$C_0^+(t;\mathcal{E}) = \frac{(\epsilon - \mu\mathcal{E}/\hbar)^2}{\Omega^2} + \frac{\Delta^2}{\Omega^2}\cos(\Omega t) - i\tanh\left(\tfrac{1}{2}\beta\hbar\Omega\right)\frac{\Delta^2}{\Omega^2}\sin(\Omega t) . \qquad (20.15)$$

The tunneling frequency $\Omega = \Omega(\mathcal{E})$ is defined in Eq. (20.8). In the second step, the correlation function (20.15) is averaged with the adiabatic weight functional given in Eq. (20.9). Thus we get in the adiabatic regime

$$C_{\mathrm{ad}}^+(t) = \frac{2}{Z_{\mathrm{ad}}}\int_{-\infty}^\infty d\mathcal{E}\, \rho(\mathcal{E})\cosh\left[\tfrac{1}{2}\beta\hbar\Omega(\mathcal{E})\right]C_0^+(t;\mathcal{E}) . \qquad (20.16)$$

---

[1]Here we assume that the two eigenstates with energies $E_\pm(\mathcal{E})$ are thermally occupied. The real part of Eq. (20.15) is the result of a standard quantum mechanical calculation for a biased TSS, and the imaginary part follows with the fluctuation-dissipation theorem [cf. Eqs. (6.63) and (6.64)].

Hence the dynamics for an adiabatic bath is reduced to quadrature. The two-state correlation function of the free system is broadened by adiabatic bath fluctuations. Due to the time-independent part in Eq. (20.15), the "adiabatic" expression (20.16) does not decay to zero at asymptotic time. To render proper decay of the correlations, we must add dynamical fluctuations [see Chapter 21].

20.1.2 Marcus theory for electron transfer

In the absence of nuclear (solvent) tunneling, the electron transfer dynamics is well understood in two different limits. In the *adiabatic limit*, the electronic coupling Δ is so large that the parameter p is of order one or smaller. When p is above one, but not extremely large, the rate is controlled by the motion of the solvent on the lower electronic surface [cf. Fig. 20.1 (right)] and well described by the classical activated rate theory [106]. It is expected to depend on the electronic coupling Δ like

$$k \propto \exp[-(F_-^* - \hbar\Delta/2)/k_{\rm B}T] \,, \tag{20.17}$$

where F_-^* is the free energy barrier for $\Delta = 0$, $F_-^* = \Lambda_{\rm cl}/4$.

In the opposite *nonadiabatic limit*, the electronic coupling is weak, $p \gg 1$. To lowest order in the coupling Δ, the transfer rate is given by the Golden Rule formula

$$k \propto \Delta^2 \exp(-\Lambda_{\rm cl}/4k_{\rm B}T) \,. \tag{20.18}$$

In the two limits, one assumes a separation of time scales. The adiabatic limit is characterized by slow fluctuations of the solvent (nuclear degree of freedom) such that the Born-Oppenheimer approximation for the electron can be used. In the nonadiabatic limit, the solvent fluctuations are assumed to be fast. Then the ET is determined by thermal excitations of the solvent to the vicinity of the crossing region which are followed by a fast transversal of this region.

The nonadiabatic regime is conveniently discussed with the adiabatic potential surfaces (20.11). For $\Delta = 0$, the free energy curves become two intersecting parabola,

$$F_\pm(\mathcal{E}) = \pm \frac{\hbar\epsilon}{2} + \frac{1}{4\Lambda_{\rm cl}}(\mu\mathcal{E} \mp \Lambda_{\rm cl})^2 - \frac{\Lambda_{\rm cl}}{4} \,. \tag{20.19}$$

which are known in the literature as the *Marcus parabola*. The parabola are the adiabatic energy functions for *diabatic* states ("diabatic" states are the eigenstates for $\Delta = 0$). In Fig. 20.2 the Marcus parabola are displayed schematically for three different values of the bias parameter ϵ. Here we choose ϵ positive. The higher local minimum in Figs. 20.2 (b) and (c) which is situated at $\mathcal{E} = \Lambda_{\rm cl}/\mu$ is the donor (D) state, and the lower local minimum situated at $\mathcal{E} = -\Lambda_{\rm cl}/\mu$ is the acceptor (A) state. Pictorially, the reorganization energy is the difference of the free energy $F_-(\mathcal{E})$ between the donor state and the acceptor state,

$$\Lambda_{\rm cl} = F_-(\mathcal{E} = \Lambda_{\rm cl}/\mu) - F_-(\mathcal{E} = -\Lambda_{\rm cl}/\mu) \,. \tag{20.20}$$

Since the nuclear degree of freedom (polarization) is slow compared to the electronic degree of freedom, the transition is determined by the *Franck-Condon principle*. This

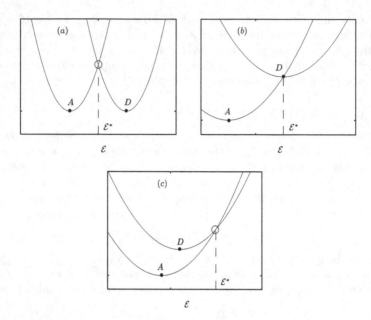

Figure 20.2: Free energy of the diabatic ($\Delta = 0$) electronic states as a function of the nuclear reaction coordinate $\mathcal{E}$ ("Marcus parabola"). The figures show the schematic dependence on the bias ϵ. (a) Symmetric case $\epsilon = 0$. (b) Activationless case $\epsilon = \Lambda_{\mathrm{cl}}/\hbar$. (c) Inverted regime $\epsilon > \Lambda_{\mathrm{cl}}/\hbar$. The intersection point of the two parabola is at $\mathcal{E}^* = \hbar\epsilon/\mu$.

principle states that transitions between electronic states occur vertically in Fig. 20.2; that is, the nuclear reaction coordinate $\mathcal{E}$ does not change during the electronic transition process. The vertical process, however, is energetically unfavorable unless the bath configuration $\mathcal{E}$ is in the *Landau-Zener* regime, $\mathcal{E} \approx \mathcal{E}^*$, where $\mathcal{E}^* = \hbar\epsilon/\mu$ is the point of intersection of the two curves. The activation (free) energies for this particular bath fluctuation relative to the D and A state are

$$
\begin{aligned}
F_D^* &\equiv F_+(\mathcal{E}^*) - F_+(\mathcal{E} = +\Lambda_{\mathrm{cl}}/\mu) = (\hbar\epsilon - \Lambda_{\mathrm{cl}})^2/4\Lambda_{\mathrm{cl}} , \\
F_A^* &\equiv F_-(\mathcal{E}^*) - F_-(\mathcal{E} = -\Lambda_{\mathrm{cl}}/\mu) = (\hbar\epsilon + \Lambda_{\mathrm{cl}})^2/4\Lambda_{\mathrm{cl}} .
\end{aligned}
\tag{20.21}
$$

In the symmetric case $\epsilon = 0$, we have $F_D^* = F_A^* = \Lambda_{\mathrm{cl}}/4$. With increasing bias the activation energy relative to the donor state becomes smaller, and therefore the $D \to A$ (forward) rate becomes larger. When the condition $\hbar\epsilon = \Lambda_{\mathrm{cl}}$ is reached, the free energy barrier has disappeared. Since at this point $F_D^* = 0$, we have activationless transfer which depends only weakly on temperature. As the bias is increased further, the activation energy grows again. For this reason, the regime $\hbar\epsilon > \Lambda_{\mathrm{cl}}$ is called the *inverted regime*. Altogether, the Marcus theory predicts that the transfer rate is a non-monotonous function of the bias. The sweeping success of Marcus' theory is founded on the discovery of this characteristic behavior in ET measurements.

In the low temperature regime, nuclear tunneling becomes relevant [106]. This collective tunneling phenomenon is generally thought to prevail in the inverted regime. However, one finds that nuclear tunneling at low T may dominate the transfer dynamics even for a symmetric system. The quantum effects depend on the low-frequency behavior of the spectral density, the coupling strength, and the cutoff frequency. These features will be discussed in some detail for the nonadiabatic regime in Section 20.2.

In a *classical* description of activated barrier crossing, the rate expression consists of the Boltzmann factor $\exp(-\beta F^*_{D/A})$ and of the pre-exponential factor or attempt frequency [cf. Eq. (10.4)]. The Boltzmann factor represents the probability for a fluctuation so that $\mathcal{E} = \mathcal{E}^*$. It has been attempted to integrate higher-order effects in the electronic coupling by using a Landau-Zener pre-exponential factor which describes the probability for tunneling on resonance $\mathcal{E} = \mathcal{E}^*$ and depends on the details of the dynamics in the Landau-Zener transition region [106, 410, 411]. Another possibility is to include terms of higher order in the electronic coupling Δ as a geometrical series [412, 413]. When the collective bath motion is slow compared with the motion of the TSS, we may insert into the formally exact series given below in Eq. (22.196) for the bath correlation function (18.44) the approximate form $Q(t) = i[\Lambda_{cl}t - \Lambda_{cl}\omega_R t^2]/\hbar + \Lambda_{cl}t^2/\hbar^2\beta$, where ω_R is a characteristic bath frequency. Then the integrations over $s_1,\ldots,s_{n-1}$ in Eq. (22.196) lead to a product of $n-1$ delta functions. As a result, the $\tau_2,\ldots,\tau_n$ integrals are trivial, and the residual τ_1 integral is Gaussian. In the end, the classical rate k^+_{cl} from the D state to the A state (forward rate), and the respective backward rate k^-_{cl}, take the Zusman-type form

$$k^\pm_{cl}(T, \epsilon) = \frac{\hbar\Delta^2}{4 + 2\pi\hbar\Delta^2/\Lambda_{cl}\omega_R} \left(\frac{\pi}{\Lambda_{cl}k_B T}\right)^{1/2} \exp\left(-\frac{(\hbar\epsilon \mp \Lambda_{cl})^2}{4\Lambda_{cl}k_B T}\right). \tag{20.22}$$

The prefactor describes dynamical recrossings of the dividing surface. The forward and backward rates obey detailed balance as required,

$$k^-_{cl}(T, \epsilon) = k^+_{cl}(T, -\epsilon) = k^+_{cl}(T, \epsilon) \exp(-\beta\hbar\epsilon). \tag{20.23}$$

The frequency scale ω_R in Eq. (20.22) depends on details of the solvent model [412]. It may be identified, e.g., with the renormalized frequency defined in Eq. (14.12). The classical forward rate has a maximum at $\hbar\epsilon = \Lambda_{cl}$, and it is *symmetric* around this maximum. At very high temperatures, the Arrhenius factor is saturated, and the rate shows a universal $T^{-1/2}$ temperature dependence. In the adiabatic ("large" Δ) limit, the Marcus parabola are split up and the transfer is described by Transition State Theory (TST) on the lower energy surface. The rate expression becomes independent of the electronic coupling. We obtain from Eq. (20.22)

$$k^+_{cl,TST}(T, \epsilon) = \frac{\omega_R}{2\pi} \left(\frac{\pi\Lambda_{cl}}{k_B T}\right)^{1/2} \exp\left(-\frac{(\hbar\epsilon - \Lambda_{cl})^2}{4\Lambda_{cl}k_B T}\right). \tag{20.24}$$

In the opposite nonadiabatic ("small" Δ) limit, the expression (20.22) reduces to the classical *Golden Rule* rate

$$k_{\text{cl,GR}}^{+}(T, \epsilon) = \frac{\hbar \Delta^2}{4} \left(\frac{\pi}{\Lambda_{\text{cl}} k_{\text{B}} T} \right)^{1/2} \exp\left(-\frac{(\hbar \epsilon - \Lambda_{\text{cl}})^2}{4 \Lambda_{\text{cl}} k_{\text{B}} T} \right) . \qquad (20.25)$$

Whether the expression (20.24) or the formula (20.25) applies depends on whether the "adiabaticity parameter"

$$g = \pi \hbar \Delta^2 / 2 \omega_{\text{R}} \Lambda_{\text{cl}} \qquad (20.26)$$

is large or small. For $g \ll 1$, we have the nonadiabatic rate expression (20.25), while for $g \gg 1$, we obtain the Δ independent adiabatic rate (20.24). The rate in the adiabatic limit is generally much larger than in the nonadiabatic limit.

Attempts for a unified treatment of the crossover region between the adiabatic and nonadiabatic limits, in which the assumed separation of time scales is violated, have been largely founded on an imaginary-time formulation and on an assumed relationship between the ET rate and an analytic continuation of the free energy (cf. Section 12.5). Among these approaches are treatments which are based on the centroid free energy method (cf. Section 12.6) put forward, e.g., by Gehlen *et al.* [414] and extended later on to complex-valued centroid coordinates by Stuchebrukhov and Song [415], and on numerical instanton methods [416].

A profound theoretical treatment of the crossover region which is free of questionable assumptions is based on the real-time path integral formulation. The respective discussion is given in Section 22.8.

20.2   Incoherent tunneling in the nonadiabatic regime

When the reduced system is ergodic (cf. Section 3.2), it approaches thermal equilibrium as time passes by. For sufficiently high temperature and/or strong coupling, the system exhibits overdamped exponential relaxation towards the equilibrium state. This regime may be studied by computation of the evolution of occupation probabilities. We shall postpone the corresponding real-time study to the next chapter. In this section, our concern is to calculate the transition rates of the spin-boson model using the free energy (imaginary-time) method. Later on, we shall confirm that the results obtained agree with those of the real-time approach.

In Section 12.1, we have given formally exact quantum mechanical rate expressions which form the basis of numerical computation. Here we confine ourselves to the so-called nonadiabatic transfer limit. This is the limit in which the tunneling matrix element Δ is small compared to characteristic vibrational frequencies and to the reorganization frequency $\Lambda_{\text{cl}}/\hbar$. Then we may employ perturbation theory in the tunneling matrix element Δ. The so-called "Golden Rule" approach to the spin-boson tunneling rate, which is widely used in chemical physics and in condensed matter physics, treats the system-environment coupling to all orders and the transfer matrix in second order. In the terminology of the extended system, the nonadiabatic rate corresponds to the single-bounce rate expression. The nonadiabatic regime is relevant, e.g., for electron transfer and for defect tunneling in solids (cf. the recent review in Ref. [378]). Systematic corrections to the nonadiabatic rate are discussed within a dynamical approach in Subsection 22.8.

20.2.1 General expressions for the nonadiabatic rate

In the nonadiabatic regime, the second term in the expression (19.32) describes the dominant tunneling contribution to the free energy connected with the right/left state. We have discussed already in Subsection 17.4 that the stationary point $\bar{\tau}$ of the integrand in Eq. (19.32) is a saddle point with the direction of steepest descent along real time. Thus, by the corresponding deformation of the integration path, the free energy (19.32) acquires an imaginary part which is

$$\mathrm{Im}\, F_{\mathrm{L/R}} = -\frac{\hbar\Delta^2}{8i} \int_{\bar{\tau}-i\infty}^{\bar{\tau}+i\infty} d\tau\, e^{\pm\,\epsilon\tau-W(\tau)} = -\frac{\hbar\Delta^2}{8} \int_{-\infty-i\bar{\tau}}^{\infty-i\bar{\tau}} dz\, e^{\pm\,i\epsilon z-Q(z)} . \qquad (20.27)$$

The analytically continued function $Q(z) = W(\tau = iz)$ is the complex pair interaction defined in Eq. (18.44). Now observe that $\mathrm{Im} F_{\mathrm{L/R}}$ is related to the forward/backward tunneling rate $k^\pm$ from the energetically higher/lower state $\sigma = \mp 1$ to the lower/higher state $\sigma = \pm 1$ by [cf. Eq (14.40) with $\kappa = 1$]

$$k^\pm = -(2/\hbar)\,\mathrm{Im}\, F_{\mathrm{L/R}} . \qquad (20.28)$$

Putting $z = t - i\bar{\tau}$, we obtain for the forward/backward rate the integral expression

$$k^\pm(\epsilon) = \frac{\Delta^2}{4} e^{\pm\epsilon\bar{\tau}} \int_{-\infty}^{\infty} dt\, e^{\pm i\epsilon t}\, e^{-Q(t-i\bar{\tau})} . \qquad (20.29)$$

Because $Q(z)$ is analytic in the strip $0 \geq \mathrm{Im}\, z > -\hbar\beta$, we may choose for $\bar{\tau}$ an arbitrary value in the interval $0 \leq \bar{\tau} < \hbar\beta$. Setting $\bar{\tau} = 0$, Eq. (20.29) takes the form

$$k^\pm(\epsilon) = \frac{\Delta^2}{4} \int_{-\infty}^{\infty} dt\, e^{\pm i\epsilon t}\, e^{-Q(t)} = \frac{\Delta^2}{2} \int_{0}^{\infty} dt\, \cos\left(\pm\epsilon t - Q''(t)\right) e^{-Q'(t)} . \qquad (20.30)$$

Another convenient choice is $\bar{\tau} = \hbar\beta/2$ in Eq. (20.29) [429]. We then have

$$k^\pm(\epsilon) = \frac{\Delta^2}{4} e^{\beta\hbar\epsilon/2} \int_{-\infty}^{\infty} dt\, e^{\pm i\epsilon t}\, e^{-X(t)} = \frac{\Delta^2}{2} e^{\pm\beta\hbar\epsilon/2} \int_{0}^{\infty} dt\, \cos(\epsilon t)\, e^{-X(t)} , \qquad (20.31)$$

where we have utilized that the function $X(t)$ is an even function of t, as we see from the defining integral expression (18.49).

The backward rate $k^-(\epsilon)$ is related to the forward rate $k^+(\epsilon)$ by

$$k^-(\epsilon) = k^+(-\epsilon) = e^{-\beta\hbar\epsilon}\, k^+(\epsilon) . \qquad (20.32)$$

The first form is the definition of the backward rate in terms of the forward rate, and the second form is obvious from Eq. (20.31). Thus, the forward and backward tunneling rates satisfy the *principle of detailed balance*. It is evident from the derivation that detailed balance formally is a consequence of the analytic properties of the function $Q(z)$ in the strip $0 \geq \operatorname{Im} z > -\hbar\beta$, and of the reflection property (18.45).

The above rate expressions are nonperturbative in the bath coupling. They correspond to "Golden Rule" in the electronic coupling and are exact in the nonadiabatic limit. The same forms are also found in the dynamical approach given below [cf. Sec. 21.3 and Subsec. 22.8]. Finally, we remark that equivalent expressions have been obtained already in the sixties [383, 129], and were rederived later on repeatedly.

20.2.2 *Probability for energy exchange: general results*

Further insight into the understanding of the quantum rate expression (20.29) is gained by rewriting it in the form

$$k^+(\epsilon) = \tfrac{1}{2}\pi\hbar\Delta^2\, P(\hbar\epsilon) , \qquad (20.33)$$

where $P(E)$ is the Fourier transform of the correlation function $e^{-Q(z)}$,

$$P(E) = \frac{1}{2\pi\hbar}\int_{-\infty}^{\infty} dz\, e^{iEz/\hbar - Q(z)} = \frac{e^{\beta E/2}}{2\pi\hbar}\int_{-\infty}^{\infty} dt\, e^{iEt/\hbar - X(t)} . \qquad (20.34)$$

In electron transfer processes, the function $P(E)$ represents the nuclear contribution to the rate. In a sense, $P(E)$ is a compact expression for the thermally averaged nuclear Franck-Condon overlap integral [410, 106]. The function $P(E)$ is equally useful to describe the influence of an electromagnetic environment in single charge tunneling [417], as discussed below in Section 20.3. It is the essence of the quantum rate expression (20.33), and satisfies a number of general properties holding without specification of the spectral density $G(\omega)$. First of all, $P(E)$ is normalized as

$$\int_{-\infty}^{\infty} dE\, P(E) = e^{-Q(0)} = 1 . \qquad (20.35)$$

Secondly, we see from Eq. (20.32) that $P(E)$ obeys the detailed balance relation

$$P(-E) = e^{-\beta E}\, P(E) . \qquad (20.36)$$

The properties (20.35) and (20.36) imply that $P(E)$ is a positive function of E. It is therefore natural to regard $P(E)$ for $E > 0$ as the spectral probability density function that the reservoir absorbs the energy E from the system during the tunneling event. Correspondingly, $P(E)$ with $E < 0$ is the probability density for emission of energy E from the reservoir. Thirdly, using the defining expression (20.34) for $P(E)$, we find the sum rule

$$\int_{-\infty}^{\infty} dE\, E P(E) = -i\hbar \left.\frac{dQ(z)}{dz}\right|_{z=0} = \Lambda_{\mathrm{cl}} . \qquad (20.37)$$

To obtain the second form, we have used the spectral representation (18.44) for $Q(z)$ and we have observed that the resulting expression is just the classical bath reorganization energy defined in Eq. (20.4). The l.h.s. is the balance for the mean energy absorbed and emitted by the environment, $\langle E \rangle_{\rm abs} - \langle E \rangle_{\rm em}$. Thus, the sum rule (20.37) expresses that the net absorbed energy is equal to the reorganization energy. A related discussion of the role of the function $P(E)$ is given in Ref. [418].

To consolidate the meaning of $P(E)$, we invoke an old technique originally introduced by Lax [383] for the study of optical line shape functions, and employed in the ET theory by Jortner [419] (see also Refs. [410, 417, 155]).

Consider first the case where the thermal reservoir consists of a single oscillator with eigenfrequency ω_0. The respective spectral density is

$$G(\omega) \;=\; \lambda_0^2\, \delta(\omega - \omega_0)\,. \tag{20.38}$$

Assume that the oscillator is in the ground state. Then the correlation function $Q(z)$ takes the form $Q(z) = \rho[\,1 - {\rm e}^{-i\omega_0 z}\,]$ where $\rho = (\lambda_0/\omega_0)^2$. Upon expanding ${\rm e}^{-Q(z)}$ in powers of ${\rm e}^{-i\omega_0 z}$, the function $P(E)$ takes the form of an infinite series of delta functions,

$$P(E) \;=\; \sum_{m=0}^{\infty} p_m\, \delta(E - m\hbar\omega_0)\,, \qquad p_m \;=\; \frac{\rho^m}{m!}\,{\rm e}^{-\rho}\,. \tag{20.39}$$

The quantities $p_1, p_2, \cdots, p_m$ are the probabilities that the oscillator absorbs $1, 2, \cdots, m$ quanta of energy $\hbar\omega_0$. The set of probabilities $\{p_m\}$ forms a normalized Poisson distribution with mean number ρ of absorbed quanta. Thus, $P(E)$ is the probability function for absorption of energy E resulting from the individual uncorrelated absorption processes .

At finite temperature, the oscillator does not only absorb, but does also emit quanta. The mean numbers of absorbed and emitted bosons are $\rho_{\rm a} = (1 + n)\rho$, and $\rho_{\rm e} = n\rho$, where $n = 1/[\exp(\beta\hbar\omega_0) - 1]$ is the Bose distribution function in thermal equilibrium. The quantities $\rho_{\rm a}$ and $\rho_{\rm e}$ are the Huang-Rhys factors [383]. With the Poissonian probabilites $p_{m,{\rm a}} = \rho_{\rm a}^m\,{\rm e}^{-\rho_{\rm a}}/m!$ for absorption of m quanta and $p_{\ell,{\rm e}} = \rho_{\rm e}^\ell\,{\rm e}^{-\rho_{\rm e}}/\ell!$ for emission of ℓ quanta, the probability for exchange of energy E of the quantum oscillator in thermal equilibrium with the TSS takes in generalization of Eq. (20.39) the form

$$P(E) \;=\; {\rm e}^{-(\rho_{\rm e}+\rho_{\rm a})} \sum_{\ell,m} \frac{\rho_{\rm e}^\ell\, \rho_{\rm a}^m}{\ell!\, m!}\, \delta\!\left(E - (m - \ell)\hbar\omega_0\right)\,. \tag{20.40}$$

Putting $k = m - \ell$ as summation variable the sum over the variable ℓ takes the form of the ascending series of the modified Bessel functions $I_k(z)$ [89]. Putting $\mu = \hbar\beta\omega_0/2$ we readily obtain

$$P(E) \;=\; {\rm e}^{-\rho\coth\mu} \sum_{k=-\infty}^{\infty} I_k(\rho/\sinh\mu)\, {\rm e}^{k\mu}\, \delta(E - k\hbar\omega_0)\,. \tag{20.41}$$

We now show that the same expression follows from the second form for $P(E)$ in Eq. (20.34). Upon using the expression (18.49) for $X(t)$ and the form (20.38) for $G(\omega)$ we get

$$e^{-X(t)} = e^{-\rho \coth \mu} \exp[\,(\rho/\sinh \mu) \cos(\omega_0 t)\,]$$

$$= e^{-\rho \coth \mu} \sum_{k=-\infty}^{\infty} I_k(\rho/\sinh \mu)\, e^{-ik\omega_0 t} \,. \tag{20.42}$$

Here the second line follows upon identifying the second exponential factor in the first line with the generating functional of the modified Bessel functions [89]. With the series expression (20.42) the time integration in Eq. (20.34) can be done easily, and the result is just the expression (20.41).

It is straightforward to generalize the discussion to two and three, and finally to infinitely many reservoir oscillators. This confirms that $P(E)$ is the probability for exchange of energy with the environment resulting from uncorrelated exchange of bosons between the reservoir and the TSS in compliance with detailed balance.

Often in nature, a system is influenced by a number of different dissipative mechanisms connected with different bath correlation functions $Q_1(z), \cdots , Q_n(z)$. Then, the correlation factor $e^{-Q(z)}$ can be factorized into the separate contributions. The function $P(E)$, which involves the totality of energy exchange processes, can now be written as a convolution of the partial probability densities $P_j(E)$ connected with the individual correlation functions $Q_j(z)$ [419],

$$P(E) = \int_{-\infty}^{\infty} dE_1 \cdots dE_{n-1}\, P_1(E - E_1) P_2(E_1 - E_2) \cdots P_n(E_{n-1}) \,. \tag{20.43}$$

Correspondingly, also the total forward/backward rate can be written as a convolution of the partial contributions modulo a factor.

In numerical computations of $P(E)$ for a given spectral density $G_{\mathrm{lf}}(\omega)$, the most difficult task is to calculate the probability function at $T = 0$, $P_0(E)$, for E near to zero since $Q_0(z) \equiv Q_{T=0}(z)$ depends algebraically on z in the regime $\omega_{\mathrm{c}}|z| \gg 1$ for $s \neq 1$, and logarithmically on z for $s = 1$. Fortunately, the computation is facilitated by employing an integral equation for $P_0(E)$. Following Ref. [420], we start out from a relation for the bath correlation function $Q_0(z)$ at zero temperature,

$$\frac{d}{dz} e^{-Q_0(z)} = -i \int_0^{\infty} d\omega\, \frac{G_{\mathrm{lf}}(\omega)}{\omega}\, e^{-i\omega z}\, e^{-Q_0(z)} \,. \tag{20.44}$$

Taking the Fourier transform and integrating by parts the resulting integral on the l.h.s., we then obtain for $P_0(E)$ the integral equation [410, b] (see also Refs. [421, 155]),

$$E\, P_0(E) = \int_0^{E} dE'\, \frac{G_{\mathrm{lf}}[\,(E - E')/\hbar\,]}{(E - E')/\hbar}\, P_0(E') \,. \tag{20.45}$$

Thus we have avoided to calculate $Q_0(t)$. Numerical integration of Eq. (20.45) yields $P_0(E)$ up to a multiplicative constant, the starting value $P_0(0)$. Subsequently, the constant can be fixed by the normalization condition (20.35).

At finite temperature, we write $Q_1(z) = Q(z) - Q_0(z)$. The direct integration of Eq. (20.34) for $P_1(E)$ is without difficulties, since $Q_1(z)$ is exponentially cut off. Finally, $P(E)$ is obtained as a convolution of $P_0(E)$ and $P_1(E)$ [cf. Eq. (20.43)].

We conclude this subsection with calculating the integral (20.34) by the method of steepest descent. Putting

$$R(z) = Q(z) - iEz/\hbar, \tag{20.46}$$

the saddle point z^* obeys the relation $R'(z^*) = 0$ (the prime denotes differentiation with respect to the argument). We find $z^* = -i\tau_B$ with $0 < \tau_B \leq \hbar\beta/2$, and

$$P(E) = \frac{1}{\hbar} \left(\frac{1}{2\pi R''(z^*)} \right)^{1/2} e^{-\beta F_{qm}} \qquad \text{with} \qquad F_{qm} = R(z^*)/\beta. \tag{20.47}$$

The quantities F_{qm} and $R''(z^*)$ are positive real. Here, F_{qm} is the "quantum activation free energy" [422] of the donor state, and $R''(z^*)$ introduces quantum effects in the pre-exponential factor. The expression (20.47) is still formal. In general, it is not possible to invert the condition $R'(z^*) = 0$ in analytic form to find the saddle point z^*. The various cases where the steepest-descent formula (20.47) for $P(E)$ is justified and explicit expressions can be given are discussed in the following subsections.

In the classical limit, $\kappa \equiv 1/\beta\hbar\omega_c \to \infty$, F_{qm} coincides with the classical activation energy F_D^* for the donor state introduced in Eq. (20.21). We then obtain

$$P(E) = \sqrt{\beta/4\pi\Lambda_{cl}}\, e^{-\beta F_D^*} \qquad \text{with} \qquad F_D^* = (E - \Lambda_{cl})^2/4\Lambda_{cl}. \tag{20.48}$$

This expression is in correspondence with the Marcus form (20.25).

20.2.3 The spectral probability density for absorption at $T = 0$

Consider the probability density for absorption of energy by the reservoir ($E > 0$) at zero temperature. For $T = 0$ the pair interaction (18.47) takes the simple form

$$Q_{T=0}(z) = 2\delta_s \frac{\Gamma(s)}{s-1} \left(\frac{\omega_c}{\omega_{ph}} \right)^{s-1} \left[1 - (1 + i\omega_c z)^{1-s} \right],$$

$$Q_{T=0}(z) = 2K\ln(1 + i\omega_c z), \qquad \text{for} \qquad s = 1. \tag{20.49}$$

Substituting Eq. (20.49) into Eq. (20.46), the saddle point condition $R'(z^*) = 0$ gives $z^* = -i\tau_B$ with $\tau_B = [\,(\Lambda_{cl}/E)^{1/s}-1\,]/\omega_c$, where Λ_{cl} is the solvation energy, Eq. (20.5). The length τ_B is the width of the bounce (kink–anti-kink pair) in imaginary time. In steepest-descent approximation, we find from Eq. (20.47)

$$P_{T=0}(E) = \frac{(\Lambda_{cl}/E)^{(1+s)/2s}}{\sqrt{2\pi s \Lambda_{cl}\hbar\omega_c}} \exp\left\{ -\frac{E}{\hbar\omega_c} + \frac{1}{s-1} \frac{\Lambda_{cl}}{\hbar\omega_c} \left[s\left(\frac{E}{\Lambda_{cl}} \right)^{\frac{s-1}{s}} - 1 \right] \right\}. \tag{20.50}$$

The expression (20.50) is asymptotically exact in the regime

$$[2\delta_s\Gamma(s)]^{1/s}(E/\hbar\omega_{ph})^{(s-1)/s} = (E/\Lambda_{cl})^{(s-1)/s}\Lambda_{cl}/\hbar\omega_c \gg (s+2)(s+1)/8s. \tag{20.51}$$

Physically, this is the regime of multi-phonon emission processes. It roughly corresponds to the regime $E \lesssim \hbar\omega_{ph}\delta_s^{1/(1-s)}$ in the sub-Ohmic case, $E \gtrsim \hbar\omega_{ph}/\delta_s^{1/(s-1)}$ in the super-Ohmic case, and to $\delta_1 \equiv K \gtrsim 1$ in the Ohmic case.

The probability for transfer of energy to the environment has a maximum at $E = E_{\max}$, where $E_{\max}$ obeys the transcendental equation

$$[\, 1 - (\Lambda_{\mathrm{cl}}/E_{\max})^{1/s}\,]\, E_{\max} + [\,(s+1)/2s\,]\, \hbar\omega_{\mathrm{c}} = 0 \,. \tag{20.52}$$

The expressions (20.50) – (20.52) hold for general s. The form of the ascending flank for E well below $E_{\max}$ depends on the particular value of s.

In the *sub-Ohmic* case, $s < 1$, the formula (20.50) is valid down to $E = 0$. The function $P(E)$ has a distinct maximum at $E = E_{\max}$. In the limit $\omega_{\mathrm{c}} \to \infty$, the position of the maximum is selfconsistently determined from Eq. (20.52) as

$$E_{\max} = [\, 2s/(1+s)\,]^{s/(1-s)}\, [\, 2\delta_s \Gamma(s)\,]^{1/(1-s)}\, \hbar\omega_{\mathrm{ph}} \,. \tag{20.53}$$

As a result of the high density of low-energy excitations, the probability of the TSS to lose only a small amount of energy is exponentially small and therefore elastic tunneling processes are absent,

$$P_{T=0}(E \to 0) \propto E^{-\frac{1+s}{2s}} \exp\left[-\frac{s}{(1-s)}\frac{\Lambda_{\mathrm{cl}}}{\hbar\omega_{\mathrm{c}}}\left(\frac{\Lambda_{\mathrm{cl}}}{E}\right)^{\frac{1-s}{s}} \right]. \tag{20.54}$$

Hence, barrier crossing is completely quenched for a symmetric system at zero temperature [with the exception of the special case (18.38)].

The descending flank of $P_{T=0}(E)$ for E above $E_{\max}$ is formed by multi-phonon emission. These processes gradually die out as E is increased further, so that the steepest descent formula (20.50) ceases to be valid. For $E \gg E_{\max}$, the influence of the environment is only weak and the probability for absorption of energy by the reservoir is determined by the one-phonon process. Putting $e^{-Q(t)} \approx 1 - Q(t)$ and substituting the representation (18.44) for $Q(t)$, we obtain[2] for $E \gg E_{\max}$ and $s < 1$

$$P_{T=0}(E) = \frac{1}{E}\frac{G_{\mathrm{lf}}(E/\hbar)}{E/\hbar} = \frac{1}{\Gamma(s)}\frac{\Lambda_{\mathrm{cl}}}{(\hbar\omega_{\mathrm{c}})^2}\left(\frac{\hbar\omega_{\mathrm{c}}}{E}\right)^{2-s} e^{-E/\hbar\omega_{\mathrm{c}}} \,. \tag{20.55}$$

From this we see that the one-phonon contribution to $P(E)$ directly reflects the spectral properties of the reservoir coupling modulo a factor E^{-2}.

The *super-Ohmic* case with $s > 2$ shows opposite behavior. Now, the low-energy tail of $P_{T=0}(E)$ ($E \ll E_{\max}$) is determined by the one-phonon process, Eq. (20.55). With increasing E, multi-phonon processes become active and form the ascending flank of $P_{T=0}(E)$. For $E \gtrsim E_{\max}$, the asymptotic multi-phonon expression (20.50) is valid. The maximum of $P_{T=0}(E)$ is situated at $E_{\max} = \Lambda_{\mathrm{cl}} - \frac{1}{2}(1+s)\hbar\omega_{\mathrm{c}}$. In the limit $E \to \infty$, $P_{T=0}(E)$ drops to zero faster than algebraically. Physically, this is a consequence of the exponential cutoff in the spectral density $G_{\mathrm{lf}}(\omega)$.

Consider next the Ohmic case. Taking in Eq. (20.50) the limit $s \to 1$, we obtain

$$P_{T=0}(E) = \sqrt{\frac{K}{\pi}}\frac{e^{2K}}{(2K)^{2K}}\frac{e^{-E/\hbar\omega_{\mathrm{c}}}}{E}\left(\frac{E}{\hbar\omega_{\mathrm{c}}}\right)^{2K} \,. \tag{20.56}$$

---

[2] At finite T, the one-phonon emission process has an additional factor $1/[\,1 - \exp(-\beta E)\,]$.

For $K > \frac{1}{2}$, the probability for absorption of energy by the reservoir shows a maximum at $E_{\max} = (2K - 1)\hbar\omega_c = \Lambda_{cl} - \hbar\omega_c$. Well below the maximum, the function $P_{T=0}(E)$ has power law form $\propto E^{2K-1}$.

Finally, we anticipate that the integral in Eq. (20.34) can be calculated exactly in the Ohmic scaling limit. With use of Eqs. (20.82) and (20.33), we find

$$P_{T=0}(E) = \frac{1}{\Gamma(2K)} \frac{e^{-E/\hbar\omega_c}}{E} \left(\frac{E}{\hbar\omega_c}\right)^{2K}. \qquad (20.57)$$

Employing Stirling's asymptotic representation for $\Gamma(2K)$, which applies for large values of K, we recover the steepest descent expression (20.56).

20.2.4 Crossover from quantum-mechanical to classical behavior

We now extend the study of $P(E)$ to finite temperature. Consider first the case $E = 0$ in the expression (20.47). For $E = 0$, the saddle point is at $z^* = -i\hbar\beta/2$. It is therefore natural to take up the representation (20.31) in which the integrand is stationary at $t = 0$. Expanding the integrand in Eq. (18.49) about $t = 0$, we find

$$X(t) = \Lambda_1\beta/4 + (\Lambda_2/\hbar^2\beta)\,t^2 + \mathcal{O}(t^4)\,, \qquad (20.58)$$

where

$$\Lambda_1 = \frac{4}{\beta} \int_0^\infty d\omega\, \frac{G_{lf}(\omega)}{\omega^2} \frac{[\cosh(\frac{1}{2}\beta\hbar\omega) - 1]}{\sinh(\frac{1}{2}\beta\hbar\omega)}\,, \qquad (20.59)$$

$$\Lambda_2 = \frac{\hbar^2\beta}{2} \int_0^\infty d\omega\, G_{lf}(\omega)\, \frac{1}{\sinh(\frac{1}{2}\beta\hbar\omega)}\,. \qquad (20.60)$$

The quantities Λ_1 and Λ_2 are quantum mechanical reorganization energies. For the form (18.46) for $G_{lf}(\omega)$ we find from Eq. (18.50) the analytical expressions

$$\Lambda_1 = 8k_BT B_s + b_s(\kappa)\, k_BT\, (T/T_{ph})^{s-1}\,, \qquad (20.61)$$

$$\Lambda_2 = a_s(\kappa)\, k_BT\, (T/T_{ph})^{s-1}\,, \qquad (20.62)$$

$$b_s(\kappa) = 16\delta_s\Gamma(s - 1)\Big\{\zeta(s - 1, 1 + \kappa) - \zeta(s - 1, \tfrac{1}{2} + \kappa)\Big\}\,, \qquad (20.63)$$

$$a_s(\kappa) = 2\delta_s\Gamma(s + 1)\zeta(s + 1, \tfrac{1}{2} + \kappa)\,, \qquad (20.64)$$

where $T_{ph} = \hbar\omega_{ph}/k_B$. The first term in Eq. (20.61) gives rise to the adiabatic dressing factor introduced in Eqs. (18.34). We see from the integral representations (20.59) and (20.60) that $\Lambda_1 \geq \Lambda_2$. In the limit $\kappa \equiv k_BT/\hbar\omega_c \to \infty$, the quantum reorganization energies Λ_1 and Λ_2 approach the classical reorganization energy,

$$\lim_{\kappa\to\infty} \Lambda_1 = \Lambda_{cl}\,, \qquad \lim_{\kappa\to\infty} \Lambda_2 = \Lambda_{cl}\,. \qquad (20.65)$$

These limiting cases can be seen either from the integral representations or by substituing the asymptotic expansion of the zeta function into Eqs. (20.61) and (20.62).

In the deep quantum regime $\kappa \ll 1$, the coefficient functions take the form

$$b_s(0) = 16\delta_s \Gamma(s-1)[\, 2 - 2^{s-1}\,] \zeta(s-1) \,, \tag{20.66}$$

$$a_s(0) = 2\delta_s \Gamma(s+1)[\, 2^{s+1} - 1\,] \zeta(s+1) \,. \tag{20.67}$$

Substituting Eq. (20.58) into Eq. (20.31), we obtain the probability for exchange of zero energy as (cf. Eq. (4.4) in Ref. [423])

$$P(0) = \sqrt{\beta/4\pi\Lambda_2}\; e^{-\beta\Lambda_1/4} \,. \tag{20.68}$$

Within the bounce picture discussed in Subsection 17.4, the formula (20.68) has the following interpretation. The bounce consists of an instanton–anti-instanton pair with (imaginary-time) distance $\hbar\beta/2$, which is half of the bounce period $\hbar\beta$. The exponent $\Lambda_1/4k_{\mathrm{B}}T$ is the interaction between the instantons,

$$\Lambda_1/4k_{\mathrm{B}}T = W(\hbar\beta/2) = Q(-i\hbar\beta/2) = X(0) \,, \tag{20.69}$$

and $\Lambda_2/k_{\mathrm{B}}T$ is directly connected with the negative eigenvalue of the breathing mode of the bounce. Hence it is a measure for the stiffness of the bounce width.

The leading correction to the formula (20.68) arises from the $\mathcal{O}(t^4)$-term in Eq. (20.58). This contribution is small when the condition

$$32\delta_s \left(\frac{T}{T_{\mathrm{ph}}}\right)^{s-1} \frac{[\,\Gamma(s+1)\,\zeta(s+1,\tfrac{1}{2}+\kappa)\,]^2}{\Gamma(s+3)\,\zeta(s+3,\tfrac{1}{2}+\kappa)} \gg 1 \tag{20.70}$$

is satisfied. Physically, this condition holds in the regime of multi-phonon processes. In the Ohmic case, Eq. (20.70) corresponds to the damping regime $K \gg 1$. In the super-Ohmic case, the condition (20.70) is met for temperatures well above T_{ph}. For a sub-Ohmic bath, the condition (20.70) is fulfilled when T is fairly below T_{ph}.

When the energy E exchanged with the environment is in the range

$$E \ll (1+2\kappa)\Lambda_2 \,, \tag{20.71}$$

the saddle point z^* is still near to $-i\hbar\beta/2$. Hence we may use again the short-time expansion (20.58) for $X(t)$. With this we then obtain an explicit expression for the SPA result (20.47),

$$P(E) = \sqrt{\beta/4\pi\Lambda_2}\, e^{-\beta F_{\mathrm{qm}}} \quad \text{with} \quad F_{\mathrm{qm}}(E) = \Lambda_1/4 - E/2 + E^2/4\Lambda_2 \,. \tag{20.72}$$

This expression includes quantum effects and satisfies the detailed balance relation (20.36). Corrections to the formula (20.72) are found to be small if we meet the conditions (20.70) and (20.71). The energy range (20.71) sensitively depends on temperature. It shrinks $\propto T^s$ as $T \to 0$, as we see from Eq. (20.62). In the classical limit, $\kappa \to \infty$, the influence of *every* bath is characterized by only one parameter, the reorganization energy Λ_{cl}. In this limit, the expression (20.72) reduces to the classical probability function (20.48), and the condition (20.71) is also met when the exchanged energy E is of the order of Λ_{cl} or larger.

On the other hand, in the quantum regime the particular spectral properties of the spectral density $G_{lf}(\omega)$ become relevant. This indicates that nuclear tunneling becomes effective. The quantum effects are captured by the deviations of the activation energies Λ_1 and Λ_2 from the classical solvation energy Λ_{cl}. We should like to remark that the above expressions are valid for general s. However, the conditions (20.70) and (20.71) depend in the quantum regime sensitively on the parameter s.

As E is increased, the stationary point z^* moves from $z^* = -i\hbar\beta/2$ towards $z^* = 0$. When the condition (20.71) is violated, the short-time expansion about $z = -i\hbar\beta/2$, Eq. (20.58), is not applicable any more, and the expression (20.72) is incorrect. When z^* is distant from $-i\hbar\beta/2$, the expression (20.47) must be evaluated numerically. However, when z^* is near to zero, which is the case when E and/or T is very large, we can analyze again the expression (20.47) in analytic form.

For E near to E_{max} and $s \geq 1$, the stationary point is found to be near $z = 0$ for all T. A consistent analysis can thus be performed upon expanding $Q(z)$ in Eq. (20.34) about $z = 0$ up to fourth order in z. We find

$$Q(z) = i\frac{\Lambda_{cl}}{\hbar}z + \frac{A_2}{2!}(\omega_c z)^2 - i\frac{A_3}{3!}(\omega_c z)^3 - \frac{A_4}{4!}(\omega_c z)^4 + \mathcal{O}[(\omega_c z)^5]. \qquad (20.73)$$

For the spectral density (18.33), the dimensionless coefficient functions read

$$A_2(\kappa) = s[1 + 2\kappa^{s+1}\zeta(s+1, 1+\kappa)]\Lambda_{cl}/\hbar\omega_c,$$

$$A_3(\kappa) = s(s+1)\Lambda_{cl}/\hbar\omega_c, \qquad (20.74)$$

$$A_4(\kappa) = s(s+1)(s+2)[1 + 2\kappa^{s+3}\zeta(s+3, 1+\kappa)]\Lambda_{cl}/\hbar\omega_c.$$

Near to the maximum we obtain the normalized Gaussian distribution

$$P(E) = \frac{1}{\sqrt{2\pi\langle E^2\rangle}}\exp\left(-\frac{(E - E_{max})^2}{2\langle E^2\rangle}\right). \qquad (20.75)$$

The position of the maximum and the width of the distribution are given by

$$E_{max} = \Lambda_{cl} - \frac{A_3}{2A_2}\hbar\omega_c,$$

$$\langle E^2\rangle \equiv 2k_B T_{eff}\Lambda_{cl} = \left(1 + \frac{A_3^2}{2A_2^3} - \frac{A_4}{2A_2^2}\right)A_2(\hbar\omega_c)^2. \qquad (20.76)$$

The expression (20.75) with (20.76) and (20.74) correctly describes $P(E)$ near to the maximum for all T. Observe that the transfer of energy is activationless at the maximum for all temperatures. In the zero temperature limit, we find from Eq. (20.76)

$$E_{max} = \Lambda_{cl} - \tfrac{1}{2}(1 + s)\hbar\omega_c, \qquad \langle E^2\rangle = s\hbar\omega_c\Lambda_{cl} - \tfrac{1}{2}(s+1)(\hbar\omega_c)^2. \qquad (20.77)$$

The resulting expression for $P(E)$ is consistent with the previous result (20.50) around the maximum. In the classical limit, $\kappa \gg 1$, E_{max} and $\langle E^2\rangle$ take the Marcus form,

$$E_{max} = \Lambda_{cl}, \qquad \langle E^2\rangle = 2k_B T\Lambda_{cl}. \qquad (20.78)$$

Thus, the position of the maximum is shifted upwards towards the reorganization energy as the temperature is raised from the quantum regime ($\kappa \ll 1$) to the classical regime ($\kappa \gg 1$). Concomitantly, the width of the Gaussian distribution is substantially broadened as temperature is increased. The effective temperature T_{eff} in Eq. (20.76) approaches T at high temperatures and becomes temperature-independent as the actual temperature decreases. The expression (20.75) with Eq. (20.76) takes consistently into account all quantum effects for E around E_{max}. It is the systematic generalization of a nonadiabatic rate formula discussed by Garg et al. [412].[3] These authors take into account the quantum effects included in the function $A_2(\kappa)$, and they disregard those described by $A_3(\kappa)$ and $A_4(\kappa)$. The functions $A_3(\kappa)$ and $A_4(\kappa)$ vanish in the classical limit, but may become relevant in the quantum regime. The treatment in Ref. [412] has several precursors [424] and has sometimes been termed "semiclassical" [424, 422].

In the classical limit, the probability $P(E)$ is symmetric about the maximum. In the quantum regime, the function $P(E)$ is asymmetric. The function $P(E)$ is enhanced above E_{max} because of the smaller width of the barrier in the inverted regime which leads to a higher tunneling probability. This behavior is confirmed, e.g., by the $T = 0$ expressions (20.50) and (20.56). The asymmetry of $P(E)$ around the maximum is reduced as the temperature and/or the damping strength are increased.

20.2.5   The Ohmic case

Consider first the Ohmic case, $s = 1$, in the scaling regime $k_{\text{B}}T$, $\hbar|\epsilon| \ll \hbar\omega_{\text{c}}$, in which the rate expression (20.29) can be calculated in analytic form. Substituting the scaling form (18.54) into Eq. (20.29), the Golden Rule rate is found to lowest order in $k_{\text{B}}T/\hbar\omega_{\text{c}}$ and in $\epsilon/\omega_{\text{c}}$ as [425, 426, 427]

$$k^+(T,\epsilon) = \frac{1}{4}\frac{\Delta^2}{\omega_{\text{c}}}\left(\frac{\hbar\omega_{\text{c}}}{2\pi k_{\text{B}}T}\right)^{1-2K}\frac{|\Gamma(K + i\hbar\epsilon/2\pi k_{\text{B}}T)|^2}{\Gamma(2K)}e^{\hbar\epsilon/2k_{\text{B}}T} . \qquad (20.79)$$

With the form (18.54), the integrand in Eq. (20.29) is singular at the origin. Naively, one would expect that the result (20.79) is only valid in the regime $K < \frac{1}{2}$. However, observe that we have calculated a contour integral which actually bypasses the singularity. Hence the expression (20.79) can be analytically continued to the regime $K \geq \frac{1}{2}$ and therefore holds without restriction on the parameter K. This is easily substantiated by use of the representation (20.31) with the form (18.55) for $X(t)$. The resulting integrand is nonsingular on the real axis for $K \geq 0$, and the respective integral is tabulated in Ref. [88]. In the end, we find the previous form (20.79).

For $K < 1$, it is convenient to absorb the ω_{c} dependence into the effective tunneling matrix element Δ_{eff} defined in Eq. (18.37). Then Eq. (20.79) is rewritten as

---

[3]The model in Ref. [412] differs in the description of the medium. The reaction coordinate is modeled by a damped harmonic oscillator leading to a spectral density of the form (3.228).

$$k^+(T,\epsilon) = \Delta_{\text{eff}} \frac{\sin(\pi K)}{2\pi} \left(\frac{\hbar\Delta_{\text{eff}}}{2\pi k_{\text{B}}T}\right)^{1-2K} |\Gamma(K + i\hbar\epsilon/2\pi k_{\text{B}}T)|^2\, e^{\hbar\epsilon/2k_{\text{B}}T}. \quad (20.80)$$

For weak damping $K \ll 1$, this expression reduces to the form

$$k^+(T,\epsilon) = \left(\frac{2\pi k_{\text{B}}T}{\hbar\Delta_{\text{eff}}}\right)^{2K} \frac{\pi K \Delta_{\text{eff}}^2}{[(2\pi K k_{\text{B}}T/\hbar)^2 + \epsilon^2]} \frac{\epsilon}{[1 - \exp(-\hbar\epsilon/k_{\text{B}}T)]}. \quad (20.81)$$

At zero temperature, we can calculate the rate exactly for arbitrary ω_{c}. Substituting the form $Q_{T=0}(z) = 2K \ln(1 + i\omega_{\text{c}}z)$ into the integral (20.34), we obtain

$$k^+(0,\epsilon) = \frac{\pi}{2\Gamma(2K)} \frac{\Delta^2}{\omega_{\text{c}}} \left(\frac{\epsilon}{\omega_{\text{c}}}\right)^{2K-1} e^{-\epsilon/\omega_{\text{c}}}, \quad (20.82)$$

whereas $k^-(0,\epsilon) = 0$, as follows with Eq. (20.32). Hence at $T = 0$, there are only transitions from the upper to the lower well, as expected. We can also calculate the asymptotic enhancement at low T for arbitrary ω_{c}. Expanding $Q(z)$ up to terms of order T^2, we formally obtain $k^+(T,\epsilon) = \exp[\,K(\pi^2/3)\partial^2/(\partial\hbar\beta\epsilon)^2\,]\,k^+(0,\epsilon)$, yielding

$$k^+(T,\epsilon) = k^+(0,\epsilon) \exp\left\{\frac{K}{3}\left(\frac{\pi}{\hbar\beta}\right)^2\left[\left(\frac{2K-1}{|\epsilon|} - \frac{1}{\omega_{\text{c}}}\right)^2 - \frac{2K-1}{\epsilon^2}\right]\right\}. \quad (20.83)$$

The finite-temperature enhancement of the rate varies as T^2. This in agreement with the corresponding result for an extended biased potential, Eq. (17.16). The width of the instanton-anti-instanton pair in steepest descent is $\bar{\tau} = 2K/\epsilon$, so that for $K \gg 1$ also the prefactor of the T^2 law corresponds with Eq. (17.16). When $\hbar\epsilon$ is chosen near to the maximum of the rate, the result (20.83) is consistent with the low-temperature expansion of the rate formula (20.75) with (20.76) for the Ohmic case.

On the other hand, in the absence of a bias ($\epsilon = 0$), Eqs. (20.79) and (20.80) give

$$\begin{aligned}
k^{\pm}(T,0) &= \frac{\sqrt{\pi}\,\Gamma(K)}{4\Gamma(K+\frac{1}{2})} \frac{\Delta^2}{\omega_{\text{c}}} \left(\frac{\pi k_{\text{B}}T}{\hbar\omega_{\text{c}}}\right)^{2K-1} \\[2mm]
&= \frac{\Delta_{\text{eff}}}{2} \frac{\Gamma(K)}{\Gamma(1-K)} \left(\frac{\hbar\Delta_{\text{eff}}}{2\pi k_{\text{B}}T}\right)^{1-2K}.
\end{aligned} \quad (20.84)$$

The forms expressed in terms of the effective tunneling coupling Δ_{eff} apply in the regime $K < 1$, while the forms with the explicit ω_{c} dependence are also valid in the regime $K > 1$. We remark that for the particular potentials discussed in Subsection 17.4 and Section 18.1 the transfer matrix element Δ can be expressed in analytic form in terms of the original potential parameters.

In the regime $K < \frac{1}{2}$, the rate (20.84) increases as temperature is decreased. This phenomenon was predicted first by Kondo and experimentally confirmed in the diffusion of charged interstitials in metals (see Ref. [379, 136], and references therein). The increase of the rate with decreasing temperature is a signature that the dynamics becomes coherent at sufficiently low temperature. The incoherent-coherent transition is discussed in Subsection 22.2.2, and the relation with the diffusion coefficient is discussed in Subsection 25.5.1.

In the regime $0 \le K < 1$, the above Golden Rule results are valid if the temperature or the bias exceeds a certain value (cf. the discussion in Subsection 22.1.1). For $K < \frac{1}{2}$, low temperatures and weak bias, the dynamics may actually be coherent and therefore cannot be simply described in terms of a single rate. For $\frac{1}{2} < K < 1$, the Golden rule approximation may break down at low T. On the other hand, for $K > 1$, the above Golden Rule rate expressions are valid down to zero temperature.

Next, consider the crossover from the quantum to the classical regime in the bias range given in Eq. (20.71). The quantum activation energies Λ_1 and Λ_2 take the form

$$\Lambda_1 = 8Kk_{\rm B}T \ln[\,\hbar\beta\omega_{\rm c}\Gamma^2(1+\kappa)/\Gamma^2(\tfrac{1}{2}+\kappa)]\,, \qquad (20.85)$$

$$\Lambda_2 = 2Kk_{\rm B}T\psi'(\tfrac{1}{2}+\kappa)\,, \qquad (20.86)$$

where $\kappa = 1/\beta\hbar\omega_{\rm c}$. With these forms, the rate formula (20.33) with (20.72) reads

$$k^+(T,\epsilon) = \frac{\hbar\beta\Delta^2}{4(\beta\hbar\omega_{\rm c})^{2K}} \sqrt{\frac{\pi\,e^{\beta\hbar\epsilon}}{2K\psi'(\tfrac{1}{2}+\kappa)}}\, \frac{\Gamma^{4K}(\tfrac{1}{2}+\kappa)}{\Gamma^{4K}(1+\kappa)} \exp\left(-\frac{(\beta\hbar\epsilon)^2}{8K\psi'(\tfrac{1}{2}+\kappa)}\right). \qquad (20.87)$$

This reduces in the quantum regime $k_{\rm B}T \ll \hbar\omega_{\rm c}$ to the form

$$k^+(T,\epsilon) = k^+(T,0)\exp\left\{\frac{\beta\hbar\epsilon}{2} - \frac{1}{K}\left(\frac{\beta\hbar\epsilon}{2\pi}\right)^2\right\}\,, \qquad (20.88)$$

$$k^+(T,0) = \frac{\sqrt{\pi}}{4\sqrt{K}}\frac{\Delta^2}{\omega_{\rm c}}\left(\frac{\pi}{\beta\hbar\omega_{\rm c}}\right)^{2K-1}. \qquad (20.89)$$

The bias dependence is correctly described by Eq. (20.88) for $\epsilon \ll K\pi k_{\rm B}T/\hbar$. In the regime $K \gtrsim 1$, we have $\Gamma(K)/\Gamma(K+\tfrac{1}{2}) \approx 1/\sqrt{K}$. Thus, in this damping regime, the steepest-descent expression (20.89) coincides with the exact formula (20.84).

In the classical regime, $k_{\rm B}T \gg \hbar\omega_{\rm c}$, Eq. (20.87) reduces to the form

$$k^+(T,\epsilon) = \frac{\hbar\Delta^2}{4}\left(\frac{\pi}{2K\hbar\omega_{\rm c}k_{\rm B}T}\right)^{1/2} \exp\left(-\frac{\hbar(\epsilon-2K\omega_{\rm c})^2}{8K\omega_{\rm c}k_{\rm B}T}\right). \qquad (20.90)$$

The general expression (20.87) describes the transition from the quantum rate expression, Eq. (20.88) with (20.89), to the classical rate, Eq. (20.90).

Finally, the rate expression near to the maximum is given by Eqs. (20.33), (20.75), and (20.76) with the coefficient functions $A_2 = 2K[1+2\kappa^2\psi'(1+\kappa)]$, $A_3 = 4K$, and $A_4 = 4K[3+\kappa^4\psi^{(3)}(1+\kappa)]$. As T is increased from $\kappa \ll 1$ to $\kappa \gg 1$, $\epsilon_{\max}$ moves from $(2K-1)\omega_{\rm c}$ to $2K\omega_{\rm c}$, and the width $\langle\epsilon^2\rangle$ expands from $(2K-1)\omega_{\rm c}^2$ to $4K\omega_{\rm c}k_{\rm B}T/\hbar$.

The above rate expressions are directly relevant for a large variety of tunneling problems involving an Ohmic heat bath, e.g., conduction electron reservoirs in metals, and resistive electromagnetic environments.

We remark that the theoretical interest in Ohmic dissipation stems, to a large part, from the power-law behaviors of the transfer rate discussed in this subsection.

20.2.6 Exact nonadiabatic rates for $K = \frac{1}{2}$ and $K = 1$

For the special value $K = \frac{1}{2}$, the nonadiabatic rate expression (20.29) can be calculated in analytic form for arbitrary value of $\kappa = 1/\beta\hbar\omega_c$. Using for $Q(z)$ the form (18.52) and closing the contour in the half-plane Im $z > 0$, we pick up the residua of the simple poles of $\Gamma(\kappa + iz/\hbar\beta)$. The poles are located at $z = i(n + \kappa)\hbar\beta$, where $n = 0, 1, 2, \cdots$, and the respective residua are proportional to $\Gamma(n + 2\kappa + 1)$. Upon using Euler's integral representation for the function $\Gamma(n + 2\kappa + 1)$ [88], the residua of the infinite sequence of poles can be summed. Eventually, the remaining integration can be performed. We readily find [428]

$$k^+(T, \epsilon; K = \tfrac{1}{2}) \; = \; \frac{\Gamma(\tfrac{1}{2} + \kappa)}{\sqrt{\pi}\,\Gamma(1 + \kappa)} \frac{e^{\beta\hbar\epsilon/2}}{[\cosh(\beta\hbar\epsilon/2)]^{1+2\kappa}} \frac{\gamma}{2} \,, \tag{20.91}$$

where $\gamma \equiv \Delta_{\rm eff}(K = \tfrac{1}{2}) = \pi\Delta^2/2\omega_c$. Note that the expression (20.91) satisfies the detailed balance condition (20.32). For the symmetric system, Eq. (20.91) becomes

$$k^+(T, 0; K = \tfrac{1}{2}) \; = \; k^-(T, 0; K = \tfrac{1}{2}) \; = \; \frac{\Gamma(\tfrac{1}{2} + \kappa)}{\sqrt{\pi}\,\Gamma(1 + \kappa)} \frac{\gamma}{2} \,. \tag{20.92}$$

In the low-temperature regime $\kappa \equiv 1/\beta\hbar\omega_c \ll 1$, the rate (20.92) becomes temperature-independent, $k_+ = k_- = \gamma/2$, which is in agreement with the expression (20.84) for $K = \frac{1}{2}$. In the opposite classical limit $\kappa \gg 1$, we find

$$k^+(T, 0; K = \tfrac{1}{2}) \; = \; \frac{\Delta^2}{4\omega_c} \left(\frac{\pi\hbar\omega_c}{k_{\rm B}T}\right)^{1/2} \propto \frac{1}{\sqrt{T}} \,. \tag{20.93}$$

In the special Ohmic case $K = \frac{1}{2}$, the solvation energy $\Lambda_{\rm cl}$ is equal to $\hbar\omega_c$, as we see from Eq. (20.6). Thus we recover the high-temperature limit of the formula (20.25).

As a function of the bias, the rate (20.91) has a maximum at $\epsilon = \epsilon_{\rm max}$, where

$$\hbar\epsilon_{\rm max} \; = \; k_{\rm B}T \, \ln(1 + \hbar\omega_c/k_{\rm B}T) \,. \tag{20.94}$$

In the classical limit, this reduces to the Marcus result $\epsilon_{\rm max} = \omega_c$. As well, the expression (20.91) reduces to the Marcus form (20.25). As T is lowered, the maximum of the rate shifts to smaller values for $\epsilon_{\rm max}$ and finally approaches $\epsilon_{\rm max} = 0$ as $T \to 0$.

Making use of the convolution property

$$k^+(\epsilon; 2K) \; = \; \frac{2}{\pi\Delta^2} \int_{-\infty}^{\infty} d\epsilon' \, k^+(\epsilon'; K) k^+(\epsilon - \epsilon'; K) \,, \tag{20.95}$$

which follows from Eqs. (20.33) and (20.43), we can obtain the exact rate in analytic form also for $K = 1$. Substituting Eq. (20.91) in Eq. (20.95), we find

$$k^+(T, \epsilon; K = 1) \; = \; \frac{\Delta^2}{2\hbar\beta\omega_c^2} \left(\frac{\Gamma(\tfrac{1}{2} + \kappa)}{\Gamma(1 + \kappa)}\right)^2 \frac{e^{\beta\hbar\epsilon/2}\mathcal{Q}_{2\kappa}[\coth(\beta\hbar\epsilon/2)]}{[\sinh(\beta\hbar\epsilon/2)]^{1+2\kappa}} \,, \tag{20.96}$$

where $Q_\nu(z)$ is the Legendre function of the second kind [88]. Upon using the property $Q_\nu(z) = z^{-\nu-1}F_\nu(z^2)$, where $F_\nu(z^2)$ is a function of z^2, it is straightforward to see that the expression (20.96) satisfies detailed balance. For zero bias, we obtain

$$k^+(T, 0; K = 1) \; = \; \frac{\pi\Delta^2}{2\hbar\beta\omega_c^2} \frac{\Gamma^4(1+2\kappa)}{\Gamma^4(1+\kappa)\Gamma(2+4\kappa)} \, , \qquad (20.97)$$

which agrees at low temperatures with the temperature dependence in Eq. (20.84) (where $K = 1$). In the high-temperature limit, we recover again the $T^{-1/2}$ Marcus law. At high temperatures, the rate expression (20.96) as a function of the bias is symmetric around the maximum which is at $\epsilon = \epsilon_{\max}$, where $\epsilon_{\max} = \Lambda_{\mathrm{cl}}/\hbar = 2\omega_c$. With decreasing temperature, the maximum is shifted to $\epsilon_{\max} = \Lambda_{\mathrm{cl}}/2\hbar$, and the rate becomes asymmetric around this value. In summary, the special cases $K = \frac{1}{2}$ and $K = 1$ show in the various limits the general characteristic features discussed above.

20.2.7   The sub-Ohmic case ($0 < s < 1$)

The sub-Ohmic case is distinguished by the high density of low-energy excitations. Hence the dissipative effects become most noticeable at low T and weak bias. The forward rate for a biased system at zero temperature is given by

$$k^+(0, \epsilon) \; = \; (\pi\hbar/2)\, \Delta^2 P_{T=0}(\hbar\epsilon) \, . \qquad (20.98)$$

The behavior of $P_{T=0}(\hbar\epsilon)$ has been discussed in Subsection 20.2.3.

Consider next the low-temperature corrections in the regime $k_{\mathrm{B}}T \ll \hbar\epsilon$. We should expect that the ascending shoulder of $P(E)$ is shifted to lower energies as the temperature is raised above zero since then the particle can also absorb energy from the reservoir. This effect can easily be studied asymptotically. Since the form (20.49) cuts off the Fourier integral (20.34) for $P(E)$ at $|z| \gtrsim 1/(\omega_{\mathrm{ph}}\delta_s^{1/(1-s)})$, the temperature-dependent term of $Q(z)$ in Eq. (18.47) may be expanded in powers of z^2,

$$\Delta Q(z) \; = \; 2\delta_s\Gamma(s+1)\zeta(s+1)\,(\hbar\beta\omega_{\mathrm{ph}})^{-(s+1)}(\omega_{\mathrm{ph}}z)^2 + \mathcal{O}[(\omega_{\mathrm{ph}}z)^4] \, . \qquad (20.99)$$

The leading temperature correction for $k_{\mathrm{B}}T \ll E_{\max}$ is given by the factor $\exp[-\Delta Q(z^*)]$ where $z^* = -i\tau_{\mathrm{B}}$ is the stationary point for $T = 0$. Since $\Delta Q(z^*)$ is negative, this term actually leads to an enhancement of the probability function and hence of the crossing rate. We readily find

$$k^+(T, \epsilon) \; = \; k^+(0, \epsilon)\exp\left\{2\delta_s\Gamma(s+1)\zeta(s+1)(\omega_{\mathrm{ph}}\tau_{\mathrm{B}})^2\Big(\frac{T}{T_{\mathrm{ph}}}\Big)^{1+s}\right\} , \qquad (20.100)$$

where we have put $\hbar\omega_{\mathrm{ph}} = k_{\mathrm{B}}T_{\mathrm{ph}}$. Note that, with Eqs. (3.72), (3.141), and (17.14), the thermal enhancement factor in Eq. (20.100) is exactly in the form discussed in Section 17.2 for a general extended metastable potential [cf. Eq. (17.12) with Eq. (17.15)]. The enhancement function $\mathcal{A}(T)$, which is the curly bracket in Eq. (20.100), varies as T^{1+s}. This is in agreement with the general conclusions reported in Section 17.2.

When s approaches one from below, the condition (20.51) reduces to the condition $K \gg 1$. Performing this limit in Eq. (20.100) with Eq. (20.50), the resulting expression is consistent with the Ohmic result (20.83) with Eq. (20.82) for large K and $\hbar\epsilon \ll \Lambda_{\text{cl}}$.

Finally, we turn to the discussion of the rate for a symmetric system. In the low-temperature regime determined by the condition (20.70) we may use the rate expression (20.68) with the activation energies (20.61) and (20.62). Thus we find

$$
k^+(T,0) = \frac{\Delta^2 e^{2|B_s|}}{4\omega_{\text{ph}}} \sqrt{\frac{\pi}{a_s(\kappa)}} \left(\frac{T_{\text{ph}}}{T}\right)^{\frac{1+s}{2}} \exp\left\{ -\frac{b_s(\kappa)}{4} \left(\frac{T_{\text{ph}}}{T}\right)^{1-s} \right\} ,
\tag{20.101}
$$

where $a_s(\kappa)$ and $b_s(\kappa)$ are defined in Eqs. (20.63) and (20.64).

In the regime $k_B T \ll \hbar\omega_c$, the coefficient functions $a_s(0)$ and $b_s(0)$ are given in Eqs. (20.66) and (20.67).[4] As T approaches zero, the rate vanishes with an essential singularity for the reasons mentioned above.

By carefully taking the limit $s \to 1^-$ in Eq. (20.101), we obtain the Ohmic results for $\epsilon = 0$, Eq. (20.87) and Eq. (20.89), respectively. The generalization to the biased case can be taken from Subsection 20.2.4.

20.2.8   The super-Ohmic case ($s > 1$)

We now study the tunneling rate for super-Ohmic spectral densities and we identify the cutoff frequency ω_c with the Debye frequency. We write $X(t)$ as in Eq. (18.49),

$$
X(t) = X_1 - X_2(t) .
\tag{20.102}
$$

The term

$$
X_1 = \int_0^\infty d\omega \, \frac{G_{\text{lf}}(\omega)}{\omega^2} \coth(\tfrac{1}{2}\beta\hbar\omega)
\tag{20.103}
$$

is the Huang-Rhys factor (18.51) [383]. For $s > 2$, the functions X_1 and $X_2(t)$ are individually nonsingular, whereas for $s \le 2$ only $X(t)$ is nonsingular.

Consider first the regime well below the Debye temperature, $T \ll T_D \equiv \hbar\omega_D/k_B$. Then we may put $\kappa = 0$ in Eq. (18.50). It is natural to absorb the adiabatic term X_1 into the definition of a renormalized tunneling matrix element. We write

$$
\tilde{\Delta} = \Delta\, e^{-X_1/2} = \Delta_{\text{eff}}\, e^{-D_s} , \qquad \Delta_{\text{eff}} = \Delta\, e^{-B_s} .
\tag{20.104}
$$

Here, Δ_{eff} is the polaron-dressed tunneling matrix element at $T = 0$ defined in Eq. (18.34) with Eq. (18.35). The Franck-Condon factor e^{-D_s} describes the reduction of the tunneling amplitude by thermal excitations of the polaron cloud. We have

$$
D_s = d_s(T/T'_{\text{ph}})^{s-1} \quad \text{with} \quad d_s = \Gamma(s-1)\zeta(s-1)/\pi^2 ,
\tag{20.105}
$$

where we have introduced for convenience the temperature scale

$$
T'_{\text{ph}} = T_{\text{ph}}/(2\delta_s\pi^2)^{1/(s-1)} .
\tag{20.106}
$$

---

[4] We remark that the corresponding factors in Eq. (6.12) of Ref. [85] are incorrect.

For $s = 3$, T'_{ph} coincides with the temperature scale used in the literature [429, 430]. Here we restrict the attention to the most relevant regime

$$\hbar \Delta_{\mathrm{eff}}/k_{\mathrm{B}} \ll T'_{\mathrm{ph}} \ll T_{\mathrm{D}} . \tag{20.107}$$

For $T \ll T'_{\mathrm{ph}}$, the dressed matrix element $\tilde{\Delta}$ is practically temperature-independent. Putting $\kappa = 0$ in Eq. (18.47), we have

$$X_2(t) = 2[\Gamma(s-1)/\pi^2] (T/T'_{\mathrm{ph}})^{s-1} \operatorname{Re} \zeta\big(s - 1, \tfrac{1}{2} + it/\hbar\beta\big) . \tag{20.108}$$

The forward tunneling rate expression (20.31) takes the form

$$k^+(T, \epsilon) = \frac{\tilde{\Delta}^2}{4} \, e^{\beta\hbar\epsilon/2} \int_{-\infty}^{\infty} dt \, e^{i\epsilon t} \, e^{X_2(t)} . \tag{20.109}$$

Expansion of the integrand in powers of $X_2(t)$ gives the multi-phonon series for the rate. The term linear in $X_2(t)$ is the one-phonon contribution. It takes the form

$$k^+(T, \epsilon) = \frac{\pi}{2} \frac{\tilde{\Delta}^2}{\epsilon^2} \frac{G_{\mathrm{lf}}(|\epsilon|)}{|1 - \exp(-\beta\hbar\epsilon)|} = \frac{1}{2\pi} \frac{\tilde{\Delta}^2}{|\epsilon|} \frac{(\hbar|\epsilon|/k_{\mathrm{B}}T'_{\mathrm{ph}})^{s-1}}{|1 - \exp(-\beta\hbar\epsilon)|} . \tag{20.110}$$

With the backward rate $k^-(T, \epsilon) = k^+(T, -\epsilon)$ subjoined the one-phonon contribution to the relaxation rate reads

$$\gamma_{\mathrm{r}} \equiv k^+ + k^- = (\pi\tilde{\Delta}^2/2\epsilon^2) \, G_{\mathrm{lf}}(|\epsilon|) \coth(\beta\hbar|\epsilon|/2) . \tag{20.111}$$

Thus, for nonzero bias and T above $\hbar\epsilon/k_{\mathrm{B}}$, the one-phonon rate is proportional to T.

For $s < 3$, the one-phonon rate diverges in the limit $\epsilon \to 0$. For $s = 3$, the forward rate remains finite in this limit,

$$k^+(T, 0) = (\hbar\tilde{\Delta}^2/2\pi k_{\mathrm{B}}T'_{\mathrm{ph}}) \, T/T'_{\mathrm{ph}} . \tag{20.112}$$

The rate varies in the regime $T \ll T'_{\mathrm{ph}}$, in which $\tilde{\Delta} \approx \Delta_{\mathrm{eff}}$, *linearly* with T [134].

The one-phonon process vanishes for $s > 3$ as $\epsilon \to 0$. Thus for a symmetric system, the leading contribution for $T \ll T'_{\mathrm{ph}}$ is the "two-phonon assisted" process which is

$$
\begin{aligned}
k^+(T, 0) &= \frac{\pi}{2} \frac{\tilde{\Delta}^2}{4} \int_0^{\infty} \frac{d\omega}{\omega^4} \frac{G_{\mathrm{lf}}^2(\omega)}{\sinh^2(\beta\hbar\omega/2)} \\
&= \frac{\Gamma(2s-3)\zeta(2s-4)}{2\pi^3} \frac{\hbar\tilde{\Delta}^2}{k_{\mathrm{B}}T_{\mathrm{ph}}} \left(\frac{T}{T_{\mathrm{ph}}}\right)^{2s-3} .
\end{aligned}
\tag{20.113}
$$

Thus in the regime $T \ll T'_{\mathrm{ph}}$, the rate varies as T^{2s-3}. This gives for the diffusion constant[5] in the case $s = 5$ the familiar T^7 law [132] in the small-polaron model [129].

At this point, we should like to remark that the present treatment relies on the assumption that the dynamics is incoherent. However, for $T \ll T'_{\mathrm{ph}}$, the dynamics is

---

[5]In the incoherent tunneling regime, the diffusion constant is proportional to the tunneling rate [cf. Subsection 25.2.1].

actually coherent and therefore the Golden Rule approach is inadequate. The relevant discussion within a dynamical approach is given in Section 22.3. There it turns out that the rate (20.111) is the rate describing relaxation of the incoherent part of $\langle\sigma_z\rangle_t$ towards the equilibrium value $\langle\sigma_z\rangle_\infty$ for $\epsilon \gg \tilde{\Delta}$ [cf. Eqs. (22.91), (22.92), and (22.95)]. Similar conclusions hold for the two- and multi-phonon rate. The discussion of the dynamics beyond the one-phonon process is given in Subsection 22.1.2.

For T of the order of T'_{ph} or larger, phonon processes of higher order contribute, and for $T \gg T'_{\text{ph}}$ the full multi-phonon series has to be taken into account. In this regime, the formula (20.72) applies. It is convenient to write the activation energies Λ_1 and Λ_2 given in Eqs. (20.61) and (20.62) in the form

$$\Lambda_1 = 4[2B_s - c_s(\kappa)(T/T'_{\text{ph}})^{s-1}]k_B T , \qquad \Lambda_2 = d_s(\kappa)(T/T'_{\text{ph}})^{s-1}k_B T/4 , \quad (20.114)$$

$$c_s(\kappa) = 2\Gamma(s-1)\{\zeta(s-1,\tfrac{1}{2}+\kappa) - \zeta(s-1;1+\kappa)\}/\pi^2 , \quad (20.115)$$

$$d_s = 4\Gamma(s+1)\zeta(s+1,\tfrac{1}{2}+\kappa)/\pi^2 . \quad (20.116)$$

We then obtain

$$k^+(T,\epsilon) = \frac{\hbar\Delta^2 e^{-2B_s}}{2k_B T'_{\text{ph}}} \sqrt{\frac{\pi}{d_s(\kappa)}} \left(\frac{T'_{\text{ph}}}{T}\right)^{(s+1)/2}$$

$$\times \exp\left\{ \frac{\hbar\epsilon}{2k_B T} + c_s(\kappa)\left(\frac{T}{T'_{\text{ph}}}\right)^{s-1} \right. \quad (20.117)$$

$$\left. - \frac{1}{d_s(\kappa)}\left(\frac{\hbar\epsilon}{k_B T'_{\text{ph}}}\right)^2 \left(\frac{T'_{\text{ph}}}{T}\right)^{s+1} \right\} .$$

The rate expression (20.117) is the result of the integral expression (20.109) in steepest descent approximation. It practically holds in the entire regime $T \gtrsim T'_{\text{ph}}$.[6] In the regime $T'_{\text{ph}} \lesssim T \ll T_D$, we have $\kappa \approx 0$, and hence the coefficients $c_s(0)$ and $d_s(0)$ are temperature independent. For zero bias, the rate varies as $k \propto T^{-(s+1)/2} \exp[c_s(0)(T/T'_{\text{ph}})^{s-1}]$. We have $c_s(0) = (2^s-4)\Gamma(s-1)\zeta(s-1)/\pi^2 > 0$ for $s > 1$. Thus the thermal polaronic effects lead to an exponential enhancement of the rate. Observe that $c_s(0)$ is regular at $s = 2$. Further, the singularity at $s = 1$ cancels a singularity in the term B_s. Thus, Λ_1 is regular at $s = 1$. At higher temperature the dependence on κ becomes relevant, and in the classical regime $\kappa \gg 1$ the expression (20.117) matches with the classical rate (20.25). In this limit, the adiabatic dressing factor e^{-2B_s} is fully compensated by thermal effects and thus is absent. In the context of the small-polaron problem, the regime $T \gg T_D$ has been studied in Refs. [129, b] and [431].

In conclusion, the rate formula (20.117) covers the entire domain of multi-phonon processes which extends from the quantum regime up to the classical regime.

---

[6]The precise conditions for the validity of the form (20.117) are given in Eqs. (20.70) and (20.71).

20.2.9 Incoherent defect tunneling in metals

The tunneling of a defect in a metal at low temperature is predominantly influenced by the interaction with conduction electrons. At higher temperature, the coupling to acoustic phonons becomes important. The coupling to electrons leads to a decrease of the rate with increasing temperature for $K < 1/2$ and $\epsilon = 0$, as we may see from Eq. (20.84). The latter coupling causes phonon-assisted exponential enhancement of the rate at higher temperature [cf. Eq. (20.117) for $\epsilon = 0$]. Thus, the concerted influences lead to a minimum of the rate as a function of temperature.

To the study of incoherent tunneling near the minimum of the rate we take as a basis the combined spectral density

$$G_{\text{lf}}(\omega) = 2K\omega\, e^{-\omega/\omega_c} + 2\delta_3 \omega_{\text{ph}}^{-2}\, \omega^3\, e^{-\omega/\omega_D}\ . \qquad (20.118)$$

Interestingly, for the spectral density (20.118), the nonadiabatic tunneling rate can be found in closed analytical form in the entire temperature regime.

Consider first the regime $T \ll T_D$, $\hbar\omega_c/k_B$. For $\kappa_{\text{el}} = \kappa_{\text{ph}} \approx 0$, we obtain from the general expression (18.50) for the pair interaction $X(t)$ the form

$$X(t) = X_{\text{el}}(t) + X_{\text{ph}}(t)\ ,$$

$$X_{\text{el}}(t) = 2K \ln\Big((\beta\hbar\omega_c/\pi)\cosh(\pi t/\hbar\beta) \Big)\ , \qquad (20.119)$$

$$X_{\text{ph}}(t) = 2B_3 + \phi/3 - \phi/\cosh^2(\pi t/\hbar\beta)\ ,$$

where

$$\phi \equiv (T/T_{\text{ph}}')^2 = 2\pi^2 \delta_3 (k_B T/\hbar\omega_{\text{ph}})^2\ . \qquad (20.120)$$

With use of the expressions (20.119) the tunneling rate (20.31) may be written as

$$k^+(T, \epsilon; K) = k_{\text{el}}^+(T, \epsilon; K)\, \mathcal{A}_{\text{ph}}(T, \epsilon)\ , \qquad (20.121)$$

where k_{el}^+ is the forward rate in the presence of the conduction electrons alone, Eq. (20.79) or Eq. (20.80), and where $\mathcal{A}_{\text{ph}}(T, \epsilon)$ represents the activation factor due to the contact with the phonon bath in the presence of the fermionic excitations. Expanding the factor $\exp[\phi/\cosh^2(\pi t/\hbar\beta)]$ in the integrand of Eq. (20.31) in a power series in $\phi/\cosh^2(\pi t/\hbar\beta)$, and integrating each term, the resulting series turns out as a generalized hypergeometric series $_2F_2(a, b; c, d; z)$ [89]. The phonon activation factor takes the form [429]

$$\mathcal{A}_{\text{ph}}(T, \epsilon) = e^{-2B_3}\, e^{-\phi/3}\, {}_2F_2(K + i\hbar\beta\epsilon/2\pi,\ K - i\hbar\beta\epsilon/2\pi;\ K,\ K + \tfrac{1}{2};\ \phi)\ . \qquad (20.122)$$

Equation (20.121) with Eq. (20.122) and Eq. (20.79) or (20.80) constitutes the exact nonadiabatic rate expression in the regime $T \ll T_D$, $\hbar\omega_c/k_B$. Implementation of the hypergeometric series in Eq. (20.121) yields the expansion of the tunneling rate in terms of the multi-phonon processes in the presence of the electronic influences.

It is instructive to consider this result more closely in various limits. For zero bias, $k_{\text{el}}(T, 0)$ is given in Eq. (20.84), and the phonon factor (20.122) reduces to a degenerate hypergeometric function (Kummer function) [89]. We then have

$$k^{\pm}(T,0;K) = \frac{\sqrt{\pi}\,\Gamma(K)}{4\Gamma(K+\frac{1}{2})} \left(\frac{\hbar\omega_c}{\pi k_B T}\right)^{1-2K} \frac{\Delta^2}{\omega_c} e^{-(2B_3+\phi/3)} \,_1F_1(K; K+\tfrac{1}{2}; \phi) . \quad (20.123)$$

In the temperature regime $T \ll T'_{\rm ph}$, we may expand the Kummer function in a power series in ϕ. This corresponds to a categorization of the rate in terms of the one-phonon process, two-phonon process, etc., but in which the Ohmic contribution is fully taken into account [432].

In the opposite regime $T \gg T'_{\rm ph}$ (but still well below the Debye temperature), we may substitute the asymptotic representation of the Kummer function for $\phi \gg 1$. Then the expression (20.123) reduces to the form

$$k^{+}(T,0;K) = \frac{1}{4\sqrt{\pi}} \frac{\hbar\Delta^2 e^{-2B_3}}{k_B T'_{\rm ph}} \left(\frac{\pi k_B T}{\hbar\omega_c}\right)^{2K} \left(\frac{T'_{\rm ph}}{T}\right)^2 \exp\left(\frac{2T^2}{3T'^2_{\rm ph}}\right), \quad (20.124)$$

where the term $e^{-\phi/3}$ is absorbed into the last exponential factor. This expression is nonperturbative in the phonon coupling. For $K = 0$, Eq. (20.124) coincides with an early result by Holstein [129, b] and by Pirc and Gosar [433].

On the other hand, for $K = 0$ and $\epsilon \neq 0$, the expression (20.121) with (20.122) and with $\tilde{\Delta}^2 = \Delta^2 e^{-(2B_3+\phi/3)}$ reduces to the form

$$k^{+}(T,\epsilon;0) = \frac{1}{2\pi} \frac{\epsilon}{1-e^{-\hbar\beta\epsilon}} \left(\frac{\hbar\tilde{\Delta}}{k_B T'_{\rm ph}}\right)^2 \,_2F_2\left(1+i\frac{\hbar\beta\epsilon}{2\pi}, 1-i\frac{\hbar\beta\epsilon}{2\pi}; 2, \frac{3}{2}; \phi\right). \quad (20.125)$$

Next, consider a symmetric system in the absence of the coupling to conduction electrons, $K = 0$. We obtain from Eq. (20.125) the multi-phonon series expansion

$$k^{+}(T,0;0) = \frac{\tilde{\Delta}}{2\pi} \frac{\hbar\tilde{\Delta}}{k_B T'_{\rm ph}} \sum_{n=1}^{\infty} \frac{\Gamma(\frac{3}{2})}{n\Gamma(n+\frac{1}{2})} \left(\frac{T}{T'_{\rm ph}}\right)^{2n-1}. \quad (20.126)$$

The terms $n = 1$ and $n = 2$ correspond to the one-phonon process, Eq. (20.112), and two-phonon process, Eq. (20.113). Alternatively, we can derive the form (20.126) from Eq. (20.123). However, to circumvent a divergence in the limit $K \to 0$, the diagonal (zero-phonon) process has to be subtracted [129, 85]. The asymptotic representation of the series (20.126) in the limit $T \gg T'_{\rm ph}$ is given in Eq. (20.124) for $K = 0$.

As the temperature is increased further, the parameter $\kappa_{\rm ph} = 1/\hbar\beta\omega_{\rm D}$ moves away from zero. Since we have for the quantity (20.60) $\Lambda_2^{(\rm ph)} \gg \Lambda_2^{(\rm el)}$ in the regime $T \gg T'_{\rm ph}$, we find for a symmetric system

$$k^{+}(T,0;K) = \frac{\hbar\Delta^2 e^{-2B_3}}{2k_B T'_{\rm ph}} \sqrt{\frac{\pi}{d_3(\kappa_{\rm ph})}} \left(\frac{\pi k_B T}{\hbar\omega_c}\right)^{2K}$$
$$\times \left(\frac{T'_{\rm ph}}{T}\right)^2 \exp\left\{c_3(\kappa_{\rm ph})\left(\frac{T}{T'_{\rm ph}}\right)^2\right\}, \quad (20.127)$$

where

$$c_3(\kappa) = 2[\zeta(2, \tfrac{1}{2} + \kappa) - \zeta(2, 1 + \kappa)]/\pi^2 , \qquad d_3(\kappa) = 24\zeta(4, \tfrac{1}{2} + \kappa)/\pi^2 . \quad (20.128)$$

The expression (20.127) describes the crossover between the multi-phonon quantum rate expression (20.124) and the classical Marcus form[7]

$$k^+(T, 0; K) = \frac{\hbar\Delta^2}{4} \left(\frac{\pi k_B T}{\hbar\omega_c}\right)^{2K} \left(\frac{\pi}{\Lambda_{cl}^{(ph)} k_B T}\right)^{1/2} \exp\left(-\frac{\Lambda_{cl}^{(ph)}}{4 k_B T}\right), \quad (20.129)$$

where $\Lambda_{cl}^{(ph)} = 2\delta_3(\omega_D/\omega_{ph})^2 \hbar\omega_D$. Here we tacitly have assumed that we are still in the scaling limit with regard to the electron bath, $k_B T \ll \hbar\omega_c$. The appropriate generalizations to the biased case and to the quantum-classical crossover with respect to the electron bath are straightforward and left to the reader.

The theory of tunneling and diffusion of light interstitials in metals has been reviewed in Ref. [381]. The cooperation of the electron bath with the phonon bath leads to a minimum of the tunneling rate as a function of temperature for $K < \tfrac{1}{2}$. The minimum has been observed for muon diffusion in Al and in Cu [100], for incoherent hydrogen tunneling in Niobium [99, 101], and for defect tunneling in mesoscopic Bi wires [104, 103, 430]. Recently, the data for jump rates of defects in mesoscopic Bi wires [103] have been analyzed by using a roughly guessed formula interpolating between the one-phonon rate and the asymptotic expression (20.124) [434].

For tunneling of hydrogen in Niobium, the minimum of the crossing rate is found for T near T_D. The corresponding rate has been studied numerically in this regime [435]. It would now be interesting to apply the formula (20.127) to this problem. One final remark is appropriate. Because of disorder in real systems, the above rate expressions should be averaged with a distribution of bias energies if a quantitative comparison of theory with experiment is attempted.

20.3 Single charge tunneling

A tunnel junction embedded in an electrical circuit forms a quantum system violating Ohm's law. The tunneling transitions of electrons through the barrier are liable to inelastic scattering processes with creation of electromagnetic modes. Thereby the current is reduced at low voltage, an effect called dynamical Coulomb blockade (DCB). The Coulomb blockade regime is most developed in the weak-tunneling limit, $R_T \gg R_K$, where R_T is the tunneling resistance, and $R_K = 2\pi\hbar/e^2$ is the resistance quantum.

Control of single electron tunneling processes by tuning of gate voltages may be achieved in the DCB regime. The common principle of single electron devices, such as turnstiles, pumps, and transistors, is to transfer electrons one by one in systems of small tunnel junctions (cf. the contributions by D. Esteve, and by D. V. Averin and K. H. Likharev in Ref. [154]). As an indispensable ground work, it is necessary to understand the DCB effect in charge tunneling through a single junction.

---

[7]Note that the effects of the electrons are still quantum mechanical.

20.3.1 Weak-tunneling regime

For weak tunneling, the wave function of an excess electron is localized near to the barrier. Then the discrete charge representation (3.223) applies and tunneling through the barrier can be treated in Golden Rule approximation which formally corresponds to incoherent tunneling in a two-state system in the nonadiabatic limit. The effects of the electromagnetic environment are in the phase-phase equilibrium correlation function $Q_\varphi(t)$ given in Eq. (3.215). We rely on the global model (3.222) introduced in Section 3.5. We consider for simplicity the case of a constant tunneling amplitude, $T_{\boldsymbol{k},\boldsymbol{k}'} = T_T$, and we assume for a normal junction a constant density of states in the electron band around the Fermi energy E_F. It is convenient to combine the relevant junction properties in the tunneling resistance R_T,

$$\frac{1}{R_T} = \frac{4\pi e^2}{\hbar} N_R(0) N_L(0) \Omega_R \Omega_L |T_T|^2 , \qquad (20.130)$$

where $N_{R/L}(0)$ and $\Omega_{R/L}$ are the density of states and the volume of the right/left electrode, respectively. We assume that the tunneling resistance R_T is large compared to the resistance quantum $R_K = 2\pi\hbar/e^2$ so that thermal equilibrium in each electrode is maintained and the tunneling term H_T, Eq. (3.220), can be treated as a perturbation. The tunneling current may be written as (we put $v = eV_a/\hbar$)

$$I(V_a) = e\left[k^+(v) - k^-(v)\right] = e\left[1 - e^{-\beta\hbar v}\right] k^+(v) , \qquad (20.131)$$

where $k^+(v)$ is the (forward) rate for tunneling from the left electrode to the right electrode, and $k^-(v)$ is the backward tunneling rate. In the second form, we have used the detailed balance relation (20.32). The calculation of the rate to second order in H_T is straightforward. The thermal average of the quasiparticle modes in the electrodes introduces the Fermi distribution function $f(\omega)$, as in Eq. (4.131) with Eq. (4.132). Switching to the frequency variables $\omega' = E_{\boldsymbol{k}'}/\hbar$ and $\omega'' = E_{\boldsymbol{k}''}/\hbar$, we obtain

$$k^+(v) = \frac{\hbar}{2\pi}\frac{R_K}{R_T} \int_{-\omega_c}^{\omega_c} d\omega' \int_{-\omega_c}^{\omega_c} d\omega'' f(\omega') f(-\omega'') P_{em}(\hbar\omega' + \hbar v - \hbar\omega'') . \qquad (20.132)$$

Since the cutoff ω_c in the electron band is, in general, the largest frequency of the problem, we eventually take the limit $\omega_c \to \infty$.

The formula (20.132) has a neat interpretation. The integrand is proportional to the spectral probability density for an occupied state at energy $\hbar\omega'$ in the left electrode and an empty state at energy $\hbar\omega''$ in the right electrode times the probability density $P_{em}(\hbar\omega' + \hbar v - \hbar\omega'')$ that the electromagnetic environment absorbs the energy $\hbar(\omega' + v - \omega'')$. The bias energy $\hbar v = eV_a$ expresses the difference of the Fermi energies in the two electrodes due to the ideal voltage source V_a. The probability density of the electromagnetic environment to absorb the energy E is given by

$$P_{em}(E) = \frac{1}{2\pi\hbar} \int_{-\infty}^{\infty} dt \, e^{iEt/\hbar - Q_{em}(t)} . \qquad (20.133)$$

For the electromagnetic environment described by the Hamiltonian (3.207) the function $Q_{em}(t)$ corresponds to the phase correlation function $Q_\varphi(t)$ which describes the time correlations of the fluctuations of the phase jump $\varphi(t)$ at the junction due to the electromagnetic environment, $Q_{em}(t) = Q_\varphi(t)$,

$$e^{-Q_{em}(t)} \equiv \left\langle e^{i\varphi(t)} e^{-i\varphi(0)} \right\rangle_\beta = e^{-\langle [\varphi(0)-\varphi(t)]\varphi(0)\rangle_\beta} . \qquad (20.134)$$

To obtain the second form, we have employed the Gaussian statistical properties of the fluctuating phase $\varphi(t)$. The phase correlation function $Q_{em}(t) \equiv \langle [\varphi(0) - \varphi(t)]\varphi(0)\rangle_\beta$ can be written as [cf. Eqs. (3.215) and (3.216)]

$$Q_{em}(t) = \int_0^\infty d\omega \, \frac{G_{em}(\omega)}{\omega^2} \left\{ \coth\left(\frac{\hbar\beta\omega}{2}\right)\left(1 - \cos(\omega t)\right) + i\sin(\omega t) \right\} \qquad (20.135)$$

with

$$G_{em}(\omega) = (e^2/\pi\hbar)\,\omega^2\tilde{\chi}''(\omega) = 2\omega \operatorname{Re} Z_t^*(\omega)/R_K . \qquad (20.136)$$

The general properties of the probability function $P_{em}(E)$ have been discussed in Subsection 20.2.2. In particular, the relation (20.36) for $P_{em}(E)$ ensures detailed balance for the rate, which we have employed already in Eq. (20.131).

The spectral density of an Ohmic electromagnetic environment is [cf. Eq. (3.225]

$$G_{em}(\omega) = \frac{2\alpha\,\omega}{1 + (\omega/\omega_R)^2} , \qquad (20.137)$$

where $\alpha = R/R_K$, $\omega_R = 1/RC = E_c/\pi\hbar\alpha$, and where $E_c = e^2/2C$ is the charging energy. For such environment, the reorganization energy Λ_{cl} or absorbed net energy (20.37) coincides with the charging energy,

$$\Lambda_{cl} = E_c . \qquad (20.138)$$

Thus, for a high-impedance environment, the junction behaves classically down to fairly low temperature. In this regime, $P_{em}(E)$ has the Gaussian form (20.25),

$$P_{em}(E) = \frac{1}{(4\pi E_c k_B T)^{1/2}} \exp\left(-\frac{(E - E_c)^2}{4E_c k_B T}\right) . \qquad (20.139)$$

With increasing quality factor Q_{qual} in the spectral density (3.228), the probability function $P_{em}(E)$ changes from the smooth form (20.139) to the resonant characteristics given in Eq. (20.41), where ω_0 corresponds to ω_L.

Next, we return to the rate expression (20.132. With the substitution $\omega'' = \omega + \omega'$, the ω'-integration in Eq. (20.132) can be done,

$$\int_{-\infty}^\infty d\omega' \, f(\omega')\, f(-\omega' - \omega) = \int_{-\infty}^\infty d\omega' \, \frac{f(\omega') - f(\omega' + \omega)}{1 - e^{-\beta\hbar\omega}} = \frac{\omega}{1 - e^{-\beta\hbar\omega}} . \qquad (20.140)$$

With this relation, the rate expression (20.132) is reduced to the single integral

$$k^+(v) = \hbar \int_{-\infty}^{\infty} d\omega\, k_0^+(\omega) P_{\text{em}}(\hbar v - \hbar\omega)\,, \qquad (20.141)$$

$$k_0^+(\omega) = \frac{1}{2\pi} \frac{R_{\text{K}}}{R_{\text{T}}} \frac{\omega}{1 - e^{-\hbar\beta\omega}}\,. \qquad (20.142)$$

For "elastic" tunneling we have $P_{\text{em}}(\hbar v - \hbar\omega) = \delta(\hbar v - \hbar\omega)$ and hence $k^+(v) = k_0^+(v)$. Hence the rate $k_0^+(v)$ describes tunneling of the fermionic quasiparticle through the barrier in the absence of coupling to the electromagnetic environment.

It is instructive to see that the tunneling rate expression (20.132) for the fermionic entity can be transformed into an expression of the form (20.30) which describes tunneling of a bosonic entity. To prepare the ground, we first observe that the detailed balance relation (20.36) is met if we write the function $P(\hbar\omega)$, Eq. (20.34), which describes the reservoir's probability density for absorption of energy $\hbar\omega$, as [436]

$$P(\hbar\omega) = D(\omega) f(-\omega) / \hbar\omega_{\text{c}}\,, \qquad (20.143)$$

where $f(\omega)$ is the Fermi function. The function $D(\omega)$ is symmetric about the Fermi energy, $D(\omega) = D(-\omega)$, and the normalization is $\int_0^{\infty} d\omega\, D(\omega) = \omega_{\text{c}}$, as follows from Eq. (20.35). As we shall see, the function $D(\omega)$ may be perceived as an effective density of states for fermionic quasiparticles.

Consider next the Fourier transform of the correlator $\exp[-Q_{\text{ohm}}(t, K)]$, where $Q_{\text{ohm}}(t; K)$ is the bath correlation function in the Ohmic scaling limit, Eq. (18.54),

$$Q_{\text{ohm}}(t; K) = 2K \ln[(\hbar\beta\omega_{\text{c}}/\pi) \sinh(\pi t/\hbar\beta)] + i\pi K \operatorname{sgn}(t)\,. \qquad (20.144)$$

The Fourier integral

$$\frac{1}{2\pi\hbar} \int_{-\infty}^{\infty} dt\, e^{-i\omega t} \exp\left[-Q_{\text{ohm}}(t; K)\right] = \frac{1}{\hbar\omega_{\text{c}}} D_{\text{ohm}}(\omega; K)\, f(\omega) \qquad (20.145)$$

can be done in analytic form [cf. Eq. (20.29) with Eq. (20.79)], yielding ($|\omega| \ll \omega_{\text{c}}$)

$$D_{\text{ohm}}(\omega; K) = \frac{1}{\Gamma(2K)} \left(\frac{\hbar\beta\omega_{\text{c}}}{2\pi}\right)^{1-2K} \frac{|\Gamma(K + i\hbar\beta\omega/2\pi)|^2}{|\Gamma(1/2 + i\hbar\beta\omega/2\pi)|^2}\,. \qquad (20.146)$$

Observing that the density of states is constant in the fermionic band $|\omega| < \omega_{\text{c}}$ for the particular value $K = \frac{1}{2}$, $D_{\text{ohm}}(\omega; \frac{1}{2}) = 1$, we can directly relate the Fermi function with the $P(E)$-function at $K = \frac{1}{2}$ discussed in Subsection 20.2.2,

$$f(\omega) = \frac{\omega_{\text{c}}}{2\pi} \int_{-\infty}^{\infty} dt\, e^{-i\omega t} \exp[-Q_{\text{ohm}}(t; K = \tfrac{1}{2})]\,. \qquad (20.147)$$

Next, we substitute the Fourier representation (20.147) both for $f(\omega')$ and for $f(-\omega'')$ and the Fourier representation (20.133) for $P_{\text{em}}(t)$ into the expression (20.132). Since the expression (20.132) is in the form of a convolution, we immediately get with use of the obvious relation $2Q_{\text{ohm}}(t, \frac{1}{2}) = Q_{\text{ohm}}(t, 1)$

$$k^+(v) = \left(\frac{\omega_{\text{c}}}{2\pi}\right)^2 \frac{R_{\text{K}}}{R_{\text{T}}} \int_{-\infty}^{\infty} dt\, e^{ivt} \exp[-Q_{\text{ohm}}(t; K = 1) - Q_{\text{em}}(t)]\,. \qquad (20.148)$$

Thus we have found a surprising result: tunneling of a fermionic particle through a barrier embedded in a conductor is like tunneling of a boson which undergoes Ohmic dissipation with Kondo parameter $K = 1$. In the presence of an electromagnetic environment, the respective correlation function $Q_{em}(t)$ is simply added to the the Ohmic correlation function.

For an Ohmic impedance, the spectral density $G_{em}(\omega)$ is given by the Drude form (20.137). The correlation function $Q_{em}(t; \alpha)$ for the algebraic cutoff at frequency ω_R differs in essence from the form (18.54), being appropriate for exponential cutoff at frequency ω_c, by an adiabatic correction

$$Q_{em}(t; \alpha) = Q_{ohm}(t; \alpha) + \Delta Q_{em}(\alpha) , \qquad (20.149)$$

$$\Delta Q_{em}(\alpha) = 2\alpha \ln(\omega_R/\omega_c) + 2\alpha\zeta_D , \qquad (20.150)$$

$$\zeta_D = \psi\left(1 + \frac{\hbar\beta\omega_R}{2\pi}\right) - \psi(1) - \ln\left(\frac{\hbar\beta\omega_R}{2\pi}\right) \xrightarrow{\hbar\beta\omega_R \gg 1} C_E . \quad (20.151)$$

The relation $\zeta_D = C_E$ holds in the scaling regime, where $C_E = 0.577\ldots$ is Euler's constant. With this, the forward tunneling rate (20.148) takes the form

$$k^+(v) = \left(\frac{\omega_c}{2\pi}\right)^2 \frac{R_K}{R_T} e^{-\Delta Q_{em}(\alpha)} \int_{-\infty}^{\infty} dt\, e^{ivt} \exp[-Q_{ohm}(t; K = 1 + \alpha)] . \quad (20.152)$$

Thus, there is an entire low-energy regime where weak tunneling of electrons coupled to a resistive electromagnetic environment behaves exactly like tunneling of bosons embedded in an Ohmic environment with Ohmic damping parameter

$$K = 1 + \alpha . \qquad (20.153)$$

One remark is in order. The expression (20.152) is the leading contribution to the rate in the weak-tunneling limit. Generalization to the full weak-tunneling power series in R_K/R_T, and to the related strong-tunneling series will be given in Chapter 28.

Alternatively, we can transfer the effects of the electromagnetic environment into an effective frequency- and temperature-dependent tunneling density. Writing

$$\frac{1}{2\pi\hbar} \int_{-\infty}^{\infty} dt\, e^{-i\omega t} e^{-[Q_{ohm}(t; K=1) + Q_{em}(t)]/2} = \frac{1}{\hbar\omega_c} D_{eff}(\omega) f(\omega) , \qquad (20.154)$$

the forward rate (20.148) is transformed into

$$k^+(v) = \frac{1}{2\pi} \frac{R_K}{R_T} \int_{-\omega_c}^{\omega_c} d\omega\, D_{eff}(\omega) D_{eff}(\omega + v) f(\omega)\, f(-\omega - v) . \qquad (20.155)$$

In this formulation, the effects of the environment are described in terms of an effective frequency-dependent tunneling density of states of the fermionic charge carrier. In the absence of the environmental coupling, $Q_{em}(t) = 0$, we have $D_{eff}(\omega) = 1$, so that $k^+(v)$ takes the form (20.142), $k^+(v) = k_0^+(v)$.

The expressions (20.148) and (20.155) are bosonic and fermionic representations, respectively, of the same physical tunneling process.

20.3.2 The current-voltage characteristics

With use of the expressions (20.141) and (20.131), the current-voltage characteristics of a single junction embedded in an electromagnetic environment is found as

$$I(V_a) = \int_{-\infty}^{\infty} dE \, \frac{1 - e^{-\beta e V_a}}{1 - e^{-\beta E}} \, P_{em}(eV_a - E) \, I_0(E/e) \,, \qquad (20.156)$$

where

$$I_0(V) = V/R_T \qquad (20.157)$$

is the Ohmic tunneling current. The expression (20.156) satisfies the plausible relation $I(-V_a) = -I(V_a)$, as follows with use of the detailed balance relation (20.36).

When the impedance of the environment at zero frequency $Z(\omega = 0) = R$ is small compared to the resistance quantum R_K, the transfer of the charge through the junction is predominantly elastic, $P_{em}(E) \approx \delta(E)$. With this we obtain the form (20.142) for the forward tunneling rate, and finally with Eq. (20.131) or directly from Eq. (20.156) the Ohmic law $I(V_a) = V_a/R_T$. This warrants the interpretation of R_T as a tunneling resistance.

In the opposite limit of a high-impedance environment, the Gaussian distribution (20.139) for $P_{em}(E)$ at low T is very narrow about $E = E_c$ and can well be approximated by $P_{em}(E) = \delta(E - E_c)$. In this limit, the energy absorbed by the environment equals the charging energy, and the current is

$$I(V_a) = \Theta(V_a - E_c/e) \left[V_a - E_c/e \right]/R_T \,. \qquad (20.158)$$

The Coulomb gap $eV_a > E_c$ is in correspondence with the energy balance (3.204).

At zero temperature, we have $P_{em}(E) = 0$ for $E < 0$ since the bath cannot supply energy anymore. In this limit, the expression (20.156) reduces to

$$I(V_a) = \Theta(V_a) \frac{1}{eR_T} \int_0^{eV_a} dE \, (eV_a - E) \, P_{em}(E) \,. \qquad (20.159)$$

Direct information about the distribution of energy absorbed by the environment is gained from the second derivative of the current with respect to the applied voltage,

$$\frac{d^2 I(V_a)}{dV_a^2} = \frac{e}{R_T} P_{em}(eV_a) \,. \qquad (20.160)$$

In the regime $eV_a \gg k_B T$ and $P_{em}(|eV_a|) \ll P_{em}(0)$ we obtain from Eq. (20.156)

$$I(V_a) = \frac{1}{eR_T} \int_{-V_a}^{V_a} dE \, (eV_a - E) \, P_{em}(E) \,. \qquad (20.161)$$

Using the sum rules (20.35) and (20.37), and assuming a resistive environment for which Λ_{cl} equals the charging energy, Eq. (20.138), we find the linear current-voltage characteristics at low temperatures in the classical regime $k_B T \ll e^2/2C < eV_a$ as

$$I(V_a) = \left[V_a - e/2C \right]/R_T \,. \qquad (20.162)$$

The shift in voltage by $e/2C$ is the manifestation of the Coulomb blockade in the classical regime.

Consider now the current-voltage characteristics in the low temperature quantum regime $k_B T \ll eV_a \ll e^2/2C$. We immediately see from the nonadiabatic rate formula (20.152) that the analytic rate expressions (20.79) – (20.84) directly apply to single-electron tunneling in a resistive environment described by the spectral density (20.137) if we employ the assignments

$$K \to 1+\alpha\,, \quad \text{and} \quad \frac{\Delta^2}{4} \to \left(\frac{\omega_c}{2\pi}\right)^2 \frac{R_K}{R_T} e^{-\Delta Q_{em}(\alpha)}\,. \tag{20.163}$$

With use of these correspondences and of the rate expression (20.82), the current-voltage characteristics in the regime $k_B T \ll eV_a \ll \hbar\omega_R$ is found to read

$$I(V_a) = \frac{e^{-2\alpha C_E}}{\Gamma(2+2\alpha)} \frac{V_a}{R_T} \left(\frac{\pi\alpha eV_a}{E_c}\right)^{2\alpha}\,, \quad \alpha = R/R_K\,. \tag{20.164}$$

We see that in the dynamical Coulomb blockade expression (20.164) the sharp Coulomb gap appearing in the classical expression (20.158) is rounded by quantum fluctuations. The dynamical Coulomb blockade shows itself in a super-linear behavior $I(V_a) \propto V_a^{1+2\alpha}$ in the regime $k_B T \ll eV_a$, or equivalently in the so-called zero-bias anomaly $dI(V_a)/dV_a \propto V_a^{2\alpha}$, instead of the voltage-independent conductance without the electromagnetic influences. In view of the substitution rule $K \to 1+\alpha$ in the correspondence, it is not surprising that the current shows the characteristic power laws which we have already encountered in the discussion of the Ohmic case in Subsection 20.2.5. The tunneling density of states associated with the Ohmic impedance is obtained from Eq. (20.146) with Eq. (20.163) at zero temperature for $\omega \ll \omega_R$ as

$$D_{em}(\omega) = e^{-\alpha C_E} \left(|\omega|/\omega_R\right)^\alpha /\Gamma(1+\alpha)\,. \tag{20.165}$$

The tunneling density of states is nonanalytic at the Fermi energy and it is thinned down near the Fermi level for $\alpha > 0$ compared with the constant tunneling density of states for $\alpha = 0$. In the expression (20.155), the zero-bias anomaly originates from the low density of states near the Fermi level. The power law in Eq. (20.164) directly reflects the nonanalytic behavior $D_{em}(\omega \to 0) \propto |\omega|^\alpha$.

We should like to remark that we have the same power-law form as in Eq. (20.164) for any other environment with a finite impedance at $\omega = 0$, where $\alpha = Z(0)/R_K$. Only the prefactor depends on the spectral properties of the impedance.

Consider next the linear conductance at finite temperature. Using the correspondences (20.163), we obtain from Eq. (20.131) with Eq. (20.84) the expression

$$\left.\frac{dI}{dV_a}\right|_{V_a=0} = \frac{e^{-2\alpha C_E}}{R_T} \frac{\sqrt{\pi}\Gamma(1+\alpha)}{2\Gamma(\frac{3}{2}+\alpha)} \left(\frac{\pi^2\alpha k_B T}{E_c}\right)^{2\alpha}\,. \tag{20.166}$$

In the high voltage regime $eV_a \gg E_c$, we may use in Eq. (20.159) for $P_{em}(E)$ the form (20.55). Upon substituting for the spectral density $G_{em}(\omega)$ the expression (20.137), we obtain the leading correction to the strict Coulomb gap form (20.158),

$$I(V_a) = \frac{1}{R_T} \left(V_a - \frac{e}{2C} + \frac{\alpha}{\pi^2} \frac{e^2}{4C^2} \frac{1}{V_a} \right) , \qquad \text{for} \qquad eV_a \gg E_c . \qquad (20.167)$$

Thus, the actual offset is smaller than the offset in Eq. (20.158).

Let us finally consider an LC transmission line described by the sub-Ohmic spectral density (3.230) with (3.231). Putting $s = \frac{1}{2}$ in the former result (20.50) and taking the limit $\omega_c \to \infty$, we find for the probability function $P_{em}(E)$ the expression[8]

$$P_{em}(E) = \sqrt{\frac{eV_c}{4\pi E^3}} \exp\left(-\frac{eV_c}{4E} \right) \qquad \text{with} \qquad V_c = \frac{4eR_0}{C_0 R_K} . \qquad (20.168)$$

The function $P_{em}(E)$ has a maximum at $E = eV_c/6$. Upon inserting this form into the current (20.159), we find exponential suppression of the current $\propto \exp(-V_c/4V_a)$ for $V_a \ll V_c$ instead of the power law suppression in the Ohmic case, Eq. (20.164). The strong suppression of the charge transfer is due to the much higher density of low-frequency excitations of the LC transmission line. Finally, we remark that at higher voltage the current characteristics is similar to the case of an Ohmic resistive environment.

20.3.3 Weak tunneling of 1D interacting electrons

In 1D quantum wires, the electron-electron interaction leads to a break-down of the Fermi liquid model. The usual quasiparticle description does not apply any more. Instead, the low-energy excitations of the correlated system are collective bosonic excitations, which can be viewed as collective density fluctuations of a harmonic fluid. The appropriate model is the Luttinger liquid model which is described in boson representation in Chapter 28. Now assume that the transport through the quantum wire is impeded by a barrier or impurity. The respective transport problem is treated below in Section 28.1. At this point, it is insightful to show the formal similarity of correlated electron-tunneling through a strong barrier with weak single-electron tunneling in the presence of a resistive electromagnetic environment. To this aim, we only have to anticipate the particular form of the equilibrium correlation function of the phase $\phi(0,t)$ of the fermion field (28.2), $4\pi\langle [\phi(0,0) - \phi(0,t)]\phi(0,0)\rangle_\beta \equiv Q(t; 1/g)$. Here, g is a dimensionless interaction constant in the Luttinger model. The case $g < 1$ corresponds to repulsive electron-electron interaction, and $g = 1$ is the Fermi liquid point. We find from Eq. (28.22) that the phase correlation function $Q(t; 1/g)$ is formally identical with the Ohmic bath correlation function $Q_{ohm}(t; K)$ given in Eq. (20.144), with $K = 1/g$,

$$Q_{ohm}(t; 1/g) \equiv \frac{2}{g} \ln \left[\frac{\beta\hbar\omega_c}{\pi} \sinh\left(\frac{\pi|t|}{\hbar\beta} \right) \right] + i\frac{\pi}{g} \text{sgn}(t) . \qquad (20.169)$$

In boson representation, the forward tunneling rate of the unit charge through the barrier is given, in analogy with Eq. (20.148), by

---

[8]The expression for V_c is misprinted in Ref. [155].

$$k^+(v) = \left(\frac{\omega_c}{2\pi}\right)^2 \frac{R_K}{R_T} \int_{-\infty}^{\infty} dt\, e^{ivt} \exp\left[-Q_{ohm}(t; 1/g)\right] , \tag{20.170}$$

where $V_a = \hbar v/e$ is the applied voltage at the impurity. Thus we find a direct formal correspondence between weak tunneling of 1D interacting electrons and weak tunneling of electrons coupled to an electrical circuit with Ohmic impedance. In the correspondence we have

$$1 + \alpha = 1/g . \tag{20.171}$$

We shall discuss below in Chapter 28, in particular in Subsection 28.1.4, that the correspondence als holds for joint tunneling of two, three and many charges, and that it is also valid in the weak-barrier, or equivalently strong-tunneling regime.

Obviously, we may reverse the previous arguments and introduce an effective tunneling density of states of the fermionic entity, $D_{ohm}(\omega; 1/g)$, as in Eq. (20.146). Thus we find in analogy with Eq. (20.155) the expression [436]

$$k^+(v) = \frac{1}{2\pi} \frac{R_K}{R_T} \int_{-\omega_c}^{\omega_c} d\omega\, D_{ohm}(\omega; 1/2g) D_{ohm}(\omega + v; 1/2g) f(\omega) f(-\omega - v) . \tag{20.172}$$

It is also easy to see from Eq. (20.131) with the form (20.170), and with use of the expression (20.145) for $Q_{ohm}(t; 1/2g) = \frac{1}{2} Q_{ohm}(t; 1/g)$, that the current may be written as

$$I(V_a) = \frac{\hbar \omega_c}{e R_T}\left(1 - e^{-\beta \hbar v}\right) f(-v) D_{ohm}(v; 1/g) , \tag{20.173}$$

where $D_{ohm}(\omega; K)$ is given in Eq. (20.146).

For $g = 1$ we have $D_{ohm}(\omega; 1) = (\omega/\omega_c)\coth(\beta\hbar\omega/2)$. With this we see that the expression (20.173) reduces at the Fermi liquid point to Ohm's law $I(V_a) = V_a/R_T$.

In a double junction system, besides sequential tunneling, higher order tunneling processes may occur in which the Coulomb barrier is bypassed by virtual occupation of the island. A theoretical description of resonant tunneling for correlated electrons using the above concept of effective tunneling densities is reported in Ref. [437].

A nonperturbative formalism in the presence of strong Coulomb interactions has been developed and applied to resonant tunneling in Ref. [438]. A review of mesoscopic electron transport is given in Ref. [439]. Electrical conduction in single-molecule circuits is discussed in a recent book by Cuevas and Scheer [440].

20.3.4   Tunneling of Cooper pairs

Tunneling of Cooper pairs through a Josephson junction is affected by the electromagnetic environment as well. For weak Josephson coupling energy, $E_J \ll E_c = 2e^2/C$, we may calculate the crossing rate for a Cooper pair in Golden Rule approximation with respect to the Josephson coupling energy. Since the tunneling entities are bosons, the rate can be written in a form analogous to Eq. (20.29) [cf. Fig. 3.3 and Subsection 3.5.3]. With the externally applied voltage V_x, the rate expression is

$$k^+ = \frac{E_J^2}{4\hbar^2} \int_{-\infty}^{\infty} dt\, e^{i2eV_x t/\hbar}\, e^{-Q_\psi(t)} . \tag{20.174}$$

The bias energy is $2eV_x$ because the charge transferred by a Cooper pair across the junction is $2e$, and the phase correlation function for Cooper pairs $Q_\psi(t)$ is given in Eq. (3.234). With the correspondences

$$\hbar\Delta \hat{=} E_J = (\hbar/2e)I_c \,, \qquad \hbar\epsilon \hat{=} 2eV_x \,, \qquad K \hat{=} \rho = R/R_Q \,, \qquad (20.175)$$

where $R_Q = 2\pi\hbar/4e^2$, and with

$$Q_\psi(t) \equiv Q_{\mathrm{ohm}}(t; K = \rho) + 2\rho\ln(\omega_R/\omega_c) + 2\rho\zeta \,, \qquad (20.176)$$

we can directly apply the results for incoherent tunneling of a boson with Ohmic friction presented in Section 20.2 to Cooper pair tunneling. The logarithmic term in Eq. (20.176) takes into account that the cut-off frequency is

$$\omega_R = 1/Z(0)C = E_c/\pi\rho\hbar \,, \qquad (20.177)$$

instead of ω_c. The quantity ζ accounts for the particular high-frequency dependence of the total impedance $Z_t^*(\omega)$ in adiabatic approximation,

$$\zeta = \zeta_D + \int_0^\infty \frac{d\omega}{\omega} \left[\frac{\mathrm{Re}\, Z_t^*(\omega)}{\rho R_Q} - \frac{1}{1 + (\pi\rho\hbar\omega/E_c)^2} \right] \,. \qquad (20.178)$$

The term ζ_D, given in Eq. (20.151), arises from the Drude form (20.137), and the integral accounts for deviation of the actual $Z_t^*(\omega)$ from this behavior.

Introducing the function $P_\psi(E)$ as the Fourier transform of $e^{-Q_\psi(t)}$ [see Subsection 20.2.2 for the physical meaning and the properties of $P_\psi(E)$], the current-voltage characteristics takes the form

$$I(V_x) = 2e(k^+ - k^-) = \frac{\pi e(E_J\, e^{-\rho\zeta})^2}{\hbar} \left[1 - e^{-2\beta eV_x} \right] P_\psi(2eV_x) \,. \qquad (20.179)$$

Using the rate expression (20.79) and the above correspondences, the current-voltage characteristics is found to read [441]

$$I(V_x) = \frac{\pi e\rho}{\hbar} \frac{(E_J\, e^{-\rho\zeta})^2}{E_c} \left(\frac{\beta E_c}{2\pi^2\rho} \right)^{1-2\rho} \frac{|\Gamma(\rho + i\beta eV_x/\pi)|^2}{\Gamma(2\rho)} \sinh(\beta eV_x) \,. \qquad (20.180)$$

This expression holds for a wide range of temperatures in the weak tunneling regime $E_J \ll E_c$. For small ρ, the supercurrent-voltage characteristics (20.180) shows a peak at voltage $V = \pi\rho/e\beta$ which becomes increasingly marked as temperature is lowered.

The current at zero temperature is found from (20.180) as

$$I(V_x) = \frac{\pi^{5/2}\,\rho}{2\,\Gamma(\rho)\Gamma(\rho + \frac{1}{2})} \left(\frac{E_J\, e^{-\rho\zeta}}{eV_x} \right)^2 \left(\frac{\pi\rho eV_x}{E_c} \right)^{2\rho} \frac{V_x}{R} \,. \qquad (20.181)$$

Hence the supercurrent exhibits the zero bias anomaly $I(V_x) \propto V_x^{2\rho-1}$ [421]. It describes suppression of the current by the Coulomb blockade effect for $\rho > 1$. Also the zero bias conductance at finite temperature exhibits power law behavior,

$$\frac{dI}{dV_x}\bigg|_{V_x=0} = \frac{1}{R_Q} \frac{\sqrt{\pi}}{2} \frac{\Gamma(\rho)}{\Gamma(\rho+\frac{1}{2})} \left(\frac{E_J \, e^{-\rho\varsigma}}{E_c/\rho\pi^2}\right)^2 \left(\frac{\beta E_c}{\rho\pi^2}\right)^{2-2\rho}. \tag{20.182}$$

For small impedance $Z(0) \ll R_Q$, we get from Eq. (20.180)

$$I(V_x) = \frac{\pi e}{\hbar} \frac{E_J^2 \, e^{-2\rho\varsigma}}{E_c} \rho^{2\rho} \left(\frac{\beta E_c}{2\pi^2}\right)^{1-2\rho} \frac{2\pi^2\rho\beta e V_x}{(\beta e V_x)^2 + (\pi\rho)^2}, \tag{20.183}$$

which reduces further in the classical limit $k_B T \gg \hbar\omega_R$ to the form

$$I(V_x) = \frac{I_c^2 \, e^{-2\rho\varsigma}}{2} \frac{Z(0)\,V_x}{V_x^2 + [\,2eZ(0)k_B T/\hbar\,]^2}. \tag{20.184}$$

At $\rho < 1$, which is the usual case, the conductance found from Eq. (20.181) diverges in the zero bias limit. Also the conductance (20.182) diverges with decreasing temperature as $T^{2\rho-2}$. This indicates onset of strong tunneling in the low-energy regime. The singularity in the weak-tunneling conductance is smoothed out by taking into account terms of higher order in E_J^2 (see Section 26). At this point, a preliminary final remark is appropriate. The case of weak Josephson coupling is related to the case of large Josephson coupling [Eq. (17.79)] by a duality symmetry. This issue is discussed in Subsection 26.3.

20.3.5 Tunneling of quasiparticles

Quasiparticle tunneling in a superconducting junction is similar to quasiparticle tunneling in a normal junction. The key difference is that the density of states of the BCS quasiparticles strongly depends on frequency, in particular in the region slightly above the gap frequency. The normalized density of states $\mathcal{N}_{qp}(\omega)$ of the BCS quasiparticle excitations is given in Eq. (4.167). Observing detailed balance for the tunneling rate, Eq. (20.32), we may write the current again in the form (20.131). In modification of Eq. (20.132), the forward rate for quasiparticle tunneling is

$$k^+(v) = \frac{\hbar}{2\pi} \frac{R_K}{R_T} \int_{-\omega_c}^{\omega_c} d\omega \int_{-\omega_c}^{\omega_c} d\omega' \, \mathcal{N}_{qp}(\omega)\mathcal{N}_{qp}(\omega') \, f(\omega)f(-\omega') \, P_{em}(\hbar\omega + \hbar v - \hbar\omega').$$

Other forms analogous to those given for quasiparticle tunneling in a normal junction are easily found. For instance, in analogy with Eq. (20.156) the quasiparticle current through the junction in the presence of the environment, $I_{qp,em}$, is related to the quasiparticle current in the absence of it, $I_{qp,0}$, by the integral relation [421]

$$I_{qp,em}(V_a, T) = \int_{-\infty}^{\infty} dE \, \frac{1 - e^{-\beta e V_a}}{1 - e^{-\beta E}} \, P_{em}(eV_a - E) \, I_{qp,0}(E/e, T). \tag{20.185}$$

The quasiparticle current in the absence of the environment reads

$$I_{qp,0}(V_a, T) = \frac{\hbar}{eR_T} \int_{-\infty}^{\infty} d\omega \, \mathcal{N}_{qp}(\omega)\mathcal{N}_{qp}(\omega + eV_a/\hbar) \, [\,f(\omega) - f(\omega + eV_a/\hbar)\,]. \tag{20.186}$$

The ω-integral can be done in analytic form in the limit $T \to 0$. The resulting expression is given in terms of the hypergeometric function $_2F_1(z)$ as

$$I_{\text{qp},0}(V_\text{a}, 0) = \Theta(V_\text{a} - V_\text{g}) \frac{V_\text{a}}{R_\text{T}} \frac{V_\text{a}}{2(V_\text{a} + V_\text{g})}$$

$$\times \left\{ B(\tfrac{1}{2}, \tfrac{1}{2})\ _2F_1(\tfrac{1}{2}, \tfrac{1}{2}; 1; z) \right. \tag{20.187}$$

$$\left. - \frac{(V_\text{a} - V_\text{g})^2}{V_\text{a}^2} B(\tfrac{3}{2}, \tfrac{1}{2})\ _2F_1(\tfrac{1}{2}, \tfrac{3}{2}; 2; z) \right\} ,$$

where the gap voltage is $V_\text{g} = 2\hbar\Delta_\text{g}/e$ and $z = (V_\text{a} - V_\text{g})^2/(V_\text{a} + V_\text{g})^2$.

The quasiparticle current $I_{\text{qp},0}(V_\text{a}, 0)$ is zero in the regime $V_\text{a} < V_\text{g}$. At the gap voltage, $V_\text{a} = V_\text{g}$, the quasiparticle current jumps from zero to the finite value

$$I_{\text{qp},0}(V_\text{g}, 0) = \frac{\pi}{4} \frac{V_\text{g}}{R_\text{T}} . \tag{20.188}$$

The quasiparticle current approaches asymptotically Ohm's law

$$I_{\text{qp},0}(V_\text{a} \gg V_\text{g}, 0) = \frac{V_\text{a}}{R_\text{T}} . \tag{20.189}$$

Consider next the effects of an electromagnetic environment. For an Ohmic spectral density $G_{\text{em}} = 2\alpha\omega$ with $\alpha = R/R_\text{K}$ we obtain from Eq. (20.185) upon using findings from Subsection 20.2.5, in particular Eq. (20.82), the anomalous threshold behavior

$$I_{\text{qp,em}}(V_\text{a}, 0) \propto \Theta(V_\text{a} - V_\text{g}) \left(V_\text{a} - V_\text{g}\right)^{2\alpha} . \tag{20.190}$$

Hence the jump (20.188) at the threshold is dissolved in fact by quantum fluctuations of the Ohmic resistor.

At voltage $V_\text{a} \gg V_\text{g}$, the quasiparticle current varies as

$$I_{\text{qp,em}}(V_\text{a}, 0) \propto V_\text{a}^{2\alpha+1} . \tag{20.191}$$

The deviation of $I_{\text{qp,em}}(V_\text{a}, 0)$ from the Ohmic law $I(V_\text{a}) \propto V_\text{a}$ is again a signature of dynamical Coulomb blockade by an Ohmic electromagnetic environment.

21. Two-state dynamics: basics and methods

Up to now, we have mainly discussed thermodynamic properties of the open two-state system. As far as dynamics is concerned, we have been limited to the study of nonadiabatic tunneling rates in the incoherent regime. We now present the real-time approach based on the method given in Chapter 5. The approach will cover the full dynamics in a unified manner for different kinds of the initial preparation and for arbitrary linear dissipation, both in the incoherent and oscillatory regime. We shall provide explicit expressions in most regions of the parameter space. Emphasis is put on the regime in which the system shows quantum coherent oscillations.

21.1   Initial preparation, expectation values, and correlations

We have discussed already in Sections 5.2 – 5.5 initial conditions, preparation functions and propagating functions for a general global system. We now deal with the specification for the open two-state system. In the sequel, we put emphasis on answering questions which are of experimental relevance.

21.1.1   Product initial state

Consider now first the case in which the initial state of the density matrix of the global system is a product initial state (pis)

$$W_{\text{pis}}(t = 0, \overline{\sigma}) = |R><R| \otimes \exp\{-\beta[H_{\text{Res}} - \tfrac{1}{2}\overline{\sigma}\mathfrak{E}(t = 0)]\}/Z_{\text{Res}} . \qquad (21.1)$$

The TSS is prepared in the eigenstate $\sigma = +1$ of σ_z (right well), and the bath is in a shifted canonical distribution. Here, H_{Res} is the bare reservoir Hamiltonian (3.3), $\mathfrak{E}(t)$ is the collective bath mode defined in Eq. (3.136), and $\overline{\sigma}$ is a control parameter for the shift of the bath in the initial state at time zero.

The influence functional tailored to the initial state (21.1) is conveniently expressed in terms of an influence functional of the Feynman-Vernon form in which the system-reservoir coupling is switched on at time t_0 where $t_0 \leq 0$. Substituting the spin path (4.67) into the influence functional (5.23) with Eq. (5.24) and observing that the last term in Eq. (5.24) does not contribute since $\sigma^2(t) = \sigma'^2(t)$, we find

$$\mathcal{F}[\sigma, \sigma'; t_0] = \exp\left\{ -\frac{1}{4}\int_{t_0}^{t} dt' \int_{t_0}^{t'} dt'' \left(\sigma(t') - \sigma'(t') \right) \right.$$
$$\left. \times \left(\mathfrak{L}(t' - t'')\,\sigma(t'') - \mathfrak{L}^*(t' - t'')\,\sigma'(t'') \right) \right\} . \qquad (21.2)$$

We have substituted $L(t) = (\hbar/q_0^2)\mathfrak{L}(t)$, and $\mathfrak{L}(t)$ is related to $Q(t)$ by $\ddot{Q}(t) = \mathfrak{L}(t)$.

Imagine that the bath is in canonical equilibrium of the bare Hamiltonian (3.3) and the system is suddenly prepared at time zero in the state $\sigma = +1$. Then, the choice $\overline{\sigma} = 0$ in Eq. (21.1) corresponds to the situation in which the system evolves out of this state before the bath has relaxed to the shifted thermal equilibrium distribution. Mathematically, this is achieved by switching on the system-bath coupling at time $t_0 = 0$. We shall refer to this case as the preparation class A henceforth. Preparation according to class A might be relevant in electron transfer reactions where a particular electronic donor state is suddenly prepared by photoinjection.[1] The corresponding influence functional is

$$\mathcal{F}_{\overline{\sigma}=0}[\sigma, \sigma'] = \mathcal{F}[\sigma, \sigma'; t_0 = 0] . \qquad (21.3)$$

The other important case is when $\overline{\sigma} = +1$ in Eq. (21.1), referred to as class B in the sequel. This product initial state is prepared by holding the system for some large

---

[1] The observability of electronic coherence in ET reactions for preparation A is studied in Refs. [442, 443], and references therein.

time in the state $\sigma = +1$, so that the environment could have come into thermal equilibrium with it. The product initial state can be arranged, for example, by applying a strong negative bias $-\hbar\epsilon_0\Theta(-t)$ with $\epsilon_0 \gg \Delta$ for all times $t < 0$, so that the system is trapped in the state $\sigma = +1$. At time zero the constraint is released, and for $t > 0$ the dynamics is governed by the spin-boson Hamiltonian (18.15). Such initial preparation is achieved, e.g., in the rf SQUID device by a suitable choice of the applied magnetic field (cf. Subsection 3.3.2). The relevant influence functional is

$$\mathcal{F}_{\bar{\sigma}}[\sigma, \sigma'] = \mathcal{F}[\sigma, \sigma'; t_0 \to -\infty] , \qquad (21.4)$$

in which the double path $\sigma(t')$, $\sigma'(t')$ is constrained for all times $t' < 0$ to the state $\sigma = \sigma' = \bar{\sigma}$. Upon using Eqs. (21.3) and (21.4), we find the relation

$$\mathcal{F}_{\bar{\sigma}}[\sigma, \sigma'] = \mathcal{F}_{\bar{\sigma}=0}[\sigma, \sigma'] \exp\left\{ i\frac{\bar{\sigma}}{2} \int_0^t dt' \left[\sigma(t') - \sigma'(t') \right] \dot{Q}''(t') \right\} , \qquad (21.5)$$

where $\dot{Q}''(t)$ is the time derivative of the imaginary part of the complex bath correlation function $Q(t)$ defined in Eq. (18.44). Thus, the effects of preparation class A can be described for the Hamiltonian (3.137) in terms of a particular time-dependent bias, $\epsilon(t) = \bar{\sigma}\dot{Q}''(t)$.

The expression (21.5) establishes the connection between the two different kinds of preparation of the thermal bath. For a strict Ohmic spectral density ($\omega_c \to \infty$), we have $\dot{Q}''(t) \propto \delta(t)$. Hence the integral in Eq. (21.5) vanishes, and the different preparation has no effect. In the general case, the effects of different preparation (class A or class B) vanishes on a time scale of order $1/\omega_c$ and therefore are relevant only in the adiabatic limit, in which ω_c is of the order of Δ or smaller.

All information on the two-state system at a later time $t > 0$ is contained in the reduced density matrix. The diagonal elements $\rho_{1,1}$ and $\rho_{-1,-1}$ are the population probabilities of the two states, and the off-diagonal elements $\rho_{-1,1}$ and $\rho_{1,-1}$ are the coherences. The reduced density matrix can be written as a linear combination of the Pauli matrices and of the unit matrix, $\rho(t) = \frac{1}{2}[\mathbf{1} + \sum_{j=x,y,z}\langle\sigma_j\rangle_t\sigma_j]$, where

$$\langle\sigma_j\rangle_t \equiv \text{tr}_{\text{Res}}\left\{ \exp[-\beta H_{\text{Res}} + \frac{1}{2}\bar{\sigma}\beta\mathfrak{E}(0)] < R|\sigma_j(t)|R > \right\}/Z_{\text{Res}} \qquad (21.6)$$

is the expectation value of σ_j, and $\sigma_j(t)$ is taken in the Heisenberg representation with respect to the full Hamiltonian H,

$$\sigma_j(t) = e^{iHt/\hbar} \sigma_j e^{-iHt/\hbar} . \qquad (21.7)$$

The specifications for the preparation classes A and B are given in Subsection 21.2.3.

The expectation values $\langle\sigma_j\rangle_t$ are related to the reduced density matrix by

$$\begin{aligned}
\langle\sigma_z\rangle_t &= \rho_{1,1}(t) - \rho_{-1,-1}(t) , \\
\langle\sigma_x\rangle_t &= \rho_{1,-1}(t) + \rho_{-1,1}(t) , \\
\langle\sigma_y\rangle_t &= i\rho_{1,-1}(t) - i\rho_{-1,1}(t) .
\end{aligned} \qquad (21.8)$$

In our subsequent studies we choose the initial condition $\rho_{\sigma,\sigma'}(t=0) = \delta_{\sigma,1}\delta_{\sigma',1}$. The quantity $\langle\sigma_z\rangle_t$ describes the difference of the populations of the two localized states for initial population of the right state. It gives immediate information about the tunneling dynamics and is most directly relevant in studies of "macroscopic quantum coherence" (MQC). The understanding of the TSS dynamics is completed by the knowledge of the coherences $\langle\sigma_x\rangle_t$ and $\langle\sigma_y\rangle_t$ [cf. Section 4.1]. We obtain from Eq. (4.6) the bound

$$\langle\sigma_x\rangle_t^2 + \langle\sigma_y\rangle_t^2 + \langle\sigma_z\rangle_t^2 \leq 1 . \tag{21.9}$$

The equality sign holds in the absence of damping. Using Eq. (21.7) and the commutation relation $(H, \sigma_z) = i\hbar\Delta\,\sigma_x$, one finds

$$\langle\sigma_y\rangle_t = -\frac{1}{\Delta}\frac{d\langle\sigma_z\rangle_t}{dt} . \tag{21.10}$$

Hence the coherence $\langle\sigma_y\rangle_t$ is proportional to the tunneling current.

In the absence of the system-bath coupling, it is straightforward to calculate the transition amplitudes and to join them together to construct the density matrix.[2] Alternatively, we may evaluate directly for zero damping the expressions for the expectation values $\langle\sigma_j\rangle_t$ given below in Subsection 21.2.3. In any event, the result is

$$\langle\sigma_z\rangle_t^{(0)} = \epsilon^2/\Delta_b^2 + (\Delta^2/\Delta_b^2)\cos(\Delta_b t) ,$$
$$\langle\sigma_x\rangle_t^{(0)} = (\epsilon\Delta/\Delta_b^2)[1 - \cos(\Delta_b t)] , \tag{21.11}$$
$$\langle\sigma_y\rangle_t^{(0)} = (\Delta/\Delta_b)\sin(\Delta_b t) .$$

Observe that these expressions saturate the bound in Eq. (21.9) for all t. The expression for $\langle\sigma_z\rangle_t^{(0)}$ is sometimes called Rabi's formula. The RDM displays oscillations with the transition frequency $\Delta_b = \sqrt{\Delta^2 + \epsilon^2}$, which is a signature of phase-coherent dynamics. For $\epsilon = 0$, we have $(H, \sigma_x) = 0$, and hence $\sigma_x(t) = \sigma_x(0)$. Therefore, $\langle\sigma_x\rangle_t^{(0)}$ is frozen up at the initial value, which is zero.

The expectation values $\langle\sigma_j\rangle_t$ ($j = x, y, z$) of the dissipative two-state system are conveniently expressed in terms of the Feynman-Vernon two-time conditional propagating function $J(\zeta, t; \zeta_0, t_0)$ introduced in Eqs. (5.11), (5.12). Here, ζ denotes one of the four states of the reduced density matrix [see Fig. 21.1]. The RDM has two diagonal states (populations), denoted by $\zeta = (\eta, 0) = \eta$, and two off-diagonal states (coherences), denoted by $\zeta = (0, \xi) = \xi$ (in the sequel, we drop the redundant zero). We now assume that the TSS starts out at time zero from the diagonal state $\eta_0 = 1$, and the reservoir's initial state may be either class A or class B, as discussed above. Propagation of the RDM under the full Hamiltonian is then given by

$$\rho_\zeta(t) = J(\zeta, t; \eta_0 = 1, 0) , \tag{21.12}$$

and the expectation values of the populations and coherences take the form

---

[2]For a discussion of the bare TSS and the fictitious spin $\frac{1}{2}$ system, we refer the reader to Ref. [444].

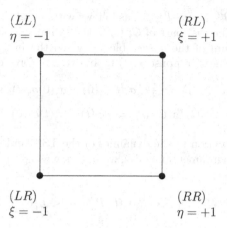

Figure 21.1: Graphical representation of the four states of the reduced density matrix. The diagonal states are labelled by $\eta = \pm 1$, and the off-diagonal states by $\xi = \pm 1$.

$$\langle \sigma_z \rangle_t = \sum_{\eta = \pm 1} \eta \, J(\eta, t; \eta_0 = 1, 0) \,, \tag{21.13}$$

$$\langle \sigma_x \rangle_t = \sum_{\xi = \pm 1} J(\xi, t; \eta_0 = 1, 0) \,, \tag{21.14}$$

$$\langle \sigma_y \rangle_t = i \sum_{\xi = \pm 1} \xi \, J(\xi, t; \eta_0 = 1, 0) \,. \tag{21.15}$$

We shall take up these expressions below in Subsection 21.2.3.

21.1.2 Thermal initial state

In many cases of interest, the system under consideration, e.g. a tunneling system in a solid, can not be prepared in a particular pure state. In the usual experimentally reproducible situation, before measurements take place, the system has relaxed to the thermal equilibrium state which is entangled with the environment. After that, it is prepared by a measurement of a certain observable, say at time zero. The measurement of the same observable at a later time t gives then direct information about the equilibrium autocorrelation function of this observable.

The equilibrium autocorrelation functions of the TSS are ($j = x, y, z$)

$$C_j^{\pm}(t) \equiv \langle \sigma_j(\pm t)\sigma_j(0) \rangle_\beta - \langle \sigma_j \rangle_\infty^2 = \operatorname{tr}\left\{ e^{-\beta H} \sigma_j(\pm t)\sigma_j(0) \right\}/Z - \langle \sigma_j \rangle_\infty^2 \,, \tag{21.16}$$

where $\sigma_j(t)$ is given in Eq. (21.7). As in Eq. (6.11), we have subtracted the equilibrium average so that $\lim_{t \to \infty} C_j^{\pm}(t) \to 0$. This is needed in order that the Fourier transform of $C_j^{\pm}(t)$ is well-defined. The subtraction term is absent for $C_y^{\pm}(t)$, since $\langle \sigma_y \rangle_\infty = 0$, as follows from Eq. (21.10). Actually, $C_x^{\pm}(t)$ can be found from $C_z^{\pm}(t)$ by differentiation,

$C_y^\pm(t-t') = (\partial^2/\partial t\,\partial t')\,C_z^\pm(t-t')/\Delta^2$, as follows with use of the commutation relation $(H, \sigma_z) = i\hbar\Delta\,\sigma_y$. The real part of $C_j^\pm(t)$ is the symmetric autocorrelation function in thermal equilibrium of the observable σ_j, while the imaginary part of $C_j^\pm(t)$ is connected with the linear response $\chi_j(t)$ to an external force coupled to σ_j. We have

$$S_j(t) \;\equiv\; \mathrm{Re}\,C_j^\pm(t) \;=\; \tfrac{1}{2}\langle\,[\,\sigma_j(t)\sigma_j(0) + \sigma_j(0)\sigma_j(t)\,]\,\rangle_\beta - \langle\sigma_j\rangle_\infty^2 \qquad (21.17)$$

$$\hbar\chi_j(t) \;\equiv\; -2\,\Theta(t)\,\mathrm{Im}\,C_j^+(t) \;=\; i\,\Theta(t)\,\langle\,[\,\sigma_j(t)\sigma_j(0) - \sigma_j(0)\sigma_j(t)\,]\,\rangle_\beta\ . \quad (21.18)$$

If we restrict the attention to the dynamics of the TSS, and disregard for a moment the average of the environmental modes, we get upon using the representations (3.129)

$$
\begin{aligned}
C_z^+(t) \;=\; \frac{1}{Z}\Big\{ & <R|\,\mathrm{e}^{-\beta H}|R> < R|\sigma_z(t)|R> \;-\; <L|\,\mathrm{e}^{-\beta H}|L> < L|\sigma_z(t)|L> \\
&+\; <R|\,\mathrm{e}^{-\beta H}|L><L|\sigma_z(t)|R> \;-\; <L|\,\mathrm{e}^{-\beta H}|R><R|\sigma_z(t)|L> \Big\} - \langle\sigma_z\rangle_\infty^2\ ,
\end{aligned}
$$

and with $C_x(t) = C_x^+(t)$

$$
\begin{aligned}
C_x(t) \;=\; \frac{1}{Z}\Big\{ & <R|\,\mathrm{e}^{-\beta H}|R> < R|\sigma_x(t)|L> \;+\; <L|\,\mathrm{e}^{-\beta H}|L> < L|\sigma_x(t)|R> \\
&+\; <R|\,\mathrm{e}^{-\beta H}|L><L|\sigma_x(t)|L> \;+\; <L|\,\mathrm{e}^{-\beta H}|R><R|\sigma_x(t)|R> \Big\} - \langle\sigma_x\rangle_\infty^2\ .
\end{aligned}
$$

Next, we take into account that the TSS is actually entangled with the environment when the first measurement at time zero takes place. To deal with this case, we proceed as follows. Suppose that at time t_p, where $t_\mathrm{p} < 0$, the system is released from the RDM state ζ_p, and then propagates under the full Hamiltonian. At time $t_0 = 0$, the system is measured to be in the state ζ_0, and eventually at later time t a second measurement is performed in which it is found in the state ζ. The corresponding three-time conditional propagating function for this sequence of events is denoted by $J(\zeta, t;\ \zeta_0, 0;\ \zeta_p, t_\mathrm{p})$. For an ergodic open system, the correlations of a dynamical variable at time t with the same (or a different) dynamical variable at time zero do not depend on the particular initial state of the reduced system chosen at time t_p, if the time of preparation t_p is displaced to the infinite past. Thus, for convenience, we may choose at time t_p in the infinite past a product initial state, in which the reservoir is in thermal equilibrium and the system in a diagonal state, say in the state $\eta_\mathrm{p} = +1$.[3] Then, before the first measurement of a TSS observable at time zero is performed, the TSS has already equilibrated with the environment and is in an entangled canonical state of the system-plus-reservoir complex. With these preliminary thoughts, we are now in the position to study equilibrium correlation functions within the standard real-time influence functional approach introduced in Section 5.6.

---

[3]This product state may be prepared, e.g., according to the preparation class A, or according to the preparation class B, which have been introduced in Subsection 21.1.1.

Consider first the correlation function $C_z^+(t)$, in which the system finally ends up in a diagonal state of the RDM. Regarding the symmetrized part, the system dwells at time zero in a diagonal state, whereas for the anti-symmetrized part it occupies at this time an off-diagonal state. Taking into account the respective weight factors of these states, we get with $S_z(t) = \operatorname{Re} C_z^{\pm}(t)$ and $\hbar \chi_z(t) = -2\Theta(t) \operatorname{Im} C_z^+(t)$

$$S_z(t) = \lim_{t_p \to -\infty} \sum_{\eta=\pm 1} \sum_{\eta_0=\pm 1} \eta \eta_0 \, J(\eta, |t|; \eta_0, 0; \eta_p, t_p) - \langle \sigma_z \rangle_\infty^2 , \qquad (21.19)$$

$$\hbar \chi_z(t) = \lim_{t_p \to -\infty} 2i\, \Theta(t) \sum_{\eta=\pm 1} \sum_{\xi_0=\pm 1} \eta \xi_0 \, J(\eta, t; \xi_0, 0; \eta_p, t_p) . \qquad (21.20)$$

The symmetric population correlations are relevant, e.g., for the description of inelastic neutron scattering from interstitials or defects in a lattice. The differential cross section of this process is given by [210]

$$\frac{\partial^2 \Sigma_{\text{inel}}}{\partial \Omega \, \partial \omega} = \frac{\Sigma_{\text{inc}}}{4\pi} \frac{k_f}{k_i} S_{\text{inel}}(\mathbf{k}, \omega) , \qquad (21.21)$$

where $\hbar \mathbf{k}$ and $\hbar \omega$ are momentum and energy transfer of the neutrons, and Σ_{inc} is the incoherent cross section of the scatterer. The scattering function, which is the dynamic structure factor for a scatterer with position $\mathbf{q}(t)$ at time t, is given by

$$S_{\text{inel}}(\mathbf{k}, \omega) = \frac{1}{2\pi} \int_{-\infty}^{\infty} dt\, e^{i\omega t} \left(\langle e^{-i\mathbf{k}\cdot\mathbf{q}(0)}\, e^{i\mathbf{k}\cdot\mathbf{q}(t)} \rangle_\beta - \left| \langle e^{i\mathbf{k}\cdot\mathbf{q}} \rangle_\beta \right|^2 \right) . \qquad (21.22)$$

For a defect tunneling between positions $\frac{1}{2}\mathbf{q}_0$ and $-\frac{1}{2}\mathbf{q}_0$, the operator $\mathbf{q}(t)$ takes the form $\mathbf{q}(t) = \frac{1}{2}\mathbf{q}_0 \sigma_z(t)$. We then find

$$S_{\text{inel}}(\mathbf{k}, \omega) = \sin^2\left(\tfrac{1}{2}\mathbf{k}\cdot\mathbf{q}_0\right) \tilde{C}_z^-(\omega)/2\pi . \qquad (21.23)$$

Here,

$$\tilde{C}_z^-(\omega) = \int_{-\infty}^{\infty} dt\, e^{i\omega t} \left(\langle \sigma_z(0)\sigma_z(t) \rangle_\beta - \langle \sigma_z \rangle_\infty^2 \right) \qquad (21.24)$$

is the Fourier transform of the pseudo-spin correlation function $C_z^-(t)$. Using the spectral relation (6.21), we may express $\tilde{C}_z^\pm(\omega)$ in terms of the symmetrized correlation function $\tilde{S}_z(\omega)$

$$\tilde{C}_z^\pm(\omega) = \frac{2}{1 + e^{\mp \beta \hbar \omega}} \tilde{S}_z(\omega) . \qquad (21.25)$$

The differential cross section for *inelastic* neutron scattering takes the form

$$\frac{\partial^2 \Sigma_{\text{inel}}}{\partial \Omega \, \partial \omega} = \frac{\Sigma_{\text{inc}}}{4\pi^2} \frac{k_f}{k_i} \sin^2\left(\frac{\mathbf{k}\cdot\mathbf{q}_0}{2}\right) \frac{\tilde{S}_z(\omega)}{1 + e^{\beta \hbar \omega}} . \qquad (21.26)$$

Hence, inelastic neutron scattering yields data directly for the spectral function $\tilde{S}_z(\omega)$.

The other important equilibrium correlation function is that of the tunneling or coherence operator σ_x. Since the polaron unitary operator (18.39) does not commute with σ_x, the correlation function of the bare coherence operator,

$$C_x(t) = \mathrm{tr}\left\{ e^{-\beta H}\, e^{iHt/\hbar}\, \sigma_x\, e^{-iHt/\hbar}\, \sigma_x \right\}/Z - \left[\mathrm{tr}\left\{ e^{-\beta H}\, \sigma_x \right\}/Z \right]^2 , \qquad (21.27)$$

differs from that of the polaron-dressed coherence operator (18.41),

$$C_x^{(\mathrm{pd})}(t) = \mathrm{tr}\left\{ e^{-\beta H}\, e^{iHt/\hbar}\, \tilde{\sigma}_x\, e^{-iHt/\hbar}\, \tilde{\sigma}_x \right\}/Z - \left[\mathrm{tr}\left\{ e^{-\beta H}\, \tilde{\sigma}_x \right\}/Z \right]^2 , \quad (21.28)$$

$$= \mathrm{tr}\left\{ e^{-\beta \tilde{H}}\, e^{i\tilde{H}t/\hbar}\, \sigma_x\, e^{-i\tilde{H}t/\hbar}\, \sigma_x \right\}/Z - \left[\mathrm{tr}\left\{ e^{-\beta \tilde{H}}\, \sigma_x \right\}/Z \right]^2 . \quad (21.29)$$

We see with use of Eq. (18.40) that the forms (21.28) and (21.29) are equivalent. The undressed and dressed coherence correlations are subject to different bath correlations, as we shall find out in Subsection 21.2.5. Due to the operation of σ_x at time zero, the TSS makes at this time a transition from an off-diagonal to a diagonal state or vice versa. Hence the coherence correlations are described in terms of a four-time propagating function $J(\zeta, t; \zeta_+, 0_+; \zeta_-, 0_-, \zeta_\mathrm{p}, t_\mathrm{p})$. The system finally ends up in an off-diagonal state. Assuming that the system is again prepared in the infinite past in a diagonal state, we get for the symmetrized part

$$\begin{aligned}
S_x(t) = \lim_{t_\mathrm{p}\to -\infty} \frac{1}{2}\Bigg\{ &\sum_{\{\xi, \xi_+, \eta_- = \pm 1\}} J(\xi, |t|; \xi_+, 0_+; \eta_-, 0_-; \eta_\mathrm{p}, t_\mathrm{p}) \\
+ &\sum_{\{\xi, \eta_+, \xi_- = \pm 1\}} J(\xi, |t|; \eta_+, 0_+; \xi_-, 0_-; \eta_\mathrm{p}, t_\mathrm{p}) \Bigg\} - \langle \sigma_x \rangle_\infty^2 ,
\end{aligned}$$

$$\tag{21.30}$$

and for the linear response function

$$\begin{aligned}
\hbar \chi_x(t) = \lim_{t_\mathrm{p}\to -\infty} i\,\Theta(t)\Bigg\{ &\sum_{\{\xi, \xi_+, \eta_- = \pm 1\}} \xi_+ \eta_-\, J(\xi, t; \xi_+, 0_+; \eta_-, 0_-; \eta_\mathrm{p}, t_\mathrm{p}) \\
+ &\sum_{\{\xi, \eta_+, \xi_- = \pm 1\}} \eta_+ \xi_-\, J(\xi, t; \eta_+, 0_+; \xi_-, 0_-; \eta_\mathrm{p}, t_\mathrm{p}) \Bigg\} .
\end{aligned}$$

$$\tag{21.31}$$

We see that both $S_x(t)$ and $\chi_x(t)$ consist of two different groups. In group A (given in the respective first line), the system hops at time zero from a diagonal to an off-diagonal state, whereas in group B (the respective second line), it hops at this time from an off-diagonal to a diagonal state. Evidently, the coherence correlations (21.30) and (21.31) are more intricate than the population correlations (21.19) and (21.20). We shall resume the discussion of the coherence correlations in Subsection 21.2.5.

21.2 Exact formal expressions for the system dynamics

21.2.1 Sojourns and blips

According to the discussion in Chapter 5, dynamical quantities are expressed in terms of double path integrals of the form

$$\int \mathcal{D}\sigma(\cdot) \int \mathcal{D}\sigma'(\cdot)\, \mathcal{A}[\sigma(\cdot)]\mathcal{A}^*[\sigma'(\cdot)]\mathcal{F}[\sigma(\cdot), \sigma'(\cdot)] \qquad (21.32)$$

with appropriately chosen boundary conditions for the spin paths $\sigma(t')$ and $\sigma'(t')$. The functional $\mathcal{A}[\sigma(\cdot)]$ is the probability amplitude for the free (undamped) TSS to follow the path $\sigma(t')$, and $\mathcal{F}[\sigma(\cdot), \sigma'(\cdot)]$ is the real-time influence functional discussed in Section 5.6. For the two-state system, the paths $\sigma(t')$ and $\sigma'(t')$ jump between the two discrete values $+1$ and -1, as depicted in Fig. 4.1. According to Eq. (5.32), it is convenient to define antisymmetric and symmetric spin paths $\xi(t')$ and $\eta(t')$,

$$\xi(t') \equiv \tfrac{1}{2}[\sigma(t') - \sigma'(t')] ; \qquad \eta(t') \equiv \tfrac{1}{2}[\sigma(t') + \sigma'(t')] , \tag{21.33}$$

satisfying the relation $\xi(t')\eta(t') = 0$.

Since the spin paths $\xi(t')$ and $\eta(t')$ are piecewise constant with sudden jumps in between, it is convenient to perform integrations by parts in the expression (21.2). Then the influence functional takes the form (apart from possible boundary terms)

$$\mathcal{F}[\sigma, \sigma'] = \exp\left(\int_{t_0}^{t} dt' \, \dot{\xi}(t') \int_{t_0}^{t'} dt'' [Q'(t' - t'')\dot{\xi}(t'') + i\, Q''(t' - t'')\dot{\eta}(t'')] \right) , \tag{21.34}$$

where $Q(t) = Q'(t) + iQ''(t)$ is the complex bath correlation function defined in Eq. (18.44). This form is especially convenient for the piecewise constant spin path with discontinuous jumps [cf. Fig. 4.1, and Eq. (21.35) below]. It is the charge representation of the influence function, whereas Eq. (21.2) is the state representation.

The double path sum can be visualized as a path sum for a single path that visits the four states of the reduced density. The combinatorial problem facing us is the sum of paths a walker can go along the edges of the square sketched in Fig. 21.1. A period the path spends in a diagonal state has been dubbed *sojourn* by Leggett et al. [85], and a period the path dwells in an off-diagonal state has been termed *blip*. During a sojourn, the function $\xi(\tau)$ is zero, whereas during a blip interval the function $\eta(\tau)$ is zero. There are two sojourn states, labelled by $\eta_j = +1$ [state (RR)] and $\eta_j = -1$ [state (LL)]. Similarly, there are two kinds of blips, and we assign the label $\xi_j = +1$ to the off-diagonal state (RL) and the label $\xi_j = -1$ to the off-diagonal state (LR).

A general sojourn-to-sojourn path along the edges of the square in Fig. 21.1 making $2n$ transitions at intermediate times t_j $(j = 1, 2, \ldots, 2n)$ is parametrized by

$$\eta^{(n)}(t') = \sum_{j=0}^{n} \eta_j [\Theta(t' - t_{2j}) - \Theta(t' - t_{2j+1})] ,$$
$$\xi^{(n)}(t') = \sum_{j=1}^{n} \xi_j [\Theta(t' - t_{2j-1}) - \Theta(t' - t_{2j})] . \tag{21.35}$$

Denoting the intervals spent in a blip state by τ_j and the intervals spent in a sojourn state by s_j, we have

$$\tau_j = t_{2j} - t_{2j-1} , \qquad s_j = t_{2j+1} - t_{2j} . \tag{21.36}$$

In the charge picture, the velocity path $\dot{\eta}^{(n)}(t')$ represents a sequence of sojourn dipoles. The dipole j has successive charges η_j and $-\eta_j$ at distance $s_j = t_{2j+1} - t_{2j}$.

Correspondingly, the velocity path $\dot{\xi}^{(n)}(t')$ represents a sequence of blip dipoles, where the dipole j has successive charges ξ_j and $-\xi_j$ at distance $\tau_j = t_{2j} - t_{2j-1}$. We see from Eq. (21.34) that the interaction between a blip and a sojourn charge is $Q''(\tau)$ and the interaction of a pair of blip charges is $Q'(\tau)$, while sojourn charges do not interact with each other. With the notation $Q_{j,k} = Q(t_j - t_k)$ we may write the interaction $\Lambda_{j,k}$ between the blip dipole k and a later blip dipole j and the interaction $X_{j,k}$ between a sojourn dipole k and a later blip dipole j compactly as

$$
\begin{aligned}
\Lambda_{j,k} &= Q'_{2j,2k-1} + Q'_{2j-1,2k} - Q'_{2j,2k} - Q'_{2j-1,2k-1} , \\
X_{j,k} &= Q''_{2j,2k+1} + Q''_{2j-1,2k} - Q''_{2j,2k} - Q''_{2j-1,2k+1} .
\end{aligned}
\tag{21.37}
$$

The charge representation of the influence functional for the double path (21.35) can now be written as

$$
\begin{aligned}
\mathcal{F}^{(n)} &= G_n H_n , \\
G_n &= \exp\left[-\sum_{j=1}^{n} Q'_{2j,2j-1} \right] \exp\left[-\sum_{j=2}^{n} \sum_{k=1}^{j-1} \xi_j \xi_k \Lambda_{j,k} \right] , \\
H_n &= \exp\left[i\sum_{j=1}^{n} \sum_{k=0}^{j-1} \xi_j \eta_k X_{j,k} \right] = \exp\left[i\sum_{k=0}^{n-1} \sum_{j=k+1}^{n} \xi_j \eta_k X_{j,k} \right] .
\end{aligned}
\tag{21.38}
$$

The blip interactions are bundled up in the real-valued function G_n. The first exponential factor contains the intrablip (intradipole) interactions of the n blips, and the second exponential factor represents the interblip correlations. The interactions between the sojourns and all subsequent blips are in the phase factor H_n.

The function G_n is a filtering function which suppresses long blips. Hence, it is favored that the system dwells in sojourn states. Physically, this is because the environment is continuously measuring σ_z and therefore suppresses occupation of blip states, and quantum interference between the eigenstates of σ_z.

It is straightforward to write down also the various weight factors resulting from the amplitude product $A[\sigma] A^*[\sigma']$ for the TSS or spin Hamiltonian (3.128). The weight to switch per unit time from a diagonal state η to an off-diagonal state ξ, or vice versa, is

$$
-i\, \xi\eta\, \Delta/2 .
\tag{21.39}
$$

Thus we have the weight factor $-i\Delta/2$ for transitions $(RL) \Longleftrightarrow (RR)$ and $(LL) \Longleftrightarrow (LR)$, and $i\Delta/2$ for transitions $(RL) \Longleftrightarrow (LL)$ and $(RR) \Longleftrightarrow (LR)$.

The weight to stay in a sojourn is unity, while the weight factor to stay in the jth blip with label ξ_j and length τ_j is $\exp(i\epsilon\xi_j\tau_j)$. The bias factors for n blip states are accumulated in the total bias factor

$$
B_n = \exp\left(i\epsilon \sum_{j=1}^{n} \xi_j \tau_j \right) .
\tag{21.40}
$$

The prescription to sum over all paths along the edges of the square sketched in Fig. 21.1 now means (i) to sum over the possible intermediate sojourn and blip states the paths with a given number of transitions can visit, (ii) to integrate over the time-ordered jump times of these paths, and (III) to sum over the possible numbers of transitions the system may take,

$$
\sum_{\text{all paths}} \cdots \rightarrow \sum_{n=0}^{\infty} \sum_{\{\eta_j = \pm 1\}} \sum_{\{\xi_j = \pm 1\}} \int_{t_n \geq t_{n-1} \cdots \geq t_1} dt_n \, dt_{n-1} \cdots dt_1 \cdots . \tag{21.41}
$$

The appropriate way to integrate correlated system-reservoir initial states at time zero, e.g. in equilibrium correlation functions, is to consider the evolution of the system since a fictitious time t_{p} at which the global system is prepared in some factorizing initial state, where t_{p} is eventually sent to the infinite past, as we have explained in Section 5.6 (cf. the contours C_{III} or C_{IV} specified in Fig. 5.4). To this end we divide the integrations over the time-ordered flip times $\{t_j\}$ into a negative and a positive time branch. It is useful to employ the compact notation

$$
\int_{t_{\text{p}}}^{t} \mathcal{D}_{k,\ell}\{t_j\} \times \cdots \equiv \int_{0}^{t} dt_{\ell+k} \cdots \int_{0}^{t_{\ell+2}} dt_{\ell+1} \int_{t_{\text{p}}}^{0} dt_{\ell} \cdots \int_{t_{\text{p}}}^{t_2} dt_1 \, \Delta^{k+\ell} \times \cdots . \tag{21.42}
$$

Here, ℓ is the number of steps in the negative time branch $t_{\text{p}} \leq t' \leq 0$ and k is the number of steps in the positive time branch, $0 \leq t' \leq t$. For convenience, we have included in the integration symbol the tunneling matrix factor. We are now ready to discuss the conditional propagating functions for the boundary conditions of interest.

21.2.2 Conditional propagating functions

We have seen in Section 21.1 that the relevant expectation values and equilibrium autocorrelation functions can be expressed in terms of two-time, three-time, and four-time conditional propagating functions.

If the expectation values $\langle \sigma_j \rangle_t$ $(j = x,\, y,\, z)$ imply a factorizing initial state of the global system at time zero, they can be expressed in terms of the two-time conditional propagating function $J(\zeta, t; \eta_0, 0)$. Here the initial state of the system is the sojourn state η_0, and the preparation of the reservoir may be conducted either according to class A or according to class B. When the final state is again a sojourn, the number of transitions the system makes is even, while it is odd, when the final state is a blip.

Upon collecting the various weight factors discussed in the previous subsection we obtain for the two-time sojourn-to-sojourn correlation function the series expression

$$
J(\eta, t; \eta_0, 0) = \delta_{\eta, \eta_0} + \eta \eta_0 \sum_{m=1}^{\infty} \frac{(-1)^m}{2^{2m}} \int_{0}^{t} \mathcal{D}_{2m,0}\{t_j\} \sum_{\{\xi_j = \pm 1\}} G_m B_m \sum_{\{\eta_j = \pm 1\}'} H_m . \tag{21.43}
$$

Correspondingly, the series for the sojourn-to-blip correlation function reads

$$J(\xi, t; \eta_0, 0) = -i\xi\eta_0 \sum_{m=1}^{\infty} \frac{(-1)^{m-1}}{2^{2m-1}} \int_0^t \mathcal{D}_{2m-1,0}\{t_j\} \sum_{\{\xi_j=\pm 1\}'} G_m B_m \sum_{\{\eta_j=\pm 1\}'} H_m . \quad (21.44)$$

The prime in $\{\eta_j = \pm 1\}'$ and $\{\xi_j = \pm 1\}'$ indicates that the boundary states are fixed as indicated in the arguments of the propagating functions.

The three-time conditional propagating function for being in sojourn states at the preparation time t_p, at time zero, and again at time t reads

$$J(\eta, t; \eta_0, 0; \eta_p, t_p) = \delta_{\eta,\eta_0}\delta_{\eta_0,\eta_p} \quad (21.45)$$

$$+ \delta_{\eta,\eta_0}\eta_0\eta_p \sum_{n=1}^{\infty} \left(-\frac{1}{4}\right)^n \int_{t_p}^t \mathcal{D}_{0,2n}\{t_j\} \sum_{\{\xi_j=\pm 1\}} G_n B_n \sum_{\{\eta_j=\pm 1\}'} H_n$$

$$+ \delta_{\eta_0,\eta_p}\eta\eta_0 \sum_{m=1}^{\infty} \left(-\frac{1}{4}\right)^m \int_{t_p}^t \mathcal{D}_{2m,0}\{t_j\} \sum_{\{\xi_j=\pm 1\}} G_m B_m \sum_{\{\eta_j=\pm 1\}'} H_m$$

$$+ \eta\eta_p \sum_{m=1}^{\infty}\sum_{n=1}^{\infty} \left(-\frac{1}{4}\right)^{m+n} \int_{t_p}^t \mathcal{D}_{2m,2n}\{t_j\} \sum_{\{\xi_j=\pm 1\}} G_{m+n} B_{m+n} \sum_{\{\eta_j=\pm 1\}'} H_{m+n} .$$

The sum over arrangements $\{\xi_j\}$ and $\{\eta_j\}$ extends over the possible values ± 1 of the intermediate states the system may take for $2n$ transitions in the interval $t_p < t' < 0$ and $2m$ transitions in the interval $0 < t' < t$. The prime in $\{\eta_j\}'$ is to indicate that the sojourns at times t_p, 0, and t are fixed as indicated in the arguments of propagating function. Correspondingly, we obtain for the case in which a blip state is occupied at time zero the series

$$J(\eta, t; \xi_0, 0; \eta_p, t_p) = \eta\eta_p \sum_{m=1}^{\infty}\sum_{n=1}^{\infty} \left(-\frac{1}{4}\right)^{n+m-1} \int_{t_p}^t \mathcal{D}_{2m-1,2n-1}\{t_j\}$$

$$\times \sum_{\{\xi_j=\pm 1\}'} G_{m+n-1} B_{m+n-1} \sum_{\{\eta_j=\pm 1\}'} H_{m+n-1} . \quad (21.46)$$

Here the sum is over all sequences of blips and sojourns which are in accordance with the constraints indicated by the arguments of the propagating function.

The discussion of the coherence correlations is given in Subsection 21.2.5.

21.2.3 The expectation values $\langle \sigma_j \rangle_t$ $(j = x, y, z)$

Consider first the population $\langle \sigma_z \rangle_t$. Inserting the series expression (21.43) for the two-time conditional propagating function into Eq. (21.13), and performing the summation over the intermediate sojourn states, $\{\eta_j = \pm 1\}'$, we obtain the exact formal series expression [85]

$$\langle \sigma_z \rangle_t = 1 + \sum_{m=1}^{\infty} (-1)^m \int_0^t \mathcal{D}_{2m,0}\{t_j\} \frac{1}{2^m} \sum_{\{\xi_j=\pm 1\}} \left(F_m^{(+)} B_m^{(s)} - F_m^{(-)} B_m^{(a)} \right) . \quad (21.47)$$

Here the bias dependence is divided up into the even and odd contributions

$$B_m^{(s)} = \cos\left(\epsilon \sum_{j=1}^{m} \xi_j \tau_j\right) ; \qquad B_m^{(a)} = \sin\left(\epsilon \sum_{j=1}^{m} \xi_j \tau_j\right) , \qquad (21.48)$$

and the effects of the environment are in the influence functions

$$F_m^{(+)} = G_m \prod_{k=0}^{m-1} \cos\left(\phi_{k,m}\right) , \quad F_m^{(-)} = G_m \sin\left(\phi_{0,m}\right) \prod_{k=1}^{m-1} \cos\left(\phi_{k,m}\right) . \qquad (21.49)$$

The term G_m describes the intra- and inter-blip correlations, and the residual factors contain the phase correlations. The bath correlations between the kth sojourn and the $m - k$ succeeding blips are combined in the phase

$$\phi_{k,m} = \sum_{j=k+1}^{m} \xi_j X_{j,k} . \qquad (21.50)$$

Finally, the sum over the labels $\{\xi_j\}$ runs over all intermediate blip states, $\{\xi_j = \pm 1\}$. For subsequent purposes, we introduce the notation

$$\langle \sigma_z \rangle_t \equiv P(t) = P_{\mathrm{R}}(t) - P_{\mathrm{L}}(t) = 2P_{\mathrm{R}}(t) - 1 , \qquad (21.51)$$

where $P_{\mathrm{R/L}}(t)$ is the occupation probability of the right/left well, respectively. With the initial condition (21.13) we have $P(0) = 1$. The function $P(t)$ may be split into the components which are symmetric (s) and antisymmetric (a) under inversion of the bias ($\epsilon \to -\epsilon$),

$$P(t) = P_{\mathrm{s}}(t) + P_{\mathrm{a}}(t) . \qquad (21.52)$$

When the damping of the system persists until time infinity, the system is ergodic and it relaxes to the thermal equilibrium state.[4] Then the part $P_{\mathrm{s}}(t)$, which is even in the bias, goes to zero as $t \to \infty$, whereas the part $P_{\mathrm{a}}(t)$, which is odd in the bias, approaches the equilibrium value $P(t \to \infty) = P_\infty = \langle \sigma_z \rangle_\infty = \langle \sigma_z \rangle^{(\mathrm{eq})}$,

$$\langle \sigma_z \rangle_\infty = P_\infty = \lim_{t \to \infty} \sum_{m=1}^{\infty} (-1)^{m-1} \int_0^t \mathcal{D}_{2m,0}\{t_j\} \frac{1}{2^m} \sum_{\{\xi_j = \pm 1\}} F_m^{(-)} B_m^{(a)} . \qquad (21.53)$$

Observe that this expression for the equilibrium distribution P_∞ is found within a dynamical nonequilibrium approach. On the other hand, the equilibrium distribution P_∞ may be expressed in terms of thermodynamic quantities. Denoting the partition functions associated with the left and right state by Z_{R} and Z_{L}, respectively, we have [cf. Eqs. (19.10) and (19.11)]

$$\langle \sigma_z \rangle_\infty = \langle \sigma_z \rangle^{(\mathrm{eq})} = (Z_{\mathrm{R}} - Z_{\mathrm{L}})/(Z_{\mathrm{R}} + Z_{\mathrm{L}}) . \qquad (21.54)$$

For an ergodic system, the expressions (21.53) and (21.54) are equivalent.

---

[4]The criteria for the spectral density $G(\omega)$ of an ergodic open system are specified in Section 3.2.

It is also straightforward to derive the series expressions for the coherences $\langle \sigma_x \rangle_t$ and $\langle \sigma_y \rangle_t$. Upon employing the series expression (21.44), the relations (21.14) and (21.15) yield [445]

$$\langle \sigma_x \rangle_t = \sum_{m=1}^{\infty} (-1)^{m-1} \int_0^t \mathcal{D}_{2m-1,0}\{t_j\}\, \frac{1}{2^m} \sum_{\{\xi_j=\pm 1\}} \xi_m \left(F_m^{(+)} B_m^{(a)} + F_m^{(-)} B_m^{(s)} \right), \quad (21.55)$$

$$\langle \sigma_y \rangle_t = \sum_{m=1}^{\infty} (-1)^{m-1} \int_0^t \mathcal{D}_{2m-1,0}\{t_j\}\, \frac{1}{2^m} \sum_{\{\xi_j=\pm 1\}} \left(F_m^{(+)} B_m^{(s)} - F_m^{(-)} B_m^{(a)} \right). \quad (21.56)$$

The expressions (21.47), (21.55) and (21.56) are exact formal expressions for the time evolution of the reduced density matrix. Observe that the expressions (21.47) and (21.56) satisfy the relation (21.10).

Let us finally comment on the different kinds of preparation discussed in Subsection 21.1.1. In the expressions (21.47), (21.55), and (21.56) the differences between class A and class B are captured by the correlations of the blips with the initial sojourn, as follows from Eq. (21.5) with Eq. (21.33). These correlations are described by the functions $X_{j,0}$ given in Eq. (21.37). For class A, the initial sojourn begins at time zero. Thus we have $X_{j,0}^{(A)} = Q_{2j,1}'' + Q_{2j-1,0}'' - Q_{2j,0}'' - Q_{2j-1,1}''$. On the other hand, for class B, the initial sojourn has length infinity. In this limit, the second and third term in $X_{j,0}$ cancel out, yielding $X_{j,0}^{(B)} = Q_{2j,1}'' - Q_{2j-1,1}''$. For an Ohmic spectral density and the scaling limit (18.57), the bath correlation function is given by Eq. (18.54). In this limit, the preparation classes A and B coincide, $X_{j,0}^{(A)} = X_{j,0}^{(B)}$. Substituting $Q''(t) = \pi K \operatorname{sgn}(t)$ into Eqs. (21.49), the influence functions simplify to the forms

$$F_m^{(+)} = [\cos(\pi K)]^m\, G_m\,; \qquad F_m^{(-)} = \xi_1 \sin(\pi K)[\cos(\pi K)]^{m-1}\, G_m\,. \quad (21.57)$$

We shall use these forms below in Section 22.6.

21.2.4   Correlation and response function of the populations

Upon inserting the series expression (21.45) into Eq. (21.19) and summing again over the intermediate sojourns, we obtain for the symmetric correlation function [220]

$$S_z(t) = S_z^{(\mathrm{unc})}(t) + R(t)\,, \quad (21.58)$$

$$S_z^{(\mathrm{unc})}(t) = P_{\mathrm{s}}(|t|) + P_\infty [P_{\mathrm{a}}(|t|) - P_\infty]\,. \quad (21.59)$$

Here, $P_{\mathrm{s,a}}(t)$ is the part of $\langle \sigma_z \rangle_t$ in Eq. (21.47) which is symmetric (antisymmetric) in the bias, and $P_\infty = P_{\mathrm{a}}(t \to \infty) = \langle \sigma_z \rangle_\infty$ is the equilibrium value of σ_z. The term $S_z^{(\mathrm{unc})}(t)$ is the contribution to $S_z(t)$ in which the entanglement of the system with the reservoir at time zero is disregarded, which formally corresponds to neglect of all bath correlations $Q(\tau)$ between the negative-time and the positive-time branch of the full spin path. If we had chosen a product initial state for the system-plus-reservoir complex at time zero (just before the first measurement takes place), then

Eq. (21.59) would be the resulting expression for the symmetrized σ_z autocorrelation function. The residual term $R(t)$ in Eq. (21.58) describes the dynamical effects at time t resulting from the system-bath correlations in the initial state at time zero,

$$
\begin{aligned}
R(t) = \ & \lim_{t_p \to -\infty} \sum_{m=1}^{\infty} \sum_{n=1}^{\infty} (-1)^{m+n-1} \int_{t_p}^{|t|} \mathcal{D}_{2m,2n}\{t_j\} \\
& \times \frac{1}{2^{m+n}} \sum_{\{\xi_j = \pm 1\}} \left[F_{m,n} B_m^{(s)} B_n^{(s)} - \left(F_{m,n} - F_m^{(-)} F_n^{(-)} \right) B_m^{(a)} B_n^{(a)} \right],
\end{aligned}
$$

where

$$
F_{m,n} = G_{m+n} \sin\left(\phi_{n,m+n}\right) \sin\left(\phi_{0,m+n}\right) \prod_{k=1\,(k \neq n)}^{m+n-1} \cos\left(\phi_{k,m+n}\right). \tag{21.60}
$$

The second term in the round bracket in Eq. (21.60) $\propto F_m^{(-)} F_n^{(-)}$ is a subtraction term in which the bath correlations are only inside of the negative and positive time branch. It is easily seen from Eqs. (21.47) and (21.53) that this term is just the counter term of the contribution $P_\infty P_a(t)$ in Eq. (21.59). We should expect that at low temperature and long time the correlation term $R(t)$ gives significant contributions. Indeed, this we shall find below. For later convenience, we also give the simplified form of Eq. (21.60) holding in the scaling limit (18.57) with (18.54),

$$
F_{m,n} = \xi_1 \xi_{n+1} [\sin(\pi K)]^2 [\cos(\pi K)]^{m+n-2} G_{m+n}. \tag{21.61}
$$

Likewise, the response function (21.20) is found with use of the expression (21.46) as

$$
\chi_z(t) = \lim_{t_p \to -\infty} \frac{4}{\hbar} \sum_{m=1}^{\infty} \sum_{n=1}^{\infty} \frac{(-1)^{m+n}}{2^{m+n}} \int_{t_p}^{t} \mathcal{D}_{2m-1,2n-1}\{t_j\} \sum_{\{\xi_j = \pm 1\}} \xi_n F_{m+n-1}^{(-)} B_{m+n-1}^{(s)},
$$

$$
\tag{21.62}
$$

where $t > 0$ and where $F_m^{(-)}$ is defined in Eq. (21.49). The peculiarity of this expression is that the TSS is in a blip state at time zero.

The static nonlinear susceptibility is defined by the relations

$$
\overline{\chi}_z \equiv \tilde{\chi}_z(\omega = 0) = \int_0^{\infty} dt\, \chi_z(t). \tag{21.63}
$$

We now substitute the form (21.62) into Eq. (21.63) and use that the integrals over the last interval at negative times ($\tau_n^- = -t_{2n-1}$) and the first interval at positive times ($\tau_n^+ = t_{2n}$) are of the form ($\tau_n = \tau_n^- + \tau_n^+$ is the full blip length)

$$
\int_0^{\infty} d\tau_n^-\, d\tau_n^+\, f(\tau_n^- + \tau_n^+) = \int_0^{\infty} d\tau_n\, \tau_n f(\tau_n).
$$

Upon rearrangement of the double series, we then obtain the series expression

$$
\overline{\chi}_z = \lim_{t \to \infty} \frac{2}{\hbar} \sum_{m=1}^{\infty} (-1)^{m-1} \int_0^{t} \mathcal{D}_{2m,0}\{t_j\} \frac{1}{2^m} \sum_{\{\xi_j \pm 1\}} F_m^{(-)} B_m^{(s)} \left(\sum_{n=1}^{m} \xi_n \tau_n \right), \tag{21.64}
$$

which expresses the static susceptibility as the asymptotic long-time limit of a dynamical quantity. Now, upon comparing the series (21.64) with the series (21.53), we see that they are related by

$$\overline{\chi}_z = \frac{2}{\hbar} \frac{\partial \langle \sigma_z \rangle_\infty}{\partial \epsilon} \,. \tag{21.65}$$

This result of the dynamical is in correspondence with the thermodynamic relation (19.13). Our subsequent study of the dynamics of the dissipative two-state system is based upon the above exact formal series expressions.

21.2.5 Correlation and response function of the coherences

The case of the coherence correlation function is more intricate than that of the population correlations for two reasons. First, the system makes an additional flip at time zero. The flip is enforced by the operation of σ_x at this time, and hence is not dynamical. Secondly, the system ends up in an off-diagonal state at time t. The final off-diagonal state gives rise to a boundary term in the influence function which exactly appears as if there would be an extra step at time t back to a diagonal state.

In the remainder of this subsection we restrict the attention to the case of Ohmic dissipation in the scaling limit (18.57) with (18.54). For the just mentioned two reasons, the influence function introduces additional bath correlation terms. These lead to an overall factor $(\Delta_{\mathrm{r}}/\Delta)^2 = (\Delta_{\mathrm{r}}/\omega_{\mathrm{c}})^{2K}$ in the series expression for $C_x(t)$. Because of the explicit dependence on ω_{c}, the function $C_x(t)$ is non-universal, and it vanishes in the scaling limit, as observed first by Guinea [446].

Universality is restored, however, by taking into account the adiabatic dynamics of the bath modes when the particle tunnels from the one localized state to the other. These effects are included in the correlation function $C_x^{(\mathrm{pd})}(t)$ of the *polaron-dressed* tunneling operator, Eq. (21.28). Since the dressed operator $\tilde{\sigma}_x$ acts in the full system-plus-reservoir space and the correlated initial state involves a particular preparation of the reservoir, the procedure of the elimination of the bath modes has to be reconsidered. The relevant analysis has been performed in Ref. [447]. In the end, one finds a modified influence functional. This, however, can be cast into the standard form at the expense of introducing modified system paths. The findings can be put in simple terms. First, the terms in Eqs. (21.30) and (21.31) with $\xi = -\xi_+ = \eta_-$ in group A, and $\xi = -\eta_+ = \xi_-$ in group B, vanish again in the scaling limit. In the residual terms, which have $\xi = \xi_+$ in group A and $\xi = \eta_+$ in group B, the system's jumps at time zero and time t actually do not give rise to bath correlations in the influence functional. In the equivalent charge picture, the corresponding charges are removed by the polaron transformation. For this reason, these terms turn out to be universal. At this point, we anticipate that the equilibrium expectation value $\langle \sigma_x \rangle_\infty$ contributing to the expression (21.30) is non-universal and therefore vanishes in the scaling limit [see also the discussion after Eq.(22.79)].

Because of the additional step the system makes (which is without a factor Δ), it is convenient to introduce the modified integration symbol [cf. Eq.(21.42)]

$$\int_{t_{\mathrm{p}}}^{t} \tilde{\mathcal{D}}_{k,\ell}\{t_j\} \times \cdots \equiv \int_{0}^{t} dt_{\ell+k+1} \cdots \int_{0}^{t_{\ell+3}} dt_{\ell+2} \int_{t_{\mathrm{p}}}^{0} dt_{\ell} \cdots \int_{t_{\mathrm{p}}}^{t_2} dt_1\, \Delta^{k+\ell} \times \cdots , \quad (21.66)$$

and the extra step is exactly at time zero, $t_{\ell+1} = 0$. For the groups A and B specified below Eq. (21.31) the system is finally in a blip state. The symmetrized correlation function $S_x(t) = \mathrm{Re}\, C_x^{(\mathrm{pd})}(t)$ is found as $S_x(t) = S_x^{\mathrm{A}}(t) + S_x^{\mathrm{B}}(t)$, where

$$S_x^{\mathrm{A}}(t) = \frac{1}{2} \sum_{m=1}^{\infty} \left(-\frac{\cos(\pi K)}{2}\right)^{m-1} \int_{-\infty}^{|t|} \tilde{\mathcal{D}}_{2m-2,0}\{t_j\} \sum_{\{\xi_j=\pm1\}_{\mathrm{A}}} G_m^{\mathrm{A}} B_m^{(\mathrm{s})} , \quad (21.67)$$

$$S_x^{\mathrm{B}}(t) = -\sum_{m=2}^{\infty}\sum_{n=1}^{\infty} \left(-\frac{\cos(\pi K)}{2}\right)^{n+m-1} \sin^2(\pi K) \int_{-\infty}^{|t|} \tilde{\mathcal{D}}_{2m-1,2n-1}\{t_j\} \quad (21.68)$$

$$\times \sum_{\substack{\{\xi_j=\pm1\}\\ \xi_n=\xi_{n+1}=-\xi_{m+n}}} \xi_{m+n}\, \xi_1\, G_{m+n}^{\mathrm{B}} B_{m+n}^{(\mathrm{s})} .$$

Correspondingly, the response function $\chi_x(t) = -(2/\hbar)\,\Theta(t)\,\mathrm{Im}\, C_x^{(\mathrm{pd})}(t)$ is found as

$$\chi_x^{\mathrm{A}}(t) = \frac{\Theta(t)}{\hbar} \sum_{m=1}^{\infty}\sum_{n=1}^{\infty} \left(-\frac{\cos(\pi K)}{2}\right)^{m+n-1} \tan(\pi K) \int_{-\infty}^{t} \tilde{\mathcal{D}}_{2m-2,2n}\{t_j\} \quad (21.69)$$

$$\times \sum_{\{\xi_j=\pm1\}_{\mathrm{A}}} \xi_{m+n}\, \xi_1\, G_{m+n}^{\mathrm{A}} B_{m+n}^{(\mathrm{s})} ,$$

$$\chi_x^{\mathrm{B}}(t) = \frac{\Theta(t)}{\hbar} \sum_{m=1}^{\infty}\sum_{n=1}^{\infty} \left(-\frac{\cos(\pi K)}{2}\right)^{m+n-1} \tan(\pi K) \int_{-\infty}^{t} \tilde{\mathcal{D}}_{2m-1,2n-1}\{t_j\} \quad (21.70)$$

$$\times \sum_{\{\xi_j=\pm1\}_{\mathrm{B}}} \left[\, \sin^2(\pi K)\, \xi_{n+1} + \cos^2(\pi K)\, \xi_{n+m}\, \right] \xi_1\, G_{m+n}^{\mathrm{B}} B_{m+n}^{(\mathrm{s})} .$$

The summations over the sojourn states are already performed. The subscripts $\{\ldots\}_{\mathrm{A}}$ and $\{\ldots\}_{\mathrm{B}}$ indicate that the blip labels are subject to the constraints

$$\xi_{m+n} = \xi_{n+1} \quad \text{(group A)}, \qquad \xi_{m+n} = -\xi_n \quad \text{(group B)} . \quad (21.71)$$

The charge interaction $Q'(t)$ is given in Eq. (18.54). The interaction factors G_{m+n}^{A} and G_{m+n}^{B} differ from the form (21.38) for G_{m+n} by the absence of the charges situated at the origin, $t' = 0$, and at the end point, $t' = t$. Just for this reason, the remaining $2m + 2n - 2$ blip charges come with the universal factor $\Delta_{\mathrm{r}}^{(2-2K)(m+n-1)}$, and there is no extra dependence on Δ and ω_{c}.

Equations (21.67) – (21.70) are exact formal expressions for the polaron-dressed coherence correlation function in the scaling limit.

21.2.6 Generalized exact master equation and integral relations

We see from the exact formal expressions for the expectation values $\langle\sigma_j\rangle_t$ that the system's transitions are correlated with each other through the influence functions

$F_m^{(\pm)}(\{t_j\})$. Notwithstanding these formidable intricacies, it is possible to describe the dynamics of the conditional populations $P(i,t;j,0)$ [i and j label diagonal states] for any N-state system in terms of a set of exact generalized master equations (GME)

$$\dot{P}(i,t;j,0) = -\sum_{k=1}^{N} \int_0^t dt' \, K(i,t;k,t') P(k,t';j,0), \qquad t > 0. \qquad (21.72)$$

Conservation of probability provides the sum rule $\sum_i K(i,t;j,t') = 0$.

Let us now consider the case of a two states, which we label with the eigenvalues ± 1 of σ_z. We put $P(\pm 1, t; 1, 0) = \frac{1}{2}[1 \pm \langle \sigma_z \rangle_t]$. Using the sum rule for the kernels, $K(\mp 1, t; \pm 1, t') + K(\pm 1, t; \pm 1, t') = 0$, and introducing linear combinations which are even and odd under bias inversion, $K_z^{(s,a)}(t,t') = K(-1,t;-1,t') \pm K(1,t;1,t')$, the GME for $\langle \sigma_z \rangle_t$ with product initial state at time zero is found as [448]

$$\frac{d\langle \sigma_z \rangle_t}{dt} = \int_0^t dt' \, [\, K_z^{(a)}(t,t') - K_z^{(s)}(t,t') \langle \sigma_z \rangle_{t'} \,]. \qquad (21.73)$$

Irreducible kernels and self-energies

The kernels are, by definition, the irreducible components in the exact formal series (21.47). Irreducibility of a kernel means that it cannot be cut into two uncorrelated pieces at an intermediate sojourn without removing bath correlations across this sojourn. We now define *irreducible* influence clusters $\tilde{F}_n^{(\pm)}$ by subtraction of the reducible components in $F_n^{(\pm)}$, which are in product form. Since the bias factor factorizes correspondingly in the subtractions, as we can see from the cluster-wise $\{\xi\}$-summations, we consider the product $F_n^{(\pm)} B_n^{(s,a)}$. For a path with n blips and time growing from right to left in each term, we find

$$\tilde{F}_n^{(\pm)} B_n^{(s,a)} \equiv F_n^{(\pm)} B_n^{(s,a)} \qquad (21.74)$$

$$-\sum_{j=2}^{n} (-1)^j \sum_{m_1, \cdots, m_j} F_{m_1}^{(+)} B_{m_1}^{(s)} \, F_{m_2}^{(+)} B_{m_2}^{(s)} \cdots F_{m_j}^{(\pm)} B_{m_j}^{(s,a)} \delta_{m_1 + \cdots + m_j, n}.$$

The inner sum is over positive integers m_j. The subtractions represent all possibilities of factorizing the influence functions into clusters. For instance, the $n = 3$ term reads

$$\tilde{F}_3^{(\pm)} B_3^{(s,a)} = F_3^{(\pm)} B_3^{(s,a)} - F_2^{(+)} B_2^{(s)} \, F_1^{(\pm)} B_1^{(s,a)} \qquad (21.75)$$

$$-F_1^{(+)} B_1^{(s)} \, F_2^{(\pm)} B_2^{(s,a)} + F_1^{(+)} B_1^{(s)} \, F_1^{(+)} B_1^{(s)} \, F_1^{(\pm)} B_1^{(s,a)}.$$

In the subtractions, the bath correlations are only inside the individual cluster factors $F_{m_j}^{(\pm)}$, and there are no bath correlations between the clusters.

The kernels are fixed by matching the iterative solution of Eq. (21.73) with the exact formal series expression (21.47). Eventually, we obtain

$$K_z^{(s,a)}(t,t') = \Delta^2 F_1^{(\pm)}(t,t')\, B_1^{(s,a)}(t,t') \tag{21.76}$$

$$+ \sum_{n=2}^{\infty}(-1)^{n-1}\frac{\Delta^{2n}}{2^n}\int_{t'}^{t}dt_{2n-1}\cdots\int_{t'}^{t_3}dt_2\sum_{\{\xi_j=\pm 1\}}\tilde{F}_n^{(\pm)}\, B_n^{(s,a)}\,.$$

The product function $\tilde{F}_n^{(\pm)} B_n^{(s,a)}$ depends on $2n$ flip times, where the first transition or flip just occurs at $t_1 = t'$, and the last one at $t_{2n} = t$. The intermediate flip times are subject to time-ordered integration.

The GME (21.73) with the kernels (21.76) is the exact dynamical equation for general (state-independent) dissipation.

The coherence $\langle\sigma_x\rangle_t$ is connected with $\langle\sigma_z\rangle_t$ by the exact integral relation

$$\langle\sigma_x\rangle_t = \int_0^t dt'[K_x^{(s)}(t,t') + K_x^{(a)}(t,t')\langle\sigma_z\rangle_{t'}]\,. \tag{21.77}$$

The kernels $K_x^{(s,a)}(t,t')$ are given again in the form of series expressions involving the modified influence functions. We find upon matching Eq. (21.77) with the exact formal series expression (21.55)

$$K_x^{(s,a)}(t,t') = \Delta\, F_1^{(\mp)}(t,t')\, B_1^{(s,a)}(t,t') \tag{21.78}$$

$$+ \sum_{n=2}^{\infty}(-1)^{n-1}\frac{\Delta^{2n-1}}{2^n}\int_{t'}^{t}dt_{2n-1}\cdots\int_{t'}^{t_3}dt_2\sum_{\{\xi_j=\pm 1\}}\xi_n\tilde{F}_n^{(\mp)}\, B_n^{(s,a)}\,.$$

It is appropriate to remark that the GME (21.73) and the integral relation (21.77) also hold when the bias and the tunneling matrix element are time-dependent. A time-dependent tunneling amplitude requires the substitution (23.5), and a time-dependent bias is taken into account with use of the modified bias factors (23.3) with (23.2).

In the absence of external driving, the kernels $K_z^{(s,a)}(t,t')$ and $K_x^{(s,a)}(t,t')$ depend only on the relative time $t - t'$. Then, the expressions (21.73) and (21.77) are convolutions which can be solved by switching to the Laplace transforms.[5] We obtain

$$\langle\sigma_z(\lambda)\rangle = \frac{1 + \hat{K}_z^{(a)}(\lambda)/\lambda}{\lambda + \hat{K}_z^{(s)}(\lambda)} = \frac{1 + \Sigma_z^{(a)}(\lambda)/\lambda}{\lambda + \lambda\Delta^2/(\lambda^2 + \epsilon^2) + \Sigma_z^{(s)}(\lambda)}\,, \tag{21.79}$$

$$\langle\sigma_x(\lambda)\rangle = \frac{1}{\lambda}\Sigma_x^{(s)}(\lambda) + \left(\epsilon\Delta/(\lambda^2 + \epsilon^2) + \Sigma_x^{(a)}(\lambda)\right)\langle\sigma_z(\lambda)\rangle\,. \tag{21.80}$$

For subsequent convenience we have introduced the self-energies $\hbar\Sigma_z^{(s,a)}(\lambda)$ and $\hbar\Delta\Sigma_x^{(s,a)}(\lambda)$. The self-energies are the Laplace transforms of the kernels (21.76) and (21.78) in which the respective kernels of the isolated system are subtracted,[6]

---

[5]We use for the Laplace transform the convention as given in Eqs. (6.33) and (6.34).

[6]The quantity $\Sigma_z^{(s,a)}(\lambda)$ has dimension frequency, whereas $\Sigma_x^{(s,a)}(\lambda)$ is dimensionless.

$$\Sigma_z^{(s)}(\lambda) \;\equiv\; \hat{K}_z^{(s)}(\lambda) - \lambda\Delta^2/(\lambda^2 + \epsilon^2) \,, \qquad \Sigma_z^{(a)}(\lambda) \;\equiv\; \hat{K}_z^{(a)}(\lambda) \,,$$

$$\Sigma_x^{(a)}(\lambda) \;\equiv\; \hat{K}_x^{(a)}(\lambda) - \epsilon\Delta/(\lambda^2 + \epsilon^2) \,, \qquad \Sigma_x^{(s)}(\lambda) \;\equiv\; \hat{K}_x^{(s)}(\lambda) \,. \tag{21.81}$$

By definition, the self-energies $\hbar\Sigma_z^{(s,a)}(\lambda)$ and $\hbar\Delta\Sigma_x^{(s,a)}(\lambda)$ are zero for vanishing coupling to the reservoir. Once the self-energies are known, the expressions (21.79) and (21.80) can then be inverted to obtain $\langle\sigma_z\rangle_t$ and $\langle\sigma_x\rangle_t$.

When the reduced system is ergodic, the expectation values relax to the respective thermal equilibrium state. The residua of the pole at $\lambda = 0$ of $\langle\sigma_z(\lambda)\rangle$ and $\langle\sigma_x(\lambda)\rangle$ give us directly $\langle\sigma_z\rangle_{t\to\infty}$ and $\langle\sigma_x\rangle_{t\to\infty}$, as the contributions of all other singularities in the complex λ-plane fade away in the course of time. Since the self-energies $\Sigma_z^{(s,a)}(\lambda)$ and $\Sigma_x^{(s,a)}(\lambda)$ have well-defined limits for $\lambda \to 0$, we find for the equilibrium values the exact formal expressions

$$\langle\sigma_z\rangle_\infty \;=\; \frac{\Sigma_z^{(a)}(0)}{\Sigma_z^{(s)}(0)} \,, \qquad \langle\sigma_x\rangle_\infty \;=\; \Sigma_x^{(s)}(0) + \left(\frac{\Delta}{\epsilon} + \Sigma_x^{(a)}(0)\right)\langle\sigma_z\rangle_\infty \,. \tag{21.82}$$

When explicit inversion of the transforms (21.79) and (21.80) is not possible, much about the dynamics can nevertheless be learned from a study of the singularities.

Despite the progress achieved by the exact results (21.79) and (21.80), the formal expressions for the kernels as well as the series expression for $C_z^{(\pm)}(t)$ obtained in the preceding section, while exact, are extremely cumbersome. Nevertheless, they can be evaluated in certain limits by analytical methods. This is discussed in the sequel. Some exact results in the regime of incoherent exponential relaxation at $T = 0$ are discussed in Subsection 22.8.1.

21.3   The noninteracting-blip approximation (NIBA)

21.3.1   Assumptions

After having stated exact formal expressions for the system's dynamics, I now turn to the discussion of the dynamics within the noninteracting-blip approximation (NIBA). This approach will turn out to be exact in various limits [85].

The simple assumption underlying the NIBA is that the average time $\langle s \rangle$ spent by the system in a diagonal or sojourn state is very large compared to the average time $\langle \tau \rangle$ spent in an off-diagonal or blip state. On this condition we may take the view that the blip gas is dilute and noninteracting. In detail, the NIBA assumption leads to two simple prescriptions regarding the sojourn-blip correlations $X_{j,k}$ and interblip correlations $\Lambda_{j,k}$ given in Eq. (21.37).

(1) Set the sojourn-blip correlations $X_{j,k}$ equal to zero when $j \neq k+1$, and put $X_{k+1,k} = Q''(t_{2k+2} - t_{2k+1})$. Thus, the correlations between a sojourn k and the subsequent blips reduce to the intrablip phase correlation $\phi_{k,m} = \xi_{k+1}Q''(\tau_{k+1})$.

(2) Set all interblip interactions $\Lambda_{j,k}$ in the factor G_n in Eq. (21.38) equal to zero.

With these specifications, the influence functional (21.38) reduces to a factorized form of intra-blip bath correlations in which the sign of the individual blip phase depends on the label of respective preceding sojourn,

$$\mathcal{F}_{\mathrm{NIBA}}^{(n)} = \prod_{j=1}^{n} \exp\left\{ -Q'(\tau_j) + i\xi_j\eta_{j-1}Q''(\tau_j) \right\}. \tag{21.83}$$

In the Ohmic scaling limit, in which the form (18.54) for $Q(t)$ applies, the assumption (1) is exact. Thus in this important case, the NIBA consists in the neglect of the interblip interactions $\Lambda_{j,k}$.

Generally, the NIBA can be justified in at least three different limits [85]:

(a) Weak-coupling and zero bias: because of the $\{\xi_j\}$-summations in Eq. (21.76), those contributions to the kernel $K_z^{(s)}(\tau)$ which are *linear* in the interblip correlations $\Lambda_{j,k}$ cancel each other. Thus, the interblip correlations of the kernel $K_z^{(s)}(\tau)$ are of second order in the coupling δ_s, whereas the intrablip correlations are of first order in δ_s. Hence the NIBA for $\langle\sigma_z\rangle_t$ is exact for zero bias in the weak damping (one-phonon exchange) limit. On the other hand, the interblip correlations of first order cancel only partially in the kernel $K_z^{(s,a)}(\tau)$ for nonzero bias, and in the kernel $K_x^{(s,a)}(\tau)$ for zero and nonzero bias. Therefore, neglect of interblip correlations in the NIBA may result in flaws. To these I will allude at the end of this subsection.

(b) The average length of a sojourn is of the order of Δ^{-1} or larger. Further, for $s > 1$ at zero temperature and for $s > 2$ at finite temperature, the numerical value $Q'(t \approx \Delta^{-1})$ differs only little from the asymptotic value $Q'(t \to \infty)$. Thus, the interblip interactions $\Lambda_{j,k}$ tend to be small compared to the intrablip interactions, independently from the partial cancellations in Eq. (21.37).

(c) Long blips are suppressed when the function $Q'(t)$ increases with t at long times. This occurs at $T = 0$ for sub-Ohmic damping, $s < 1$, and at finite temperatures for $s < 2$.[7] Thus, the NIBA is justified (1) in the sub-Ohmic case at all T, (2) in the Ohmic case for large damping and/or high temperature, and (3) in the super-Ohmic case with $s < 2$ for high temperatures. Long blips are also suppressed, independently of the form of $Q(t)$, when the bias is very large.

Thus, the NIBA gives consistent results in three physically quite different regimes.

The picture we now have is that of a noninteracting gas of blips caged in an interval of length t. All influences of the environment are contained in the intrablip interactions which are fully taken into account.[8] In the absence of interblip correlations, the irreducible influence functions $\tilde{F}_n(\pm)$ in Eq. (21.74) are zero for $n > 1$. Thus, within

---

[7]At $T = 0$, we have $Q'(t \to \infty) \propto t^{1-s}$, whereas for finite temperatures $Q'(t \to \infty) \propto t^{2-s}$, as follows from Eq. (18.47).

[8]A simple derivation of the NIBA in the Heisenberg picture is given in Ref. [449].

the NIBA, the kernels given in Subsection 21.2.6 are determined by the one-blip contribution, which is the term of lowest order in Δ in Eq. (21.76) and in Eq. (21.78), respectively. Switching over to the self-energies (21.81), we have in the NIBA[9]

$$
\Sigma_z^{(s)}(\lambda) \;=\; \Delta^2 \int_0^{\infty} d\tau\, e^{-\lambda\tau} \cos(\epsilon\tau) \left(e^{-Q'(\tau)} \cos[Q''(\tau)] - 1 \right) ,
$$

$$
\Sigma_z^{(a)}(\lambda) \;=\; \Delta^2 \int_0^{\infty} d\tau\, e^{-\lambda\tau} \sin(\epsilon\tau)\, e^{-Q'(\tau)} \sin[Q''(\tau)] ,
$$

$$
\Sigma_x^{(s)}(\lambda) \;=\; \Delta \int_0^{\infty} d\tau\, e^{-\lambda\tau} \cos(\epsilon\tau)\, e^{-Q'(\tau)} \sin[Q''(\tau)] ,
$$
(21.84)

$$
\Sigma_x^{(a)}(\lambda) \;=\; \Delta \int_0^{\infty} d\tau\, e^{-\lambda\tau} \sin(\epsilon\tau) \left(e^{-Q'(\tau)} \cos[Q''(\tau)] - 1 \right) .
$$

If we use these forms in Eqs. (21.79) and (21.80), Laplace inversion of the resulting expressions gives the evolution of the damped TSS in the NIBA. One finds that the NIBA gives a good qualitative (and often quantitative) account of the dynamics in the most interesting time regime of several Δ_r^{-1}. However, it does not correctly describe at low T the long-time behavior of $\langle\sigma_j\rangle_t$ and of $C_j^{\pm}(t)$ $(j=x,y,z)$.

21.3.2 Limitations

Before we embark on the discussion of $\langle\sigma_j\rangle_t$ and $C_z^{\pm}(t)$ for particular spectral densities, we point out two flaws of the NIBA which occur at low temperature in the asymptotic time regime. First, since blip-sojourn and blip-blip interactions are absent, the function $R(t)$ in Eq. (21.58), which describes the effects of system-bath correlations in the initial state, is zero in the NIBA. Hence the function $S_z(t)$ reduces to the function $S_z^{(\mathrm{unc})}(t)$ given in Eq. (21.59), in which the components $P_{\mathrm{s,a}}(t)$ and $P_{\infty} = \langle\sigma_z\rangle_{\infty}$ are the respective expressions in the NIBA. As a result of the missing system-bath correlations in the initial state, the algebraic long-time tails of $S_z(t)$ at $T = 0$ (cf. Subsection 22.7) are disregarded in the NIBA. Secondly, the equilibrium values $\langle\sigma_z\rangle_{\infty}$ and $\langle\sigma_x\rangle_{\infty}$ for $\epsilon \neq 0$ are qualitatively incorrect at low temperature. In the NIBA we have

$$
\Sigma_z^{(s)}(0) \;=\; k^+ + k^- , \qquad \Sigma_z^{(a)}(0) \;=\; k^+ - k^- ,
$$
(21.85)

where $k^{\pm}$ are the nonadiabatic rate expressions discussed in Subsection 20.2.1. We find from Eq. (21.82) upon using the detailed balance relation (20.32)

$$
\langle\sigma_z\rangle_{\infty} \;=\; \tanh\left(\frac{\hbar\epsilon}{2k_{\mathrm{B}}T}\right) ,
$$
(21.86)

$$
\langle\sigma_x\rangle_{\infty} \;=\; \frac{\Delta}{\epsilon} \tanh\left(\frac{\hbar\epsilon}{2k_{\mathrm{B}}T}\right) .
$$
(21.87)

---

[9]The generalization for time-dependent driving is studied in Section 23.1.

Hence the NIBA predicts symmetry breaking and strict localization in the lower well at zero temperature, even when the bias is infinitesimal, $\langle\sigma_z\rangle_{\infty\,|T=0} = \mathrm{sgn}\,(\epsilon)$. Furthermore, we get from Eq. (21.87) $\langle\sigma_x\rangle_{\infty\,|T=0} = \Delta/|\epsilon|$. Unfortunately, this expression violates the bound $|\langle\sigma_x\rangle_t| \leq 1$ in the regime $|\epsilon| < \Delta$.

To find the correct weak-damping forms for $\langle\sigma_z\rangle_\infty$ and $\langle\sigma_x\rangle_\infty$ we follow a simple thought. First, assuming that the eigenstates of H_{TSS} are thermally occupied, we get

$$\langle\sigma_z\rangle_\infty = \frac{1}{Z}\left[<R|e^{-\beta H_{\mathrm{TSS}}}|R> - <L|e^{-\beta H_{\mathrm{TSS}}}|L>\right],$$

$$\langle\sigma_x\rangle_\infty = \frac{1}{Z}\left[<R|e^{-\beta H_{\mathrm{TSS}}}|L> + <L|e^{-\beta H_{\mathrm{TSS}}}|R>\right].$$

Next, we write the localized states $|R>$ and $|L>$ in terms of the eigenstates $|g>$ and $|e>$ of the Hamiltonian H_{TSS}. With use of the relations (3.130) we then obtain

$$\langle\sigma_z\rangle_\infty = \frac{\epsilon}{\Delta_{\mathrm{b}}}\tanh\left(\frac{\hbar\Delta_{\mathrm{b}}}{2k_{\mathrm{B}}T}\right) = \cos\varphi\,\langle\tau_z\rangle_\beta\,, \tag{21.88}$$

$$\langle\sigma_x\rangle_\infty = \frac{\Delta}{\Delta_{\mathrm{b}}}\tanh\left(\frac{\hbar\Delta_{\mathrm{b}}}{2k_{\mathrm{B}}T}\right) = \sin\varphi\,\langle\tau_z\rangle_\beta\,, \tag{21.89}$$

where $\langle\tau_z\rangle_\beta$ is the thermal equilibrium value in the eigenbasis of H_{TSS},

$$\langle\tau_z\rangle_\beta = \tanh(\beta\hbar\Delta_{\mathrm{b}}/2)\,, \tag{21.90}$$

and where $\tan\varphi = \Delta/\epsilon$. The expression (21.88) shows that the particle is not confined in reality to one of the two wells at zero temperature, $\langle\sigma_z\rangle_{\infty\,|T=0} = \epsilon/\Delta_{\mathrm{b}}$, and the expression (21.89) satisfies the bound $|\langle\sigma_x\rangle_{\infty\,|T=0}| \leq 1$.

This observation shows that the NIBA fails for a biased system in the regime $k_{\mathrm{B}}T \lesssim \hbar\Delta_{\mathrm{b}}$. The appropriate treatment of this regime within the dynamical approach is given in Section 22.3.

The expressions (21.86) and (21.87) match the exact weak-damping expressions (21.88) and (21.89), respectively, in the temperature regime $T \gtrsim \hbar\Delta_{\mathrm{b}}/k_{\mathrm{B}}$. This points to the fact that the NIBA gives the behavior of both the expectation values $\langle\sigma_j\rangle_t$ and the equilibrium correlation function $C_z^\pm(t)$ correctly for all times at temperatures above $\hbar\Delta_{\mathrm{b}}/k_{\mathrm{B}}$.[10]

21.4 The interacting-blip chain approximation (IBCA)

A systematic improvement on the NIBA consists in taking into account, besides the intra-blip correlations, the nearest-neighbor correlations between blips and the full phase correlations between neighboring sojourn-blip pairs. Diagrammatically, we

---

[10]The attentive reader may have noticed that the argument has been given for a weakly damped system. However, since any system behaves more classically as the damping is increased, this condition should be valid for any damping strength.

then have a chain of blips in which the nearest-neighbor interblip correlations are fully included. A pictorial description illustrating the contribution of three blips to $\langle\sigma_z\rangle_t$ is sketched in Fig. 21.2. Because of the chain-like structure, this approximation has been dubbed "interacting-blip chain approximation" (IBCA) [450].

In the IBCA, the bath influence function $\mathcal{F}^{(n)}$ in Eq. (21.38) is approximated as

$$\mathcal{F}_{\text{IBCA}}^{(n)} = \exp\left\{ -\sum_{j=1}^{n} \left(Q'_{2j,2j-1} - i\xi_j\eta_{j-1}X_{j,j-1}\right) - \sum_{j=2}^{n} \xi_j\xi_{j-1}\Lambda_{j,j-1}\right\}. \quad (21.91)$$

The nearest-neighbor blip-blip and sojourn-blip interactions read

$$\begin{aligned}
\Lambda_{j,j-1} &= Q'_{2j,2j-3} + Q'_{2j-1,2j-2} - Q'_{2j,2j-2} - Q'_{2j-1,2j-3}\,, \\
X_{j,j-1} &= Q''_{2j,2j-1} + Q''_{2j-1,2j-2} - Q''_{2j,2j-2}\,.
\end{aligned} \quad (21.92)$$

Next, our task is to establish the dynamical equations for the reduced density matrix in the IBCA. To this end, we introduce the conditional probabilities per unit time $R_+(t;\tau)$ and $R_-(t;\tau)$ for the particle to be released from the sojourn state $\eta_0 = +1$ at time zero and to hop at time $t - \tau$ into the final blip state $\xi_f = +1$ and $\xi_f = -1$, respectively, and afterwards to remain there until time t. In addition, we define a kernel matrix $Y_{\xi,\xi'}(\tau_2, s_1, \tau_1)$ representing the possible elementary blip-sojourn-blip processes. The kernel describes a two-step transition from the blip state ξ' which has been visited for a period τ_1 via an intermediate sojourn state to the blip state ξ which is visited for a period τ_2. The time spent in the sojourn state is s_1. The kernel represents the intrablip interaction of the last blip and the interactions of this blip with the preceding blip and with the intermediate sojourn. Summation over the two possible intermediate sojourn states yields

$$Y_{\xi,\xi'}(\tau_2, s_1, \tau_1) = -\tfrac{1}{2}\,\xi\xi'\Delta^2 \exp[\,-Q'(\tau_2) - \xi\xi'\Lambda_{2,1}\,]\cos X_{2,1}\,. \quad (21.93)$$

Another element of the dynamical equations is the bias phase factor which takes into account the influences of the deterministic forces. For a stay in the blip state $\xi = \pm 1$, say lasting from time $t - \tau$ until time t, the bias phase factor is $B_\pm(\tau) = \exp(\pm i\epsilon\tau)$, and in the case of a time dependent bias [cf. Chapter 23]

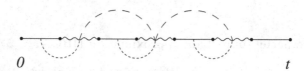

0 t

Figure 21.2: Three-blip contribution to $\langle\sigma_z\rangle_t$ in the IBCA. The solid line represents a sojourn and the curly line a blip interval. The blip-blip correlations $\Lambda_{j,j-1}$ are symbolically sketched by a dashed curve and the sojourn-blip correlations by a dotted curve. The intrablip interactions and the individual interactions in $\Lambda_{j,j-1}$ are not displayed.

$$B_{\pm}(t, \tau) = \exp\left(\pm i \int_{t-\tau}^{t} dt'\, \epsilon(t')\right). \tag{21.94}$$

In numerical computation, a major difficulty arises from the fact that the bias factor (21.94) depends on absolute times, the initial time $t-\tau$ and the final time t of the blip. When the preparation of the initial state belongs to class B [cf. Subsection 21.1.1], the initial sojourn has length infinity. Hence in this case, the first sojourn-blip pair plays a special role. The corresponding amplitude depends on the blip length τ alone and is represented by the factor

$$A_{\pm}(\tau) = \mp i\,(\Delta/2)\exp[-Q'(\tau)\pm iQ''(\tau)]. \tag{21.95}$$

The iteration of the elementary sojourn-blip sequence (21.93) generates all paths the system can go. The sum over all possibilities of stringing together sojourn-blip sequences can be combined in a set of integral equations. Piecing together the elements (21.93) – (21.95), we obtain the dynamical equations in the IBCA as

$$\begin{aligned}
R_{\pm}(t; \tau) &= B_{\pm}(t, \tau)\Big[A_{\pm}(\tau) \\
&\quad + \sum_{k=\pm} \int_{0}^{t-\tau} ds \int_{0}^{t-\tau-s} d\tau'\, Y_{\pm,k}(\tau, s, \tau')\, R_{k}(t-\tau-s; \tau')\Big].
\end{aligned} \tag{21.96}$$

Because of the bath correlations, the coupled integral equations (21.96) are not in the form of convolutions, even not when the bias is static.

Integration of the conditional probabilities $R_{\pm}(t; \tau)$ over the length τ of the final blip state gives the off-diagonal elements of the RDM at time t. We have

$$\langle\sigma_{x}\rangle_{t} = \int_{0}^{t} d\tau\,[R_{+}(t; \tau) + R_{-}(t; \tau)], \tag{21.97}$$

$$\langle\sigma_{y}\rangle_{t} = i\int_{0}^{t} d\tau\,[R_{+}(t; \tau) - R_{-}(t; \tau)]. \tag{21.98}$$

Finally, the population $\langle\sigma_{z}\rangle_{t}$ results from integration of Eq. (21.98),

$$\langle\sigma_{z}\rangle_{t} = 1 - \Delta\int_{0}^{t} dt'\,\langle\sigma_{y}\rangle_{t'}, \tag{21.99}$$

as follows from Eq. (21.10). The integration over t' takes into account that the final step back to the diagonal may occur at any time t' in the interval $0 \leq t' \leq t$.

In conclusion, the dynamical problem of finding the reduced density matrix is solved up to quadratures once the conditional quantities $R_{\pm}(t'; \tau)$ are known in the interval $0 \leq \tau \leq t' \leq t$. An efficient numerical algorithm consists in solving (21.96) by iteration on an equidistant grid in time [450].

The method presented here differs from the iterative solution of the GME (21.73) (cf. Ref. [448]). To include nearest-neighbor blip correlations in the GME, the respective kernel has to be considered at least in order Δ^4. However, iteration of the GME does not lead to linked blip clusters of higher order than those included in the kernel. Furthermore, the GME (21.73) is in the form of a convolution in the absence of time-dependent deterministic forces, while the dynamical equation (21.96) is generally in non-convolutive form.

As the nearest-neighbor blip correlations constitute the most relevant corrections to the NIBA, the IBCA is a valid approximation for longer propagation times than the NIBA. The IBCA is most suitable for moderate-to-strong damping.

Systematic improvement of the IBCA is possible along two lines of development. For weak-to-moderate damping, we may insert, in analogy with the proceeding in Subsection 22.3, all possible tunneling events of the undamped system into the intervals of the chain-links. For higher damping, the first step to do would be to include all next-to-nearest-neighbor interblip correlations. The relevant kernels $Y_{\xi,\xi',\xi''}$ would then depend on three blip labels and on five time intervals, namely the lengths of three blips and of two sojourns in between. Upon book-keeping more and more time intervals in the kernels, the range in which the bath correlations are taken into account exactly is systematically enlarged. The corresponding generalization of the numerical algorithm is clear, but the numerical costs increase drastically with each step.

22. Two-state dynamics: sundry topics

22.1 Symmetric TSS in the NIBA

22.1.1 Ohmic scaling limit

In this subsection I shall discuss first the symmetric TSS ($\epsilon = 0$) in the Ohmic scaling limit within the NIBA. Subsequently, the results will be compared with exact results known from conformal field theory. With the form (18.54) for the bath correlation function $Q(\tau)$, the resulting integral over the blip length τ in Eq. (21.84) can be evaluated in analytic form. The kernel

$$\hat{K}_z^{(s)}(\lambda) \equiv g(\lambda) = \Delta^2 \cos(\pi K) \int_0^\infty d\tau \, e^{-\lambda \tau} \, e^{-Q'(\tau)} , \qquad (22.1)$$

is found to read

$$g(\lambda) = \Delta_{\text{eff}} \left(\frac{\hbar \beta \Delta_{\text{eff}}}{2\pi} \right)^{1-2K} \frac{h(\lambda)}{K + \hbar\beta\lambda/2\pi} ; \quad h(\lambda) = \frac{\Gamma(1 + K + \hbar\beta\lambda/2\pi)}{\Gamma(1 - K + \hbar\beta\lambda/2\pi)} , \qquad (22.2)$$

where Δ_{eff} is the effective tunnel matrix element defined in Eq. (18.37). The Laplace transform of the population $\langle \sigma_z(\lambda) \rangle$ and the spectral function $\tilde{S}_z(\omega)$ of the symmetric equilibrium correlation function [cf. Eq. (21.26)] read

$$\langle \sigma_z(\lambda) \rangle = \frac{1}{\lambda + g(\lambda)} , \qquad \tilde{S}_z(\omega) = \text{Re} \, \frac{2}{-i\omega + g(\lambda = -i\omega)} . \qquad (22.3)$$

Consider now first the dynamics of the symmetric TSS at zero temperature. Taking in Eq. (22.2) the limit $T \to 0$, we get $g(\lambda) = \Delta_{\text{eff}}(\Delta_{\text{eff}}/\lambda)^{1-2K}$. Thus we have

$$\langle \sigma_z(\lambda) \rangle = \frac{1}{\lambda + \Delta_{\text{eff}}(\Delta_{\text{eff}}/\lambda)^{1-2K}} . \tag{22.4}$$

This is the Laplace transform of the Mittag-Leffler function $E_\nu(z)$ [426, a], of which the integral representation is given in Eq. (7.50) [256]. We have

$$\langle \sigma_z \rangle_t = E_{2-2K}[-(\Delta_{\text{eff}}t)^{2-2K}] = \sum_{m=0}^{\infty} \frac{(-1)^m}{\Gamma[1 + (2 - 2K)m]} (\Delta_{\text{eff}}t)^{(2-2K)m} . \tag{22.5}$$

As $K \to 0$, we recover from Eq. (22.5) the persistent oscillation $\langle \sigma_z \rangle_t = \cos(\Delta t)$, whereas for $K = 1/2$ we get pure relaxation $\langle \sigma_z \rangle_t = \exp[-(\pi\Delta^2/2\omega_c)t]$. Here we have employed the relation $\Delta_{\text{eff}}(K = \frac{1}{2}) = \pi\Delta^2/2\omega_c$, which follows from Eq. (18.37).

Since the NIBA is exact in order Δ^2, the expression (22.5) correctly describes the leading time dependence at short times,

$$\langle \sigma_z \rangle_t = 1 - (\Delta_{\text{eff}}t)^{2-2K}/\Gamma(3 - 2K) + \mathcal{O}[(\Delta_{\text{eff}}t)^{4-4K}] , \qquad \Delta_{\text{eff}}t \ll 1 . \tag{22.6}$$

The transform (22.4) is very useful in understanding the behavior of $\langle \sigma_z \rangle_t$. For $0 < K < \frac{1}{2}$, the expression (22.4) has the following singularities:

(1) A branch point at $\lambda = 0$. The complex λ-plane is cut along the negative real axis, and in the cut plane the integrand is single-valued.

(2) A complex conjugate pair of simple poles on the principal sheet at

$$\lambda_{1/2} \equiv -\gamma \pm i\Omega = \Delta_{\text{eff}} \exp[\pm i \tfrac{\pi}{2(1-K)}] . \tag{22.7}$$

Thus we may divide up $P(t) = \langle \sigma_z \rangle_t$ into a coherent and an incoherent part,

$$P(t) = P_{\text{coh}}(t) + P_{\text{inc}}(t) . \tag{22.8}$$

The cut gives a negative contribution $P_{\text{inc}}(0) = -K/(1-K)$ to the initial value $P(0) = 1$, and it yields the asymptotic behavior as $\Delta_{\text{eff}}t \to \infty$,

$$P_{\text{inc}}(t) = \sum_{n=1}^{\infty} \frac{(-1)^{n-1}}{\Gamma[1 - (2 - 2K)n]} \frac{1}{(\Delta_{\text{eff}}t)^{(2-2K)n}} . \tag{22.9}$$

The leading contribution is negative and decays as $1/(\Delta_{\text{eff}}t)^{2-2K}$. For $K = 0$ and $K = \frac{1}{2}$, the branch point is absent and hence $P_{\text{inc}}(t) = 0$.

The residues of the pair of complex conjugate poles give the coherent part

$$P_{\text{coh}}(t) = \frac{1}{1 - K} \cos(\Omega t)\, e^{-\gamma t} . \tag{22.10}$$

The oscillation frequency Ω and the decoherence rate γ are given by

$$\Omega = \Delta_{\text{eff}} \cos\left[\frac{\pi K}{2(1-K)}\right], \quad \text{and} \quad \gamma = \Delta_{\text{eff}} \sin\left[\frac{\pi K}{2(1-K)}\right]. \tag{22.11}$$

The quality factor of the oscillation Q at $T = 0$ is a function of K and it is independent of the frequency scale Δ_{eff}. The expressions (22.11) give

$$Q \equiv \Omega/\gamma = \cot\left[\frac{\pi K}{2(1-K)}\right]. \tag{22.12}$$

Recently, it was argued by Lesage and Saleur (LS) [451] by using a generalization of the boundary conditions changing operators of conformal field theory [452] that the leading long-time behavior of $P(t)$ is a damped oscillation as in Eq. (22.10). They found that the quality factor of the oscillation in the regime $0 < K < \frac{1}{2}$ is exactly given by the form (22.12), which proved that the quality factor found in in the NIBA is exact. However the LS frequency scale differs from the NIBA scale Δ_{eff}. In our notation, the expressions for Ω and γ obtained by Lesage and Saleur read

$$\Omega = \Delta_{\text{LS}} \cos\left[\frac{\pi K}{2(1-K)}\right], \quad \text{and} \quad \gamma = \Delta_{\text{LS}} \sin\left[\frac{\pi K}{2(1-K)}\right]. \tag{22.13}$$

The exact frequency scale Δ_{LS} is related to the NIBA scale Δ_{eff}, which is defined in Eq. (18.37), by

$$\Delta_{\text{LS}} = \sin\left[\frac{\pi K}{2(1-K)}\right] a(K) \Delta_{\text{eff}}, \tag{22.14}$$

where

$$a(K) = \frac{\Gamma\left(\frac{K}{2(1-K)}\right)}{\sqrt{\pi} \Gamma\left(\frac{1}{2(1-K)}\right)} \left(\frac{\Gamma(\frac{1}{2}+K)\Gamma(1-K)}{\sqrt{\pi}}\right)^{\frac{1}{2(1-K)}}. \tag{22.15}$$

The limiting behavior of (22.14), as $K \to 0$, is $\Delta_{\text{LS}} = \Delta_{\text{eff}}[1 + \mathcal{O}(K^2)]$. Similarly, as $K \to \frac{1}{2}$, we have $\Delta_{\text{LS}} = \Delta_{\text{eff}}[1 + \mathcal{O}(1-2K)]$. Thus, the frequency scale Δ_{LS} coincides with the NIBA frequency scale Δ_{eff} in Eq. (22.11) in these limits. For intermediate values of K, the scale Δ_{LS} differs from Δ_{eff} by only a few percent; e.g., for $K = \frac{1}{4}$ we find from Eq. (22.14) $\Delta_{\text{LS}} \approx 1.038\,\Delta_{\text{eff}}$.

For $\frac{1}{2} < K < 1$, the poles of the Laplace transform (22.4) are not anymore on the principal sheet, so that $P(t)$ in the NIBA is fully determined by the branch-cut contribution[1] $P_{\text{inc}}(t)$, which is a completely monotonous function of t. However, the power law form (22.9) is qualitatively incorrect. As opposed to this sluggish decay predicted in the NIBA, the nonequilibrium correlation function $P(t)$ decays in reality exponentially fast [cf. the discussion in Subsection 22.7.1].

At finite temperature, the branch point of $\langle \sigma_z(\lambda) \rangle_{T=0}$ is resolved into an infinite sequence of poles on the negative real λ-axis at $\lambda_n = -(2\pi/\hbar\beta)n - v_n$ ($n = 1, 2, \ldots$), where $-K \leq (\hbar\beta/2\pi)v_n \leq K$. Hence the incoherent part is a series of exponentially decaying contributions, $P_{\text{inc}}(t) = \sum_{n=1}^{\infty} A_n \exp(\lambda_n t)$. The amplitude A_n vanishes linearly with K as $K \to 0$, so that $P_{\text{inc}}(t)$ can be disregarded for weak damping. The coherent dynamics is determined by a pair of complex conjugate simple poles of $\langle \sigma_z(\lambda) \rangle$ so that $P_{\text{coh}}(t)$ once again takes the form (22.10). For $K \ll 1$ and $0 \leq k_{\text{B}}T \lesssim \hbar\Delta_{\text{eff}}$,

---

[1]Since there is no coherent contribution for $K > 1/2$ (in the limit $\Delta_{\text{eff}}/\omega_c \to 0$) at $T = 0$, it is unlikely that there would be any one for $T > 0$ in the damping regime $K > 1/2$.

we can determine the position of the poles iteratively from Eq. (22.3). We find for the frequency Ω and the decoherence rate γ the forms[2]

$$\Omega(T) = \Delta_{\text{eff}} \left\{ 1 + K \left[\operatorname{Re} \psi(i\hbar\beta\Delta_{\text{eff}}/2\pi) - \ln(\hbar\beta\Delta_{\text{eff}}/2\pi) \right] \right\} ,$$

$$\gamma(T) = \tfrac{1}{2}\pi K \Delta_{\text{eff}} \coth(\hbar\beta\Delta_{\text{eff}}/2) .$$

(22.16)

The decoherence rate is in Korringa form which is consistent at $T = 0$ with Eq. (22.11) for $K \ll 1$. The same temperature dependence of the level splitting and the relaxation rate has been observed in a variety of physical systems involving electron-hole excitations, such as interstitials in metals [101] and rare impurities in metals [453]. The generalization of Eq. (22.16) to the biased case is given in Section 22.3.

For intermediate to high temperatures, $k_{\text{B}}T \gtrsim \hbar\Delta_{\text{eff}}$, and general K, we may expand the function $h(\lambda)$ given in Eq. (22.2) about $\lambda = 0$ up to terms of order λ^2. Then the condition $\lambda + g(\lambda) = 0$ yields a quadratic equation for the poles of $\langle \sigma_z(\lambda) \rangle$. Introducing the dimensionless quantities

$$x = (\hbar\beta/2\pi)\,\lambda , \qquad u = (\hbar\beta\Delta_{\text{eff}}/2\pi)^{2K-2} h(0) ,$$

(22.17)

the quadratic equation for x reads

$$(1 - g_2 u)\, x^2 + (K + g_1 u)\, x + u = 0 ,$$

(22.18)

$$g_1 = K^{-1} - \pi \cot(\pi K) , \qquad g_2 = \tfrac{1}{2}\left[\psi'(1-K) - \psi'(1+K) - g_1^2 \right] .$$

(22.19)

The roots of Eq. (22.18) are complex conjugate in the coherent and real in the incoherent regime. The temperature $T^*(K)$ at which the transition between the two "phases" occurs is found from Eq. (22.18) as [454]

$$T^*(K) = \frac{\hbar\Delta_{\text{eff}}}{k_{\text{B}}} \left\{ \frac{\Gamma(K)}{K\Gamma(1-K)} \left[1 + \pi K \cot(\pi K) + 2\sqrt{W(K)} \right] \right\}^{1/2(1-K)} ,$$

(22.20)

where $W(K) = \pi K \cot(\pi K) - K^2 g_2(K)$. In the entire regime $0 \leq K \leq \tfrac{1}{2}$, the formula (22.20) has a relative error of less than 0.4% as compared with the bound calculated numerically from Eq. (22.3) by Garg [455, 85]. The bound for the coherence regime monotonously increases when K is decreased. For $K \ll 1$, we find from Eq. (22.20)

$$T^*(K) = \left[(2\pi)^K/\pi K \right]^{1/(1-K)} \hbar\Delta_{\text{r}}/k_{\text{B}} \approx \hbar\Delta_{\text{r}}/k_{\text{B}}\pi K .$$

(22.21)

The transition temperature varies inversely with K for weak damping. The formula (22.20) yields for $K = \tfrac{1}{2}$ the exact result $T^* = \hbar\Delta_{\text{eff}}/k_{\text{B}}\pi$.

We refrain from giving results in the NIBA for a biased TSS at low temperature because of the weakness of the NIBA in this parameter regime, as marked out in the preceding subsection. In the two subsequent sections I outline the systematic analysis of the biased TSS in the limit of memoryless noise correlations relevant at elevated temperature and in the limit of weak-coupling and low temperature.

---

[2]We remark again that the NIBA is exact for a symmetric system in the weak-damping limit. The expression for $\Omega(T)$ coincides with the former result (19.51) obtained from a rather sophisticated imaginary-time calculation ($\epsilon = 0$).

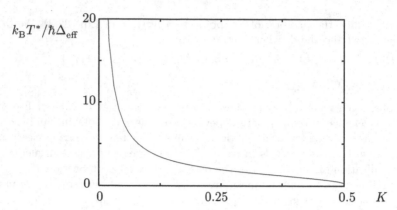

Figure 22.1: The temperature $T^*(K)$ for crossover from the coherent to the incoherent phase is shown as a function of K.

22.1.2   The super-Ohmic case

We now study the behavior of $\langle\sigma_z\rangle_t$ for a symmetric system with a super-Ohmic spectral density, Eq. (18.46) with $s > 2$. Our analysis is based on the expression[3]

$$\langle\sigma_z(\lambda)\rangle \;=\; \frac{\lambda}{\lambda^2 + \tilde{\Delta}^2 + \lambda\Sigma_z(\lambda)}\,, \tag{22.22}$$

where we have introduced the polaron dressed tunneling matrix element $\tilde{\Delta}$ defined in Eq. (20.104). The self-energy in the NIBA, Eq. (21.84), takes the form

$$\hbar\Sigma_z^{(\mathrm{blip})}(\lambda) \;=\; \hbar\tilde{\Delta}^2 \int_0^\infty d\tau\, e^{-\lambda\tau}\left(e^{-[Q'(\tau)-X_1]}\cos[\,Q''(\tau)\,] - 1\right). \tag{22.23}$$

For low-to-moderate temperatures, we expect to find the poles of Eq. (22.22) near $\pm i\tilde{\Delta}$. Upon iteration, we find the pole condition

$$\lambda^2 + \lambda\Sigma_z^{(\mathrm{blip})}(\pm i\tilde{\Delta}) + \tilde{\Delta}^2 \;=\; 0\,. \tag{22.24}$$

The shift of the oscillation frequency in leading order of $\Sigma_z^{(\mathrm{blip})}$ is determined by the imaginary part of the self-energy.[4] However, since generally $\mathrm{Im}\,\Sigma_z^{(\mathrm{blip})}(\pm i\tilde{\Delta}) \ll \mathrm{Re}\,\Sigma_z^{(\mathrm{blip})}(\pm i\tilde{\Delta})$ for $s > 2$ [458], here we disregard the imaginary part of the self-energy.

We now put $\mathrm{Re}\,\Sigma_z^{(\mathrm{blip})}(\pm i\tilde{\Delta}) \equiv \Upsilon$. With a manipulation similar to that which has led us from Eq. (20.30) to the expression (20.31), we find

$$\Upsilon \;=\; \tilde{\Delta}^2 \cosh\left(\tfrac{1}{2}\beta\hbar\tilde{\Delta}\right) \int_0^\infty dt\, \cos(\tilde{\Delta}t)\left(e^{X_2(t)} - 1\right), \tag{22.25}$$

---

[3]In this subsection we put $\Sigma_z^{(s)}(\lambda) = \Sigma_z(\lambda)$.

[4]The shift due to the one-phonon process is given in Eq. (22.91) [cf. also Eq. (19.42)].

where $X_2(t)$ is given in Eqs. (20.102) and (20.108). Expansion of the integrand in powers of $X_2(t)$ gives the multi-phonon series for the rate Υ. For $s = 3$, we obtain a hypergeometric series [cf. Eq. (20.125)],

$$\Upsilon = \frac{\tilde{\Delta}}{2\pi} \coth\left(\tfrac{1}{2}\beta\hbar\tilde{\Delta}\right) \left(\frac{\hbar\tilde{\Delta}}{k_B T'_{ph}}\right)^2 {}_2F_2\left(1 + i\frac{\beta\hbar\tilde{\Delta}}{2\pi}, 1 - i\frac{\beta\hbar\tilde{\Delta}}{2\pi}; 2, \frac{3}{2}; \phi\right), \quad (22.26)$$

where ϕ and T'_{ph} are given in Eq. (20.120). It is convenient to write $\Upsilon = \Upsilon_+ + \Upsilon_-$, where Υ_+ captures the even and Υ_- the odd multi-phonon processes. At temperatures $T \lesssim \hbar\tilde{\Delta}/k_B$, only the one-phonon process is relevant because of the condition (20.107),

$$\Upsilon_- = (\tilde{\Delta}/2\pi)(\hbar\tilde{\Delta}/k_B T'_{ph})^2 \coth(\tfrac{1}{2}\beta\hbar\tilde{\Delta}), \qquad \Upsilon_+ \ll \Upsilon_-. \quad (22.27)$$

At higher temperatures, $\hbar\tilde{\Delta}/k_B \ll T \ll T_D$, we find from Eq. (22.26) the expansions in terms of even and odd multi-phonon processes,

$$\Upsilon_+ = \frac{\tilde{\Delta}}{\pi} \frac{\hbar\tilde{\Delta}}{k_B T} \sum_{n=1}^{\infty} \frac{\Gamma(\tfrac{3}{2})}{2n\Gamma(2n + \tfrac{1}{2})} \left(\frac{T}{T'_{ph}}\right)^{4n}, \quad (22.28)$$

$$\Upsilon_- = \frac{\tilde{\Delta}}{\pi} \frac{\hbar\tilde{\Delta}}{k_B T'_{ph}} \frac{T}{T'_{ph}} \sum_{n=1}^{\infty} \frac{\Gamma(\tfrac{3}{2})}{(2n-1)\Gamma(2n - \tfrac{1}{2})} \left(\frac{T}{T'_{ph}}\right)^{4n-4}. \quad (22.29)$$

With growing temperature, higher multi-phonon processes become increasingly relevant, and for T above T'_{ph}, but still well below the Debye temperature, we may use the asymptotic expression for the multi-phonon series. We find from Eq.(22.25) by steepest descent for general s [cf. Eq. (20.117) for $\kappa = 0$ and $\epsilon = 0$]

$$\Upsilon_\pm = \tfrac{1}{2}\Upsilon = \frac{\hbar\Delta^2 e^{-2B_s}}{2k_B T'_{ph}} \sqrt{\frac{\pi}{d_s(0)}} \left(\frac{T'_{ph}}{T}\right)^{(s+1)/2} \exp\left\{c_s(0)\left(\frac{T}{T'_{ph}}\right)^{s-1}\right\}. \quad (22.30)$$

For $s = 3$, this form reduces to

$$\Upsilon_\pm = \tfrac{1}{2}\Upsilon = \frac{1}{4\sqrt{\pi}} \frac{\hbar\Delta^2 e^{-2B_3}}{k_B T'_{ph}} \left(\frac{T'_{ph}}{T}\right)^2 \exp\left(\frac{2T^2}{3T'^2_{ph}}\right). \quad (22.31)$$

In the under-damped regime, the poles of Eq. (22.22) are at $\lambda_{1/2} = -\gamma \pm i\Omega$ with

$$\gamma = \tfrac{1}{2}\Upsilon, \qquad \Omega = \sqrt{\tilde{\Delta}^2 - \tfrac{1}{4}\Upsilon^2}, \quad (22.32)$$

and the populations undergo damped oscillations

$$\langle\sigma_z\rangle_t = \cos(\Omega t - \phi)e^{-\gamma t}/\cos\phi, \qquad \phi = \arctan(\gamma/\Omega). \quad (22.33)$$

The temperature T^*, at which the transition from underdamped to overdamped motion occurs, is determined by the equality $\Upsilon = 2\tilde{\Delta}$, as follows from Eq. (22.32). With Eqs. (20.104) and (22.30), we obtain for T^* the transcendental equation

$$T^* = T'_{\text{ph}} \left[\frac{1}{[c_s(0) + d_s]} \ln \left(\frac{\Delta \, e^{-B_s}}{A_s(T^*)} \right) \right]^{1/(s-1)}, \qquad (22.34)$$

where $A_s(T)$ is the preexponential factor in the expression (22.30).

For T above T^*, we have [observe the subtle difference to Eq. (22.56)],

$$\langle \sigma_z \rangle_t = \frac{1}{(\gamma_1 - \gamma_2)} \left(\gamma_1 \, e^{-\gamma_1 t} - \gamma_2 \, e^{-\gamma_2 t} \right), \qquad \gamma_{1/2} = \tfrac{1}{2} \Upsilon \pm \sqrt{\tfrac{1}{4} \Upsilon^2 - \tilde{\Delta}^2} \,. \quad (22.35)$$

For T far above T^*, we have $\Upsilon \gg 2\tilde{\Delta}$. Then the amplitude of the second term is small compared to that of the first term, and we obtain single exponential decay

$$\langle \sigma_z \rangle_t = e^{-\Upsilon t} \,. \qquad (22.36)$$

Until now, our study of the super-Ohmic case has been restricted to the NIBA. Recently, an approximate summation of the self-energy $\hbar\Sigma_z(\lambda)$ in all orders of $\tilde{\Delta}^2$ has been put forward [458]. The approximation consists in disregarding all bath correlations except for the intra-sojourn correlations in the terms $n \geq 2$ in the series (21.76). This converts the series into a convolution. Switching to the Laplace transform, each blip-sojourn sequence in the terms $n \geq 2$ contributes a factor $-\Sigma_z^{(\text{soj})}(\lambda)/\lambda$, where the factor $1/\lambda$ originates from the blip interval, and $\hbar\Sigma_z^{(\text{soj})}(\lambda)$ is the self-energy contribution from a single sojourn. The resulting expression is a geometrical series which is summed to

$$\Sigma_z(\lambda) = \Sigma_z^{(\text{blip})}(\lambda) - \frac{\tilde{\Delta}^2}{\lambda} \frac{\Sigma_z^{(\text{soj})}(\lambda)/\lambda}{1 + \Sigma_z^{(\text{soj})}(\lambda)/\lambda} \,. \qquad (22.37)$$

Because of the $\{\xi_j\}$-summation in Eq. (21.76), the odd multi-phonon contributions to $\Sigma_z^{(\text{soj})}(\lambda)$ cancel out, and we obtain

$$\Sigma_z^{(\text{soj})}(\lambda) = \tilde{\Delta}^2 \int_0^\infty ds \, e^{-\lambda s} \left(\cosh[\, Q'(s) - X_1 \,] \cos[\, Q''(s) \,] - 1 \right) . \qquad (22.38)$$

Substituting the expression (22.37) into Eq. (22.22), we find

$$\langle \sigma_z(\lambda) \rangle = \frac{\lambda + \Sigma_z^{(\text{soj})}(\lambda)}{[\lambda + \Sigma_z^{(\text{soj})}(\lambda)][\lambda + \Sigma_z^{(\text{blip})}(\lambda)] + \tilde{\Delta}^2} \,. \qquad (22.39)$$

The iteration of the pole condition in Eq. (22.39) leads to the substitutions $\Sigma_z^{(\text{soj})}(\lambda) \rightarrow \Upsilon_+$, $\Sigma_z^{(\text{blip})}(\lambda) \rightarrow \Upsilon_+ + \Upsilon_-$. With these we have

$$\langle \sigma_z(\lambda) \rangle = \frac{\lambda + \Upsilon_+}{[\lambda + \Upsilon_+][\lambda + \Upsilon_+ + \Upsilon_-] + \tilde{\Delta}^2} \,. \qquad (22.40)$$

For $k_{\text{B}} T \lesssim \hbar\tilde{\Delta}$, the partial rates Υ_+ and Υ_- are given in Eq. (22.27), and for higher temperature in Eqs. (22.28) and (22.29).[5]

---

[5]For $k_{\text{B}} T \lesssim \hbar\tilde{\Delta}$, a possible generalization is to insert in the blip interval in Eq. (22.23) and in the sojourn interval in Eq. (22.38) the corresponding propagators of the free system [cf. the discussion in Subsections 19.1.6, 19.1.7, and 22.3.1]. The iteration of the pole condition results in the substitutions $\Sigma_z^{(\text{S})}(\lambda) \rightarrow \tfrac{1}{2}(1+p)\Upsilon_+$, $\Sigma_z^{(\text{B})}(\lambda) \rightarrow \tfrac{1}{2}(1+p)\Upsilon_+ + \Upsilon_-$, where $p = \hbar\beta\tilde{\Delta}/\sinh(\hbar\beta\tilde{\Delta})$ [458]. Since the quantity p is different from 1 only for $T \lesssim \hbar\tilde{\Delta}/k_{\text{B}}$, we then have $T \ll T'_{\text{ph}}$ [cf. Eq. (20.107)], and hence $\Upsilon_+ \ll \Upsilon_-$. This shows that the expression (22.40) is also valid in the regime $T \lesssim \hbar\tilde{\Delta}/k_{\text{B}}$.

The expression (22.40) describes again underdamped oscillations for $T < T^*$,

$$\langle \sigma_z \rangle_t = \cos(\Omega t - \phi)\, e^{-\gamma t} / \cos(\phi) \,, \qquad \phi = \arctan(\Upsilon_-/2\Omega) \,, \qquad (22.41)$$

with

$$\gamma = \Upsilon_+ + \tfrac{1}{2}\Upsilon_- \,, \qquad \Omega = \tilde{\Delta}\sqrt{1 - (\Upsilon_-/2\tilde{\Delta})^2} \,. \qquad (22.42)$$

This result differs from the NIBA result (22.32), (22.33) by a factor 2 in the *even* multi-phonon contributions to the decoherence rate γ, and also the phase shifts differ.

In the temperature regime $T_D \gg T \gg T'_{\rm ph}$, the asymptotic expression of the multi-phonon series (22.28) and (22.29) is given for general s in Eq. (22.30), and for $s = 3$ in Eq. (22.31). The oscillation of $\langle \sigma_z \rangle_t$ vanishes when the condition $\Upsilon_- = 2\tilde{\Delta}$ is reached. The transcendental equation for the transition temperature T^* is of the form (22.34) with an extra factor 2 in the argument of the logarithm.

In the regime $T > T^*$ we have

$$\langle \sigma_z \rangle_t = \frac{\gamma_1 - \tfrac{1}{2}\Upsilon}{\gamma_1 - \gamma_2}\, e^{-\gamma_1 t} + \frac{\tfrac{1}{2}\Upsilon - \gamma_2}{\gamma_1 - \gamma_2}\, e^{-\gamma_2 t} \,, \qquad \gamma_{1/2} = \tfrac{3}{4}\Upsilon \pm \sqrt{\left(\tfrac{1}{4}\Upsilon\right)^2 - \tilde{\Delta}^2} \,. \quad (22.43)$$

For T well above T^*, the rate γ_1 is near Υ, whereas γ_2 approaches $\tfrac{1}{2}\Upsilon$. Hence the residuum of the second relaxation contribution becomes negligibly small, and we obtain the NIBA result (22.36). As the temperature is increased further, the crossover to the classical Marcus regime takes place. This has been discussed already in some detail in Subsection 20.2.8.

In Eq. (22.37) a special interblip correlation term is summed in all orders in $\tilde{\Delta}^2$. The treatment of interblip correlations is not systematic. Nevertheless, the characteristic features of the resulting dynamics are found to be in qualitative agreement with the NIBA results. There is only a marginal difference in the crossover temperature T^*. The major difference around the transition temperature is an additional factor 2 in the even multi-phonon contribution Υ_+ to the rate.

22.2 White-noise regime

22.2.1 Power spectrum of the stochastic force

With a view to the regimes of weak damping and moderate-to-high temperature, it is useful to introduce a number of notions and definitions. First, let us define a normalized power spectrum of the stochastic force $\xi(t)$ with dimension frequency as

$$S_\xi(\omega) \equiv \frac{1}{2}\left(\frac{q_0}{\hbar}\right)^2 \tilde{\mathcal{X}}(\omega) = \frac{\pi}{2}\, G_{\rm lf}(|\omega|) \coth\left(\frac{\beta \hbar |\omega|}{2}\right) \,. \qquad (22.44)$$

Here $\tilde{\mathcal{X}}(\omega)$ is given in Eq. (2.34). The second form follows with use of Eqs. (3.26) and (3.141). The relation (22.44) is a ramification of the fluctuation-dissipation theorem. In the white noise limit

$$\langle \xi(t)\xi(0) \rangle = 2M\gamma k_{\rm B}T\, \delta(t) \,, \qquad (22.45)$$

the power spectrum $S_\xi(\omega)$ becomes independent of the frequency ω,

$$S_{\xi,\mathrm{wn}} \equiv \vartheta = 2\pi K k_\mathrm{B} T/\hbar \,, \tag{22.46}$$

which is a thermal frequency scaled with the Ohmic damping parameter K. For $K \ll 1$ the Markovian or white-noise regime applies at elevated temperature

$$T_\mathrm{b} \lesssim T \ll \hbar\omega_\mathrm{c}/k_\mathrm{B} \,, \qquad\qquad T_\mathrm{b} \equiv \hbar(\Delta_\mathrm{eff}^2 + \epsilon^2)^{1/2}/k_\mathrm{B} \,. \tag{22.47}$$

Next, we observe that in the temperature range $T \gtrsim T_\mathrm{b}$ the relevant time domain of the pair interaction $Q(t)$ is $|t| \gg \hbar\beta/\pi$. In this time regime the Ohmic scaling form (18.54) of the pair interaction reduces to the "white noise" expression

$$Q_\mathrm{wn}(t) = Q_\mathrm{adia} + \vartheta\,|t| + i\,\pi K\,\mathrm{sgn}(t) \,,$$

$$Q_\mathrm{adia} \equiv \int_{\nu_1}^{\omega_\mathrm{c}} d\omega\, \frac{G_\mathrm{ohm}(\omega)}{\omega^2} = 2K\ln\left(\frac{\hbar\omega_\mathrm{c}}{2\pi k_\mathrm{B} T}\right) . \tag{22.48}$$

Recalling the relation $\ddot{Q}(t) = q_0^2 L(t)/\hbar$ and Eq. (5.40) we see that the expression (22.48) is consistent with the memoryless force correlation (2.2). This shows that the pair interaction $Q_\mathrm{wn}(t)$ establishes the Markovian dynamics of the open system in the charge representation (21.34). The term Q_adia is the adiabatic contribution of the bath modes which dresses the tunneling matrix element as in Eq. (18.30). The dynamical (second and third) terms in $Q_\mathrm{wn}(t)$ are due to the modes with $\omega < \nu_1 = 2\pi/\hbar\beta$.

22.2.2 Symmetric Ohmic TSS at moderate-to-high temperature

In this subsection, we study the case of weak Ohmic dissipation, $K \ll 1$, in the temperature regime (22.47). The temperature regime beneath, $0 \le T \lesssim T_\mathrm{b}$, is discussed below in Section 22.3. The case $K \ll 1$ applies, e.g., to interstitial tunneling in metals. In addition, the weak-damping case is of fundamental importance with regard to "Macroscopic Quantum Coherence" (MQC) experiments since the dynamics might be coherent in the accessible temperature regime.

The argument of this subsection is based on the observation that in the regime (22.47) the characteristic fluctuations of the collective bath mode $\mathfrak{E}(t) = q_0\xi(t)$ are virtually memory-less. Then, the pair correlation function takes the form $Q_\mathrm{wn}(t)$ given in Eq. (22.48). Use of the form (22.48) instead of the full expression (18.54) in the breathing mode integral for $\Sigma_z^{(\mathrm{s})}(\lambda)$ in Eq. (21.84) is equivalent to the substitution $h(\lambda) \to h(0)$ in Eq. (22.2). This is a consistent approximation when the relevant poles $\lambda_{1/2}$ of $\langle\sigma_z(\lambda)\rangle$ satisfy the condition $\hbar\beta|\lambda_{1/2}| \ll 1$. The condition is fulfilled indeed in the regime (22.47). This verifies the presented reasoning.

With the white-noise form (22.48) for the pair correlation function the interblip interactions cancel out exactly. Thus we have $\Lambda_{jk} = 0$, and hence the noninteracting-blip assumption is *exact* in the regime (22.47).

It is convenient to absorb the adiabatic bath contribution Q_adia into an effective temperature-dependent tunneling matrix element,

$$\Delta_T \equiv \Delta \, e^{-Q_{\text{adia}}/2} \; = \; \Delta_r \, (2\pi k_B T / \hbar \Delta_r)^K \; = \; \Delta_{\text{eff}} \, (2\pi k_B T / \hbar \Delta_{\text{eff}})^K \, \sqrt{h(0)} \, . \quad (22.49)$$

In the latter equality we have used $h(0) = \Gamma(1+K)/\Gamma(1-K)$ and we have disregarded terms of order K^2.

With these preliminaries, it is now obvious that the kernel (22.1) reduces in the white-noise limit to the simple form

$$\hat{K}_z^{(s)}(\lambda) \; = \; \Delta_T^2 / [\lambda + \vartheta] \; = \; \Delta_T^2 / [\lambda + 2\pi K / \hbar \beta] \, . \quad (22.50)$$

Let us consider first the *symmetric* two-state system. Inserting Eq. (22.50) into the expression (21.79) and setting $\epsilon = 0$, we obtain[6]

$$\langle \sigma_z(\lambda) \rangle \; = \; \frac{\lambda + \vartheta}{\lambda(\lambda + \vartheta) + \Delta_T^2} \; = \; \frac{\lambda - \lambda_1 - \lambda_2}{(\lambda - \lambda_1)(\lambda - \lambda_2)} \, . \quad (22.51)$$

The positions $\lambda_{1/2}$ of the poles of $\langle \sigma_z(\lambda) \rangle$ are again complex-conjugate for $T < T^*$, and real-valued for $T > T^*$. The transition temperature $T^*(K)$ separating the coherent from the incoherent "phase" is given in Eq. (22.21). For small K, it varies with $1/K$, $T^*(K) = \hbar \Delta_r / k_B \pi K$. In the coherent regime $\hbar \Delta_{\text{eff}} / k_B \lesssim T \leq T^*(K)$, we have $\lambda_{1/2} = -\gamma \pm i\Omega$. Hence the population $\langle \sigma_z \rangle_\beta$ undertakes damped oscillations,

$$\langle \sigma_z \rangle_t \; = \; \frac{\cos(\Omega t - \phi)}{\cos \phi} \, e^{-\gamma t} \, , \qquad \phi = \arctan(\gamma / \Omega) \, . \quad (22.52)$$

The corresponding spectral function $\tilde{S}_z(\omega)$, which is the Fourier transform of the symmetrized correlation function $S_z(t) = \text{Re} \, C_z^\pm(t)$, is a superposition of two inelastic lines of Lorentzian shape centered at $\omega = \pm\Omega(T)$ with line width $2\gamma(T)$,

$$\tilde{S}_z(\omega) = \frac{\gamma + (\Omega - \omega) \tan \phi}{(\omega - \Omega)^2 + \gamma^2} + \frac{\gamma + (\Omega + \omega) \tan \phi}{(\omega + \Omega)^2 + \gamma^2} \, . \quad (22.53)$$

The parameters of the equivalent damped oscillator in the regime $T \leq T^*$ are

$$\Omega(T) \; = \; \Delta_T \sqrt{1 - (T/T^*)^{2-2K}} \; ; \qquad \gamma(T) \; = \; \vartheta/2 = \pi K \, k_B T / \hbar \, . \quad (22.54)$$

Together with the findings in Subsection 22.1.1 we now have a complete picture of the unbiased Ohmic TSS. In the regime $T \lesssim \hbar \Delta_r / k_B$ the expressions (22.16) for $\Omega(T)$ and $\gamma(T)$ apply. They smoothly match near $T = \hbar \Delta_r / k_B$ with the expressions (22.54). The oscillation frequency increases $\propto T^2$ at low temperatures,[7] as we can see from Eq. (22.16). For $T \gtrsim \hbar \Delta_r / k_B$ the form (22.54) applies. Observing that $\Delta_T \propto T^K$ one finds that the oscillation frequency $\Omega(T)$ has a maximum at $T = K^{1/(2-2K)} T^*$. Above this temperature, $\Omega(T)$ decreases monotonously and approaches zero for $T = T^*$. The coherent quantum oscillations "dephase" on the time scale $1/\gamma(T)$ which decreases inversely with T.

---

[6]The pole condition is the weak-damping limit of Eq. (22.18). The generalization to larger values of K is analogous to Eq. (22.18). Only the coefficients of the algebraic equation are modified.

[7]The T^2-law for $\Omega(T)$ gives a linear contribution to the specific heat (cf. Subsection 19.2.1).

For T above T^*, the two poles of $\langle \sigma_z(\lambda) \rangle$ in Eq. (22.51) are situated on the real negative λ-axis at $\lambda = -\gamma_1$ and $\lambda = -\gamma_2$. The rates are

$$\gamma_{1,2} = \pi K/\hbar\beta \pm \sqrt{(\pi K/\hbar\beta)^2 - \Delta_T^2} \, , \tag{22.55}$$

and instead of the oscillatory behavior (22.52) we have pure bi-exponential relaxation

$$\langle \sigma_z \rangle_t = \frac{\gamma_1}{\gamma_1 - \gamma_2} e^{-\gamma_2 t} + \frac{\gamma_2}{\gamma_2 - \gamma_1} e^{-\gamma_1 t} \, . \tag{22.56}$$

As T is increased further, the poles move in opposite direction along the negative real λ-axis. The rate γ_1 increases with temperature and attains linear dependence on T in the regime $T \gg T^*$, $\gamma_1 = 2\pi K k_B T/\hbar$. In contrast, the rate γ_2 decreases with increasing temperature and takes for $T \gg T^*(K)$ the asymptotic form

$$\gamma_2 \equiv \gamma_r(T) = \frac{1}{K} \Delta_r \left(\frac{\hbar \Delta_r}{2\pi k_B T} \right)^{1-2K} . \tag{22.57}$$

Fairly above T^*, the weight of the pole λ_1 is very small, so that only the relaxation pole moving towards the origin, $\lambda = -\gamma_r$, is relevant. Then, the spectral function $\tilde{S}_z(\omega)$ is described by a quasi-elastic Lorentzian peak, and $\langle \sigma_z \rangle_t$ decays exponentially,

$$\tilde{S}_z(\omega) = \frac{2\gamma_r}{\omega^2 + \gamma_r^2} \, , \qquad \langle \sigma_z \rangle_t = e^{-\gamma_r t} \, . \tag{22.58}$$

The decay rate γ_r is the sum of the the forward and backward "Golden Rule" tunneling rate, and it is consistent with form (20.84) for $K \ll 1$. Since the tunneling rate varies with temperature as T^{2K-1}, the width of the Lorentzian line shrinks with increasing temperature when $K < 1/2$. The anomalous T^{2K-1} law has been predicted first by Kondo [379, 136]. According to this characteristic feature, the temperature regime well above T^* is called the Kondo regime.

22.2.3   Biased Ohmic TSS at moderate-to-high temperature

The analysis of the dynamics in the white-noise regime is easily extended to the *biased* system. With Eq. (22.2) we have

$$\begin{aligned} \hat{K}_z^{(s)}(\lambda) &= \tfrac{1}{2}[g(\lambda + i\epsilon) + g(\lambda - i\epsilon)] \, , \\ \hat{K}_z^{(a)}(\lambda) &= i\pi K \tfrac{1}{2}[g(\lambda + i\epsilon) - g(\lambda - i\epsilon)] \, , \end{aligned} \tag{22.59}$$

and similar forms hold in the NIBA for $\hat{K}_x^{(s,a)}(\lambda)$. Next, we insert form (22.48) in Eq. (21.84). This amounts to putting $h(0)$ instead of $h(\lambda)$ in Eq. (22.2). Eventually, we find [445]

$$\begin{aligned} \hat{K}_z^{(s)}(\lambda) &= \Delta_T^2(\lambda + \vartheta)/[(\lambda + \vartheta)^2 + \epsilon^2] \, , \\ \hat{K}_z^{(a)}(\lambda) &= \Delta_T^2 \pi K \epsilon/[(\lambda + \vartheta)^2 + \epsilon^2] \, , \\ \hat{K}_x^{(s)}(\lambda) &= (\pi K/\Delta)\hat{K}_z^{(s)}(\lambda) \, , \\ \hat{K}_x^{(a)}(\lambda) &= \hat{K}_z^{(a)}(\lambda)/(\pi K \Delta) \, . \end{aligned} \tag{22.60}$$

With these expressions, the expectation values (21.79) and (21.80) are found as

$$\langle \sigma_z(\lambda) \rangle = \frac{1}{\lambda} \left(1 + \frac{(\pi K \epsilon - \lambda - \vartheta) \Delta_T^2}{N(\lambda)} \right) , \tag{22.61}$$

$$\langle \sigma_x(\lambda) \rangle = \hat{K}_x^{(s)}(\lambda)/\lambda + \hat{K}_x^{(a)}(\lambda) \langle \sigma_z(\lambda) \rangle , \tag{22.62}$$

where

$$N(\lambda) = \lambda \left[(\lambda + \vartheta)^2 + \epsilon^2 \right] + (\lambda + \vartheta) \Delta_T^2 . \tag{22.63}$$

The dynamics is determined by four singularities in the complex λ-plane: a simple pole at $\lambda = 0$, the residuum of which represents the equilibrium value,

$$\langle \sigma_z \rangle_\infty = \frac{\hbar \epsilon}{2 k_B T} ; \quad \langle \sigma_x \rangle_\infty = \frac{\Delta_T}{\Delta} \frac{\hbar \Delta_T}{2 k_B T} , \tag{22.64}$$

and three simple poles which are located at the zeros of the cubic equation $N(\lambda) = 0$. At quick glance, it seems that there occur additional singularities in $\langle \sigma_x(\lambda) \rangle$, which are in the kernels $\hat{K}_x^{(s,a)}(\lambda)$ resulting from the zeros of the equation $(\lambda + \vartheta)^2 + \epsilon^2 = 0$. However, the associated residua of the two terms in Eq. (22.62) cancel each other.

The characteristics of the cubic equation $N(\lambda) = 0$ is as follows [454]. For a bias in the range $|\epsilon| < \epsilon_c$ where $\epsilon_c = \Delta_T/\sqrt{8}$, we may distinguish three temperature regimes with qualitatively different behaviors. For $T < T_1$ and $T > T_2$, where $T_2 > T_1$, one of the zeros of $N(\lambda)$ is negative real, and the other two are complex conjugate. The former gives rise to exponential relaxation and the latter to damped oscillations,

$$\lambda_1 = -\gamma_r, \quad \lambda_{2/3} = -\gamma \pm i\Omega . \tag{22.65}$$

In the regime $T_1 < T < T_2$, all zeros are negative real. Thus in the intermediate regime, the system shows over-damped relaxation. The transition temperatures are

$$T_{2/1} = \frac{1}{4\sqrt{2\pi K k_B}} \frac{\hbar}{|\epsilon|} \sqrt{\Delta_T^4 + 20\epsilon^2 \Delta_T^2 - 8\epsilon^4 \pm \Delta_T (\Delta_T^2 - 8\epsilon^2)^{3/2}} . \tag{22.66}$$

At the critical bias strength $|\epsilon| = \epsilon_c \equiv \Delta_T/\sqrt{8}$, the transition temperatures T_1 and T_2 coincide, $T_1(\epsilon_c) = T_2(\epsilon_c) = 3\sqrt{3/2}\,\hbar\Delta_T/(4\pi K k_B)$.

For $|\epsilon| \ll \epsilon_c$, we may expand the expression (22.66). We then obtain $T_1 = [\hbar(\Delta_T - \epsilon^2/2\Delta_T)/\pi K + \mathcal{O}(\epsilon^4)]/k_B$, and $T_2 = [\hbar\Delta_T^2/4\pi K|\epsilon| + \hbar|\epsilon|/2\pi K + \mathcal{O}(|\epsilon|^3)]/k_B$.

When $|\epsilon|$ exceeds the critical bias ϵ_c, there is one real root and two complex conjugate ones for all temperatures, as given in Eq. (22.65).

The Vieta relations for γ_r, γ and Ω read $(T \gtrsim T_b)$

$$\gamma_r + 2\gamma = 2\vartheta , \tag{22.67}$$

$$\gamma_r (\gamma^2 + \Omega^2) = \Delta_T^2 \vartheta , \tag{22.68}$$

$$\gamma^2 + 2\gamma\gamma_r + \Omega^2 = \Delta_T^2 + \epsilon^2 + \vartheta^2 . \tag{22.69}$$

We see from the Vieta relations that in the Markov regime decoherence and relaxation depend via ϑ on the product of damping strength and temperature, apart from the dependence of Δ_T on K and T resulting from the polaronic cloud.

For temperatures well below T_1, but above T_b, we find

$$
\gamma_{\mathrm{r}} = \frac{\Delta_T^2}{\Delta_T^2 + \epsilon^2}\,\vartheta - \frac{\Delta_T^2 \epsilon^4}{(\Delta_T^2 + \epsilon^2)^2}\,\vartheta^3 + \mathcal{O}(\vartheta^5)\,,
$$

$$
\gamma = \frac{\Delta_T^2 + 2\epsilon^2}{2(\Delta_T^2 + \epsilon^2)}\,\vartheta + \frac{\Delta_T^2 \epsilon^4}{2(\Delta_T^2 + \epsilon^2)^2}\,\vartheta^3 + \mathcal{O}(\vartheta^5)\,, \tag{22.70}
$$

$$
\Omega^2 = \Delta_T^2 + \epsilon^2 - \frac{\Delta_T^2(\Delta_T^2 + 4\epsilon^2)}{4(\Delta_T^2 + \epsilon^2)^2}\,\vartheta^2 + \mathcal{O}(\vartheta^4)\,.
$$

In this temperature regime, the dynamics is dominated by the system's Hamiltonian, and the environmental coupling is a perturbation. In leading order (one-phonon exchange), the relaxation rate γ_{r} and the decoherence rate γ grow linearly with ϑ,

$$
\gamma_{\mathrm{r}} = \sin^2\varphi\,\vartheta\,, \qquad \text{and} \qquad \gamma = \tfrac{1}{2}\gamma_{\mathrm{r}} + \cos^2\varphi\,\vartheta\,, \tag{22.71}
$$

where $\tan\varphi = \Delta_T/\epsilon$.[8] The trigonometric factors $\sin^2\varphi$ and $\cos^2\varphi$ determine the weights of the transverse and longitudinal coupling in the rates, respectively. On the other hand, the oscillation frequency decreases with growing ϑ. Since damping strength K and temperature T are combined in the scaled thermal frequency ϑ given in Eq. (22.46), perturbative treatment of the damping breaks down even for small K when T is sufficiently large. As temperature is raised, multi-phonon exchange becomes increasingly important.

In the high-temperature regime T well above T_2 we obtain from the cubic equation

$$
\gamma_{\mathrm{r}} = \frac{\Delta_T^2}{\vartheta} + \Delta_T^2(\Delta_T^2 - \epsilon^2)\,\frac{1}{\vartheta^3} + \mathcal{O}(1/\vartheta^5)\,,
$$

$$
\gamma = \vartheta - \frac{\Delta_T^2}{2\vartheta} - \frac{\Delta_T^2(\Delta_T^2 - \epsilon^2)}{2}\,\frac{1}{\vartheta^3} + \mathcal{O}(1/\vartheta^5)\,, \tag{22.72}
$$

$$
\Omega^2 = \epsilon^2 + \frac{\Delta_T^2(4\epsilon^2 - \Delta_T^2)}{4\epsilon^2}\,\frac{1}{\vartheta^2} + \mathcal{O}(1/\vartheta^4)\,.
$$

In this regime, the dynamics is dominated by the noise force. Now, the dephasing and relaxation rates show converse trends. The dephasing rate γ continues growing linearly with ϑ in leading order of ϑ, whereas the relaxation rate γ_{r} drops inversely with ϑ. In the strict sense, the relaxation rate shows Kondo-like behavior $\propto T^{2K-1}$. As a result, decoherence is much faster than relaxation. In addition, the oscillation frequency approaches the bias frequency as T is increased.

The location of the crossover between the system-controlled behavior (22.70) and the noise-controlled behavior (22.72) is determined by the zeros of the discriminant of the cubic equation (22.63).

---

[8]In difference to Eq. (3.131) here we define φ in terms of the dressed tunneling coupling Δ_T.

Laplace inversion of the expression (22.61) for $\langle \sigma_z(\lambda) \rangle$ yields in the time regime

$$\langle \sigma_z \rangle_t = a_1 e^{-\gamma_r t} + [(1 - a_1 - \langle \sigma_z \rangle_\infty) \cos \Omega t + a_2 \sin \Omega t] e^{-\gamma t} + \langle \sigma_z \rangle_\infty ,$$

with the initial condition $\langle \sigma_z \rangle_{t=0} = 1$ and $\langle \dot{\sigma}_z \rangle_{t=0} = 0$. The amplitudes are given by

$$
\begin{aligned}
a_1 &= [\Omega^2 + \gamma^2 - \Delta_T^2 - (\Omega^2 + \gamma^2)\langle \sigma_z \rangle_\infty]/D , \\
a_2 &= [(\gamma_r - \gamma)a_1 + \gamma(1 - \langle \sigma_z \rangle_\infty)]/\Omega ,
\end{aligned}
\tag{22.73}
$$

where $D = \Omega^2 + (\gamma - \gamma_r)^2$. Similarly, we obtain from Eq. (22.62)

$$\langle \sigma_x \rangle_t = b_1 e^{-\gamma_r t} + [-(b_1 + \langle \sigma_x \rangle_\infty) \cos \Omega t + b_2 \sin \Omega t] e^{-\gamma t} + \langle \sigma_x \rangle_\infty , \tag{22.74}$$

with the initial conditions $\langle \sigma_x \rangle_{t=0} = 0$ and $\langle \dot{\sigma}_x \rangle_{t=0} = \pi K \Delta_T^2/\Delta$. The amplitudes b_1 and b_2 are found to read

$$
\begin{aligned}
b_1 &= \left(\frac{\epsilon \Delta_T^2}{\Delta} - \frac{\epsilon^2(\gamma^2 + \Omega^2)}{\gamma^2 + \Omega^2 - \Delta_T^2}\langle \sigma_x \rangle_\infty \right)\frac{1}{D} , \\
b_2 &= \left(\pi K \frac{\Delta_T^2}{\Delta} + (\gamma_r - \gamma)b_1 - \gamma \langle \sigma_x \rangle_\infty \right)\frac{1}{\Omega} .
\end{aligned}
\tag{22.75}
$$

The expressions (22.70) – (22.75) describe the dynamics of the expectation values $\langle \sigma_z \rangle_t$ and $\langle \sigma_x \rangle_t$ for $K \ll 1$ in the Markov or white-noise regime $T \gtrsim T_b$.

Consider next the spin correlation function. In the NIBA, the correlation term $R(t)$ in Eq. (21.58) vanishes. We find from Eq. (21.59) with Eqs. (21.79) and (21.81)

$$\tilde{S}_z(\omega) = 2 \operatorname{Re} \frac{-i\omega - \langle \sigma_z \rangle_\infty \hat{K}_z^{(a)}(-i\omega)}{-i\omega [-i\omega + \hat{K}_z^{(s)}(-i\omega)]} , \tag{22.76}$$

where the kernels are given in Eq. (22.59), and $\langle \sigma_z \rangle_\infty$ in Eq. (21.86). The same form has been found in Ref. [456] using a perturbative Liouville relaxation method. The dynamical susceptibility $\tilde{\chi}_z(\omega)$ in the NIBA is given below in Eq. (23.18). We have already discussed after Eq. (21.84) that the NIBA gives inconsistent results for a biased system at low T and small frequencies (long times). For instance, the weight of the quasi-elastic line at the origin described by Eq. (22.76) is $(\epsilon/\Omega)^2 - \langle \sigma_z \rangle_\infty^2$. With the NIBA form (21.86) for $\langle \sigma_z \rangle_\infty$, the weight becomes negative at sufficiently low T, which is unphysical.

For temperatures in the Markov regime (22.47), the NIBA is correct for all frequencies. Substituting the NIBA kernels (22.60) into Eq. (22.76) and decomposing the resulting expression into partial fractions, we obtain the spectral spin correlation function as a superposition of three Lorentzians [we use $\langle \sigma_z \rangle_\infty = (\epsilon/\Omega) \tanh(\beta \hbar \Omega/2)$],

$$\tilde{S}_z(\omega) = \sin^2 \varphi \sum_{\zeta = \pm 1} \frac{\gamma + (\Omega - \zeta\omega)\tan \phi}{(\omega - \zeta\Omega)^2 + \gamma^2} + \frac{\cos^2 \varphi}{\cosh^2(\frac{1}{2}\beta\hbar\Omega)}\frac{2\gamma_r}{\omega^2 + \gamma_r^2} , \tag{22.77}$$

where $\cos \varphi = \epsilon/\Omega$, and $\tan \phi = \gamma/\Omega + (\gamma_r/\Omega)(\epsilon^2 - \langle \sigma_z \rangle_\infty^2 \Omega^2)/(\Omega^2 - \epsilon^2)$. The inelastic peaks of width γ centered at $\omega = \pm\Omega$ are signatures of coherent tunneling at frequency

Ω with dephasing time $1/\gamma$. The quasi-elastic peak at $\omega = 0$ represents incoherent relaxation to the thermal equilibrium state on the time scale $1/\gamma_r$.

For completeness, it should be mentioned that, in difference to $\tilde{S}_z(\omega)$ the spectral function $\tilde{C}_z^{\pm}(\omega)$ has additional weight factors $\kappa_{\text{inel}}^{\pm}(\zeta) = 1 \pm \zeta \tanh(\frac{1}{2}\beta\hbar\Omega)$ for the inelastic peaks at $\omega = \zeta\Omega$, as follows from the relation (21.25). These factors fix the relative magnitudes of the absorption and emission lines. Disregarding the terms $\propto \tan\phi$ we have

$$\tilde{C}_z^{\pm}(\omega) = \sin^2\varphi \sum_{\zeta=\pm 1} \frac{[1 \pm \zeta \tanh(\frac{1}{2}\beta\hbar\Omega)]\gamma}{(\omega - \zeta\Omega)^2 + \gamma^2} + \frac{\cos^2\varphi}{\cosh^2(\frac{1}{2}\beta\hbar\Omega)} \frac{2\gamma_r}{\omega^2 + \gamma_r^2}. \qquad (22.78)$$

The form (22.78) with perturbative (one-phonon) expressions for γ and γ_r are also found with the Redfield theory [32] presented in Subsection 2.3.1.

The inelastic peaks of $\tilde{S}_z(\omega)$ and $\tilde{C}_z^{\pm}(\omega)$ merge near $T = T_1$ with the quasi-elastic peak. In the temperature regime well above T_2 we have a single Lorentzian [426]

$$\tilde{C}_z^{\pm}(\omega) = \tilde{S}_z(\omega) = \frac{2\gamma_r}{\omega^2 + \gamma_r^2}. \qquad (22.79)$$

The line width γ_r given in Eq. (22.72). In the Kondo regime $T \gg T_2$, the width γ_r decreases as temperature is increased.

Finally, one remark concerning the scaling limit (18.57) is appropriate. We see from the above results that the expectation value $\langle\sigma_z\rangle_t$ is universal in the sense specified in Subsection 18.2.2. In contrast, the expectation values $\langle\sigma_x\rangle_t$ and $\langle\sigma_y\rangle_t$ are non-universal since they have an overall factor $\Delta_r/\Delta \propto (\Delta_r/\omega_c)^K$. Hence, the off-diagonal elements of the reduced density matrix vanish in the scaling limit (18.57).

22.3 Weak quantum noise in the biased TSS

For weak damping and temperature in the range of $T_b = \hbar(\Delta_{\text{eff}}^2 + \epsilon^2)^{1/2}/k_B$ or smaller, the Markov assumption (22.45) and the corresponding form (22.48) for $Q(\tau)$ is not valid, and the justification of the noninteracting-blip approximation is questionable. For a symmetric system, the NIBA is found to be inconsistent for the coherence $\langle\sigma_x\rangle_t$, as we have remarked already in item (a) in Section 21.3, and after Eq. (21.87). For nonzero bias, the NIBA breaks down for the same reason at temperatures below T_b for the population $\langle\sigma_z\rangle_t$ and the coherence $\langle\sigma_x\rangle_t$, as well as for the correlation functions. In all these cases, the interblip correlations are already relevant for the one-phonon process. This is confirmed by the subsequent study. The bias energy is important for tunneling systems in crystals with higher defect concentration because of strain fields and in amorphous materials, e.g., in dielectric and metallic glasses.

We now consider the regime $T \lesssim T_b = \hbar(\Delta_{\text{eff}}^2 + \epsilon^2)^{1/2}/k_B$ for weak damping. Here Δ_{eff} is the effective tunneling coupling which is given in Eq. (18.34) for a super-Ohmic, and in Eq. (18.37) for an Ohmic bath, and which is equal to Δ in the sub-Ohmic case. We shall treat the interblip correlations systematically to first order in the coupling strength. To be general, we assume that the spectral density $G_{\mathrm{lf}}(\omega)$ has the power-law form (18.33).

22.3.1 The one-boson self-energy

Interestingly enough, the series (21.76) and (21.78) for the kernels $K_z^{(\mathrm{s,a})}(t,t')$ and $K_x^{(\mathrm{s,a})}(t,t')$ can be summed in analytic form for weak damping with a strategy similar to that presented in Subsection 19.1.6[457, 448, 445]. We proceed by considering the blip correlations to linear order in $\Lambda_{j,k}$ in the series expression (21.76). Switching to the Laplace transform, the self energy $\hbar\Sigma_z^{(\mathrm{s})}(\lambda)$ is found to read

$$
\begin{aligned}
\Sigma_z^{(\mathrm{s})}(\lambda) \;=\; & \Delta_{\mathrm{eff}}^2 \int_0^\infty d\tau\, e^{-\lambda\tau} \cos(\epsilon\tau)\big[\,q - Q'\,\big] \\
& - \lambda \sum_{n=2}^\infty (-\Delta_{\mathrm{eff}}^2)^n \prod_{j=1}^n \left(\int_0^\infty d\tau_j\, ds_j\, e^{-\lambda(s_j+\tau_j)} \right) \\
& \times \Lambda_{n1} \sin(\epsilon\tau_1)\sin(\epsilon\tau_n)\prod_{k=2}^{n-1}\cos(\epsilon\tau_k) \,.
\end{aligned}
\tag{22.80}
$$

Here we have included the full (nonperturbative) adiabatic dressing of the tunneling coupling. The first term is the self-energy in the NIBA, and q is a counterterm so that the oscillation frequency at $T=0$ will emerge as $\Delta_{\mathrm{b}} = \sqrt{\Delta_{\mathrm{eff}}^2 + \epsilon^2}$. The residual term represents the contribution of the one-boson interblip correlations. In diagrammatic language, the series is the sum of all *irreducible* multi-blip diagrams in which the first blip interacts with the last blip via one-boson exchange. The insertion of noninteracting blips in between accounts for all intermediate uncorrelated tunneling events. The series of the insertions can be summed and yields [cf. Eq. (21.11)]

$$
P_0(s) \;=\; \langle \sigma_z \rangle_s^{(0)} \;=\; \frac{\epsilon^2}{\Delta_{\mathrm{b}}^2} + \frac{\Delta_{\mathrm{eff}}^2}{\Delta_{\mathrm{b}}^2}\cos(\Delta_{\mathrm{b}}s)\,, \qquad \Delta_{\mathrm{b}} = \sqrt{\Delta_{\mathrm{eff}}^2 + \epsilon^2}\,,
\tag{22.81}
$$

where s is the interval between the first and the last blip. Thus we get

$$
\begin{aligned}
\Sigma_z^{(\mathrm{s})}(\lambda) \;=\; & \Delta_{\mathrm{eff}}^2 \int_0^\infty d\tau\, e^{-\lambda\tau} \cos(\epsilon\tau)\big[\,q - Q'(\tau)\,\big] \\
& - \Delta_{\mathrm{eff}}^4 \int_0^\infty d\tau_2\, ds\, d\tau_1\, e^{-\lambda(\tau_2+s+\tau_1)} \sin(\epsilon\tau_2)\,P_0(s)\sin(\epsilon\tau_1) \\
& \times \big[\, Q'(\tau_1+\tau_2+s) + Q'(s) - Q'(\tau_1+s) - Q'(\tau_2+s)\,\big]\,.
\end{aligned}
\tag{22.82}
$$

Upon using the spectral representation (18.44) and interchanging the frequency integral and the time integrals [457], we get

$$
\Sigma_z^{(\mathrm{s})}(\lambda) \;=\; -\frac{2\Delta_{\mathrm{eff}}^2}{\Delta_{\mathrm{b}}^2}\,\frac{\Delta_{\mathrm{eff}}^2\lambda^3 u(\lambda) + \epsilon^2(\lambda^2 + \Delta_{\mathrm{b}}^2)\,\mathrm{Re}\,\{(\lambda - i\Delta_{\mathrm{b}})u(\lambda + i\Delta_{\mathrm{b}})\}}{(\lambda^2 + \epsilon^2)^2}\,.
\tag{22.83}
$$

where the function $u(z)$ is given by

$$u(z) = \frac{1}{2} \int_0^\infty d\omega \frac{G_{\mathrm{lf}}(\omega)}{\omega^2 + z^2} \left[\coth(\tfrac{1}{2}\beta\hbar\omega) - 1 \right] . \tag{22.84}$$

The function $u(z)$ is related to the function $v(y)$ occurring in the imaginary-time approach [cf. Eq. (19.43)] by analytic continuation, $u(z) = v(y = -iz)$.

Similarly, the selfenergy $\Sigma_z^{(\mathrm{a})}(\lambda)$ represents the correlations between the first sojourn and the last blip with insertion of all possible tunneling events in between. The corresponding expression is

$$\begin{aligned}
\Sigma_z^{(\mathrm{a})}(\lambda) = {} & \Delta_{\mathrm{eff}}^2 \int_0^\infty d\tau\, e^{-\lambda\tau} \sin(\epsilon\tau)\, Q''(\tau) \\
& - \Delta_{\mathrm{eff}}^4 \int_0^\infty d\tau_2\, ds\, d\tau_1\, e^{-\lambda(\tau_2+s+\tau_1)} \sin(\epsilon\tau_2)\, P_0(s) \cos(\epsilon\tau_1) \qquad (22.85) \\
& \times \left[Q''(\tau_1 + \tau_2 + s) - Q''(\tau_1 + s) \right] .
\end{aligned}$$

Use of the spectral representation (18.44) for $Q''(\tau)$ gives

$$\Sigma_z^{(\mathrm{a})}(\lambda) = i\,\epsilon\, \frac{\Delta_{\mathrm{eff}}^2(\lambda^2 + \Delta_{\mathrm{b}}^2)}{\Delta_{\mathrm{b}}(\lambda^2 + \epsilon^2)} \left[w(\lambda + i\Delta_{\mathrm{b}}) - w(\lambda - i\Delta_{\mathrm{b}}) \right] , \tag{22.86}$$

where

$$w(z) = \frac{1}{2} \int_0^\infty d\omega \frac{G_{\mathrm{lf}}(\omega)}{\omega(\omega^2 + z^2)} . \tag{22.87}$$

Corresponding expressions are found from Eq. (21.78) for the selfenergies $\Sigma_x^{(\mathrm{s,a})}(\lambda)$,

$$\begin{aligned}
\Sigma_x^{(\mathrm{s})}(\lambda) = {} & \frac{\Delta_{\mathrm{eff}}^2}{\Delta} \int_0^\infty d\tau\, e^{-\lambda\tau} \cos(\epsilon\tau)\, Q''(\tau) \\
& - \frac{\Delta_{\mathrm{eff}}^4}{\Delta} \int_0^\infty d\tau_2\, ds\, d\tau_1\, e^{-\lambda(\tau_2+s+\tau_1)} \cos(\epsilon\tau_2)\, P_0(s) \cos(\epsilon\tau_1) \qquad (22.88) \\
& \times \left[Q''(\tau_1 + \tau_2 + s) - Q''(\tau_1 + s) \right] ,
\end{aligned}$$

$$\begin{aligned}
\Sigma_x^{(\mathrm{a})}(\lambda) = {} & \frac{\Delta_{\mathrm{eff}}^2}{\Delta} \int_0^\infty d\tau\, e^{-\lambda\tau} \sin(\epsilon\tau) \left(q - Q'(\tau) \right) \\
& + \frac{\Delta_{\mathrm{eff}}^4}{\Delta} \int_0^\infty d\tau_2\, ds\, d\tau_1\, e^{-\lambda(\tau_2+s+\tau_1)} \cos(\epsilon\tau_2)\, P_0(s) \sin(\epsilon\tau_1) \qquad (22.89) \\
& \times \left[Q'(\tau_1 + \tau_2 + s) + Q'(s) - Q'(\tau_1 + s) - Q'(\tau_2 + s) \right] .
\end{aligned}$$

We conclude with the remark that these expressions for the self-energies $\Sigma_x^{(\mathrm{s,a})}(\lambda)$ can be cast in forms which are similar to the expressions (22.86) and (22.83).

22.3.2 Populations and coherences (super-Ohmic and Ohmic)

To determine the dynamics, we substitute the expressions (22.83) and (22.86) into Eq. (21.79) and study the singularities of $\langle \sigma_z(\lambda) \rangle$. First, we see that the self-energy term shifts the two complex-conjugate poles from the imaginary axis into the left half-plane to $\lambda = -\gamma \pm i\Omega$. These poles render damped oscillations in the time domain, and inelastic spectral lines in the frequency regime. Furthermore, there are additional poles at $\lambda = -\gamma_r$ and at $\lambda = 0$. The former is the relaxation pole. It contributes in the time regime a term describing exponential relaxation with the rate γ_r, and in frequency space a quasi-elastic peak of width γ_r centered at $\omega = 0$. The residuum of the latter pole is the equilibrium value $\langle \sigma_z \rangle_\infty$. It turns out as[9]

$$\langle \sigma_z \rangle_\infty = \frac{\Sigma_z^{(a)}(0)}{\Sigma_z^{(s)}(0)} = \frac{\epsilon}{\Delta_b} \tanh\left(\frac{\hbar \Delta_b}{2 k_B T}\right) = \cos\varphi\, \langle \tau_z \rangle_\beta \,, \tag{22.90}$$

which is the weak-damping form anticipated already in Eq. (21.88).

With the one-boson exchange contribution taken into account, the oscillation frequency Ω, relaxation rate γ_r and dephasing rate γ read

$$\Omega^2 = \Delta_b^2[1 - 2\sin^2\varphi\, \mathrm{Re}\, u(i\Delta_b)] \,,$$

$$\gamma_r = \frac{\Delta_{\mathrm{eff}}^2}{\Delta_b^2} S_\xi(\Delta_b) = \sin^2\varphi\, S_\xi(\Delta_b) \,, \tag{22.91}$$

$$\gamma = \frac{1}{2}\gamma_r + \frac{\epsilon^2}{\Delta_b^2} S_\xi(0) = \frac{1}{2}\gamma_r + \cos^2\varphi\, \gamma_{\mathrm{pd}} \,.$$

Here $S_\xi(\omega)$ is the power spectrum of the random force introduced in Subsection 22.2.1,

$$S_\xi(\omega) = \frac{\pi}{2} G_{\mathrm{lf}}(\omega) \coth\left(\frac{\hbar\omega}{2 k_B T}\right) \,. \tag{22.92}$$

The factors $\cos^2\varphi$ and $\sin^2\varphi$ allocate the weights of the longitudinal ($\propto \cos\varphi$) and transverse ($\propto \sin\varphi$) coupling in the Hamiltonian (3.138). The relaxation rate γ_r represents the inverse time scale for relaxation of diagonal states of the RDM to the thermal equilibrium state. It is proportional to $\sin^2\varphi$ and to the spectral power of the random force at the level splitting frequency. The decoherence rate γ characterizes the decay of off-diagonal elements of the RDM and is the inverse time scale for loss of phase coherence between the two states. It is a combination of the relaxation contribution $\frac{1}{2}\gamma_r$ and the "pure-dephasing" contribution due to longitudinal noise at zero frequency

$$\gamma_{\mathrm{pd}} = S_\xi(0) = \begin{cases} \dfrac{\pi}{2} \lim\limits_{\omega \to 0} G_{\mathrm{lf}}(\omega) \,, & T = 0 \\[2ex] \pi \dfrac{k_B T}{\hbar} \lim\limits_{\omega \to 0} \dfrac{G_{\mathrm{lf}}(\omega)}{\omega} \,, & T > 0 \end{cases} \,. \tag{22.93}$$

---

[9]In this subsection we define φ in terms of the dressed tunneling amplitude, $\tan\varphi = \Delta_{\mathrm{eff}}/\epsilon$.

The longitudinal coupling $\propto \cos\varphi$ leads to fluctuations of the eigenenergies which results in a random phase between the two eigenstates. Since no energy is exchanged, the power spectrum at zero frequency appears in the pure dephasing rate. This expression is only meaningful if the noise power does not diverge in the limit $\omega \to 0$ (see the discussion in Subsection 22.4). The level splitting $\hbar\Omega$ is renormalized due to transverse noise. The shift of the level splitting is the analogue of the Lamb shift.

For Ohmic dissipation, we obtain the expressions

$$
\begin{aligned}
\Omega^2 &= \Delta_{\rm b}^2 \left\{ 1 + 2K\sin^2\varphi\,[\,{\rm Re}\,\psi(i\hbar\Delta_{\rm b}/2\pi k_{\rm B}T) - \ln(\hbar\Delta_{\rm b}/2\pi k_{\rm B}T)]\right\}, \\
\gamma_{\rm r} &= \pi K \sin^2\varphi \coth(\hbar\Delta_{\rm b}/2k_{\rm B}T)\,\Delta_{\rm b}, \\
\gamma &= \tfrac{1}{2}\gamma_{\rm r} + \cos^2\varphi\,\gamma_{\rm pd}, \\
\gamma_{\rm pd} &= \vartheta = 2\pi K k_{\rm B}T/\hbar.
\end{aligned}
\tag{22.94}
$$

Evidently, the expression for Ω^2 coincides with the results (19.42) and (19.51) of the imaginary-time approach. Furthermore, in the temperature range $k_{\rm B}T \gtrsim \hbar\Delta_{\rm b}$, the results for γ and $\gamma_{\rm r}$ smoothly map on those of the Markov regime given in Eq. (22.70). In summary, these results together with those presented in Subsection 22.2.3 yield a faithful quantitative account of the dynamics of the biased Ohmic TSS in the entire temperature regime.

In the sub-Ohmic regime $0 < s < 1$ the pure dephasing rate $\gamma_{\rm pd}$ vanishes for zero temperature, whereas it is infinity for finite T. These drastically different behaviors become clear from the discussion of the exactly treatable pure dephasing regime given in Section 22.4.

The transform (21.79) with (22.86) is easily inverted to obtain $\langle\sigma_z\rangle_t$. We find[10]

$$
\begin{aligned}
\langle\sigma_z\rangle_t &= [\epsilon^2/\Omega^2 - \langle\sigma_z\rangle_\infty]\,{\rm e}^{-\gamma_{\rm r}t} + \langle\sigma_z\rangle_\infty \\
&\quad + (\Delta_{\rm eff}^2/\Omega^2)\cos(\Omega t)\,{\rm e}^{-\gamma t} \\
&\quad + [\,(\gamma_{\rm r}\epsilon^2 + \gamma\Delta_{\rm eff}^2)/\Omega^3 - \gamma_{\rm r}\langle\sigma_z\rangle_\infty/\Omega\,]\sin\Omega t\,{\rm e}^{-\gamma t}.
\end{aligned}
\tag{22.95}
$$

The spectral spin correlation function $\tilde{S}_z(\omega)$ is found in the form (22.77) in which $\langle\sigma_z\rangle_\infty$ is given by Eq. (22.90). With the expression (22.90), the weight of the quasi-elastic peak is $(\epsilon^2/\Omega^2)[\,1 - \tanh^2(\hbar\Omega/2k_{\rm B}T)\,]$, which is positive down to $T = 0$. Thus the NIBA flaw that the weight of the quasi-elastic peak becomes negative at low T, which is discussed below Eq. (22.76), is dissolved by the interblip correlations.

Following similar lines, it is also straightforward to calculate $\langle\sigma_x\rangle_t$. We obtain

$$
\langle\sigma_x\rangle_t = \left[\frac{\epsilon\Delta_{\rm eff}^2}{\Delta\Omega^2} - \langle\sigma_x\rangle_\infty\right]{\rm e}^{-\gamma_{\rm r}t} + \left[-\frac{\epsilon\Delta_{\rm eff}^2}{\Delta\Omega^2}\cos\Omega t + b_2\sin\Omega t\right]{\rm e}^{-\gamma t} + \langle\sigma_x\rangle_\infty, \tag{22.96}
$$

where

$$
b_2 = (\Delta_{\rm eff}^2/\Delta\Omega)[\,\pi K + \epsilon\,(\gamma_{\rm r} - \gamma)/\Omega^2\,] - \gamma_{\rm r}\langle\sigma_x\rangle_\infty/\Omega. \tag{22.97}
$$

---

[10]We deliberately substitute Ω for $\Delta_{\rm b}$ in the amplitudes.

The equilibrium value is

$$\langle \sigma_x \rangle_\infty = \frac{\Delta_{\text{eff}}^2}{\Delta \Delta_b} \tanh \left(\frac{\hbar \Delta_b}{2 k_B T} \right) \approx \frac{\Delta_{\text{eff}}^2}{\Delta \Omega} \tanh \left(\frac{\hbar \Omega}{2 k_B T} \right). \qquad (22.98)$$

The above expressions for $\langle \sigma_z \rangle_t$ and $\langle \sigma_x \rangle_t$ match in the Ohmic case with the expressions (22.70) – (22.75) at temperatures $T \gtrsim T_b$. Thus, we have found in the Ohmic case for $K \ll 1$ analytic expressions for the dynamics in the entire temperature range of interest. Interestingly, the dynamical approach presented here yields the correct expressions for $\langle \sigma_z \rangle_\infty$ and $\langle \sigma_x \rangle_\infty$. They coincide with the forms (21.88) and (21.89) found by a simple thermostatic consideration.

An insightful quantity is the difference of the populations of the ground state and the excited state. In the eigenbasis of H_{TSS} we have $\langle \tau_z \rangle \equiv \langle |g><g| \rangle - \langle |e><e| \rangle$. By means of the rotations (3.133) and (3.130) we may relate $\langle \tau_z \rangle_t$ to $\langle \sigma_x \rangle_t$ and $\langle \sigma_z \rangle_t$ as

$$\langle \tau_z \rangle_t = \sin \varphi \, \langle \sigma_x \rangle_t + \cos \varphi \, \langle \sigma_z \rangle_t. \qquad (22.99)$$

With the above results for the expectation values $\langle \sigma_x \rangle_t$ and $\langle \sigma_z \rangle_t$ we get upon disregarding terms of order γ_r / Ω

$$\langle \tau_z \rangle_t = \left(\epsilon / \Omega - \langle \tau_z \rangle_\beta \right) e^{-\gamma_r t} + \langle \tau_z \rangle_\beta. \qquad (22.100)$$

This expression points out again the role of the rate γ_r. It describes relaxation from the initial value ϵ / Ω to the equilibrium value $\langle \tau_z \rangle_\beta = \tanh(\hbar \Omega / 2 k_B T)$.

Finally, let us take a look at the decay of the off-diagonal elements or coherences of the TSS density matrix in energy representation. These are the expectation values of the flip operators $\tau_\pm$. The latter are expressed in terms of the σ-matrices by the relation (3.135). With the expressions (22.95), (22.96) and (21.10) we find

$$\langle \tau_\pm \rangle_t = \frac{1}{2} \left(\cos \varphi \, \langle \sigma_x \rangle_t \pm i \, \langle \sigma_y \rangle_t - \sin \varphi \, \langle \sigma_z \rangle_t \right) = -\frac{\Delta}{2\Omega} e^{\mp i \Omega t} e^{-\gamma t}. \qquad (22.101)$$

In the latter form we have disregarded amplitudes of order γ / Ω and γ_r / Ω. Observe that in the result (22.101) relaxation terms $\propto e^{-\gamma_r t}$ have dropped out. This shows again clearly that $1/\gamma$ is the time scale for dephasing.

The analysis of the dynamics can be generalized to the case of time-dependent external fields coupled to σ_z and to σ_x by employing the substitutions given in Subsection 23.1.1. The dynamics of $\langle \sigma_z \rangle_t$ under influence of a monochromatic high-frequency field coupled to σ_z has been studied in Ref. [448]. Alternative approaches based on second-order perturbation in the TSS-bath coupling have been frequently employed in this parameter regime. In these treatments, the adiabatic renormalization of the bare tunneling matrix element is usually disregarded. For a discussion of $\langle \sigma_z \rangle_t$ and $\langle \tau_z \rangle_t$ under monochromatic low-frequency driving, we refer to Refs. [459, 460].

22.4 Pure dephasing

The dynamics induced by the spin-boson Hamiltonian (18.15) is easily solved if we disregard the tunneling term. For $\Delta = 0$, we have $\tau_j = \sigma_j$, where $j = x, y, z$, and the observables of interest are the coherences. These are the expectation values of the flip operators $\sigma_\pm = \frac{1}{2}(\sigma_x \pm i\,\sigma_y)$,

$$\langle \sigma_\pm \rangle_t \;=\; \mathrm{tr}\,[\sigma_\pm(t)\hat\rho(0)]\,. \tag{22.102}$$

In the absence of tunneling, the operator $\sigma_\pm(t)$ obeys the equation of motion

$$\dot\sigma_\pm(t) \;=\; (i/\hbar)[H_{\mathrm{SB}},\sigma_\pm(t)]_{\Delta=0} \;=\; \mp i\,[\epsilon + \mathfrak{E}(t)/\hbar]\sigma_\pm(t)\,, \tag{22.103}$$

where $\mathfrak{E}(t) = \sum_\alpha \hbar\lambda_\alpha[b_\alpha\,e^{-i\omega_\alpha t} + b_\alpha^\dagger\,e^{i\omega_\alpha t}]$ is the polarization energy. Hence we have

$$\langle \sigma_\pm \rangle_t \;=\; \langle \sigma_\pm \rangle_0\,e^{\mp i\epsilon t}\left\langle \exp\left\{\mp\frac{i}{\hbar}\int_0^t dt'\,\mathfrak{E}(t')\right\}\right\rangle$$

$$\tag{22.104}$$

$$\;=\; \langle \sigma_\pm \rangle)_0\,e^{\mp i\epsilon t}\,\exp\left\{-\frac{1}{\hbar^2}\int_0^t dt'\int_0^{t'} dt''\langle\,\mathfrak{E}(t')\mathfrak{E}(t'')\rangle\right\}\,.$$

The second form holds for Gaussian statistics of the fluctuating polarization energy $\mathfrak{E}(t)$. With the thermal averages (3.39) and the definition (18.16) for the spectral density $G(\omega)$ one finds that, apart from the minus sign, the exponent is just the spectral representation (18.44) of the pair correlation function $Q(t)$ discussed in Section 18.2. Hence the off-diagonal elements of the RDM decay exactly as

$$\langle \sigma_\pm \rangle_t \;=\; \langle \sigma_\pm \rangle_0\,e^{\mp i\epsilon t}\,e^{-Q(t)}\,. \tag{22.105}$$

An analytic expression for $Q(t)$ for general spectral bath parameter s, general T and all times is given in Eq. (18.47). The corresponding zero temperature expression is stated in Eq. (20.49).

The loss of coherence process is characterized by three different time regimes:
(i) the core regime $0 < t < 1/\omega_c$, in which $Q(t)$ depends on the particular choice of the cutoff in the spectral density $G(\omega)$.
(ii) the quantum regime $1/\omega_c \ll t \ll \hbar\beta$, in which $Q(t)$ is essentially determined by the ground state of the reservoir $[\coth(\beta\hbar\omega/2) \approx 1]$, i.e, $Q(t) = Q_{T=0}(t)$.
(iii) the thermal regime $t \gg \hbar\beta$, in which $Q(t)$ is dominated by the classical state of the reservoir $[\coth(\beta\hbar\omega/2) \approx 2/\beta\hbar\omega]$. From this it is clear that $Q(t)$ in the thermal regime has an additional factor t as compared with $Q_{T=0}(t)$. Hence the loss of coherence is always faster in the time regime $t > \hbar\beta$ in relation to the loss in the regime $t < \hbar\beta$ [461]. At zero temperature, the quantum regime persists until time infinity.

Consider now first the case of Ohmic friction, $s = 1$. With the Ohmic form (20.49) for the pair correlation function $Q(t)$ we find in the quantum regime, including the initial slippage in the core regime,

$$|\langle \sigma_\pm \rangle_t| \;=\; |\langle \sigma_\pm \rangle_0|\,\frac{\cos[2K\arctan(\omega_c t)]}{[1 + (\omega_c t)^2]^K}\,, \qquad 0 \le t \ll \hbar\beta \tag{22.106}$$

with the limiting power-law drop

$$|\langle \sigma_\pm \rangle_t| \;=\; |\langle \sigma_\pm \rangle_0|\,\cos(\pi K)\,(\omega_c t)^{-2K}\,, \qquad 1/\omega_c \ll t \ll \hbar\beta\,, \tag{22.107}$$

The asymptotic behavior is characterized by exponential decay, as we infer from Eq. (22.48) with Eq. (22.46),

$$|\langle\sigma_{\pm}\rangle_t| = |\langle\sigma_{\pm}\rangle_0| \cos(\pi K) e^{-Q_{\text{adia}}} e^{-\gamma_{\text{pd}}t}, \qquad t \gg \hbar\beta. \qquad (22.108)$$

Here $\gamma_{\text{pd}} = S_\xi(0) = \vartheta$ is the pure dephasing rate introduced in Eq. (22.94). Thus the off-diagonal states of the RDM decay algebraically in the quantum regime, and the power of the decay law depends on the coupling strength. As $T \to 0$, the pure dephasing rate γ_{pd} drops to zero. As we now see from Eq. (22.107), the off-diagonal states dephase anyhow, but with slower than exponential, viz. algebraic drop.

In the super-Ohmic range $s > 1$ we find from Eq. (18.44) the limiting forms

$$\operatorname{Re} Q(t) = Q_{\text{adia}} - 2\delta_s \frac{\Gamma(s)\sin(\frac{1}{2}\pi s)}{s-1} \frac{1}{(\omega_{\text{ph}}t)^{s-1}}, \qquad \omega_c^{-1} \ll t \ll \hbar\beta, \quad (22.109)$$

$$\operatorname{Re} Q(t) = Q_{\text{adia}} + 4\delta_s \frac{\Gamma(s-2)\cos(\frac{1}{2}\pi s)}{\beta\hbar\omega_{\text{ph}}} (\omega_{\text{ph}}t)^{2-s}, \qquad t \gg \hbar\beta, \quad (22.110)$$

where $Q_{\text{adia}} = 2\delta_s \left[\Gamma(s)/(s-1)\right](\omega_c/\omega_{\text{ph}})^{s-1}$.

The reduction of the off-diagonal states $\propto e^{-Q_{\text{adia}}}$ by adiabatic dressing is quickly built up in the initial core regime $t \lesssim 1/\omega_c$. From the viewpoint of quantum state engineering, this is already a substantial loss of phase coherence. The drop depends strongly on the cutoff frequency ω_c. In the subsequent time regime $1/\omega_c < t < \hbar\beta$ coherent oscillations persist in which the amplitude roughly maintains a plateau value.

In the range $1 < s < 2$ there is finally in the thermal regime $t \gg \hbar\beta$ decay $\propto \exp[-\text{const} \times T(\omega_{\text{ph}}t)^{2-s}]$. In contrast, in the range $s > 2$ the plateau regime persists in practice until time infinity. The preliminary result (22.93) for the pure dephasing rate γ_{pd} is consistent with these findings. Since for $s > 1$ the decay is actually slower than exponential, γ_{pd} vanishes in this regime.

Let us finally discuss the sub-Ohmic regime $0 < s < 1$. In the limit $\omega_c \to \infty$ the core regime becomes arbitrarily narrow, and there is no reduction of the coherences in this regime. In the subsequent time regimes we have

$$\operatorname{Re} Q(t) = 2\delta_s \frac{\Gamma(s)\sin(\frac{1}{2}\pi s)}{1-s} (\omega_{\text{ph}}t)^{1-s}, \qquad 0 \ll t \ll \hbar\beta, \quad (22.111)$$

$$\operatorname{Re} Q(t) = 4\delta_s \frac{\Gamma(s-2)\cos(\frac{1}{2}\pi s)}{\beta\hbar\omega_{\text{ph}}} (\omega_{\text{ph}}t)^{2-s}, \qquad t \gg \hbar\beta. \quad (22.112)$$

The expressions (22.93) are consistent with these time behaviors. The pure dephasing rate γ_{pd} vanishes for $T = 0$, since the actual decay $\propto \exp[-\text{const} \times (\omega_{\text{ph}}t)^{1-s}]$ in the quantum regime is slower than exponential decay $\propto e^{-\gamma_{\text{pd}}t}$ would be. On the other hand, the pure dephasing rate γ_{pd} diverges for finite T, because the decay $\propto \exp[-\text{const} \times (T/\omega_{\text{ph}})(\omega_{\text{ph}}t)^{2-s}]$ in the thermal regime is actually faster than decay under $e^{-\gamma_{\text{pd}}t}$.

22.5 $1/f$ noise and decoherence

Experiments with superconducting qubits showed that the relaxation dynamics is dominated by Ohmic high frequency noise (ω near Ω), while the dominant source for dephasing at low T is slow flicker noise with roughly a $1/f$ power spectrum [117, 118]. The $1/f$ noise may have different origin depending on the system. A widely-used assumption is that the qubit is affected by impurities which essentially act as an ensemble of two-level systems [116]. The discrete nature of a TLS environment was observed before, e.g., in metals and in single-electron devices. There, the random or coherent switching causes conductance fluctuations and $1/f$ current noise. In charge qubits, the discrete fluctuators add to the fluctuations of the gate charge regulating the qubit.

Basically, a TLS environment is non-Gaussian. Therefore in general the effects of it can not simply be described in terms of the pair correlation function $Q(t)$. Rather the full qubit-TLS dynamics may be relevant which becomes apparent in the transient qubit dynamics and in saturation effects [462, 463]. Gaussian statistics is found to hold definitely only in the limit of very weak qubit-TLS coupling. Slow flicker noise becomes effective when there is a whole ensemble of thermal or non-thermal fluctuators with a roughly log-uniform distribution of flip rates [116, 464].

Let us now take a closer look to the effects of a TLS environment in Gaussian approximation. We start with the observation that the TLS coupling induces in the qubit Hamiltonian (3.138) the random polarization energy

$$\mathfrak{E}(t) \; = \; \sum_\alpha v_\alpha \, \sigma_{z,\alpha}(t) \; . \tag{22.113}$$

The power spectrum of the related noise, normalized as in Eq. (22.44), then is

$$S_{\text{TLS}}(\omega) \; = \; \frac{1}{2\hbar^2} \sum_\alpha v_\alpha^2 \tilde{C}_z^+(\omega) \; , \tag{22.114}$$

where $\tilde{C}_z^+(\omega)$ is the Fourier transform of the σ_z autocorrelation function $C_z^+(t)$ which is defined in Eq. (21.16). We now consider the different cases the slow noise comes from a randomly switching and from a coherent TLS environment.

22.5.1 $1/f$ noise from fluctuating background charges

One possibility is that the two-level systems are fluctuating background charges (BC) switching randomly between two states [116]. In the low-frequency regime of interest the spectral correlation function of the randomly switching BCs is given by the quasi-elastic Lorentzian (22.79). The power spectrum resulting from the back ground charges then is

$$S_{\text{BC}}(\omega) \; = \; \frac{1}{2\hbar^2} \sum_\alpha v_\alpha^2 \frac{2\gamma_{\text{r},\alpha}}{\omega^2 + \gamma_{\text{r},\alpha}^2} \; , \tag{22.115}$$

where $\gamma_{\text{r},\alpha}$ is the switching rate of two-state fluctuator α. Since in general there are many fluctuators, one must average over the distribution of coupling strengths and switching rates,

$$S_{\mathrm{BC}}(\omega) \;\propto\; \int dv\, d\gamma_{\mathrm r}\, P(v,\gamma_{\mathrm r})\, v^2\, \frac{2\gamma_{\mathrm r}}{\omega^2 + \gamma_{\mathrm r}^2} \;. \tag{22.116}$$

Various forms for the distribution function $P(v,\gamma_{\mathrm r})$ with uncorrelated v and $\gamma_{\mathrm r}$ have been analyzed in Refs. [464, 465]. Generally one should distinguish between results for specific samples and those averaged over a statistical ensemble of samples. This is particularly important when the qubit is influenced by a small number of BCs and hence there are sizeable sample-to-sample fluctuations. There is also the possibility that the bulk of the weights in v and in $\gamma_{\mathrm r}$ come from different BCs.

Consider now the simple case in which the switching rates are dispersed with a smooth distribution on a log scale, as in the standard tunneling model of glasses [119, 120, 121]. Putting $\gamma_{\mathrm r} \propto e^{-\ell/\ell_0}$, where ℓ may be thought of as the tunneling distance or any power of it, and assuming uniform distribution of ℓ over a range substantially larger than ℓ_0, we obtain precisely $1/f$ noise,

$$S_{\mathrm{BC}}(\omega) \;\propto\; \overline{v^2} \int d\ell\, \frac{2\gamma_{\mathrm r}(\ell)}{\omega^2 + \gamma_{\mathrm r}^2(\ell)} \;\propto\; \overline{v^2} \int \frac{d\gamma_{\mathrm r}}{\gamma_{\mathrm r}}\, \frac{2\gamma_{\mathrm r}}{\omega^2 + \gamma_{\mathrm r}^2} \;\propto\; \frac{\overline{v^2}}{|\omega|} \;. \tag{22.117}$$

Here $\overline{v^2}$ is the average squared coupling. This finding may now be compared with the power spectrum resulting from an oscillator bath. We obtain from Eq. (22.44) in the regime $\omega \ll k_{\mathrm B}T/\hbar$

$$S_\xi(\omega) = \frac{\pi k_{\mathrm B}T}{\hbar}\, \frac{G_{\mathrm{lf}}(|\omega|)}{|\omega|} \;. \tag{22.118}$$

From this we see that the expression (22.118) yields the $1/f$ power spectrum (22.117) precisely in the sub-Ohmic limiting case $s = 0$ in Eq. (18.46). The respective spectral density of the coupling is constant,

$$G_{\mathrm{lf}}(\omega) \;\;\to\;\; G_{\mathrm{BC}}(\omega) = 2\delta_0 \omega_{\mathrm{ph}}\, \Theta(\omega - \omega_{\mathrm{ir}}) \;. \tag{22.119}$$

Here we have subjoined an intrinsic low-frequency cut-off ω_{ir} for the $1/f$ noise spectrum. Since the slow $1/f$ noise sources are largely out of equilibrium, the bath temperature in Eq. (22.118) should be considered as a fit parameter $T_{\mathrm b}$. Upon introducing a suitable frequency scale $\omega_{1/\mathrm f}$, we then may write

$$S_{\mathrm{BC}}(\omega) = \frac{\omega_{1/\mathrm f}^2}{|\omega|}\, \Theta(\omega - \omega_{\mathrm{ir}}) \quad \text{with} \quad \omega_{1/\mathrm f} = \sqrt{2\pi\delta_0\, \omega_{\mathrm{ph}} k_{\mathrm B} T_{\mathrm b}/\hbar} \;. \tag{22.120}$$

22.5.2 $1/f$ noise from coherent two-level systems

The case in which the noise comes from a set of coherent two level systems is equally interesting. Now the dynamics of the dissipative TSS is embodied in the form (22.78) of the spectral spin correlation function $\tilde{C}_z^+(\omega)$. In addition, we take a continuous distribution $P(v,\epsilon,\Delta)$ of the coupling parameter v and the TSS parameters ϵ and Δ. At low T the emission line at $\omega = -\Omega$ is irrelevant since it is exponentially suppressed as $e^{-\beta\hbar\Omega}$. Then the power spectrum takes the form

$$S_{\mathrm{TLS}}(\omega) \;\propto\; \int dv\, d\epsilon\, d\Delta\, P(v, \epsilon, \Delta)$$

$$\times \left(\frac{2\gamma \sin^2 \varphi}{(\omega - \Omega)^2 + \gamma^2} + \frac{\cos^2 \varphi}{\cosh^2(\tfrac{1}{2}\beta\hbar\Omega)} \frac{2\gamma_{\mathrm{r}}}{\omega^2 + \gamma_{\mathrm{r}}^2} \right), \qquad (22.121)$$

where $\Omega = \sqrt{\Delta^2 + \epsilon^2}$. It is natural to assume that the distribution of v is uncorrelated with that of Δ and v. When the width γ_{r} of the quasielastic peak is small compared to the Rabi frequency Ω, the terms in Eq. (22.121) divide into low-frequency noise resulting from the quasielastic peak and high-frequency noise stemming from the inelastic peak.

If we now assume a log-uniform distribution of level splittings Δ or switching rates $\gamma_{\mathrm{r}} \propto \Delta^2$, the low-frequency behavior resulting from Eq. (22.121) is again $1/f$ noise as in Eq. (22.117). It is self-evident to choose again a log uniform distribution of level splittings, $P_\Delta(\Delta) \propto 1/\Delta$, for the inelastic contribution in Eq. (22.121). Taking the integral over Δ, we then obtain

$$S_{\mathrm{TLS}}(\omega) \;\propto\; \frac{1}{\omega} \int_0^\omega d\epsilon\, P_\epsilon(\epsilon)\,. \qquad (22.122)$$

The choice $P_\epsilon(\epsilon) \propto \epsilon$ readily yields Ohmic power spectrum $S_{\mathrm{TLS}}(\omega) \propto \omega$ at high frequency. Most interestingly, the integral over ϵ in the low-frequency contribution $S(\omega) \propto 1/|\omega|$ with the same assumption $P_\epsilon(\epsilon) \propto \epsilon$ yields a weight factor T^2.

In sum, with the distribution $P(\epsilon, \Delta) \propto \epsilon/\Delta$ the expression (22.121) simultaneously yields the limiting behaviors [118]

$$S_{\mathrm{TLS}}(\omega) = \begin{cases} a \left(\dfrac{k_{\mathrm{B}} T}{\hbar} \right)^2 \dfrac{1}{\omega} & \text{for} \quad 0 < \omega \ll \Omega\,, \\[2ex] a\,\omega & \text{for} \quad \omega \approx \Omega\,. \end{cases} \qquad (22.123)$$

These different spectral power-law forms with the same prefactor a intersect at $\omega = k_{\mathrm{B}} T/\hbar$. A connection between the low-frequency $1/f$ noise power relevant for dephasing and the Ohmic high-frequency noise power responsible for relaxation was observed in experiments with Josephson devices [468]. The above T^2 dependence of the strength of the $1/f$ noise is consistent with these experiments. This indicates relevance of a coherent TLS environment in Josephson qubit devices.

There are still open questions about the statistics of the low-frequency noise, and whether the average over the TLS parameters is universal or dependent on the particular sample.

22.5.3 Decoherence from $1/f$ noise

Consider now the transient coherent dynamics of the dissipative TSS in which the power spectrum of the BC ensemble is characterized by the form (22.120). In this case the expression (22.93) for the pure dephasing rate is not valid. In fact, the decay of coherences is given in generalization of Eq. (22.105) by the expression

$$\langle \sigma_\pm \rangle_t = \langle \sigma_\pm \rangle_0 \, e^{\mp i\,\Omega t} \, e^{-\gamma_r t/2} \, e^{-\cos^2\varphi \, Q(t)} \,. \tag{22.124}$$

In one-boson exchange approximation, the relaxation rate reads $\gamma_r = \sin^2\varphi \, S_{BC}(\Omega)$. With the power spectrum (22.120) the bath correlation function (18.44) in the time regime $\hbar\beta \ll t \ll 1/\omega_{ir}$ takes the form

$$\mathrm{Re}\, Q(t) = (\omega_{1/f} t)^2 \, |\ln(\omega_{ir} t)|/\pi \,. \tag{22.125}$$

For longitudinal noise, $\varphi = 0$, the expression (22.124) with (22.125) yields for the coherence decay quadratic time-dependence in the exponent (see also Ref. [466]),

$$|\langle \sigma_\pm \rangle_t| = |\langle \sigma_\pm \rangle_0| \, e^{-(\gamma^* t)^2} \qquad \text{with} \qquad \gamma^* = \omega_{1/f} \sqrt{\ln(\omega_{1/f}/\omega_{ir})/\pi} \,. \tag{22.126}$$

In contrast, transverse noise, $\varphi = \pi/2$, leads to exponential decay

$$|\langle \sigma_\pm \rangle_t| = |\langle \sigma_\pm \rangle_0| \, e^{-\gamma t} \qquad \text{with} \qquad \gamma = \kappa \omega_{1/f}^2/\Omega \,. \tag{22.127}$$

The one-phonon exchange contribution gives $\gamma = \frac{1}{2}\gamma_r = \frac{1}{2}\omega_{1/f}^2/\Omega$, and thus $\kappa = \frac{1}{2}$. The analysis of multi-phonon contributions to the rate shows that these are sensitive to the divergent low-frequency power spectrum and thus may dominate. It was found with a diagrammatic self-energy method that by consideration of higher-order contributions the factor κ is actually moved with logarithmic accuracy to [467]

$$\kappa \approx \ln(\omega_{1/f}^2/\omega_{ir}\Omega)/\pi \,. \tag{22.128}$$

Hence the ratio γ/γ_r is actually much larger than $\frac{1}{2}$. In the experiments reported in Ref. [123] for γ/γ_r the numerical value $\kappa \approx 3$ was found.

22.6 The Ohmic TSS at and close to the Toulouse point

For the special case $K = \frac{1}{2}$, the Ohmic TSS can be mapped on the Toulouse model, as we have discussed already in Subsection 19.2.2. The equations of motion for the imaginary-time Green's function of the d-level are solved without difficulties and yield the expression (19.77). The relevant equilibrium correlation functions can be expressed in terms of this function, as we shall see below in Subsection 22.6.6. Alternatively, we can directly sum the exact series expressions given in Subsections 21.2.3 and 21.2.4 [219]. As an advantage over the fermionic approach, the latter method facilitates the calculation of nonequilibrium conditional probabilities.

22.6.1 Grand-canonical sums of collapsed blips and sojourns

The reason that the path summation is possible in analytic form in the particular case $K = \frac{1}{2}$ can most easily be understood using the concept of *collapsed* blips and *collapsed* sojourns. In the corresponding charge picture, neighboring charges of opposite sign form *collapsed* dipoles which have vanishing dipole moment and hence are not interacting with other charges. The computation is done by putting $K = \frac{1}{2} - \kappa$

and eventually taking the limit $\kappa \to 0$. In the scaling limit (18.57) with (18.54), the bath influence factors $F_m^{(\pm)}$ and $F_{m,n}$ are given in Eqs. (21.57) and (21.61). The central point now is that the factor $\cos(\pi K) = \sin(\pi \kappa)$ occurring in the series expressions given in Subsection 21.2.3 is zero as $\kappa \to 0$. To obtain a non-vanishing contribution, each $\cos(\pi K)$-factor must come along with a factor which diverges at $K = \frac{1}{2}$, so that the combined expression is finite as $K \to \frac{1}{2}$. The singularity required is provided by the short-distance singular behavior of the intra-dipole interaction in the breathing mode integral of a dipole, $\lim_{\tau \to 0} e^{-Q'(\tau)} \approx (\omega_c \tau)^{-1+2\kappa}$. The dipole length τ may represent either a sojourn or a blip interval. The finite contribution of a blip (sojourn) dipole together with the associated $\cos(\pi K)$-factor is

$$I_{b(s)}(K = \tfrac{1}{2}) \equiv \lim_{\kappa \to 0} \frac{\Delta^2}{2} \cos[\pi(\tfrac{1}{2} - \kappa)] \int_0^\infty d\tau \, e^{-Q'(\tau)} B_{b(s)}(\tau) f_{b(s)}(\tau) \,. \qquad (22.129)$$

The bias factor for a blip dipole is $B_b(\tau) = e^{\pm i \epsilon \tau}$, and is unity for a sojourn dipole, $B_s(\tau) = 1$. The function $f_{b(s)}(\tau)$ represents the interaction factor of the blip (sojourn) dipole with the other charges. Observing that a blip (sojourn) dipole with zero dipole length does not interact with other charges, we have $f_{b(s)}(0) = 1$. Thus we obtain

$$I_{b(s)}(K = \tfrac{1}{2}) = \lim_{\kappa \to 0} \frac{\Delta^2}{2\omega_c} B_{b(s)}(0) f_{b(s)}(0) \sin(\pi \kappa) \Gamma(2\kappa) = \frac{\pi}{4} \frac{\Delta^2}{\omega_c} = \frac{\gamma}{2} \,. \qquad (22.130)$$

From this we see that a collapsed blip or sojourn dipole depends neither on the bias strength nor on temperature. The expression $I(K = \tfrac{1}{2})$ is half the effective frequency $\Delta_{\text{eff}}(K = \tfrac{1}{2}) = \gamma$ introduced in Eq. (19.62).

Of particular importance now is that collapsed dipoles are noninteracting. Therefore, the grand-canonical sum of these entities can be performed without difficulty.

During a wide sojourn interval the system may make visits of duration zero to either of the two blip states. We refer to these "blitz" visits of blip states as collapsed blips (CB). All the possibilities of these short visits within a wide sojourn interval form a grand-canonical ensemble of these entities. Taking into account the minus sign associated with each factor Δ^2, and the multiplicity of the possible intermediate blip and sojourn states,[11] the grand-canonical sum of noninteracting collapsed blips in the sojourn interval of length s adds up to the exponential factor

$$U_{\text{CB}}(s) = e^{-\gamma s} \,, \qquad (22.131)$$

which we refer to as CB form factor. Equivalently, during an extended stay in a blip state the system may make any number of visits of duration zero to a sojourn state. The ensemble of these short visits may be viewed as a noninteracting gas of collapsed sojourns (CS). The grand-canonical sum of these entities in the blip interval of length τ yields the CS form factor

---

[11]After a short visit to a sojourn state, the system must return to the same blip state, whereas after a short visit to a blip state, the system can return to the same or to the other sojourn state. This is the reason for the multiplicity factor 2 in the exponent of the CB form factor.

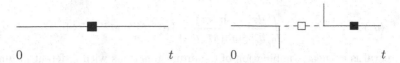

Figure 22.2: The diagrams for $P_s(t)$ (left) and $P_a(t)$ (right). The full (dashed) lines represent sojourns (blips). A full box represents the insertion of a CB form factor U_{CB} within a sojourn interval, and the empty box stands for a CS form factor U_{CS} inserted in a blip interval. The downward and upward spikes symbolize the remaining two solitary charges forming an extended blip dipole.

$$U_{CS}(\tau) = e^{-\gamma\tau/2} , \qquad (22.132)$$

where we have taken into acount that the multiplicity factor for collapsed sojourns is one. With these preliminaries, the evaluation of the path or charge sum is facilitated substantially.

22.6.2 The expectation value $\langle\sigma_z\rangle_t$ for $K = \frac{1}{2}$

Consider now first the symmetric bias contribution $P_s(t)$ to $\langle\sigma_z\rangle_t$ in Eq. (21.47). Because the influence factor $F_m^{(+)}$ in Eq. (21.57) has $m\,\cos(\pi K)$ factors, there are m collapsed blips. Hence all blips are collapsed ones, and $P_s(t)$ is represented by a single wide sojourn of length t which is dressed by a CB form factor, as sketched diagrammatically in Fig. 22.2 (left). Thus we obtain exponential decay with the temperature- and bias-independent relaxation rate $\gamma = \pi\Delta^2/2\omega_c$,

$$P_s(t) = U_{CB}(t) = e^{-\gamma t} . \qquad (22.133)$$

Consider next the antisymmetric contribution $P_a(t)$ to $\langle\sigma_z\rangle_t$. We see from Eq. (21.57) that the influence function $F_m^{(-)}$ has one $\cos(\pi K)$-factor less than $F_m^{(+)}$. Hence there is a single wide blip. Because of the factor ξ_1 in $F_m^{(-)}$ and summation over $\xi_1 = \pm 1$ is implied, we see from the product $F_m^{(-)}B_m^{(a)}$ in Eq. (21.47) with Eq. (21.48) that the first blip, has the bias factor $\sin(\xi_1\epsilon\tau_1)$. Thus, the first blip is the wide one. The picture we now have is as sketched in Fig. 22.2 (right). The contribution $P_a(t)$ is pictorially represented by two wide sojourns of lengths s_1 and s_2 which are separated by a single wide blip of length $\tau_1 = \tau$. The first sojourn is bare, whereas the subsequent blip is dressed by a CS form factor and the terminative sojourn is dressed by a CB form factor. The intervals of these three components add up to the total time, $s_1 + \tau + s_2 = t$. Joining the various pieces together, and taking into account the interaction between the remaining two solitary charges, we obtain

$$P_a(t) = \Delta^2 \int_0^\infty ds_1\, d\tau\, ds_2\, \delta(t - s_1 - \tau - s_2)\sin(\epsilon\tau)\, e^{-Q'(\tau)}\, e^{-\gamma\tau/2}\, e^{-\gamma s_2} . \qquad (22.134)$$

Using Eqs. (22.133), (22.134), and (18.54), we finally get

$$\langle\sigma_z\rangle_t = e^{-\gamma t} + 2\int_0^t \frac{d\tau}{\hbar\beta}\frac{\sin(\epsilon\tau)}{\sinh(\pi\tau/\hbar\beta)}\left[e^{-\gamma\tau/2} - e^{-\gamma t}\,e^{\gamma\tau/2}\right]. \qquad (22.135)$$

The integral is a linear combination of digamma functions with different arguments.

Consider now first the case $T = 0$. At long times, $\sqrt{\gamma^2/4 + \epsilon^2}\,t \gg 1$, we find from Eq. (22.135) that the equilibrium value is approached as

$$\langle\sigma_z\rangle_t \approx \langle\sigma_z\rangle_\infty - \frac{8}{\pi}\frac{\gamma^2}{\gamma^2 + 4\epsilon^2}\frac{\sin(\epsilon t)\,e^{-\gamma t/2}}{\gamma t}, \qquad (22.136)$$

and the expectation value in thermal equilibrium is

$$\langle\sigma_z\rangle_\infty = (2/\pi)\arctan(2\epsilon/\gamma). \qquad (22.137)$$

Thus, there is a finite occupation of the higher (left) state at zero temperature.

At low temperature, $0 < k_B T \ll \hbar\gamma$, the leading asymptotic behavior is

$$\langle\sigma_z\rangle_t \approx \langle\sigma_z\rangle_\infty - \frac{16}{\hbar\beta\gamma}\frac{\gamma^2}{\gamma^2 + 4\epsilon^2}\sin(\epsilon t)\,e^{-(\gamma+\nu_1)t/2}, \qquad (22.138)$$

where $\nu_1 = 2\pi/\hbar\beta$ is the lowest Matsubara frequency.

The equilibrium value reached at asymptotic times is found from Eq. (22.135) in analytic form for all T and ϵ,

$$\langle\sigma_z\rangle_\infty = 2\int_0^\infty \frac{d\tau}{\hbar\beta}\frac{\sin(\epsilon\tau)\,e^{-\gamma\tau/2}}{\sinh(\pi\tau/\hbar\beta)} = \frac{2}{\pi}\,\mathrm{Im}\,\psi\left(\frac{1}{2} + \frac{\hbar\gamma}{4\pi k_B T} + i\frac{\hbar\epsilon}{2\pi k_B T}\right), \qquad (22.139)$$

where $\psi(z)$ is the digamma function.

From these results for $K = \frac{1}{2}$, we can draw two conclusions. First, a bias leads to an increase of the quality factor of the oscillation, whereas the amplitude is decreased. Second, with increasing temperature, the quality factor of the oscillation decreases, whereas the amplitude is increased. Similar behavior has been found in numerical studies also for $K \neq \frac{1}{2}$ [407].

As we have dicussed at the end of Subsection 22.2.3, the expectation values $\langle\sigma_x\rangle_t$ and $\langle\sigma_y\rangle_t$ are non-universal. Hence they are zero in the universality limit (18.57).

22.6.3 The case $K = \frac{1}{2} - \kappa$; coherent-incoherent crossover

For a symmetric system at zero temperature, the NIBA expression (22.5) gives for $K = \frac{1}{2} - \kappa$ with $|\kappa| \ll 1$ a coherent contribution $P_{\mathrm{coh}}(t)$ of the form (22.10), where

$$\Omega = 2\pi\kappa\,\Delta_{\mathrm{eff}}, \qquad \gamma = \Delta_{\mathrm{eff}}, \qquad (22.140)$$

and an incoherent part

$$P_{\mathrm{inc}}(t) = -2\kappa/(\Delta_{\mathrm{eff}}t)^{1+2\kappa}. \qquad (22.141)$$

Thus, NIBA predicts the coherent-incoherent transition to occur at $K = \frac{1}{2}$.

To investigate the transition beyond NIBA, we now follow Ref. [469] and perform a systematic expansion of K about the exactly solvable case $K = \frac{1}{2}$. For $K = \frac{1}{2} - \kappa$, the blip length is finite, and within every blip one has to take into account all sequences of collapsed sojourns. This crucially modifies the kernel $K_z^{(s)}(\lambda)$ in the limit $\lambda \to 0$, as compared with the NIBA. The grand-canonical sum of collapsed sojourns within a blip of length τ simply gives a factor $\exp(-\Delta_{\text{eff}}\tau/2)$, as explained in the preceding subsection. Thus we have

$$K_z^{(s)}(\lambda) = \Delta^2 \pi\kappa \int_0^\infty d\tau \, \frac{e^{-(\lambda+\Delta_{\text{eff}}/2)\tau}}{(\omega_{\text{c}}\tau)^{1-2\kappa}} = \Delta_{\text{eff}} \left[\frac{\lambda}{\Delta_{\text{eff}}} + \frac{1}{2} \right]^{-2\kappa}. \tag{22.142}$$

With the regularized kernel (22.142), the pole condition reads

$$\lambda[\, 1/2 + \lambda/\Delta_{\text{eff}}\,]^{2\kappa} = -\Delta_{\text{eff}}. \tag{22.143}$$

This gives for the oscillation frequency and the decay rate the expressions

$$\Omega = 2\pi\kappa\,\Delta_{\text{eff}}[\, 1+\mathcal{O}(\kappa)\,], \qquad \gamma = \Delta_{\text{eff}}[\, 1+\mathcal{O}(\kappa)\,], \tag{22.144}$$

which coincide in the explicitly given orders with the NIBA expressions (22.140). From this we conclude that the value $K = \frac{1}{2}$ is exactly the critical damping strength where the coherent-incoherent transition occurs at zero temperature in the limit $\Delta/\omega_{\text{c}} \to 0$, as already predicted by the NIBA.

The kernel (22.142) shifts the branch point of $\langle \sigma_s(\lambda) \rangle$ from $\lambda = 0$ to $\lambda = -\Delta_{\text{eff}}/2$ and thereby removes the spurious algebraic long-time tail (22.141). For $\kappa > 0$, the leading cut contribution at times $t \gg \Delta_{\text{eff}}^{-1}$ is given by

$$P_{\text{inc}}(t) = -2\kappa \, \frac{\exp(-\Delta_{\text{eff}}t/2)}{(\Delta_{\text{eff}}t)^{1+2\kappa}}, \tag{22.145}$$

while for $\kappa < 0$ we find

$$P_{\text{inc}}(t) = 8|\kappa| \, \frac{\exp(-\Delta_{\text{eff}}t/2)}{(\Delta_{\text{eff}}t)^{1+2|\kappa|}}. \tag{22.146}$$

Thus the unphysical algebraic long-time tail (22.141) predicted by the NIBA is actually suppressed by an exponential decay factor. It is straightforward to see that for $\kappa > 0$ ($K < \frac{1}{2}$) the cut contribution is negative for all times, while it becomes positive for $\kappa < 0$ ($K > \frac{1}{2}$). For $K \geq \frac{1}{2}$, the expectation value $\langle \sigma_z \rangle_t$ is decaying monotonously from the initial value $+1$ to zero, which means that the dynamics is fully incoherent. In marked contrast to the NIBA result (22.141), the power of the algebraic decay factor does not depend on the sign of κ.

22.6.4 Equilibrium σ_z autocorrelation function

We begin the discussion of the equilibrium correlation functions by considering the symmetric σ_z autocorrelation function. Starting out from the expression (21.58) with Eq. (21.60), and gathering up the results achieved in Subsection 22.6.2, we find

$$S_z^{(\text{unc})}(t) = e^{-\gamma|t|} + \langle \sigma_z \rangle_\infty [\, P_{\text{a}}(|t|) - \langle \sigma_z \rangle_\infty \,], \tag{22.147}$$

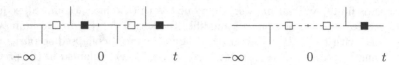

Figure 22.3: The left diagram represents the correlation contribution $R(t)$ to $S_z(t)$, and the right diagram pictorially gives the response function $\chi_z(t)$. The symbols are as in Fig. 22.2.

where $P_{\mathrm{a}}(t)$ and $\langle\sigma_z\rangle_\infty$ are given in Eqs. (22.134) and (22.139), respectively.

Because of the factor $\xi_1\xi_{n+1}\sin^2(\pi K)$ in the expression (21.61) for $F_{m,n}$, the correlation term $R(t)$ consists of two extended blips, the one being the first blip in the negative-time branch, and the other the first blip in the positive-time branch. The interval of each of the two blips is filled with a gas of collapsed sojourns which result in form factors U_{CS} as explained in Subsection 22.6.1. The two extended blips are followed in the respective remaining interval of the negative- and positive-time branches by grand-canonical sums of collapsed noninteracting blips, yielding form factors U_{CB}. The resulting correlation term $R(t)$ is pictorially sketched in Fig. 22.3 (left).

For $K = \frac{1}{2}$, the blip interaction factor for two blips of type ξ and $\pm\xi$ with flip times t_1, t_2 $(t_2 > t_1)$ and t_3, t_4 $(t_4 > t_3)$, respectively, can be decomposed as [cf. the similar decomposition (19.59) for imaginary times]

$$
\begin{aligned}
G_2(\xi,\xi) &= \mathrm{e}^{-Q'_{21}}\,\mathrm{e}^{-Q'_{43}} + \mathrm{e}^{-Q'_{32}}\,\mathrm{e}^{-Q'_{41}}\,, \\
G_2(\xi,-\xi) &= \mathrm{e}^{-Q'_{21}}\,\mathrm{e}^{-Q'_{43}} - \mathrm{e}^{-Q'_{31}}\,\mathrm{e}^{-Q'_{42}}\,.
\end{aligned}
\tag{22.148}
$$

We find that the contributions from the first term in $G_2(\xi,\xi)$ and in $G_2(\xi,-\xi)$ cancel (after accomplishment of the ξ summation) the subtraction term in Eq. (21.60). Piecing together these observations, we find for the correlation term $R(t)$ the form

$$
\begin{aligned}
R(t) &= -\frac{\Delta^4}{2}\int_0^\infty d\tau_1\,d\tau_2\,ds_1\,ds_2\,\Theta(|t|-\tau_2-s_2)\,\mathrm{e}^{-\gamma(\tau_1+\tau_2)/2}\,\mathrm{e}^{-\gamma s_1}\,\mathrm{e}^{-\gamma(|t|-\tau_2-s_2)} \\
&\quad \times \Big\{\cos[\,\epsilon(\tau_1+\tau_2)\,]\,\mathrm{e}^{-Q'(s_1+s_2)}\,\mathrm{e}^{-Q'(\tau_1+s_1+s_2+\tau_2)} \\
&\quad\quad + \cos[\,\epsilon(\tau_1-\tau_2)\,]\,\mathrm{e}^{-Q'(\tau_1+s_1+s_2)}\,\mathrm{e}^{-Q'(s_1+s_2+\tau_2)}\Big\}\,.
\end{aligned}
\tag{22.149}
$$

Here, τ_1 and τ_2 are the lengths of the two wide blips. The length of the intermediate sojourn is s_1+s_2, where s_1 is the interval until time zero and s_2 is the residual interval in the positive-time branch. The pair interactions $Q'(\tau)$ convey correlations between the negative-time and the positive-time branch.

The expression (22.149) is readily evaluated by (i) introducing the arguments of the function $Q'(\tau)$ as new integration variables and (ii) by performing the other integrations. Combining the resulting form for $R(t)$ with the expression (22.147), we find in the end for the symmetrized correlation function the analytic form

$$
S_z(t) = \mathrm{e}^{-\gamma|t|} - F_1^2(t) - F_2^2(t)\,,
\tag{22.150}
$$

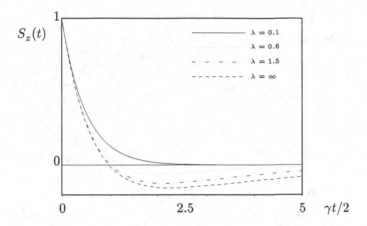

Figure 22.4: The formula (22.150) for $S_z(t)$ is plotted for zero bias as a function of $\gamma t/2$ for different values of the scaled inverse temperature $\lambda = \hbar\gamma/2\pi k_B T$. In the high temperature case $\lambda = 0.1$, the curve for $S_z(t)$ cannot be resolved from the function $\langle \sigma_z \rangle_t = e^{-\gamma t}$.

$$F_1(t) = \frac{\Delta^2}{2\gamma} \int_0^\infty d\tau \, \sin(\epsilon\tau) \, e^{-Q'(\tau)} \left[e^{-\gamma|t-\tau|/2} + e^{-\gamma(t+\tau)/2} \right] ,$$
$$F_2(t) = \frac{\Delta^2}{2\gamma} \int_0^\infty d\tau \, \cos(\epsilon\tau) \, e^{-Q'(\tau)} \left[e^{-\gamma|t-\tau|/2} - e^{-\gamma(t+\tau)/2} \right] .$$
(22.151)

The result (22.150) with Eq. (22.151) is exact for all T and ϵ, and it holds for all t. For $\gamma t \gg 1$ and $T = 0$, the functions $F_1(t)$ and $F_2(t)$ are asymptotically given by

$$F_1(t) \approx \frac{4}{\pi} \frac{\gamma^2}{\gamma^2 + 4\epsilon^2} \frac{\sin(\epsilon t)}{\gamma t} , \qquad F_2(t) \approx \frac{4}{\pi} \frac{\gamma^2}{\gamma^2 + 4\epsilon^2} \frac{\cos(\epsilon t)}{\gamma t} .$$
(22.152)

Consider now first the unbiased case. Insertion of the expression (22.152) in Eq. (22.150) yields for the symmetric system at $T = 0$ at asymptotic time the algebraic tail

$$S_z(t) = - \left(\frac{4}{\pi\gamma} \right)^2 \frac{1}{t^2} .$$
(22.153)

At very low T, the algebraic law (22.153) holds in the intermediate time region $\gamma^{-1} \ll |t| \ll \hbar\beta$, while in the asymptotic regime $|t| \gg \hbar\beta$ one has exponential decay with the rate given by the lowest Matsubara frequency $\nu_1 = 2\pi/\hbar\beta$ [219],

$$S_z(t) = - (8k_B T/\hbar\gamma)^2 e^{-\nu_1|t|} .$$
(22.154)

The behavior of $S_z(t)$ is shown for different temperatures in Fig. 22.4.

For nonzero bias and short to intermediate times $t \lesssim 1/\gamma$, the correlation function $S_z(t)$ is given by Eq. (22.147). At $T = 0$, we find for t near $1/\gamma$ damped oscillations,

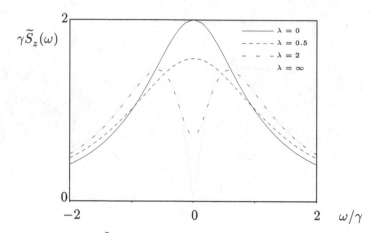

Figure 22.5: The function $\gamma \tilde{S}_z(\omega)$ is shown for $\epsilon = 0$ for a set of values of the scaled inverse temperature $\lambda = \hbar\gamma/2\pi k_{\mathrm{B}}T$.

$$S_z(t) \approx -\frac{8}{\pi} \frac{\gamma^2}{\gamma^2 + 4\epsilon^2} \langle \sigma_z \rangle_\infty \frac{\sin(\epsilon t)\, e^{-\gamma|t|/2}}{\gamma t}. \tag{22.155}$$

From this we draw two conclusions. First, a bias can induce a transition from incoherent relaxation to damped coherent oscillation. Numerical simulations indicate that the transition generally occurs in the regime $\frac{1}{2} < K < 1$ when the bias exceeds a critical value [407]. Secondly, the quality factors for the bias-induced oscillations of $\langle \sigma_z \rangle_t$ and $S_z(t)$ are the same. This observation holds generally, as we can infer from the form (22.147) which rules the behavior of $S_z(t)$ in the time domain $t \lesssim 1/\Delta_{\mathrm{eff}}$.

At zero temperature, the decay of the correlations at asymptotic times $t \gg 1/\gamma$ is dominated by the correlation term $R(t)$ in Eq. (21.58). We find from Eq. (22.150) with Eq. (22.151) that $S_z(t)$ drops algebraically to zero in this regime,

$$S_z(t) = -\left(\frac{\hbar\overline{\chi}_z}{2}\right)^2 \frac{1}{t^2}, \qquad \overline{\chi}_z = \frac{8}{\pi\hbar\gamma} \frac{\gamma^2}{\gamma^2 + 4\epsilon^2}. \tag{22.156}$$

Here we have introduced for convenience the static susceptibility $\overline{\chi}_z$ at $T = 0$, which may be calculated either from Eq. (21.65) with Eq. (22.137), or alternatively from Eq. (19.87). The form (22.156) is in correspondence with the behavior (6.86) for $s = 1$ at the Toulouse point $\delta_1 = K = \frac{1}{2}$.

The effects of the correlations in the initial state also clearly emerge in frequency space. The expression (22.150) may be Fourier-transformed in analytic form for general T and ϵ. We obtain

$$\tilde{S}_z(\omega) = \coth\left(\frac{\hbar\omega}{2k_{\mathrm{B}}T}\right) \frac{2}{\pi} \frac{\gamma}{\omega[\omega^2 + \gamma^2]} \left(\omega \Phi''(\omega) - \gamma \Phi'(\omega)\right), \tag{22.157}$$

where $\Phi'(\omega)$ and $\Phi''(\omega)$ are the real and imaginary parts of the complex function

$$\Phi(\omega) = \psi(x_+) + \psi(x_-) - \psi(x_+ - i\hbar\beta\omega/2\pi) - \psi(x_- - i\hbar\beta\omega/2\pi) \,. \qquad (22.158)$$

Here $\psi(x)$ is the digamma function, and $x_\pm = \frac{1}{2} + (\frac{1}{2}\gamma \pm i\epsilon)\hbar\beta/2\pi$. At zero temperature we have

$$
\begin{aligned}
\Phi'(\omega) &= -\tfrac{1}{2}\ln\{[\gamma^2 + 4(\omega + \epsilon)^2][\gamma^2 + 4(\omega - \epsilon)^2]/[\gamma^2 + 4\epsilon^2]^2\} \,, \\
\Phi''(\omega) &= \arctan[2(\omega + \epsilon)/\gamma] + \arctan[2(\omega - \epsilon)/\gamma] \,.
\end{aligned}
\qquad (22.159)
$$

Because of the algebraic decay $\propto t^{-2}$ in Eq. (22.156), the low-frequency behavior at zero temperature is non-analytic. We find from Eq. (22.157)

$$\lim_{\omega \to 0} \tilde{S}_z(\omega)/|\omega| = \pi(\hbar\overline{\chi}_z/2)^2 \,. \qquad (22.160)$$

This is the Shiba relation (6.86) for $s = 1$ and $\delta_1 = K = \frac{1}{2}$.

While the uncorrelated part of the spectral function is temperature-independent for zero bias and is given by the Lorentzian form $\tilde{S}_z^{(\mathrm{unc})}(\omega) = 2\gamma/(\omega^2 + \gamma^2)$, the expression (22.157) is drastically different at $T = 0$ in the low-frequency regime. The curve for $\lambda = \infty$ in Fig. 22.5 illustrates the nonanalytic behavior disclosed in Eq. (22.160). As T is increased, the trough of $\tilde{S}_z(\omega)$ is leveled off. At high temperatures where $\lambda \ll 1$, the difference between $\tilde{S}_z(\omega)$ and $\tilde{S}_z^{(\mathrm{unc})}(\omega)$ becomes negligibly small.

Next, we study the response function $\chi_z(t)$. The analysis again proceeds by employing the concept of collapsed dipoles. First, we recognize from the formally exact expression for $\chi_z(t)$, Eq. (21.62), that the system is in a blip state at time zero. Secondly, we see from the form (21.57) for the influence factor $F_m^{(-)}$ that there is only one extended blip for $K = \frac{1}{2}$. It is then clear that the system hops into the extended blip state in the negative time branch and finishes off this state only at a positive time. Within the extended blip, there are insertions of collapsed sojourns, resulting in a CS form factor U_{CS}. The extended blip is followed by an extended sojourn with insertions of collapsed blips. These lead to a CB form factor U_{CB}. The respective diagram is shown in Fig. 22.3 (right). In mathematical terms, we have

$$
\begin{aligned}
\chi_z(t) = \; &\frac{2}{\hbar}\,\Theta(t)\,\Delta^2 \int_0^\infty d\tau_1\,d\tau_2\,ds_2\,\delta(t - \tau_2 - s_2) \\
&\times \cos[\epsilon(\tau_1 + \tau_2)]\,e^{-Q'(\tau_1 + \tau_2)}\,e^{-\gamma(\tau_1 + \tau_2)/2}\,e^{-\gamma s_2}\,,
\end{aligned}
\qquad (22.161)
$$

where $\tau_1 + \tau_2$ is the overall length of the blip and s_2 is the remaining sojourn length. Introducing $\tau = \tau_1 + \tau_2$ as a new integration variable and performing the other integrals, we find for the response function the exact analytic expression

$$\chi_z(t) = (4/\hbar)\,\Theta(t)\,F_2(t)\,e^{-\gamma t/2}\,, \qquad (22.162)$$

where the function $F_2(t)$ is defined in Eq. (22.151).

Employing the form (22.152) for $F_2(t)$, we find that the response function at $T = 0$ decays asymptotically as

$$\chi_z(t) \approx \frac{16}{\pi\hbar}\frac{\gamma^2}{\gamma^2 + 4\epsilon^2}\frac{\cos(\epsilon t)\,e^{-\gamma t/2}}{\gamma t}\,, \qquad \gamma t \gg 1\,. \qquad (22.163)$$

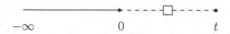

Figure 22.6: The diagram describing $S_x(t)$ with symbols as in Fig. 22.2. The bullets mark transitions which are free of bath correlations because of the modified influence functional.

Fourier transformation of Eq. (22.162) yields an analytic expression for the dynamical susceptibility at all T and ϵ,

$$\tilde{\chi}_z(\omega) \; = \; \frac{2}{\pi} \frac{1}{\hbar\omega} \frac{\gamma}{\omega + i\gamma} \, \Phi(\omega) \, , \tag{22.164}$$

where $\Phi(\omega)$ is the complex function defined in Eq. (22.158).

Two remarks are appropriate. First, we see that the independently derived expressions (22.157) and (22.164) satisfy the fluctuation-dissipation theorem

$$\tilde{S}_z(\omega) \; = \; \hbar \coth(\beta\hbar\omega/2)\, \tilde{\chi}_z''(\omega) \, . \tag{22.165}$$

Secondly, the static susceptibility $\overline{\chi}_z = \tilde{\chi}_z(\omega = 0)$ found from Eq. (22.164) coincides with the result (19.87) obtained within a thermodynamic approach and normalized as in Eq. (19.13). The expression is also in agreement with the result found from the dispersion relation

$$\overline{\chi}_z \; = \; \frac{2}{\pi} \int_0^\infty d\omega \, \frac{\tilde{\chi}_z''(\omega)}{\omega} \, . \tag{22.166}$$

22.6.5   Equilibrium σ_x autocorrelation function

With the concept of collapsed blips and collapsed sojourns explained in Subsection 22.6.1, it is possible to evaluate also the series expressions (21.67) – (21.70) for the dressed coherence correlation function in the scaling limit for $K = \frac{1}{2}$.

Consider first the symmetrized correlation function. Again we assign a factor $\cos(\pi K)$ to each collapsing dipole. We then find that in each term of the series (21.68) there is a surplus of one $\cos(\pi K)$ factor. As a result, the contribution $S_x^B(t)$ vanishes as $K \to \frac{1}{2}$. In the contribution $S_x^A(t)$, the system dwells in the initial sojourn state η throughout the negative-time branch. At time zero it then hops into the blip state $\xi = -\eta$. Afterwards, it stays there until time t except for flash visits of sojourn states, which altogether sum up to a CS form factor $e^{-\gamma t/2}$. Piecing together the bias factor of the blip state with the CS form factor, we find the damped oscillation

$$S_x(t) \; = \; \cos(\epsilon t)\, e^{-\gamma|t|/2} \, . \tag{22.167}$$

The contribution $S_x^A(t)$ is sketched diagrammatically in Fig. 22.6. Since all charges merge into collapsed dipoles, there are no excess charges, and hence $S_x(t)$ does not depend on temperature.

Next, we study the response function $\chi_x(t)$. The contribution of group A is sketched in Fig. 22.7 (left). In the negative-time branch, there is an extended blip state which is followed by an extended sojourn state, both being dressed with CS and CB form factor, respectively. At time zero, the system hops back into a blip state and stays there until time t. The extended blip state is again dressed by a CS form factor. In mathematical terms we have

$$\chi_x^A(t) = \Theta(t)\frac{2}{\hbar}\sin(\epsilon t)\,e^{-\gamma t/2}\Delta^2\int_0^\infty d\tau\int_0^\infty ds\,\sin(\epsilon\tau)\,e^{-Q'(\tau)}\,e^{-\gamma\tau/2}\,e^{-\gamma s}\,. \qquad (22.168)$$

Now, the double integral times the factor Δ^2 is just $\langle\sigma_z\rangle_\infty = P_a(t\to\infty)$ given in Eq. (22.139), as follows with Eq. (22.134). In the end, we find [219]

$$\chi_x^A(t) = (2/\hbar)\,\langle\sigma_z\rangle_\infty\,\sin(\epsilon t)\,e^{-\gamma t/2}\,. \qquad (22.169)$$

Thus we have again exponential decay. This is, because each extended interval, except for the first sojourn, is cut off exponentially by the respective form factor.

Now we turn to the contributions of group B. Because of the $\cos^2(\pi K)$-factor in the second term in the square bracket of Eq. (21.70), the respective contribution vanishes $\propto (K-\frac{1}{2})^2$ as $K\to\frac{1}{2}$, whereas the first term in the square bracket is nonzero in this limit. In this contribution, the system hops from a sojourn into a blip at negative time $-\tau$ and stays there until time zero, where it returns to a sojourn state. At time s, it hops again into a blip state and dwells in this state until time t. Again, each extended blip interval is dressed with a CS form factor, as discussed above. However, because of the factor ξ_{n+1} in the first term of the square bracket in Eq. (21.70) and the summation over the values ± 1, the extended sojourn in the positive-time branch is free of collapsed blips. The extended dipole of length $\tau + s$ introduces correlations between the negative- and positive-time branches as depicted in Fig. 22.7 (right). These features are condensed in the formula

$$\chi_x^B(t) = \Theta(t)\frac{\Delta^2}{\hbar}\int_0^\infty d\tau\int_0^t ds\,e^{-\gamma(t+\tau-s)/2}\,e^{-S(\tau+s)}\cos[\epsilon(t-\tau-s)]\,. \qquad (22.170)$$

Introducing the dipole length $\tau + s$ as a new integration variable, performing the other integrations, and combining the resulting expression with Eq. (22.169), we find

$$\chi_x(t) = (2/\hbar)\big[\sin(\epsilon t)F_1(t) + \cos(\epsilon t)F_2(t)\big]\,, \qquad (22.171)$$

where the functions $F_1(t)$ and $F_2(t)$ have been given already in Eq. (22.151). The function $\chi_x(t)$ describes the linear response of the system to a variation of the tunneling splitting Δ. Using Eq. (22.152), we find asymptotically algebraic decay,

$$\chi_x(t) \approx \frac{8}{\pi\hbar}\frac{\gamma^2}{\gamma^2+4\epsilon^2}\frac{1}{\gamma t} \qquad \text{for} \qquad t\gg 1/\gamma\,. \qquad (22.172)$$

Next, consider the spectral properties. Taking the Fourier transform of $S_x(t)$ given in Eq. (22.167), we obtain

$$\tilde{S}_x(\omega) = \gamma\frac{\omega^2+\epsilon^2+\gamma^2/4}{(\omega^2+\epsilon^2+\gamma^2/4)^2-4\epsilon^2\omega^2}\,. \qquad (22.173)$$

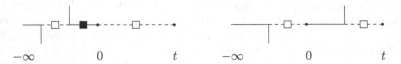

Figure 22.7: The contribution of group A (left) and group B (right) to $\chi_x(t)$. The symbols are as in Fig. 22.2. Each diagram has only one extended dipole.

On the other hand, we may calculate $\tilde{\chi}_x''(\omega)$ from the expression (22.171). We then discover that the resulting expression is in accordance with the fluctuation-dissipation theorem

$$\hbar\tilde{\chi}_x''(\omega) = \tanh(\beta\hbar\omega/2)\,\tilde{S}_x(\omega)\,. \tag{22.174}$$

Finally, the real part of the dynamical susceptibility is found in the unbiased case as

$$\tilde{\chi}_x'(\omega) = \frac{8\gamma}{\pi\hbar}\frac{1}{\omega^2+\gamma^2/4}\,\mathrm{Re}\left\{\psi\left(\frac{1}{2}+\frac{\hbar\gamma}{4\pi k_{\mathrm B}T}\right) - \psi\left(\frac{1}{2}+i\,\frac{\hbar\omega}{2\pi k_{\mathrm B}T}\right)\right\}\,. \tag{22.175}$$

From this we find that the static susceptibility $\overline{\chi}_x = \tilde{\chi}_x(\omega=0)$ diverges logarithmically for zero bias as temperature goes to zero.

22.6.6 Correlation functions in the Toulouse model

Additional insights are gained by calculating the equilibrium tunneling and coherence correlation functions in the fermionic model (19.66). The equivalence relations of the TSS operators with the fermionic operators of the d level are

$$\sigma_z = 2d^\dagger d - 1\,, \qquad \sigma_x = d^\dagger + d\,, \qquad \sigma_y = i(d - d^\dagger)\,. \tag{22.176}$$

To proceed, we rewrite the Matsubara sum (19.77) for $G^{(\mathrm E)}(\tau)$ as a contour integral. Analytic continuation to real time gives for the Green function $G(t) \equiv \langle T_t d^\dagger(t)d(0)\rangle_\beta$ of the d-level the expression

$$G(\pm t) = G^{(\mathrm E)}(\tau=\pm it) = \tfrac{1}{2}\mathrm{e}^{\mp i\epsilon t}\left[F_1(t)\mp iF_2(t)\pm\mathrm{e}^{-\gamma|t|/2}\right]\,, \tag{22.177}$$

where the functions $F_1(t)$ and $F_2(t)$ are defined by the integral representations

$$
\begin{aligned}
F_1(t) &= \frac{\gamma}{2\pi}\int_{-\infty}^{\infty}d\omega\,\frac{\cos(\omega t)}{\omega^2+\gamma^2/4}\tanh\left(\frac{\hbar(\omega+\epsilon)}{2k_{\mathrm B}T}\right)\,,\\
F_2(t) &= \frac{\gamma}{2\pi}\int_{-\infty}^{\infty}d\omega\,\frac{\sin(\omega t)}{\omega^2+\gamma^2/4}\tanh\left(\frac{\hbar(\omega+\epsilon)}{2k_{\mathrm B}T}\right)\,.
\end{aligned}
\tag{22.178}
$$

The key relation linking the fermionic representation to the bosonic one is

$$\tanh(\hbar\omega/2k_{\mathrm B}T) = (2\omega_{\mathrm c}/\pi)\int_0^{\infty}d\tau\,\sin(\omega\tau)\,\mathrm{e}^{-Q'_{K=1/2}(\tau)}\,, \tag{22.179}$$

where $Q'_{K=1/2}(\tau)$ is the function (18.54) for $K = \frac{1}{2}$. With use of the relation (22.179), the "fermionic" expressions (22.178) can directly be transformed into the bosonic representations (22.151). Thus we find as an important intermediate result that the expressions (22.151) and (22.178) are different integral representations of the same functions. Next, we insert the form (22.176) for σ_z into Eq. (21.16), use Eq. (21.18) for the index $j = z$, and observe that the two-particle Green function factorizes into products of one-particle Green functions. We readily obtain

$$S_z(t) = 1 + 4\,\mathrm{Re}\left\{G^2(0^+) - G(0^+) - G(t)G(-t)\right\},$$

$$\chi_z(t) = (8/\hbar)\Theta(t)\,\mathrm{Im}\left\{G(t)G(-t)\right\}. \tag{22.180}$$

With the forms (22.177) for $G(\pm t)$ we recover the earlier results for $S_z(t)$ and $\chi_z(t)$, Eqs. (22.150) and (22.162).

In the fermionic representation, it is straightforward to calculate also the symmetric σ_j autocorrelation function $S_j(t) = \mathrm{Re}\,\langle\sigma_j(t)\sigma_j(0)\rangle_\beta$ and the response function $\chi_j(t) = -(2/\hbar)\Theta(t)\,\mathrm{Im}\,\langle\sigma_j(t)\sigma_j(0)\rangle_\beta$ for $j = x,\,y$. By virtue of the relations (22.176) we find

$$S_y(t) = S_x(t), \quad \text{and} \quad \chi_y(t) = \chi_x(t). \tag{22.181}$$

The resulting expressions are

$$S_x(t) = \mathrm{Re}\left\{G(t) - G(-t)\right\} = \cos(\epsilon t)\,e^{-\gamma|t|/2}, \tag{22.182}$$

$$\chi_x(t) = -\frac{2}{\hbar}\Theta(t)\mathrm{Im}\left\{G(t) - G(-t)\right\} = \frac{2}{\hbar}\Theta(t)\left[\sin(\epsilon t)F_1(t) + \cos(\epsilon t)F_2(t)\right].$$

These expressions coincide with the analogous correlation functions of the spin-boson model, Eqs. (22.167) and (22.171). Because σ_x in Eq. (22.176) is the bare tunneling operator, it is evident that we have worked in the representation (21.29). Thus, the Toulouse Hamiltonian (19.66), and the resonant-level Hamiltonian (19.106), directly correspond to the polaron-transformed Hamiltonian $\tilde{H}$ given in Eq. (18.40). This concludes the discussion of the exactly solvable case $K = \frac{1}{2}$.

22.7 Long-time behavior at $T = 0$ for $K < 1$: general discussion

We have seen in the previous section that in the special case $K = \frac{1}{2}$ the grand-canonical sum of collapsed dipoles within an extended dipole gives rise to a dipole form factor. For K different from $\frac{1}{2}$ the dipoles do not collapse anymore and therefore they are interacting with each other. For this reason, the grand-canonical sum of extended dipoles can not be performed in analytic form any more. We argue that in the limit of long times the alternating series of these entities nevertheless add up to an effective form factor which cuts off the respective time interval at a length of order $1/\Delta_{\mathrm{eff}}$. With these preliminaries, the asymptotic dynamics can be understood by means of two rules:

(I) Every time interval which is free of a form factor for $K = \frac{1}{2}$ is free of a form factor also for $K \neq \frac{1}{2}$. We shall refer to such intervals as bare intervals.

(II) The totality of charge arrangement placed between bare intervals form a grand-canonical ensemble (charge cluster) which effectively acts as a form factor. The form factor restricts the respective interval to a length of order $1/\Delta_{\mathrm{eff}}$.

Rule (I) applies to blip states in the series expressions given in Subsections 21.2.4 and 21.2.5 which have weight factors ξ_j. For instance in the expression (21.61), these special blips are the first blip in the negative- and the first blip in the positive-time branch. For these blips, the $\{\xi_j\}$ summations lead to cancellations among the interactions stretching over the respective preceding sojourn interval. As a result, these sojourns stay bare. We now apply these rules to correlation functions of the TSS.

22.7.1 The populations

The noninteracting-blip approximation does not describe the long-time dynamics of the population difference $\langle \sigma_z \rangle_t$ correctly. In the Ohmic case for $K < 1$ and in the absence of a bias, the expectation value $\langle \sigma_z \rangle_t$ is found to decay asymptotically as $\propto (\Delta_{\mathrm{eff}} t)^{-(2-2K)}$, as follows from Eq. (22.9). The algebraic decay in the NIBA originates from a branch point of the Laplace-transformed kernel $\hat{K}_z^{(s)}(\lambda) \propto \lambda^{2K-1}$ at $\lambda = 0$ [cf. Eq. (22.4)]. The asymptotic behavior is changed qualitatively by the interblip correlations, as we have shown for the particular case $K = \frac{1}{2} - \kappa$ in Subsection 22.6.3. It is difficult to work out the effects of the interblip correlations quantitatively for general K since many different frequency scales act in combination. This phenomenon is well-known from the closely related Kondo problem in the antiferromagnetic sector. The resummation of the alternating series (21.76) of blip sequences results in kernels $\hat{K}_z^{(s,a)}(\lambda)$ which are regular at $\lambda = 0$ (see Subsection 22.8.1). Thus, $\langle \sigma_z \rangle_t$, or the envelope function of it, approaches the equilibrium value exponentially fast, as follows by Laplace inversion of Eq. (21.79). Further discussion of relaxation and decoherence, and several exact analytical results at $T = 0$ are given in Section 22.8.

The exponential decay towards the equilibrium distribution $\langle \sigma_z \rangle_\infty$ also follows immediately from the above two rules. As in the diagrams in Fig. 22.2, there is (at long time) only a single cluster both for $P_s(t)$ and $P_a(t)$, and hence exponential decay.

22.7.2 The population correlations and Shiba relation

With rules (I) and (II), we see from Fig 22.3 (right) that the response function $\chi_z(t)$ is diagrammatically represented by a single neutral cluster surrounding the origin of the time axis. Hence, $\chi_z(t)$ decays exponentially at long times.

Similarly, the contribution $S_z^{(\mathrm{unc})}(t)$ to $S_z(t)$ given in Eq. (21.59) decays exponentially, as follows with the results obtained in Subsection 22.7.1. Employing rules (I, II) to the diagram in Fig. 22.3 (left), we find that the correlation term $R(t)$ is represented by two neutral clusters, the one in the negative time branch near the origin, the other in the positive time branch near t. Therefore, at asymptotic times, the clusters are separated by an interval of length t. Since the interval is bare, the clusters are interacting with the unscreened dipole-dipole interaction which in the Ohmic case is $\ddot{Q}'(t) = -2K/t^2$. Hence we should expect that $S_z(t)$ decays as $1/t^2$ for Ohmic dissipation.

To put this argument in concrete form, we expand the blip interaction factor G_{n+m} in Eq. (21.61) under the assumption that the effective length of the two clusters is small compared to the interval t between the clusters. We then have

$$F_{m,n} \rightarrow F_m^{(-)}(\{\tau_k\}) F_n^{(-)}(\{\tau_j\}) \left(1 - \ddot{Q}'(t) \sum_{j=1}^{n} \sum_{k=n+1}^{n+m} \xi_j \xi_k \tau_j \tau_k \right), \qquad (22.183)$$

where $F_n^{(-)}$ and $F_m^{(-)}$ are the influence functions (21.57) for the clusters with n blips in the negative and m blips in the positive time branch. Next, we insert Eq. (22.183) into Eq. (21.60). Because of the $\{\xi_j\}$ summations, those terms which are odd in the blip labels ξ_j for the individual time branches cancel out. The remaining contribution can be written in the symmetric form

$$R(t) = \ddot{Q}'(t) \lim_{t_q \to \infty} \sum_{m=1}^{\infty} (-1)^m \int_0^{t_q} \mathcal{D}_{2m,0}\{t_j\} \frac{1}{2m} \sum_{\{\xi_j = \pm 1\}} B_m^{(s)} F_m^{(-)} \left(\sum_{k=1}^{m} \xi_k \tau_k \right)$$

$$\times \lim_{t_p \to -\infty} \sum_{n=1}^{\infty} (-1)^n \int_{t_p}^{0} \mathcal{D}_{0,2n}\{t_j\} \frac{1}{2n} \sum_{\{\xi_j = \pm 1\}} B_n^{(s)} F_n^{(-)} \left(\sum_{\ell=1}^{n} \xi_\ell \tau_\ell \right).$$

This expression describes two neutral clusters which are interacting with each other through the dipole-dipole interaction $\ddot{Q}'(t)$. Next, we observe that the series expansion for each cluster can be identified with the expression (21.64) for the static suscepti- bility $\overline{\chi}_z$ at $T = 0$ (apart from a missing factor $2/\hbar$). With this identification, and provided that the zero temperature limit of the expression (21.64) is well-defined, we obtain for $S_z(t)$ the exact asymptotic behavior [cf. Eq. (6.86) with s=1 and $\delta_1 = K$]

$$S_z(t) = \left(\frac{\hbar \overline{\chi}_z}{2} \right)^2 \ddot{Q}'_{T=0}(t) = - 2K \left(\frac{\hbar \overline{\chi}_z}{2} \right)^2 \frac{1}{t^2}. \qquad (22.184)$$

This relation holds for any bias and for any K smaller than one. Interestingly, the bias and the damping parameter K enter into (22.184) only implicitly through the static zero temperature susceptibility, except for the extra factor K. For $K > 1$, the Kondo temperature $T_K \equiv 2/(\pi k_B \overline{\chi}_z)$ is zero in the scaling limit, and the asymp- totic expansion discussed here does not exist. In this damping regime, the high- temperature expansion is valid down to $T = 0$. In the absence of a bias, one obtains from Eq. (22.90) in the zero damping limit $\overline{\chi}_z \rightarrow 2/\hbar \Delta$, while for $K = 1/2$ one has $\overline{\chi}_z = 8/(\pi \hbar \gamma)$ [cf. Eq.(22.156)], where $\gamma = \pi \Delta^2/2\omega_c$. From the correspondence with the Kondo model [141, 85], we find in the damping regime $1 - K \ll 1$ with Eq. (19.105) $\overline{\chi}_z = [2(1 - K)]^{K/(1-K)}/(\hbar \Delta_r)$, where $\Delta_r = (\Delta/\omega_c)^{K/(1-K)} \Delta$.

We remark that the zero temperature behavior (22.184) is also valid at finite temperatures in the time region $\hbar \overline{\chi}_z \ll t \ll \hbar \beta$. At times $t \gg \hbar \beta$, the correlation function $C(t)$ approaches the equilibrium value exponentially fast with a rate given by the smallest Matsubara frequency $\nu_1 = 2\pi/\hbar\beta$.

From Eq. (22.184) we may also infer the behavior in frequency space near $\omega = 0$, known as the *Shiba relation*,[12]

$$\lim_{\omega\to 0} \tilde{S}_z(\omega \to 0)/|\omega| = 2\pi K\big(\hbar\overline{\chi}_z/2\big)^2 , \tag{22.185}$$

or equivalently

$$\lim_{\omega\to 0} \hbar\tilde{\chi}_z''(\omega)/\omega = 2\pi K\big(\hbar\overline{\chi}_z/2\big)^2 . \tag{22.186}$$

These relations are analogous to a relation that has been proven by Shiba [249] for the Anderson model. While Shiba's derivation is essentially based upon a particle number conservation law, the proof given here is based upon the Coulomb gas representation in complex time. With the oscillator-spin correspondence $\chi_{n,0} = \overline{\chi}_z$, the relations (22.185) and (22.186) are in correspondence with the expressions (6.87).

Generalized Shiba relation for non-Ohmic spectral density

The exact formal expression (21.64) for the static susceptibility is well-defined in the limit $T \to 0$ also in the super-Ohmic case $s > 1$. This we may infer, for instance, indirectly from the expression (21.65) with (22.90). In the sub-Ohmic weak-damping regime (18.38), the renormalized tunneling matrix element Δ_r at $T = 0$ is nonzero, as follows from Eq. (18.32). Hence the static susceptibility at zero temperature is nonzero as well. As a result, the argumentation leading to the first equality of Eq. (22.184), $S_z(t) = (\hbar\overline{\chi}_z/2)^2 \ddot{Q}'(t)$, applies for these regimes, just as for the Ohmic case. We then find with the expression (20.49) for $Q(t)$ algebraic decay $\propto t^{-1-s}$ at asymptotic time,

$$S_z(t) = -2\delta_s\Gamma(1 + s)\sin(\pi s/2)\omega_\mathrm{ph}^{1-s}(\hbar\overline{\chi}_z/2)^2 |t|^{-1-s} , \tag{22.187}$$

and alternatively in frequency space

$$\lim_{\omega\to 0^\pm} \tilde{S}_z(\omega)/|\omega|^s = \lim_{\omega\to 0^\pm} \hbar\,\mathrm{sgn}\,(\omega)\tilde{\chi}_z''(\omega)/|\omega|^s = 2\pi\delta_s\omega_\mathrm{ph}^{1-s}(\hbar\overline{\chi}_z/2)^2 . \tag{22.188}$$

The Shiba relation (22.185), and the respective non-Ohmic generalization (22.187) and (22.188) have been derived first in Ref. [220]. Recently, this relation has been corroborated by diverse other methods. It has been verified numerically by employing the correspondence with the anisotropic Kondo model [470, 471], and by numerical integration of the flow equations resulting from a continuous sequence of infinitesimal unitary transformations [391, 392]. The relation has been confirmed also analytically by using nonperturbative methods derived from a Bethe ansatz [472].

22.7.3   The coherence correlation function

Finally, consider the asymptotic dynamics of the coherence correlation function at $T = 0$. Using rules (I) and (II), one finds that the contribution of group A has

---

[12]For the normalization of $\overline{\chi}_z$, see Eq. (19.13) and footnote, and Eq. (19.104).

a single neutral cluster surrounding the origin of the time axis. Hence, both $S_x^A(t)$ and $\chi_x^A(t)$ decay exponentially fast. In group B, we have two clusters with opposite charges ± 1, one in each time branch. Since in each branch the initial sojourn is free of insertions, the two clusters are situated near the origin and near t. Since the initial sojourn interval in the positive-time branch is bare, the two clusters interact with the unscreened charge-charge interaction $e^{-Q'(t)} \propto t^{-2K}$. Evidently, this interaction directly determines the algebraic long-time tails of $S_x(t)$ and $\chi_x(t)$ for $0 < K < 1$,

$$S_x(t) = c_S(K)\, t^{-2K}, \qquad \chi_x(t) = c_\chi(K)\, t^{-2K}. \tag{22.189}$$

Thus, the coherence correlations decay with a power law in which the exponent depends on the damping strength.

Taking the Fourier transforms of $S_x(t)$ and $\chi_x(t)$, we find with the asymptotic behaviors (22.189) the relation

$$\lim_{\omega \to 0^+} \tilde{S}_x(\omega)/\tilde{\chi}_x(\omega) = 2 \tan(\pi K) c_S(K)/c_\chi(K). \tag{22.190}$$

The ratio $c_S(K)/c_\chi(K)$ can be fixed by matching the relation (22.190) with the fluctuation-dissipation theorem (22.174) in the same limit. We then obtain in the time regime

$$\lim_{t \to \infty} S_x(t)/\chi_x(t) = \tfrac{1}{2}\hbar \cot(\pi K). \tag{22.101}$$

For $K = \frac{1}{2}$, the function $\chi_x(t)$ decays asymptotically as $1/t$, whereas $S_x(t)$ decays exponentially, as we can see from Eqs. (22.172) and (22.167). These behaviors are consistent with the relation (22.191) since the $\cot(\pi K)$-factor is zero at $K = \frac{1}{2}$.

The law (22.189) leads to differing behaviors of the linear static susceptibility for K below and above $\frac{1}{2}$. For $K < \frac{1}{2}$, the slow decay of $\chi_x(t)$ implies that the linear static susceptibility diverges algebraically, $\overline{\chi}_x^{(\ell)} \propto T^{2K-1}$, as $T \to 0$, which indicates that the response at zero temperature is actually nonlinear. On the other hand, for $K > \frac{1}{2}$, the decay of $\chi_x(t)$ is sufficiently fast, so that the linear static susceptibility is finite at $T = 0$. Recently, these properties have been confirmed numerically [473].

In summary, the decay of the population correlation function $S_z(t)$ at long times is determined by the unscreened dipole interaction, whereas the decay of the coherence correlation function $S_x(t)$ is determined by the unscreened charge interaction.

22.8 From weak to strong tunneling: relaxation and decoherence

22.8.1 Incoherent tunneling beyond the nonadiabatic limit

In the Markov limit the generalized master equation (GME) (21.73) simplifies to the rate equation[13]

---

[13]Strictly speaking, the total transfer rate $k = k^+ + k^-$ is given by the smallest real solution of the equation $k - \hat{K}_z^{(s)}(\lambda = -k) = 0$. When k is small compared to the other frequency scales of the system, this relation reduces to the form (22.193). In the coherent regime, Eq. (22.192) is not valid (see Eq. (22.95)). In this regime, the rate $k \hat{=} \gamma_r$ describes the decay of the incoherent part of $\langle \sigma_z \rangle_t$.

$$d\langle\sigma_z\rangle_t/dt = k^+ - k^- - (k^+ + k^-)\langle\sigma_z\rangle_t,\qquad(22.192)$$

where $k^\pm$ is the forward/backward rate from the state $\sigma = \pm 1$ to the state $\sigma = \mp 1$,

$$k^\pm = \frac{1}{2}\int_0^\infty d\tau\left[K_z^{(s)}(\tau) \pm K_z^{(a)}(\tau)\right] = \frac{1}{2}\left[\hat{K}_z^{(s)}(\lambda = 0) \pm \hat{K}_z^{(a)}(\lambda = 0)\right].\quad(22.193)$$

Equation (22.192) describes the population dynamics in the incoherent regime. The solution of this equation with initial value $\langle\sigma_z\rangle_0$ is

$$\langle\sigma_z\rangle_t = \langle\sigma_z\rangle_\infty + \left[\langle\sigma_z\rangle_0 - \langle\sigma_z\rangle_\infty\right]e^{-(k^+ + k^-)t}.\qquad(22.194)$$

The equilibrium value

$$\langle\sigma_z\rangle_\infty = \frac{k^+ - k^-}{k^+ + k^-}\qquad(22.195)$$

is in agreement with the exact formal expression given in Eq. (21.82).

Writing the series (21.76) for the kernels of the GME in terms of the blip and sojourn intervals (21.36) and switching to the Laplace transform, we obtain the series expansion for the forward/backward rate as

$$k^\pm = \sum_{n=1}^\infty (-1)^{n-1}\frac{\Delta^{2n}}{2^{n+1}}\int_0^\infty d\tau_n\left[\prod_{j=1}^{n-1}ds_j\,d\tau_j\right]\sum_{\{\xi_j=\pm 1\}}\left(\tilde{F}_n^{(+)}B_n^{(s)} \pm \tilde{F}_n^{(-)}B_n^{(a)}\right).\quad(22.196)$$

The rate expression is in the form of a series in the number of bounces (blips). The $n=1$ term is the nonadabatic rate, which has been discussed in Sec. 20.2. The terms with $n > 1$ represent the adiabatic corrections. They depend significantly on the inter-blip correlations and would vanish if the inter-blip interactions were disregarded. In the strong-tunneling regime, the full series may be relevant.

A graph-theoretic approach to the computation of adiabatic corrections in the real-time path integral formulation has been given by Stockburger and Mak [474]. The linked cluster sum considered by these authors is the graphical equivalent of the series expression (22.196) with the irreducible influence functions $\tilde{F}_n^{(\pm)}$. They presented an approximate summation of the multi-blip contributions under the assumption that the separation of blips is larger than $1/\omega_c$. The numerical studies show two competing adiabatic effects. At high temperatures, the rate is reduced (compared to the nonadiabatic case) because of correlated re-crossings of the Landau-Zener region. For low enough temperatures, the rate is enhanced above the nonadiabatic rate because of an effective reduction of the reaction barrier by the electronic coupling and by nuclear tunneling. The exact summation of the series expressions (21.76) and (22.196) in the weak-damping limit is given above in Section 22.3.

Exact solution in analytic form at $T = 0$

Interestingly, in the Ohmic scaling limit at $T = 0$, the incoherent rate can be determined in analytic form in all orders of Δ. The analysis of the Coulomb gas representation (22.196), which is performed in Ref. [475], reveals a close relationship between the rate contribution of order Δ^{2n} in the TSS with the rate k_n^+ from site 0 to site n of the same order in the Schmid model, (see Section 26.6). With the explicit form of the rate k_n^+ given below in Eq. (26.133), one then obtains the weak-tunneling series of the TSS (forward) rate in the analytic form (see Ref. [475] for details)

$$k^+(\epsilon) = \frac{\epsilon}{2\sqrt{\pi}} \sum_{m=1}^{\infty} \frac{1}{m!} \frac{\Gamma(Km)\,[\,1 - \cos(2\pi Km)\,]}{\Gamma[\frac{3}{2} + (K-1)m]} \left(\frac{\epsilon_{\mathrm{sb}}}{\epsilon}\right)^{(2-2K)m} . \tag{22.197}$$

The Kondo-like frequency scale ϵ_{sb} is related to the parameters of the TSS by

$$\epsilon_{\mathrm{sb}}^{2-2K} = \frac{\Gamma^2(1-K)}{2^{2K}} \frac{\Delta^2}{\omega_{\mathrm{c}}^{2K}} . \tag{22.198}$$

For rational K, the series (22.197) can be written as linear combination of hypergeometric functions. In the regime $K \gtrsim 1$, the weak-tunneling series (22.197) is absolutely converging when the bias is small/large enough. The leading term is the nonadiabatic or golden rule rate (20.82). Expressed in terms of ϵ_{sb}, the nonadiabatic rate takes the form

$$k_{\mathrm{gr}}^+(\epsilon) = \sin^2(\pi K) \frac{\Gamma(K)}{\Gamma(\frac{1}{2} + K)} \frac{\epsilon}{\sqrt{\pi}} \left(\frac{\epsilon_{\mathrm{sb}}}{\epsilon}\right)^{2-2K} . \tag{22.199}$$

At $K = \frac{1}{2}$, the terms with $m > 1$ are zero. Hence the golden rule rate $k_{\mathrm{gr}}^+(\epsilon)$ is the exact rate at the Toulouse point. This is in agreement with the expression (22.133).

Upon performing transformations analogously to those executed below in Subsection 26.5.3, the series (22.197) can be transformed into the integral representation

$$k^+(\epsilon) = \mathrm{Re}\, \frac{\epsilon}{2\pi i} \int_C \frac{dz}{z} \left\{ \sqrt{z - 1 - z^K u_2} - \sqrt{z - 1 - z^K u_1} \right\} , \tag{22.200}$$

where $u_1 = (\epsilon/\epsilon_{\mathrm{sb}})^{2K-2}$ and $u_2 = e^{i2\pi K} u_1$. The contour C starts at the origin, encircles the branch point in counter-clockwise sense, and returns to the origin. The representation (22.200) converges not only in the regime where the series (22.197) does, but for all values of K and of ϵ/ϵ_0, . This allows us to determine the asymptotic series.

Consider first the regime $K < 1$. Upon changing variable z to $y = z^{1-K}/u_1$ and $y = z^{1-K}/u_2$, and expanding the ensuing integrands in powers of $\epsilon/\epsilon_{\mathrm{sb}}$, we get the asymptotic (strong-tunneling) series [475],

$$k^+(\epsilon) = \frac{\epsilon_{\mathrm{sb}}}{2\sqrt{\pi}} \sum_{n=0}^{\infty} b_n(K) \frac{1}{n!} \frac{\Gamma[(\frac{1}{2} - n)\frac{K}{1-K}]}{(\frac{1}{2} - n)\Gamma[(\frac{1}{2} - n)\frac{1}{1-K}]} \left(\frac{\epsilon}{\epsilon_{\mathrm{sb}}}\right)^{2n} . \tag{22.201}$$

The function $b_n(K)$ is given by

$$b_n(K) = \begin{cases} 2\sin^2\left[\dfrac{\pi K}{2(1-K)}(1-2n)\right] , & \text{for} \quad K < \frac{1}{3} , \\[4mm] 1 , & \text{for} \quad \frac{1}{3} < K < 1 . \end{cases} \tag{22.202}$$

For very weak damping $K \ll 1$, both the Taylor series (22.197) and the asymptotic series (22.201) can be summed to the simple form ($\epsilon_{\mathrm{sb}} = \Delta_{\mathrm{eff}}$)

$$k^+(\epsilon) \; = \; \pi K \epsilon_{sb}^2 \big/ \sqrt{\epsilon_{sb}^2 + \epsilon^2} \,, \tag{22.203}$$

which agrees with the earlier result (22.94), $k^+ = \gamma_{\mathrm{r}}(T = 0)$.

In the zero bias limit, the asymptotic series (22.201) reduces to the $n = 0$-term. This yields for the forward rate of the unbiased TSS the expression

$$k^+(0) \; = \; \frac{b_0(K)}{\sqrt{\pi}} \, \frac{\Gamma[\frac{K}{2(1-K)}]}{\Gamma[\frac{1}{2(1-K)}]} \, \epsilon_{sb} \,. \tag{22.204}$$

In the narrow regime $1 - K \ll 1$, we get from Eq. (22.204)

$$k^+(0) \; = \; \sqrt{\frac{2}{\pi}(1-K)} \; \epsilon_{sb} \quad \text{with} \quad \epsilon_{sb} = \left(\tfrac{1}{2(1-K)}\right)^{\frac{1}{1-K}} \Delta_{\mathrm{r}} = \frac{2}{\pi} \frac{k_{\mathrm{B}} T_{\mathrm{K}}}{\hbar} \,. \tag{22.205}$$

The second form relates ϵ_{sb} to the Kondo temperature T_{K} of the anisotropic Kondo model in the corresponding regime $\rho J_\perp \ll \rho J_\parallel$ [see Eq. (19.105)].

Consider next the asymptotic expansion for $K > 1$ and large bias. This may be derived from (22.200) upon changing from variable z to $t = e^{-i\pi} u_{1/2} z^K$. We then find in the section $K = p + \kappa$, where $p = 1, 2, \cdots$, and $0 \le \kappa < 1$,

$$k^+(\epsilon) \; = \; \frac{\epsilon}{\sqrt{\pi}} \sum_{m=1}^{\infty} \frac{(-1)^m}{m!} \, \frac{\Gamma(\frac{m}{K}) \sin[\frac{1+p}{K} m\pi] \sin(\frac{p}{K} m\pi)}{K \, \Gamma[\frac{3}{2} + (\frac{1}{K} - 1)m]} \left(\frac{\epsilon}{\epsilon_{sb}}\right)^{(2/K-2)m} \,, \tag{22.206}$$

From this we see that the rate is zero for a symmetric system. This confirms that there is a transition to self-trapping at $K = 1$ in the Ohmic scaling limit (18.57), as is known also from the anisotropic Kondo model. The localization transition has been discussed by Chakravarty [142], Bray and Moore [143], and by Hakim et al. [144].

In Fig. 22.8 (left) the normalized rate k^+/k_{gr}^+ is plotted for various values of K in the regime $K < 1$. The horizontal line represents the particular case $K = \frac{1}{2}$. For $K < \frac{1}{4}$, the full rate is always lower than the golden rule rate. Hence the numerous multi-bounce contributions interfere destructively in this regime. For $\frac{1}{2} < K < 1$, the multi-bounce contributions interfere constructively for all x so that the full rate is always above the golden rule rate. In the regime $\frac{1}{4} < K < \frac{1}{2}$, the rate goes through a maximum as tunneling is increased, and finally falls below the golden rule rate. This reflects constructive interference at small and intermediate $x = (\Delta_{\mathrm{r}}/\epsilon)^{1-K}$, and destructive interference at large x.

Fig. 22.8 (right) shows plots of the normalized rate k^+/k_{gr}^+ for $K > 1$. The normalized rate goes through a maximum which is shifted to higher x when K is increased. At large enough x, the rate k^+ falls below the golden rule rate. Hence there is constructive interference of the tunneling terms at small and intermediate x, and destructive interference in the strong-tunneling regime.

22.8.2 Decoherence at zero temperature: analytic results

One might guess from the known special cases that there is in the spin-boson model a close relationship between the relaxation rate $k^+(\epsilon)$ and the decoherence rate $\gamma(\epsilon)$.

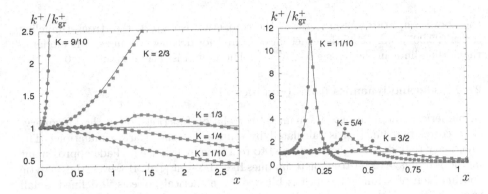

Figure 22.8: The scaled rate k^+/k_{gr}^+ is plotted versus $x = (\Delta_r/\epsilon)^{1-K} = (\epsilon/\omega_c)^K \Delta/\epsilon$ for various values of K, $K < 1$ (left) and $K > 1$ (right). The circles are calculated from the weak-tunneling series, and the squares from the asymptotic strong-tunneling series. The full curve is the respective hypergeometric function expression.

The former describes relaxation to the ground state of the TSS, and the latter loss of phase coherence, which reveals itself in the damping of the oscillatory dynamics as given, e.g., in the expression (22.95). In the weak damping limit, we simply have $\gamma(\epsilon) = \frac{1}{2}k^+(\epsilon)$, as we can see from the last relation in Eq. (22.94). In Ref. [475] it was conjectured that the strong-tunneling series for the decoherence rate at zero temperature in the damping regime $0 < K \le \frac{1}{2}$ reads

$$\gamma(\epsilon) = \frac{\epsilon_{sb}}{2\sqrt{\pi}} \sum_{n=0}^{\infty} \tilde{b}_n(K) \frac{1}{n!} \frac{\Gamma[(\frac{1}{2}-n)\frac{K}{1-K}]}{(\frac{1}{2}-n)\Gamma[(\frac{1}{2}-n)\frac{1}{1-K}]} \left(\frac{\epsilon}{\epsilon_{sb}}\right)^{2n} . \qquad (22.207)$$

The function $\tilde{b}_n(K)$ is given by [note the distinction to Eq. (22.202)]

$$\tilde{b}_n(K) = \sin^2\left[\frac{\pi K}{2(1-K)}(1-2n)\right], \qquad 0 < K \le \frac{1}{2} . \qquad (22.208)$$

The series (22.207) is in agreement with the exact expressions for the decoherence rate known in special cases.

A significant cheque of the expression (22.207) is the asymptotic strong-tunneling limit $\epsilon \to 0$, in which we get

$$\gamma(0) = \frac{1}{\sqrt{\pi}} \sin^2\left[\frac{\pi K}{2(1-K)}\right] \frac{\Gamma[\frac{K}{2(1-K)}]}{\Gamma[\frac{1}{2(1-K)}]} \epsilon_{sb} . \qquad (22.209)$$

The decoherence rate of the unbiased Ohmic TSS has been calculated within the framework of integrable QFT [451]. The exact result is given above in Eqs. (22.13)-(22.15) and it coincides with the expression (22.209).

The series (22.207) matches also the known exact expressions for the decoherence rate in the regimes $K \ll 1$ and K close to $\frac{1}{2}$. These congruences strongly support the above conjecture. Assuming that the conjecture is correct, the strong-tunneling series (22.207) represents the exact lower bound for decoherence in the SB model in the scaling limit in the regime $0 < K \leq \frac{1}{2}$. The bound is saturated at $T = 0$.

22.9 Thermodynamics from dynamics

In numerical computations of dynamical correlation functions, often the imaginary-time correlation function is calculated by standard Monte Carlo simulations. The imaginary-time data are then continued to real time by using, e.g., Padé approximant methods or image reconstruction techniques like Max-Ent [476]. However, the analytic continuation of a function which is blurred by numerical noise is ill-defined: small errors in the input data can lead to exponentially enhanced errors in the output data.

Here we wish to disseminate that sometimes the opposite proceeding, namely computation of thermodynamics from the asymptotic dynamics may have considerable advantages.

In the standard thermodynamic approach, first of all the partition function is computed, which then provides the basis to calculate the static susceptibility, specific heat, etc. This proceeding has two major disadvantages. First, it is difficult to perform approximations in the Coulomb gas representation for the partition function which preserve the symmetry of the imaginary-time interaction, $W(\tau) = W(\hbar\beta - \tau)$. Secondly, the fugacity expansion for Z is often found to converge badly. Remarkably, the fugacity expansion for $\ln Z$ or the free energy (cumulant expansion) usually converges much faster [cf. the discussion in Ref. [477] for an unordered Coulomb gas].

Instead of using the imaginary-time approach for the partition function Z, we can find the thermodynamic properties of an ergodic open system within a dynamical approach by studying the equilibrium state which the system takes at asymptotic time. To proceed, we first note that the equilibrium distribution $\langle \sigma_z \rangle_\infty$ is related to the partition function Z by $\langle \sigma_z \rangle_\infty = (2/\hbar\beta)\partial \ln Z/\partial\epsilon = -(2/\hbar)\partial F/\partial\epsilon$ [cf. Eq. (19.12)]. Thus, we directly obtain the free energy if we integrate the formal expression for $\langle \sigma_z \rangle_\infty$ with respect to the bias. Since the integration constant does not depend on Δ and β, it may be set equal to zero without any restriction. It is just in this way that we directly obtain an exact formal series expression for $\ln Z$ or the free energy.[14] We now briefly sketch this approach, and then use it to show that the damped system has a surprising universal behavior at low temperature (see also [220]).

Based on the formal solution (21.53) for $\langle \sigma_z \rangle_\infty$, we obtain upon integration with respect to the bias the series expansion of the free energy in the form

$$F(T, \epsilon) \;=\; \lim_{t\to\infty} \frac{\hbar}{2} \sum_{m=1}^{\infty} (-1)^{m-1} \int_0^t \mathcal{D}_{2m,0}\{t_j\} \; \frac{1}{2^m} \sum_{\{\xi_j\}} F_m^{(-)} A_m^{(s)} \,, \qquad (22.210)$$

---

[14]We recall that the imaginary-time approach directly leads to a series expression for the partition function Z (cf. Chapter 19).

where

$$A_m^{(s)} = \cos\left(\epsilon \sum_{j=1}^{m} \xi_j \tau_j\right) \Big/ \sum_{j=1}^{m} \xi_j \tau_j . \tag{22.211}$$

We are not allowed to interchange in Eq. (22.210) $\lim_{t \to \infty}$ with the summation of the series since the limit $t \to \infty$ would give a divergent result for each individual term.

Secondly, we use the fact that the static (nonlinear) susceptibility $\overline{\chi}_z$ is related to $\langle \sigma_z \rangle_\infty$ by $\overline{\chi}_z = (2/\hbar)\partial\langle \sigma_z \rangle_\infty/\partial\epsilon$ [cf. Eq. (19.13)]. Using this relation, we obtain from the series expression (21.53) for $\langle \sigma_z \rangle_\infty$ the exact formal solution for $\overline{\chi}_z$ in the form

$$\overline{\chi}_z(T, \epsilon) = \lim_{t \to \infty} \frac{2}{\hbar} \sum_{m=1}^{\infty} (-1)^{m-1} \int_0^t \mathcal{D}_{2m,0}\{t_j\} \frac{1}{2^m} \sum_{\{\xi_j\}} F_m^{(-)} C_m^{(s)} , \tag{22.212}$$

where

$$C_m^{(s)} = \cos\left(\epsilon \sum_{j=1}^{m} \xi_j \tau_j\right) \sum_{j=1}^{m} \xi_j \tau_j . \tag{22.213}$$

Let us now determine the asymptotic low temperature expansion of F and $\overline{\chi}_z$. First, we observe that the temperature enters into the expressions (22.210) and (22.212) only through the pair interaction $Q'(\tau)$. Secondly, we write $Q'(\tau) = Q'_0(\tau) + Q'_1(\tau)$, where $Q'_0(\tau)$ is the pair interaction for zero temperature, and $Q'_1(\tau)$ is the finite temperature contribution. The leading correction at low T is easily computed for the form (18.46) of $G_{\text{lf}}(\omega)$ from Eq. (18.47), and is

$$Q'_1(\tau) = \kappa_s (\hbar\beta\omega_{\text{ph}})^{1-s}(\tau/\hbar\beta)^2\left\{1 + \mathcal{O}[(\tau/\hbar\beta)^2]\right\} , \tag{22.214}$$

where

$$\kappa_s = 2\delta_s \Gamma(1+s)\zeta(1+s) , \tag{22.215}$$

and where $\Gamma(z)$ and $\zeta(z)$ are the gamma and Riemann zeta function. With this expression for $Q'_1(\tau)$, the low temperature expansion of the blip correlation factor G_m defined in Eq. (21.38) takes the form

$$G_m = G_m^{(0)}\left\{1 - \kappa_s \omega_{\text{ph}}^{1-s}\left(\frac{k_{\text{B}}T}{\hbar}\right)^{1+s}\left(\sum_{j=1}^{m} \xi_j \tau_j\right)^2 + \mathcal{O}(T^{3+s})\right\} , \tag{22.216}$$

where $G_m^{(0)}$ is the blip interaction factor at zero temperature. Inserting now (15.96) into the expressions (15.92) and (15.93), and observing that the resulting powers of $\sum_j \xi_j \tau_j$ can be generated by differentiations with respect to the bias, we immediately find for F and $\overline{\chi}_z$ the low temperature expansions

$$F(T, \epsilon) = F(0, \epsilon) - \frac{1}{4}\kappa_s \overline{\chi}_z(0, \epsilon)\left(\hbar\omega_{\text{ph}}\right)^{1-s}(k_{\text{B}}T)^{1+s} + \mathcal{O}(T^{3+s}) , \tag{22.217}$$

$$\overline{\chi}_z(T, \epsilon) = \overline{\chi}_z(0, \epsilon) + \kappa_s \overline{\chi}_z''(0, \epsilon)\omega_{\text{ph}}^{1-s}(k_{\text{B}}T/\hbar)^{1+s} + \mathcal{O}(T^{3+s}) . \tag{22.218}$$

Here, $F(0, \epsilon$ and $\overline{\chi}_z(0, \epsilon)$ are the free energy and static nonlinear susceptibility at zero temperature and $\overline{\chi}''_z = \partial^2 \overline{\chi}_z/(\partial \epsilon)^2$. The expressions (22.217) and (22.218) are the asymptotic expansions of F and $\overline{\chi}_z$ for a two-state system described by a spectral density of the form (18.46) under the limitations discussed below. The leading temperature dependence is given by the power law T^{1+s}. The prefactor depends on the factor γ_s and on $\overline{\chi}_z^{(0)}$. Notice that the bias enters only through the nonlinear susceptibility at zero temperature. Observing now that the specific heat is defined by

$$c(T)/k_{\rm B} = -k_{\rm B}^{-1} T \partial^2 F/(\partial T)^2 \,, \qquad (22.219)$$

we immediately see that the leading dependence of $c(T)$ on temperature is T^s. Thus it is natural to adjust the definition of the Wilson ratio, Eq. (19.53), to non-Ohmic spectral density $(s \neq 1)$. As anticipated in Eq. (6.121), the generalized Wilson ratio is

$$R_s \equiv \lim_{T \to 0} \frac{4c(T)/k_{\rm B}}{\overline{\chi}_z \, (\hbar \omega_{\rm ph})^{1-s}(k_{\rm B} T)^s} \,, \qquad (22.220)$$

where $\overline{\chi}_z$ is the nonlinear static susceptibility. We then obtain from Eq. (22.217)

$$R_s = 2s\Gamma(2+s)\zeta(1+s)\,\delta_s \,. \qquad (22.221)$$

These results for the free energy, the specific heat, and the Wilson ratio are generally valid provided that the system has a nonzero static susceptibility at zero temperature. This generally holds for arbitrary power s when the system is *biased* (nonlinear static susceptibility). The linear susceptibility is finite in the super-Ohmic case for all δ_s, and in the Ohmic case for K in the regime $0 < K < 1$. Thus in these cases the relation (22.221) is also valid for the unbiased system. For $s = 1$ and $K > 1$, and in the sub-Ohmic case $s < 1$ and any $\delta_s > 0$, the limit $\lim_{\epsilon \to 0} \overline{\chi}_z(T = 0, \epsilon)$ is divergent. Hence in these parameter regimes, the above analysis is not applicable. Instead of that, in these regimes the high temperature expansions for $\overline{\chi}(T, \epsilon = 0)$ and $F(T)$, which are the expansions in powers of Δ^2, Eq. (22.210) and Eq. (22.212), are valid down to $T = 0$. For an Ohmic bath, the Wilson ratio simplifies to

$$R_1 = 2K\pi^2/3 \,. \qquad (22.222)$$

Using the correspondence relation (19.102) for K, we see that the Wilson ratio (22.222) of the Ohmic TSS coincides with the Wilson ratio (19.104) of the s-d model. The Wilson ratio has been studied numerically for the s-d model, and excellent agreement with the formula (22.222) has been found [471].

23. The driven two-state system

With the advance of laser technology and the possibility for experimental time resolution in the sub-picosecond regime, there has been growing interest in the study of the dynamics of quantum systems that are driven by strong time-dependent external fields. Quantum dynamics of explicitly time-dependent Hamiltonians shows a variety of novel effects, such as the phenomenon of coherent destruction of tunneling [478], stabilization of localized states which would otherwise decay [479], and the possibility of controlling the tunneling dynamics with pulsed monochromatic light [480, a] [481] and sinusoidal fields [480, b] [481]. For a review, see Refs. [482, 483]. Here we restrict the attention to a two-state system which is simultaneously exposed to a fluctuating force by the surroundings and to deterministic time-dependent forces.

23.1 Time-dependent external fields

We generalize the Hamiltonian (18.15) of the global system by taking into account the interaction with external time-dependent fields. We assume that the external fields couple to the system's operators σ_z and σ_x only, and have no effect on the bath.

23.1.1 Diagonal and off-diagonal driving

The coupling of an external field to σ_z describes, e.g., an electric field coupled to the dipole moment of the TSS, or in the rf SQUID device discussed in Subsection 3.3.2 a time-dependent magnet flux threading the ring. This coupling leads to a temporal modulation of the bias which is superimposed on the static bias. This entails in the Hamiltonian (18.15) the substitution

$$\epsilon \quad \Longrightarrow \quad \epsilon(t) = \epsilon_0 + \epsilon_1(t) . \tag{23.1}$$

Here, ϵ_0 is the bias frequency related to the intrinsic static strain field, and $\epsilon_1(t)$ is a bias modulation due to the externally applied time-dependent force. Regarding electron transfer in a solvent, it is conceivable to control charge tunneling by application of strong continuous laser fields. For charge transfer in nano-structured devices, we may think of regulating the dynamics by turning on microwave irradiation or a high-frequency voltage. In pump-probe set-ups, the relevant system is subject to a pulse-shaped driving force. The modulation of the bias energy leads to modified bias phases. The accumulated phases for a path with a sequence of m blips is given by

$$\varphi_m = \sum_{j=1}^{m} \xi_j \vartheta(t_{2j}, t_{2j-1}) , \quad \text{where} \quad \vartheta(t_2, t_1) = \int_{t_1}^{t_2} dt' \, \epsilon(t') . \tag{23.2}$$

The bias factors replacing the forms (21.48) are expressed in terms of φ_m as

$$B_m^{(s)} = \cos(\varphi_m) , \qquad B_m^{(a)} = \sin(\varphi_m) . \tag{23.3}$$

Secondly, a "quadrupole"-like time-dependent coupling induces a temporal variation of the barrier opacity. A barrier modulation can be realized, e.g., in a superconducting loop with two Josephson junctions [484]. The quadrupole coupling leads to a *multiplicative* modification of the tunneling coupling. For harmonic pulsation of the barrier, we have in the Hamiltonian (18.15) the substitution [485]

$$\Delta \quad \Longrightarrow \quad \Delta(t) = \Delta \exp[\mu \cos(\nu t)] , \tag{23.4}$$

where ν is the angular frequency and μ a suitable dimensionless amplitude.

23.1.2 Exact formal solution

In the exact formal series expressions obtained in Chapter 21, the time-dependent tunneling matrix element is accounted for by the substitution [e.g., in Eq. (21.42)]

$$\Delta^m \quad \longrightarrow \quad \Delta_m(\{t_j\}) = \prod_{j=1}^{m} \Delta(t_j) . \tag{23.5}$$

In the non-convolutive case, the Laplace transform of the master equation (21.73) for $P(t) = \langle \sigma_z \rangle_t$ takes the form[1]

$$\lambda \hat{P}(\lambda) = 1 + \int_0^\infty dt' \, e^{-\lambda t'} \left[\hat{K}_\lambda^{(a)}(t') - \hat{K}_\lambda^{(s)}(t') P(t') \right] \tag{23.6}$$

with the kernel $\hat{K}_\lambda^{(s,a)}(t') = \int_0^\infty d\tau \, e^{-\lambda \tau} K_z^{(s,a)}(t' + \tau, t')$.

Consider now first the case of a periodic bias modulation with period $\mathcal{T} = 2\pi/\omega$,

$$\epsilon(t) = \epsilon(t + 2\pi n/\omega) . \tag{23.7}$$

Then the kernel $\hat{K}_\lambda^{(s,a)}(t)$ is time-periodic and may be written as Fourier series,

$$\hat{K}_\lambda^{(s,a)}(t) = \sum_{m=-\infty}^{\infty} k_m^{(s,a)}(\lambda) \, e^{-i m \omega t} . \tag{23.8}$$

With employment of Eq. (23.8) in Eq. (23.6), one obtains the solution of Eq. (23.6) in recursive form [486]. The resulting population $P(t)$ consists of a transient part and a part representing persistent oscillations. The transient behavior is determined by the zeros of the characteristic equation

$$\lambda + i m \omega + k_0^{(s)}(\lambda + i m \omega) = 0 . \tag{23.9}$$

The solution of Eq. (23.9) for $m = 0$, denoted by λ_0, yields the primary transient. The time scale for the decay of this transient is set by the smallest real negative part of λ_0. All the other solutions of Eq. (23.9) for $m = \pm 1, \pm 2, \ldots$, are obtained from λ_0 by shifts with a multiple of the driving frequency in imaginary direction, $\lambda_m = \lambda_0 - i m \omega$. The respective transient contributions decay on the same time scale as the primary transient.

After the transient terms have died out, $P(t)$ oscillates persistently with period $\mathcal{T}$. The asymptotic regime is described by the Fourier series [486] (see also Ref. [487])

$$\lim_{t \to \infty} P(t) = P^{(as)}(t) = \sum_{m=-\infty}^{\infty} p_m \, e^{-i m \omega t} . \tag{23.10}$$

The Fourier coefficient p_m is found from Eq. (23.6) as the residuum of the pole of $\hat{P}(\lambda)$ at $\lambda = -i m \omega$. The coefficients obey the recursive relations

---

[1]We restrict the attention to the discussion of $\langle \sigma_z \rangle_t$. The corresponding study of $\langle \sigma_x \rangle_t$ is without difficulty and left to the reader.

$$p_0 = \frac{k_0^{(a)}(0)}{k_0^{(s)}(0)} - \sum_{m \neq 0} \frac{k_m^{(s)}(0)}{k_0^{(s)}(0)} p_m ,$$

$$p_m = \frac{1}{-i\,m\omega} \left(k_m^{(a)}(-i\,m\omega) - \sum_n k_{m-n}^{(s)}(-i\,m\omega) p_n \right) .$$

(23.11)

For monochromatic modulation of the bias

$$\epsilon(t) = \epsilon_0 + \epsilon_1 \cos(\omega t) ,$$ (23.12)

the kernel coefficients $k_{2m}^{(a)}(\lambda)$ and $k_{2m+1}^{(s)}(\lambda)$ are zero if $\epsilon_0 = 0$. With this property, we find from Eq. (23.11) the selection rule[2]

$$p_{2m} = 0 .$$ (23.13)

Thus in the absence of a static bias only terms with odd multiples of the fundamental frequency contribute to the dynamics of $P^{(as)}(t)$.

In the general case, the kernels $k_m^{(s,a)}(\lambda)$ are power series expressions in Δ^2. In the NIBA they are of order Δ^2 and take the simple form

$$k_m^{(s)}(\lambda) = \Delta^2 \int_0^\infty d\tau\, e^{-\lambda\tau} e^{-Q'(\tau)} \cos[Q''(\tau)] A_m^{(s)}(\tau) ,$$

$$k_m^{(a)}(\lambda) = \Delta^2 \int_0^\infty d\tau\, e^{-\lambda\tau} e^{-Q'(\tau)} \sin[Q''(\tau)] A_m^{(a)}(\tau) .$$

(23.14)

with

$$A_{2m}^{(s)}(\tau) = (-1)^m e^{-i\,m\omega\tau} \cos(\epsilon_0\tau) J_{2m}\left(\frac{2\epsilon_1}{\omega} \sin\frac{\omega\tau}{2}\right) ,$$

$$A_{2m}^{(a)}(\tau) = (-1)^m e^{-i\,m\omega\tau} \sin(\epsilon_0\tau) J_{2m}\left(\frac{2\epsilon_1}{\omega} \sin\frac{\omega\tau}{2}\right) ,$$

(23.15)

$$A_{2m+1}^{(s)}(\tau) = (-1)^{m+1} e^{-i\,(m+1/2)\omega\tau} \sin(\epsilon_0\tau) J_{|2m+1|}\left(\frac{2\epsilon_1}{\omega} \sin\frac{\omega\tau}{2}\right) ,$$

$$A_{2m+1}^{(a)}(\tau) = (-1)^m e^{-i\,(m+1/2)\omega\tau} \cos(\epsilon_0\tau) J_{|2m+1|}\left(\frac{2\epsilon_1}{\omega} \sin\frac{\omega\tau}{2}\right) ,$$

(23.16)

where $J_n(z)$ is a Bessel function of the first kind. We see from these forms that the oscillating bias suppresses long blips. Thus, in the regimes in which the NIBA is appropriate in the absence of the oscillating field, it is even more qualified in the presence of time-dependent driving. One can now use these forms to calculate the dynamics of $P(t)$ numerically in the NIBA [483].

---

[2] $\langle\sigma_y\rangle_t^{(as)}$ obeys the same selection rule [cf. Eq. (21.10)], and $\langle\sigma_x\rangle_t^{(as)}$ has only even harmonics.

23.1.3 Linear response

When the driving amplitude ϵ_1 is small compared to the driving frequency ω, we may expand the kernels (23.14) in a power series in ϵ_1. Within linear response, the relevant kernels are of order unity and of order ϵ_1. We then find in the Fourier series (23.10) the limitation $|m| \leq 1$. The term $k_0^{(s,a)}(\lambda)$ is independent of ϵ_1, and $k_{\pm 1}^{(s,a)}(\lambda) = \mathcal{O}(\epsilon_1)$. As a result, $P^{(as)}(t)$ behaves as

$$P^{(as)}(t) = P_\infty + \hbar\epsilon_1 \left[\tilde{\chi}(\omega)\, e^{-i\omega t} + \tilde{\chi}(-\omega)\, e^{i\omega t} \right]/4 , \qquad (23.17)$$

where $P_\infty = \lim_{\epsilon_1 \to 0} k_0^{(a)}(0)/k_0^{(s)}(0) = \tanh(\frac{1}{2}\hbar\beta\epsilon_0)$ [cf. Eq. (21.86)]. The evaluation of the recursive relations (23.11) for $|m| \leq 1$ gives

$$\tilde{\chi}(\omega) = \lim_{\epsilon_1 \to 0} \frac{4}{\hbar\epsilon_1 \left[-i\,\omega + k_0^{(s)}(-i\,\omega)\right]} \left[k_1^{(a)}(-i\,\omega) - P_\infty k_1^{(s)}(-i\,\omega)\right] . \qquad (23.18)$$

This is the dynamical susceptibility in the NIBA. The reader may easily convince himself that the expression (23.18) is directly connected with the spectral correlation function (22.76) by the fluctuation-dissipation theorem (22.165).

23.1.4 The Ohmic case with Kondo parameter $K = \frac{1}{2}$

For an Ohmic heat bath with damping strength $K = \frac{1}{2}$ and $\hbar\beta\omega_c \gg 1$, the dynamics can be solved exactly, up to quadratures, in the presence of diagonal and off-diagonal driving. In generalization of the expression (22.135), we obtain [487, 485]

$$P(t) = \exp\left(-\int_0^t d\tau\,\gamma(\tau)\right) + P_a(t) , \qquad \gamma(\tau) = \frac{\pi}{2}\frac{\Delta^2(\tau)}{\omega_c} ,$$

$$P_a(t) = \int_0^t dt_2\,\Delta(t_2) \exp\left(-\int_{t_2}^t d\tau\,\gamma(\tau)\right) \qquad (23.19)$$

$$\times \int_0^{t_2} dt_1\,\Delta(t_1)\sin[\vartheta(t_2,t_1)]\, e^{-Q'(t_2-t_1)} \exp\left(-\frac{1}{2}\int_{t_1}^{t_2} d\tau'\,\gamma(\tau')\right) ,$$

where $\vartheta(t_2,t_1)$ is given in Eq. (23.2). The asymptotic dynamics is described by $P_a(t)$.

Consider now the case of time-independent tunneling coupling in some more detail. We find from Eq. (23.19) for the Fourier amplitudes p_m the exact expression

$$p_m(\omega,\epsilon_1) = \frac{\Delta^2}{-i\,m\omega + \gamma} \int_0^\infty d\tau\, e^{i\,m\omega\tau - \gamma\tau/2 - Q'(\tau)} A_m^{(a)}(\tau) . \qquad (23.20)$$

The spectral amplitude of the fundamental frequency, $\eta_1(\omega,\epsilon_1) = 4\pi|p_1(\omega,\epsilon_1)/\hbar\epsilon_1|^2$ [cf. Eq. (23.40)], is plotted versus ω for various temperatures in Fig. 23.1.[3] At high temperature, $\eta_1(\omega)$ is peaked at $\omega = 0$, and there is only little structure in the frequency dependence. As the temperature is decreased, resonances are formed at fractional values of the static bias, $\omega = \epsilon_0/n$ $(n = 1, 2, \dots)$. At these frequencies, the driving-induced coherent motion is amplified.

---

[3]Figs. 23.1 – 23.4 are by courtesy of M. Grifoni, P. Hänggi and L. Hartmann.

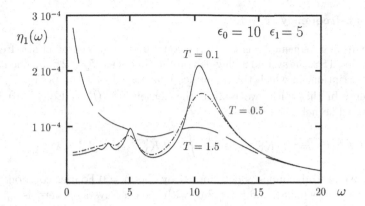

Figure 23.1: The spectral amplitude η_1 is plotted versus ω for different temperatures. See text for details. Frequencies and temperature are in units of $\gamma = \pi\Delta^2/2\omega_c$.

23.2 Markovian regime

When the system's characteristic motion is slow on the time scale τ on which the kernels $K_z^{(s,a)}(t, t-\tau)$ decay, the incoherent dynamics at long times is described by a time-local master equation. The GME (21.73) simplifies to the form

$$\dot{P}(t) = \mathcal{M}_z^{(a)}(t) - \mathcal{M}_z^{(s)}(t)P(t) , \qquad (23.21)$$

with

$$\mathcal{M}_z^{(s)}(t) = \int_0^\infty d\tau\, K_z^{(s)}(t, t-\tau) = \Delta^2 \int_0^\infty d\tau\, e^{-Q'(\tau)} \cos[Q''(\tau)] \cos[\vartheta(t, t-\tau)] ,$$

$$\mathcal{M}_z^{(a)}(t) = \int_0^\infty d\tau\, K_z^{(a)}(t, t-\tau) = \Delta^2 \int_0^\infty d\tau\, e^{-Q'(\tau)} \sin[Q''(\tau)] \sin[\vartheta(t, t-\tau)] .$$

The respective second forms are the NIBA expressions, and $\vartheta(t_2, t_1)$ is given in Eq. (23.2). The master equation (23.21) can be solved directly. The solution with initial value $P(0) = 1$ is

$$P(t) = \exp\left\{-\int_0^t dt'\, \mathcal{M}_z^{(s)}(t')\right\} + \int_0^t dt'\, \mathcal{M}_z^{(a)}(t') \exp\left\{-\int_{t'}^t dt''\, \mathcal{M}_z^{(s)}(t'')\right\} .$$

In the adiabatic limit we may put $\vartheta(t, t-\tau) = \epsilon(t)\,\tau$. With this form, the kernels are found as linear combinations of the forward/backward Golden Rule rates $k^\pm$ introduced in Subsection 20.2.1,

$$\mathcal{M}_z^{(s,a)}(t) = k^+[\epsilon(t)] \pm k^-[\epsilon(t)] , \qquad (23.22)$$

In these expressions the time-dependence is solely in the bias $\epsilon(t)$. The adiabatic rates obey detailed balance with respect to the current bias $\epsilon(t)$.

23.3 High-frequency regime

When the driving frequency ω is very large compared to the characteristic frequencies of the damped TSS, the system is sluggish on the time scale $\mathcal{T}_\omega = 2\pi/\omega$. Then a coarse-grained description, in which the dynamics is averaged over the period $\mathcal{T}_\omega = 2\pi/\omega$, is expedient. In this spirit, we replace the kernel $\hat{K}_\lambda^{(s,a)}(t)$ in Eq. (23.6) with the time-averaged kernel

$$\hat{K}_\lambda^{(s,a)}(t) \;\Longrightarrow\; \langle \hat{K}_\lambda^{(s,a)} \rangle \;\equiv\; \frac{1}{\mathcal{T}_\omega}\int_0^{\mathcal{T}_\omega} dt\, \hat{K}_\lambda^{(s,a)}(t) \;=\; k_0^{(s,a)}(\lambda)\,. \tag{23.23}$$

Then the averaged dynamics is determined by the $m=0$ Fourier component of the kernels in the series (23.8), and Eq. (23.6) with the time-averaged kernels $k_0^{(s,a)}(\lambda)$ can be easily solved for $\hat{P}(\lambda)$. The resulting expression is analogous in form to Eq. (21.79),

$$\hat{P}(\lambda) \;=\; \frac{1 + k_0^{(a)}(\lambda)/\lambda}{\lambda + k_0^{(s)}(\lambda)}\,. \tag{23.24}$$

This expression establishes the averaged transient and the long-time behavior.

The characteristics of the transient dynamics is determined by the zeros of the equation $\lambda + k_0^{(s)}(\lambda) = 0$. With use of the decomposition

$$J_0(2z\sin\alpha) \;=\; \sum_{n=-\infty}^{\infty} J_n^2(z)\cos(2n\alpha) \tag{23.25}$$

in the expression for $k_0^{(s)}(\lambda)$ given in Eq. (23.14) with (23.15), the pole condition of Eq. (23.24) is found to read

$$\lambda + \frac{1}{2}\sum_{n=-\infty}^{\infty} J_n^2\!\left(\frac{\epsilon_1}{\omega}\right)\Big\{ g[\lambda + i\,(n\omega + \epsilon_0)] + g[\lambda - i\,(n\omega + \epsilon_0)]\Big\} \;=\; 0\,, \tag{23.26}$$

where the kernel $g(\lambda)$ is given in Eq. (22.1).

In the incoherent regime, we keep the relaxation pole at $\lambda = -\gamma_r \equiv -k_0^{(s)}(\lambda = 0)$ and disregard all other poles since their residua are very small. Thus we obtain

$$P(t) \;=\; P_\infty + [\,1 - P_\infty\,]\,e^{-\gamma_r t} \tag{23.27}$$

where $P_\infty = k_0^{(a)}(0)/k_0^{(s)}(0)$ is the time-averaged equilibrium value. The relaxation rate is given by the series[4]

$$\gamma_r(\epsilon_0; \omega, \epsilon_1) \;=\; \sum_{n=-\infty}^{\infty} J_n^2\!\left(\frac{\epsilon_1}{\omega}\right)\Big[k^+(\epsilon_0 + n\omega) + k^-(\epsilon_0 + n\omega)\Big]\,. \tag{23.28}$$

---

[4]Representations of the form (23.28) for transport quantities have a long tradition since the pioneering work by Tien and Gordon [488].

The formula (23.28) describes the inclusive relaxation rate γ_r as a sum over all possible decay channels. The individual channels represent tunneling processes with simultaneous emission and absorption of a fixed number of quanta with fundamental frequency ω, and the weight factor for n quanta is $J_n^2(\epsilon_1/\omega)$. The partial rates in Eq. (23.28) are given in terms of the relaxation rates $k^{\pm}$ for a static bias $\epsilon_n = \epsilon_0 + n\omega$ for the dissipative mechanism under consideration. These rates have been discussed in Section 20.2. The inclusive relaxation rate γ_r sensitively depends on the parameters of the driving field. We see in Fig. 23.2 that the rate is enhanced compared to the static case when the resonance condition $\epsilon_0 = \pm n\omega$ is met and when the weight does not fall on a zero of the respective Bessel function. Upon tuning ϵ_1/ω to a zero of one of the Bessel functions in Eq. (23.28), the corresponding decay channel is missing.

The striking dependence of γ_r on the external field parameters may be utilized, e.g., to control electron transfer rates in chemical reaction processes.

The fast ac field leads to an asymptotic population $P_\infty = k_0^{(a)}(0)/k_0^{(s)}(0)$ which may differ drastically from the static case, $P_\infty = \tanh(\hbar\beta\epsilon_0/2)$. Even inverted population can be reached by means of a suitable choice of the parameters ϵ_1 and ω, as predicted, e.g., in the works in Ref. [489].

Next, consider a symmetric Ohmic TSS. Using for $g(\lambda)$ the NIBA form (22.2), the pole condition (23.26) can be written in the regime $\omega \gg 2\pi K k_B T/\hbar$ as [485]

$$\lambda + g(\lambda)\left\{ J_0^2(\epsilon_1/\omega) + \left(\frac{2\pi K + \hbar\beta\lambda}{\hbar\beta\omega}\right)^2 \sum_{n \neq 0} \frac{J_n^2(\epsilon_1/\omega)}{n^2} \right\} = 0 \,. \qquad (23.29)$$

The first term in the curly bracket dominates for large ω when the ratio ϵ_1/ω is sufficiently distant from a zero of the J_0-Bessel function. Then the pole condition simply reads

$$\lambda + g(\lambda)J_0^2(\epsilon_1/\omega) = 0 \,. \qquad (23.30)$$

Thus the results presented in Subsections 22.2.2 and 22.2.3 directly apply to the present case if we replace Δ by a driving-renormalized tunneling matrix element,

$$\Delta \longrightarrow J_0(\epsilon_1/\omega)\,\Delta \,. \qquad (23.31)$$

There is weakly-damped coherent dynamics in the transient regime when $K \ll 1$.

The coherent dynamics is suppressed when ϵ_1/ω is tuned to a zero of the Bessel function $J_0(\epsilon_1/\omega)$. This phenomenon of an undamped or weakly damped driven TSS is termed "coherent destruction of tunneling" [490, 491]. At the destruction point $J_0(\epsilon_1/\omega) = 0$, the system relaxes incoherently with the rate

$$\gamma_r = g(0)\left(\frac{2\pi K}{\hbar\beta\omega}\right)^2 \sum_{n \neq 0} \frac{J_n^2(\epsilon_1/\omega)}{n^2} \,. \qquad (23.32)$$

Consider now the coherent regime for $K \ll 1$ and $\epsilon_1/\omega \ll 1$. In this case we may perform a systematic weak-damping study as explained in Section 22.3. For large ω, the leading effects of the driving force are taken into account by expanding the

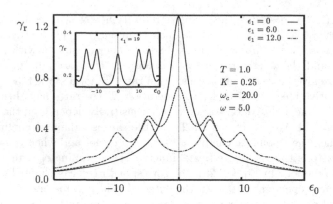

Figure 23.2: The inclusive rate γ_r is shown as a function of ϵ_0 for different amplitudes ϵ_1. The frequency parameters are given in units of $\Delta/2$. In the inset, the amplitude ϵ_1 is tuned on a zero of the $n = 1$ Bessel function, $J_1(\epsilon_1/\omega) = 0$ so that the first side band is absent.

Bessel functions in Eq. (23.26) up to terms of order $(\epsilon_1/\omega)^2$. In this approximation, the *undamped* driven system is characterized by three bias frequencies $\mu_1 = \epsilon_0$, $\mu_2 = \epsilon_0 + \omega$, and $\mu_3 = \epsilon_0 - \omega$ and three tunneling frequencies ν_j. The squared ν_j are the solutions of a cubic equation in ν^2,

$$\nu^6 - a_4\nu^4 + a_2\nu^2 - a_0 = 0 , \tag{23.33}$$

$$a_0 = \mu_1^2\mu_2^2\mu_3^2 + [\Delta_1^2\mu_2^2\mu_3^2 + \text{cycl.}] , \qquad a_2 = \Delta_1^2(\mu_2^2 + \mu_3^2) + \mu_2^2\mu_3^2 + \text{cycl.} ,$$

$$a_4 = \mu_1^2 + \Delta_1^2 + \text{cycl.} , \qquad \Delta_1^2 = (1 - \epsilon_1^2/2\omega^2)\Delta^2 , \qquad \Delta_2^2 = \Delta_3^2 = (\epsilon_1^2/2\omega^2)\Delta^2 .$$

The undamped system performs a superposition of coherent oscillations

$$P_{\text{undamped}}(t) = p_0 + \sum_{j=1}^{3} p_j \cos(\nu_j t) , \tag{23.34}$$

with the amplitudes (p_2, p_3 cycl.)

$$p_1 = \frac{(\nu_1^2 - \mu_1^2)(\nu_1^2 - \mu_2^2)(\nu_1^2 - \mu_3^2)}{\nu_1^2(\nu_1^2 - \nu_2^2)(\nu_1^2 - \nu_3^2)} , \qquad p_0 = \prod_{j=1}^{3} \frac{\mu_j^2}{\nu_j^2} . \tag{23.35}$$

For weak damping, the shift of the poles can be calculated as explained in Section 22.3. Disregarding irrelevant frequency shifts, one finds

$$P(t) = \sum_{j=1}^{3} p_j \cos(\nu_j t)\, e^{-\gamma_j t} + (p_0 - P_\infty)\, e^{-\gamma_0 t} + P_\infty . \tag{23.36}$$

The rates are linear combinations of one-phonon emission and absorption processes [cf. Eq. (22.91)] for the possible transition frequencies.[5] One finds again by comparison of the exact weak-damping expressions with the NIBA results that the NIBA disregards the frequency shifts in the transition frequencies and amplitudes of the one-phonon processes resulting from the tunneling coupling.

One final remark is in order. On a closer look, the steady state exhibits oscillations about the mean value P_∞ given in Eq.(23.27). We find from Eq. (23.11) for large ω

$$P^{(\mathrm{as})}(t) = P_\infty + \sum_{m \neq 0} \frac{1}{-i\,m\omega} \left(k_m^{(\mathrm{a})}(-i\,m\omega) - k_m^{(\mathrm{s})}(-i\,m\omega)P_\infty \right) e^{-i\,m\omega t} . \qquad (23.37)$$

With increasing driving-frequency, the oscillations of $P^{(\mathrm{as})}(t)$ about P_∞ become suppressed since the $m \neq 0$ Fourier components get less important.

23.4 Quantum stochastic resonance

The stochastic resonance phenomenon refers to the amplification of the response to an applied periodic signal at a certain optimal value of the noise strength and is a cooperative effect of friction, noise, and periodic driving in a bistable system. This phenomenon has been dubbed *stochastic resonance* (SR) since classically the maximal output signal occurs when the thermal hopping rate is in resonance with the frequency of the driving force. It has been argued that this phenomenon is of fundamental importance in biological evolution. Several comprehensive reviews on classical stochastic resonance are available [492]. Qualitatively new signatures of stochastic resonance appear in the quantum regime [493, 494].

When the condition (3.127) is met, the double well system is well described by the discrete two-state system. Then the quantity of interest in SR is the power spectrum

$$S(\nu) = \int_{-\infty}^{\infty} d\tau\, e^{i\nu\tau}\, \overline{C}^{(\mathrm{as})}(\tau) , \qquad (23.38)$$

where $\overline{C}^{(\mathrm{as})}(\tau)$ is the time-averaged steady-state population correlation function,

$$\overline{C}^{(\mathrm{as})}(\tau) \equiv \lim_{\tau \to \infty} \frac{\omega}{2\pi}\, \mathrm{Re} \int_0^{2\pi/\omega} dt\, \langle \sigma_z(t+\tau)\sigma_z(t) \rangle = \sum_{m=-\infty}^{\infty} |p_m(\omega,\epsilon_1)|^2\, e^{-i\,m\omega\tau} .$$

The set of amplitudes $\{p_m(\omega,\epsilon_1)\}$ obeys the recursion relations (23.11). The power spectrum takes the form

$$S(\nu) = 2\pi \sum_{m=-\infty}^{\infty} |p_m(\omega,\epsilon_1)|^2\, \delta(\nu - m\omega) . \qquad (23.39)$$

A quantitative study of the power amplitude

---

[5]The various rate expressions are given in Ref. [448].

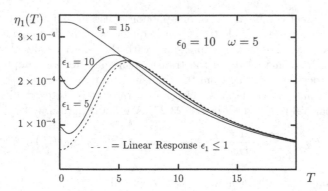

Figure 23.3: Depiction of QSR in the deep quantum regime for Ohmic damping $K = \frac{1}{2}$. The power amplitude η_1 of the fundamental frequency is plotted as a function of temperature for different driving amplitudes. See text for details. The units are the same as in Fig. 23.1.

$$\eta_m(\omega, \epsilon_1) \;=\; 4\pi \, |p_m(\omega, \epsilon_1)/\hbar\epsilon_1|^2 \tag{23.40}$$

can be performed by first calculating numerically the kernels $k_m^{(s,a)}(\lambda)$ for the environment of interest and subsequently solving the recursive relations (23.11). In the sequel, we confine ourselves to an Ohmic coupling with a large cutoff.

In classical SR, the spectral amplification is maximal for a *symmetric* system [492], whereas in the deep quantum regime and $K < 1$, QSR is only effective in the presence of a static bias. One finds that QSR is most striking when the static bias ϵ_0 is larger than the driving amplitude ϵ_1. For $K = \frac{1}{2}$, the power amplitudes are given by the expression (23.20). In Fig. 23.3, the power amplitude $\eta_1(T)$ for $K = \frac{1}{2}$ is plotted as a function of temperature for different driving amplitudes. When $\epsilon_1 > \epsilon_0$, the power amplitude decreases monotonously with increasing T. As ϵ_1 is decreased, a minimum at low T followed by a QSR maximum at $T \approx \hbar\omega/k_B$ is formed. For $\epsilon_1 < 5\gamma$, the QSR can be studied within linear response theory.

In the linear response regime, we find from Eqs. (23.17) and (23.40) the relation $p_1 = -\hbar\epsilon_1\tilde{\chi}(\omega)/4$. In the regime $\hbar\omega < 2\pi\alpha k_B T$ and $k_B T > \hbar\Delta$ and/or $K > 1$, the NIBA form (23.18) is correct. From this we obtain in the Lorentzian approximation for the quasi-elastic peak centered at $\omega = 0$ the expression

$$\eta_1(\omega) \;=\; \frac{\pi}{(2k_B T)^2} \, \frac{1}{\cosh^4(\hbar\epsilon_0/2k_B T)} \, \frac{\gamma_r^2}{\omega^2 + \gamma_r^2} \,, \tag{23.41}$$

where $\gamma_r = k^+ + k^- = [1 + \exp(-\hbar\beta\epsilon)]k^+$ is the width, and where $k^+(T, \epsilon_0)$ is the forward tunneling rate (20.79). Since the quasi-elastic peak reflects exponential relaxation, it is not surprising that the same form is also found in the classical case, e.g. for the bistable double well, in which γ_r is the thermal relaxation rate [492].

For large amplitude ϵ_1 of the driving force, the response of higher harmonics may become significant. One finds from the recursive relations (23.11) that quan-

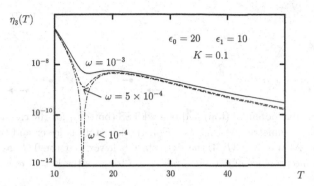

Figure 23.4: Noise-induced suppression of higher harmonics is illustrated for the third power amplitude η_3. Frequencies and temperature are given in units of Δ_{eff}.

tum noise can enhance or suppress higher harmonics in $P^{(\text{as})}(t)$. In Fig. 23.4, the temperature dependence of the power amplitude η_3 is shown for different ω. As the driving frequency is decreased, noise-induced suppression of this amplitude occurs at a temperature where the fundamental power amplitude η_1 has a maximum.

23.5 Driving-induced symmetry breaking

As a final neat example we show that driving-induced localization can occur when both the bias and the tunneling coupling energy are harmonically modulated [495].

Consider the cooperative effect of monochromatic fields modulating both the bias of a TSS with zero static bias and the tunneling coupling,

$$
\begin{aligned}
\epsilon(t) &= \epsilon_1 \sin(\omega t)\,, \\
\Delta(t) &= \Delta \exp[\mu \sin(\nu t)]\,.
\end{aligned}
\tag{23.42}
$$

If either the bias or the coupling energy is modulated, the left-right symmetry is dynamically broken. However, on average over a period, one finds asymptotically again equal occupation of both states.

When both parameters are modulated with commensurable frequencies,

$$
\omega = m\Omega \quad \text{and} \quad \nu = n\Omega \quad \text{with} \quad m, n \text{ integer}\,,
\tag{23.43}
$$

the TSS Hamiltonian has a discrete time translation symmetry. Interestingly, when n and m are odd, the left-right symmetry is broken even on average. The symmetry breaking is maximal when $n = m$, i.e., $\nu = \omega$. This case is pictorially sketched and qualitatively explained in Fig. 23.5. We expect from the illustrative presentation that at long time the occupation of the left state is preferred on average.

The averaged dynamics is again established by the expression (23.24) for $\hat{P}(\lambda)$ in which the kernels $k_0^{(\text{s,a})}(\lambda)$ must be calculated for time-periodic bias and tunneling

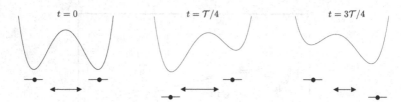

Figure 23.5: Bistable potential (top) and related TSS (bottom) for the case $\nu = \omega$. At time $t = 0$, the TSS is symmetric. At time $t = \mathcal{T}/4$, the left state is lower and the TC has the maximum value. At time $t = 3\mathcal{T}/4$, the right state is lower, but the TC has the minimum value. Thus, in the presence of damping, relaxation towards the left state is preferred.

coupling. For simplicity, we now restrict the attention to the averaged equilibrium state

$$P_\infty = \frac{k_0^{(a)}(0)}{k_0^{(s)}(0)} = \frac{\gamma^+ - \gamma^-}{\gamma^+ + \gamma^-} \, . \tag{23.44}$$

Here, $\gamma^\pm$ are the inclusive forward/backward tunneling rates. We start out from the time-dependent NIBA kernel

$$K_z^{(s,a)}(t, t - \tau) = \Delta(t)\Delta(t - \tau) \, e^{-Q'(\tau)} \left\{ \begin{array}{l} \cos[Q''(\tau)] \cos \vartheta(t, t - \tau) \\[2mm] \sin[Q''(\tau)] \sin \vartheta(t, t - \tau) \end{array} \right. , \tag{23.45}$$

with the time-dependent bias phase and tunneling term

$$e^{\pm i\vartheta(t, t - \tau)} = e^{\mp i(\epsilon_1/\omega) \cos(\omega t)} \, e^{\pm i(\epsilon_1/\omega) \cos[\omega(t - \tau)]} \, , \tag{23.46}$$

$$\Delta(t)\Delta(t - \tau) = e^{\mu \sin(\omega t)} \, e^{\mu \sin[\omega(t - \tau)]} \, , \tag{23.47}$$

respectively. The time-averaged Laplace transform of the kernel (23.45)

$$k_0^{(s,a)}(\lambda) = \frac{\omega}{2\pi} \int_0^{2\pi/\omega} dt \int_0^\infty d\tau \, e^{-\lambda \tau} \, K_z^{(s,a)}(t, t - \tau) \, . \tag{23.48}$$

can be calculated in analytic form by expanding the bias phase term (23.46) in a double series in J-Bessel functions, and the oscillatory tunneling coupling (23.47) in a double series in I-Bessel functions. Readily we obtain the inclusive rates $\gamma^\pm$ as

$$\gamma^\pm = \sum_{\ell,m,n=-\infty}^{\infty} I_\ell(\mu) I_m(\mu) J_n(\epsilon_1/\omega) J_{\ell+n-m}(\epsilon_1/\omega) \, k^\pm[(n - m)\omega] \, , \tag{23.49}$$

where $k^\pm(\epsilon)$ is the exclusive forward/backward rate for a static bias ϵ given in Eq. (20.30). The expression (23.49) is the generalization of the earlier expression (23.28) to the case of an additional periodic modulation of the tunneling coupling. It reveals that the forward/backward symmetry is dynamically broken, $\gamma^+ \neq \gamma^-$, when both the driving amplitude of the bias ϵ_1 and the modulation amplitude of the tunneling coupling μ are nonzero.

PART V

THE DISSIPATIVE MULTI-STATE SYSTEM

A quantum Brownian particle in a multi-well potential coupled to a dissipative environment is archetypal for many problems in physics and chemistry. Examples include super-ionic conductors, atoms on surfaces, and interstitials in dielectrics and metals. Also the current-voltage characteristics of a Josephson junction, and charge transport in a quantum wire hindered by impurity scattering are described by this model. At high temperature, the particle moves forward or backward from well to well by incoherent tunneling or thermally activated transitions. As the temperature is lowered, coherent tunneling across many wells may become significant, and the competing different tunneling paths may interfere constructively or destructively. The model has a profound and powerful duality symmetry between the weak-binding and strong-binding representation. In view of the broad area of applications, the understanding of quantum transport in multi-well systems is a central issue.

24. Quantum Brownian particle in a washboard potential

24.1 Introduction

In the preceding part, I have considered the dynamics of a damped quantum system which is effectively restricted to a two-dimensional Hilbert space. Many of the concepts and approximation schemes developed in Part IV can be generalized for a dissipative system with N tight-binding sites. A system with three sites, $N = 3$, for instance, is of particular interest for the study of the ultrafast primary electron transfer in bacterial photosynthesis [496]. In many systems in chemical and biological physics, the transfer of a particle or a charge from a donor to an acceptor state occurs via a bridge of few or many intermediate tight-binding states. The RDM for a discrete N-site system can be visualized as a square lattice with $N \times N$ lattice sites. The path sum for the conditional propagating function (5.12) covers all paths on this lattice for given boundary sites. Each path consists of a sequence of segments in which the path dwells for some time on a lattice site and then undertakes a sudden flip to a neighboring site. A general path can be divided into a sequence of clusters: each cluster is a path section between two successive visits of a diagonal state. A natural generalization of the noninteracting-blip approximation to the case of N states is the "noninteracting-cluster" approximation (NICA) developed in Ref. [497]. In the NICA, the intra-cluster correlations are taken into account while the inter-cluster correlations are disregarded. The infinite series resulting from the iteration of the kernels can be cast into master equations for the populations [see Eq. (25.21)]. The kernels are given as the sum of all clusters with the same final diagonal state relative to the initial diagonal state.

Quantum Brownian motion (QBM) in a periodic potential is a key model for many transport phenomena in condensed matter [498]. Early work was mainly concerned with the significance of polaronic effects [132] – [134]. Recent interest has been focussed on the influence of frequency-independent damping. For instance, the electron-hole drag of charged particles in metals, as well as quasiparticle tunneling in Josephson junctions [499, 152] give rise to Ohmic dissipation (cf. Subsections 4.2.8 and 4.2.10). A duality symmetry between the weak- and strong-binding representations of the QBM model was put forward by Schmid [145]. The duality in the QBM model corresponds to the charge-phase duality in the Josephson junction model [152]. Transport of charge through impurities in quantum wires [500] and tunneling of edge currents through constrictions in fractional quantum Hall devices [501] are also described by the QBM model. The collective excitations of the correlated fermions away from the barrier manifest themselves as density fluctuations. In a theoretical description they are represented by a Luttinger harmonic liquid, and they corresponds to the thermal reservoir in the QBM model. Short-range electron interaction is equivalent to Ohmic damping in the related QBM model, whereas unscreened long-range Coulomb repulsion [502] corresponds to a sub-Ohmic reservoir coupling [503]. Many other physical and chemical systems involve transport of charge through barriers under Ohmic and super-Ohmic dissipation [85, 175].

24.2   Weak- and tight-binding representation

Consider a quantum Brownian particle moving in a tilted corrugated potential. The dynamics is described by the translational-invariant global model (3.11) with a bilinear coordinate coupling of the particle to a bath of harmonic oscillators,

$$H_{\mathrm{WB}} = \frac{P^2}{2M} + V(X,t) + \sum_\alpha \left[\frac{p_\alpha^2}{2m_\alpha} + \frac{m_\alpha \omega_\alpha^2}{2} \left(x_\alpha - \frac{c_\alpha}{m_\alpha \omega_\alpha^2} X \right)^2 \right]. \qquad (24.1)$$

The label WB indicates the "weak binding" representation. We choose a trigonometric form for the corrugation with period X_0 and a global tilting force $F = 2\pi V_{\mathrm{tilt}}/X_0 = \hbar \epsilon_{\mathrm{WB}}/X_0$. To study the particle's response in thermal equilibrium, we assume that this force is acting only for $t > 0$. We then have [cf. Eq.(17.43)]

$$V(X,t) = -V_0 \cos\left(2\pi X/X_0\right) - \Theta(t) F X . \qquad (24.2)$$

There are many different physical situations in which an underlying discrete translational symmetry assigns a periodic potential. In real systems, there is often complexity due to perturbations of the periodic order by impurities or by local strain fields.

Clearly, the model (24.1) with (24.2) provides an idealized description which shows already interesting nontrivial behaviors. Complexity may potentially be included by taking coupling and bias parameters as random or noisy.

In the tight-binding regime $k_B T/\hbar\omega_0$, $FX_0/\hbar\omega_0 \ll 1 \ll V_0/\hbar\omega_0$, where $\hbar\omega_0$ is the frequency of small oscillations about the potential minima, only the lowest state in each well is occupied. Then the potential system is effectively reduced to a single-band tight-binding lattice. The Hilbert space of the isolated system is spanned by the (localized) ground states of the individual wells. These states are coupled by a transfer matrix element Δ which represents quantum mechanical overlap of nearest-neighbor localized states. The tight-binding (TB) model with transfer matrix element Δ, lattice constant q_0, and translational-invariant coupling is

$$
\begin{aligned}
H_{\mathrm{TB}} = & -\frac{\hbar\Delta}{2}\left(e^{iq_0p/\hbar} + e^{-iq_0p/\hbar}\right) - \Theta(t)Fq \\
& + \sum_\alpha \left[\frac{\pi_\alpha^2}{2M_\alpha} + \frac{M_\alpha\Omega_\alpha^2}{2}\left(u_\alpha - \frac{d_\alpha}{M_\alpha\Omega_\alpha^2}q\right)^2\right],
\end{aligned}
\tag{24.3}
$$

with

$$
\begin{aligned}
q &= q_0 \sum_n n\, a_n^\dagger a_n = q_0 \sum_n n\, |n\rangle\langle n|\,, \\
e^{iq_0p/\hbar} &= \sum_n a_n^\dagger a_{n+1} = \sum_n |n\rangle\langle n+1|\,.
\end{aligned}
\tag{24.4}
$$

Here we have given two equivalent operator forms for the position and translation operators. The operators $a_n^\dagger$ and a_n create and annihilate a particle at the site n, and $|n\rangle$ denotes the localized state at this site. In the sequel, we express the tilting force F in terms of the potential drop $\hbar\epsilon$ between neighboring sites,

$$
F = \hbar\epsilon/q_0\,,
\tag{24.5}
$$

As far as we are interested in properties of the WB and TB particle, the set of coupling constants $\{c_\alpha\}$ and $\{u_\alpha\}$, and the parameters of the reservoirs are relevant only via the respective spectral densities of the coupling. We define the spectral densities for the two models as in Eq. (3.24),

$$
J_{\mathrm{WB}}(\omega) = \frac{\pi}{2}\sum_\alpha \frac{c_\alpha^2}{m_\alpha\omega_\alpha}\delta(\omega - \omega_\alpha), \qquad J_{\mathrm{TB}}(\omega) = \frac{\pi}{2}\sum_\alpha \frac{d_\alpha^2}{M_\alpha\Omega_\alpha}\delta(\omega - \Omega_\alpha)\,.
\tag{24.6}
$$

At this point I anticipate that the tight-binding and weak-binding models are related by a duality symmetry which becomes an exact self-duality in the Ohmic scaling limit. The respective discussion is given in Chapter 26. Explicit computations are conveniently performed in the discrete TB representation. Subsequently, the results can easily be transferred, upon using the duality transformation, to the dual weak-binding model.

25. Multi-state dynamics

25.1   Quantum transport and quantum-statistical fluctuations

The reduced density matrix describes all properties pertaining to the Brownian particle. In the TB limit, the originally continuous coordinate q is discrete, $q = nq_0$
($n = 0, \pm1, \pm2, \cdots$). Correspondingly, the density matrix is discrete and conveniently
labelled with integer indices which number the wells. We now study the dynamics for
two different kinds of initial states.

25.1.1   Product initial state

The first kind is a product initial state (pis) in which the state of the system and the
thermal state of the reservoir are in factorized form as in Eq. (5.10). Suppose that
the particle was prepared to start out at time zero from the site $n = 0$ of the discrete
lattice. The dynamical quantity of interest is then the probability $P_n(t)$ for finding
the particle at site n at a later time $t > 0$. To formulate the evolution of $P_n(t)$, we
use again the real-time influence functional method for a product initial state of the
system-plus-reservoir complex discussed in Section 5.2. The populations $P_n(t)$ are the
diagonal elements of the reduced density matrix. They can be written as a double
path integral for the propagating function,

$$P_n(t) = J(nn, t; 00, 0) = \int \mathcal{D}q(\cdot) \int \mathcal{D}q'(\cdot)\, \mathcal{A}[q(\cdot)]\mathcal{A}^*[q'(\cdot)]\mathcal{F}[q(\cdot), q'(\cdot)] . \quad (25.1)$$

Here, $q(\cdot)$, $q'(\cdot)$ are discontinuous paths propagating vertically and horizontally in
steps of length q_0 in the (q, q')-plane [see Eq. (25.27)]. The paths obey the boundary
conditions

$$q(0) = q'(0) = 0 , \quad \text{and} \quad q(t) = q'(t) = n\, q_0 . \quad (25.2)$$

The functional $\mathcal{A}[q(\cdot)]$ is the probability amplitude that the particle propagates along
the path $q(\cdot)$, and $\int \mathcal{D}q(\cdot) \int \mathcal{D}q'(\cdot)$ means summation of all paths on the TB lattice
with boundary points (25.2). Again, all the influences of the coupling to the reservoir
are contained in the influence functional $\mathcal{F}[q(\cdot), q'(\cdot)]$ which conveys self-interactions
of the paths $q(\cdot)$ and $q'(\cdot)$ and interactions between the paths $q(\cdot)$ and $q'(\cdot)$.

25.1.2   Characteristic functions of moments and cumulants

The Fourier transform of the populations is the central dynamical quantity,

$$\mathcal{Z}(\rho, t) = \sum_{n=-\infty}^{\infty} e^{i\rho n q_0} P_n(t) = \left\langle e^{i\rho q(t)} \right\rangle = \sum_{m=0}^{\infty} \frac{(i\rho)^m}{m!} \left\langle q^m(t) \right\rangle . \quad (25.3)$$

The *characteristic function* $\mathcal{Z}(\rho, t)$ is the *moment generating function* (MGF), where
ρ is the counting field. The Nth derivative of the MGF at $\rho{=}0$ gives the Nth moment

$$\langle q^N(t) \rangle \equiv q_0^N \sum_{n=-\infty}^{\infty} n^N P_n(t) = \left(-i \frac{\partial}{\partial \rho} \right)^N \mathcal{Z}(\rho, t) \Big|_{\rho=0}. \qquad (25.4)$$

The reducible moments of $P_n(t)$ can be written in terms of the irreducible moments or cumulants $\langle q^N(t) \rangle_c$ as

$$\langle q(t) \rangle = \langle q(t) \rangle_c,$$
$$\langle q^2(t) \rangle = \langle q(t) \rangle_c^2 + \langle q^2(t) \rangle_c, \qquad (25.5)$$
$$\langle q^3(t) \rangle = \langle q(t) \rangle_c^3 + 3\langle q^2(t) \rangle_c \langle q(t) \rangle_c + \langle q^3(t) \rangle_c.$$

The cumulant generating function (CGF) $\ln \mathcal{Z}(\rho, t)$ yields the cumulant expansion

$$\ln \mathcal{Z}(\rho, t) = \sum_{m=1}^{\infty} \frac{(i\rho)^m}{m!} \langle q^m(t) \rangle_c. \qquad (25.6)$$

Evidently, the Nth cumulant can be found from the CGF by differentiation,

$$\langle q^N(t) \rangle_c = \left(-i \frac{\partial}{\partial \rho} \right)^N \ln \mathcal{Z}(\rho, t) \Big|_{\rho=0}. \qquad (25.7)$$

The cumulant generating function $\ln \mathcal{Z}(\rho, t)$ carries all statistical properties of the quantum transport process. The transport dynamics is referred to as diffusive when $\ln \mathcal{Z}(\rho, t)$, and hence all irreducible moments or cumulants, grow linearly with t at long time. This is the case when the dynamics takes place by incoherent tunneling events between the system's diagonal states (see Section 25.2).

The first cumulant gives the average current. From this one may deduce the mobility or conductance. The second cumulant yields the diffusion constant or, in connection with charge transport, the (nonequilibrium) dc current noise. All higher cumulants are zero when the statistics is Gaussian. The third cumulant, referred to as skewness, gives information about the leading asymmetric deviation from the Gaussian distribution. Experimentally, it can be discriminated from Gaussian noise by inverting the current. The fourth cumulant, called curtosis or sharpness, is a measure for the flatness of the distribution compared to the standard distribution. When the curtosis is positive, the distribution is sharp, and when it is negative, the distribution is flat.

25.1.3 Thermal initial state and correlation functions

The second kind of initial state is a thermal initial state: the system is prepared by letting the particle equilibrate with the reservoir, and then at time $t = 0$ a measurement of the observable of interest of the particle is performed. The measurement leads to a reduction of the canonical density operator according to $\hat{W}_0 = \hat{P} \hat{W}_\beta \hat{P}$, where $\hat{W}_\beta$ is the equilibrium density operator of the global system, and the operator $\hat{P}$ projects onto the measured interval of the observable in question. Upon measuring the same observable again at a later time $t > 0$, we obtain information about the equilibrium autocorrelations of this observable. Here we restrict our attention to position correlation functions. These can be deduced from the generating function

$$\mathcal{Z}_{\mathrm{th}}(\rho, \kappa, \mu; t) \;\equiv\; \left\langle e^{i\rho q(0)}\, e^{i\kappa q(t)}\, e^{i\mu q(0)} \right\rangle_{\beta}, \tag{25.8}$$

where $\langle \cdots \rangle_{\beta}$ means thermal average of the states of the global system. We are mainly interested in the evolution of the average position $\langle q(t) \rangle_{\beta}$, the mean square displacement

$$D_{\mathrm{th}}(t) \;\equiv\; \left\langle [\, q(t) - q(0)\,]^{2} \right\rangle_{\beta}, \tag{25.9}$$

and the antisymmetrized correlation function

$$A(t) \;\equiv\; \frac{1}{2i} \left\langle [\, q(t)q(0) - q(0)q(t)\,] \right\rangle_{\beta}\Big|_{F=0}. \tag{25.10}$$

These functions are directly found from $\mathcal{Z}_{\mathrm{th}}$ upon differentiation,

$$
\begin{aligned}
\langle q(t) \rangle_{\beta} &= \left. -i \frac{\partial}{\partial \kappa} \mathcal{Z}_{\mathrm{th}} \right|_{\rho=\kappa=\mu=0}, \\[4pt]
D_{\mathrm{th}}(t) &= \left[-\frac{\partial^{2}}{\partial \kappa^{2}} - \frac{\partial}{\partial \rho}\frac{\partial}{\partial \mu} + \frac{\partial}{\partial \kappa}\frac{\partial}{\partial \mu} + \frac{\partial}{\partial \kappa}\frac{\partial}{\partial \rho} \right] \mathcal{Z}_{\mathrm{th}} \bigg|_{\rho=\kappa=\mu=0}, \\[4pt]
A(t) &= \frac{i}{2}\left[\frac{\partial}{\partial \kappa}\frac{\partial}{\partial \mu} - \frac{\partial}{\partial \kappa}\frac{\partial}{\partial \rho} \right] \mathcal{Z}_{\mathrm{th}} \bigg|_{\rho=\kappa=\mu=0,\, F=0}.
\end{aligned} \tag{25.11}
$$

The generating function $\mathcal{Z}_{\mathrm{th}}$ can be expressed in terms of three-time conditional propagating functions. For an ergodic open system, these can be written as real-time path integrals which represent propagation since the infinite past (see Section 5.6 and Subsection 21.2.2). Suppose that the particle has been prepared at some negative time t_{p} in the diagonal state $(0,0)$. Then the particle will have relaxed at time zero to the thermal equilibrium state if we send t_{p} to the infinite past. Performing now at time $t = 0$ a measurement of the observable of interest, the desired thermal correlated initial state is prepared. We have

$$\mathcal{Z}_{\mathrm{th}}(\rho, \kappa, \mu; t) \;=\; \lim_{t_{\mathrm{p}} \to -\infty} \sum_{n,m,r} J(n\,n, t;\, m\,r, 0;\, 0\,0, t_{\mathrm{p}})\, e^{iq_{0}(\rho m + \kappa n + \mu r)}. \tag{25.12}$$

The propagating function may be expressed as a double path integral of the form (25.1) in which the paths on the (q, q')-lattice are constrained as

$$
\begin{aligned}
q(t_{\mathrm{p}}) &= 0, & q(0) &= m\,q_{0}, & q(t) &= n\,q_{0}, \\
q'(t_{\mathrm{p}}) &= 0, & q'(0) &= r\,q_{0}, & q'(t) &= n\,q_{0}.
\end{aligned}
$$

The effects of different initial preparation are now easily visualized. For the product initial state discussed in Subsection 25.1.1, the dynamics of the particle at negative times is quenched. For the thermal initial state discussed here, the particle undergoes dynamics since the infinite past, and the effects of the correlations at time zero are represented by the interactions between the negative- and positive-time branch.

25.2 Poissonian quantum transport

25.2.1 Incoherent nearest-neighbor transitions (weak tunneling)

If the temperature is high and/or the bath coupling is strong, the particle tunnels incoherently between neighboring TB states.[1] This implies that the quantum particle occupies a diagonal state of the RDM after every second transition it makes. Hence the paths on the (q, q')-lattice are restricted to visits of sites of a tridiagonal matrix. Each path of a walker on the RDM consists of sequences of three elementary two-step (order Δ^2) processes from a diagonal to a diagonal state:

(1) The walker may step forward to the next diagonal state

(2) The walker may walk back to the preceding diagonal state

(3) The walker may return to the same diagonal state

The elementary processes (1) and (2) have each two variants, whereas the third has four variants, as counted from the number of possible intermediate off-diagonal states. On the assumption that the above elementary processes are noninteracting and thus statistically independent, the nearest-neighbor tunneling transitions are incoherent. We denote the weights of the double jumps per unit time by w^+, w^-, and w, respectively. The weights are of order Δ^2. They are directly related to the non-adiabatic forward and backward tunneling rates in the two-state system discussed in Section 20.2. We have $w^+ = k^+$, $w^- = k^-$, and $w = -k$, where $k = k^+ + k^-$. As the walker finally reaches the diagonal state (nn), there is an excess of n k^+-transitions over the k^--transitions. Thus the grand-canonical sum of the statistically independent elementary two-step processes is

$$P_n(t) = \sum_{j,\ell,m=0}^{\infty} \delta_{j,\ell+n} \frac{1}{j!\,\ell!\,m!} (k^+)^j (k^-)^\ell (-k)^m \, t^{j+\ell+m} , \qquad (25.13)$$

which can be summed upon using the detailed balance relation (20.32) to the form

$$P_n(t) = \exp(n\beta\hbar\epsilon/2 - kt) \, I_{|n|}[\, kt/\cosh(\beta\hbar\epsilon/2)] , \qquad (25.14)$$

where $I_n(z)$ is a modified Bessel function. A simple interpretation of this result may be obtained by observing that the nearest-neighbor hopping dynamics is alternatively described in terms of the master equation [425, 498]

$$\dot{P}_n(t) = k^+ P_{n-1}(t) + k^- P_{n+1}(t) - k P_n(t) . \qquad (25.15)$$

Upon using a recursion formula of the modified Bessel function, it is straightforward to see that the expression (25.14) solves the master equation.

Next, we discover that the Fourier series (25.3) with the populations (25.14) can be summed up to the generating function of the modified Bessel functions. The resulting expression for $\mathcal{Z}(\rho, t)$ is

---

[1]The precise conditions for this weak-tunneling regime are discussed (in connection with the dissipative two-state system) in Chapters 20 and 21, and in Sections 26.4 and 26.5.

$$\mathcal{Z}(\rho, t) = \exp\left\{\left[\cos(\rho q_0) - 1\right] k\, t + i \sin(\rho q_0)[\, k^+ - k^-]\, t \right\} . \tag{25.16}$$

The linear time-dependence in the exponent tells us that the dynamics is purely diffusive. With this form, we find from Eq. (25.7) the first and second cumulants as

$$\langle q(t) \rangle_{\mathrm{c}} = q_0 \left[k^+(\epsilon) - k^-(\epsilon) \right] t , \qquad \langle q^2(t) \rangle_{\mathrm{c}} = q_0^2\, k(\epsilon)\, t . \tag{25.17}$$

If the forward and backward rate are related by detailed balance [see Eq. (20.32)], we have $k^+(\epsilon) - k^-(\epsilon) = \tanh(\beta \hbar \epsilon/2)\, k(\epsilon)$.

Next, we employ the definining expressions for the nonlinear mobility $\mu(T, \epsilon)$ and the diffusion coefficient $D(T, \epsilon)$,

$$\mu = \frac{q_0}{\hbar \epsilon} \lim_{t \to \infty} \frac{\langle q(t) \rangle}{t} , \qquad D = \lim_{t \to \infty} \frac{\langle q^2(t) \rangle_{\mathrm{c}}}{2t} . \tag{25.18}$$

With these definitions, we finally find from the forms (25.17) and the defining expressions (25.18) the nonlinear mobility and the diffusion coefficient as

$$\mu(T, \epsilon) = \frac{\tanh(\beta \hbar \epsilon/2)}{\hbar \epsilon}\, q_0^2\, k(T, \epsilon) , \tag{25.19}$$

$$D(T, \epsilon) = \frac{q_0^2}{2}\, k(T, \epsilon) = k_{\mathrm{B}} T \frac{\beta \hbar \epsilon/2}{\tanh(\beta \hbar \epsilon/2)}\, \mu(T, \epsilon) . \tag{25.20}$$

The nonadiabatic forward/backward tunneling rates $k^{\pm}(T, \epsilon)$ in a TB double-well system have been discussed in Section 20.2. With these results we then get explicit expressions for $\mu(T, \epsilon)$ and $D(T, \epsilon)$. The resulting expression for the nonlinear mobility in the Ohmic scaling limit is given below in Subsection 25.5.1.

25.2.2 The general case (strong tunneling)

We now generalize the discussion to a TB model in which the transport of mass or charge between the sites $n = 0, \pm 1, \pm 2, \cdots$ takes place via direct forward and backward transitions by ℓ sites, $\ell = 1, 2, \cdots$. We denote the respective transitions weights (transition "rates") by $k_\ell^{\pm}$. Assuming statistically independent transitions, the dynamics of the population probability $P_n(t)$ of site n is governed by the master equation

$$\dot{P}_n(t) = \sum_{\ell=1}^{\infty} \left[k_\ell^+ P_{n-\ell}(t) + k_\ell^- P_{n+\ell}(t) - (k_\ell^+ + k_\ell^-) P_n(t) \right] . \tag{25.21}$$

The moment generating function defined in Eq. (25.3) is found from this equation as

$$\mathcal{Z}(\rho, t) = \prod_{n=1}^{\infty} \exp\left\{ t(\, e^{i\rho q_0 n} - 1)k_n^+ + t(\, e^{-i\rho q_0 n} - 1)k_n^- \right\} . \tag{25.22}$$

The cumulants are obtained from the cumulant generating function $\ln \mathcal{Z}(\rho, t)$ by differentiation. With use of Eq. (25.7) we get

$$\langle q^N(t) \rangle_c = q_0^N \sum_{n=1}^{\infty} n^N [\, k_n^+ + (-1)^N k_n^- \,] \, t \,. \tag{25.23}$$

The characteristic function is conveniently written in terms of partial forward/backward currents $I_n^{\pm} = n k_n^{\pm}$ as

$$\mathcal{Z}(\rho, t) = \prod_{n=1}^{\infty} Z_n^+(\rho, t) \, Z_n^-(\rho, t) \,, \tag{25.24}$$

where

$$Z_n^{\pm}(\rho, t) = \sum_{\ell=0}^{\infty} \mathrm{e}^{\pm i \rho q_0 n \ell} \frac{(t I_n^{\pm}/n)^{\ell}}{\ell!} \, \mathrm{e}^{-t I_n^{\pm}/n} = \exp[\, t(\mathrm{e}^{\pm i \rho q_0 n} - 1) I_n^{\pm}/n \,] \,. \tag{25.25}$$

The physical meaning of this expression is elucidated as follows. Suppose that the mass or charge transferred per unit time in forward direction were the results of a Poisson process for particles of unit mass propagating via uncorrelated nearest-neighbor forward transitions k_1^+ contributing a current $I_1^+ = k_1^+$, plus a Poisson process of uncorrelated forward moves via next-to-nearest-neighbor transitions contributing a current $I_2^+ = 2k_2^+$, etc. Suppose also that the total backward current were the result of independent Poisson processes with partial backward currents $I_n^- = n k_n^-$, $n = 1, 2, \cdots$. The final form of the characteristic function would then be the expression (25.24) with (25.25).

The transport model (25.21) finds application, e.g., to charge transport through a weak link or through a quantum impurity in a 1D quantum wire, and to tunneling of edge currents in the fractional quantum Hall regime [cf. Chapter 28]. In these cases, the interpretation would be that the expression (25.24) with (25.25) describes independent Poisson processes for particles of charge one going across the impurity's barrier in forward/backward direction, contributing a current $I_1^{\pm}$, plus a Poisson process for particles of charge two (or for joint transport of a pair of particles of charge one) contributing a forward/backward current $I_2^{\pm}$, etc.

When the tunneling entities are coupled to a thermal reservoir, then the forward and backward transition weights are connected by detailed balance. Assuming that the potential drop per lattice period in forward direction is $\hbar\epsilon$, we have

$$k_n^-(\epsilon) = \mathrm{e}^{-n\beta\hbar\epsilon} k_n^+(\epsilon) \,. \tag{25.26}$$

The nonequilibrium relation between diffusion and mobility (25.20) does not hold when different pairs of transfer rates, say $k_1^{\pm}$, $k_2^{\pm}$, $\cdots$, contribute to the transport process, since they have different detailed balance factors, $\mathrm{e}^{-\hbar\beta\epsilon}$, $\mathrm{e}^{-2\hbar\beta\epsilon}$, $\cdots$.

In a classical transport process, the transition weights $k_n^{\pm}$, $n = 1, 2, \cdots$ have the usual meaning of rates, since they are all positive. When the Poisson processes come about quantum mechanically, the sign of transition weights can be negative (see, e.g., explicit results in Section 26.6, in which the sign of $k_n^{\pm}$ alternates as a function of n).

This does not spoil conservation of probability, since the master equation (25.21) provides $\sum_n \dot{P}_n(t) = 0$ regardles of the particular form chosen for the set $\{k_n^\pm\}$. Nevertheless, each population $P_n(t)$, $n = 1, 2, \cdots$, must be non-negative at any time.

Below we shall see that the quantum mechanical transition weights $k_n^\pm(\epsilon)$ can be given in analytic form for particular models in ample regions of the parameter space.

25.3 Exact formal expressions for the system dynamics

In the tight-binding model (24.3), the quantum particle makes sudden transitions between neighboring states of the density matrix with a probability amplitude $\pm i\Delta/2$ per unit time. The sequence of states of the density matrix visited in succession can be visualized in terms of a paths of a walker on an infinite square lattice spanned by the sites of the discrete (q, q') double path. The walker starts on some site on the principal diagonal, say at $(0,0)$, and then randomly makes horizontal and vertical steps along the q- and q'-axis, respectively. At each site of the square lattice there are four possible directions to continue. Return to a site on the principal diagonal is possible with an even number of steps.

A TB path on the square lattice is conveniently parametrized in terms of charges $u_j = \pm 1$ and $v_j = \pm 1$. A path with k steps in $q(\tau)$ and ℓ steps in $q'(\tau)$ is then given by

$$q^{(k)}(\tau) = q_0 \sum_{j=1}^{k} u_j \Theta(\tau - t_j) , \qquad q'^{(\ell)}(\tau) = q_0 \sum_{i=1}^{\ell} v_i \Theta(\tau - t'_i) . \qquad (25.27)$$

A double path from the state $(0,0)$ to the state (n,n) is subject to the constraints

$$\sum_{j=1}^{k} u_j = n , \qquad \sum_{i=1}^{\ell} v_i = n . \qquad (25.28)$$

The environmental coupling manifests itself in complex-valued interactions between the charges. The influence action is conveniently split into the self-interactions of the paths $q(\tau)$ and $q'(\tau)$ denoted by Φ_1 and Φ_1^*, and into the interaction between the paths $q(\tau)$ and $q'(\tau)$, denoted by Φ_2. Upon inserting the tight-binding paths (25.27) into Eq. (5.23) with (5.24), the influence function then takes the form

$$\mathcal{F}[q^{(k)}, q'^{(\ell)}] = \exp\left(\Phi_1[q^{(k)}] + \Phi_1^*[q'^{(\ell)}] + \Phi_2[q^{(k)}, q'^{(\ell)}] \right) , \qquad (25.29)$$

with the interactions

$$\Phi_1[q^{(k)}] = \sum_{j=2}^{k} \sum_{i=1}^{j-1} u_j u_i Q(t_j - t_i) ,$$

$$\Phi_1^*[q'^{(\ell)}] = \sum_{j=2}^{\ell} \sum_{i=1}^{j-1} v_j v_i Q^*(t'_j - t'_i) , \qquad (25.30)$$

$$\Phi_2[q^{(k)}, q'^{(\ell)}] = -\sum_{j=1}^{\ell} \sum_{i=1}^{k} v_j u_i Q(t'_j - t_i) ,$$

where $Q(t)$ is the pair correlation function introduced in Section 18.2.

For computational purposes it is again useful to introduce symmetric and anti-symmetric spin paths,

$$\eta(\tau) \equiv [q(\tau) + q'(\tau)]/q_0 , \qquad \xi(\tau) = [q(\tau) - q'(\tau)]/q_0 . \tag{25.31}$$

The path $\eta(\tau)$ describes propagation on the (q, q')-lattice in direction parallel to the principal diagonal, whereas the path $\xi(\tau)$ measures moves in perpendicular direction. A general path with $2m$ time-ordered steps on the square lattice, which is released at time t_p from the diagonal state $\xi = 0$, $\eta = 0$, is represented as

$$\eta^{(2m)}(\tau) = \sum_{k=1}^{2m} \eta_k \, \theta(\tau - t_k) , \qquad \xi^{(2m)}(\tau) = \sum_{k=1}^{2m} \xi_k \, \theta(\tau - t_k) , \tag{25.32}$$

where $t_k > t_p$ for all k. The charges $\xi_k = \pm 1$ and $\eta_k = \pm 1$ label the four possibilities to move at time t_k instantaneously from a site to a neighboring site. Every individual u- or v-charge is associated with a $\{\eta, \xi\}$-pair. We have the correspondences

$$\begin{aligned} u_j = \pm 1 \quad &\longleftrightarrow \quad \{\eta_j, \xi_j\} = \{\pm 1, \pm 1\} , \\ v_k = \pm 1 \quad &\longleftrightarrow \quad \{\eta_k, \xi_k\} = \{\pm 1, \mp 1\} . \end{aligned} \tag{25.33}$$

In time-ordered (η, ξ)-representation, the influence function (25.29) for $2m$ steps reads

$$\mathcal{F}^{(2m)} = G_{2m} H_{2m} , \qquad \text{where} \qquad \begin{aligned} G_n &= \exp\left[\sum_{j=2}^{n} \sum_{k=1}^{j-1} \xi_j \xi_k Q'(t_j - t_k) \right] , \\ H_n &= \exp\left[i \sum_{k=1}^{n-1} \eta_k \chi_{k,n} \right] , \end{aligned} \tag{25.34}$$

with the influence phase

$$\chi_{k,n} = \sum_{j=k+1}^{n} \xi_j Q''(t_j - t_k) . \tag{25.35}$$

The filter function G_{2m} is the charge representation of the noise action factor $e^{-S^{(N)}/\hbar}$ in Eq. (5.23) with (5.33). It suppresses far and long excursions away from the diagonal of the RDM. The phase factor H_{2m} is the corresponding friction action factor $e^{-i S^{(F)}/\hbar}$.

For nonzero tilting force at time $t > 0$ in the Hamiltonian (24.3) , $F = \Theta(t)\hbar\epsilon/q_0$, the path (25.32) at positive time is weighted with a bias phase factor. For k transitions at negative times and $j = 2m - k$ transitions at positive times, the bias factor reads

$$B_{j,k} = \exp(i\,\varphi_{j,k}) \qquad \varphi_{j,k} = \epsilon \int_0^t d\tau \, \xi^{(k+j)}(\tau) = \epsilon \sum_{i=k+1}^{k+j} \xi_i(t - t_i) . \tag{25.36}$$

With these preliminaries, it is straightforward to write down the path sum for all dynamical quantities of interest.

One remark is in order. We shall see that the (η, ξ)-representation is expedient in actual computations. On the other hand, the (q, q')-representation (25.29) is advantageous if one aims at derivation of fluctuation-dissipation theorems, e.g., detailed balance and the Einstein relation (see Subsection 25.4.2).

25.3.1 Product initial state

Consider now first the case of a product initial state of the global system. The path sum $\int \mathcal{D}q \int \mathcal{D}q'$ in Eq. (25.1) is expressed for the discrete system as follows. First, $P_n(t)$ is represented as an expansion in even numbers of steps or tunneling transitions. Secondly, in a given order of steps, say $2m$, we have to sum over all possible arrangements of labels $\{\xi_j = \pm 1\}$ and $\{\eta_j = \pm 1\}$ which fulfill the boundary conditions (25.2). This entails the constraints

$$\sum_{j=1}^{2m} \eta_j = 2n \,, \qquad\qquad \sum_{j=1}^{2m} \xi_j = 0 \,. \tag{25.37}$$

Thirdly, we must take into account that the individual steps may occur at any time. To save space, we compactly write time-ordered integrations for ℓ steps in the negative time-branch and k steps in the positive-time branch as

$$\int_{t_{\rm p}}^{t} \mathcal{D}_{k,\ell}\{t_j\} \times \cdots \equiv \int_{0}^{t} dt_{\ell+k} \cdots \int_{0}^{t_{\ell+2}} dt_{\ell+1} \int_{t_{\rm p}}^{0} dt_{\ell} \cdots \int_{t_{\rm p}}^{t_2} dt_1 \, \Delta^{k+\ell} \times \cdots \,, \tag{25.38}$$

where $t_{\rm p} < 0$. We have included in the definition the product of the tunneling amplitudes. The various components are combined to yield for $P_n(t)$ the form

$$P_n(t) = \sum_{m=|n|}^{\infty} (-1)^{m-n} \frac{1}{2^{2m}} \int_0^t \mathcal{D}_{2m,0}\{t_j\} \sum_{\{\xi_j\}'} B_{2m,0} G_{2m} \sum_{\{\eta_j\}'} H_{2m} \,. \tag{25.39}$$

The prime in $\{\xi_j\}'$ and $\{\chi_j\}'$ denotes summation under the constraints (25.37). With this form, the characteristic function (25.3) for a product initial state (pis) reads

$$\mathcal{Z}_{\rm pis}(\rho,t) = \sum_{m=0}^{\infty} (-1)^m \int_0^t \mathcal{D}_{2m,0}\{t_j\} \sum_{\{\xi_j\}'} B_{2m,0} G_{2m} F_{2m}(\rho) \,. \tag{25.40}$$

The weighted average over the populations $P_n(t)$ is in the function

$$
\begin{aligned}
F_{2m}(\rho) &= \frac{1}{2^{2m}} \sum_{n=-m}^{+m} (-1)^n \, e^{i\rho n q_0} \sum_{\{\eta_j\}'} H_{2m} \\
&= \sin\left(\frac{\rho q_0}{2}\right) \prod_{j=1}^{2m-1} \sin\left(\frac{\rho q_0}{2} + \chi_{j,2m}\right) \,.
\end{aligned} \tag{25.41}
$$

To obtain the second form, we have used the Fourier representation for the Kronecker delta which brings in the constraint $\sum_j \eta_j = 2n$, and then performed the sum over n and $\{\eta_j = \pm 1\}$. Observe that every individual path together with its charge-conjugate counter part gives a real contribution to the generating function.

It is now easy to find the expressions for the moments of the probability distribution. For the most interesting cases $N = 1$ and $N = 2$ we obtain using Eq. (25.4)

$$\langle q(t)\rangle_{\mathrm{pis}} = q_0 \sum_{m=1}^{\infty} \int_0^t \mathcal{D}_{2m,0}\{t_j\} \sum_{\{\xi_j\}'} b_{2m,0} B_{2m,0} \, G_{2m} \,, \qquad (25.42)$$

$$\langle q^2(t)\rangle_{\mathrm{pis}} = q_0^2 \sum_{m=1}^{\infty} \int_0^t \mathcal{D}_{2m,0}\{t_j\} \sum_{\{\xi_j\}'} c_{2m,0} B_{2m,0} \, G_{2m} \,, \qquad (25.43)$$

where

$$b_{2m,0} = \frac{i}{2} (-1)^{m-1} \prod_{\ell=1}^{2m-1} \sin(\chi_{\ell,2m}) \,, \qquad (25.44)$$

$$c_{2m,0} = \frac{1}{2} (-1)^{m-1} \sum_{k=1}^{2m-1} \cos(\chi_{k,2m}) \prod_{\ell=1,\ell\neq k}^{2m-1} \sin(\chi_{\ell,2m}) \,. \qquad (25.45)$$

Eqs. (25.39) – (25.45) are exact expressions appropriate to a product initial state. The correlations resulting from the friction action are in the phase functions $\chi_{k,2m}$. The correlations due to the noise action are captured by the filter function G_{2m}.

We also give the corresponding Laplace transforms. These are conveniently expressed in terms of the intervals the system spends in a particular state, $\tau_\ell = t_{\ell+1} - t_\ell$. We introduce for the corresponding integrations the compact symbol

$$\int_0^{\infty} \hat{\mathcal{D}}_{k,0}(\lambda, \{\tau_\ell\}) \times \cdots \equiv \Delta^k \int_0^{\infty} \prod_{\ell=1}^{k-1} d\tau_\ell \, e^{-\lambda\tau_\ell} \times \cdots \,, \qquad \tau_\ell = t_{\ell+1} - t_\ell \,. \quad (25.46)$$

The Laplace transforms of $\langle q(t)\rangle_{\mathrm{pis}}$ and $\langle q^2(t)\rangle_{\mathrm{pis}}$ are then given by

$$\langle \hat{q}(\lambda)\rangle_{\mathrm{pis}} = i\frac{q_0}{\lambda^2} \sum_{m=1}^{\infty} \int_0^{\infty} \hat{\mathcal{D}}_{2m,0}(\lambda, \{\tau_\ell\}) \sum_{\{\xi_j\}'} b_{2m,0} G_{2m} \sin \varphi_{2m,0} \,, \quad (25.47)$$

$$\langle \hat{q}^2(\lambda)\rangle_{\mathrm{pis}} = \frac{q_0^2}{\lambda^2} \sum_{m=1}^{\infty} \int_0^{\infty} \hat{\mathcal{D}}_{2m,0}(\lambda, \{\tau_\ell\}) \sum_{\{\xi_j\}'} c_{2m,0} G_{2m} \cos \varphi_{2m,0} \,. \quad (25.48)$$

The bias phase $\varphi_{2m,0}$ is conveniently expressed in terms of the cumulated charge

$$p_\ell = \sum_{k=1}^{\ell} \xi_k = -\sum_{k=\ell+1}^{2m} \xi_k \,. \qquad (25.49)$$

The second form is obtained with use of the constraint (25.37). The cumulated charge p_ℓ measures how far the system is off-diagonal after ℓ transitions. The meaning of the charges $\{\xi_j\}$ and the cumulated charges $\{p_j\}$ is illustrated in Fig. 25.1. With the cumulated charges $\{p_j\}$ and with $\sum_i \xi_i = 0$, the bias phase (25.36) can be written as

$$\varphi_{j,k} = \epsilon\, p_k(t_{k+1} - t) + \epsilon \sum_{\ell=k+1}^{k+j-1} p_\ell \tau_\ell \,, \qquad \varphi_{2m,0} = \epsilon \sum_{\ell=1}^{2m-1} p_\ell \tau_\ell \,. \quad (25.50)$$

The diffusive dynamics to be reached at long time can now be traced from the residua of the singularities at $\lambda = 0$ in the expressions (25.47) and (25.48).

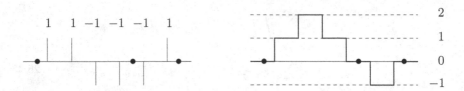

1   1 −1 −1 −1   1

Figure 25.1: A sequence of six steps falling into two clusters separated by a sojourn (•) is sketched. In the charge picture (left), the spikes or charges $\{\xi_j = \pm1\}$ give the direction of the moves perpendicular to the diagonal of the RDM. In the respective state representation (right), the cumulated charge p_j measures distance from the main diagonal after j steps.

25.3.2   Thermal initial state

To derive the dynamical expressions for a thermal initial state, we follow the discussion in Section 5.6 and Subsection 21.2.2. We then obtain the propagating function, which is a constituent of the expression (25.12), in the form

$$
J(n\,n, t; m\,r, 0; 0\,0, t_{\mathrm{p}}) \;=\; \sum_{\substack{k=|m|+|r| \\ j=|n-m|+|n-r|}}^{\infty} (-1)^n \left(\frac{i}{2}\right)^{j+k}
$$

$$
\times \int_{t_{\mathrm{p}}}^{t} \mathcal{D}_{j,k}\{t_\ell\} \sum_{\{\xi_\ell\}''} B_{j,k} G_{j+k} \sum_{\{\eta_\ell\}''} H_{j+k} \; . \tag{25.51}
$$

The double prime in $\{\xi_\ell\}''$ and $\{\eta_\ell\}''$ is to indicate the constraints

$$
\sum_{\ell=1}^{k} \eta_\ell \;=\; m+r\,, \qquad \sum_{\ell=k+1}^{k+j} \eta_\ell \;=\; 2n-(m+r)\,,
$$

$$
\sum_{\ell=1}^{k} \xi_\ell \;=\; m-r\,, \qquad \sum_{\ell=k+1}^{k+j} \xi_\ell \;=\; r-m\,.
$$

The summations $\{\xi_\ell = \pm1\}''$ and $\{\eta_\ell = \pm1\}''$ create all possible paths on the (q, q')-lattice to go in k steps in the negative time interval $-|t_{\mathrm{p}}| < t' < 0$ from $(0,0)$ to (m, r) and in j steps at positive times $0 < t' < t$ from (m, r) to (n, n). Substituting the form (25.51) into Eq. (25.12), the generating function $\mathcal{Z}_{\mathrm{th}}$ takes the form

$$
\mathcal{Z}_{\mathrm{th}}(\rho, \kappa, \mu; t) \;=\; \sum_{j=0}^{\infty} \sum_{k=0}^{\infty} \int_{-\infty}^{t} \mathcal{D}_{j,k}\{t_\ell\} \sum_{\{\xi_\ell\}'} B_{j,k} G_{j+k} F_{j,k}(\rho, \kappa, \mu)\,, \tag{25.52}
$$

where the prime in $\{\xi_\ell\}'$ denotes again summation over all charge sequences obeying the neutrality condition (25.37). The function $F_{j,k}(\rho, \kappa, \mu)$ is given by

$$F_{j,k}(\rho, \kappa, \mu) = \left(\frac{1}{2}\right)^{j+k} \sum_{n,m,r=-\infty}^{+\infty} (-1)^{(j+k-2n)/2} \, e^{iq_0(n\kappa+m\rho+r\mu)}$$

$$\times \, \Theta(k - |m| - |r|) \, \theta(j - |n - m| - |n - r|) \sum_{\{n_\ell\}''} H_{j+k} \, . \tag{25.53}$$

Following the strategy which has led us to the second form in Eq. (25.41), it is again easy to perform the constrained sums. The resulting expression is

$$F_{j,k}(\rho, \kappa, \mu) = i^{j+k} \, \exp\left(i\frac{q_0}{2}(\rho - \mu)p_k\right) \sin\left(\frac{\kappa q_0}{2}\right)$$

$$\times \prod_{\ell=1}^{k} \sin\left(\frac{q_0}{2}(\kappa + \rho + \mu) + \chi_{\ell,j+k}\right) \tag{25.54}$$

$$\times \prod_{m=k+1}^{j+k-1} \sin\left(\frac{q_0}{2}\kappa + \chi_{m,j+k}\right) \, .$$

Finally, performing the differentiations of $\mathcal{Z}_{\text{th}}$ given in Eq. (25.11), we find for $t > 0$

$$\langle q(t) \rangle_{\text{th}} = \langle q(t) \rangle_{\text{pis}} + R_1(t) \, , \tag{25.55}$$

$$D_{\text{th}}(t) = \langle q^2(t) \rangle_{\text{pis}} + R_2(t) \, . \tag{25.56}$$

The effects of the different initial conditions for $\langle \cdots \rangle_{\text{pis}}$ and $\langle \cdots \rangle_{\text{th}}$ are described by

$$R_1(t) = q_0 \sum_{j=1}^{\infty} \sum_{k=1}^{\infty} \int_{-\infty}^{t} \mathcal{D}_{j,k}\{t_\ell\} \sum_{\{\xi_\ell\}'} b_{j,k} B_{j,k} G_{j+k} \, , \tag{25.57}$$

$$R_2(t) = q_0^2 \sum_{j=1}^{\infty} \sum_{k=1}^{\infty} \int_{-\infty}^{|t|} \mathcal{D}_{j,k}\{t_\ell\} \sum_{\{\xi_\ell\}'} c_{j,k} B_{j,k} G_{j+k} \, . \tag{25.58}$$

The coefficients $b_{j,k}$ and $c_{j,k}$ are given by

$$b_{j,k} = i\frac{1}{2}(-1)^{(j+k-2)/2} \prod_{\ell=1}^{j+k-1} \sin(\chi_{\ell,j+k}) \, ,$$

$$c_{j,k} = \frac{1}{2}(-1)^{(j+k-2)/2} \prod_{\ell=1}^{k} \sin(\chi_{\ell,j+k}) \tag{25.59}$$

$$\times \left[\sum_{i=k+1}^{j+k-1} \cos(\chi_{i,j+k}) \prod_{m=k+1,m\neq i}^{j+k-1} \sin(\chi_{m,j+k})\right] \, .$$

Further, the response function $\chi(t)$, which is related to the antisymmetric correlation function $A(t)$ by the relation (6.15), is found as

$$\chi(t) \;=\; \frac{i}{\hbar} q_0^2\, \Theta(t) \sum_{j=1}^{\infty} \sum_{k=1}^{\infty} \int_{-\infty}^{t} \mathcal{D}_{j,k}\{t_\ell\} \sum_{\{\xi_\ell\}'} p_k b_{j,k} G_{j+k} \cos \varphi_{j,k}\,. \tag{25.60}$$

Finally, consider the Laplace transforms of $R_1(t)$ and $R_{2,+}(t) = \Theta(t)\, R_2(t)$. In generalization of the expression (25.46), we define

$$\int_0^{\infty} \hat{\mathcal{D}}_{j,k}(\lambda, \{\tau_\ell\}) \times \cdots \equiv \Delta^{j+k} \prod_{i=k+1}^{k+j-1} \int_0^{\infty} d\tau_i\, e^{-\lambda \tau_i} \prod_{\ell=1}^{k-1} \int_0^{\infty} d\tau_\ell\, e^{-0^+ \tau_\ell} \times \cdots\,. \tag{25.61}$$

The additional two intervals not included in Eq. (25.61) are

$$r_k = -t_k\,; \qquad s_k = t_{k+1}\,. \tag{25.62}$$

The interval r_k is the time passed between the last step in the negativ-time branch and time zero, and s_k is the interval between time zero and the first step at positive time. The Laplace transforms of the functions $R_1(t)$ and $R_{2,+}(t)$ are

$$
\begin{aligned}
\hat{R}_1(\lambda) &= i \frac{q_0}{\lambda} \sum_{j=1}^{\infty} \sum_{k=1}^{\infty} \int_0^{\infty} \hat{\mathcal{D}}_{j,k}(\lambda, \{\tau_\ell\}) \\
&\quad \times \int_0^{\infty} dr_k\, ds_k\, e^{-(\lambda s_k + 0^+ r_k)} \sum_{\{\xi_\ell\}'} b_{j,k} G_{j+k} \sin \varphi_{j,k}\,, \\[2mm]
\hat{R}_{2,+}(\lambda) &= \frac{q_0^2}{\lambda} \sum_{j=1}^{\infty} \sum_{k=1}^{\infty} \int_0^{\infty} \hat{\mathcal{D}}_{j,k}(\lambda, \{\tau_\ell\}) \\
&\quad \times \int_0^{\infty} dr_k\, ds_k\, e^{-(\lambda s_k + 0^+ r_k)} \sum_{\{\xi_\ell\}'} c_{j,k} G_{j+k} \cos \varphi_{j,k}\,,
\end{aligned}
\tag{25.63}
$$

and the Laplace transform of the response function for zero bias is

$$
\begin{aligned}
\hat{\chi}(\lambda) &= i \frac{q_0^2}{\hbar \lambda} \sum_{j=1}^{\infty} \sum_{k=1}^{\infty} \int_0^{\infty} \hat{\mathcal{D}}_{j,k}(\lambda, \{\tau_\ell\}) \\
&\quad \times \int_0^{\infty} dr_k\, ds_k\, e^{-(\lambda s_k + 0^+ r_k)} \sum_{\{\xi_\ell\}'} p_k b_{j,k} G_{j+k}\,.
\end{aligned}
\tag{25.64}
$$

The functions $R_1(t)$ and $R_{2,+}(t)$ describe the effects of the correlated initial state in the dynamics of $\langle q(t)\rangle_{\mathrm{th}}$ and $D_{\mathrm{th}}(t)$.

25.4 Mobility and Diffusion

25.4.1 Exact formal series expressions for transport coefficients

The mobility and the diffusion coefficient are found from the limits $\lim_{\lambda \to 0} \lambda^2 \langle \hat{q}(\lambda)\rangle_{\mathrm{pis}}$ and $\lim_{\lambda \to 0} \lambda^2 \langle \hat{q}^2(\lambda)\rangle_{\mathrm{pis}}$ in the expressions (25.47) and (25.48).

In the charge picture, an off-diagonal (blip) interval with cumulated charge p_k separates a sequence of charges with total charge zero into two sections with charge $\pm p_k$. When the blip length τ_k is large compared to the effective length of the sections, the blip interaction factor is $\exp\left[-p_k^2 Q'(\tau_k)\right]$. For the form (18.46) of the spectral density $G_{lf}(\omega)$, the function $Q(\tau)$ behaves for nonzero T at asymptotic time as

$$Q'(\tau) \propto \tau^{2-s}, \qquad Q''(\tau) \propto \tau^{1-s}, \qquad \text{for} \qquad \tau \gg 1/\omega_c,\, \hbar\beta, \qquad (25.65)$$

as we see from Eq. (18.47). Hence the integral over the length of a blip is convergent even in the limit $\lambda \to 0$ when $s < 2$. For $s \geq 2$ and $\lambda = 0$, the interaction between charged clusters can not suppress long blips anymore, and the integral is only convergent when the biasing force F is nonzero.

For a sojourn state $k = 2j$, the cumulated charge p_{2j} is zero because the charge clusters to the left and right are neutral. Since the dipole-dipole interaction $\propto \ddot{Q}'(\tau)$ drops to zero as τ^{-s}, it can not provide convergence of the integral over the sojourn interval τ_{2j} in the limit $\lambda \to 0$. But inspection of the phase factor $b_{2m,0}$ shows that just this factor furnishes convergence of the τ_{2j}-integral when $s \geq 1$. Assuming convergence of the blip and sojourn integrals within a neutral charge cluster we then obtain at long time

$$\lim_{t\to\infty} \langle q(t) \rangle_{\mathrm{pis}} = F\mu t + q_\infty, \qquad (25.66)$$

where we have added the next to leading order term. The quantity μ is the mobility. It is called nonlinear mobility when it depends on the bias, $\mu = \mu(T, \epsilon)$, and linear mobility in the linear response regime, $\mu_{\mathrm{lin}}(T) = \mu(T, 0)$. We obtain the mobility from Eq. (25.47) in the form of a series in even numbers of transitions,

$$\mu(\epsilon, T) = \lim_{\lambda\to 0^+} i\frac{q_0^2}{\hbar\epsilon} \sum_{m=1}^{\infty} \int_0^\infty \hat{D}_{2m,0}(\lambda, \{\tau_\ell\}) \sum_{\{\xi_l\}'} b_{2m,0} G_{2m} \sin\left(\epsilon \sum_{j=1}^{2m-1} p_j\tau_j\right). \qquad (25.67)$$

This series expression is formally exact. With the above arguments we see that the nonlinear mobility is well defined for $s \geq 1$ and the linear mobility for $T > 0$ in the range $2 > s \geq 1$. At zero temperature, $Q'(\tau)$ approaches a constant as $\tau \to \infty$ for $s > 1$. Thus the linear mobility becomes divergent as $T \to 0$ also in the super-Ohmic regime $1 < s < 2$. This indicates that the mean position moves faster than linear in t. The sub-Ohmic case is special. It will be discussed below in Chapter 29.

Consider next the second moment $\langle \hat{q}^2(\lambda) \rangle_{\mathrm{pis}}$ for the unbiased washboard ($\epsilon = 0$). The phase factor (25.45) is nonzero for a sojourn $k = 2j$, $p_{2j} = 0$, because of the term $\cos(\chi_{2j,2m})$. In addition, the interactions in the factor G_{2m} across the sojourn are not strong enough for $s \geq 1$ to regularize the integral over the sojourn length. Thus the sojourn's breathing integral seems to diverge in the limit $\lambda \to 0$. However one finds that the divergent contributions cancel out in the sum of all paths which visit the sojourn $2j$. As a result, the limiting expression $\lim_{\lambda\to 0} \lambda^2 \langle \hat{q}^2(\lambda) \rangle_{\mathrm{pis}}$ in the regime $1 \leq s < 2$ and $T > 0$ is a constant. This entails

$$\lim_{t\to\infty} \langle q^2(t) \rangle_{\mathrm{pis}} / t = 2D. \qquad (25.68)$$

The exact formal weak-tunneling series for the diffusion coefficient is

$$D(T,0) = \lim_{\lambda \to 0^+} \frac{q_0^2}{2} \sum_{m=1}^{\infty} \int_0^{\infty} \hat{D}_{2m,0}(\lambda, \{\tau_\ell\}) \sum_{\{\xi_j\}'} c_{2m,0} G_{2m} . \tag{25.69}$$

In conclusion, the TB model shows diffusive behavior in the regime $1 < s < 2$. This is different from free Brownian motion which is superdiffusive in this regime, as we have discussed in Section 7.4.

25.4.2 Einstein relation

The Einstein relation links the diffusion coefficient to the linear mobility,

$$D(T,0) = k_{\mathrm{B}} T \, \mu_{\mathrm{lin}}(T) . \tag{25.70}$$

As emphasized in the preceeding subsection, the static transport coefficients D and μ_{lin} in this relation are well-defined in the parameter regime $1 \leq s < 2$. We have seen already in Section 7.3 for the case of free Brownian motion that the relation (25.70) is a special version of the fluctuation-dissipation theorem. The relation was proven rigorously for Brownian motion in a corrugated potential in the classical regime [504]. There have been raised concerns about the general validity of the Einstein relation in the quantum regime [218]. The questions are related to possible problems with the definition of μ_{lin} since the stationary state in which the particle moves with constant velocity is not normalizable and therefore the standard Green-Kubo method might not be applicable. On the other hand, if one defines the mobility as in Eq. (25.66), it is not obvious a priori whether the Einstein relation is valid in all orders in Δ^2.

To prove the relation (25.70) for the TB model, we rewrite μ_{lin} and D as moments of the distribution function, and return to the parametrization (25.27). We have

$$\mu_{\mathrm{lin}} = q_0^2 \sum_n n X_n , \qquad D = \tfrac{1}{2} q_0^2 \sum_n n^2 Y_n , \tag{25.71}$$

$$X_n = \sum_{k,\ell=|n|}^{\infty} i^{\ell-k} \left(\frac{\Delta}{2}\right)^{\ell+k} \int_0^{\infty} \prod_{j=1}^{k-1} d\rho_j \prod_{i=1}^{\ell-1} d\rho_i' \int_{-\infty}^{\infty} d\tau \sum_{\{u_k, v_\ell\}'} \frac{i\varphi_{k,\ell}}{\hbar\epsilon} \mathcal{F}[q^{(k)}, q'^{(\ell)}] ,$$

$$Y_n = \sum_{k,\ell=|n|}^{\infty} i^{\ell-k} \left(\frac{\Delta}{2}\right)^{\ell+k} \int_0^{\infty} \prod_{j=1}^{k-1} d\rho_j \prod_{i=1}^{\ell-1} d\rho_i' \int_{-\infty}^{\infty} d\tau \sum_{\{u_k, v_\ell\}'} \mathcal{F}[q^{(k)}, q'^{(\ell)}] ,$$

where $\rho_j = t_{j+1} - t_j$ and $\rho_i' = t_{i+1}' - t_i'$ are the intervals in the paths $q(\tau)$ and $q'(\tau)$, and $\tau = t_1' - t_k$. The bias phase is $\varphi_{k,\ell} = \epsilon (\sum_i v_i t_i' - \sum_j u_j t_j)$. The ρ- and ρ'-integrals preserve charge ordering in the individual sets $\{u_k\}$ and $\{v_\ell\}$, respectively. The unbounded τ-integral introduces all possibilities of mixing up the set of charges $\{u_k\}$ with the set of charges $\{v_\ell\}$.

To proceed, we assign to the double path $\{q^{(k)}(\tau), q'^{(\ell)}(\tau)\}$ a conjugate double path $\{q^{(\ell)}(\tau), q'^{(k)}(\tau)\}$ in which the charges are in reverse order. Since the function

$Q(z)$ is analytic in the strip $0 \geq \operatorname{Im} z > -\hbar\beta$, the τ-integration contour can be translated parallely in this strip as explained below Eq. (20.29).[2] If we perform a translation of the contour by $-i\hbar\beta$, $\tilde{\tau} = \tau - i\hbar\beta$. we may use the reflection symmetry (18.45) of $Q(z)$. As a result, the influence function $\tilde{\mathcal{F}}$ calculated for a conjugate pair of double paths on the shifted contour $\tilde{\tau}$ has the property

$$\tilde{\mathcal{F}}[q^{(k)}, q'^{(\ell)}] + \tilde{\mathcal{F}}[q^{(\ell)}, q'^{(k)}] = \mathcal{F}^*[q^{(k)}, q'^{(\ell)}] + \mathcal{F}^*[q^{(\ell)}, q'^{(k)}], \qquad (25.72)$$

and the bias phase for the shifted contour is related to that for the original contour by

$$\tilde{\varphi}_{k,\ell} = \varphi_{k,\ell} - i\beta\hbar\epsilon n. \qquad (25.73)$$

Upon writing X_n for the shifted contour, then employing the relations (25.72) and (25.73), and comparing the resulting expression with the above form for Y_n, we find for $1 \leq s < 2$ the relation [360]

$$X_n = -X_n + n\beta Y_n. \qquad (25.74)$$

This relation is valid in all orders in Δ^2. Substituting Eq. (25.74) into the expressions (25.71), we find the Einstein relation (25.70) (see also the discussion in Ref. [505]).

Since a conjugate pair of double paths visit the same sites, the relation (25.74) and hence the Einstein relation holds also in the presence of disorder.

25.5 The Ohmic case

We now consider the mobility for the Ohmic spectral density in various limits. We base the treatment on the scaling form (18.54) for the pair interaction with the usual dimensionless damping parameter introduced by Kondo [cf. Eq. (18.21)]

$$K = \frac{\eta q_0^2}{2\pi\hbar}. \qquad (25.75)$$

With the scaling form (18.54) we find from (25.67) the normalized TB mobility as

$$\frac{\mu_{\mathrm{TB}}(T, \epsilon)}{\mu_0} = \frac{\pi K}{\epsilon} \sum_{m=1}^{\infty} (-1)^{m-1} \Delta^{2m} \int_0^\infty d\tau_1 \, d\tau_2 \cdots d\tau_{2m-1}$$

$$\times \sum_{\{\xi_j = \pm 1\}'} G_{2m}\left(\{\tau_j\}; K, \{\xi_j\}\right) \sin\left(\sum_{i=1}^{2m-1} \epsilon\, p_i \tau_i\right) \prod_{k=1}^{2m-1} \sin(\pi K p_k), \qquad (25.76)$$

with the interaction factor (we put $\tau_{jk} \equiv t_j - t_k = \sum_{\ell=k}^{j-1} \tau_\ell$)

$$G_{2m}\left(\{\tau_j\}; K, \{\xi_j\}\right) = \exp\left(2K \sum_{j>k=1}^{2m} \xi_j \xi_k \ln\left[(\beta\hbar\omega_c/\pi)\sinh(\pi\tau_{jk}/\hbar\beta)\right]\right). \qquad (25.77)$$

Here, $\mu_0 = 1/\eta = q_0^2/2\pi\hbar K$ is the mobility of a free Brownian particle [cf. Eq. (7.27)]. Due to the product of sine factors, every double-path contributing to the mobility does not return to the diagonal at intermediate times, i.e., the affiliated cumulated charges p_k have all the same sign. In charge picture, the form (25.76) is the grand-canonical sum of all charge sequences which can not be separated into neutral sub-clusters.

---

[2]In the absence of a bias, $\exp[-Q(z)]$ must vanish for $|z| \to \infty$. This gives the bound $s < 2$.

25.5.1 Weak-tunneling regime

The $m = 1$ term of the series (25.76) can be calculated by direct integration. Alternatively, it can be found by combining previous results. With the form (18.54) for $Q(t)$, the nonadiabatic forward tunneling rate k^+ takes the form (20.79). With this expression, and with $k = (1+e^{-\beta\hbar\epsilon})k^+$, we then obtain the mobility from Eq. (25.19). In either case we find

$$\frac{\mu_{\mathrm{TB}}(T,\epsilon)}{\mu_0} = \pi^2 K \frac{|\Gamma(K + i\hbar\epsilon/2\pi k_{\mathrm{B}}T)|^2}{\Gamma(2K)} \frac{\sinh(\hbar\epsilon/2k_{\mathrm{B}}T)}{\hbar\epsilon/2k_{\mathrm{B}}T} \left(\frac{\hbar\Delta_{\mathrm{r}}}{2\pi k_{\mathrm{B}}T}\right)^{2-2K}. \tag{25.78}$$

From this we find for the nonlinear mobility at zero temperature

$$\frac{\mu^{(\mathrm{TB})}(0,\epsilon)}{\mu_0} = \frac{\pi^2 K}{\Gamma(2K)} \left(\frac{\Delta_{\mathrm{r}}}{\epsilon}\right)^{2-2K}, \tag{25.79}$$

and for the linear mobility at finite temperature

$$\frac{\mu_{\mathrm{TB,\,lin}}(T,0)}{\mu_0} = \frac{\pi^2\sqrt{\pi}}{2}\frac{\Gamma(1+K)}{\Gamma(1/2+K)}\left(\frac{\hbar\Delta_{\mathrm{r}}}{\pi k_{\mathrm{B}}T}\right)^{2-2K}. \tag{25.80}$$

These expressions describe the mobility for an Ohmic environment in the nearest-neighbor tunneling regime discussed in Subsection 25.2.1. The diffusion coefficient $D(T,\epsilon)$ in order Δ^2 can be found with use of Eq. (25.78) in the expression (25.20). Combining the expressions (25.20) and (25.80), the diffusion coefficient reads

$$D(T,0) = \mu_0\hbar\Delta_{\mathrm{r}}\frac{\pi\sqrt{\pi}\,\Gamma(1+K)}{2\Gamma(\frac{1}{2}+K)}\left(\frac{\hbar\Delta_{\mathrm{r}}}{\pi k_{\mathrm{B}}T}\right)^{1-2K}. \tag{25.81}$$

Thus we have Kondo behavior $D(T,0) \propto T^{2K-1}$ [cf. the discussion below Eq. (20.84)].

25.5.2 Weak-damping limit

When the Kondo parameter is very small, $K \ll 1$, and the bias is low, paths which include visits of far off-diagonal states of the RDM and long dwell times in these states give the leading contributions in the series (25.76). In the regime $\tau \gg \hbar\beta$, the noise correlation function $Q'(\tau)$ takes the "white noise" form [cf. Eq. (22.48)]

$$Q'(\tau) = 2K\left[\pi|\tau|/\hbar\beta + \ln(\beta\hbar\omega_c/2\pi)\right]. \tag{25.82}$$

With the correlation function (25.82), the combined filter function of the noise filter (25.34) and the bias filter $e^{i\varphi_{2m,0}}$ takes product form and provides an exponential filter for each individual time interval,

$$G_{2m}\,e^{i\varphi_{2m,0}} = \left(\frac{2\pi}{\beta\hbar\omega_c}\right)^{2Km}\prod_{k=1}^{2m-1}\exp\left\{-\left(2\pi K\frac{p_k}{\hbar\beta} - i\,\epsilon\right)p_k\tau_k\right\}. \tag{25.83}$$

In the parameter regime $\beta\hbar\epsilon \ll 2\pi p K$, where p is a characteristic cumulated charge, the path measure is dominated by the noise filter function G_{2m}. In the opposite regime $\beta\hbar\epsilon \gg 2\pi pK$, the filter function G_{2m} is ineffective, and the path measure is dominated by the bias factor $e^{i\varphi_{2m,0}}$ instead.

With the product form (25.83) all the time integrals in the Laplace representation $\langle \hat{q}^N(\lambda)\rangle_{\mathrm{pis}}$ are uncorrelated, and each integral is a simple definite integral of an exponential function. Eventually, the mobility is found from Eq. (25.76) as

$$\frac{\mu_{\mathrm{TB}}(T,\epsilon)}{\mu_0} = -\frac{2\pi K}{\beta\hbar\epsilon}\sum_{m=1}^{\infty}(-u)^m \sum_{\{\xi_j=\pm 1\}'} \mathrm{Im}\prod_{k=1}^{2m-1}\frac{\sin(\pi K p_k)}{\pi K p_k[p_k - i\,\beta\hbar\epsilon/2\pi K]}\,, \qquad (25.84)$$

where

$$u \equiv (\beta\hbar\Delta_T/2)^2\,, \qquad \Delta_T = \Delta_{\mathrm{r}}\,(2\pi k_{\mathrm{B}}T/\hbar\Delta_{\mathrm{r}})^K\,. \qquad (25.85)$$

The adiabatic Franck-Condon factor $(2\pi/\beta\hbar\omega_{\mathrm{c}})^K$ in Eq. (25.83) is included in the frequency scale Δ_T [cf. Eq. (22.49)]. The series (25.84) resembles the grand-canonical partition function of a one-dimensional Coulomb gas [506]. The summation over all arrangements of charges $\xi_j = \pm 1$ satisfying the neutrality condition (25.37) leads to the continued fraction expression [506, 499]

$$\frac{\mu_{\mathrm{TB}}(T,\epsilon)}{\mu_0} = \frac{4\pi K}{\beta\hbar\epsilon}\,\mathrm{Im}\,\cfrac{a_1 u}{1 + \cfrac{a_1 a_2 u}{1 + \cfrac{a_2 a_3 u}{1 + \cdots}}} \qquad (25.86)$$

with coefficients

$$a_n = \frac{\sin(\pi K n)}{\pi K n}\,\frac{1}{n - i\,\beta\hbar\epsilon/2\pi K}\,. \qquad (25.87)$$

For $K \ll 1$, the sine function in Eq. (25.87) may be linearized. This yields the form

$$a_n = 1/(n - i\nu) \qquad \text{with} \qquad \nu \equiv \beta\hbar\epsilon/2\pi K\,. \qquad (25.88)$$

With this form, the expression (25.86) matches the continued fraction (9.1.73) in Ref. [89] which can be written in terms of modified Bessel functions $I_\mu(z)$ with complex order μ [507]. With adjustment of the parameters, we obtain the mobility in analytic form for all T and ϵ as

$$\frac{\mu_{\mathrm{TB}}(T,\epsilon)}{\mu_0} = \frac{\beta\hbar\Delta_T}{\nu}\,\mathrm{Im}\left(\frac{I_{1-i\nu}(\beta\hbar\Delta_T)}{I_{-i\nu}(\beta\hbar\Delta_T)}\right)\,. \qquad (25.89)$$

Employing recursion relations of the modified Bessel function, this expression can be rewritten as

$$\frac{\mu_{\mathrm{TB}}(T,\epsilon)}{\mu_0} = 1 - \frac{\sinh(\pi\nu)}{\pi\nu}\,\frac{1}{|I_{i\nu}(\beta\hbar\Delta_T)|^2}\,. \qquad (25.90)$$

Upon performing the limit $\epsilon \to 0$ either in Eq. (25.89) or in Eq. (25.90), the linear mobility is found as

$$\frac{\mu_{\mathrm{TB,\,lin}}(T)}{\mu_0} = 1 - \frac{1}{I_0^2(\beta\hbar\Delta_T)}\,. \qquad (25.91)$$

From this we see that the linear mobility at zero temperature coincides with the mobility of a free Brownian particle, $\mu_{\text{lin}}^{(\text{TB})}(T=0) = \mu_0$.

Consider next the nonlinear mobility at $T = 0$. Now, thermal noise is absent and quantum noise is so weak in the limit $K \to 0$ that the only effect of the filter function G_{2m} in the series expression (25.76) is to regularize the time integrals, $\lim_{\kappa \to 0^+} \int_0^\infty d\tau_j \, e^{-(\kappa - i\,\epsilon p_j)\tau_j}$, $j = 1, \cdots, 2m-1$. Again the time integrals are simple. In addition, we may linearize the sine factors in Eq. (25.84), $\sin(\pi K p_k) \to \pi K p_k$. Then the cumulated charges $\{p_j\}$ drop out. As a result, for fixed number of charges in a neutral cluster, every individual sequence of charges yields the same contribution. Altogether, the perturbative series for the nonlinear mobility takes the form

$$\frac{\mu_{\text{TB}}(0, \epsilon)}{\mu_0} = \sum_{m=1}^{\infty} \left(\frac{\pi K \Delta}{\epsilon}\right)^{2m} \sum_{\{\xi_j = \pm 1\}'} 1 \,. \tag{25.92}$$

The number of possibilities for $2m$ charges with total charge zero to form a single irreducible cluster, i.e. all $p_j \neq 0$, is $2^{2m-1}\Gamma(m-1/2)/\sqrt{\pi}\Gamma(m+1)$. Thus we find

$$\frac{\mu_{\text{TB}}(0, \epsilon)}{\mu_0} = \frac{1}{2\sqrt{\pi}} \sum_{m=1}^{\infty} \frac{\Gamma(m-1/2)}{\Gamma(m+1)} \left(\frac{2\pi K \Delta}{\epsilon}\right)^{2m} \,. \tag{25.93}$$

This series is convergent in the regime

$$\epsilon > \epsilon_0 \equiv 2\pi K \Delta \,, \tag{25.94}$$

and is summed to the square root expression

$$\frac{\mu_{\text{TB}}(0, \epsilon)}{\mu_0} = 1 - \sqrt{1 - (\epsilon_0/\epsilon)^2} \,. \tag{25.95}$$

Two remarks seem expedient. First, the expression (25.95) is also found from the continued fraction expression (25.86) if we substitute $a_n = i/\nu$ and put $u = (\hbar\beta\Delta/2)^2$. Secondly, in the zero temperature limit of the expression (25.89), or of the expression (25.90), we may employ the uniform asymptotic expansion of the modified Bessel function for large complex order [cf. Eq. (9.7.7) in Ref. [89]]. Taking into account the asymptotically leading term and disregarding the adiabatic Franck-Condon dressing factor, we then arrive again at the expression (25.95).

25.6 Exact solution in the Ohmic scaling limit at $K = \frac{1}{2}$

The quantum transport problem of the Ohmic TB particle for the particular case $K = \frac{1}{2}$ can be analytically solved in two different ways. In the first, one employs in the Coulomb gas representation of the TB model presented in Section 25.3 the concept of collapsed dipoles, which has been introduced in Subsection 22.6.1. In the second way, one passes on to a representation of noninteracting fermionic quasiparticles by following the treatment given in Subsections 19.2.2 and 22.6.6.

25.6.1   Current and mobility

In the Ohmic scaling limit, the TB series (25.76) for the mobility and the series (25.47) for $\langle \hat{q}(\lambda) \rangle_{\text{pis}}$ can be calculated in analytic form for the coupling value $K = \frac{1}{2}$ using the concept of collapsed dipoles developed in Subsection 22.6.1 [508, 360]. The calculation is done by putting $K = \frac{1}{2} - \kappa$ and eventually taking the limit $\kappa \to 0$. In this limit, the phase factors $\sin(\pi K p_k)$ are reduced to factors $1, 2\pi\kappa, -1, -4\pi\kappa, \cdots$ for $p_k = 1, 2, 3, 4, \cdots$. On the other hand, the breathing mode integral of a dipole diverges as $1/\kappa$ in the limit $\kappa \to 0$. The singularity originates again from the $\tau^{-1+2\kappa}$ short-distance behavior of the intra-dipole interaction $\exp[-Q'(\tau)]$. Altogether, only those charge sequences, in which the occurring κ-factors are compensated by $1/\kappa$-singularities, contribute to the mobility. These are just the formations which have cumulated charge sequence $p_k = 1, 2, 1, 2, \cdots, 2, 1$ and $p_k = -1, -2, -1, -2, \cdots, -2, -1$, or pictorially

$$
\begin{aligned}
+ \ (+-) \ (+-) \ \cdots \ (+-) \ (+-) \ - \ , \\
- \ (-+) \ (-+) \ \cdots \ (-+) \ (-+) \ + \ .
\end{aligned}
\tag{25.96}
$$

Here, the round brackets indicate the charge pairs which form collapsed dipoles. Thus, to the path sum for the mobility only paths contribute which are limited to the pentadiagonal matrix of the (q, q')-lattice.

For convenience in Subsection 26.2.1, we now use the frequency scale $\tilde{\epsilon}_0$ introduced for general K in Eq. (26.119). At $K = \frac{1}{2}$, we have

$$
\tilde{\epsilon}_0 = 2\gamma = \pi\Delta^2/\omega_c \, ,
\tag{25.97}
$$

where $\gamma = \Delta_{\text{eff}}(K = \frac{1}{2})$ is the frequency scale familiar from the spin-boson system at the Toulouse point $K = \frac{1}{2}$ (see Subsection 19.2.2 and Section 22.6).

The grand-canonical sum of collapsed noninteracting dipoles in the interval τ between the two peripheral charges yields the form factor (see Subsection 22.6.1)

$$
U_{\text{CD}}(\tau) = e^{-\tilde{\epsilon}_0\tau} \, .
\tag{25.98}
$$

The exponent in Eq. (25.98) differs from the exponent in the CB form factor (22.131) by a factor two. This factor stems from the phase factor $\sin(2\pi K) = 2\pi\kappa$ of the collapsed dipole compared with the blip phase factor $\cos(\pi K) = \pi\kappa$ of the collapsed blip in Eq. (22.129). From these considerations we see that the dressed extended $+-$ and $-+$ dipoles depicted in (25.96) determine the mean particle current and hence also the mobility, $\langle I \rangle = (\hbar\epsilon/q_0)\,\mu = q_0(\epsilon/\pi)\,\mu/\mu_0$. The breathing mode integrals of the two extended dipoles readily yields

$$
\langle I \rangle = \lim_{t \to \infty} \langle q(t) \rangle_c/t = q_0\,\Delta^2 \int_0^\infty d\tau\, e^{-Q'(\tau) - \tilde{\epsilon}_0\tau} \sin(\epsilon\tau) \, ,
\tag{25.99}
$$

which is in essence an integral representation of the digamma function,

$$\langle I \rangle = q_0 \frac{\tilde{\epsilon}_0}{\pi} \operatorname{Im} \psi \left(\frac{1}{2} + \frac{\beta \hbar \tilde{\epsilon}_0}{2\pi} + i \frac{\beta \hbar \epsilon}{2\pi} \right) . \tag{25.100}$$

This yields for the normalized nonlinear mobility the analytic expression

$$\frac{\mu}{\mu_0} = \frac{\tilde{\epsilon}_0}{\epsilon} \operatorname{Im} \psi \left(\frac{1}{2} + \frac{\beta \hbar \tilde{\epsilon}_0}{2\pi} + i \frac{\beta \hbar \epsilon}{2\pi} \right) . \tag{25.101}$$

At high temperature, $\hbar \tilde{\epsilon}_0 / 2\pi k_B T \ll 1$, we get in agreement with expression (25.78)

$$\frac{\mu}{\mu_0} = \frac{\pi \tilde{\epsilon}_0}{2\epsilon} \tanh \left(\frac{\beta \hbar \epsilon}{2} \right) . \tag{25.102}$$

In the opposite limit $T \to 0$, we obtain the equivalent forms

$$\frac{\mu}{\mu_0} = \frac{\tilde{\epsilon}_0}{\epsilon} \arctan \left(\frac{\epsilon}{\tilde{\epsilon}_0} \right) = \frac{\tilde{\epsilon}_0}{\epsilon} \left\{ \frac{\pi}{2} - \arctan \left(\frac{\tilde{\epsilon}_0}{\epsilon} \right) \right\} . \tag{25.103}$$

Thus we find, as in the limit $K \to 0$ in Subsection 25.5.2, $\mu_{\text{lin}}(T = 0) = \mu_0$. We shall see in Subsection 26.4.1 that this relation generally holds in the regime $K < 1$.

Fermionic representation

It has been realized by Guinea that the TB model (24.3) in the Ohmic scaling limit at $K = \frac{1}{2}$ can be mapped on free fermions [509] (see also Ref. [500], Sec. VIII). The correspondence is analogous to that of the spin-boson model, which maps at $K = \frac{1}{2}$ on the Toulouse point of the anisotropic Kondo model (cf. Subsection 19.4.1) and on the resonance level model with vanishing repulsive contact interaction (cf. Subsection 19.4.2). The equivalence of the bosonic Coulomb gas representation (presented in Subsection 25.3.1) at $K = \frac{1}{2}$ with the respective fermionic representation directly follows from an integral representation of the charge interaction factor,

$$e^{-Q'(\tau, K = \frac{1}{2})} = \frac{i}{2\omega_c} \int_{-\infty}^{\infty} d\omega \, \tanh \left(\frac{\beta \hbar \omega}{2} \right) e^{-i\omega\tau} . \tag{25.104}$$

With use of this representation the integral expression (25.99) for the current can be transformed into the frequency integral

$$\langle I \rangle = 2q_0 \int_0^{\infty} \frac{d\omega}{2\pi} \, \mathcal{T}(\omega) [\mathcal{N}_+(\omega, \epsilon) - \mathcal{N}_-(\omega, \epsilon)] , \tag{25.105}$$

where

$$\mathcal{N}_{\pm}(\omega, \epsilon) = f(\omega \mp \epsilon) [1 - f(\omega \pm \epsilon)] , \tag{25.106}$$

and where $f(\omega) = 1/(e^{\beta \hbar \omega} + 1)$ is the Fermi function,

The interpretation of this expression is straightforward. The quantity $\mathcal{N}_+(\omega, \epsilon)$ represents the spectral weight for right-moving fermions for an occupied state on the left side at frequency $\omega - \epsilon$ and an empty state on the right side of the barrier at frequency $\omega + \epsilon$. Accordingly, the quantity $\mathcal{N}_-(\omega, \epsilon)$ represents the spectral weight for left-moving fermions for an occupied state on the right side at frequency $\omega + \epsilon$ and an empty state on the left side of the barrier at frequency $\omega - \epsilon$.

The functions $\mathcal{T}(\omega)$ and $\mathcal{R}(\omega)$ represent the spectral transmission and reflection probability of the fermionic entity at frequency ω, $\mathcal{T}(\omega) + \mathcal{R}(\omega) = 1$,

$$\mathcal{T}(\omega) = \frac{\tilde{\epsilon}_0^2}{\tilde{\epsilon}_0^2 + \omega^2}, \qquad \mathcal{R}(\omega) = \frac{\omega^2}{\tilde{\epsilon}_0^2 + \omega^2}. \qquad (25.107)$$

The expression (25.105) describes the net current across the barrier.

It is convenient to introduce the linear combinations

$$F_1(\omega, \epsilon) \equiv \mathcal{N}_+(\omega, \epsilon) - \mathcal{N}_-(\omega, \epsilon) = f(\omega - \epsilon) - f(\omega + \epsilon),$$
$$F_2(\omega, \epsilon) \equiv \mathcal{N}_+(\omega, \epsilon) + \mathcal{N}_-(\omega, \epsilon) = \coth(\hbar\epsilon/k_B T) F_1(\omega, \epsilon). \qquad (25.108)$$

In the second line, we have used the explicit form of the Fermi function $f(\omega)$. In terms of these functions, the first cumulant (25.105) may be written as

$$\langle I \rangle = 2q_0 \int_0^\infty \frac{d\omega}{2\pi} \mathcal{T}(\omega) F_1(\omega, \epsilon) = q_0 \frac{\epsilon}{\pi} - 2q_0 \int_0^\infty \frac{d\omega}{2\pi} \mathcal{R}(\omega) F_1(\omega, \epsilon). \qquad (25.109)$$

The first form expresses the current in terms of the transmission through the barrier. The first term in the second form is the maximum current, which appears in the absence of the barrier. The second term diminishes the maximum current because of the possibility that the particle is reflected at the barrier.

One remark about small deviations from $K = \frac{1}{2}$ is in order. For $K = \frac{1}{2} - \kappa$, $|\kappa| \ll 1$, a leading log summation of all diagrams contributing to the current can be performed [510]. The resulting expression for the mobility is again of the form (25.101), though with a bias- and temperature-dependent renormalization of the frequency scale,

$$\tilde{\epsilon}_0 \rightarrow \tilde{\epsilon}(\epsilon, T) = \bar{\epsilon} \left[(\epsilon/\bar{\epsilon})^2 + (2\pi k_B T/\hbar\bar{\epsilon})^2 \right]^{-\kappa} \qquad (25.110)$$

with $\bar{\epsilon} = \tilde{\epsilon}_0 (\omega_c/\tilde{\epsilon}_0)^{2\kappa}$. We remark that the expression (25.101) with (25.110) is consistent with the scaling forms at $K = \frac{1}{2} - \kappa$ of the weak- and strong-tunneling expansions of the mobility discussed below in Section 26.4.

25.6.2 Diffusion and skewness

The calculation of the series (25.48) for the second moment $\langle \hat{q}^2(\lambda) \rangle_{\text{pis}}$ is more intricate because of the cosine factor in the phase term (25.45). This factor allows the particle to make a single virtual visit of diagonal state at some intermediate time. The combined analysis of zeros from phase factors and short-distance singularities from the charge interaction shows that the paths which contribute to the second cumulant are limited to the heptadiagonal matrix of the (q, q')-lattice. In the equivalent charge picture, the charge sequences are restricted to configurations in which the cumulated charges p_j are in the range $|p_j| \leq 3$.

For zero bias, the resulting "bosonic" integral expression at long times reads [360]

$$\langle q^2(t) \rangle_c = t\, q_0^2\, \frac{\tilde{\epsilon}_0}{2} \left\{ 1 - \frac{4\tilde{\epsilon}_0\, \omega_c^2}{\pi^2} \int_0^{\infty} d\tau_1\, d\tau_2\, d\tau_3\; \mathrm{e}^{-\tilde{\epsilon}_0(\tau_1+\tau_3)}\; \mathrm{e}^{-Q'(\tau_1+\tau_2)-Q'(\tau_2+\tau_3)} \right\} .$$

Evaluation of the triple integral yields for the diffusion coefficient the analytic form

$$D(T) = (\hbar\tilde{\epsilon}_0/2\pi)\, \psi'\left(1/2 + \hbar\tilde{\epsilon}_0/2\pi k_{\mathrm{B}}T\right) . \qquad (25.111)$$

It is evident that the linear mobility resulting from Eq. (25.101) and the diffusion coefficient (25.111) satisfy the Einstein relation (25.70), $D(T) = k_{\mathrm{B}}T\mu_{\mathrm{lin}}(T)$.

The diffusion coefficient vanishes for zero bias and $T = 0$. This indicates that the dispersion is sub-diffusive. We see in Subsection 25.7.2 that the second moment for zero bias and zero temperature spreads out at long time only logarithmically with time, $\langle q^2(t \to \infty) \rangle = (2/\pi^2)q_0^2 \ln(\tilde{\epsilon}_0 t)$, and that the logarithmic spread of the position dispersion at long time generally holds for all translational-invariant systems with Ohmic dissipation which have nonzero linear mobility at $T = 0$ (see also Subsection 7.3.2). A study of the diffusion coefficient in the Coulomb gas representation for the biased case is reported in Ref. [511].

The bosonic forms of the cumulants, extracted from the formally exact series (25.40) [e.g. Eq. (25.43)], may be transformed with use of the integral representation (25.104) into fermionic representation. The resulting expressions for the second cumulant (diffusion) and third cumulant (skewness) in the biased case are

$$\langle q^2(t) \rangle_c = t\,(2q_0)^2 \int_0^{\infty} \frac{d\omega}{2\pi} \left[\mathcal{T}(\omega)F_2(\omega,\epsilon) - \mathcal{T}^2(\omega)F_1^2(\omega,\epsilon) \right], \qquad (25.112)$$

$$\langle q^3(t) \rangle_c = t\,(2q_0)^3 \int_0^{\infty} \frac{d\omega}{2\pi} \mathcal{T}(\omega)F_1(\omega,\epsilon) \qquad (25.113)$$

$$\times \left[1 - 3\mathcal{T}(\omega)F_2(\omega,\epsilon) + 2\mathcal{T}^2(\omega)F_1^2(\omega,\epsilon) \right].$$

Upon employing the expressions (25.108) and (25.107), the cumulants (25.112) and (25.113) can be expressed in terms of cumulants of lower order. One finds

$$\langle q^2(t) \rangle_c = q_0 \left\{ \coth(\beta\hbar\epsilon)\,\tilde{\epsilon}_0 \frac{d}{d\tilde{\epsilon}_0} + \left(2 - \tilde{\epsilon}_0 \frac{d}{d\tilde{\epsilon}_0}\right)\frac{1}{\hbar\beta}\frac{d}{d\epsilon} \right\} \langle q(t) \rangle_c , \qquad (25.114)$$

$$\langle q^3(t) \rangle_c = q_0 \left\{ \coth(\beta\hbar\epsilon)\,\tilde{\epsilon}_0 \frac{d}{d\tilde{\epsilon}_0} + \left(2 - \frac{1}{2}\tilde{\epsilon}_0 \frac{d}{d\tilde{\epsilon}_0}\right)\frac{1}{\hbar\beta}\frac{d}{d\epsilon} \right\} \langle q^2(t) \rangle_c \qquad (25.115)$$

$$- q_0^2 \left\{ \coth(\beta\hbar\epsilon)\frac{1}{\hbar\beta}\frac{d}{d\epsilon} + \frac{1}{\sinh^2(\beta\hbar\epsilon)} \right\}\tilde{\epsilon}_0 \frac{d}{d\tilde{\epsilon}_0} \langle q(t) \rangle_c .$$

With use of the analytic expression (25.100) for the current, it is now straightforward to derive analytic expressions for diffusion and skewness.

25.7 The effects of a thermal initial state

25.7.1 Mean position and variance

We have seen in Subsection 25.3.2 that the difference between $\langle q(t) \rangle_{\mathrm{th}}$ and $\langle q(t) \rangle_{\mathrm{pis}}$ is described by the function $R_1(t)$ given in Eq. (25.57). Under conditions specified in Subsection 25.4.1 all the integrals in the expression for $\hat{R}_1(\lambda)$ in Eq. (25.63) are convergent in the limit $\lambda \to 0$. Hence, because of the extra $1/\lambda$-factor, the function $R_1(t)$ approaches a constant at asymptotic time, $R_1(t \to \infty) = R_{1,\infty}$. As a result, the leading asymptotic behavior of the mean values $\langle q(t) \rangle_{\mathrm{th}}$ and $\langle q(t) \rangle_{\mathrm{pis}} \propto t$ is exactly the same for both kinds of initial conditions, while sub-leading terms are different,

$$\lim_{t \to \infty} \left[\langle q(t) \rangle_{\mathrm{th}} - \langle q(t) \rangle_{\mathrm{pis}} \right] = R_{1,\infty} . \tag{25.116}$$

In the presence of drift due to a bias, the variances are defined by

$$\sigma_{\mathrm{pis}}^2(t) \equiv \langle q^2(t) \rangle_{\mathrm{pis}} - \langle q(t) \rangle_{\mathrm{pis}}^2 , \quad \text{and} \quad \sigma_{\mathrm{th}}^2(t) \equiv D_{\mathrm{th}}(t) - \langle q(t) \rangle_{\mathrm{th}}^2 . \tag{25.117}$$

Consider the term of order Δ^{2m} in Eq. (25.48) with (25.45). The leading contribution to $\langle \hat{q}^2(\lambda) \rangle_{\mathrm{pis}}$ in the limit $\lambda \to 0$ comes from the terms in which the factor $\cos(\chi_{k,2m})$ in Eq. (25.45) is associated with a sojourn interval τ_k ($p_k = 0$). The respective τ_k-integration yields a factor $1/\lambda$. The two branches to the left and right of this interval are easily identified as terms of the series expansion of $\lambda \langle \hat{q}(\lambda) \rangle_{\mathrm{pis}} = F\mu/\lambda + q_\infty + \mathcal{O}(\lambda)$. Besides, the diffusion contribution $2D/\lambda^2$ to $\langle \hat{q}^2(\lambda) \rangle_{\mathrm{pis}}$ results from the summands in which the phase factor $\cos(\chi_{k,2m})$ falls on a blip state ($p_k \neq 0$). In the end we find

$$\lim_{t \to \infty} \langle q^2(t) \rangle_{\mathrm{pis}} = (F\mu t)^2 + 2(F\mu q_\infty + D)t . \tag{25.118}$$

With this form and with the subtraction of $\langle q(t) \rangle_{\mathrm{pis}}^2$, we obtain for the variance in the diffusive regime

$$\lim_{t \to \infty} \sigma_{\mathrm{pis}}^2(t) = 2Dt . \tag{25.119}$$

The effects of the different initial conditions are in the function

$$\sigma_{\mathrm{th}}^2(t) - \sigma_{\mathrm{pis}}^2(t) = R_2(t) - R_1^2(t) - 2\langle q(t) \rangle_{\mathrm{pis}} R_1(t) , \tag{25.120}$$

as follows from Eqs. (25.117), (25.55), and (25.56). In the limit $\lambda \to 0$, the leading contribution in order Δ^{i+j} to $\hat{R}_2(\lambda)$ comes again from terms with the factor $\cos(\chi_{k,i+j})$ in which τ_k is a sojourn. The branch to the left of this interval can be identified with a contribution of $\hat{R}_1(\lambda)$, and the branch to the right is a term of the series expression for $\langle \hat{q}(\lambda) \rangle_{\mathrm{pis}}$. Readily we find that in the combined expression $R_2(t) - 2\langle q(t) \rangle_{\mathrm{pis}} R_1(t)$ the asymptotically leading term of $R_2(t)$, which grows linearly with t, is cancelled. The remaining term approaches asymptotically a constant value,

$$\lim_{t \to \infty} \left[R_2(t) - 2\langle q(t) \rangle_{\mathrm{pis}} R_1(t) \right] = R_{2,\infty} . \tag{25.121}$$

Therefore, the leading behavior at long times of $\sigma_{\text{th}}^2(t)$ coincides with that of $\sigma_{\text{pis}}^2(t)$, Eq. (25.119), whereas sub-leading terms are different,

$$\lim_{t\to\infty} \left[\sigma_{\text{th}}^2(t) - \sigma_{\text{pis}}^2(t) \right] \;=\; R_{2,\infty} - R_{1,\infty}^2 \;. \tag{25.122}$$

Thus in the diffusive regime, the variances $\sigma_{\text{th}}^2(t)$ and $\sigma_{\text{pis}}^2(t)$ differ by a constant at asymptotic time.

25.7.2   Linear response

In this subsection, we focus the attention on the linear response to an external force $F = \hbar\epsilon/q_0$ switched on at time $t = 0$. One may think that the definition of the mobility within Kubo's linear-response theory is problematic [218] since the stationary state at asymptotic times is not normalizable. We now draw our attention to this question.

The Kubo formalism yields for the mean velocity at time t the expression

$$\bar{v}_{\text{lin}}(t) \;=\; -(2\epsilon/q_0)\, \Theta(t)\, A(t) \;=\; \frac{\hbar\epsilon}{q_0}\, \chi(t)\;. \tag{25.123}$$

Here, $A(t)$ is the antisymmetric equilibrium correlation function (25.10) at zero bias, and $\chi(t)$ is the response function. Alternatively, we may define the mean velocity $\bar{v}_{\text{lin}}(t)$ as

$$\bar{v}_{\text{lin}}(t) \;=\; \epsilon \lim_{\epsilon\to 0} \left[\langle \dot{q}(t) \rangle_{\text{th}} / \epsilon \right]\;. \tag{25.124}$$

It is now interesting to see, using Eqs. (25.55) and (25.60), whether the two expressions coincide. Equating the expressions (25.123) and (25.124) and switching to the Laplace transforms, we obtain

$$\lim_{\epsilon\to 0} \left[\langle \hat{q}(\lambda) \rangle_{\text{pis}} + \hat{R}_1(\lambda) \right]/\epsilon \;=\; (\hbar/q_0)\, \hat{\chi}(\lambda)/\lambda\;. \tag{25.125}$$

To prove the validity of this relation, we use for the time integrals occurring in the series expression for $\hat{R}_1(\lambda)$ and $\hat{\chi}(\lambda)$ the integral identities

$$\int_0^\infty dr_k\, ds_k\, e^{-\lambda s_k} f(r_k + s_k) \;=\; \int_0^\infty d\tau\, f(\tau)\frac{1}{\lambda}\left(1 - e^{-\lambda\tau} \right),$$

$$\int_0^\infty dr_k\, ds_k\, s_k\, e^{-\lambda s_k} f(r_k + s_k) \;=\; \int_0^\infty d\tau\, f(\tau)\left(\frac{1}{\lambda^2}\left(1 - e^{-\lambda\tau} \right) - \frac{\tau}{\lambda}e^{-\lambda\tau} \right),$$

where $\tau = r_k + s_k$, and where the factor $f(\tau)$ contains the interactions between the two sets of charges to the left and right of the interval τ. With these forms, it is straightforward to see that the relation (25.125) is satisfied for all λ. Thus, the expressions (25.123) and (25.124) agree.

Using Eqs. (25.66) and (25.125), we find that the *linear* mobility is found from the absorptive part of the dynamical susceptibility as

$$\mu_{\text{lin}} \;=\; \lim_{\omega \to 0} \omega \operatorname{Im} \tilde{\chi}(\omega) \;=\; \lim_{\omega \to 0} \omega \operatorname{Im} \int_0^{\infty} dt\, \chi(t)\, e^{i\omega t}\,. \tag{25.126}$$

Consider next the mean square displacement. In the absence of the bias, $\epsilon = 0$, the variance $\sigma_{\text{th}}^2(t)$ coincides with the equilibrium correlation function $D_{\text{th},+}(t)$. With use of the expression (7.23) for the Laplace transform and with Eq. (25.126) we get

$$\lim_{\lambda \to 0} \lambda^2\, \hat{D}_{\text{th},+}(\lambda) \;=\; 2\, k_{\text{B}} T \lim_{\omega \to 0} \omega \operatorname{Im} \tilde{\chi}(\omega) \;=\; 2\, k_{\text{B}} T\, \mu_{\text{lin}}\,. \tag{25.127}$$

On the other hand, we obtain in the diffusive regime with the findings of the preceding subsection

$$\lim_{\lambda \to 0} \lambda^2\, \hat{D}_{\text{th},+}(\lambda) \;=\; \lim_{\lambda \to 0} \lambda^2\, \langle \hat{q}^2(\lambda) \rangle_{\text{pis}} \;=\; 2D\,. \tag{25.128}$$

Equating Eq. (25.127) with Eq. (25.128), we arrive at the Einstein relation (25.70).

At zero temperature, we find from Eq. (7.23)

$$\lim_{\lambda \to 0} \hat{D}_{\text{th},+}(\lambda) \;=\; - (2\hbar/\pi)\, \mu_{\text{lin},0}\, \ln(\lambda t_0)/\lambda\,, \tag{25.129}$$

and hence

$$\lim_{t \to \infty} D_{\text{th}}(t) \;=\; (2\hbar/\pi)\mu_{\text{lin},0}\, \ln\big(t/t_0\big)\,, \tag{25.130}$$

where $\mu_{\text{lin},0} = \mu_{\text{lin}}(T = 0)$, and t_0 is a reference time. Here we have assumed that the infrared limit in Eq. (25.126) at $T = 0$, and hence the linear mobility $\mu_{\text{lin},0}$, is nonzero. Actually this is the case for the TB model in the regime $K < 1$, as we shall see below in Subsection 26.4.1. Since we have not referred to a particular system in the derivation, the logarithmic spread of $D_{\text{th}}(t)$ at long time holds at zero temperature for any system of which the linear mobility at $T = 0$ is nonzero.

We can also study the spread of the second moment at long time directly from the exact formal expression (25.48). As already noted in Subsection 25.4.1, the leading contribution in order Δ^{2m} for $T = 0$ in the limit $\lambda \to 0$ comes again from the terms in which the factor $\cos(\chi_{2k,2m})$ in the coefficient $c_{2m,0}$, Eq. (25.45), falls on a sojourn, i.e., $p_{2k} = 0$. This sojourn divides the charge sequence into two neutral clusters, and the average sojourn length $\bar{\tau}_{2k}$ is very large compared to the effective lengths of the clusters. On this assumption, we have $\cos(\chi_{2k,2m}) = 1$ and thus

$$c_{2m,0} G_{2m}^{(0)} \;=\; (-1)^{m-1} \frac{1}{2} \sum_{k=1}^{m-1} \prod_{\ell=1,\ell \neq 2k}^{2m-1} \sin\big(\chi_{\ell,2m}\big)\, G_{2m-2k}^{(0)} G_{2k}^{(0)}$$

$$\times \left(1 - \ddot{Q}'(\tau_{2k}) \sum_{i=1}^{2k-1} p_i \tau_i \sum_{j=2k+1}^{2m-1} p_j \tau_j \right)\,. \tag{25.131}$$

The terms $G_{2k}^{(0)}$ and $G_{2m-2k}^{(0)}$ are the interaction factors (25.34) at $T = 0$ for the two clusters without their mutual interaction across the interval τ_{2k}. The second term in the square bracket represents the dipole-dipole interaction between the two neutral clusters. Higher-order multipole interactions are negligible for a long sojourn.

Inserting the first term into the expression (25.48), we obtain the first term in Eq. (25.118). This term vanishes for zero bias. The additional contribution to $\langle \hat{q}^2(\lambda) \rangle_{\mathrm{pis}}$ stemming from the second (dipole-dipole interaction) term is found in product form as well. Performing the summation over all possible charge sequences on both sides of the interval τ_k, we find that the two distinguished branches for $\epsilon = 0$ can be identified with the series (25.67) for the zero temperature linear mobility $\mu_{\mathrm{lin},0}$, and the remaining factor represents the dipole-dipole interaction between the two neutral branches. We obtain

$$
\begin{aligned}
\lim_{\lambda \to 0} \lambda^2 \langle \hat{q}^2(\lambda) \rangle_{\mathrm{pis}} &= -2 \left(\frac{\hbar}{q_0} \mu_{\mathrm{lin},0} \right)^2 \int_{\tau_{\min}}^{\infty} d\tau \, \ddot{Q}'(\tau) \, e^{-\lambda \tau} \\
&= -\frac{2\hbar}{\pi} \frac{\mu_{\mathrm{lin},0}^2}{\mu_0} \lambda \, \ln(\lambda t_0') \, .
\end{aligned}
\tag{25.132}
$$

Here we have put $\epsilon = 0$ and used $\ddot{Q}'(\tau) = -2K/\tau^2$, and we have extracted the λ-dependence of the integral from the region $\tau \gg \tau_{\min}$. Hence we have

$$
\lim_{t \to \infty} \langle q^2(t) \rangle_{\mathrm{pis}} = (2\hbar/\pi) \, (\mu_{\mathrm{lin},0}^2/\mu_0) \, \ln \left(t/t_0' \right) \, .
\tag{25.133}
$$

Since the Ohmic system is ergodic [cf. Section 3.2], the prefactor of the logarithmic law in Eqs. (25.130) must coincide with that in Eq. (25.133). As a surprising consequence of this requirement, we find that the linear mobility at $T = 0$ must coincide with the mobility in the absence of the cosine potential,

$$
\mu_{\mathrm{lin},0} = \mu_0 \, .
\tag{25.134}
$$

Thus in the zero temperature limit the TB potential is renormalized to zero. This result is in correspondence with the findings in Subsection 26.4.1.

25.7.3 The exactly solvable case $K = \frac{1}{2}$

In the Ohmic scaling limit at the Toulouse point $K = \frac{1}{2}$, the function $\langle q(t) \rangle_{\mathrm{th}}$ and the correlation functions $D_{\mathrm{th}}(t)$ and $\chi(t)$ can be calculated in analytic form [511] using again the concept of collapsed blips explained in Subsection 22.6.1. The Laplace transform of the response function $\chi(t)$ is found as

$$
\hat{\chi}(\lambda) = -(q_0^2 \tilde{\epsilon}_0 / 2\pi\hbar) \hat{\Psi}(\lambda)
\tag{25.135}
$$

with the function

$$
\begin{aligned}
\hat{\Psi}(\lambda) &\equiv -\frac{2\omega_c}{\lambda} \int_0^{\infty} d\tau \, ds \, e^{-\lambda s} \, e^{-\tilde{\epsilon}_0(\tau+s)} \, e^{-Q'(\tau+s)} \\
&= \frac{2}{\lambda^2} \left\{ \psi \left(\frac{1}{2} + \frac{\beta\hbar\tilde{\epsilon}_0}{2\pi} \right) - \psi \left(\frac{1}{2} + \frac{\beta\hbar(\tilde{\epsilon}_0 + \lambda)}{2\pi} \right) \right\} \, ,
\end{aligned}
\tag{25.136}
$$

where $\tilde{\epsilon}_0 = 2\gamma = \pi\Delta^2/\omega_c$, and where $\psi(z)$ is the digamma function. We see from the integral expression that $\hat{\chi}(\lambda)$ is represented by a single extended dipole of length $\tau+s$. Here τ is the length in the negative and s the length in the positive time branch. The factors $e^{-\tilde{\epsilon}_0\tau}$ and $e^{-\tilde{\epsilon}_0 s}$ represent grand-canonical sums of collapsed dipoles in the intervals τ and s, respectively. The dynamical susceptibility is found from Eq. (25.135) as $\tilde{\chi}(\omega) = -(q_0^2\tilde{\epsilon}_0/2\pi\hbar)\,\hat{\Psi}(-i\omega)$. This yields

$$
\begin{aligned}
\mathrm{Im}\,\tilde{\chi}(\omega) &= \frac{q_0^2}{\pi\hbar}\frac{\tilde{\epsilon}_0\omega_c}{\omega^2}\int_0^\infty d\rho\, e^{-Q'(\rho)-\tilde{\epsilon}_0\rho}\sin(\omega\rho) \\
&= -\frac{q_0^2}{\pi\hbar}\frac{\tilde{\epsilon}_0}{\omega^2}\,\mathrm{Im}\,\psi\Big(\frac{1}{2}+\frac{\beta\hbar(\tilde{\epsilon}_0-i\omega)}{2\pi}\Big)\,.
\end{aligned}
\tag{25.137}
$$

The calculation of $\hat{D}_{\mathrm{th}}(\lambda)$ is more difficult than that of $\hat{\chi}(\lambda)$ since there are additional contributions from paths which cross the diagonal (cf. Ref. [511]). The resulting integral expression in Laplace space for $\epsilon = 0$ is

$$
\hat{D}_{\mathrm{th},+}(\lambda) = \frac{q_0^2\tilde{\epsilon}_0\omega_c}{\pi^2\lambda}\int_{-\infty}^{\infty}d\omega'\,\frac{\coth(\beta\hbar\omega'/2)}{\omega'^2+\lambda^2}\int_0^\infty d\rho\,e^{-Q'(\rho)-\tilde{\epsilon}_0\rho}\sin(\omega'\rho)\,.
\tag{25.138}
$$

With the integral expression (25.137) the Laplace transform $\hat{D}_{\mathrm{th},+}(\lambda)$ can be cast into the form (7.23), which is consistent with the fluctuation-dissipation theorem. In fact, we obtain from Eq. (25.138) with (25.137) in Fourierspace

$$
\tilde{D}_{\mathrm{th}}(\omega) = -2\hbar\coth(\beta\hbar\omega/2)\,\mathrm{Im}\,\tilde{\chi}(\omega)\,.
\tag{25.139}
$$

Finally, it may be interesting for the reader to compare the results of this subsection with those obtained in Subsection 22.6.4 for the σ_z-autocorrelation function of the TSS at the Toulouse point.

26. Duality symmetry

Duality is by now an old concept. In electromagnetism, the simplest form of duality is the invariance of the source free Maxwell equations under interchange of electric and magnetic field, $B \to E$, $E \to -B$. In the presence of sources, the product of electric and magnetic charges obeys the Dirac quantization condition. In general, duality maps a theory with strong coupling to one with weak coupling. Thus, if a duality symmetry exists, we can study the strong-coupling regime via the perturbative analysis of the weak-coupling regime. A theory is self-dual when there is an exact map between the strong-coupling sector and the weak-coupling sector of the same theory.

26.1 Duality for general spectral density

The model of the Brownian particle in a cosine potential contains a duality symmetry within itself. The weak-binding representation of this model can be mapped exactly on its tight-binding representation, and vice versa. Schmid [145] demonstrated in an imaginary-time path sum approach by comparing the perturbative series in the corrugation strength V_0 with the perturbative series in the tunneling matrix element Δ (multi-kink expansion) that the equilibrium density matrix at $T = 0$ is self-dual in the Ohmic scaling limit.[1] The analysis by Schmid led to a transformation for the dc mobility in which diffusive and localized behavior are interchanged. The duality was generalized by Fisher and Zwerger in Ref. [498] to real time and finite temperature. They showed within the influence functional method that in the duality transformation the strict Ohmic spectral density in the WB model (24.1) is mapped onto an Ohmic spectral density with a Drude cut-off in the dual TB model (24.3).

About ten years later, it was shown that the Hamiltonians (24.1) and (24.3) are exactly related by a duality transformation for general density $J(\omega)$ with power $s < 2$ at low frequencies [512]. In the duality, the power s maps on the power $2 - s$ and vice versa. Hence super-Ohmic and sub-Ohmic friction are interchanged, while Ohmic friction maps on Ohmic, but with the coupling strength transformed into the inverse of it. This matter is discussed in the following subsection.

We shall expound in Section 26.3 that a corresponding duality symmetry holds between the charge and phase representation for a Josephson junction. An interesting correspondence between impurity scattering in a Tomonaga-Luttinger liquid and Brownian motion in a periodic potential is discussed in Chapter 28. In this case, there is an exact self-duality map between weak and strong backscattering.

26.1.1   The map between the TB and WB Hamiltonian

We begin with demonstrating that the Hamiltonians (24.1) and (24.3) can directly be mapped onto each other. For simplicity, we confine ourselves to the case $F = 0$.[2] In the first step, we perform the canonical transformation

$$
\begin{aligned}
p_\alpha &\to -m_\alpha \omega_\alpha x_\alpha, & x_\alpha &\to p_\alpha/m_\alpha \omega_\alpha + pc_\alpha/\kappa m_\alpha \omega_\alpha^2, \\
X &\to p/\kappa, & P &\to -\kappa q + \sum_\alpha c_\alpha x_\alpha/\omega_\alpha.
\end{aligned}
\tag{26.1}
$$

To map the cosine potential in Eq. (24.1) on the hopping term in Eq. (24.3), we choose

$$
V_0 = \hbar\Delta, \qquad\qquad \kappa = 2\pi\hbar/X_0 q_0. \tag{26.2}
$$

With these transformations the Hamiltonian (24.1) takes the form

$$
\begin{aligned}
H &= -\hbar\Delta \cos\left(\frac{q_0 p}{\hbar}\right) + \frac{\kappa^2 q^2}{2M} - \frac{\kappa q}{M} \sum_\alpha \frac{c_\alpha x_\alpha}{\omega_\alpha} + H_{\mathrm{R}}, \\
H_{\mathrm{R}} &= \sum_\alpha \left(\frac{p_\alpha^2}{2m_\alpha} + \frac{m_\alpha \omega_\alpha^2 x_\alpha^2}{2}\right) + \frac{1}{2M}\left(\sum_\alpha \frac{c_\alpha x_\alpha}{\omega_\alpha}\right)^2.
\end{aligned}
\tag{26.3}
$$

---

[1]The universality or scaling limit is discussed in Subsection 18.2.2.

[2]The case $F \neq 0$ is more intricate since the duality transformation changes a force coupled to a coordinate into a force coupled to a momentum. In the Ohmic scaling limit, the final result is as discussed in Subsection 26.1.3.

Now, in the transformed reservoir Hamiltonian H_R the modes of the environment are coupled with each other. In the next step, we diagonalize H_R. This leads to new canonical variables π_α and u_α, and new parameters M_α and Ω_α. By the transformation, the interaction term $-(\kappa q/M)\sum_\alpha c_\alpha x_\alpha/\omega_\alpha$ is converted into $-q\sum_\alpha d_\alpha u_\alpha$. The mapping is completed by imposing spatial invariance on the system-bath coupling, which requires that the q^2 term in Eq. (26.3) becomes the counter term of the model (24.3). This condition is implemented by the constraint

$$\kappa^2/M \ = \ \sum_\alpha d_\alpha^2/M_\alpha\Omega_\alpha^2 \,. \tag{26.4}$$

A relation between $J_{WB}(\omega)$ and $J_{TB}(\omega)$ is found with manipulations of the dynamical matrix $A_{\alpha\beta}$ of the reservoir Hamiltonian H_R. We have

$$A_{\alpha\beta}(\omega) \ = \ B_{\alpha\beta}(\omega) + c_\alpha c_\beta/M\omega_\alpha\omega_\beta \,, \qquad B_{\alpha\beta}(\omega) \ = \ \delta_{\alpha\beta}m_\alpha(\omega_\alpha^2 - \omega^2) \,. \tag{26.5}$$

Since the coupling term of the bath modes is in the form of an exterior vector product, the ratio of the determinants of the matrices $\boldsymbol{A}(\omega)$ and $\boldsymbol{B}(\omega)$ gives

$$\frac{\det \boldsymbol{A}(\omega)}{\det \boldsymbol{B}(\omega)} \ = \ 1 + \frac{1}{M}\sum_\alpha \frac{c_\alpha^2}{m_\alpha\omega_\alpha^2(\omega_\alpha^2 - \omega^2)} \ = \ 1 + i\frac{\tilde\gamma_{WB}(\omega)}{\omega} \,. \tag{26.6}$$

In the last form, we have introduced the spectral damping function $\tilde\gamma(\omega)$ for the weak binding model. In the continuum limit, we have

$$\tilde\gamma(\omega) = \lim_{\varepsilon\to 0}\frac{-i\omega}{M}\frac{2}{\pi}\int_0^\infty d\omega' \frac{J(\omega')}{\omega'(\omega'^2 - \omega^2 - i\varepsilon\,\mathrm{sgn}\,\omega)} \,. \tag{26.7}$$

On the other hand, we may resolve $A_{\alpha\beta}$ for $B_{\alpha\beta}$ and then switch to the unitarily equivalent form in which $\boldsymbol{A}$ is diagonal. We then get

$$\tilde B_{\alpha\beta}(\omega) \ = \ \tilde A_{\alpha\beta}(\omega) - (M/\kappa^2)\,d_\alpha d_\beta \,, \qquad \tilde A_{\alpha\beta}(\omega) \ = \ \delta_{\alpha\beta}M_\alpha(\Omega_\alpha^2 - \omega^2) \,, \tag{26.8}$$

$$\frac{\det \boldsymbol{B}(\omega)}{\det \boldsymbol{A}(\omega)} \ = \ 1 - \frac{M}{\kappa^2}\sum_\alpha \frac{d_\alpha^2}{M_\alpha(\Omega_\alpha^2 - \omega^2)} \ = \ -i\,\frac{M^2}{\kappa^2}\,\omega\tilde\gamma_{TB}(\omega) \,. \tag{26.9}$$

To obtain the second form, we have used Eq. (26.4) and we have introduced the damping function $\tilde\gamma_{TB}(\omega)$ of the TB model. Combining Eq. (26.6) with Eq. (26.9), we find an exact relation between the damping functions of the two models [512],

$$\tilde\gamma_{TB}(\omega)\,[\tilde\gamma_{WB}(\omega) - i\,\omega] \ = \ \kappa^2/M^2 \,. \tag{26.10}$$

Using $J(\omega) = M\omega\,\mathrm{Re}\,\tilde\gamma(\omega)$, we then have

$$J_{WB}(\omega) \ = \ (\kappa^2/M^2)J_{TB}(\omega)/|\tilde\gamma_{TB}(\omega)|^2 \,, \tag{26.11}$$

$$J_{TB}(\omega) \ = \ (\kappa^2/M^2)J_{WB}(\omega)/|\tilde\gamma_{WB}(\omega) - i\,\omega|^2 \,. \tag{26.12}$$

Thus, the spectral density of the one model can be calculated for any form of the spectral density of the other model.

Consider now the WB model with the spectral power-law form

$$J_{\mathrm{WB}}(\omega) = M\gamma_s\omega_{\mathrm{ph}}(\omega/\omega_{\mathrm{ph}})^s\,, \qquad 0 < s < 2\,. \tag{26.13}$$

Here we have introduced for $s \neq 1$ a phononic reference frequency ω_{ph}, as in Eq. (3.72). The corresponding spectral damping function is [cf. Eq. (3.83)]

$$\tilde{\gamma}_{\mathrm{WB}}(\omega) \equiv \tilde{\gamma}(\omega, s) = \lambda_s(-i\,\omega/\omega_{\mathrm{ph}})^{s-1}\,, \qquad \lambda_s \equiv \gamma_s/\sin(\pi s/2)\,. \tag{26.14}$$

Let us now check the consistency of the transformation. We see from Eq. (26.9) that the constraint (26.4) is satisfied if $\omega\tilde{\gamma}_{\mathrm{TB}}(\omega)$ vanishes in the limit $\omega \to 0$. This in turn implies that $\lim_{\omega\to 0}\tilde{\gamma}_{\mathrm{WB}}(\omega)/\omega$ must diverge, which is the case in the parameter range $0 < s < 2$, as we see with use of Eq. (26.14). For $s \geq 2$, we have at low frequency in leading order $\tilde{\gamma}_{\mathrm{WB}}(\omega) \propto \omega$, as we catch from Eq. (3.83). This term leads to mass renormalization [87], not to friction, as we have discussed in Subsection 3.1.7. Moreover, with the form $\tilde{\gamma}(\omega) \propto \omega$ the condition (26.4) cannot be satisfied.

Thus we have shown that there is an exact duality between the weak corrugation model (24.1) and the tight-binding model (24.3) in the regime $s < 2$. In the mapping, the continuous coordinate X in the WB model is identified, up to a scale factor $1/\kappa$, with the quasimomentum p in the dual TB model. Hence the duality is a sort of a Fourier transformation between real and momentum space [499]. A non-zero system-bath coupling is essential since otherwise the scale factor $1/\kappa$ is infinity. Strictly speaking, the mapping holds under the condition $\lim_{\omega\to 0}\omega^2/J(\omega) \to 0$. There are no restrictions on the form of $J(\omega)$ at finite frequencies except those enforced on physical grounds. There follows from Eq. (26.11) or Eq. (26.12) that the spectral density $J_{\mathrm{WB}}(\omega) \propto \omega^s$ of the weak corrugation model maps on the spectral density $J_{\mathrm{TB}}(\omega \to 0) \propto \omega^{2-s}$ of the dual TB model. Thus the power s in the spectral density is mapped on the power $2 - s$. This means that sub-Ohmic and super-Ohmic friction are interchanged in the transformation, while Ohmic friction is mapped on Ohmic.

So far, we have considered the scales X_0 and q_0 as free parameters. In order to make the relation (26.10) as simple as possible, we now make the specific choice

$$M\lambda_s X_0 q_0/2\pi\hbar \equiv M\lambda_s/\kappa = 1\,. \tag{26.15}$$

Then the standard dimensionless coupling parameters $K_{\mathrm{WB}} \equiv M\lambda_s X_0^2/2\pi\hbar$ and $K_{\mathrm{TB}} \equiv M\lambda_s q_0^2/2\pi\hbar$ are related by

$$X_0/q_0 = K_{\mathrm{WB}} = 1/K_{\mathrm{TB}}\,. \tag{26.16}$$

Upon inserting the expressions (26.13) and (26.14) into Eq. (26.12), the spectral density of the dual TB model obtains the power law ω^{2-s} with an algebraic cut-off

$$J_{\mathrm{TB}}(\omega) = M\lambda_s^2\,\frac{\omega_{\mathrm{ph}}}{\gamma_s}\,\frac{(\omega/\omega_{\mathrm{ph}})^{2-s}}{1 + \left[\cot(\pi s/2) - (\omega_{\mathrm{ph}}/\gamma_s)(\omega/\omega_{\mathrm{ph}})^{2-s}\right]^2}\,. \tag{26.17}$$

In the Ohmic case, $s = 1$, the density $J_{\mathrm{TB}}(\omega)$ reduces to the Drude form [498]

$$J_{\mathrm{TB}}(\omega) \; = \; \frac{M\gamma\omega}{1 + (\omega/\omega_c)^2} \quad \text{with} \quad \omega_c = \gamma \,. \tag{26.18}$$

Let us briefly pause and dwell on the issue why $J_{\mathrm{TB}}(\omega)$ has a soft cut-off while the cutoff is missing in $J_{\mathrm{WB}}(\omega)$. For simplicity, we consider the Ohmic case $s = 1$. First, we observe that the trajectories in the TB lattice are step-like paths as given in Eq. (25.27) or in Eq. (25.32), while those in the cosine potential are rounded on the timescale γ^{-1}. The rounding of the Heaviside step function is described by the function $h(\tau) = \Theta_{\mathrm{WB}}(\tau)$, which is a particular solution of the differential equation $-\ddot{h}(\tau) + \gamma\dot{h}(\tau) = \gamma\delta(\tau)$. The smeared Θ-function is [498]

$$\Theta_{\mathrm{WB}}(\tau) \; = \; e^{\gamma t}\,\Theta(-\tau) + \Theta(\tau) \,. \tag{26.19}$$

In the WB model, the $(\eta,\,\xi)$-path analogous to Eq. (25.32) is smoothed out by replacing the Heaviside function $\Theta(\tau)$ with the smeared function $\Theta_{\mathrm{WB}}(\tau)$. It is straightforward to see that the influence function (25.34) for the smoothed $(\eta,\,\xi)$-path with the straight spectral density $J_{\mathrm{WB}}(\omega) = M\gamma\omega$ is the same as that for the steplike path (25.32), but with the modified spectral density (26.18), which has a soft cutoff at frequency γ. In other words, the algebraic cutoff in the spectral density $J_{\mathrm{TB}}(\omega)$ is due to the sharp tight-binding trajectory (25.27) instead of the smoothed path in the WB model. Formally, the two influence functions are reconciled by the change of the spectral densities which enter the respective kernels $Q(\tau)$, as given by the relations (26.11) and (26.12). Physically, the inertia of the particle with a finite mass provides a natural cutoff in the coupling to high-frequency oscillators of the reservoir.

When the cutoff frequency is sent to infinity, the spectral damping function dual to Eq. (26.14) becomes symmetric in form, as we see from Eq. (26.10) with (26.15),

$$\hat{\gamma}_{\mathrm{TB}}(\omega) \; = \; \tilde{\gamma}(\omega, 2-s) \; = \; \lambda_s(-i\,\omega/\omega_{\mathrm{ph}})^{1-s} \,. \tag{26.20}$$

We shall use these forms in Chapter 29 in which we study quantum transport for sub- and super-Ohmic friction both in the TB and WB model.

26.1.2 Frequency-dependent linear mobility

We now turn to a study of the linear ac mobilities of the WB and TB model. First, we observe that by the canonical transformation (26.1) a coordinate autocorrelation function of the WB model is transformed into a momentum autocorrelation function of the associated TB model. Thus we have for the linear mobility of the WB model

$$\tilde{\mu}_{\mathrm{WB}}(\omega) \; \equiv \; -i\omega\frac{i}{\hbar}\int_0^\infty dt\, e^{i\omega t}\,\langle [X(t), X(0)]\rangle_\beta \; = \; -i\omega\tilde{\Pi}_{\mathrm{TB}}(\omega)/\kappa^2 \,, \tag{26.21}$$

where $\tilde{\Pi}_{\mathrm{TB}}(\omega)$ is the Fourier transform of the retarded momentum response function of the TB model, $\Pi_{\mathrm{TB}}(t) = (i/\hbar)\,\Theta(t)\langle [p(t), p(0)]\rangle_\beta$. On the other hand, the ac mobility of the TB model is related to the Fourier transform $\tilde{Y}_{\mathrm{TB}}(\omega)$ of the respective retarded coordinate response function $Y_{\mathrm{TB}}(t) = (i/\hbar)\Theta(t)\langle [q(t), q(0)]\rangle_\beta$ by

$$\tilde{\mu}_{\mathrm{TB}}(\omega) \; = \; -i\,\omega\tilde{Y}_{\mathrm{TB}}(\omega) \,. \tag{26.22}$$

To connect $\tilde{\Pi}_{\text{TB}}(\omega)$ with $\tilde{Y}_{\text{TB}}(\omega)$, we use the equations of motion resulting from the Hamiltonian (24.3). We then obtain the relation

$$\omega^2 \tilde{\Pi}_{\text{TB}}(\omega) = iM\omega \tilde{\gamma}_{\text{TB}}(\omega) - M^2\omega^2 \tilde{\gamma}_{\text{TB}}^2(\omega) \tilde{Y}_{\text{TB}}(\omega) \; . \tag{26.23}$$

Using this form, and Eqs. (26.21), (26.22) and (26.10), we find the exact relations

$$\tilde{\mu}_{\text{WB}}(\omega, V_0/\hbar) = \frac{1}{M[\tilde{\gamma}_{\text{WB}}(\omega) - i\omega]} - \frac{\tilde{\gamma}_{\text{TB}}(\omega)}{\tilde{\gamma}_{\text{WB}}(\omega) - i\omega} \tilde{\mu}_{\text{TB}}(\omega, \Delta) \; , \tag{26.24}$$

$$\tilde{\mu}_{\text{TB}}(\omega, \Delta) = \frac{1}{M\tilde{\gamma}_{\text{TB}}(\omega)} - \frac{\tilde{\gamma}_{\text{WB}}(\omega) - i\omega}{\tilde{\gamma}_{\text{TB}}(\omega)} \tilde{\mu}_{\text{WB}}(\omega, V_0/\hbar) \; . \tag{26.25}$$

The expressions (26.10) – (26.12), (26.24) and (26.25) are the central duality relations. Most remarkably, they hold for spectral densities of the coupling with low-frequency power s in the range $0 < s < 2$. At higher frequency the only restrictions are those placed for physical reasons. Here, the correspondence is shown for the frequency-dependent linear mobility. Clearly, the mapping can be extended to different kinds of dynamical variables.

In the low-frequency regime $\omega/\tilde{\gamma}(\omega, s) \ll 1$, the relations (26.24) and (26.25) become symmetric. With use of Eqs. (26.14) and (26.20) we obtain the concise form

$$M\tilde{\gamma}(\omega, s)\tilde{\mu}_{\text{WB}}(\omega, V_0/\hbar; s) \; = \; 1 - M\tilde{\gamma}(\omega, 2-s)\tilde{\mu}_{\text{TB}}(\omega, \Delta; 2-s) \; . \tag{26.26}$$

In the Ohmic scaling limit we have $\tilde{\gamma}_{\text{WB}}(\omega) = \tilde{\gamma}_{\text{TB}}(\omega) = \gamma$, and the mobility of the free Brownian particle is $\mu_0 = 1/M\gamma$. Then the duality relation reads

$$\tilde{\mu}_{\text{WB}}(\omega, V_0/\hbar; K) \; = \; \mu_0 - \tilde{\mu}_{\text{TB}}(\omega, \Delta; 1/K) \; . \tag{26.27}$$

In this form, the mobilities are still labelled with WB or TB because of the dependence on $V_0/\hbar$ or Δ, respectively. We shall see in Section 26.4 that there is a universal frequency scale ϵ_0 with which the mobility loses differentiation between WB and TB model. With the frequency scale ϵ_0, the self-duality relation takes the concise form

$$\tilde{\mu}(\omega, \tilde{\epsilon}_0, K) \; = \; \mu_0 - \tilde{\mu}(\omega, \epsilon_0, 1/K) \; . \tag{26.28}$$

This relation equally applies within the TB or WB model, and between these models.

26.1.3 Nonlinear static mobility

In the presence of an external constant force F, the potential drop per lattice constant X_0 in the continuous WB potential (24.2) is $\hbar\epsilon_{\text{WB}} = FX_0$, whereas the potential drop in the dual TB model per lattice constant q_0 is $\hbar\epsilon_{\text{TB}} = Fq_0$. Using the relation (26.16), the energy drop in the two models is related by $\epsilon_{\text{WB}} = \epsilon_{\text{TB}}/K_{\text{TB}}$. Thus we find in generalization of Eq. (26.27) that the nonlinear static mobilities in the Ohmic scaling limit are related by

$$\mu_{\rm WB}(V_0/\hbar, T,\, \epsilon,\, K) \; = \; \mu_0 - \mu_{\rm TB}(\Delta, T,\, \epsilon/K,\, 1/K)\,, \qquad (26.29)$$

which holds for all T and ϵ. Again we may eliminate the explicit dependence on the TB and WB frequency scales Δ and $V_0/\hbar$ by introducing the universal frequency scale $\tilde{\epsilon}_0$ given in Eq. (26.119),

$$\mu(\tilde{\epsilon}_0, T,\, \epsilon,\, K) \; = \; \mu_0 - \mu(\tilde{\epsilon}_0, T,\, \epsilon/K,\, 1/K)\,. \qquad (26.30)$$

At this point, one remark is expedient. The self-duality relation (26.30) is free of explicit dependence on the particular model. In the mapping, both the Ohmic viscosity η and the biasing force stay constant. Because the lattice constants transform as $X_0 \leftrightarrow q_0/K$, where $K = \eta q_0^2/2\pi\hbar$, we have in the duality $K \leftrightarrow 1/K$ and $\epsilon \leftrightarrow \epsilon/K$. The mapping described by the self-duality relation (26.30) is discussed in Section 26.4.

For $T = 0$, it is convenient to eliminate the change of the bias $\epsilon \leftrightarrow \epsilon/K$ by employing the modified frequency scale ϵ_0 given in Eqs. (26.117) and (26.118), as explained in Subsection 26.5.1. With the scaled bias $v = \epsilon/\epsilon_0$, we then have

$$\mu(v, K) \; = \; \mu_0 - \mu(v, 1/K)\,. \qquad (26.31)$$

As example, we now translate the TB expression (25.90), which holds for $K \ll 1$, into the mobility of the WB model. The duality relation $\omega_c = \gamma$ maps the scaling limit of the TB model on the high-friction limit of the WB model, $\hbar\gamma \gg V_0,\, \hbar\epsilon_{\rm WB}$. We see from the definition $K_{\rm WB} = M\gamma X_0^2/2\pi\hbar$ and the duality $K_{\rm WB} = 1/K_{\rm TB}$ that the condition $K_{\rm TB} \to 0$ corresponds to the classical limit in the dual WB model.

Using the self-duality relation (26.29), the mobility of the WB model in the classical limit $K_{\rm WB} \to \infty$ is found from the expression (25.90) to read ($\epsilon = \epsilon_{\rm WB} = FX_0/\hbar$)

$$\frac{\mu_{\rm WB}(T, \epsilon)}{\mu_0} \; = \; \frac{\sinh(\beta\hbar\epsilon/2)}{\beta\hbar\epsilon/2}\, \frac{1}{|I_{i\beta\hbar\epsilon/2\pi}(\beta V_0)|^2}\,. \qquad (26.32)$$

This expression reduces in the special cases $\epsilon = 0$ and $T = 0$, respectively, to

$$\frac{\mu_{\rm WB}(T, 0)}{\mu_0} \; = \; \frac{1}{I_0^2(\beta V_0)}\,, \qquad (26.33)$$

$$\frac{\mu_{\rm WB}(0, \epsilon)}{\mu_0} \; = \; \Theta(\epsilon - \epsilon_0)\, \sqrt{1 - (\epsilon_0/\epsilon)^2}\,. \qquad (26.34)$$

In the classical limit, the Kondo frequency scale ϵ_0 is related to the corrugation strength V_0 of the weak-binding model by

$$\epsilon_0 \; = \; 2\pi V_0/\hbar\,. \qquad (26.35)$$

To our satisfaction, the results (26.32) - (26.34) coincide with solutions of the Smoluchowski diffusion equation (11.19) in corresponding limits (see Ref. [304]). For $\epsilon \leq \epsilon_0$ or equivalently $V_{\rm tilt} \equiv FX_0/2\pi \leq V_0$, the mobility is zero in the classical limit because the creeping motion of the overdamped particle comes to a stop at a point with zero slope and cannot move anymore since there are neither quantal nor thermal fluctuations. For $V_{\rm tilt} > V_0$, the potential (24.2) is sloping downwards everywhere, so that the particle is not trapped. This results in a nonzero mobility at zero temperature in the classical limit for $V_{\text{tilt}} > V_0$.

As already emphasized, the mobility is most easily calculated in the TB representation. A major advantage of the duality now is that one can shift the numerical problem of calculating or simulating the dynamics of the WB model to the discrete TB model. For instance in quantum Monte Carlo simulations of the dynamics, the discrete variables of the latter model significantly reduce the relevant configuration space subject to Monte Carlo sampling compared to the continuous model, and meaningful real-time simulations in the interesting non-perturbative low-temperature regime are possible [514]. In many interesting cases, e. g. charge transfer in chemical reactions, the density $J_{\text{WB}}(\omega)$ is not in the simple power-law form (26.13), and also the band width ω_c may be of the same order as the other frequencies. Nevertheless, the associated density $J_{\text{TB}}(\omega)$ is given by (26.12). Therefore, one may take advantage of performing numerical simulations for the equivalent TB model. Since the mapping is exact, the entire regime from quantum tunneling to thermal hopping across the barrier can be studied in the equivalent TB model.

In conclusion, we have discussed an exact duality symmetry between quantum Brownian motion in a continuous cosine potential and in a discrete tight-binding lattice. Because the tight-binding approximation ignores excited states at each lattice site, one might think that a general mapping between the two models is impossible. However, the TB model is fully sensitive to all aspects of both the quantum and the thermal hopping dynamics of the continuous model. In fact, the tight-binding model is reconciled with the continuous model by taking into account the alteration of the spectral density.

26.2   Self-duality in the exactly solvable cases $K = \frac{1}{2}$ and $K = 2$

In Section 25.6 we have discussed the three lowest cumulants of the tight-binding transport model (24.3) in the Ohmic scaling regime at the Toulouse point $K = \frac{1}{2}$. Now, we first extend the discussion to the full counting statistics. Then we demonstrate, upon employing self-duality, that the resulting cumulant generating function can be transformed into the CGF of the TB model at $K = 2$. Finally, we show that the CGFs also describe the full-counting statistics of the WB model (24.1) in corresponding regimes.

26.2.1   Full counting statistics at $K = \frac{1}{2}$

In Subsection 25.6.1 we have seen that the current of the bosonic transport problem at $K = \frac{1}{2}$ can be mapped on one of free fermions with the transmission probability (25.107). Accordingly, it is quite natural to conjecture that the cumulant generating function is given by the Levitov-Lesovik formula [515]

$$\ln \mathcal{Z}(\rho, t) = t \int_0^\infty \frac{d\omega}{2\pi} \ln\left\{1 + \mathcal{T}(\omega)\left[N_+(\omega, \epsilon)\left(e^{i2q_0\rho} - 1\right)\right.\right.$$
$$\left.\left. + N_-(\omega, \epsilon)\left(e^{-i2q_0\rho} - 1\right)\right]\right\} \tag{26.36}$$

with the spectral transmission probability at frequency ω [see Eq. (25.107)]

$$\mathcal{T}(\omega) = \frac{\tilde{\epsilon}_0^2}{\tilde{\epsilon}_0^2 + \omega^2} . \tag{26.37}$$

In fact, the CGF (26.36) reproduces the cumulants $\langle q(t) \rangle_c$, $\langle q^2(t) \rangle_c$ and $\langle q^3(t) \rangle_c$ as given in Eqs. (25.109), (25.112) and (25.113).

In the remainder of this subsection, we consider transport and noise at zero temperature. Now, up-hill particle flow is absent, $N_+(\omega, \epsilon) = \Theta(\epsilon - |\omega|)$, and $N_-(\omega, \epsilon) = 0$. In this limit, the CGF (26.36) yields for the cumulants the concise form

$$\langle q^N(t) \rangle_c = t \, 2q_0^N \int_0^\epsilon \frac{d\omega}{2\pi} \mathcal{T}_N(\omega) . \tag{26.38}$$

Here
$$\mathcal{T}_1(\omega) = \mathcal{T}(\omega) ,$$
$$\mathcal{T}_2(\omega) = 2\mathcal{T}(\omega)[1 - \mathcal{T}(\omega)] , \tag{26.39}$$
$$\mathcal{T}_3(\omega) = 4\mathcal{T}(\omega)[1 - \mathcal{T}(\omega)][1 - 2\mathcal{T}(\omega)] ,$$

etc. The function $\mathcal{T}_N(\omega)$ can be generated by differentiation,

$$\mathcal{T}_N(\omega) = [\tilde{\epsilon}_0(d/d\tilde{\epsilon}_0)]^{N-1}\mathcal{T}(\omega) . \tag{26.40}$$

This allows to express every single cumulant in terms of derivatives of the current,

$$\langle q^N(t) \rangle_c = \left(q_0 \tilde{\epsilon}_0 \frac{d}{d\tilde{\epsilon}_0}\right)^{N-1} \langle q(t) \rangle_c = t \left(q_0 \tilde{\epsilon}_0 \frac{d}{d\tilde{\epsilon}_0}\right)^{N-1} \langle I \rangle . \tag{26.41}$$

On the other hand, the CGF may be written in the tight-binding form (25.22). At zero temperature, backward flow is absent. We then have

$$\ln \mathcal{Z}(\rho, t) = t \sum_{n=1}^\infty \left(e^{i\rho q_0 n} - 1\right) k_n^+ , \tag{26.42}$$

and the cumulants take the form [cf. Eq. (25.23)]

$$\langle q^N(t) \rangle_c = t \, q_0^N \sum_{n=1}^\infty n^N k_n^+ . \tag{26.43}$$

It is immediately obvious, that the form (26.43) is consistent with the expression (26.41) for arbitrary N if and only if the transition weight k_n^+ is of order $\tilde{\epsilon}_0^n$. This specific property allows us to extract all the transition weights k_n^+ from the power series of one of the cumulants, e.g. from the perturbative series of the mobility. The power series of the second form in Eq. (25.103) yields for the transition weight k_n^+ at zero temperature the analytic expression

$$k_n^+ = \frac{\epsilon}{\pi} \frac{(-1)^{n-1}}{n} \frac{\sin[\frac{\pi}{2}(n-1)]}{n-1} \left(\frac{\tilde{\epsilon}_0}{\epsilon}\right)^n . \qquad (26.44)$$

We see that the k_n^+ is zero for all odd n apart from the case $n = 1$. We have $k_1^+ = \tilde{\epsilon}_0/2$, and for even n

$$k_{2n}^+ = \frac{\epsilon}{\pi} \frac{(-1)^n}{2n} \frac{1}{2n-1} \left(\frac{\tilde{\epsilon}_0}{\epsilon}\right)^{2n} . \qquad (26.45)$$

Hence the sign of the transition weights is not strictly positive. This is the interference effect addressed in Subsection 25.2.2 and discussed in greater depth in Section 26.6.

The expressions given here have lost explicit dependence on the particular representation. In application to the TB model (24.3), the length q_0 is the lattice constant, ϵ is the bias frequency and the renormalized tunneling amplitude is [cf. Eq. (25.97)]

$$\tilde{\epsilon}_0 = \pi \Delta^2 / \omega_c . \qquad (26.46)$$

Passage to the WB representation (24.1) with lattice period X_0 and bias frequency ϵ_{WB} implies the substitutions $q_0 \to X_0/2$ and $\epsilon \to \epsilon_{\text{WB}}/2$, as follows from the relation (26.16). Further, we see from the relation (26.119) at $K = \frac{1}{2}$ that the dressed tunneling amplitude $\tilde{\epsilon}_0$ is expressed in terms of the WB parameters as

$$\tilde{\epsilon}_0 = \frac{1}{2\pi} \frac{\omega_c^2}{V_0/\hbar} . \qquad (26.47)$$

With these assignments, the above expressions for the CGF and the cumulants are equally valid both for the WB and the TB model at $K = \frac{1}{2}$. The perturbative series (26.43) likewise represents the weak-tunneling series of the TB model, and the strong-backscattering series of the WB model.

26.2.2 Full counting statistics at $K = 2$

In the preceding subsection, we have seen that the full counting statistics at the Toulouse point $K = \frac{1}{2}$ is available in analytic form. Equipped with this solution, the case $K = 2$ can now be tackled without splendid calculation upon employing self-duality of the model under consideration. The dual mapping $K = \frac{1}{2} \leftrightarrow K = 2$ in the TB representation implies

$$\tilde{\epsilon}_0 \leftrightarrow \tilde{\epsilon}_0 , \qquad \text{and} \qquad \epsilon \leftrightarrow \epsilon/2 , \qquad (26.48)$$

as follows from the relation (26.30).

Current and mobility:

Employing the mapping (26.30) for $K = \frac{1}{2}$, we find from the expression (25.100) that the normalized mobility at $K = 2$ takes the form

$$\frac{\mu}{\mu_0} = 1 - \frac{2\tilde{\epsilon}_0}{\epsilon} \operatorname{Im} \psi\left(\frac{1}{2} + \frac{\beta\hbar\tilde{\epsilon}_0}{2\pi} + i\frac{\beta\hbar\epsilon}{4\pi}\right), \tag{26.49}$$

and the mean current is

$$\langle I \rangle = q_0 \frac{\epsilon}{4\pi} - q_0 \frac{\tilde{\epsilon}_0}{2\pi} \operatorname{Im} \psi\left(\frac{1}{2} + \frac{\beta\hbar\tilde{\epsilon}_0}{2\pi} + i\frac{\beta\hbar\epsilon}{4\pi}\right). \tag{26.50}$$

Here, q_0 and $\epsilon = Fq_0/\hbar$ are the lattice constant and bias frequency in the TB representation (24.3). The renormalized scattering amplitude $\tilde{\epsilon}_0$ is connected with the TB parameters Δ and ω_c by the relation

$$\tilde{\epsilon}_0 = \frac{1}{2\pi} \frac{\omega_c^2}{\Delta}, \tag{26.51}$$

as follows from Eq. (26.119) at $K = 2$. Because $\tilde{\epsilon}_0 \propto 1/\Delta$, the weak-tunneling series of current and mobility emerge from the asymptotic expansion of the digamma function in the expressions (26.50) and (26.49). The leading term of the weak tunneling series resulting from the expression (26.49) coincides indeed with the weak-tunneling expression (25.78) at $K = 2$,

$$\frac{\mu}{\mu_0} = \frac{\pi^2}{3} \frac{\Delta^2}{\omega_c^4} \left(\epsilon^2 + \left(\frac{2\pi}{\hbar\beta}\right)^2\right). \tag{26.52}$$

Passage to the WB model implies the substitutions $q_0 \to 2X_0$ and $\epsilon \to 2\epsilon_{\mathrm{WB}}$, and the relation of the dressed amplitude with the WB parameters is

$$\tilde{\epsilon}_0 = \pi(V_0/\hbar\omega_c)^2\,\omega_c. \tag{26.53}$$

With these assignments, we may switch easily between the dual representations.

In the limit $T \to 0$, the expression (26.50) reduces to the equivalent forms

$$\langle I \rangle = q_0 \frac{\epsilon}{4\pi} - q_0 \frac{\tilde{\epsilon}_0}{2\pi} \arctan\left(\frac{\epsilon}{2\tilde{\epsilon}_0}\right) = q_0 \frac{\epsilon}{4\pi} - q_0 \frac{\tilde{\epsilon}_0}{4} + q_0 \frac{\tilde{\epsilon}_0}{2\pi} \arctan\left(\frac{2\tilde{\epsilon}_0}{\epsilon}\right). \tag{26.54}$$

The Taylor expansion of the first form yields the weak-tunneling series of the current for the TB model, and the strong-backscattering series for the WB model. The Taylor expansion of the second form gives the strong-tunneling and weak-backscsattering series of the TB and WB model, respectively.

By taking up the line pursued in Subsection 25.6.1, we may transform the expression (26.50) into fermionic representation. The resulting forms are

$$\langle I \rangle = q_0 \frac{\epsilon}{4\pi} - q_0 \int_0^\infty \frac{d\omega}{2\pi} \overline{\mathcal{R}}(\omega)\, F_1(\omega, \epsilon/2) = q_0 \int_0^\infty \frac{d\omega}{2\pi} \overline{\mathcal{T}}(\omega)\, F_1(\omega, \epsilon/2), \tag{26.55}$$

where $\overline{\mathcal{T}}(\omega)$ and $\overline{\mathcal{R}}$ are the spectral transition and reflexion probabilities at $K = 2$,

$$\overline{\mathcal{T}}(\omega) = \frac{\omega^2}{\omega^2 + \tilde{\epsilon}_0^2}, \qquad \overline{\mathcal{R}}(\omega) = \frac{\tilde{\epsilon}_0^2}{\omega^2 + \tilde{\epsilon}_0^2}. \tag{26.56}$$

Comparison of the expressions (26.55) with the forms (25.109) shows that in the mapping $K = \frac{1}{2} \leftrightarrow K = 2$ transmission and reflexion perform role reversal

$$\mathcal{T}(\omega) = \overline{\mathcal{R}}(\omega) \, , \qquad \text{and} \qquad \mathcal{R}(\omega) = \overline{\mathcal{T}}(\omega) \, . \tag{26.57}$$

The meaning of the expression (26.55) is as in the case $K = \frac{1}{2}$. The first term in the first form is the maximum current in the absence of the barrier. The second term diminishes the current because of the possibility that the particle is reflected at the barrier. The second form expresses the current in terms of the transmission through the barrier.

Full counting statistics:

With the relations (26.48) and the substitution $\mathcal{T}(\omega) \to \overline{\mathcal{T}}(\omega)$, we may also deduce the cumulant generating function at $K = 2$ from the CGF expression at $K = \frac{1}{2}$, Eq. (26.36) . The resulting expression is

$$\ln \mathcal{Z}(\rho, t) \;=\; t \int_0^\infty \frac{d\omega}{2\pi} \ln \Big\{ 1 + \overline{\mathcal{T}}(\omega) \left[\mathcal{N}_+(\omega, \epsilon/2) \left(e^{iq_0\rho} - 1 \right) \right. \tag{26.58}$$
$$\left. + \, \mathcal{N}_-(\omega, \epsilon/2) \left(e^{-iq_0\rho} - 1 \right) \right] \Big\} \, .$$

The CGF yields for the second and third cumulant the fermionic frequency integrals analogous to the expressions (25.112) and (25.113),

$$\langle q^2(t) \rangle_c \;=\; t q_0^2 \int_0^\infty \frac{d\omega}{2\pi} \left[\overline{\mathcal{T}}(\omega) F_2(\omega, \epsilon/2) - \overline{\mathcal{T}}^2(\omega) F_1^2(\omega, \epsilon/2) \right] ,$$

$$\langle q^3(t) \rangle_c \;=\; t q_0^3 \int_0^\infty \frac{d\omega}{2\pi} \, \overline{\mathcal{T}}(\omega) F_1(\omega, \epsilon/2) \tag{26.59}$$
$$\times \left[1 - 3\overline{\mathcal{T}}(\omega) F_2(\omega, \epsilon/2) + 2\overline{\mathcal{T}}^2(\omega) F_1^2(\omega, \epsilon/2) \right] .$$

Again, these forms can be rewritten in terms of derivatives of lower-order cumulants,

$$\langle q^2(t) \rangle_c \;=\; \frac{q_0}{2} \Big\{ - \coth\left(\tfrac{1}{2}\beta\hbar\epsilon \right) \tilde{\epsilon}_0 \frac{d}{d\tilde{\epsilon}_0} + \Big(2 + \tilde{\epsilon}_0 \frac{d}{d\tilde{\epsilon}_0} \Big) \frac{2}{\hbar\beta} \frac{d}{d\epsilon} \Big\} \langle q(t) \rangle_c \, , \tag{26.60}$$

$$\langle q^3(t) \rangle_c \;=\; \frac{q_0}{2} \Big\{ - \coth\left(\tfrac{1}{2}\beta\hbar\epsilon \right) \tilde{\epsilon}_0 \frac{d}{d\tilde{\epsilon}_0} + \Big(2 + \frac{\tilde{\epsilon}_0}{2} \frac{d}{d\tilde{\epsilon}_0} \Big) \frac{2}{\hbar\beta} \frac{d}{d\epsilon} \Big\} \langle q^2(t) \rangle_c \tag{26.61}$$

$$+ \frac{q_0^2}{4} \Big\{ \coth\left(\tfrac{1}{2}\beta\hbar\epsilon \right) \frac{2}{\hbar\beta} \frac{d}{d\epsilon} + \frac{1}{\sinh^2(\tfrac{1}{2}\beta\hbar\epsilon)} \Big\} \tilde{\epsilon}_0 \frac{d}{d\tilde{\epsilon}_0} \langle q(t) \rangle_c \, .$$

At zero temperature, we have $\mathcal{N}_+(\omega) = \Theta(\epsilon - |\omega|)$ and $\mathcal{N}_-(\omega) = 0$. We then obtain from Eq. (26.58) a relation which expresses cumulants of higher order again in terms of derivatives of the mean current,

$$\langle q^N(t) \rangle_{\rm c} = t \left(-\frac{q_0}{2} \tilde{\epsilon}_0 \frac{d}{d\tilde{\epsilon}_0} \right)^{N-1} \langle I \rangle . \tag{26.62}$$

The weak-tunneling representations of the CGF and the cumulants can be written again in the forms (26.42) and (26.43), respectively, but now the transition weight k_n^+ is

$$k_n^+ = \frac{\epsilon}{4\pi} \frac{(-1)^{n-1}}{n(2n+1)} \left(\frac{\epsilon}{2\tilde{\epsilon}_0} \right)^{2n} . \tag{26.63}$$

Like in the case $K = \frac{1}{2}$, Eq. (26.45), the sign of the transition weight k_n^+ alternates.

We conclude the discussion of the special cases $K = \frac{1}{2}$ and $K = 2$ with a prospect. First we anticipate that the full counting statistics at $T = 0$ can be found in analytic form for general K (see Section 26.6). The functional relations (26.41) and (26.62) are consistent with the general functional relations (26.132). Similarly, the rate expressions (26.44) and (26.63) are special cases of the general rate expression (26.133). The indefinite sign of the transition weights is a signature of quantum interference of the different decay channels.

26.3 Duality and supercurrent in Josephson junctions

26.3.1 Charge-phase duality

In Subsection 3.5.3 we have introduced the model for the Josephson junction in the charge representation which is appropriate for weak Josephson coupling, $E_{\rm J} \ll E_{\rm c}$. In this limit, the charge on the junction is nearly sharp, and the phase fluctuations are described by the correlation function (3.234).

In the opposite limit of large Josephson coupling, $E_{\rm J} \gg E_{\rm c}$, the jump of the phase at the junction is nearly sharp. The relevant Hamiltonian is

$$H = -U_0 \cos \left(\frac{2\pi \bar{Q}}{2e} \right) + \sum_\alpha \left[\frac{\bar{q}_\alpha^2}{2\bar{C}_\alpha} + \left(\frac{\hbar}{2e} \right)^2 \frac{1}{2\bar{L}_\alpha} (\bar{\psi} - \bar{\psi}_\alpha)^2 \right] , \tag{26.64}$$

with

$$\bar{\psi} = 2\pi \sum_n n a_n^\dagger a_n = 2\pi \sum_n n |n\rangle \langle n| ,$$

$$e^{i 2\pi \bar{Q}/2e} = \sum_n a_n^\dagger a_{n+1} = |n\rangle \langle n+1| . \tag{26.65}$$

Here we have introduced a basis of discrete phase states. The Hamiltonians (3.233) and (26.64) correspond to the WB and TB Hamiltonians, Eq. (24.1) and Eq. (24.3), respectively. Following the discussion given in Subsection 26.1.1, we now show that the Hamiltonian (3.233) can be transformed into the form (26.64). First, we put $E_{\rm J} \to U_0$, and we perform the canonical transformation [we put $\lambda_\alpha = (\hbar/2e) \sqrt{C_\alpha/L_\alpha}$],

$$q_\alpha \;\to\; -\lambda_\alpha \psi_\alpha\,, \qquad \psi_\alpha \;\to\; q_\alpha/\lambda_\alpha + (\pi/e)\,\bar{Q}\,,$$

$$\psi \;\to\; \pi\bar{Q}/e\,, \qquad Q \;\to\; -(e/\pi)\,\bar{\psi} + \sum_\alpha \lambda_\alpha \psi_\alpha\,. \tag{26.66}$$

In the transformed reservoir Hamiltonian H_R the electromagnetic modes are coupled,

$$H \;=\; -U_0 \cos\left(\frac{2\pi\bar{Q}}{2e}\right) + \left(\frac{e}{\pi}\right)^2 \frac{\bar{\psi}^2}{2C} - \frac{e\bar{\psi}}{\pi C}\sum_\alpha \lambda_\alpha \psi_\alpha + H_\mathrm{R}\,,$$

$$H_\mathrm{R} \;=\; \sum_\alpha \left[\frac{q_\alpha^2}{2C_\alpha} + \left(\frac{\hbar}{2e}\right)^2 \frac{\psi_\alpha^2}{2L_\alpha}\right] + \frac{1}{2C}\left(\sum_\alpha \lambda_\alpha \psi_\alpha\right)^2\,. \tag{26.67}$$

The mapping is completed by equating the $\bar{\psi}^2$ terms in Eqs. (26.67) and (26.64),

$$\left(\frac{e}{\pi}\right)^2 \frac{1}{C} \;=\; \left(\frac{\hbar}{2e}\right)^2 \sum_\alpha \frac{1}{\bar{L}_\alpha}\,, \tag{26.68}$$

and by switching to the diagonal basis of the bath modes. The ratio of the determinants of the dynamical matrices $\boldsymbol{A}(\omega)$ and $\boldsymbol{B}(\omega)$,

$$B_{\alpha\beta}(\omega) \;=\; \delta_{\alpha\beta} C_\alpha \,(\hbar/2e)^2 \,(\omega_\alpha^2 - \omega^2)\,, \qquad A_{\alpha\beta}(\omega) \;=\; B_{\alpha\beta}(\omega) + \lambda_\alpha \lambda_\beta/C\,, \tag{26.69}$$

where $\omega_\alpha = 1/\sqrt{L_\alpha C_\alpha}$, reads

$$\frac{\det \boldsymbol{A}(\omega)}{\det \boldsymbol{B}(\omega)} \;=\; 1 + \frac{1}{C}\sum_\alpha \frac{1}{L_\alpha}\frac{1}{\omega_\alpha^2 - \omega^2} \;=\; \frac{-i\omega C + Y^*(\omega)}{-i\omega C}\,. \tag{26.70}$$

In the second form we have introduced the admittance $Y(\omega)$ of the weak-coupling model (3.211). On the other hand, in the eigen basis of the matrix $\boldsymbol{A}(\omega)$ we have

$$\bar{A}_{\alpha\beta} \;=\; \delta_{\alpha\beta}\bar{C}_\alpha \left(\frac{\hbar}{2e}\right)^2 (\bar{\omega}_\alpha^2 - \omega^2)\,,$$

$$\bar{B}_{\alpha\beta} \;=\; \bar{A}_{\alpha\beta} - \frac{\pi^2 C}{e^2}\left(\frac{\hbar}{2e}\right)^2 \bar{\lambda}_\alpha \bar{\omega}_\alpha \bar{\lambda}_\beta \bar{\omega}_\beta\,. \tag{26.71}$$

In this representation we obtain with use of Eq. (26.68)

$$\frac{\det \boldsymbol{B}(\omega)}{\det \boldsymbol{A}(\omega)} \;=\; -i\omega C\left(\frac{\pi\hbar}{2e^2}\right)^2 \sum_\alpha \frac{1}{\bar{L}_\alpha}\frac{-i\omega}{\bar{\omega}_\alpha^2 - \omega^2} \;=\; -i\omega C\, R_\mathrm{Q}^2 \bar{Y}^*(\omega)\,, \tag{26.72}$$

where $R_\mathrm{Q} = h/4e^2$ is the resistance quantum for Cooper pairs. Equating the inverse of Eq. (26.72) with the expression (26.70), we find a simple relation between the admittance $Y(\omega)$ of the model for weak Josephson coupling and the admittance $\bar{Y}(\omega)$ of the strong-coupling model,

$$\bar{Y}(\omega)\,[\,i\omega C + Y(\omega)\,] \;=\; 1/R_\mathrm{Q}^2\,. \tag{26.73}$$

Using this correspondence, we can express quantitites of the one model in terms of those of the other. The phase-charge duality transformations are [152, 155]

$$\psi \longleftrightarrow \pi Q/e , \qquad \frac{1}{R_Q[\,i\omega C + Y(\omega)\,]} \longleftrightarrow R_Q Y(\omega) . \qquad (26.74)$$

The phase autocorrelation function $\langle [\,\psi(0) - \psi(t)\,]\psi(0)\rangle_\beta$ takes the form (3.234). The spectral density of the coupling reads [cf. Eq. (3.235) with (3.214)]

$$G_\psi(\omega) = 2\omega R_Q \operatorname{Re} \bar{Y}(\omega) = 2\omega \operatorname{Re} \frac{1}{R_Q[\,i\omega C + Y(\omega)\,]} . \qquad (26.75)$$

In the strong-coupling limit $E_J \gg E_c$, the influence of the environment is described by the charge autocorrelation function $Q_Q(t) \equiv \langle [\,Q(0) - Q(t)\,]Q(0)\rangle_\beta$. Using the correspondence (26.74), we find the function $Q_Q(t)$ in the form (3.234) with the spectral density

$$G_Q(\omega) = 2(e/\pi)^2 R_Q \operatorname{Re} Y(\omega)\,\omega . \qquad (26.76)$$

In the Ohmic scaling limit, we then obtain with

$$\rho = R/R_Q \qquad (26.77)$$

the forms

$$G_\psi(\omega) = 2\rho\omega , \qquad \text{and} \qquad G_Q(\omega) = (e/\pi)^2 (2/\rho)\,\omega , \qquad (26.78)$$

which provide exact self-duality between the charge and phase representation. These results can directly be transferred to electron tunneling through a normal junction with the substitution $2e \to e$ and $R_Q \to R_K$. The respective spectral densities are in correspondence with the earlier results (3.212) with (3.213) and (3.215) with (3.216).

Nano-electronic devices based on low-capacitance Josephson junctions appear to be suitable for large-scale integration and physical realization of quantum bits. In the regime $E_c \gg E_J$, the Coulomb blockade effect allows the controll of individual charges [114, 516]. Single- and two-bit operations may be performed upon applying sequences of gate voltages. When the gate voltage is chosen such that neighboring charge states are degenerate, the system reduces to two states which are weakly coupled by E_J [cf. Subsection 3.3.1]. One may also think of quantum-logic Josephson elements with flux states instead of charge states. The effects of the electromagnetic environment are described by the phase correlation function (3.234) with (3.235).

In Subsection 20.3.4 we have studied the supercurrent through an ultrasmall Josephson junction with Josephson coupling energy E_J in the weak-tunneling limit, in which $I \propto E_J^2$. In the region $\rho < 1$, we found divergence of the expression (20.180) for $\rho < 1$ in the low-energy regime $k_B T$, $eV_x \ll E_J(E_J/E_c)^{\rho/(1-\rho)}$. This indicates that tunneling processes of higher order in E_J become relevant in this parameter regime. As we shall see, the higher-order terms smooth out the singularity in question.

Interestingly enough, in the particular cases $\rho \ll 1$, $\rho = \frac{1}{2}$ and $\rho = 2$, the tunneling dynamics of Cooper pairs can be solved in analytic form in all orders in E_J.

26.3.2 Supercurrent-voltage characteristics for $\rho \ll 1$

By virtue of the correspondence relations (20.175) between tunneling of bosons under Ohmic friction and tunneling of Cooper pairs in a resistive electromagnetic environment, we can translate the Ivanchenko-Zil'berman expression (25.89) into the supercurrent-voltage characteristics of a small capacitance Josephson junction [513],

$$I(V_x) = \frac{V_x}{R} \frac{\beta E_J^*}{\nu} \, \mathrm{Im} \left(\frac{I_{1-i\nu}(\beta E_J^*)}{I_{-i\nu}(\beta E_J^*)} \right) \tag{26.79}$$

where $\nu = \beta e V_x / \pi \rho$. Furthermore, E_J^* is the equivalent of $\hbar \Delta_T$ defined in Eq. (25.85),

$$E_J^* = E_J \, e^{-\rho \zeta} \left(\frac{2\pi^2 \rho}{\beta E_c} \right)^{\rho} , \tag{26.80}$$

where ζ is given in Eqs. (20.178) and (20.151). The formula (26.79) can be rewritten as [see Eqs. (25.89) and (25.90)]

$$I(V_x) = \frac{V_x}{R} \left(1 - \frac{\sinh(\pi \nu)}{\pi \nu} \frac{1}{|I_{i\nu}(\beta E_J^*)|^2} \right) . \tag{26.81}$$

The supercurrent-voltage characteristics (26.79) or (26.81) covers the entire regime ranging from weak to strong Cooper pair tunneling. It shows a peak at small voltage, as explained below Eq. (20.180).

Upon expanding the expression (26.79) to order E_J^2, we recover the weak-tunneling expression (20.183). Furthermore, we obtain from the expression (26.81) the linear superconductance at general T as

$$G_{\mathrm{lin}}(T) = \frac{1}{R} \left[1 - \frac{1}{I_0^2(\beta E_J^*)} \right] , \tag{26.82}$$

and the supercurrent-voltage characteristics at zero temperature as

$$I(V_x) = \Theta(V_x - RI_c) \frac{V_x}{R} \left[1 - \sqrt{1 - (RI_c/V_x)^2} \right] , \tag{26.83}$$

where we have used the relation $\beta E_J / \nu = RI_c / V_x$. These results hold for small capacitance Josephson junctions in the domain $R \ll R_Q$.

By employment of duality it is straightforward to translate the expression (26.81) into the I-V characteristics of a large capacitance Josephson junction ($E_J \gg E_c$) coupled to a very strong resistive environment, $R/R_Q \gg 1$. The relevant Hamiltonian is given in Eq. (26.64), and the spectral density weighting the charge fluctuations is $G_Q(\omega) = (e/\pi)^2 2(R_Q/R) \, \omega$. The changeover to the latter case is analogous to the passage from the expression (25.90) to the form (26.32). In the classical limit $R/R_Q \to \infty$, the dc voltage drop V at the current-biased junction with bias current $I_{\mathrm{ext}} = I_b$ due to the backscattering current is (we put $\sigma = e\beta RI_b/\pi$) [517]

$$V(I_b) = RI_b \left(1 - \frac{\sinh(\pi \sigma)}{\pi \sigma} \frac{1}{|I_{i\sigma}(\beta U_0)|^2} \right) . \tag{26.84}$$

As $T \to 0$, this reduces to

$$V(I_b) = \Theta(I_b - \pi U_0/eR) \, RI_b \left[1 - \sqrt{1 - (\pi U_0/eRI_b)^2} \right] . \tag{26.85}$$

26.3.3 Supercurrent-voltage characteristics at $\rho = \frac{1}{2}$

With the spadework set out in Section 25.6, it is only a small step to give the supercurrent-voltage characteristics for the case $\rho = \frac{1}{2}$, or equivalently $R = \pi\hbar/4e^2$. Employing the correspondence relations (20.175) - (20.177), the renormalized tunneling amplitude (26.46) is expressed in terms of the device parameters as

$$\tilde{\epsilon}_0 = \frac{\pi^2}{2\hbar} \frac{E_J^2\, e^{-\varsigma}}{E_c} = \frac{e\pi}{\hbar} \frac{E_J\, e^{-\varsigma}}{E_c}\, RI_c \,. \tag{26.86}$$

Putting $q_0 \hat{=} 2e$, the expression (25.100) transforms itself into the supercurrent

$$I(V_x) = I_c \frac{\hbar\tilde{\epsilon}_0}{\pi E_J} \operatorname{Im} \psi\!\left(\frac{1}{2} + \frac{\beta\hbar\tilde{\epsilon}_0}{2\pi} + i\frac{\beta e V_x}{\pi} \right). \tag{26.87}$$

At zero temperature, the current-voltage characteristics takes the form [518]

$$I(V_x) = I_c \frac{\hbar\tilde{\epsilon}_0}{\pi E_J} \arctan\!\left(\frac{2e V_x}{\hbar\tilde{\epsilon}_0} \right) = I_c \frac{\hbar\tilde{\epsilon}_0}{2E_J} - I_c \frac{\hbar\tilde{\epsilon}_0}{\pi E_J} \arctan\!\left(\frac{\hbar\tilde{\epsilon}_0}{2e V_x} \right). \tag{26.88}$$

In contrast to the regime $\rho < \frac{1}{2}$, this expression does not show a peak structure anymore. The first term in the second form is the plateau value. It is equal to the Coulomb blockade expression (20.181) at $\rho = \frac{1}{2}$. This term is the leading one for weak Cooper pair tunneling, i.e., when the bias energy $2e V_x$ is large. The opposite regime of strong Cooper pair tunneling is captured, when the bias energy $2e V_x$ is small compared to $\hbar\tilde{\epsilon}_0$.

26.3.4 Supercurrent-voltage characteristics at $\rho = 2$

At $\rho = 2$, which corresponds to $R = \pi\hbar/e^2$, the renormalized frequency (26.51) with the correspondence relations (20.175) - (20.177) takes the form

$$\tilde{\epsilon}_0 = \frac{e}{\hbar} \left(\frac{E_c\, e^\varsigma}{4\pi^2 E_J} \right)^2 RI_c \,. \tag{26.89}$$

The supercurrent-voltage characteristics dual to the expression (26.87) is

$$I(V_x) = \frac{V_x}{R} - \left(\frac{E_c\, e^\varsigma}{4\pi^2 E_J} \right)^2 I_c \operatorname{Im} \psi\!\left(\frac{1}{2} + \frac{\beta\hbar\tilde{\epsilon}_0}{2\pi} + i\frac{\beta e V_x}{2\pi} \right). \tag{26.90}$$

In the zero temperature limit, this expressions simplifies to [518]

$$I(V_x) = \frac{V_x}{R} - \left(\frac{E_c\, e^\varsigma}{4\pi^2 E_J} \right)^2 I_c \arctan\!\left[\left(\frac{4\pi^2 E_J}{E_c\, e^\varsigma} \right)^2 \frac{V_x}{RI_c} \right]. \tag{26.91}$$

The perturbative expansion in V_x yields the weak-tunneling series. The leading term $\propto V_x^3$ coincides with the Coulomb blockade expression (20.181) at $\rho = 2$.

In the opposite limit of very large V_x, we have strong Cooper pair tunneling. The respective weak-backscattering series may be deduced from the complementary supercurrent-voltage relation

$$I(V_x) = \frac{V_x}{R} - \frac{\pi}{2}\left(\frac{E_c\, e^\varsigma}{4\pi^2 E_J}\right)^2 I_c + \left(\frac{E_c\, e^\varsigma}{4\pi^2 E_J}\right)^2 I_c \arctan\left(\left(\frac{E_c\, e^\varsigma}{4\pi^2 E_J}\right)^2 \frac{RI_c}{V_x}\right). \quad (26.92)$$

The leading backscattering term turns out to be independent of the applied voltage.

With the correspondence relations (20.175) - (20.177), it is easy to translate the expressions for higher order cumulants and the CGF, which for $K = \frac{1}{2}$ are given in Eqs. (25.114), (25.115), and (26.36), and for $K = 2$ in Eqs. (26.60), (26.61) and (26.58), into the corresponding expressions for Cooper pair tunneling at $\rho = \frac{1}{2}$ and at $\rho = 2$.

26.4   Self-duality in the Ohmic scaling limit

In the WB model (24.1), the scaling limit (18.57) corresponds to the high-friction (Smoluchowski) limit $\gamma = \omega_c \gg V_0/\hbar,\ \epsilon_{WB}$. In the absence of the potential, the mobility is μ_0. Actually, the mobility is reduced because of backscattering of the quantum particle in the corrugated potential. The backscattering contribution as perturbative series in the corrugation strength V_0 is analogous in form to the expression (25.76). We have [145, 498]

$$\frac{\mu_{WB}(T, \epsilon_{WB})}{\mu_0} = 1 - \frac{\pi K_{WB}}{\epsilon_{WB}} \sum_{m=1}^{\infty} (-1)^{m-1} \left(\frac{V_0}{\hbar}\right)^{2m}$$

$$\times \int_0^\infty d\tau_1\, d\tau_2 \cdots d\tau_{2m-1} \sum_{\{\xi_j\}'} G_{2m}\left(\{\tau_j\}; K_{WB}, \{\xi_j\}\right) \quad (26.93)$$

$$\times \sin\left(\sum_{i=1}^{2m-1} \epsilon_{WB}\, p_i \tau_i\right) \prod_{k=1}^{2m-1} \sin(\pi K_{WB} p_k)\,.$$

The interaction factor G_{2m} is given in Eq. (25.77). On the other hand, the TB series for the mobility in the Ohmic scaling limit is given by the series (25.76). With the expansion (26.93) and the expansion (26.93) for the TB model, the duality relation (26.29) is easily verified.

26.4.1   Linear mobility at finite T

We see from the TB series (25.76) by simply counting dimensions that, for zero bias, T is combined with the frequencies Δ and ω_c in the form $\Delta^2 T^{2K-2} \omega_c^{-2K}$. On the other hand, taking into account the mapping (26.16), $K_{WB} = 1/K$, we find from the WB series (26.93) that V_0 is combined with T and ω_c as $V_0^2 T^{2/K-2} \omega_c^{-2/K}$. From this we see that there is a Kondo-type temperature which is related to the parameters of the TB and WB model according to

$$k_B T_0 \;=\; a_{\mathrm{TB}}(K)\,(\Delta/\omega_c)^{K/(1-K)}\,\hbar\Delta \;=\; a_{\mathrm{WB}}(K)(V_0/\hbar\omega_c)^{1/(K-1)}V_0\,, \qquad (26.94)$$

where the function $a_{\mathrm{TB}}(K)$ depends only on K, and where

$$a_{\mathrm{WB}}(K) \;=\; a_{\mathrm{TB}}(1/K)\,. \qquad (26.95)$$

Upon introducing the Kondo-scaled temperature

$$\vartheta \;=\; T/T_0\,, \qquad (26.96)$$

the TB series (25.76) and the WB series (26.93) for the dimensionless linear mobility $\mathcal{M}_{\mathrm{lin}}(\vartheta, K) \equiv \mu_{\mathrm{lin}}/\mu_0$ can be rewritten in the self-dual scaling forms

$$\mathcal{M}_{\mathrm{lin}}(\vartheta, K) \;=\; \sum_{m=1}^{\infty} d_m(K)\,\vartheta^{2(K-1)m}\,, \qquad (26.97)$$

$$\mathcal{M}_{\mathrm{lin}}(\vartheta, K) \;=\; 1 - \sum_{m=1}^{\infty} d_m(1/K)\,\vartheta^{2(1/K-1)m}\,. \qquad (26.98)$$

In the Coulomb gas representation (25.76), the dimensionless coefficient $d_m(K)$ is given in terms of a $(2m-1)$-fold integral.

Self-duality means that the expressions (26.97) and (26.98) are the Taylor and the asymptotic expansion of the same mobility function. By the self-duality transformation $K \to 1/K$, the two expansions are interchanged. Equating Eq. (26.97) with Eq. (26.98), we obtain the nonperturbative self-duality

$$\mathcal{M}_{\mathrm{lin}}(\vartheta, K) \;=\; 1 - \mathcal{M}_{\mathrm{lin}}(\vartheta, 1/K)\,, \qquad (26.99)$$

which is valid for all values of K. By this relation, the Taylor expansion is mapped on the asymptotic expansion and vice versa. With the series expressions (26.97) and (26.98), the mapping holds term by term. Observe that the expressions (26.97) - (26.99) have lost explicit dependence on the TB and on the WB model. The series expressions (26.97) and (26.98) describe the same physical system in complementary regions of the parameter space.

The radius of convergence of the power series (26.97) and (26.98) is conveniently expressed in terms of the temperature scale

$$T_{\mathrm{cr}}(K) \;\equiv\; \lim_{m\to\infty} |d_m(K)|^{1/[2(1-K)m]} T_0 \;=\; T_{\mathrm{cr}}(1/K)\,. \qquad (26.100)$$

The equality is a consequence of self-duality [cf. the validation of the corresponding relation (26.116) with the expression (26.115)]. The scale T_{cr} is a crossover temperature analogous to the Kondo temperature in Kondo models. For $K < 1$, the series

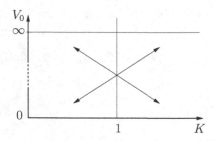
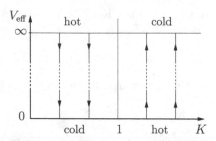

Figure 26.1: Regarding the TB lattice as the large barrier limit of the WB model, both models can be represented in a single diagram of V_0 versus K. The duality maps a WB model with small V_0 and Kondo parameter $K \lesssim 1$ to a TB model with small Δ (large V_0) and $K \gtrsim 1$ (left diagram). In the right diagram, the flow of the effective barrier height with decreasing temperature is sketched for the linear mobility. For $K < 1$ the system flows towards a vanishing barrier whereas it becomes localized for $K > 1$ at zero temperature. For the nonlinear mobility at $T = 0$, the flows as a function of the bias are similar.

(26.97) absolutely converges for $T > T_{\mathrm{cr}}$, and the series (26.98) for $T < T_{\mathrm{cr}}$. For $K > 1$, the regions of convergence of these two series are interchanged.

For $K < 1$, we have the following picture holding both for the TB and WB model. At $T \gg T_{\mathrm{cr}}$, we are in the perturbative regime of the series (26.97), the $m = 1$ term being the leading one. As T is lowered, the barrier gets effectively weaker so that higher-order terms of the series become increasingly important. Physically, this means that coherent tunneling through two, three, $\cdots$, many barriers give relevant contributions to the mobility. At $T = T_{\mathrm{cr}}$, the region of convergence of the series (26.97) is left. For $T < T_{\mathrm{cr}}$, the asymptotic series (26.98) applies. As temperature is lowered further, convergence of the series is improved. At $T = 0$, all terms $m \geq 1$ have dropped to zero, and we are left with $\mu_{\mathrm{lin}} = \mu_0$. Thus by cooling down the system from high to zero temperature, the originally strong barrier gradually fades away.

In the characteristic behavior for $K > 1$, high and low temperatures are interchanged. Thus, as T approaches zero, the barrier becomes infinitely high and the particle is localized. The pertinent flows of the effective barrier height are depicted in Fig. 26.1 (right).

26.4.2 Nonlinear mobility at $T = 0$

Similar scaling behavior is found for the nonlinear mobility at $T = 0$. In this case, the bias takes the role of temperature. By counting dimensions in the series expressions (25.76) and (26.93), we see that the Kondo energy scale is given in terms of the parameters of the TB and WB model as

$$\epsilon_0 = b_{\mathrm{TB}}(K)\,\Delta^{1/(1-K)}\omega_c^{-K/(1-K)} = b_{\mathrm{WB}}(K)(V_0/\hbar)^{K/(K-1)}\omega_c^{-1/(K-1)} \ . \quad (26.101)$$

It is now convenient to introduce the Kondo-scaled dimensionless bias frequency

$$v = \epsilon/\epsilon_0 . \tag{26.102}$$

Since v is chosen invariant, while ϵ maps on ϵ/K in the duality transfromation (26.29), we must have in difference to the relation (26.95)

$$b_{\mathrm{WB}}(K) = K\, b_{\mathrm{TB}}(1/K) . \tag{26.103}$$

Explicit expressions for the prefactors $b_{\mathrm{TB}}(K)$ and $b_{\mathrm{WB}}(K)$, which satisfy the relation (26.103), can be extracted from the relations given below in Eqs. (26.117) and (26.118).

With the scaled bias v, the TB series (25.76) for the dimensionless nonlinear mobility at zero temperature, $\mathcal{M}(v, K) \equiv \mu(T = 0, \epsilon, K)/\mu_0$, takes the scaling form

$$\mathcal{M}(v, K) = \sum_{m=1}^{\infty} c_m(K)\, v^{2(K-1)m} . \tag{26.104}$$

Likewise, the WB series (26.98) at $T = 0$ can be written as

$$\mathcal{M}(v, K) = 1 - \sum_{m=1}^{\infty} c_m(1/K)\, v^{2(1/K-1)m} . \tag{26.105}$$

Thus we find again nonperturbative self-duality, but now of the form

$$\mathcal{M}(v, K) = 1 - \mathcal{M}(v, 1/K) . \tag{26.106}$$

In the Coulomb gas representation, the coefficient $c_m(K)$ is again given in terms of a $(2m-1)$-fold integral. The convergence radius of the expansions (26.104) and (26.105) is determined by the critical frequency

$$\epsilon_{\mathrm{cr}}(K) \equiv \lim_{m \to \infty} \left| c_m(K) \right|^{1/[2(1-K)m]} \epsilon_0 = \epsilon_{\mathrm{cr}}(1/K) . \tag{26.107}$$

For $K < 1$, the series (26.104) converges absolutely for $\epsilon > \epsilon_{\mathrm{cr}}$, wheras the series (26.105) converges for $\epsilon < \epsilon_{\mathrm{cr}}$. Again for $K > 1$, the regions of convergence are interchanged.

The dual series expressions (26.104) and (26.105) are again the Taylor and asymptotic expansion of the same physical quantity, and vice versa, depending on whether K is smaller or larger than one and whether the bias ϵ is larger or smaller than ϵ_{cr}. They hold both for the TB and WB model. Upon utilizing the self-duality property of the model, we determine in Subsection 26.5.1 the coefficients $c_m(K)$ for general K. There, we also relate the dressed frequencies ϵ_0 and ϵ_{cr} to the parameters of the original WB and TB model. Finally, the relation (26.101) will permit to express the parameters of the one model in terms of those of the other [see Eq. (26.120)].

26.5   Exact scaling function at $T = 0$ for arbitrary K

26.5.1   Construction of the self-dual scaling solution

The dual expansions (26.104) and (26.105) can be derived from the contour integral representation [519]

$$\mathcal{M}(v, K) = \frac{1}{2\pi i} \int_C dz\, P(z) F(z)\, v^{2z} . \qquad (26.108)$$

Here we design the function $P(z)$ such that it provides the analytic structure of the power series (26.104) and (26.105),

$$P(z) = \frac{1}{z} \Gamma\left(1 + \frac{zK}{K-1}\right) \Gamma\left(1 + \frac{z}{1-K}\right) . \qquad (26.109)$$

The function $P(z)$ is analytic over the entire complex plane save for the points

$$
\begin{aligned}
z &= 0 , \\
z &= z_n^{(1)} \equiv (1/K - 1)n , & n &= 1, 2, \ldots , \qquad (26.110) \\
z &= z_m^{(2)} \equiv (K - 1)m , & m &= 1, 2, \ldots ,
\end{aligned}
$$

where it possesses simple poles. The function $F(z)$ serves as an adaptive function. We require that $F(z)$ is an analytic function on the entire complex z-plane. The contour C starts at infinity, then circles the origin such that the pole at $z = 0$ and the set of poles $\{z_n^{(1)}\}$ lie to the left, and the set of poles $\{z_m^{(2)}\}$ lie to the right of the integration path, and finally returns to infinity (cf. Fig. 26.2). We close the contour, in which we require that the integrand in Eq. (26.108) tends to zero faster than $1/|z|$ on the chosen semicircle. In the regime $(\epsilon_{\text{cr}}/\epsilon)^{2(1-K)} < 1$, the contour is closed such that the set of poles $\{z_m^{(2)}\}$ is circled. The resulting series expression is in the form (26.104), and the coefficients are given by

$$c_m(K) = (-1)^{m-1} \Gamma(Km+1)\, F[(K-1)m]/\Gamma(m+1) . \qquad (26.111)$$

For $(\epsilon_{\text{cr}}/\epsilon)^{2(1-K)} > 1$, the contour is closed in the opposite direction of rotation, so that the pole at $z = 0$ and the set of poles $\{z_n^{(1)}\}$ are circled. We find upon comparing the corresponding residua with the terms of the series (26.105) that $F(z)$ is independent of K at the origin, $F(0) = 1$, and the coefficients with $m \geq 1$ are given by Eq. (26.111) with $1/K$ substituted for K.

The entire analytic structure of the integral representation (26.108) required by the self-duality property is carried by the function $P(z)$. The only additional requirement imposed by self-duality is that the yet unknown function $F(z)$ is invariant under the substitution $K \to 1/K$. It is tempting to assume that self-duality at zero temperature is provided by a "minimal theory" in the sense that the whole dependence on the coupling consistant K is determined by the analytic properties of the function $P(z)$. Thus, in a minimal theory, the function $F(z)$ does not depend on K at all. Under this condition it is possible to determine $F(z)$ provided that the series (26.104)

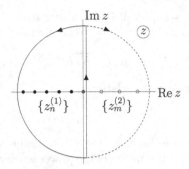

Figure 26.2: Sketch of the integration path $\mathcal{C}$ in the contour integral (26.108) for $K > 1$. Closing the contour in the counter-clockwise and clockwise sense gives the series (26.105) and (26.104), respectively. The respective poles and contour for $K < 1$ ensue by reflection at the origin.

or (26.105) is known in analytic form for a particular value of K different from unity. Fortunately, the TB mobility is known in analytic form for the case $K = \frac{1}{2}$ and in the limit $K \to 0$ as discussed in Subsections 25.5.2 and 25.6.1. The latter case is equivalent to the classical limit $K \to \infty$ in the WB model. Indeed, anyone of these special cases may be employed in order to determine the function $F(z)$.

To obtain $F(z)$, we equate the terms of the series (25.93) with those of the series (26.104) for $K = 0$ and use Eq. (26.111). Alternatively, we may match the power series of (26.34) with the series (26.105) for $K \to \infty$. In both ways, we unambiguously get

$$F(z) = \frac{\Gamma(\frac{3}{2})}{\Gamma(z + \frac{3}{2})} . \tag{26.112}$$

The matching function (26.112) is an analytic function in the entire z-plane, as postulated. Thus, we have established in fact an exact contour integral representation of the scaling function for general K at zero temperature [519],

$$\mathcal{M}(v, K) = \frac{1}{2\pi i} \int_{\mathcal{C}} dz \, \frac{1}{z} \Gamma\left(1 + \frac{zK}{K-1}\right) \Gamma\left(1 + \frac{z}{1-K}\right) \frac{\Gamma(\frac{3}{2})}{\Gamma(z + \frac{3}{2})} v^{2z} , \tag{26.113}$$

which is a typical Mellin-Barnes integral [363].

With the form (26.112) for $F(z)$, the series coefficient (26.111) is found to read

$$c_m(K) = \frac{(-1)^{m-1}}{m!} \frac{\sqrt{\pi}}{2} \frac{\Gamma(Km+1)}{\Gamma[(K-1)m + \frac{3}{2}]} . \tag{26.114}$$

The series expressions (26.104) and (26.105) with (26.114) constitute the exact solution for the mobility for general values of the parameters K and v at zero temperature.

Readily the crossover scale (26.107) delimiting the regions of convergence of the expansions (26.104) and (26.105) is found to be given by

$$\epsilon_{\mathrm{cr}}(K) \; = \; \sqrt{|1-K|}\, K^{K/[2(1-K)]}\, \epsilon_0 \; . \qquad (26.115)$$

This form satisfies for fixed ϵ_0 the relation

$$\epsilon_{\mathrm{cr}}(K) \; = \; \epsilon_{\mathrm{cr}}(1/K) \; . \qquad (26.116)$$

For instance, we find in the limit $K \to \infty$ the radius $\epsilon_{\mathrm{cr}} = 2\pi V_0/\hbar$. This is in agreement with the convergence radius of the power series resulting from Eq. (26.34).

So far we have discussed the nonlinear mobility detached from the particular model. To relate the frequency scale ϵ_0 of the scaling solution to the parameters of the WB model, we equate the $m = 1$ term of the series (26.105) with the $m = 1$ term of the series (26.93) at zero temperature. We then obtain

$$\epsilon_0^{2-2/K} \; = \; K^{2-2/K} 2^{2-2/K} [\pi/\Gamma(1/K)]^2 (V_0/\hbar)^2 \omega_c^{-2/K} \; . \qquad (26.117)$$

Likewise, we can express ϵ_0 in terms of the parameters of the TB model. Equating the $m = 1$ term of the series (26.104) with the expression (25.79), we find

$$\epsilon_0^{2-2K} \; = \; 2^{2-2K} [\pi/\Gamma(K)]^2 \Delta^2 \omega_c^{-2K} \; . \qquad (26.118)$$

The contour integral (26.113) with (26.117) and with (26.118) is an exact representation of the nonlinear dc-mobility of the WB model and TB model, respectively, holding at $T = 0$ in the scaling limit for general bias ϵ and general friction K.

The expressions (26.117) and (26.118) are symmetric under substitution $K \to 1/K$ except for the factor $K^{2-2/K}$ in (26.117). This factor appears because of the correspondence $\epsilon \leftrightarrow K\epsilon$ in the duality relation (26.29) [see relation (26.103)].

A renormalized frequency scale $\tilde{\epsilon}_0$, which is symmetric with respect to the TB and WB model, is $\tilde{\epsilon}_0 = a\, K^{1/(K-1)} \epsilon_0$, where a is an arbitrary positive constant. With the convenient choice $a = 1/8$ we obtain

$$\tilde{\epsilon}_0^{2-2K} \; = \; 2^{4K-4} \frac{\pi^2}{\Gamma^2(1+K)} \frac{\Delta^2}{\omega_c^{2K}} \; , \qquad \tilde{\epsilon}_0^{2-\frac{2}{K}} \; = \; 2^{\frac{4}{K}-4} \frac{\pi^2}{\Gamma^2(1+\frac{1}{K})} \frac{(V_0/\hbar)^2}{\omega_c^{2/K}} \; . \qquad (26.119)$$

At this point we wish to emphasize the following conclusions [519]:

(1) With use of the relations (26.117) and (26.118) [or equivalently the relations (26.119)] and elimination of $\tilde{\epsilon}_0$, we can express Δ in terms of V_0,

$$\Delta \; = \; [\Gamma(1+K)][\Gamma(1+1/K)]^K (\omega_c/\pi)^{1+K} (V_0/\hbar)^{-K} \; . \qquad (26.120)$$

Resolution of this relation for V_0 gives the same functional form except that $1/K$ is substituted for K. It is interesting to compare the expression (26.120) with the corresponding relation (17.72) obtained by use of the bounce method for large K. Employing Stirling's asymptotic formula for $\Gamma(K)$ and identifying the cutoff ω_c with the damping frequency γ, the expression (26.120) indeed coincides with the result (17.72) of the bounce method. However the form (26.120) is universal since it holds for all K. Interestingly enough, here we have found the single-bounce (and all multi-bounce contributions) to the TB mobility solely by use of duality and analytic properties, without calculating bounce actions and fiddly fluctuation determinants.

(2) Besides duality and analytic properties we have used in the construction of the scaling solution only knowledge of the classical mobility. From this we infer that the entire quantum regime of this transport problem is fully determined by self-duality.

(3) One should expect that also the scaling function $\mathcal{M}_{\text{lin}}(\vartheta, K)$ for the *linear* mobility can be found from the above "minimal" assumptions. However, a formidable practical complication arises: an ansatz equivalent to (26.108) does not work since one finds from the $m = 1$ term that the respective adaptive function $F(z)$ is non-analytic in the complex z-plane.

(4) The existence of self-duality is a direct consequence of integrability in the equivalent boundary sine-Gordon model [520]. Integrability determines the low-energy properties of this model and shows itself in the above scaling behavior.

26.5.2   Supercurrent-voltage characteristics at $T = 0$ for arbitrary ρ

We have seen in Section 26.3 that the charge-phase duality in the Josephson junction dynamics is an exact self-duality for strictly Ohmic damping. Thus, with the correspondence relations (20.175) - (20.177), the findings of the previous subsection can directly be applied to Cooper pair tunneling in a resistive environment. The supercurrent-voltage characteristics at $T = 0$ and arbitrary $\rho = R/R_{\text{Q}}$ takes the scaling-invariant form [518]

$$I(V_{\text{x}}) = \mathcal{M}(v, \rho)\frac{V_{\text{x}}}{R} . \tag{26.121}$$

Here, $\mathcal{M}(v, \rho)$ is the scaling function just discussed, but now it represents the normalized conductance of the Josephson contact. The various energy scales of the system are aggregated in the scaling variable

$$v = \frac{eV_{\text{x}}}{\pi E_{\text{J}}}\left[\Gamma(\rho)\left(\frac{e^{\varsigma}}{\pi^2\rho}\frac{E_{\text{c}}}{E_{\text{J}}}\right)^{\rho}\right]^{\frac{1}{1-\rho}} . \tag{26.122}$$

The current (26.121) with the leading term of the weak Cooper-pair-tunneling series (26.104) represents the Coulomb blockade expression (20.181). On the other hand, the current with the leading terms of the dual strong-tunneling series (26.105) for $\rho \ll 1$ agrees with the expression (17.79).

26.5.3   Connection with Seiberg-Witten theory

It is possible to establish an interesting connection of the duality with supersymmetric Seiberg-Witten theory [521]. Consider the series (26.105) with (26.114) and express

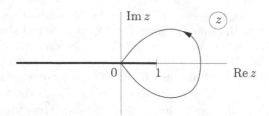

Figure 26.3: The path of integration C in Eq. (26.123) starts at the origin, circles around the branch point at $z = 1$ in the counter-clockwise sense, and returns to the origin.

the ratio of gamma functions in the coefficient $c_m(1/K)$ as a contour integral taken along the path in the complex z-plane as sketched in Fig. 26.3 [522],

$$\frac{\Gamma(x)}{\Gamma(x+y)} = \Gamma(1-y)\frac{1}{2\pi i}\int_C dz\, z^{x-1}\,(z-1)^{y-1}\,. \tag{26.123}$$

Since the series (26.105) is absolutely convergent within the circle of convergence, we can interchange the order of integration and summation. This yields

$$\mathcal{M}(v,K) = \frac{1}{4\sqrt{\pi}i}\int_C dz\,\frac{1}{\sqrt{z-1}}\left\{\sum_{n=0}^{\infty}\frac{(-1)^n}{n!}\Gamma(n+\tfrac{1}{2})\left(\frac{z^{1/K}v^{2/K-2}}{z-1}\right)^n\right\}\,. \tag{26.124}$$

The series in the curly bracket can be summed to a square root. With the substitution $z = t/v^2$, we finally obtain for the scaling function the integral representation [520]

$$\mathcal{M}(v,K) = \frac{1}{4vi}\int_{C'} dt\,\frac{1}{\sqrt{t+t^{1/K}-v^2}}\,. \tag{26.125}$$

The expression (26.125) is analytic in K and therefore holds for all K. One can now prove with this form directly the self-duality relation (26.106) without recourse to the series expansions. With the substitution $t^K = x$ in Eq. (26.125), we obtain

$$M(v,1/K) = \frac{1}{4vi}\int_{C'} dx\,\frac{1}{\sqrt{x+x^{1/K}-v^2}}\left[\left(\frac{x^{1/K-1}}{K}+1\right)-1\right]\,, \tag{26.126}$$

where C' changes accordingly. The integrand with the round bracket term is a total derivative which simply yields unity in Eq. (26.99), whereas the residual integral contribution is $-\mathcal{M}(v,K)$. Thus the contour integral (26.125) satisfies the self-duality relation (26.106).

The representation (26.125) bears resemblance with representations for mass gaps in SU(2) supersymmetric Yang-Mills theory [521]. The parameter v corresponds to an order parameter related to the expectation value of the Higgs field. A geometrical picture in terms of tori for the form (26.125) is discussed in Ref. [523].

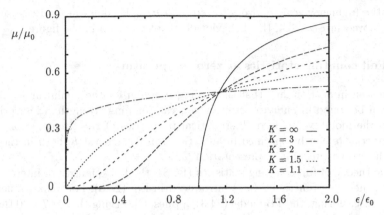

Figure 26.4: The normalized mobility μ/μ_0 at $T = 0$ is plotted as a function of ϵ/ϵ_0 for different K. In the regime $\epsilon < \epsilon_{cr}$ and $K > 1$, the weak-tunneling (TB) series (26.104) with coefficient (26.114) converges, whereas for $\epsilon > \epsilon_{cr}$ and $K > 1$ the strong-tunneling (WB) series (26.105) with coefficient (26.114) converges. The curve for $K = \infty$ shows the square root singularity of the classical case, Eq. (26.34). For finite K, the mobility is nonzero in the bias regime $\epsilon/\epsilon_0 < 1$ due to quantum mechanical tunneling.

26.5.4 Special limits

At $K = \frac{1}{2}$, the renormalized frequency scale is $\epsilon_0 = 2\tilde{\epsilon}_0 = 2\pi\Delta^2/\omega_c$. The scaling function $\mathcal{M}(v, K = \frac{1}{2})$ reproduces previous results given in Subsection 25.6.1. Namely, for $K = \frac{1}{2}$ the series (26.105) is summed to the first form given in Eq. (25.103), wheras the dual series (26.104) gives the second form in Eq. (25.103). In the narrow regime $K = 1/2 - \kappa$, with $|\kappa| \ll 1$, the scaling function is consistent with the former result (25.110) at $T = 0$.

For $K = 1 + \kappa$ with $|\kappa| \ll 1$, the series (26.105) is summed to the expression

$$\mu_{\mathrm{WB}}(\epsilon, K = 1 + \kappa)/\mu_0 = 1/[1 + (\pi V_0/\hbar\omega_c)^2 (2\omega_c/\epsilon)^{2\kappa}], \qquad (26.127)$$

where we have employed the relation (26.117). From this we see that the normalized mobility in the zero-bias limit is unity for $K < 1$ and zero for $K > 1$. The form (26.127) agrees with a result obtained by a leading-log summation in a related impurity scattering problem (discussed below in Section 28.1) with Luttinger parameter $g = 1 - \kappa$, where $g = 1/K$ [524].

The normalized mobility $\mu(\epsilon)/\mu_0$ is plotted in Fig. 26.4 as a function of ϵ/ϵ_0 for different values of the damping parameter K. The mobility shows a smooth transition from the square root singular behavior (26.34) in the classical limit $K \to \infty$ via the kink-like shape $\mu/\mu_0 = 1 - (\epsilon_0/2\epsilon)\arctan(2\epsilon/\epsilon_0)$ at $K = 2$ to the constant behavior (26.127) at $K = 1$. The curves for $K < 1$ are found from the self-duality relation (26.106). All curves cross the line $\mu/\mu_0 = 1/2$ in the interval $\epsilon/\epsilon_0 = 2/\sqrt{3} \pm 0.01$. As

$K \to 1$, the frequency scale ϵ_0 drops to zero and μ/μ_0 becomes unity. Because of the scale employed in Fig. 26.4, this behavior does not show up in the figure.

26.6   Full counting statistics at zero temperature

We have seen in Section 26.5 that the scaling function for the nonlinear mobility at $T = 0$ can be found in analytic form for general damping strength K. Surprisingly, not only the mobility but also all statistical properties of the transport process at zero temperature can be specified in analytic form for general K both in the weak- and in the strong-tunneling representation [525].

In the theory of full counting statistics (FCS), the key quantity of interest is the moment generating function $\mathcal{Z}(\rho, t)$ introduced in Eq. (25.3). For weak tunneling or strong backscattering, the TB model (24.3) applies. Observing that at $T = 0$ there are no transition from the energetically lower to the higher state, $k_n^- = 0$, the expression (25.22) reduces to

$$\ln \mathcal{Z}(\rho, t) = t \sum_{n=1}^{\infty} \left(e^{i\rho q_0 n} - 1 \right) k_n^+ . \qquad (26.128)$$

This form can be discovered indeed from a systematic cluster decomposition of the series expression (25.40) in the limit of very large t. The clusters are irreducible path segments between diagonal states of the RDM.[3] In Laplace space, an irreducible cluster becomes independent of the Laplace variable λ, as $\lambda \to 0$. By definition, the clusters are noninteracting, and therefore each time interval which separates neighboring clusters gives a factor $1/\lambda$. Path segments with intermediate visit of a diagonal state have a reducible component which factorizes into two clusters of lower order times a factor $1/\lambda$. After subtraction of the reducible component an irreducible part remains. The transition rates $k_n^{\pm}$ can be identified as the sum of all irreducible clusters which interpolate between the (arbitrary) diagonal state m and the diagonal state $m \pm n$. The clusters are formally given in terms of a power series in the number of tunneling transitions. In the Ohmic scaling limit at $T = 0$, the tunneling matrix element Δ is merged with the bias ϵ and the cut-off ω_c into the dimensionless expansion parameter

$$u = \frac{\Delta}{\epsilon} \left(\frac{\epsilon}{\omega_c} \right)^K \propto v^{K-1} . \qquad (26.129)$$

The rate k_n^+ is given in terms of the series

$$k_n^+ = \sum_{\ell=n}^{\infty} k_{n,\ell}^+ \qquad \text{with} \qquad k_{n,\ell}^+ = R_{n,\ell}^+ u^{2l} . \qquad (26.130)$$

The partial rate $k_{n,\ell}^+ \propto \Delta^{2\ell}$ covers all irreducible contributions with 2ℓ moves, or 2ℓ charges in the charge representation.

---

[3]The notion "irreducibility" is explained in Subsection 21.2.6.

The irreducible Coulomb integrals have been analyzed in the zero-temperature limit in Ref. [475]. It was found that there are formidable cancellations among the individual Coulomb integrals of the same order: only those charge sequences contribute to the partial rates in which the charges $\{u_j\}$ and $\{v_j\}$ introduced in Eq. (25.27) are all positive. As a result, the partial rate $k_{n,\ell}^+$ is zero when $\ell > n$. Accordingly, the rate k_n^+ is fully determined by the direct paths from well 0 to well n which have $2n$ moves, i.e., the minimal number of moves required in order to advance by n wells,

$$k_n^+ = k_{n,n}^+ \propto \Delta^{2n} . \tag{26.131}$$

Observing that the cumulants (25.23) are defined as moments and using the property (26.131), we immediately obtain the simple cumulant relation

$$\langle q^N(t)\rangle_c/t = \left(q_0 \frac{\Delta}{2}\frac{\partial}{\partial\Delta}\right)^{N-1}\langle I\rangle = \left(q_0 \frac{v}{2(K-1)}\frac{\partial}{\partial v}\right)^{N-1}\langle I\rangle , \tag{26.132}$$

which helps us to reduce the calculation of cumulants to the calculus of derivatives of the current $\langle I\rangle$ with respect to the tunneling coupling Δ. In addition, the central property (26.131) allows us to determine the transition rates k_n^+ unambiguously from one of the cumulants, e.g., from the current. Equating the series for the current $\langle I\rangle = \langle q(t)\rangle/t = q_0 \sum_n n k_n^+$ with the expression $\langle I\rangle = (\epsilon q_0/2\pi K)\mathcal{M}(v, K)$, where $\mathcal{M}(v, K)$ is given by the series (26.104), the rate to advance n sites is found as

$$k_n^+ = \frac{(-1)^{n-1}}{n!}\frac{\Gamma(\frac{3}{2})\Gamma(Kn)}{\Gamma[\frac{3}{2}+(K-1)n]}\frac{\epsilon}{2\pi}\left(\frac{\epsilon}{\epsilon_0}\right)^{(2K-2)n} . \tag{26.133}$$

The expression (26.128) with (26.133) is the weak-tunneling series representation of the full counting statistics for general dissipation parameter K.

When the Poissonian transport process is classical, all the individual transition rates k_n^+ in Eq. (26.128) are positive. The subtle point now is that the rates (26.133) are not. Certainly, the first contribution k_1^+ represents indeed a classical Poisson process for all K (in reality, it is a quantum mechanical tunneling process); but the coherent tunneling through two barriers with transition weight k_2^+ comes with a negative sign when $K > \frac{1}{4}$, which is an effect of quantum coherence. Similar behavior is found for tunneling transition with $n > 2$, depending on the particular value of K.

In the opposite limit of weak backscattering or strong tunneling, the series expression for $\ln \mathcal{Z}(\rho, t)$ reads

$$\ln \mathcal{Z}(\rho, t) = t\left(i\rho q_0 \frac{\epsilon}{2\pi K} + \sum_{n=1}^{\infty}\left(e^{-i\rho q_0 n/K} - 1\right)\tilde{k}_n^+\right) , \tag{26.134}$$

where

$$\tilde{k}_n^+ = \frac{(-1)^{n-1}}{n!}\frac{\Gamma(\frac{3}{2})\Gamma(n/K)}{\Gamma[\frac{3}{2}+(1/K-1)n]}\frac{\epsilon}{2\pi K}\left(\frac{\epsilon}{\epsilon_0}\right)^{(2/K-2)n} . \tag{26.135}$$

It is immediately obvious from these forms together with the relation (26.117) that the cumulant relation in the strong-tunneling representation is

$$\langle q^N(t)\rangle_{\mathrm{c}}/t = \left(-\frac{q_0}{K}\frac{V_0}{2}\frac{\partial}{\partial V_0}\right)^{N-1}\langle I\rangle = \left(-q_0\frac{v}{2(1-K)}\frac{\partial}{\partial v}\right)^{N-1}\langle I\rangle \ . \quad (26.136)$$

The strong-tunneling series (26.134) is quite similar to the weak-tunneling series (26.128), but there are subtle differences. The first term in the expression (26.134) represents the current in the absence of the barrier. The exponential factor $e^{-i\rho q_0 n/K}$ indicates that now we have tunneling over distance q_0/K and multiples thereof, and the minus sign in the exponent means that the tunneling contribution diminishes the current instead of building it up as in the weak-tunneling limit. Using the reflection formula for gamma functions, one finds that the sign of the tunneling rate $\tilde{k}_n^+$ is the one of $\cos(n\pi/K)$ when $K > 2$. Therefore, the perception of a classical Poisson process for the tunneling dynamics is quite appropriate for transition weights with modest n, when K is large. As K goes to infinity, all the transition weights become positive and hence amenable to classical interpretation but the quantum fluctuations $\langle q^n(t)\rangle_{\mathrm{c}}$ with $n > 1$ disappear, as one reaches the classical limit [cf. Eq. (26.34)],

$$\ln \mathcal{Z}(\rho,t) = it\rho\frac{q_0\,\epsilon}{2\pi K}\left[1 - \sum_{n=1}^{\infty}\frac{\Gamma(n-\frac{1}{2})}{2\sqrt{\pi}n!}\left(\frac{\epsilon_0}{\epsilon}\right)^{2n}\right] = it\rho\frac{q_0\,\epsilon}{2\pi K}\sqrt{1-\left(\frac{\epsilon_0}{\epsilon}\right)^2} \ . \quad (26.137)$$

In the classical limit, the Kondo scale ϵ_0 is related to the corrugation strength V_0 of the weak-binding model (24.1) with (24.2) by the relation $\epsilon_0 = 2\pi V_0/\hbar$.

26.7 Low temperature behavior of the characteristic function

For general damping strength K and general temperature, the evaluation of the Coulomb gas integral representation given in Subsection 25.3.1 is not possible. To make progress, one has to employ thermodynamic Bethe ansatz techniques to the related BSG model [526, 527]. In Ref. [528] a general procedure for the calculation of cumulants of any order is sketched and the third moment is explicitly calculated.

Interestingly, the leading low-temperature contribution to the zero temperature cumulant generating function can be calculated in analytic form. To this aim, we work in the Coulomb gas representation. In the first step, we perform the low temperature expansion of the charge interaction,

$$Q'(\tau) = Q'_0(\tau) + K\frac{\pi^2}{3}\left(\frac{\tau}{\hbar\beta}\right)^2 + \mathcal{O}\left[\left(\frac{\tau}{\hbar\beta}\right)^4\right], \quad (26.138)$$

where $Q'_0(\tau) = 2K\ln(\omega_c\tau)$ is the interaction at $T = 0$. Next, we determine the leading low-temperature dependence of the charge interaction factor G_{2m} given in Eq. (25.34). Upon substituting the form (26.138), we obtain

$$G_{2m} = G_{2m}^{(0)}\left\{1 - K\frac{\pi^2}{3}\left(\frac{k_{\mathrm{B}}T}{\hbar}\right)^2\left(\sum_{j=1}^{2m-1}p_j\tau_j\right)^2 + \mathcal{O}(T^4)\right\}, \quad (26.139)$$

where $G_{2m}^{(0)}$ is the full interaction factor at $T = 0$. The crucial point now is that the sum in the curly bracket in Eq. (26.139) can be expressed in terms of the bias phase $\varphi_{2m,0}$ given in Eq. (25.50) as $\sum_j p_j \tau_j = \varphi_{2m,0}/\epsilon$. Hence the square of this annoying sum can be generated by taking the second derivative of the bias factor $B_{2m,0} = e^{i\varphi_{2m,0}}$ with respect to the bias. Consequently, the asymptotic low-temperature expansion of the term $G_{2m} B_{2m,0}$ can be written as

$$G_{2m} B_{2m,0} = G_{2m}^{(0)} \left\{ 1 + K \frac{\pi^2}{3} \left(\frac{k_B T}{\hbar} \right)^2 \frac{\partial^2}{\partial \epsilon^2} + \mathcal{O}(T^4) \right\} B_{2m,0} . \qquad (26.140)$$

Upon employing the relation (26.140), we find that the T^2- contribution to the cumulant generating function can be universally written in terms of the second derivative of the zero temperature expression with respect to the bias. Examination of subleading contributions $\propto T^4$, T^6, $\cdots$ yields that they are nonuniversal, and therefore they can not be expressed in simple terms.

Substituting the form (26.140) into the series expression (25.40), we readily obtain the cumulant generating function in the TB representation in the form (26.128), in which the rate k_n^+ is given by

$$k_n^+(\epsilon, T) = \left\{ 1 + K \frac{\pi^2}{3} \left(\frac{k_B T}{\hbar} \right)^2 \frac{\partial^2}{\partial \epsilon^2} + \mathcal{O}(T^4) \right\} k_n^+(\epsilon, 0) . \qquad (26.141)$$

Observing from the form (26.133) that $k_n^+(\epsilon, 0) \propto \epsilon^{(2K-2)n+1}$, we readily get

$$k_n^+(\epsilon, T) = \left\{ 1 + \frac{\pi^2}{3} K (2K - 2)n \left[1 + (2K - 2)n \right] \left(\frac{k_B T}{\hbar \epsilon} \right)^2 \right\} k_n^+(\epsilon, 0) , \qquad (26.142)$$

where terms of order T^4 are disregarded.

The dual weak-backscattering or strong-tunneling representation of the CGF is found in generalization of the expression (26.134) as

$$\ln \mathcal{Z}(\rho, t) = \tau \left[i\rho \frac{q_0 \epsilon}{2\pi K} - \rho^2 \frac{q_0^2 k_B T}{2\pi K \hbar} + \sum_{n=1}^{\infty} \left(e^{-i\rho n q_0 / K} - 1 \right) \tilde{k}_n^+(\epsilon, T) \right] , \qquad (26.143)$$

where

$$\tilde{k}_n^+(\epsilon, T) = \left\{ 1 + \frac{\pi^2}{3} \frac{(2 - 2K)n \left[K + (2 - 2K)n \right]}{K} \left(\frac{k_B T}{\hbar \epsilon} \right)^2 \right\} \tilde{k}_n^+(\epsilon, 0) . \qquad (26.144)$$

The expressions (26.128) with (26.142) and (26.143) with (26.144) are the dual series representations of the cumulant generating function $\mathcal{Z}(\rho, t)$. It is straightforward to derive from these forms the weak- and strong-tunneling expansions of all cumulants. Observe that the second term in Eq. (26.143) gives the remaining contribution to the second cumulant in the asymptotic limit in which the corrugation is fully irrelevant.

The T^2-contribution to the full counting statistics is a distinctive signature of Ohmic dissipation inherent in the quantum transport process. A related phenomenon is the universal T^2-behavior observed at low temperatures in other open quantum systems subject to Ohmic dissipation, e.g. the T^2-enhancement in macroscopic quantum tunneling (cf. Section 17.2). It is also the origin of the universal Wilson ratio occurring e.g. in Kondo systems and in the related Ohmic two-state system (cf. Subsection 19.2.1 and Section 22.9). Other examples are the T^2 enhancement of the free energy and static susceptibility of the two-state system [cf. Eqs. (22.217) and (22.218) for $s = 1$]. In these cases, the prefactor is determined by the susceptibility at zero temperature and its second derivative, respectively. The physical origin of the T^2 corrections is the low-frequency thermal noise of Ohmic dissipation [359].

We finally remark that with the same strategy one may calculate also finite temperature corrections in the non-Ohmic case $s \neq 1$, e.g. for the nonlinear mobility. It is straightforward to see that in this case the power-law of the leading thermal enhancement is T^{1+s} [220].

27. Twisted partition function and nonlinear mobility

The grand-canonical sum of the real-time Coulomb gas (25.76) for general temperature, bias and Kondo parameter K is a problem that is to hard for analytical solution. There has been proposed a route in which the problem is solved crabwise [477, 530]. It was shown that the grand-canonical sum for the Coulomb gas in imaginary time can be solved in analytic form for twisted partition functions. From these one then obtains the nonlinear mobility with a conjectured relation which involves analytical continuation of winding numbers to the real physical bias. Here we show how the method works. In addition, we confirm results of this approach obtained in particular regions of the parameter space. The regime $K \ll 1$ for general T and ϵ, and the regime $T = 0$ for general ϵ and K have been studied in Ref. [531]. Here we present the method for the TB model (24.3). Translation to the WB model (24.1) is straightforward along the lines given above.

27.1   Solving the imaginary-time Coulomb gas with Jack polynomials

The perturbative series of the partition function is equivalent to the grand-canonical sum of a one-dimensional gas of positive and negative unit charges with overall neutrality. The charges represent the forward and backward moves in the TB lattice. While in the imaginary-time Coulomb gas for the two-state system the charges are lined up alternatingly in succession, in the multi-state TB system with infinitely many states the charges are unordered. In the term of order Δ^{2n}, we have $2n$ charges, and we must integrate over the positions $\{\tau_i'\}$ and $\{\tau_j\}$ of the n negative and n positive charges, respectively, in the period of length $\hbar\beta$. The effects of the environmental coupling are carried by the exponential charge interaction factor $e^{\pm W(\tau)}$, where the pair interaction $W(\tau)$ is given in Eq. (18.56). The grand-canonical series of the TB partition function is

$$
\begin{aligned}
\mathcal{Z}_{\mathrm{TB}} &= 1 + \sum_{n=1}^{\infty} \frac{(\Delta/2)^{2n}}{(n!)^2} \int_0^{\hbar\beta} \prod_{i=1}^{n} \mathrm{d}\tau_i \int_0^{\hbar\beta} \prod_{j=1}^{n} \mathrm{d}\tau_j' \, \exp\left\{ - \epsilon \sum_{k=1}^{n} (\tau_k - \tau_k') \right\} \\
&\times \exp\left\{ \sum_{m>k=1}^{n} [W(\tau_m - \tau_k) + W(\tau_m' - \tau_k')] - \sum_{m,k=1}^{n} W(|\tau_m - \tau_k'|) \right\}.
\end{aligned}
$$

(27.1)

The change of integration variables, $u_i = 2\pi\,\tau_i/\hbar\beta$, maps the Coulomb gas (27.1) with (18.56) on the unit circle. The perturbative series of the partition function (27.1) may be written as a power series in the effective dimensionless fugacity

$$
x = \frac{\hbar\beta\Delta}{2} \left(\frac{2\pi}{\hbar\beta\omega_c} \right)^K.
$$

(27.2)

The resulting series for the partition function is

$$
\mathcal{Z}_{\mathrm{TB}}(x,p) = 1 + \sum_{n=1}^{\infty} x^{2n} \mathcal{I}_{2n}(p),
$$

(27.3)

with the $2n$-fold Coulomb gas integral

$$
\begin{aligned}
\mathcal{I}_{2n}(p) &= \frac{2^{-2Kn}}{(n!)^2} \int_0^{2\pi} \prod_{i=1}^{n} \left(\frac{\mathrm{d}u_i}{2\pi} \frac{\mathrm{d}u_i'}{2\pi} \right) \mathrm{e}^{\mathrm{i}p \sum_i (u_i - u_i')} \\
&\times \left| \frac{\prod_{i<j} \sin(\frac{u_i - u_j}{2}) \sin(\frac{u_i' - u_j'}{2})}{\prod_{i,j} \sin(\frac{u_i - u_j'}{2})} \right|^{2K},
\end{aligned}
$$

(27.4)

where p represents the analytically continued bias,

$$
p = \mathrm{i}q \qquad \text{with} \qquad q = \frac{\hbar\epsilon}{2\pi k_{\mathrm{B}} T}.
$$

(27.5)

The integrand is regular in the parameter regime $K < \frac{1}{2}$. Changing integration variables $z_i = \mathrm{e}^{\mathrm{i}u_i}$ and $z_i' = \mathrm{e}^{\mathrm{i}u_i'}$ we get

$$
\begin{aligned}
\mathcal{I}_{2n}(p) &= \frac{1}{(n!)^2} \oint \prod_i \left(\frac{\mathrm{d}z_i}{2\mathrm{i}\pi z_i} \frac{\mathrm{d}z_i'}{2\mathrm{i}\pi z_i'} \right) \left(\frac{z_1 \cdots z_n}{z_1' \cdots z_n'} \right)^p \\
&\times \frac{[\Delta(z)\overline{\Delta(z)}]^K [\Delta(z')\overline{\Delta(z')}]^K}{\prod_{i,k}[(1 - z_i \bar{z}_k')(1 - z_k' \bar{z}_i)]^K},
\end{aligned}
$$

(27.6)

where $\Delta(z) = \prod_{i<k}(z_i - z_k)$ is the n-variable Vandermonde determinant.

The multiple integrals in (27.6) are regular in the regime $K < \frac{1}{2}$. Unfortunately, they can not be evaluated for general K and for general complex p. To advance, we must put K rational and p integer. Integer p may be regarded as a winding number due to a magnetic charge located at the origin. For rational K, we may expand the integrand in terms of Jack polynomials [477, 532]

$$\prod_{i,j} \frac{1}{(1 - r_i s_j)^K} = \sum_\lambda b_\lambda(K) P_\lambda(r, K)\, P_\lambda(s, K)\,. \tag{27.7}$$

Here, the function $P_\lambda(r, K)$ is a symmetric polynomial in the set of variables $(r_1, r_2, \cdots, r_n)$, and $\lambda = (\lambda_1, \lambda_2, \cdots, \lambda_n)$ is a partition of an integer with order $\lambda_1 \le \lambda_2 \le \cdots \le \lambda_n$. For positive integer p, we have

$$(z_1 \cdots z_n)^p\, P_\lambda(z, K) = P_{\lambda+p}(z, K)\,, \tag{27.8}$$

where $\lambda+p$ means the partition λ where p columns of length n have been added in the respective Young tableau. The multiple integrals in (27.6) can be executed by use of the orthogonality relation of the Jack polynomials, since it involves the Vandermonde determinant. In the end, one arrives at the totally ordered n-fold series expression

$$\mathcal{I}_{2n}(p) = \sum_{m_n=0}^{\infty} \sum_{m_{n-1}=0}^{m_n} \cdots \sum_{m_1=0}^{m_2} \prod_{j=1}^{n} e_j(m_j) \tag{27.9}$$

with

$$e_j(m) = \frac{1}{\Gamma^2(K)} \frac{\Gamma(jK + m)}{\Gamma(1 - K + jK + m)} \frac{\Gamma(jK + p + m)}{\Gamma(1 - K + jK + p + m)}\,. \tag{27.10}$$

It has been suggested in Ref. [530] to use (27.3) with (27.9) and (27.10) to define $\mathcal{Z}_{\mathrm{TB}}(x, p)$ for complex p. Furthermore, it has been conjectured that this is the unique analytic continuation. Below, we show that the conjecture holds in various limiting cases where analytic expressions are also available by different methods. Especially informative is the limit $|p| \to \infty$ in which the free energy is found to have the proper low-temperature behavior in all orders of Δ.

Consider now the cumulant expansion

$$\mathcal{F}_{\mathrm{TB}}(x, p) = -k_{\mathrm{B}}T \ln \mathcal{Z}_{\mathrm{TB}}(x, p) = -k_{\mathrm{B}}T \sum_{n=1}^{\infty} x^{2n}\, \mathcal{C}_{2n}(p)\,. \tag{27.11}$$

The first three cumulant coefficients are

$$\begin{aligned}
\mathcal{C}_2(p) &= \mathcal{I}_2(p)\,, \\
\mathcal{C}_4(p) &= \mathcal{I}_4(p) - \tfrac{1}{2}\mathcal{I}_2^2(p)\,, \\
\mathcal{C}_6(p) &= \mathcal{I}_6(p) - \mathcal{I}_2(p)\,\mathcal{I}_4(p) + \tfrac{1}{3}\mathcal{I}_2^3(p)\,.
\end{aligned} \tag{27.12}$$

For the analysis of the limit $|p| \to \infty$ in the cumulant expressions, it will be pivotal to rearrange the multiple sums in $\mathcal{C}_{2n}(p)$ into totally ordered sums as in (27.9). This leads to the split-up

$$\mathcal{C}_{2n}(p) = \sum_{m=1}^{n} \mathcal{C}_{2n}^{(m)}(p)\,, \tag{27.13}$$

where $\mathcal{C}_{2n}^{(m)}(p)$ is an m-fold ordered sum. The first cumulant is

$$\mathcal{C}_2(p) = \sum_{j=0}^{\infty} e_1(j) . \tag{27.14}$$

The second cumulant has two contributions,

$$
\begin{aligned}
\mathcal{C}_4^{(1)}(p) &= \frac{1}{2} \sum_{j=0}^{\infty} e_1^2(j) , \\
\mathcal{C}_4^{(2)}(p) &= \sum_{j=0}^{\infty} \sum_{k=0}^{j} \big[e_2(j) - e_1(j) \big] e_1(k) .
\end{aligned}
\tag{27.15}
$$

The multiple integral in Eq. (27.4) diverges at short distances when $K \geq 1/2$. Correspondingly, the multiple series (27.9) for the coefficient $\mathcal{I}_{2n}(p)$ diverges in the regime $K \geq 1/2$ for all n. Interestingly, the coefficient $\mathcal{C}_{2n}(p)$ in the cumulant series (27.11) is regular for K-values even when $\mathcal{I}_{2n}(p)$ is singular. This is because some of the divergences in the Coulomb integrals are cancelled when taking the connected part $\mathcal{C}_{2n}(p)$. The analysis of the ordered m-fold sums $\mathcal{C}_{2n}^{(m)}(p)$ in (27.13) shows that the cumulant coefficient $\mathcal{C}_{2n}(p)$ is nonsingular in the range $0 < K < 1 - \frac{1}{2n}$.

27.2 Nonlinear mobility

In Ref. [530] a conjecture was made which relates the nonlinear conductance directly to the partition function $\mathcal{Z}(x,p)$ or to the free energy $\mathcal{F}(x,p)$. The conjecture for the conductance of the TB model (24.3) is

$$
\begin{aligned}
\mathcal{M}_{\mathrm{TB}}(x,q) &= \frac{K}{2p} x \frac{\mathrm{d}}{\mathrm{d}x} \ln \left(\frac{\mathcal{Z}_{\mathrm{TB}}(x,-p)}{\mathcal{Z}_{\mathrm{TB}}(x,p)} \right) \bigg|_{p=iq} \\
&= \frac{K}{2p} \frac{x}{k_{\mathrm{B}}T} \frac{\mathrm{d}}{\mathrm{d}x} \big(\mathcal{F}_{\mathrm{TB}}(x,p) - \mathcal{F}_{\mathrm{TB}}(x,-p) \big) \bigg|_{p=iq} ,
\end{aligned}
\tag{27.16}
$$

where $q = \beta\hbar\epsilon/2\pi$. The conjecture is based on the fact that the free energy is real only for p integer, and that the continuation of $C_{2n}(p)$ is not even in p, so that the free energy acquires an imaginary part when p is complex.

Use of the perturbative cumulant expansion (27.11) in the relation (27.16) yields the weak-tunneling series of the conductance,

$$\mathcal{M}_{\mathrm{TB}}(x,q) = \sum_{n=1}^{\infty} \mathcal{M}_n(x,q) . \tag{27.17}$$

The perturbative tunneling contribution of order x^{2n} to the nonlinear mobility is

$$\mathcal{M}_n(x,q) = \frac{K}{iq} n \big[\mathcal{C}_{2n}(-iq) - \mathcal{C}_{2n}(iq) \big] x^{2n} . \tag{27.18}$$

We now study the conjecture in diverse regimes of the parameter space.

27.2.1 Strong barrier limit

The series that defines $C_2(p)$ can be summed in analytic form,

$$
\begin{aligned}
C_2(p) &= \frac{1}{\Gamma^2(K)} \sum_{j=0}^{\infty} \frac{\Gamma(K+j)\,\Gamma(K+p+j)}{\Gamma(1+j)\,\Gamma(1+p+j)} \\
&= \frac{\sin(\pi K)}{\pi} \frac{\Gamma(1-2K)\Gamma(K+p)}{\Gamma(1-K+p)} .
\end{aligned}
\tag{27.19}
$$

With this form the perturbative mobility (27.18) in order Δ^2 is found as

$$
\mathcal{M}_1 = K \left(\frac{\beta\hbar\Delta}{2}\right)^2 \left(\frac{\beta\hbar\omega_c}{2\pi}\right)^{-2K} \frac{\sinh(\beta\hbar\epsilon/2)}{\beta\hbar\epsilon/2} \frac{|\Gamma(K+i\,\beta\hbar\epsilon/2\pi)|^2}{\Gamma(2K)} .
\tag{27.20}
$$

The expression (27.20) coincides with the result (25.78) obtained above in the non-equilibrium real-time approach.

Let us study the equivalence in some detail. The $m=1$-term of the series (25.75) can be written as

$$
\mathcal{M}_1 = \frac{2K}{p}\, x^2 \sin(\pi K) \left[J_1(K,-p) - J_1(K,p) \right] ,
\tag{27.21}
$$

where $J_1(p)$ is the real-time noise correlation integral

$$
J_1(K,p) = \frac{1}{\pi}\int_0^{\infty} du\, \frac{e^{-2pu}}{(2\sinh u)^{2K}} = \frac{\Gamma(1-2K)\,\Gamma(K+p)}{2\pi\,\Gamma(1-K+p)} .
\tag{27.22}
$$

Eq. (27.21) with (27.22) is in agreement with the expression (27.20).

On the other hand, the conjectured relation (27.16) with (27.3) and (27.4) yields for the mobility in order x^2

$$
\mathcal{M}_1 = \frac{K}{p}\, x^2 \left[H_1(K,-p) - H_1(K,p) \right] ,
\tag{27.23}
$$

where $H_1(K,p)$ is the imaginary-time noise integral[1]

$$
H_1(K,p) = \frac{1}{\pi}\int_0^{\pi} dv\, \frac{e^{2ipv}}{(2\sin v)^{2K}} .
\tag{27.24}
$$

The analytically continued integrand of (27.22), $e^{-2pz}/(\sinh z)^{2K}$, is free of singularities in the half-strip $0 < \mathrm{Re}\,z < \infty$, $0 > \mathrm{Im}\,z > -\pi$ of the complex plane $z = u - i\,v$. Accordingly, the closed contour integral along the edges of the half-strip vanishes. This entails that the noise integrals $H_1(K,p)$ and $J_1(K,p)$ are related by

$$
H_1(K,p) = 2\,e^{ip\pi} \sin[\pi(p+K)]\, J_1(K,p) .
\tag{27.25}
$$

For integer p, this relation reduces to

---

[1] The imaginary-time noise integral $H_1(K,p)$ coincides for integer p with the cumulant $C_2(p)$.

$$H_1(K,p) \; = \; 2\sin(\pi K)\, J_1(K,p)\,, \qquad p \in \text{integers}\,. \tag{27.26}$$

Thus the expressions (27.23) and (27.21) are equal for integer p. However, it is clear that the result (27.21) of the real-time calculation is valid for complex p. From this we can now infer that the expression (27.23), which is based on the thermodynamic method and has been calculated for integer p, yields the correct mobility in order x^2 by analytical continuation to the physical bias $p = i\,\beta\hbar\epsilon/2\pi$.

Unfortunately, the real-time approach is not practicable for calculation of terms $\mathcal{M}_n(x,q)$ with $n \geq 2$ in the series (27.17), since the relevant multiple Coulomb integrals for charges distributed on the forward/backward paths cannot be carried out in analytic form. The loophole now is to take the imaginary time-route: First one calculates the free energy for integer p, as outlined in Section 27.1. Then one employs the conjectured relation (27.16) and performs analytical continuation to complex p. Let us now see whether this line of argument is valid beyond the term $\mathcal{M}_1(x,q)$.

27.2.2 The case $K \ll 1$

In the regime $K \ll 1$, the leading contribution to the coefficient $\mathcal{I}_{2n}(p)$ in the multiple series (27.9) is the term with $m_1 = m_2 = \cdots = m_n = 0$. In addition, the term $c_j(0)$ given in (27.10) reduces to the form $e_j(0) = 1/[\,j\,(j+p/K)\,]$. With these truncations the series coefficient $\mathcal{I}_{2n}(p)$ takes the concise form

$$\mathcal{I}_{2n}(p) \; = \; \prod_{j=1}^{n} e_j(0) \; = \; \frac{1}{n!}\,\frac{\Gamma(1+p/K)}{\Gamma(1+n+p/K)}\,. \tag{27.27}$$

The resulting perturbative series (27.3) of the partition function can be summed in closed form [531],

$$Z_{\mathrm{TB}}(x,p) = \frac{\Gamma(1+p/K)}{x^{p/K}} \sum_{n=0}^{\infty} \frac{x^{2n+p/K}}{n!\,\Gamma(1+n+p/K)} = \frac{\Gamma(1+p/K)}{x^{p/K}} I_{p/K}(2x)\,. \tag{27.28}$$

The function $I_\nu(z)$ is a modified Bessel function with index ν. With functional relations of the Bessel functions [89] the tunneling susceptibility $d\mathcal{F}/dx$ is found as

$$\frac{d}{dx}\mathcal{F}_{\mathrm{TB}}(x,p) \; = \; -k_{\mathrm{B}}T\,\frac{I_{1+p/K}(2x)}{I_{p/K}(2x)}\,. \tag{27.29}$$

With this form the conjecture (27.16) yields for the nonlinear mobility the expression

$$\mathcal{M}(x,q) \; = \; \frac{K}{q}\,2x\,\mathrm{Im}\,\frac{I_{1-iq/K}(2x)}{I_{-iq/K}(2x)}\,, \qquad q = \frac{\hbar\epsilon}{2\pi\,k_{\mathrm{B}}T}\,, \tag{27.30}$$

which is in agreement with the result of an elaborate real-time calculation, Eq. (25.89).

27.2.3 The limit $T \to 0$

The limit $T \to 0$ constitutes a crucial proof whether the analytic continuation from integer p to complex p in the expression (27.3) with (27.9) is correct for general K in all orders of x [531].

As $|p| \to \infty$, the ordered sums in $\mathcal{I}_{2n}(p)$ turn into ordered integrals. With the substitution $k \to p\,u$, we have the mapping

$$\sum_k \cdots \to p \int du \cdots ,\tag{27.31}$$

and the asymptotic expansion of $e_j(pu)$ yields

$$\lim_{|p| \to \infty} e_j(pu) = \frac{p^{2K-2}}{\Gamma^2(K)} \left[u\,(1+u) \right]^{K-1} \left\{ 1 + \mathcal{O}\left(\frac{1}{p}\right) \right\} .\tag{27.32}$$

With the relations (27.31) and (27.32) one finds that $\mathcal{I}_{2n}(p)$ behaves asymptotically as $p^{(2K-1)n}$. One finds from Eqs. (27.12) and (27.15) that there are systematic cancellations of the $n-1$ leading orders in $1/p$ when one passes from $\mathcal{I}_{2n}(p)$ to the connected part $\mathcal{C}_{2n}(p)$, and hence $\lim_{|p| \to \infty} \mathcal{C}_{2n}(p)/\mathcal{I}_{2n}(p) \propto p^{1-n}$. The cumulant in leading order in $1/p$ is found to read

$$\mathcal{C}_{2n}(p) = \frac{(-1)^{n-1}}{n!} \frac{\Gamma(nK)}{\Gamma^{2n}(K)} \frac{\Gamma(2n-1-2nK)}{\Gamma(n-nK)} p^{2(K-1)n+1} \left\{ 1 + \mathcal{O}\left(\frac{1}{p}\right) \right\} .\tag{27.33}$$

The fraction $c_n(K) = \Gamma(2n-1-2nK)/\Gamma(n-nK)$ in the cumulant $\mathcal{C}_{2n}(p)$ has simple poles at $K = 1 - \frac{1}{2n} + \frac{m}{n}$, where $m = 0, 1, 2, \cdots$.

Let us next perform the analytic continuation $p = i\,q$ in the expression (27.18) for $\mathcal{M}_n(x, q)$. We immediately see that the continuation produces a trigonometric factor $\cos[\,(\pi(1-K)n\,]$. Importantly, the simple zeros of this factor just compensate all the simple poles of $c_n(K)$. Thus in the end the perturbative conductance $\mathcal{M}_n(x, q)$ turns out to be regular for all $K > 0$,

$$\mathcal{M}_n(x, q) = \frac{(-1)^{n-1}}{n!} \frac{\Gamma(\tfrac{3}{2})\Gamma(1+nK)}{\Gamma^{2n}(K)\Gamma(\tfrac{3}{2} - n + nK)} \left(\frac{q}{2}\right)^{2(K-1)n} x^{2n} .\tag{27.34}$$

A crucial point now is that temperature cancels out in the expression (27.34),

$$(q/2)^{K-1} x = \Gamma(K)(\epsilon/\epsilon_0)^{K-1} .\tag{27.35}$$

Here ϵ_0 is the universal frequency scale given in Eq. (26.118). With the relation (27.35) the weak-tunneling series of the nonlinear mobility (27.17) with (27.34) becomes exactly the scaling solution (26.104) with (26.114).

Let us finally address the question whether the imaginary-time approach can also render the full-counting statistics at finite temperature. Evidently, the analytic continuation rule in Eq. (27.16) cannot provide information about individual tunneling rates. Rather we find with use of the partial rate expansion (26.130) for the terms of order x^{2n}

$$\mathcal{M}_n = \frac{K}{q}\frac{\hbar}{k_{\mathrm{B}}T} \sum_{m=1}^{n} \frac{m}{n} \left(k_{m,n}^{+} - k_{m,n}^{-}\right) ,$$

$$\mathrm{Im}\,\mathcal{F}_n(x, i\,q) = k_{\mathrm{B}}Tx^{2n}\,\mathrm{Im}\,\mathcal{C}_{2n}(-i\,q) = \frac{\hbar}{2}\sum_{m=1}^{n}\frac{m}{n}(k_{m,n}^{+} - k_{m,n}^{-}) .$$

(27.36)

The latter relation shows that the analytically continued perturbative series of the free energy yields for each term, say $\mathcal{F}_n$, a particular linear combination of partial rates of same order x^{2n}. On the other hand, the full-counting statistics at finite T requires in order x^{2n} knowledge of all the individual partial rates $k_{m,n}^{\pm}$, $m = 1, 2, \cdots, n$. Thus, on the route towards complete information of transport and noise one has to overcome the hurdle of handling the real-time Coulomb gas.

28. Charge transport in quantum impurity systems

The physics of interacting particles in one dimension is drastically different from the usual physics of interacting particles in two or three dimensions. The theoretical methods and techniques relevant to quantum physics in one dimension have been carefully reviewed in compendia by Gogolin, Nersesyan, and Tsvelik [533], and by Giamarchi [534]. Signatures of many-body correlations have attracted a great deal of interest in recent years. Many investigations have been concentrated on one-dimensional (1D) electron systems, in which the usual Fermi liquid behavior is destroyed by the interaction. The generic features of many 1D interacting fermion systems are well described in terms of the Tomanaga-Luttinger liquid (TLL) model [535, 401]. In the TLL model, all effects of the electron-electron interaction are captured by a dimensionless parameter g. A sensitive experimental probe of a Luttinger liquid state is the tunneling conductance through a point contact in a 1D quantum wire, as observed by Kane and Fisher [500]. Of interest are also the dc nonequilibrium current noise [536] and higher cumulants. The generic model is a quantum impurity embedded in a Luttinger liquid environment (QI-TLL model). Tunneling of edge currents in the fractional quantum Hall (FQH) regime provides another realization of a Luttinger phase. As shown by Wen [501], the edge state excitations are described by a (chiral) Luttinger liquid with Luttinger parameter $g = \nu$, where ν is the fractional filling parameter.

28.1 Generic models for transmission of charge through barriers

In this section, we discuss the conductance of 1D interacting spinless electrons in the presence of a barrier. First, we consider a pure interacting electron gas and give the two-terminal conductance. Then we discuss transport through a single barrier.

We approach this problem perturbatively in two limits: a very weak barrier (strong tunneling) and a very large barrier (weak tunneling). We find that the QI-TLL model is closely related to the model of an Ohmic Brownian particle in a tilted washboard potential, discussed in the preceding sections. As we shall see in Subsection 28.1.4, the model is also equivalent to a one-channel coherent conductor in a resistive electromagnetic environment.

28.1.1   The Tomonaga-Luttinger liquid

The characteristic feature of a Fermi liquid is the discontinuity of the momentum density of states at the Fermi surface. In one dimension, electron-electron interaction is so strong that this discontinuity is dissolved and replaced by a power law essential singularity at the Fermi points. The ground-state of the interacting 1D electron gas is a Tomonaga-Luttinger liquid which is distinguished by a gapless collective sound mode [401]. At low energies and long wave lengths, the strength of the interaction of spinless electrons is characterized by a single dimensionless parameter g.

The low-energy modes of the 1D interacting electron liquid are conveniently treated in the framework of bosonization [535, 401, 533]. This approach is appropriate for low temperatures where only excitation near the Fermi points are relevant. The creation operator for spinless fermions can equivalently be expressed in terms of boson phase fields $\theta(x)$ and $\phi(x)$, which obey the equal-time commutation relation

$$[\,\phi(x,t),\theta(x',t)\,] \; = \; -(i/2)\,\mathrm{sgn}(x-x')\;. \qquad (28.1)$$

Thus $\partial_x\phi(x)$ is the canonically conjugate momentum density to $\theta(x)$, and $\partial_x\theta(x)$ is conjugate to $\phi(x)$. In terms of these bosonic fields, the fermion field operator may be written as

$$\psi^\dagger(x) \; \propto \; \sum_{n\,\mathrm{odd}} \exp\{in[k_\mathrm{F}x + \sqrt{\pi}\theta(x)]\}\exp[i\sqrt{\pi}\phi(x)]\;. \qquad (28.2)$$

At long wavelengths, only the terms $n = \pm 1$ are important, corresponding to the right- and left-moving parts of the electron field. The boson representation of the electron density operator is then given by

$$\rho(x) \; = \; \frac{k_\mathrm{F}}{\pi} + \frac{1}{\sqrt{\pi}}\partial_x\theta(x) + \frac{k_\mathrm{F}}{\pi}\cos[2k_\mathrm{F}x + 2\sqrt{\pi}\theta(x)]\;, \qquad (28.3)$$

where k_F is the Fermi momentum. The first term is the background charge, the second term represents the density fluctuations of the right- and left-movers, and the last term describes interference between right- and left-movers.

The interaction between these bosons emerges from a combination of the hard-core condition of the bosons and the electron-electron interaction of the original fermions. Taking short-range electron-electron interactions and disregarding backscattering, the TLL liquid is described by the generic harmonic Hamiltonian

$$H_\mathrm{L} \; = \; \frac{\hbar v}{2}\int dx\,[(\partial_x\theta)^2/g + g(\partial_x\phi)^2]\;, \qquad (28.4)$$

where g is the interaction parameter, and $v = v_F/g$ is the sound velocity. For the noninteracting Fermi gas, we have $g = 1$, and the case $g < 1$ corresponds to repulsive interaction. We assume a sharp cutoff ω_c in the band width for the linear dispersion relation implicit in H_L and take ω_c as the largest frequency of the problem.

By switching from the Hamiltonian to the Lagrangian in either θ or ϕ, we get two equivalent representations of the Luttinger liquid. The respective Euclidean actions read

$$S_L^{(E)} = \begin{cases} \dfrac{\hbar v}{2g} \displaystyle\int dx\, d\tau \left((\partial_x \theta)^2 + \dfrac{1}{v^2}(\partial_\tau \theta)^2 \right), \\ \dfrac{\hbar v g}{2} \displaystyle\int dx\, d\tau \left((\partial_x \phi)^2 + \dfrac{1}{v^2}(\partial_\tau \phi)^2 \right). \end{cases} \tag{28.5}$$

It is well-known that the "two-terminal" conductance for a single-channel Fermi liquid is $G_0(g = 1) = e^2/2\pi\hbar$. For interacting electrons, the conductance may be calculated from the current-current correlation function in the zero frequency limit, where the current is $J = ie\dot{\theta}/\sqrt{\pi}$. One then finds that the conductance is renormalized by the electron-electron interaction [500],[1]

$$G_0(g) = g\frac{e^2}{2\pi\hbar}. \tag{28.6}$$

Thus we may regard the Luttinger parameter g as a dimensionless measure of the conductance of the pure Luttinger liquid.

28.1.2 Charge transport through a single weak barrier

The weak impurity is modelled by a barrier Hamiltonian $H_{sc} = \int dx\, V(x)\psi^\dagger(x)\psi(x)$, where $V(x)$ is a scattering potential [500]. Omitting multiple-electron backscattering processes, one finds for a short-ranged impurity potential, which is suppposed to be centered at $x = 0$, the form

$$H_{sc} = -V_0 \cos(2\sqrt{\pi}\,\bar{\theta}), \tag{28.7}$$

where $\bar{\theta}(t) = \theta(x = 0, t)$, and where V_0 is the Fourier transform of $V(x)$ at momentum $2k_F$. The scattering Hamiltonian originates from the interference between left- and right-movers and represents $2k_F$-backscattering. An applied voltage drop V_a at the impurity gives rise to the contribution

$$H_V = eV_a\bar{\theta}/\sqrt{\pi}. \tag{28.8}$$

The weak barrier or θ model is then given by

$$H_\theta = H_L + H_{sc} + H_V. \tag{28.9}$$

In the model (28.9), the nonlinear tunneling degree of freedom $\bar{\theta}(t)$ is coupled to a harmonic field which represents the modes in the leads away from $x = 0$. This is the

---

[1]The conductance considered here is a low-frequency microwave conductance. A two-terminal setup with reservoirs held at fixed chemical potentials would lead to modifications [537].

convenient starting point if one wishes to study backscattering off a weak impurity, in particular with regard to computations of conductance and statistical fluctuations.

Since the backscattering term H_{sc} only acts at $x = 0$, we may integrate out fluctuations of $\theta(x)$ for all x away from the origin. This can be done without approximation because of the harmonic nature of the pure Luttinger liquid action (28.5). With the Fourier ansatz $\theta(x,\tau) = (1/\hbar\beta) \sum_n \theta(x,\nu_n) e^{i\nu_n\tau}$, where $\nu_n = (2\pi/\hbar\beta)n$, the Euclidean action is minimized when $\theta(x,\nu_n) = \bar\theta_n \exp(-|\nu_n x|/v)$. Here, $\bar\theta_n$ is the Fourier coefficient of $\bar\theta(\tau)$. The resulting influence action [500] includes all effects of the modes in the leads on the tunneling degree of freedom $\bar\theta(\tau)$,

$$S^{(E)}_{\text{infl}}[\bar\theta]/\hbar \;=\; \frac{1}{g}\frac{1}{\hbar\beta} \sum_n |\nu_n|\,|\bar\theta_n|^2 \,. \tag{28.10}$$

It is convenient to consider the nonlinear conductance given in units of the conductance $G_0(g)$ of a pure 1D quantum wire,

$$\mathcal{G}(T,V_{\text{a}},g) \;=\; \frac{I(T,V_{\text{a}},g)}{V_{\text{a}}G_0(g)} \,. \tag{28.11}$$

For a weak barrier, the θ-representation (28.7) is appropriate. Then the conductance is determined by

$$\mathcal{G}_\theta(T,V_{\text{a}},g) \;=\; \lim_{t\to\infty} \frac{e}{\sqrt{\pi}V_{\text{a}}G_0(g)}\,\big\langle\, \dot{\bar\theta}(t) \,\big\rangle_\beta \,, \tag{28.12}$$

where $\langle\cdots\rangle_\beta$ denotes the thermal average over all modes of H_{L} away from the impurity. As we have just seen, this means average with the weight function $\exp\{-S_{\text{infl}}[\bar\theta]/\hbar\}$.

An analytical expression for $\mathcal{G}_\theta$ can be derived as follows. First, we expand the formal path integral expression into a power series in V_0^2. Then, in each term of the series, we integrate out the Gaussian field $\theta(x,\tau)$ away from $x = 0$. The resulting expression has the form of a statistical, grand-canonical ensemble of interacting discrete charges, analogous to the corresponding expressions given in Section 25.3. Because of the analogy of the interaction term (28.7) with the interaction (24.2), the charge conditions are as specified in Subsection 25.3.1. As a result of the elimination of the harmonic modes in the leads, the charges are interacting with each other. The pair interaction in imaginary time $W_\theta(\tau)$ is related to the correlator of $\bar\theta(\tau)$ by

$$\Big\langle\, \mathcal{T}\, e^{i2\sqrt{\pi}[\bar\theta(\tau)-\bar\theta(0)]} \,\Big\rangle_\beta \;=\; e^{-W_\theta(\tau)} \,. \tag{28.13}$$

Here, the thermal average $\langle\cdots\rangle$ denotes again average with the weight function $\exp[-S^{(E)}_{\text{infl}}[\bar\theta]/\hbar]$. Upon completing the square, we get

$$W_\theta(\tau) \;=\; g\,\frac{2\pi}{\hbar\beta} \sum_{n\neq0} \frac{1}{|\nu_n|} \big[1 - e^{i\nu_n\tau}\big] \,. \tag{28.14}$$

We see from the expression (4.72) with (3.56), or from Eqs. (4.207) - (4.209), that $W_\theta(\tau)$ is the charge interaction for an Ohmic spectral density $G_\theta(\omega) = 2g\,\omega$. The analytically continued real-time charge interaction $Q_\theta(t) = W_\theta(\tau = it)$ has the integral representation (18.44) and takes the Ohmic scaling form [see Eq. (18.54)]

$$Q_\theta(t) = 2g \ln \left[\frac{\hbar \beta \omega_c}{\pi} \sinh \left(\frac{\pi |t|}{\hbar \beta} \right) \right] + i \pi g \, \mathrm{sgn}(t) \,. \tag{28.15}$$

The strong-tunneling series for the normalized conductance is found to read

$$\mathcal{G}_\theta(T, V_a, g) = 1 - \frac{\pi \hbar}{e V_a} \sum_{m=1}^{\infty} (-1)^{m-1} \left(\frac{V_0}{\hbar} \right)^{2m} \int_0^\infty d\tau_1 \, d\tau_2 \cdots d\tau_{2m-1} \tag{28.16}$$

$$\times \sum_{\{\xi_j\}'} G_{2m}(\{\tau_j\}; g, \{\xi_j\}) \sin \left(\sum_{i=1}^{2m-1} g e V_a \, p_i \tau_i / \hbar \right) \prod_{k=1}^{2m-1} \sin(\pi g p_k) \,,$$

where p_i is the cumulated charge defined in Eq. (25.49). The interaction factor $G_{2m}(\{\tau_j\}; g, \{\xi_j\})$ is defined in Eq. (25.77).

28.1.3 Charge transport through a single strong barrier

In the opposite limit of a large barrier or a weak link, we have in zeroth approximation two disconnected semi-infinite Luttinger leads. These are described in terms of independent boson fields. To account for electron tunneling through the barrier, a tunneling Hamiltonian describing punctual tunneling is added,

$$H_T \propto -\hbar \Delta [\psi^\dagger(x = 0^+)\psi(x = 0^-) + \mathrm{h.c.}] \tag{28.17}$$

This term depends on the phase jump $\bar\phi \equiv \frac{1}{2}[\phi(x = 0^+) - \phi(x = 0^-)]$ of the ϕ-field at the point-like barrier. Hence the tunneling Hamiltonian may be written as

$$H_T = -\hbar \Delta \cos[2\sqrt{\pi}\bar\phi + eV_a t/\hbar] \,. \tag{28.18}$$

A voltage drop at the barrier induces a jump of the chemical potential. The jump due to the applied voltage induces a phase shift. This we have included in the expression (28.18). With addition of the terms describing the harmonic liquid in the left $(-)$ and right $(+)$ leads, we arrive at the Hamiltonian of the weak link problem (ϕ-model)

$$H_\phi = H_{L,+} + H_{L,-} + H_T \,. \tag{28.19}$$

This model describes again a nonlinear tunneling degree of freedom coupled to a harmonic field. We may again integrate out the harmonic modes away from the barrier, this time in the ϕ-representation of the lead action (28.5). Proceeding as in the previous subsection, the Euclidean influence action of the tunneling mode $\bar\phi$ resulting from the Luttinger liquid environment takes the form

$$S_{\mathrm{infl}}^{(E)}[\bar\phi]/\hbar = g \frac{1}{\hbar \beta} \sum_n |\nu_n| |\bar\phi_n|^2 \,. \tag{28.20}$$

Because of the substitution $g \to 1/g$, when switching from the θ- to the ϕ-representation, the action (28.20) is the dual of the action (28.10).

With use of the Heisenberg equation of motion for $\bar\theta(t)$, we may express the normalized conductance (28.12) in terms of the ϕ-field. The resulting expression is

$$\mathcal{G}_\phi(T, V_{\mathrm{a}}, g) = \lim_{t \to \infty} \frac{e\Delta}{V_{\mathrm{a}} G_0(g)} \left\langle \sin[\, 2\sqrt{\pi}\bar{\phi}(t) + eV_{\mathrm{a}}t/\hbar\,] \right\rangle_\beta , \qquad (28.21)$$

where $\langle \cdots \rangle_\beta$ now means average with the weight function $\exp[-S_{\mathrm{infl}}^{(\mathrm{E})}[\bar{\phi}]/\hbar]$.

The proceeding is as in the preceeding subsection. We formally expand the path integral in powers of Δ^2 and, in view of the formal similarity of the expression (28.18) with (28.7), we introduce again a charge representation. The effects of the Luttinger liquid modes are included in the charge interaction, which is again a pair correlation function, but now for $\bar{\phi}(t)$. Following the lines (28.13) – (28.15), we get for the real-time correlator

$$\left\langle \mathcal{T} \, e^{i2\sqrt{\pi}[\bar{\phi}(t) - \bar{\phi}(0)]} \right\rangle_\beta = \exp\left\{ -\frac{2}{g} \ln\left[\frac{\hbar\beta\omega_c}{\pi} \sinh\left(\frac{\pi|t|}{\hbar\beta}\right) \right] - i\frac{\pi}{g} \operatorname{sgn}(t) \right\} . \qquad (28.22)$$

The weak-tunneling series expansion for the normalized conductance is readily found as the dual of the series (28.16)

$$\mathcal{G}_\phi(T, V_{\mathrm{a}}, g) = \frac{\pi\hbar}{geV_{\mathrm{a}}} \sum_{m=1}^{\infty} (-1)^{m-1} \Delta^{2m} \int_0^\infty d\tau_1 \, d\tau_2 \cdots d\tau_{2m-1} \qquad (28.23)$$

$$\times \sum_{\{\xi_j\}'} G_{2m}(\{\tau_j\}; 1/g, \{\xi_j\}) \sin\left(\sum_i^{2m-1} eV_{\mathrm{a}} p_i \tau_i/\hbar \right) \prod_{k=1}^{2m-1} \sin(\pi p_k/g) .$$

28.1.4 Coherent conductor in an Ohmic environment

A mesoscopic conductor embedded in an electromagnetic environment forms a quantum system violating Ohm's law. The electrons in the conductor induce electromagnetic modes in the electrical circuit and therefore undergo inelastic scattering processes. By this, the current at low voltage is reduced, as we have discussed already for weak tunneling in Subsection 20.3.1. This picture of dynamical Coulomb blockade changes in the opposite limit of a good conductor. The description of tunneling of discrete charges is then no longer valid. The Luttinger parameter of the coherent conductor in the absence of the electromagnetic environment is $g = 1$. According to convenience, we may use either the θ- or the ϕ-representation of the action (28.5).

As we have seen already in Section 20.3 in the weak-tunneling limit, an Ohmic environment can well simulate the electronic interactions in a coherent conductor. We now fully extend the analogy of a one-channel conductor in an Ohmic environment to impurity scattering in a Luttinger liquid (see Ref. [538]).

Let us take the Hamiltonian (3.222) as a starting-point. For spinless electrons and a constant real tunneling amplitude T_{T}, it can be written as

$$H_{\mathrm{wt}} = H_1 + H_2 + T_{\mathrm{T}} \left[\psi_2^\dagger(0^+) \psi_1(0^-) \, e^{-i\varphi(t)} + \text{h.c.} \right] + H_{\mathrm{env}}(\mathcal{Q}, \varphi) . \qquad (28.24)$$

The label "wt" stands for *weak tunneling*, and $H_{1,2}$ is the electronic part for the left/right electrode with the voltage drop V_{a} across the barrier included in H_1, as specified in Eq. (3.221). The term $H_{\rm env}$ is supposed to describe a resistive environment with impedance R, and V_a is the applied voltage drop at the barrier. The last term couples the phase fluctuations $\varphi(t)$ induced by the impedance to the local electronic fields $\psi_{1,2}(0)$ at the edges of the barrier.

Since the barrier is point-like, electron tunneling is effectively one-dimensional. Therefore, the electronic part can be bosonized, as discussed in Subsection 28.1.1. The tunneling term in Eq. (28.24) can thus be written as [cf. Eq. (28.18)]

$$H_{\rm I}' = -\tfrac{1}{2}\hbar\Delta'\left[e^{i(2\sqrt{\pi}\bar{\phi}-\varphi)} + {\rm h.c.}\right], \qquad (28.25)$$

where $\Delta' \propto T_{\rm T}$. The relation between Δ' and the original tunneling amplitude $T_{\rm T}$ can be obtained from the expressions (20.163) and (20.130).

The autocorrelation functions for the electromagnetic phase $\varphi(t)$ and the electronic phase jump at the barrier $2\sqrt{\pi}\bar{\phi}(t)$ are [cf. Eq. (28.22) and Eq. (20.134) with Eq. (20.149)]

$$\langle\,[\varphi(0) - \varphi(t)]\,\varphi(0)\,\rangle_\beta = Q_{\rm ohm}(t;\alpha) + \Delta Q_{\rm em}(\alpha)\,, \qquad (28.26)$$

$$4\pi\langle\,[\bar{\phi}(0) - \bar{\phi}(t)]\,\bar{\phi}(0)\,\rangle_\beta = Q_{\rm ohm}(t;1)\,. \qquad (28.27)$$

Here $Q_{\rm ohm}(t;g)$ is given in (20.144) and $\Delta Q_{\rm em}(\alpha)$ in (20.149). Since $\bar{\phi}(t)$ and $\varphi(t)$ commute and the Hamiltonian $H_1 + H_2 + H_{\rm env}$ is quadratic in the combined phase $\chi = 2\sqrt{\pi}\bar{\phi} + \varphi$, the equilibrium correlation function of the auxiliary phase $\chi(t)$ is

$$\langle\,[\chi(0) - \chi(t)]\,\chi(0)\,\rangle_\beta = Q_{\rm ohm}(t;1+\alpha) + \Delta Q_{\rm em}(\alpha)\,. \qquad (28.28)$$

The constant term $\Delta Q_{\rm em}(\alpha)$ is conveniently absorbed into a dressed tunneling amplitude, $\Delta_{\rm d} = \Delta' e^{-\Delta Q_{\rm em}(\alpha)/2}$.

Thus, the Hamiltonian $H_{\rm wt}$ is equivalent to the impurity Hamiltonian H_ϕ, Eq. (28.19), if we identify $\Delta_{\rm d}$ with Δ and put

$$1 + \alpha = 1/g\,. \qquad (28.29)$$

The correspondence holds in all orders of Δ^2 in the series expression (28.23). The contribution of lowest order to the current has been discussed in Subsection 20.3.2.

We may also think of the *dual* limit of weak backscattering off the scattering potential $H_{\rm sc}$ given in bosonized form in Eq. (28.7). The electromagnetic environment causes a fluctuating potential drop $\hbar\dot{\varphi}(t)$ at the barrier which is coupled to the scattering mode $\bar{\theta}$ as in Eq. (28.8). Thus, the Hamiltonian of the coherent conductor plus electromagnetic environment takes the form

$$H_{\rm st} = H_{L,g=1} + H_{\rm sc} + H_{\rm V} + \hbar\dot{\varphi}\,\bar{\theta}/\sqrt{\pi} + H_{\rm env}(\mathcal{Q},\varphi)\,. \qquad (28.30)$$

The last term represents the electromagnetic environment as given in Eq. (3.207). It is straightforward to integrate out the modes $\{\varphi_\alpha\}$ of the electromagnetic environment.

The resulting Euclidean action in Fourier representation for an Ohmic environment, $\hat{Y}(|\nu_n|) = 1/R$, reads [cf. Eq. (4.204)]

$$S_{\text{em}}^{(\text{E})}[\bar{\theta},\varphi]/\hbar \;=\; \frac{1}{\hbar\beta}\sum_n \left\{ \frac{1}{4\pi}\frac{R_\text{K}}{R}|\nu_n||\varphi_n|^2 - i\nu_n\varphi_n\frac{\bar{\theta}_n}{\sqrt{\pi}} \right\}. \qquad (28.31)$$

In the next step, we also integrate out the phase $\varphi(t)$. Upon completing the square, we find that the effects of the electromagnetic environment on the tunneling degree of freedom $\bar{\theta}$ are described by the Euclidean influence action ($\alpha = R/R_\text{K}$)

$$S_{\text{infl, em}}^{(\text{E})}[\bar{\theta}]/\hbar \;=\; \alpha\,\frac{1}{\hbar\beta}\sum_n |\nu_n|\,|\bar{\theta}_n|^2. \qquad (28.32)$$

The Fermi liquid fluctuations in the leads ($g = 1$) give rise to the action

$$S_{\text{infl, el}}^{(\text{E})}[\bar{\theta}/]/\hbar \;=\; \frac{1}{\hbar\beta}\sum_n |\nu_n||\bar{\theta}_n|^2, \qquad (28.33)$$

as we can read off from the expression (28.10). In the sum of both terms, the resistive environment leads to a renormalization of the electronic modes, $|\nu_n| \rightarrow (1+\alpha)|\nu_n|$.

Thus, there is again a formal equivalence to backscattering off an impurity in TLL, this time in the weak-backscattering limit. The relation of the effective coupling parameter $1+\alpha$ to the fictitious TLL parameter g is the same as that found in the tunneling regime, Eq. (28.29), $1+\alpha = 1/g$. The correspondence of the model (28.30) with the QI-TLL model (28.9) is again on the level of the effective action.

28.1.5   Equivalence with quantum transport in a washboard potential

After having established a formal equivalence between the models of a quantum impurity in a TLL and a coherent one-channel conductor in a resistive environment, we now discuss the correspondence of the QI-TLL model with that of a Brownian particle in a tilted cosine potential.

The model described by H_θ, Eq. (28.9), directly corresponds to the weak-binding Brownian particle model (24.1). The correspondence can be shown by employing canonical transformations and by a study of the equations of motion of the coordinate and momentum autocorrelation functions (cf. the discussion in Subsections 26.1.1 and 26.1.2). The tunneling degree of freedom $\bar{\theta}$ corresponds to $\sqrt{\pi}X/X_0$. Ohmic damping of the mode $\bar{\theta}$ is provided by excitation of the TLL liquid away from the barrier, and the parameter g is related to the viscosity η by $g = \eta X_0^2/2\pi\hbar$. In the correspondence, the cutoff frequency ω_c of the liquid modes is identified with η/M. The equivalence becomes exact when the force of inertia is negligibly small compared with the friction force, i.e., when ω_c is the largest frequency of the problem. For $k_\text{B}T \ll V_0$ and $eV_\text{a} \ll V_0$, this means $\eta^2 \gg MV_{WB}''(0)$, or equivalently $\hbar\omega_c \gg 2\pi g V_0$. As a result of the mapping, the nonlinear conductance in the θ-model is directly related to the nonlinear mobility in the WB model.

Similarly, the high-barrier ϕ-model (28.19) is equivalent to the TB model (24.3), as follows again by unitary transformations.

Alternatively, the correspondences of the θ- and ϕ-model with the dissipative WB and TB model can also be seen directly from the exact formal series expressions for the conductance and mobility, respectively. The series (28.16) agrees with the series (26.93) for the WB mobility if we employ the correspondences

$$g \leftrightarrow K_{\mathrm{WB}} = 1/K , \quad \text{and} \quad eV_{\mathrm{a}}/\hbar \leftrightarrow \epsilon_{\mathrm{WB}} . \tag{28.34}$$

Likewise, the series expression (28.23) for the weak-link conductance concurs with the series (25.76) for the mobility of the TB model under the assignments

$$1/g \leftrightarrow K , \quad \text{and} \quad eV_{\mathrm{a}}/\hbar \leftrightarrow \epsilon . \tag{28.35}$$

The mapping relations (28.34) and (28.35) are valid on the level of the effective actions. This implies that they hold not only for the conductance but also for all statistical fluctuations.

28.2 Self-duality between weak and strong tunneling

The results of the preceding subsection can be condensed into an important correspondence between the normalized mobility of the Brownian particle and the normalized conductance for the QI-TLL model,

$$\mathcal{G}(T, V_{\mathrm{a}}, g) = \mu(T, \epsilon = eV_{\mathrm{a}}/\hbar, K = 1/g)/\mu_0 , \tag{28.36}$$

which relates the θ-model to the WB model, and the ϕ-model to the TB model. Upon employing the self-duality relation (26.29) between the TB and WB representation we immediately infer from the correspondence (28.36) that the weak-link conductance $\mathcal{G}_\phi$ is related to the weak-barrier conductance $\mathcal{G}_\theta$ by

$$\mathcal{G}_\theta(T, V_{\mathrm{a}}, g) = 1 - \mathcal{G}_\phi(T, gV_{\mathrm{a}}, 1/g) . \tag{28.37}$$

For repulsive electron interaction, the Luttinger parameter g is restricted to the domain $g < 1$, whereas in the related Brownian particle model the parameter regime is $0 < K < \infty$. The entire expansions around weak and strong backscattering are related to each other term by term by the correspondence (28.37). Upon introducing a Kondo temperature T_0 or a Kondo frequency ϵ_0, as in Section 26.4, the self-duality relations for $\mathcal{G} = \mathcal{M}$ may be cast into the form (26.99) in the linear response regime, and into the form (26.106) in the zero temperature limit. These forms have lost explicit dependence on the ϕ- or θ-model and therefore equally hold for both models.

With use of the correspondence (28.36), the analytical results obtained in Chapter 26.4 for the Brownian particle can now immediately be transferred to charge transport across a quantum impurity in a Luttinger liquid. The exact scaling solutions at $T = 0$, Eqs. (26.104) and (26.105) with Eq. (26.114), agree with expressions derived by Fendley *et al.* [526]. These authors utilized a suitable basis of interacting quasiparticles in which the model is integrable, and they employed sophisticated thermodynamic Bethe-ansatz (TBA) technology to calculate the non-Fermi distribution function and the density of states of the quasiparticles. These quantities then determine the conductance by a Boltzmann-type rate expression. The Kondo frequency ϵ_0 directly corresponds to the temperature scale T'_B used in Ref. [526], $T'_B = \hbar\epsilon_0/k_B$. The simple derivation given here in Section 26.5 sheds additional light on the underlying symmetries of these models.

28.3   Full counting statistics of charge transfer

The full counting statistics of charge transport through an impurity is encapsulated in the moment generating function $\chi(\rho, t)$, which is the Fourier transform of the probability distribution $P(Q, t)$ of the charge Q crossing the impurity during time t, $\chi(\rho, t) = \sum_Q e^{i\rho Q} P(Q, t)$, where ρ is the counting field. The function $\chi(\rho, t)$ generates moments of the charge $Q_t = \int_0^t dt' \, I(t')$ transferred during time t,

$$\chi(\rho, t) \;=\; \sum_k \frac{(i\rho)^k}{k!} \langle Q_t^k \rangle \;=\; \exp\left\{ t \sum_k \frac{(i\rho)^k}{k!} \langle \delta^k Q \rangle \right\}. \qquad (28.38)$$

Here, $t \langle \delta^k Q \rangle$ is the kth cumulant of the distribution, and $I(t)$ is the time-dependent current through the scattering region.

28.3.1   *Charge transport at low temperature for arbitrary g*

An important limiting case is the zero temperature regime in which the strong- and weak-backscattering expansions of all cumulants can be found in analytic form [525]. There are clear physical pictures in these different limits.

For strong backscattering or weak tunneling, the ϕ-model (28.19) applies, which is the equivalent of the TB model (24.3) in the Ohmic scaling limit. Here, the true ground state is that of two completely disconnected leads. Then, evidently, only quasiparticles with integer unit charge, i.e. electrons, may tunnel through the barrier between the leads. Since at $T = 0$ there are no transition from the energetically lower to the higher lead, we have in correspondence with the expression (26.128)

$$\ln \chi(\rho, t) \;=\; t \sum_{n=1}^{\infty} \left(e^{i\rho n e} - 1 \right) \frac{I_n^+}{n}, \qquad (28.39)$$

where $I_n^+ = n k_n^+$. The physical meaning of this expression is quite illuminating. Suppose that k_n^+ is the probability per unit time to transfer a particle of charge ne through the impurity barrier. Then the charge transferred in the time interval t is the result of a Poisson process for particles of charge e crossing the barrier, contributing a current I_1^+, plus a Poisson process for particles of charge $2e$ contributing a current I_2^+, etc. All these Poisson processes are represented by $\ln \chi(\rho, t)$.

Observing that the correspondence of the TB model (24.3) with the ϕ-model (28.19) does hold not only for the conductance but for all cumulants, we immediately get from the expression (26.133) for the partial current I_n^+ at $T = 0$

$$I_n^+(\epsilon, 0) = \frac{(-1)^{n-1}}{\Gamma(n)} \frac{\Gamma(\frac{3}{2})\Gamma(n/g)}{\Gamma[\frac{3}{2} + (1/g - 1)n]} \frac{\epsilon}{2\pi} \left(\frac{\epsilon}{\epsilon_0}\right)^{(2/g-2)n}. \qquad (28.40)$$

This yields for the first cumulant or mean current the expression

$$\langle \delta Q \rangle = \langle I(V_a) \rangle = e \sum_{n=1}^{\infty} I_n(eV_a/\hbar, 0). \qquad (28.41)$$

While in a classical Poisson process all the partial currents I_n^+ in Eq. (28.39) are positive, the subtle point now is that the partial currents (28.40) are not. Certainly, the first contribution I_1^+ is indeed a Poisson process for the tunneling of electrons; but the joint tunneling of pairs of electrons (and of multiples thereof), $n = 2, 4, \cdots$, comes with a negative sign, which is an effect of quantum interference.

Following the arguments given in Section 26.7, we find that the expression (28.39) also holds at low temperature, in which the universal leading low-temperature contribution is included in the partial current I_n^+,

$$I_n^+(\epsilon, T) = \left\{ 1 + \frac{\pi^2}{3g} \frac{2 - 2g}{g} n \left[1 + \frac{2 - 2g}{g} n \right] \left(\frac{k_B T}{\hbar \epsilon}\right)^2 \right\} I_n^+(\epsilon, 0). \qquad (28.42)$$

On the other hand, in the opposite limit of strong-tunneling, a collective state between the edges is formed with elementary excitations of charge ge. The resulting expression for $\ln \chi(\rho, t)$ at low temperatures reads

$$\ln \chi(\rho, t) = t \left(i\rho g e \frac{\epsilon}{2\pi} - \frac{\rho^2}{2} \frac{g e^2 k_B T}{\pi \hbar} + \sum_{n=1}^{\infty} \left(e^{-i\rho n g e} - 1 \right) \frac{\tilde{I}_n^+}{n} \right), \qquad (28.43)$$

where

$$\tilde{I}_n^+(\epsilon, T) = \left\{ 1 + \frac{\pi^2}{3g} (2g - 2) n [1 + (2g - 2) n] \left(\frac{k_B T}{\hbar \epsilon}\right)^2 \right\} \tilde{I}_n^+(\epsilon, 0) \qquad (28.44)$$

with

$$\tilde{I}_n^+(\epsilon, 0) = \frac{(-1)^{n-1}}{\Gamma(n)} \frac{\Gamma(\frac{3}{2})\Gamma(ng)}{\Gamma[\frac{3}{2} + (g - 1)n]} \frac{g e}{2\pi} \left(\frac{\epsilon}{\epsilon_0}\right)^{(2g-2)n}. \qquad (28.45)$$

The form (28.43) is quite similar to Eq. (28.39), but there are subtle differences. The first two terms in the expression (28.43) represent the current and noise of fractionally charged quasiparticles in the absence of the barrier. The exponential factor $e^{-i\rho n g e}$ indicates that now we have tunneling of quasiparticles of charge ge and of multiples thereof, and the minus sign in the exponent means that the tunneling diminishes the current instead of building it up as in the strong-backscattering limit. Since the sign of the partial current $\tilde{I}_n^+$ is the one of $\cos(ng\pi)$ for $g < \frac{1}{2}$, the perception of clusters of quasiparticles with fractional charge ge tunneling independently with a classical Poisson process is quite appropriate for bundles with modest n, when g is small. The mean current at zero temperature is

$$\langle \delta Q \rangle \; = \; \langle I \rangle \; = \; g \frac{V_a}{R_K} - ge \sum_{n=1}^{\infty} \tilde{I}_n^+(eV_a/\hbar, 0) \, . \tag{28.46}$$

As g goes to zero, all the partial currents become positive, but the quantum fluctuations $\langle \delta^n Q \rangle$ with $n > 1$ disappear at $T = 0$, as one reaches the classical limit,

$$
\begin{aligned}
\ln \chi(\rho, t) \; &= \; i \rho t \, ge \frac{\epsilon}{2\pi} \left[1 - \sum_{n=1}^{\infty} \frac{\Gamma(n - \frac{1}{2})}{2\sqrt{\pi} n!} \left(\frac{\epsilon_0}{\epsilon} \right)^{2n} \right] \\
&= \; i \rho t \, ge \frac{\epsilon}{2\pi} \sqrt{1 - \left(\frac{\epsilon_0}{\epsilon} \right)^2} \, ,
\end{aligned}
\tag{28.47}
$$

where $\epsilon_0 = 2\pi V_0/\hbar$. The first term is the current in the absence of the barrier, and the second is the backscattering contribution made up of partial backscattering currents.

The T^2-contribution to the cumulants vanishes at $g = 1$. Hence T^2-variation of the current and of its fluctuations is a distinctive signature of 1D interacting electrons.

Finally, we remark that cumulant relations analogous to the expressions (26.132) and (26.136) hold. The corresponding relation in the weak-tunneling representation is

$$\langle \delta^N Q \rangle \; = \; \left(e \frac{\Delta}{2} \frac{\partial}{\partial \Delta} \right)^{N-1} \langle \delta Q \rangle \, , \tag{28.48}$$

whereas in the strong-tunneling representation

$$\langle \delta^N Q \rangle \; = \; \left(- ge \frac{V_0}{2} \frac{\partial}{\partial V_0} \right)^{N-1} \langle \delta Q \rangle \, . \tag{28.49}$$

These relations help us to reduce every single cumulant to a calculation of derivatives of the current $\langle \delta Q \rangle$ with respect to the tunneling coupling Δ or to the corrugation strength V_0, respectively. They are in agreement with the findings from the integrable approach to the BSG model [525].

In the case of a fractional quantum Hall bar with $g = \nu = \frac{1}{3}, \frac{1}{5}, \cdots$, the collective excitations are Laughlin quasiparticles and the impurity corresponds to a point contact. In the weak-backscattering limit, Laughlin quasiparticles with fractional charge νe are tunneling. In the strong-backscattering limit, the system consists of two different Hall devices which are weakly coupled by the interaction (28.18), and only electrons can tunnel. Strict duality means that the entire expansions around weak and strong backscattering are related. In the FQHE system, the crossover from weak to strong backscattering comes along with a crossover from Laughlin quasiparticle tunneling to electron tunneling.

For general g and general temperature, the analytical calculation of higher-order Coulomb integrals in the bosonic representation is not possible. Then one may resort to thermodynamic Bethe ansatz techniques in the related BSG model [526, 527]. In Ref. [528] a general scheme for the calculation of cumulants of any order is proposed.

28.3.2 Full counting statistics at $g = \frac{1}{2}$ and general temperature

In the weak-tunneling limit (strong impurity), the mean current at $T = 0$ is $\propto V_a^{2/g-1}$, as we see from the leading term of the series (28.41). On the other hand, the current I_B that is backscattered from a weak barrier (weak impurity) varies as V_a^{2g-1}. Hence the backscattering current in leading order becomes independent of the applied voltage at $g = \frac{1}{2}$. This is a sign that $g = \frac{1}{2}$ is a special point. By virtue of the mapping (28.29), the case $g = \frac{1}{2}$ corresponds to a coherent conductor (e.g. a quantum dot) with series resistance $R = R_K$ or $\alpha = 1$. An open point contact is a realization of such series resistance, as discussed in Ref. [539].

Employing the nonequilibrium Keldysh formalism and refermionization techniques, which map a Luttinger liquid at $g = \frac{1}{2}$ on noninteracting fermions, the cumulant generating function has been calculated in Ref. [539]. With a transformation set out in Appendix C of Ref. [540], the expression of the CGF given in Ref. [539] can be converted into a form analogous to Eq. (26.58),

$$
\begin{aligned}
\ln \chi(\rho, t) = \; & t \int_0^\infty \frac{d\omega}{2\pi} \ln \Big\{ 1 + \overline{\mathcal{T}}(\omega) \left[\mathcal{N}_+(\omega, eV_a/2\hbar) \left(e^{i e \rho} - 1 \right) \right. \\
& \left. + \mathcal{N}_-(\omega, eV_a/2\hbar) \left(e^{-i e \rho} - 1 \right) \right] \Big\},
\end{aligned}
\tag{28.50}
$$

with the spectral transition probability $\overline{\mathcal{T}}(\omega) = \omega^2/(\omega^2 + \tilde{\epsilon}_0^2)$. The renormalized frequency $\tilde{\epsilon}_0$ is related to the bare impurity strength V_0 by [see the relation (26.119) at $g = \frac{1}{K} = \frac{1}{2}$ or the relation (26.53)]

$$
\tilde{\epsilon}_0 = \pi \frac{(V_0/\hbar)^2}{\omega_c} .
\tag{28.51}
$$

The functions $\mathcal{N}_\pm(\omega, \epsilon)$ and relations of them are given in Eqs. (25.106) and (25.108).

The CGF (28.50) of the QI-TLL model (28.9) at $g = \frac{1}{2}$ also describes the FCS of a coherent conductor in series with an Ohmic resistor of resistance $R = R_K$. The expression (28.50) is in agreement with the CGF (26.58) of the quantum Brownian particle model (24.1) at $K = 2$, and with the CGF of Cooper pair tunneling in a resistive environment with resistance $R = 2R_Q$. Finally, the leading cumulants are easily found from the expressions (26.59) - (26.61).

29. Quantum transport for sub- and super-Ohmic friction

We now consider the quantum transport of a particle in the cosine potential for sub- and super-Ohmic friction. The analysis is firstly done for the tight-binding model (24.3). The results obtained are then transferred with use of the duality relations developed in Section 26.1 to the weak-binding model (24.1).

29.1   Tight-binding representation

We will now use the exact formal expressions worked out in Section 25.3 in order to determine transport properties of the TB model (24.3) for the spectral density (3.72) with the power s in the range $0 < s < 2$, $s \neq 1$. The charge sequences contributing to the series expression (25.47) for $\langle \hat{q}(\lambda) \rangle$ form strings of clusters separated by sojourns.[1]

In the regime $s < 2$ and nonzero temperature, a long blip interval τ_k with cumulated charge p_k is exponentially suppressed by the charge interaction factor $\exp[-p_k^2 Q'(\tau_k)]$, as discussed above Eq. (25.65). On the other hand, the interaction between two neighboring clusters with sets of charge intervals $\{\tau_j\}$ and $\{\tau_\ell'\}$ separated by a sojourn of length ρ, where ρ is large compared to the cluster lengths, is

$$\exp\left\{ - \left(\sum_j p_j \tau_j \right) \left(\sum_\ell p_\ell \tau_\ell' \right) \ddot{Q}'(\rho) \right\} . \tag{29.1}$$

Since $\ddot{Q}'(\rho \gg \hbar\beta) \propto \rho^{-s}$, as follows from Eq. (25.65), the interaction factor (29.1) cannot bind the clusters together. Hence long sojourns are not inhibited by this interaction. The picture we thus have is as follows. In the limit $\lambda \to 0$, the series (25.47) for $\langle \hat{q}(\lambda, s) \rangle$ is the grand-canonical sum of a dilute gas of narrow charge clusters with total charge zero, with long distances between the clusters. Let us refer to the cluster contribution as kernel $\hat{k}(\lambda, s)$ and to the sojourn contribution as $\hat{f}(\lambda, s)$. As $\lambda \to 0$, the series (25.47) for the first moment to linear order in the bias forms into an initial cluster $\hat{k}_{\text{ini}}(\lambda, s)$ which is followed by a geometrical series of sojourn-cluster pairs $-\hat{f}(\lambda, s)\hat{k}(\lambda, s)$, with a minus sign subjoined [360]. In detail we have

$$\lim_{\lambda \to 0} \langle \hat{q}(\lambda, s) \rangle_{\text{lin}} = \lim_{\lambda \to 0} \frac{q_0 \epsilon}{2} \frac{1}{\lambda} \hat{k}_{\text{ini}}(\lambda, s) \frac{1}{1 + \hat{f}(\lambda, s)\hat{k}(\lambda, s)} \frac{1}{\lambda} . \tag{29.2}$$

For the dilute gas of narrow clusters, the influence phase term $b_{2m,0}$, which is defined in Eq. (25.44), may be expanded about the individual sojourn lengths in a power series of the charge intervals of the subsequent cluster. Now, since $\dot{Q}''(\rho) \propto \rho^{-s}$ as $\rho \to \infty$, only the linear term is relevant for the asymptotic dynamics. By this, a sojourn of length ρ together with its subsequent neutral cluster, say consisting of 2ℓ charges, is equipped with a factor

$$B_{\text{soj, cl}}^{(2\ell)} = \dot{Q}''(\rho) \left(\sum_{j=1}^{2\ell-1} p_j \tau_j \right) (-1)^{\ell-1} \prod_{i=1}^{2\ell-1} \sin(\chi_{i,2\ell}) . \tag{29.3}$$

Here, the phase $\chi_{i,2\ell}$ is given in Eq. (25.35), $\{\tau_j\}$ are the charge intervals of the cluster, and $\{p_j\}$ are the respective cumulated charges. The sum in the round bracket originates from the expansion of the sojourn-cluster interaction. The product term captures the intra-cluster interactions resulting from the friction action $\mathcal{S}^{(\text{F})}[r, y]$ given in Eq. (5.33).

---

[1]By definition, a cluster cannot be separated into neutral subclusters, i.e., the cumulated charges $\{p_j\}$ of a cluster are all nonzero. See also the exemplification in Fig. 25.1.

On the other hand, the initial cluster, say consisting of $2n$ charges, is free of sojourn-cluster interactions and therefore equipped with the influence phase or friction action term times a sign term

$$B_{\rm cl}^{(2n)} = (-1)^{n-1} \prod_{i=1}^{2n-1} \sin(\chi_{i,2n}) \,. \tag{29.4}$$

The factor $B_{\rm soj, cl}^{(2\ell)}$ is even under charge conjugation, whereas $B_{\rm cl}^{(2n)}$ is odd. Next we observe that the bias factor $\varphi_{2m,0}$ given in Eq. (25.50) is odd under charge conjugation. Thus it does not contribute to a sojourn-cluster pair, whereas it furnishes the term (29.4) of the initial cluster with the factor $\sum_{j=1}^{2n-1} p_j \tau_j$. As an important result we thus find that the initial kernel coincides with the subsequent kernels,

$$\hat{k}_{\rm ini}(\lambda, s) = \hat{k}(\lambda, s) \,. \tag{29.5}$$

Altogether the kernel is given by the cluster sum

$$\hat{k}(\lambda, s) = \sum_{m=1}^{\infty} (-1)^{m-1} \int_0^{\infty} \hat{\mathcal{D}}_{2m,0}(\lambda, \{\tau_\ell\}) \sum_{\{\xi_j\}'} \Big(\sum_{j=1}^{2m-1} p_j \tau_j \Big) G_{2m} \prod_{i=1}^{2m-1} \sin(\chi_{i,2m}) \,, \tag{29.6}$$

where the sum $\sum_{\{\xi_j\}'}$ is constrained as specified in the previous footnote.

With the explicit form gathered from Eq. (18.47)

$$Q''(\rho) = -2\delta_s \Gamma(s-1) \cos(\pi s/2)(\omega_{\rm ph}\rho)^{1-s} \,, \tag{29.7}$$

the relevant integral over the sojourn length is

$$\hat{f}(\lambda, s) \equiv \int_0^{\infty} d\rho \, \dot{Q}''(\rho) \, e^{-\lambda\rho} = [\, \pi\delta_s / \sin(\pi s/2) \,](\lambda/\omega_{\rm ph})^{s-1} \,. \tag{29.8}$$

With the relation $\delta_s = M\gamma_s q_0^2 / 2\pi\hbar$ we may express the sojourn contribution $\hat{f}(\lambda, s)$ in terms of the spectral damping function [cf. Eq. (3.83)]

$$\hat{\gamma}(\lambda, s) = \frac{\gamma_s}{\sin(\frac{1}{2}\pi s)} \Big(\frac{\lambda}{\omega_{\rm ph}} \Big)^{s-1} \,, \qquad \tilde{\gamma}(\omega, s) = \frac{\gamma_s}{\sin(\frac{1}{2}\pi s)} \Big(\frac{-i\,\omega}{\omega_{\rm ph}} \Big)^{s-1} \,. \tag{29.9}$$

We have

$$\hat{f}(\lambda, s) = \frac{Mq_0^2}{2\hbar} \, \hat{\gamma}(\lambda, s) \,, \qquad \tilde{f}(\omega, s) = \frac{Mq_0^2}{2\hbar} \, \tilde{\gamma}(\omega, s) \,. \tag{29.10}$$

With this we may write the expression (29.2) as

$$\lim_{\lambda \to 0} \langle \hat{q}(\lambda, s) \rangle_{\rm lin} = \lim_{\lambda \to 0} \frac{\hbar\epsilon}{q_0} \frac{1}{M\hat{\gamma}(\lambda, s)} \frac{1}{\lambda^2} \frac{\hat{f}(\lambda, s)\hat{k}(\lambda, s)}{1 + \hat{f}(\lambda, s)\hat{k}(\lambda, s)} \,. \tag{29.11}$$

Using the relations $\hat{\mu}_{\rm TB}(\lambda, s) = (q_0/\hbar\epsilon)\lambda^2 \langle \hat{q}(\lambda, s) \rangle_{\rm lin}$, we then find the frequency-dependent linear mobility as

$$\tilde{\mu}_{\rm TB}(\omega, s) = \frac{1}{M\tilde{\gamma}(\omega, s)} \frac{\tilde{f}(\omega, s)\tilde{k}(\omega, s)}{1 + \tilde{f}(\omega, s)\tilde{k}(\omega, s)} \,. \tag{29.12}$$

These expressions apply at low frequency both in the sub-Ohmic regime $0 < s < 1$ and in the super-Ohmic regime $1 < s < 2$.

29.1.1   Sub-Ohmic friction

We see from Eq. (29.8) that $\hat{f}(\lambda, s)$ diverges in the sub-Ohmic regime $s < 1$ as $\lambda \to 0$. On the other hand, the integrals in Eq. (29.6) are convergent even in the limit $\lambda \to 0$ because of the charge interaction factor G_m given in (25.34). Hence for $T > 0$ and $s < 1$ the perturbative series (29.6) of the kernel is well-defined as $\lambda \to 0$. Thus, Eq. (29.11) may be expanded for small λ in inverse powers of $\hat{f}(\lambda, s)$,

$$\langle \hat{q}(\lambda, s) \rangle_{\text{lin}} = \frac{\hbar \epsilon}{M q_0} \frac{1}{\lambda^2 \hat{\gamma}(\lambda, s)} \left\{ 1 - \frac{1}{\hat{k}(0, s) \hat{f}(\lambda, s)} + \mathcal{O}[\hat{f}^{-2}(\lambda, s)] \right\}. \tag{29.13}$$

Performing the inverse Laplace transform, we readily obtain the asymptotic expansion

$$\langle q(t) \rangle_{\text{lin}} = \frac{\hbar \sin(\frac{1}{2}\pi s)}{M q_0 \gamma_s \Gamma(1+s)} \frac{\epsilon}{\omega_{\text{ph}}} (\omega_{\text{ph}} t)^s$$
$$\times \left\{ 1 - \frac{\sin(\frac{1}{2}\pi s)}{\pi \delta_s \hat{k}(0, s)} \frac{\Gamma(1+s)}{\Gamma(2s)} (\omega_{\text{ph}} t)^{s-1} + \mathcal{O}[(\omega_{\text{ph}} t)^{2s-2}] \right\}. \tag{29.14}$$

This result is quite remarkable. First, it is nonperturbative with respect to the kernel. In fact, the dependence of $\langle q(t) \rangle_{\text{lin}}$ on temperature and on parameters of the TB lattice is fading away as time approaches the asymptotic regime. Secondly, there is sluggish subdiffusive motion at long time, and the asymptotic motion in the TB lattice is fully equivalent, including the prefactor, to subdiffusive free Brownian motion, as one might have guessed from the duality relation (26.26). Namely, using the relation (7.22), which in the present case, solved for $\chi(t)$, is

$$\chi(t) = (q_0/\hbar \epsilon) \langle \dot{q}(t) \rangle_{\text{lin}}, \tag{29.15}$$

the response function takes the form

$$\chi(t) = \frac{\sin(\frac{1}{2}\pi s)}{M \gamma_s \Gamma(s)} (\omega_{\text{ph}} t)^{s-1}. \tag{29.16}$$

This form coincides with that of the sub-Ohmic free Brownian particle, Eq. (7.54).

The direct calculation of the second moment for asymptotic time is more difficult because of the $\cos(\chi_{k,2m})$-term in Eq. (25.45). We may recall, however, that preparation effects are irrelevant for ergodic systems at asymptotic time, and therefore $\lim_{t \to \infty} \langle q^2(t) \rangle_{\text{pis}}/D_{\text{th}}(t) = 1$. With this, and with use of the relations (7.21) and (29.15), we readily find at finite T and asymptotic time

$$\langle q^2(t) \rangle_{\epsilon=0} = \frac{2 q_0}{\beta \hbar \epsilon} \langle q(t) \rangle_{\text{lin}} = \frac{2}{\beta} \int_0^t dt' \, \chi(t'). \tag{29.17}$$

With the form (29.16), it is now straight to specify $\langle q^2(t) \rangle_{\epsilon=0}$ at asymptotic time.

The linear ac mobility with expansion analogous to Eq. (29.13) takes the form

$$\tilde{\mu}_{\text{TB}}(\omega, s) = \frac{1}{M \tilde{\gamma}(\omega, s)} \left\{ 1 - \frac{\sin(\frac{1}{2}\pi s)}{\pi \delta_s \tilde{k}(0, s)} \left(\frac{-i\omega}{\omega_{\text{ph}}} \right)^{1-s} + \mathcal{O}(\omega^{2-2s}) \right\}. \tag{29.18}$$

The first term is the spectral mobility of a free sub-Ohmic particle, Eq. (7.58).

As temperature is decreased, the effective transfer matrix element of the sub-Ohmic TB model scales to lower values for fixed ω, and at zero temperature it vanishes with an essential singularity $\propto \exp(-\text{const}/\omega^{s/(1-s)})$, as $\omega \to 0$ [cf. Eq. (20.50)]. Hence the kernel $\hat{k}(\lambda)$, and as a consequence of this, also the mobility $\tilde{\mu}_{\text{TB}}(\omega, s)$ drop to zero faster than any power of ω as $\omega \to 0$. This entails that the particle becomes strictly localized in this limit.

29.1.2 Super-Ohmic friction

Consider next the TB model (24.3) for super-Ohmic friction, $1 < s < 2$. We see from Eq. (25.65) that the interaction $Q'(\tau)$ also increases with τ in the regime $T > 0$ and $1 < s < 2$. Hence the charge interaction binds again the charges together and thus the kernel $\hat{k}(\lambda, s)$ remains finite as $\lambda \to 0$. On the other hand, the sojourn factor $\hat{f}(\lambda, s)$ drops to zero in this limit, as we see from Eq. (29.8). Hence, the asymptotic dynamics is determined by the power series in $\hat{f}(\lambda, s)$ of the expression (29.2). Turning to the time regime, we then have

$$\langle q(t) \rangle_{\text{lin}} = \frac{q_0 \hat{k}(0, s)}{2} \epsilon t \left\{ 1 - \frac{\pi \delta_s \hat{k}(0, s)(\omega_{\text{ph}} t)^{1-s}}{\sin(\frac{1}{2}\pi s)\Gamma(3 - s)} + \mathcal{O}[(\omega_{\text{ph}} t)^{-1}, (\omega_{\text{ph}} t)^{2-2s}] \right\}. \quad (29.19)$$

Thus we find diffusive behavior at long time. The corresponding linear ac mobility is frequency-independent in leading order,

$$\tilde{\mu}_{\text{TB}}(\omega, s) = \frac{q_0^2}{2\hbar} \hat{k}(0, s) \left\{ 1 - \hat{k}(0, s) \frac{\pi \delta_s}{\sin(\frac{1}{2}\pi s)} \left(\frac{-i\,\omega}{\omega_{\text{ph}}} \right)^{s-1} \right\}. \quad (29.20)$$

The kernel $\hat{k}(0, s)$ is a perturbative series in Δ^2 depending on temperature.

At zero temperature, the charge interaction approaches asymptotically a constant, $Q'(\tau) - Q'(\infty) \propto \tau^{1-s}$, as we see from Eq. (20.49). It therefore cannot bind the charges anymore. Hence the kernel $\hat{k}(\lambda, s)$ diverges for $T = 0$ in the limit $\lambda \to 0$. We then find from the expression (29.2) that the asymptotic dynamics is given by the first term in Eq. (29.13). Thus we have asymptotically superdiffusive behavior, $\langle q(t) \rangle_{\text{lin}} \propto t^s$. The frequency-dependent mobility coincides with the leading term in Eq. (29.18),

$$\tilde{\mu}_{\text{TB}}(\omega, s, T = 0) = 1/M\tilde{\gamma}(\omega, s), \quad (29.21)$$

but now this form holds in the regime $1 < s < 2$ and $T = 0$. In this regime, the tight-binding lattice is again renormalized to zero, and the mobility is that of a superdiffusive free Brownian particle.

29.2 Weak-binding representation

With use of the duality relation (26.26), and with Eq. (3.83) we directly obtain from the TB expression (29.12) the linear ac mobility of the WB model (24.1) as

$$\tilde{\mu}_{\mathrm{WB}}(\omega,s) = \frac{1}{M\tilde{\gamma}(\omega,s)}\frac{1}{1+\tilde{f}(\omega,2-s)\tilde{k}(\omega,2-s)}. \tag{29.22}$$

Based on this expression it is straight to establish the WB mobility in the various regimes.

29.2.1   Super-Ohmic friction

In the super-Ohmic regime $1 < s < 2$, the function $\tilde{f}(\omega,2-s)$ diverges as $\omega \to 0$. As a result, we find from Eq. (29.22) at low frequency diffusive behavior,

$$\tilde{\mu}_{\mathrm{WB}}(\omega,s) = \frac{\sin^2(\frac{1}{2}\pi s)}{M\gamma_s\pi\delta_s\tilde{k}(0,2-s)}\left\{1 - \frac{\sin(\frac{1}{2}\pi s)}{\pi\delta_s\tilde{k}(0,2-s)}\left(\frac{-i\,\omega}{\omega_{\mathrm{ph}}}\right)^{s-1}\right\}. \tag{29.23}$$

Here we have included the next-to-leading order term. In difference to the mobility for the TB model in the regime $1 < s < 2$ given in Eq. (29.20), the expression (29.23) is nonperturbative in the corrugation strength of the potential.

As $T \to 0$, the kernel $\tilde{k}(0,2-s)$ drops to zero with an essential singularity as $\omega \to 0$, as specified below Eq. (29.18). Hence we find from Eq. (29.22) superdiffusive free Brownian motion, $\tilde{\mu}_{\mathrm{WB}}(\omega \to 0) = 1/M\tilde{\gamma}(\omega,s)$, and the potential is renormalized to zero.

29.2.2   Sub-Ohmic friction

Consider finally the sub-Ohmic regime $s < 1$. Since now the sojourn term $\tilde{f}(\omega,2-s)$ approaches zero as $\omega \to 0$, the mobility (29.22) is perturbative in the kernel,

$$\tilde{\mu}_{\mathrm{WB}}(\omega,s) = \frac{1}{M\tilde{\gamma}(\omega,s)}\left\{1 - \frac{\pi\delta_s\tilde{k}(0,2-s)}{\sin(\frac{1}{2}\pi s)}\left(\frac{-i\,\omega}{\omega_{\mathrm{ph}}}\right)^{1-s}\right\}. \tag{29.24}$$

This result is in agreement with the findings in Ref. [529].

For $T \to 0$, the kernel $\tilde{k}(\omega,2-s)$ diverges in the limit $\omega \to 0$, as argued in the paragraph below Eq. (29.20). As a result, the ac mobility for $T = 0$ and low frequency is nonperturbative in the kernel, and it is reduced compared to the mobility at finite temperature. Quantitative conclusion requires a more detailed analysis.

Considering the fact that the system becomes increasingly sensitive to its low-frequency properties as T is decreased, the transport characteristic found in this chapter is fully consistent with physical intuition. For sub-Ohmic damping and decreasing temperature, the mobility is progressively suppressed at constant low frequency in both models. This is due to the divergence of the spectral damping function $\tilde{\gamma}(\omega)$ as $\omega \to 0$. On the contrary, for super-Ohmic damping the mobility is enhanced with decreasing temperature in both models since now the spectral damping function approaches zero as $\omega \to 0$.

Bibliography

[1] G. Baym, *Lectures on Quantum Mechanics* (Benjamin, Reading, 1969).

[2] D. Chandler, *Introduction to Modern Statistical Mechanics* (Oxford University Press, New York, 1987).

[3] R. P. Feynman and A. R. Hibbs, *Quantum Mechanics and Path Integrals* (Mc Graw-Hill, New York, 1965).

[4] R. P. Feynman, *Statistical Mechanics* (Benjamin, Reading, Mass., 1972).

[5] L. S. Schulman, *Techniques and Applications of Path Integration* (Wiley, 1981).

[6] H. Kleinert, *Path Integrals in Quantum Mechanics, Statistics, Polymer Physics, and Financial Markets*, 3rd edition (World Scientific, Singapore, 2004).

[7] S. Chandrasekhar, Rev. Mod. Phys. **15**, 1 (1943).

[8] M. S. Green, J. Chem. Phys. **20**, 1281 (1952);
R. Kubo, Rep. Progr. Phys. (London) **29**, 255 (1966).

[9] P. Caldirola, Il Nuovo Cim. **18**, 393 (1941).

[10] E. Kanai, Progr. Theor. Phys. **3**, 440 (1948).

[11] W. H. Louisell, *Quantum Statistical Properties of Radiation* (Wiley, New York, 1973).

[12] D. Schuch, Phys. Rev. A **55**, 935 (1997).

[13] H. Dekker, Phys. Rev. A **16**, 2116 (1977).

[14] M. D. Kostin, J. Chem. Phys. **57**, 3589 (1972).

[15] K. Yasue, Ann. Phys. (N.Y.) **114**, 479 (1978).

[16] E. Nelson, Phys. Rev. **150**, 1079 (1966).

[17] S. Nakajima, Progr. Theor. Phys. **20**, 948 (1958).

[18] R. Zwanzig, J. Chem. Phys. **33**, 1338 (1960);
R. Zwanzig, in: *Lectures in Theoretical Physics* (Boulder), Vol. 3, ed. by W. E. Brittin, B. W. Downs, and J. Down (Interscience, New York, 1961).

[19] J. Prigogine and P. Resibois, Physica **27**, 629 (1961).

[20] J. R. Senitzky, Phys. Rev. **119**, 670 (1960).

[21] G. W. Ford, M. Kac, and P. Mazur, J. Math. Phys. **6**, 504 (1965).

[22] H. Mori, Progr. Theor. Phys. **33**, 423 (1965).

[23] F. Haake, in *Quantum Statistics in Optics and Solid State Physics*, Springer Tracts in Modern Physics, Vol. 66, ed. by G. Höhler (Springer, Berlin, 1973).

[24] H. Haken, Rev. Mod. Phys. **47**, 67 (1975).

[25] H. Spohn, Rev. Mod. Phys. **52**, 569 (1980).

[26] H. Dekker, Phys. Rep. **80**, 1 (1981).

[27] H. Grabert, *Projection Operator Techniques in Nonequilibrium Statistical Mechanics*, Springer Tracts in Modern Physics, Vol. 95 (Springer, Berlin, 1982).

[28] K. Blum, *Density Matrix Theory and Applications* (Plenum Press, 1981).

[29] P. Talkner, Ann. Phys. (N.Y.) **167**, 390 (1986).

[30] R. Alicki and K. Lendi, in *Quantum Dynamical Semigroups and Applications*, Lecture Notes in Physics Vol. 286, ed. H. Araki *et al.* (Springer, Berlin, 1987).

[31] C. W. Gardiner, *Quantum Noise* (Springer, Berlin, 1991).

[32] A.G. Redfield, IBM J. Res. Dev. **1**, 19 (1957); Adv. Magn. Reson. **1**, 1 (1965).

[33] G. C. Schatz and M. A. Ratner, *Quantum Mechanics in Chemistry* (Prentice Hall, Englewood Ciffs, New Jersey, 1993).

[34] M. Mehring, *Principles of High-Resolution NMR in Solids* (Springer, 1983).

[35] R. R. Ernst, G. Bodenhausen, and A. Wokaun, *Principles of Nuclear Magnetic Resonance in One and Two Dimensions* (Clarendon Press, Oxford, 1990).

[36] C. P. Slichter, *Principles of Magnetic Resonance* (Springer, Berlin, 1990).

[37] L. Allen and Z. H. Eberly, *Optical Resonance and Two-Level Atoms* (Wiley, New York, 1975).

[38] Y. R. Shen, *The Principles of Nonlinear Optics* (Wiley, New York, 1984).

[39] J. M. Jean, J. Chem. Phys **101**, 10464 (1994);
A. K. Felts, W. T. Pollard, and R. A. Friesner, J. Phys. Chem. **99**, 2929 (1995);
J. M. Jean and G. R. Fleming, J. Chem. Phys. **103**, 2092 (1995).

[40] O. Kühn, V. May, and M. Schreiber, J. Chem. Phys.**101**, 10404 (1994).

[41] V. Sidis, Adv. Chem. Phys. **82**, 73 (1992).

[42] T. Pacher, L. S. Cederbaum, and H. Köppel, Adv. Chem. Phys. **84**, 293 (1993).

[43] W. Domcke and G. Stock, Adv. Chem. Phys. **100**, 1 (1997).

[44] G. Lindblad, Commun. Math. Phys. **48**, 119 (1976).

[45] A. Isar *et al.*, Int. J. Mod. Phys. E **3**, 635 (1994).

[46] Sh. Gao, Phys. Rev. Lett. **79**, 3101 (1997).

[47] a.: Ph. Pechukas, Phys. Rev. Lett. **73**, 1060 (1994);
b.: A. Suárez, R. Silbey, and I. Oppenheim, J. Chem. Phys. **97**, 5101 (1992);
c.: W. J. Munro and C. W. Gardiner, Phys. Rev. A **53**, 2633 (1996).

[48] R. Karrlein and H. Grabert, Phys. Rev. E **55**, 153 (1997).

[49] E. Fick and G. Sauermann, *The Quantum Statistics of Dynamic Processes*, Springer Series in Solid-State Sciences, Vol. 88 (Springer, Berlin, 1990).

[50] C. W. Gardiner, IBM J. Res. Dev. **32**, 127 (1988).

[51] G. W. Ford and M. Kac, J. Stat. Phys. **46**, 803 (1987).

[52] G. W. Ford. J. T. Lewis, and R. F. O'Connell, Phys. Rev. A **37**, 4419 (1988).

[53] R. Benguria and M. Kac, Phys. Rev. Lett. **46**, 1 (1981).

[54] A. Schmid, J. Low Temp. Phys. **49**, 609 (1982).

[55] U. Eckern, W. Lehr, A. Menzel-Dorwarth, F. Pelzer, and A. Schmid,
J. Stat. Phys. **59**, 885 (1990).

[56] R. H. Koch, D. J. van Harlingen, and J. Clarke, Phys. Rev. Lett. **45**, 2132
(1980); *ibid.* **47**, 1216 (1981).

[57] D. Giulini, E. Joos, C. Kiefer, J. Kupsch, I.-O. Stamatescu, and H. D. Zeh, *Decoherence and the Appearance of a Classical World in Quantum Theory* (Springer,
Berlin, 1996).

[58] M. H. A. Davis, *Markov Models and Optimization* (Chapman, London, 1993).

[59] N. Gisin, Phys. Rev. Lett. **52**, 1657 (1984).

[60] N. Gisin and I. C. Percival, J. Phys. A **25**, 5677 (1992);
I. C. Percival, Proc. R. Soc. London A **447**, 189 (1994).

[61] I. Percival, *Quantum State Diffusion* (Cambridge University Press, Cambridge,
1998).

[62] H. P. Breuer and F. Petruccione, J. Phys. A: Math. Gen. **31**, 33 (1998).

[63] N. G. van Kampen, *Stochastic Processes in Physics and Chemistry* (North-Holland, Amsterdam, 1992).

[64] L. Diósi, J. Phys. Math. Gen. **21**, 2885 (1988).

[65] J. Dalibard, Y. Castin, and K. Mølmer, Phys Rev. Lett. **68**, 580 (1992).

[66] R. Dum, P. Zoller, and H. Ritsch, Phys. Rev. A **45**, 4879 (1992);
C. W. Gardiner, A. S. Parkins, and P. Zoller, *ibid.* A **46**, 4363 (1992);
R. Dum, A. S. Parkins, P. Zoller, and C. W. Gardiner, *ibid.* A **46**, 4382 (1992).

[67] H. J. Carmichael, S. Singh, R. Vyas, and P. R. Rice, Phys. Rev. A **39**, 1200
(1989);
H. J. Carmichael, *An Open System Approach to Quantum Optics* (Springer,
Berlin, 1993).

[68] M. Naraschewski and A. Schenzle, Z. Physik A **25**, 5677 (1992).

[69] K. Mølmer, Y. Castin, and J. Dalibard, J. Opt. Soc. Am. B **10**, 524 (1993).

[70] H. M. Wiseman and G. J. Milburn, Phys. Rev A **47**, 1652 (1992).

[71] H. P. Breuer and F. Petruccione, Phys. Rev. Lett. **74**, 3788 (1995);
Phys. Rev. E **51**, 4041 (1995); *ibid.* E **52**, 428 (1995).

[72] H.-P. Breuer and F. Petruccione, *The Theory of Open Quantum Systems* (Oxford University Press, Oxford, 2002).

[73] R. J. Cook, *Quantum Jumps* in Progress in Optics, Vol. XXVIII, ed. by E. Wolf (Elsevier, Amsterdam, 1990).

[74] W. Nagourney, J. Sandberg, and H. Dehmelt, Phys. Rev. Lett.**56**, 2797 (1986); Th. Sauter, W. Neuhauser, R. Blatt, and P. E. Toschek, Phys. Rev. Lett. **57**, 1699 (1986); Th. Basché, S. Kummer, and C. Bräuchle, Nature **373**, 132 (1995).

[75] P. Ullersma, Physica (Utrecht) **32**, 27, 56, 74, 90 (1966).

[76] R. Zwanzig, J. Stat. Phys. **9**, 215 (1973).

[77] A. O. Caldeira and A. J. Leggett, Phys. Rev. Lett. **46**, 211 (1981); and Ann. Phys. (N.Y.) **149**, 374 (1983); *ibid.* **153**, 445(E) (1983).

[78] K. H. Stevens, J. Phys. C **16**, 5765 (1983).

[79] V. Ambegaokar and U. Eckern, Z. Physik B **69**, 399 (1987).

[80] V. B. Magalinskiǐ, Sov. Phys.–JETP **9**, 1381 (1959).

[81] R. J. Rubin, J. Math. Phys. **1**, 309 (1960); *ibid.* **2**, 373 (1961).

[82] A. J. Leggett, Phys. Rev. B **30**, 1208 (1984).

[83] H. Grabert and U. Weiss, Z. Physik B **56**, 171 (1984).

[84] P. Hänggi, in *Stochastic Dynamics*, Lecture Notes in Physics, Vol.**484**, ed. by L. Schimansky-Geier and Th. Pöschel (Springer, Berlin, 1997), p. 15.

[85] A. J. Leggett, S. Chakravarty, A. T. Dorsey, M. P. A. Fisher, A. Garg, and W. Zwerger, Rev. Mod. Phys. **59**, 1 (1987); *ibid.* **67**, 725 (1995) [erratum].

[86] J.-D. Bao, J. Stat. Phys. **114**, 503 (2004).

[87] H. Grabert, P. Schramm, and G.-L. Ingold, Phys. Rep. **168**, 115 (1988); P. Schramm and H. Grabert, J. Stat. Phys. **49**, 767 (1987).

[88] I. S. Gradshteyn and I. M. Ryzhik, *Tables of Integrals, Series and Products* (Academic Press, London, 1965).

[89] M. Abramowitz and I. Stegun, *Handbook of Mathematical Functions* (Dover, New York, 1971).

[90] A. I. Saichev and G. M. Zaslavsky, Chaos **7**, 753 (1997).

[91] R. Metzler and J. Klafter, Phys. Rep. **339**, 1 (2000).

[92] E. Lutz, Phys. Rev. E **64**, 051106 (2001).

[93] R. J. Rubin, Phys. Rev. **131**, 964 (1963).

[94] M. Maekawa and K. Wada, Phys. Lett. A **80**, 293 (1980).

[95] A.V. Mokshin, R. M. Yulmetyev, and P. Hänggi, New J. Phys. **7**, 9 (2005); Phys. Rev. Lett. **95**, 200601 (2005).

[96] H. Spohn, *Dynamics of Charged Particles and Their Radiation Field* (Cambridge University Press, Cambridge, 2004).

[97] G. W. Ford, J. T. Lewis, and R. F. O'Connell, Phys. Rev. Lett. **55**, 2273 (1985).

[98] J.-D. Bao, P. Hänggi, and Yi-Zh. Zhuo, Phys. Rev. E **72**, 061107 (2005).

[99] H. Wipf, D. Steinbinder, K. Neumaier, P. Gutsmiedl, A. Magerl, and A. J. Dianoux, Europhys. Lett. **4**, 1379 (1989);
D. Steinbinder, H. Wipf, A. Magerl, A. D. Dianoux, and K.Neumaier, *ibid.* **6**, 535 (1988); *ibid.* **16**, 211 (1991).
For a review see H. Grabert and H. Wipf, in: Festkörperprobleme/Advances in Solid State Physics, Vol. 30, p. 1, ed. by U. Rössler (Vieweg, Braunschweig, 1990).

[100] G. M. Luke *et al.*, Phys. Rev. B **43**, 3284 (1991);
O. Hartmann *et al.*, Hyperfine Interactions **64**, 641 (1990), and references therein; I. S. Anderson, Phys. Rev. Lett. **65**, 1439 (1990).

[101] *Hydrogen in Metals III*, Topics in Applied Physics, Vol. 73, ed. by H. Wipf (Springer, Berlin, 1997).

[102] A. Würger, *From Coherent Tunneling to Relaxation*, Springer Tracts in Modern Physics, Vol. 135 (Springer, Berlin, 1997).

[103] K. Chun and N. O. Birge, Phys. Rev. B **48**, 11500 (1993); B **54**, 4629 (1996).

[104] B. Golding, N. M. Zimmermann, and S. N. Coppersmith, Phys. Rev. Lett. **68**, 998 (1992).

[105] R. Marcus, J. Chem. Phys. **24**, 966 (1956).

[106] R. A. Marcus and N. Sutin, Biochim. Biophys. Acta **811**, 265 (1985).

[107] S. Coleman, Phys. Rev. D **15**, 2929 (1977); S. Coleman, in *The Whys of Subnuclear Physics*, ed. by A. Zichichi (Plenum, New York, 1979), p. 805.

[108] U. Weiss and W. Häffner, Phys. Rev. D **27**, 2916 (1983).

[109] A. Barone and G. Paterno, *Physics and Application of the Josephson Effect* (Wiley, New York, 1982).

[110] J. R. Friedman *et al.*, Nature (London) **406**, 43 (2000).

[111] C. H. van der Wal *et. al.*, Science **290**, 773 (2000).

[112] J. E. Mooij *et al.*, Science **285**, 1036 (1999).

[113] M. V. Feigelman *et al.*, J. Low Temp. Phys. **118**, 805 (2000).

[114] Yu. Makhlin, G. Schön, and A. Shnirman, Rev. Mod. Phys. **73**, 357 (2001).

[115] Y. Nakamura, Yu. A. Pashkin, and J. S. Tsai, Nature **398**, 786 (1999).

[116] E. Paladino, L. Faoro, G. Falci, R. Fazio, Phys. Rev. Lett. **88**, 228304 (2002).

[117] G. Ithier, E. Collin, P. Joyez, P. J. Meeson, D. Vion, D. Esteve, F. Chiarello, A. Shnirman, Yu. Makhlin, J. Schriefl, and G. Schön, Phys. Rev. B **72**, 134 519 (2005).

[118] A. Shnirman, G. Schön, I. Martin, and Yu. Makhlin, Phys. Rev. Lett. **94**, 127 002 (2005).

[119] P. W. Anderson, B. I. Halperin, and C. M. Varma, Phil. Mag. **25**, 1 (1972).

[120] W. A. Phillips, J. Low Temp. Phys. **7**, 351 (1972).

[121] J. L. Black, in *Glassy Metals I*, Topics in Applied Physics, Vol. 46, ed. by H.-J. Güntherodt and H. Beck (Springer, Berlin, 1981).

[122] Yu. Makhlin, and A. Shnirman, Phys. Rev. Lett. **92**, 178301 (2004).

[123] D. Vion *et al.*, Science **296**, 886 (2002).

[124] J. M. Martinis *et al.*, Phys. Rev. Lett. **89**, 117901 (2002).

[125] I. Chiorescu, Y. Nakamura, C. Harmans, J. E. Mooij, Science **299**, 1869 (2003).

[126] A. Wallraff *et al.*, Nature **431**, 162 (2004).

[127] A. O. Niskanen, K. Harrabi, F. Yoshihara, Y. Nakamura, S. Lloyd, and J. S. Tsai, Science **316**, 723 (2007).

[128] J. Yamashita and T. Kurosawa, J. Chem. Solids **5**, 34 (1958).

[129] a.: T. Holstein, Ann. Phys. (N.Y.) **8**, 325 (1959); b.: *ibid.* **8**, 343 (1959).

[130] H. B. Shore and L. M. Sander, Phys. Rev. B **12**, 1546 (1975).

[131] M. Wagner, J. Phys. A **18**, 1915 (1986).

[132] C. P. Flynn and A. M. Stoneham, Phys. Rev. B **1**, 3966 (1970).

[133] Y. Kagan and M. I. Klinger, J. Phys. C **7**, 2791 (1974).

[134] H. Teichler and A. Seeger, Phys. Lett. **82** A, 91 (1981).

[135] H. Fröhlich, Adv. Phys. **3**, 325 (1954).

[136] J. Kondo, in *Fermi Surface Effects*, Vol. 77 of Springer Series in Solid State Sciences, eds. J. Kondo and A. Yoshimori (Springer, Berlin, 1988).

[137] T. Regelmann, L. Schimmele, and A. Seeger, Z. Physik B **95**, 441 (1994).

[138] G. D. Mahan, Many-Particle Physics (Plenum Press, New York, 1981).

[139] G. D. Mahan, in *Fermi Surface Effects*, Vol. 77 of Springer Series in Solid State Sciences, eds. J. Kondo and A. Yoshimori (Springer, Berlin, 1988).

[140] K. Ohtaka and Y. Tanabe, Rev. Mod. Phys. **62**, 929 (1990).

[141] A. M. Tsvelik and P. B. Wiegmann, Adv. Phys. **32**, 453 (1983).

[142] S. Chakravarty, Phys. Rev. Lett. **49**, 681 (1982).

[143] A. J. Bray and M. A. Moore, Phys. Rev. Lett. **49**, 1546 (1982).

[144] V. Hakim, A. Muramatsu, and F. Guinea, Phys. Rev. B **30**, 464 (1984).

[145] A. Schmid, Phys. Rev. Lett. **51**, 1506 (1983).

[146] S. Bulgadaev, Sov. Phys.–JETP Lett. **39**, 314 (1985).

[147] F. Guinea, V. Hakim, and A. Muramatsu, Phys. Rev. Lett. **54**, 263 (1985).

[148] P. G. de Gennes, *Superconductivity of Metals and Alloys* (Wesley, New York, 1989).

[149] D. J. Scalapino, in *Superconductivity*, Vol. 1, ed. by R. D. Parks (Marcel Dekker, New York, 1969).

[150] a.: V. Ambegaokar, U. Eckern, and G. Schön, Phys. Rev. Lett. **48**, 1745 (1982); b.: U. Eckern, G. Schön, and V. Ambegaokar, Phys. Rev. B **30**, 6419 (1984).

[151] A. I. Larkin and Yu. N. Ovchinnikov, Phys. Rev. B **28**, 6281 (1983).

[152] G. Schön and A. D. Zaikin, Phys. Rep. **198**, 237 (1990).

[153] Special Issue on *Single Charge Tunneling*, Z. Physik B **85** (3), 317-468 (1991).

[154] *Single Charge Tunneling*, ed. by H. Grabert and M. H. Devoret, NATO ASI Series B: Physics Vol. 294 (Plenum Press, New York, 1992).

[155] G.-L. Ingold and Yu. V. Nazarov, in Ref. [154], p. 21-107.

[156] D. V. Averin and K. K. Likharev, J. Low Temp. Phys. **62**, 345 (1986).

[157] *Quantum Tunneling of Magnetization - QTM 94*, ed. by L. Gunther and B. Barbara (Kluwer, Dordrecht, 1995).

[158] P. C. E. Stamp in Ref. [159].

[159] *Tunneling in Complex Systems*, Proceedings from the Institute for Nuclear Theory, Vol. 5, ed. by S. Tomsovic (World Scientific, Singapore, 1998).

[160] J. L. van Hemmen and A. Suto, Europhys. Lett. **1**, 481 (1986); and in Ref. [157].

[161] M. Enz and R. Schilling, J. Phys. C **19**, L 711 and 1765 (1986); and in Ref. [157].

[162] D. Gatteschi, R. Sessoli, and J. Villain, *Molecular Nanomagnets* (Oxford University Press, Oxford, 2006).

[163] I. S. Tupitsyn, N. V. Prokof'ev and P. C. E. Stamp, Intl. J. Mod. Phys. B **11**, 2901 (1997).

[164] A. O. Caldeira, A. H. Castro Neto, and T. O. de Carvalho, Phys. Rev. B **48**, 13 974 (1993).

[165] G. Mahler and V. A. Weberruß, *Quantum Networks* (Springer, Berlin, 1998).

[166] J. Allinger and U. Weiss, Z. Physik B **98**, 289 (1995).

[167] D. Cohen, Phys. Rev. E **55**, 1422 (1997); Phys. Rev. Lett. **78**, 2878 (1997).

[168] S. Doniach and E. H. Sondheimer, *Green's Functions for Solid State Physicists* (Benjamin, Reading, 1974).

[169] J. P. Sethna, Phys. Rev. B **24**, 698 (1981); *ibid.* B **25**, 5050 (1982).

[170] A. A. Louis and J. P. Sethna, Phys. Rev. Lett. **74**, 1363 (1995).

[171] H. Sugimoto and Y. Fukai, Phys. Rev. B **22**, 670 (1980);
A. Klamt and H. Teichler, Phys. Stat. Sol. (B) **134**, 103 (1986).

[172] A. Sumi and Y. Toyozawa, J. Phys. Soc. Jpn. **35**, 137 (1973).

[173] F. M. Peeters and J. T. Devreese, Phys. Rev. B **32**, 3515 (1985).

[174] B. Gerlach and H. Löwen, Rev. Mod. Phys. **63**, 63 (1991).

[175] Yu. Kagan, J. Low Temp. Phys. **87**, 525 (1992).

[176] *Quantum Tunnelling in Condensed Media*, ed. by Yu. Kagan and A. J. Leggett (Elsevier Publishers, Amsterdam, 1992).

[177] Yu. Kagan and N. V. Prokov'ev, in Ref. [176].

[178] F. Napoli, M. Sassetti, and U. Weiss, Physica B **202**, 80 (1994).

[179] J. A. Stroscio and D. M. Eigler, Science **254**, 1319 (1991).

[180] D. M. Eigler, C. P. Lutz, and W. E. Rudge, Nature **352**, 600 (1991).

[181] M.F. Crommie *et al.*, Nature (London) **363** 524 (1993); Science **262**, 218 (1993).

[182] J. E. Artacho and L. M. Falicov, Phys. Rev. B **47**, 1190 (1993).

[183] M. Sassetti, E. Galleani d'Agliano, and F. Napoli, Physica B **154**, 359 (1989).

[184] E. G. d'Agliano, P. Kumar, W. Schaich, H. Suhl, Phys. Rev. B **11**, 2122 (1975).

[185] L.-D. Chang and S. Chakravarty, Phys. Rev. B **31**, 154 (1985).

[186] P. Nozières and C. De Dominicis, Phys. Rev. **178**, 1097 (1969).

[187] F. Guinea, V. Hakim, and A. Muramatsu, Phys. Rev. B **32**, 4410 (1985).

[188] F. Sols and F. Guinea, Phys. Rev. B **36**, 7775 (1987).

[189] K. Schönhammer, Phys. Rev. B **43**, 11323 (1991).

[190] P. W. Anderson, Phys. Rev. Lett. **18**, 1049 (1967).

[191] K. Yamada and K. Yosida, Progr. Theor. Phys. **68**, 1504 (1982);
K. Yamada, A. Sakurai, S. Miyazima, and H. S. Wang, Progr. Theor. Phys. **75**, 1030 (1986).

[192] A. Oguchi and K. Yosida, Progr. Theor. Phys. **75**, 1048 (1986);
T. Kitamura, A. Oguchi and K. Yosida, Progr. Theor. Phys. **78**, 583 (1987).

[193] N. R. Wertheimer, Phys. Rev. **147**, 255 (1966).

[194] G. Schön and A. D. Zaikin, Phys. Rev. B **40**, 5231 (1989).

[195] F. Guinea and G. Schön, Physica B **152**, 165 (1988).

[196] F. W. J. Hekking, L. I. Glazman, K. A. Matveev, and R. I. Shekhter, Phys. Rev. Lett. **70**, 4138 (1993).

[197] F. W. J. Hekking and Yu V. Nazarov, Phys. Rev. Lett. **71**, 1625 (1993).

[198] E. Pollak, Chem. Phys. Lett. **127**, 178 (1986).

[199] W. P. Schleich, *Quantum Optics in Phase Space* (Wiley-VCH, 2001).

[200] H. Weyl, Z. Physik **46**, 1 (1927).

[201] M. Hillery, R. F. O'Connell, M. O. Scully, and E. P. Wigner, Phys. Rep. **106**, 122 (1984).

[202] E. Wigner, Phys. Rev. **40**, 749 (1932).

[203] U. Weiss, Z. Physik B **30**, 429 (1978).

[204] R. P. Feynman and F. L. Vernon, Ann. Phys. (N.Y.) **24**, 118 (1963).

[205] A. O. Caldeira and A. J. Leggett, Physica **121 A**, 587 (1983).

[206] A. Stern, Y. Aharonov, and Y. Imry, Phys. Rev. A **41**, 3436 (1990).

[207] D. Loss and K. Mullen, Phys. Rev. B **43**, 13 252 (1991).

[208] B. d'Espagnat, *Conceptual Foundations of Quantum Mechanics* (Benjamin, Reading, 1976).

[209] P. Grigolini, *Quantum Mechanical Irreversibility and Measurement* (World Scientific, 1993).

[210] S. Dattagupta, *Relaxation Phenomena in Condensed Matter Physics* (Academic Press, New York, 1987).

[211] J. Schwinger, J. Math. Phys. **2**, 407 (1961).

[212] L. P. Kadanoff and G. Baym, *Quantum Statistical Mechanics* (Benjamin, 1962).

[213] L. V. Keldysh, Sov. Phys.–JETP **20**, 1018 (1965).

[214] A. Kamenev, *Many-body theory of nonequilibrium systems*, in *Nanophysics: coherence and transport*, ed. by H. Bouchiat et al. (Elsevier, Amsterdam, 2005), [arXiv: condmat/0412296v2 (2005)].

[215] K. C. Chou, Z. B. Su, B. L. Hao, and L. Yu, Phys. Rep. **118**, 1 (1985).

[216] J. Rammer and H. Smith, Rev. Mod. Phys. **58**, 323 (1986).

[217] J. Rammer, *Quantum Field Theory of Nonequilibrium States* (Cambridge University Press, Cambridge, 2007).

[218] Y.-C. Chen, J. L. Lebowitz, and C. Liverani, Phys. Rev. B **40**, 4664 (1989).

[219] M. Sassetti and U. Weiss, Phys. Rev. A **41**, 5383 (1990).

[220] M. Sassetti and U. Weiss, Phys. Rev. Lett. **65**, 2262 (1990).

[221] H. Svensmark and K. Flensberg, Phys. Rev. A **47**, R23 (1993).

[222] K. S. Chow, D. A. Browne, and V. Ambegaokar, Phys. Rev. B **37**, 1624 (1988); K. S. Chow and V. Ambegaokar, Phys. Rev. B **38**, 11 168 (1988).

[223] D. S. Golubev, J. König, H. Schoeller, G. Schön, and A. D. Zaikin, Phys. Rev. B **56**, 15 782 (1997).

[224] W. T. Strunz, L. Diósi, and N. Gisin, Phys. Rev. Lett. **82**, 1801 (1999).

[225] J. T. Stockburger and C. H. Mak, Phys. Rev. Lett. **80**, 2657 (1998); J. Chem. Phys. **110**, 4983 (1999).

[226] J. T. Stockburger and H. Grabert, Phys. Rev. Lett. **88**, 170407 (2002).

[227] J. T. Stockburger, Chem. Phys. **296**, 159 (2003).

[228] J .H. Van Vleck, J. Math. Natl. Acad. Sci. U.S.A. **14**, 178 (1928).

[229] M. C. Gutzwiller, *Chaos in Classical and Quantum Mechanics*, Interdisciplinary Applied Mathematics, Vol. 1 (Springer, Berlin, 1990).

[230] M. C. Gutzwiller, J. Math. Phys. **8**, 1979 (1967).

[231] E. J. Heller, J. Chem. Phys. **75**, 2923 (1981).

[232] M. F. Herman and E. Kluk, Chem. Phys. **91**, 27 (1984).

[233] K. G. Kay, Chem. Phys. **322**, 3 (2006); G. Hochman and K. G. Kay, Phys. Rev. A **73**, 064102 (2006).

[234] W. Koch, F. Großmann, J. T. Stockburger, and J. Ankerhold, Phys. Rev. Lett. **100**, 230402 (2008).

[235] S. Zhang and E. Pollak, Phys. Rev. Lett. **91**, 190201 (2003).

[236] H. Grabert, U. Weiss, and P. Talkner, Z. Physik B **55**, 87 (1984).

[237] F. Haake and R. Reibold, Phys. Rev. A **32**, 2462 (1985).

[238] R. Jung, G.-L. Ingold, and H. Grabert, Phys. Rev. A **32**, 2510 (1985).

[239] H. Metiu and G. Schön, Phys. Rev. Lett. **53**, 13 (1984).

[240] C. Aslangul, N. Pottier, and D. Saint-James, J. Stat. Phys. **40**, 167 (1985).

[241] P. S. Riseborough, P. Hänggi, and U. Weiss, Phys. Rev. A **31**, 471 (1985).

[242] K. Lindenberg and B. J. West, Phys. Rev. A **30**, 568 (1984).

[243] B. L. Hu, J. P. Paz, and Y. Zhang, Phys. Rev. D **45**, 2843 (1992).

[244] A. Einstein, Ann. Phys. (Leipzig) **17**, 549 (1905).

[245] J. B. Johnson, Phys. Rev. **32**, 97 (1928).

[246] H. Nyquist, Phys. Rev. **32**, 110 (1928).

[247] H. B. Callen and T. A. Welton, Phys. Rev. **83**, 34 (1951).

[248] P. Talkner, Z. Physik B **41**, 365 (1981).

[249] H. Shiba, Progr. Theor. Phys. **54**, 967 (1975).

[250] G. W. Ford, J. T. Lewis, and R. F. O'Connell, Ann. of Physics **185**, 270 (1988).

[251] A. Hanke and W. Zwerger, Phys. Rev. E **52**, 6875 (1995).

[252] S. A. Adelman, J. Chem. Phys. **64**, 124 (1976).

[253] H. A. Kramers, Physica (Utrecht) **7**, 284 (1940).

[254] A. Sandulescu and H. Scutaru, Ann. Phys. (N.Y.) **173**, 277 (1987).

[255] V. Hakim and V. Ambegaokar, Phys. Rev. A **32**, 423 (1985).

[256] A. Erdélyi, *Higher Transcendental Functions*, Vol. 3 (McGraw-Hill, New York, 1955).

[257] P. Hänggi, G.-L. Ingold, and P. Talkner, New. J. Phys. **10**, 115008 (2008).

[258] R. P. Feynman, Phys. Rev. **97**, 660 (1955).

[259] R. Giachetti and V. Tognetti, Phys. Rev. Lett. **55**, 912 (1985); Phys. Rev. B **33**, 7647 (1986).

[260] R. P. Feynman and H. Kleinert, Phys. Rev. A **34**, 5080 (1986).

[261] H. Leschke, in *Path Summations: Achievements and Goals*, ed. by S. Lundquist *et al.* (World Scientific, Singapore, 1987).

[262] W. Janke, in *Path Integrals from meV to MeV*, ed. by V. Sa-yakanit *et al.* (World Scientific, Singapore, 1989).

[263] R. Giachetti, V. Tognetti, and R. Vaia, in *Path Summations: Achievements and Goals*, ed. by S. Lundquist *et al.* (World Scientific, Singapore, 1987).

[264] A. Cuccoli, V. Tognetti, and R. Vaia, in *Quantum Fluctuations in Mesoscopic and Macroscopic Systems*, ed. by H. A. Cerdeira, F. Guinea Lopez, and U. Weiss (World Scientific, Singapore, 1991).

[265] R. Giachetti and V. Tognetti, Phys. Rev. A **36**, 5512 (1987); R. Giachetti, V. Tognetti, R. Vaia, Phys. Rev. A **37**, 2165 (1988); A **38**, 1521, 1638 (1988).

[266] G. Falci, R. Fazio, and G. Giaquinta, Europhys. Lett. **14**, 145 (1991); S. Kim and M. Y. Choi, Phys. Rev. B **42**, 80 (1990).

[267] A. Cuccoli *et al.*, Phys. Rev. A **45**, 8418 (1992); A. Cuccoli, R. Giachetti, V. Tognetti, R. Vaia, and P. Verrucchi, J. Phys.: Condens. Matter **7**, 7891 (1995).

[268] H. Kleinert, Phys. Lett. A **174**, 332 (1992).

[269] H. Kleinert, W. Kürzinger, and A. Pelster, J. Phys. A **31**, 8307 (1998).

[270] A. Cuccoli, A. Rossi, V. Tognetti, and R. Vaia, Phys. Rev. E **55**, 4849 (1997).

[271] D. M. Larsen, Phys. Rev. B **32**, 2657 (1985), *ibid.* B **33**, 799 (1986); S. N. Gorshkov, A. V. Zabrodin, C. Rodriguez, V. K. Fedyanin, Theor. Math. Phys. **62**, 205 (1985); K. M. Broderix, N. Heldt, H. Leschke, Z. Physik B **66**, 507 (1987).

[272] J. T. Devreese and F. Brosens, Phys. Rev. B **45**, 6459 (1992).

[273] W. H. Zurek, Phys. Today **44** (10), 36 (1991).

[274] S. Chakravarty and A. Schmid, Phys. Rep. **140**, 193 (1986).

[275] S. Washburn and R. A. Webb, Adv. Phys. **35**, 375 (1986).

[276] P. Mohanty, E. M. Q. Jariwala, R. A. Webb, Phys. Rev. Lett. **78**, 3366 (1997).

[277] P. Mohanty and R. A. Webb, Phys. Rev. B **55**, R13452 (1997).

[278] L. Saminadayar, P. Mohanty, R. A. Webb, P. Degiovanni, and C. Bäuerle, Physica E **40**, 12 (2007).

[279] I. L. Aleiner, B. L. Altshuler, and M. E. Gershenson, Waves in Random Media **9**, 201 (1999); Phys. Rev. Lett. **82**, 3190 (1999); I. L. Aleiner, B. L. Altshuler, and M. G. Vavilov, J. Low Temp. Phys. **126**, 1377 (2002).

[280] D.S. Golubev and A. D. Zaikin, Phys. Rev. B **59**, 9195 (1999).

[281] D. S. Golubev and A. D. Zaikin, Phys. Rev. B **62**, 14061 (2000).

[282] D. S. Golubev, A. D. Zaikin, G. Schön, J. Low Temp. Phys. **126**, 1355 (2002).

[283] J. von Delft, Intl. J. Mod. Phys. B **22**, 727 (2008).

[284] F. Marquardt, J. von Delft, R. A. Smith, and V. Ambegaokar, Phys. Rev. B **76**, 195331 (2007).

[285] D. S. Golubev and A. D. Zaikin, J. Phys.: Conf. Ser. **129**, 012016 (2008).

[286] F. Guinea, Phys. Rev. B **65**, 205317 (2002).

[287] B. Altshuler, A. G. Aronov, and D. E. Khmelnitskii, J. Phys. C **15**, 7367 (1982).

[288] D. Cohen, J. von Delft, F. Marquardt, and Y. Imry, Phys. Rev. B **80**, 245410 (2009).

[289] D. Cohen, J. Phys. A **31**, 8199 (1998); D. Cohen and Y. Imry, Phys. Rev. B **59**, 11143 (1999).

[290] D. Golubev and A. D. Zaikin, Phys. Rev. Lett. **81**, 1074 (1998).

[291] R. Schuster *et al.*, Nature (London) **385**, 417 (1997).

[292] I. L. Aleiner, N. S. Wingreen, and Y. Meir, Phys. Rev. Lett. **79**, 3740 (1997).

[293] K. A. Eriksen, P. Hedegård, and H. Bruus, Phys. Rev. B **64**, 195327 (2001).

[294] S. Arrhenius, Z. Phys. Chem. (Leipzig) **4**, 226 (1889).

[295] A. J. Leggett, Contemp. Phys. **25**, 583 (1984).

[296] A. J. Leggett, in *Directions in Condensed Matter Physics*, Vol. 1, ed. by G. Grinstein and G. Mazenko (World Scientific, Singapore, 1986), p. 187.

[297] H. Grabert, P. Olschowski, and U. Weiss, Phys. Rev. B **36**, 1931 (1987).

[298] P. Hänggi, P. Talkner, and M. Borkovec, Rev. Mod. Phys. **62**, 251 (1990).

[299] J. Ankerhold, *Quantum Tunneling in Complex Systems*, Springer Tracts in Modern Physics, Vol. 224 (Springer Verlag, Berlin, 2007).

[300] A. N. Cleland, J. M. Martinis, and J. Clarke, Phys. Rev. B **37**, 5950 (1988).

[301] D. M. Brink, J. M. Neto, and H. A. Weidenmüller, Phys. Lett. B **80**, 170 (1979).

[302] K. Möhring and U. Smilansky, Nucl. Phys. A **338**, 227 (1980).

[303] D. Emin and T. Holstein, Ann. Phys. (N.Y.) **53**, 439 (1969).

[304] H. Risken, *The Fokker-Planck Equation* (Springer Verlag, Berlin, 1984).

[305] F. Hund, Z. Physik **43**, 805 (1927).

[306] J. R. Oppenheimer, Phys. Rev. **31**, 80 (1928).

[307] G. Gamow, Z. Physik **51**, 204 (1928).

[308] R. W. Gurney and E. U. Condon, Nature (London) **122**, 439 (1928).

[309] E. P. Wigner, Z. Phys. Chem. B **19**, 203 (1932).

[310] W. H. Miller, J. Chem. Phys. **62**, 1899 (1975).

[311] W. H. Miller, S. D. Schwartz, J. W. Tromp, J. Chem. Phys. **79**, 4889 (1983).

[312] W. H. Miller, J. Chem. Phys. **61**, 1823 (1974).

[313] T. Yamamoto, J. Chem. Phys. **33**, 281 (1960).

[314] E. Pollak and J.-L. Liao, J. Chem. Phys. **108**, 2733 (1998);
       G. Gershinsky and E. Pollak, J. Chem. Phys. **108**, 2756 (1998).

[315] F. Matzkies and U. Manthe, J. Chem. Phys. **106**, 2646 (1997).

[316] W. H. Thompson and W. H. Miller, J. Chem. Phys. **102**, 7409 (1995); *ibid.*
       106, 142 (1997).

[317] F. J. Mc Lafferty and Ph. Pechukas, Chem. Phys. Lett. **27**, 511 (1974).

[318] F. Haake, *Quantum Signatures of Chaos* (Springer, Berlin, 2nd edition, 2000).

[319] I. Affleck, Phys. Rev. Lett. **46**, 388 (1981).

[320] J. S. Langer, Ann. Phys. (N.Y.) **41**, 108 (1967); *ibid.* **54**, 258 (1969).

[321] J. S. Langer, in *Systems far from Equilibrium*, Lecture Notes in Physics,
       Vol. 132, ed. by L. Garrido (Springer, Berlin, 1980), p. 12.

[322] C. G. Callan and S. Coleman, Phys. Rev. D **16**, 1762 (1977).

[323] M. Stone, Phys. Lett. **67B**, 186 (1977).

[324] T. Nakamura, A. Ottewill, and S. Takagi, Ann. Phys. (N.Y.) **260**, 9 (1997).

[325] R. F. Dashen, B. Hasslacher, and A. Neveu, Phys. Rev. D **10**, 4114 (1974).

[326] R. P. Bell, *The Tunnel Effect in Chemistry* (Chapman and Hall, London, 1980).

[327] V. I. Goldanskii, Dokl. Acad. Nauk SSSR **124**, 1261 (1959); **127**, 1037 (1959).

[328] P. Reimann, M. Grifoni, and P. Hänggi, Phys. Rev. Lett. **79**, 10 (1997).

[329] S. Keshavamurthy and W. H. Miller, Chem. Phys. Lett. **218**, 189 (1994).

[330] N. T. Maitra and E. J. Heller, Phys. Rev. Lett. **78**, 3035 (1997).

[331] M. J. Gillan, J. Phys. C **20**, 3621 (1987);
       see also P. G. Wolynes, J. Chem. Phys. **87**, 6559 (1987).

[332] G. A. Voth, D. Chandler, and W. H. Miller, J. Chem. Phys. **91**, 7749 (1989).

[333] J. Cao and G. A. Voth, J. Chem. Phys. **105**, 6856 (1996).

[334] D. Makarov and M. Topaler, Phys. Rev. E **52**, 178 (1995).

[335] M. C. Gutzwiller, J. Math. Phys. **12**, 343 (1971);
M. C. Gutzwiller, Physica D (Utrecht) **5**, 183 (1982).

[336] T. Banks, C. M. Bender, and T. T. Wu, Phys. Rev. D **8**, 3346 (1973).

[337] R. F. Grote and J. T. Hynes, J. Chem. Phys. **73**, 2715 (1980);
P. Hänggi and F. Mojtabai, Phys. Rev. A **29**, 1168 (1982).

[338] P. G. Wolynes, Phys. Rev. Lett. **47**, 968 (1981); see also:
V. I. Mel'nikov and S. V. Meshkov, Sov. Phys.–JETP Lett. **60**, 38 (1983).

[339] H. Grabert, U. Weiss, and P. Hänggi, Phys. Rev. Lett. **52**, 2193 (1984).

[340] H. Grabert and U. Weiss, Phys. Rev. Lett. **53**, 1787 (1984).

[341] A. I. Larkin and Yu. N. Ovchinnikov, Sov. Phys.–JETP Lett. **37**, 382 (1983).

[342] A. I. Larkin and Yu. N. Ovchinnikov, Sov. Phys.–JETP **59**, 420 (1984).

[343] A. M. Levine, M. Shapiro, and E. Pollak, J. Chem. Phys. **88**, 1959 (1988).

[344] H. Grabert, Phys. Rev. Lett. **61**, 1683 (1988).

[345] E. Pollak, H. Grabert, and P. Hänggi, J. Chem. Phys. **91**, 4073 (1989).

[346] V. I. Mel'nikov and S. V. Meshkov, J. Chem. Phys. **85**, 1018 (1986).

[347] E. Hershkovitz and E. Pollak, J. Chem. Phys. **106**, 7678 (1997).

[348] P. Hänggi, H. Grabert, G.-L. Ingold, and U. Weiss, Phys. Rev. Lett. **55**, 761 (1985).
For an experimental confirmation of Eq. (15.16) see J. B. Bouchaud, E. Cohen de Lara, and R. Kahn, Europhys. Lett. **17**, 583 (1992).

[349] J. Ankerhold, Ph. Pechukas, and H. Grabert, Phys. Rev. Lett. **87**, 086802 (2001).

[350] W. T. Coffey, Yu. P. Kalmykov, S. V. Titov, and B. P. Mulligan, J. Phys. A: Math. Theor. **40**, F91 (2007); *ibid.* **40**, 12505 (2007).

[351] L. Machura *et al.*, Phys. Rev. E **70**, 031107 (2004);
J. Luczka, R. Rudnicki, and P. Hänggi, Physica A 351, 60 (2005).

[352] D. Waxman and A. J. Leggett, Phys. Rev. B **32**, 4450 (1985).

[353] P. Hänggi and W. Hontscha, J. Chem. Phys. **88**, 4094 (1988); Ber. Bunsenges. Phys. Chem. **95**, 379 (1991).

[354] A. Schmid, Ann. Phys. (N.Y.) **170**, 333 (1986).

[355] U. Eckern and A. Schmid, in Ref. [176].

[356] H. Grabert, P. Olschowski, and U. Weiss, Z. Physik B **68**, 193 (1987).

[357] E. Freidkin, P. S. Riseborough, and P. Hänggi, Z. Physik B **64**, 237 (1986).

[358] H. Grabert and U. Weiss, Z. Physik B **56**, 171 (1984).

[359] J. M. Martinis and H. Grabert, Phys. Rev. B **38**, 2371 (1988).

[360] U. Weiss, M. Sassetti, Th. Negele, M. Wollensak, Z. Physik B **84**, 471 (1991).

[361] S. Washburn, R. A. Webb, R. F. Voss, and S. M. Faris, Phys. Rev. Lett. **54**, 2712 (1985);
D. B. Schwartz, B. Sen, C. N. Archie, and J. E. Lukens, **55**, 1547 (1985);
A. N. Cleland, J. M. Martinis, J. Clarke, Phys. Rev. B **36**, 58 (1987).

[362] S. E. Korshunov, Sov. Phys.–JETP **65**, 1025 (1987).

[363] A. Erdélyi, *Higher Transcendental Functions*, Vol. 1 (McGraw-Hill, New York, 1955).

[364] L.-D. Chang and S. Chakravarty, Phys. Rev. B **29**, 130 (1984); *ibid*. B **30**, 1566(E) (1984).

[365] M. H. Devoret, D. Esteve, C. Urbina, J. Martinis, A. N. Cleland, and J. Clarke, in Ref. [176].

[366] S. Takagi, *Macroscopic Quantum Tunneling* (Cambridge University Press, Cambridge, 2002).

[367] G. Careri and G. Consolini, Ber. Bunsenges. Phys. Chem. **95**, 376 (1991).

[368] W. Kleemann, V. Schönknecht, D. Sommer, Phys. Rev. Lett. **66**, 762 (1991).

[369] F. Bruni, G. Consolini, and G. Careri, J. Chem. Phys. **99**, 538 (1993).

[370] W. Wernsdorfer and R. Sessoli, Science **284**, 133 (1999).

[371] M. N. Leuenberger and D. Loss, Nature **410**, 789 (2001).

[372] B. Golding, J. E. Graebner, A. B. Kane, and J. L. Black, Phys. Rev. Lett. **41**, 1487 (1978).

[373] J. L. Black and P. Fulde, Phys. Rev. Lett. **43**, 453 (1979).

[374] G. Weiss, W. Arnold, K. Dransfeld, and H.-J. Güntherodt, Solid State Comm. **33**, 111 (1980).

[375] J. Stockburger, U. Weiss, and R. Görlich, Z. Physik B **84**, 457 (1991).

[376] J. Stockburger, M. Grifoni, M. Sassetti, U. Weiss, Z. Physik B **94**, 447 (1994).

[377] P. Esquinazi, R. König, and F. Pobell, Z. Physik B **87**, 305 (1992).

[378] *Tunneling Systems in Amorphous and Crystalline Solids*, ed. by P. Esquinazi (Springer Verlag, Berlin, 1998).

[379] J. Kondo, Physica **125** B, 279 (1984); *ibid*. **126** B, 377 (1984).

[380] G. Cannelli, R. Cantelli, F. Cordero, and F. Trequattrini, in Ref. [378].

[381] H. Grabert and H. R. Schober, in Ref. [101].

[382] C. D. Tesche, Ann. N. Y. Acad. Sci. **480**, 36 (1986);
S. Chakravarty, *ibid*. **480**, 25 (1986).

[383] K. Huang and A. Rhys, Proc. Roy. Soc. A **204**, 406 (1950);
M. Lax, J. Chem. Phys. **20**, 1752 (1952);
R. Kubo and Y. Toyozawa, Progr. Theor. Phys. **13**, 160 (1955);
V. G. Levich and R. R. Dogonadze, Coll. Czech. Chem. Comm. **26**, 193 (1961).
See also V. G. Levich in *Advances in Electrochemistry and Electrochemical Engineering*, ed. by P. Delahay and C. W. Tobias, Vol. 4, 249 (Interscience, 1965).

[384] H. Spohn and R. Dümcke, J. Stat. Phys. **41**, 389 (1985).

[385] U. Weiss, H. Grabert, P. Hänggi, P. Riseborough, Phys. Rev. B **35**, 9535 (1987).

[386] S. Chakravarty and S. Kivelson, Phys. Rev. B **32**, 76 (1985).

[387] A. T. Dorsey, M. P. A. Fisher, and M. Wartak, Phys. Rev. A **33**, 1117 (1986).

[388] R. Silbey and R. A. Harris, J. Chem. Phys. **80**, 2615 (1983); J. Phys. Chem. **93**, 7062 (1989).

[389] F. Wegner, Ann. Physik (Leipzig) **3**, 77 (1994).

[390] S. Kehrein, *The Flow Equation Approach to Many-Particle Systems*, Springer Tracts in Modern Physics, Vol. 217 (Springer Verlag, Berlin, 2006).

[391] S. K. Kehrein, A. Mielke, and P. Neu, Z. Physik B **99**, 269 (1996).

[392] S. K. Kehrein and A. Mielke, Ann. Physik (Leipzig) **6**, 90 (1997); J. Stat. Phys. **90**, 889 (1997).

[393] S. K. Kehrein and A. Mielke, Phys. Lett. A **219**, 313 (1996).

[394] R. Görlich and U. Weiss, Phys. Rev. B **38**, 5254 (1988).

[395] R. Görlich and U. Weiss, Il Nuovo Cim. **11** D, 123 (1989).

[396] P. B. Vigmann and A. M. Finkel'steïn, Sov. Phys.–JETP **48**, 102 (1978).

[397] P. Nozières, J. Low Temp. Phys. **17**, 31 (1974).

[398] P. Schlottmann, J. Magn. Mater. **7**, 72 (1978); Phys. Rev. B **25**, 4805 (1982).

[399] P. W. Anderson, Phys. Rev. Lett. **18**, 1049 (1967).

[400] K. D. Schotte and U. Schotte, Phys. Rev. **182**, (1969);
K. Schönhammer, Z. Phys. B **45**, 23 (1981).

[401] V. J. Emery, in *Highly Conducting One-Dimensional Solids*, ed. by J. T. Devreese *et al.* (Plenum, New York, 1979);
F. D. M. Haldane, Phys. Rev. Lett. **47**, 1840 (1981).

[402] D. L. Cox and A. Zawadowski, Adv. Phys. **47**, 599 (1998).

[403] G. Yuval and P. W. Anderson, Phys. Rev. B **1**, 1522 (1970).

[404] P. W. Anderson and G. Yuval, J. Phys. C **4**, 607 (1971).

[405] J. Cardy, Journ. of Physics A 14, 1407 (1981).

[406] S. Chakravarty and J. Rudnick, Phys. Rev. Lett. **75**, 501 (1995).

[407] K. Völker, Phys. Rev. B **58**, 1862 (1998).

[408] H. Spohn, Comm. Math. Phys. **123**, 277 (1989).

[409] B. Carmeli and D. Chandler, J. Chem. Phys. **82**, 3400 (1985);
D. Chandler in *Liquids, Freezing and Glass Transition*, ed. by D. Levesque,
J. P. Hansen, and J. Zinn-Justin (Elsevier Science Publishers, 1990).

[410] a.: J. Ulstrup, *Charge Transfer in Condensed Media* (Springer, 1979);
b.: B. Fain, *Theory of Rate Processes in Condensed Media* (Springer, 1980).

[411] P. Ao and J. Rammer, Phys. Rev. Lett. **62**, 3004 (1989).

[412] A. Garg, J. N. Onuchic, and V. Ambegaokar, J. Chem. Phys. **83**, 4491 (1985).

[413] I. Rips and J. Jortner, J. Chem. Phys. **87**, 2090 (1987);
M. Sparpaglione and S. Mukamel, J. Chem. Phys. **88**, 3263 (1987).

[414] J. N. Gehlen and D. Chandler, J. Chem. Phys. **97**, 4958 (1992); J. N. Gehlen,
D. Chandler, H. J. Kim, and J. T. Hynes, J. Phys. Chem. **96**, 1748 (1992).

[415] X. Song and A. A. Stuchebrukhov, J. Chem. Phys. **99**, 969 (1993);
A. A. Stuchebrukhov and X. Song, J. Chem. Phys. **101**, 9354 (1994).

[416] J. Cao, C. Minichino, and G. A. Voth, J. Chem. Phys. **103**, 1391 (1995).

[417] M. H. Devoret, D. Esteve, H. Grabert, G.-L. Ingold, H. Pothier, and C. Urbina,
Phys. Rev. Lett. **64**, 1824 (1990); Physica B **165 & 166**, 977 (1990);
S. M. Girvin, L. I. Glazman, M. Jonson, D. R. Penn, and M. D. Stiles, Phys.
Rev. Lett. **64**, 3183 (1990).

[418] R. Bruinsma and P. Bak, Phys. Rev. Lett. **56**, 420 (1986).

[419] J. Jortner, J. Chem. Phys. **64**, 4860 (1975).

[420] P. Minnhagen, Phys. Lett. A **56**, 327 (1976).

[421] G. Falci, V. Bubanja, and G. Schön, Z. Physik B **85**, 451 (1991).

[422] K. Ando, J. Chem. Phys. **106**, 116 (1997).

[423] J. S. Bader, R. A. Kuharski, and D. Chandler, J. Chem. Phys. **93**, 230 (1990).

[424] P. Siders and R. A. Marcus, J. Am. Chem. Soc. **103**, 741 (1981).

[425] U. Weiss and H. Grabert, Phys. Lett. **108A**, 63 (1985).

[426] a.: H. Grabert and U. Weiss, Phys. Rev. Lett. **54**, 1605 (1985).
b.: M. P. A. Fisher and A. T. Dorsey, Phys. Rev. Lett. **54**, 1609 (1985).

[427] C. Aslangul, N. Poitier, and D. Saint-James, J. Phys. (Paris) **47**, 1671 (1986).

[428] R. Egger, C. H. Mak, and U. Weiss, J. Chem. Phys. **100**, 2651 (1994).

[429] H. Grabert, Phys. Rev. B **46**, 12753 (1992).

[430] A. Würger, in Ref. [378]; Phys. Rev. Lett. **78**, 1759 (1997).

[431] Q. Niu, J. Stat. Phys. **65**, 317 (1991).

[432] H. Grabert, U. Weiss, and H. R. Schober, Hyperfine Interactions **31**, 147 (1986).

[433] R. Pirc and P. Gosar, Phys. Kondens. Mater. **9**, 377 (1969).

[434] A. Würger, Physics Letters A **236**, 571 (1998).

[435] A. Würger, Solid State Comm. **106**, 63 (1998).

[436] M. Sassetti and U. Weiss, Europhys. Lett. **27**, 311 (1994).

[437] M. Sassetti, F. Napoli, and U. Weiss, Phys. Rev. B **52**, 11 213 (1995).

[438] H. Schoeller and G. Schön, Phys. Rev. B **50**, 18 436 (1994);
       J. König, J. Schmid, H. Schoeller, and G. Schön, Phys. Rev. B **54**, 16 820 (1996);
       H. Schoeller, in Ref. [439].

[439] *Mesoscopic Electron Transport*, ed. by L. L. Sohn, L. P. Kouwenhoven, and
       G. Schön, NATO ASI Series E, Vol. 345 (Kluwer, Dordrecht, 1997).

[440] J. C. Cuevas and E. Scheer, *Molecular Electronics*, Series in Nanoscience and
       Nanotechnology – Vol. 1 (World Scientific, Singapore, 2010).

[441] G.-L. Ingold, H. Grabert, and U. Eberhardt, Phys. Rev. B **50**, 395 (1994).

[442] R. D. Coalson, D. G. Evans, and A. Nitzan, J. Chem. Phys. **101**, 436 (1994).

[443] A. Lucke, C. H. Mak, R. Egger, J. Ankerhold, J. Stockburger, and H. Grabert,
       J. Chem. Phys. **107**, 8397 (1997).

[444] C. Cohen-Tannoudji, B. Diu, and F. Laloë, *Quantum Mechanics*, Vol. 1 (Wiley,
       New York).

[445] M. Grifoni, M. Winterstetter, and U. Weiss, Phys. Rev. E **56**, 334 (1997).

[446] F. Guinea, Phys. Rev. B **32**, 4486 (1985).

[447] G. Lang, E. Paladino, and U. Weiss, Europhys. Lett. **43**, 117 (1998); Phys. Rev.
       E **58**, 4288 (1998).

[448] M. Grifoni, M. Sassetti, and U. Weiss, Phys. Rev. E **53**, R2033 (1996).

[449] H. Dekker, Phys. Rev. A **35**, 1436 (1987).

[450] M. Winterstetter and U. Weiss, Chem. Phys. **217**, 155 (1997).

[451] F. Lesage and H. Saleur, Phys. Rev. Lett. **80** 4370 (1998).

[452] J. Cardy, Nucl. Phys. B **324**, 581 (1989);
       I. Affleck and A. W. W. Ludwig, Nucl. Phys. B **360**, 641 (1991); *ibid.* B **428**,
       545 (1994).

[453] P. Fulde and I. Peschel, Adv. Phys. **21**, 1 (1972).

[454] U. Weiss and H. Grabert, Europhys. Lett. **2**, 667 (1986);
       U. Weiss, H. Grabert, and S. Linkwitz, J. Low Temp. Phys. **68**, 213 (1987).

[455] A. Garg, Phys. Rev. B **32**, 4746 (1985).

[456] S. Dattagupta, H. Grabert, and R. Jung, J. Phys.: Cond. Mat. **1**, 1405 (1989).

[457] U. Weiss and M. Wollensak, Phys. Rev. Letters **62**, 1663 (1989);
       R. Görlich, M. Sassetti, and U. Weiss, Europhys. Lett. **10**, 507 (1989).

[458] A. Würger, Phys. Rev. B **57**, 347 (1998).

[459] D. A. Parshin, Z. Physik B **91**, 367 (1993).

[460] J. Stockburger, M. Grifoni, and M. Sassetti, Phys. Rev. B **51**, 2835 (1995).

[461] W. G. Unruh, Phys. Rev. A **51**, 992 (1995).

[462] E. Paladino, M. Sassetti, G. Falci, and U. Weiss, Phys. Rev. B **77**, 041303(RC) (2008).

[463] P. Nägele, G. Campagnano, and U. Weiss, New. Journ. of Physics **10**, 115010 (2008);
P. Nägele and U. Weiss, Physica E **42**, 622 (2010).

[464] Y. M. Galperin, B. L. Altshuler, J. Bergli, and D. V. Shantsev, Phys. Rev. Lett. **96**, 097009 (2006);
Y. M. Galperin, B. L. Altshuler, and D. V. Shantsev, ArXiv:cond-mat/0312490v1 (2003).

[465] J. Schriefl, Yu. Makhlin, A. Shnirman, and G. Schön, New. Journ. of Physics **8**, 1 (2006).

[466] G. Falci, A. D'Arrigo, A. Mastellone, and E. Paladino, Phys. Rev. Lett. **94**, 167002 (2005).

[467] Yu. Makhlin, G. Schön, and A Shnirman, in *New Directions in Mesoscopic Physics (Towards Nanoscience)*, R. Fazio, V. F. Gantmakher, and Y. Imry (Eds.) (Kluwer Academic Publishers, 2003) [ArXiv:cond-mat/0309049v1].

[468] O. Astafiev, Yu. A. Pashkin, Y. Nakamura, T. Yamamoto, and J. S. Tsai, Phys. Rev. Lett. **93**, 267007 (2004).

[469] R. Egger, H. Grabert, and U. Weiss, Phys. Rev. E **55**, R3809 (1997).

[470] T. A. Costi and C. Kieffer, Phys. Rev. Lett. **76**, 1683 (1996).

[471] T. A. Costi, Phys. Rev. B **55**, 3003 (1997); Phys. Rev. Lett. **80**, 1038 (1998).

[472] F. Lesage and H. Saleur, Nucl. Phys. B **490**, 543 (1997).

[473] S. P. Strong, Phys. Rev. E **55**, 6636 (1997).

[474] J. T. Stockburger and C. H. Mak, J. Chem. Phys. **105**, 8126 (1996).

[475] H. Baur, A. Fubini, and U. Weiss, Phys. Rev. B **70**, 024302 (2004).

[476] R. N. Silver, D. S. Sivia, and J. E. Gubernatis in: *Quantum Simulations of Condensed Matter Phenomena*, ed. by J. D. Doll and J. E. Gubernatis (World Scientific, Singapore, 1990).

[477] P. Fendley, F. Lesage, and H. Saleur, J. Stat. Phys. **79**, 799 (1995).

[478] F. Grossmann, T. Dittrich, and P. Hänggi, Phys. Rev. Lett. **67**, 516 (1991).

[479] N. Makri, J. Chem. Phys. **106**, 2286 (1997); N. Makri and L. Wei, Phys. Rev. E **55**, 2475 (1997).

[480] a.: D. G. Evans, R. D. Coalson, H. J. Kim, and Yu. Dakhnovskii, Phys. Rev. Lett. **75**, 3649 (1995).
b.: M. Morillo and R. I. Cukier, Phys. Rev. B **54**, 13 962 (1996).

[481] M. Grifoni, L. Hartmann, and P. Hänggi, Chem. Phys. **217**, 167 (1997).

[482] Special issue on *Dynamics of Driven Quantum Systems*, ed. by W. Domcke, P. Hänggi, and D. Tannor, Chem. Phys. **217** (2,3), 117-416 (1997).

[483] M. Grifoni and P. Hänggi, *Driven Quantum Tunneling*, Phys. Rep. **304**, 229 (1998).

[484] S. Han, J. Lapointe, and J. E. Lukens, Phys. Rev. Lett. **66**, 810 (1991); Phys. Rev. B **46**, 6338 (1992).

[485] M. Grifoni, Phys. Rev. E **54**, R3086 (1996).

[486] M. Grifoni, M. Sassetti, P. Hänggi, and U. Weiss, Phys. Rev. E **52**, 3596 (1995).

[487] M. Grifoni, M. Sassetti, J. Stockburger, U. Weiss, Phys. Rev. E **48**, 3497 (1993).

[488] P. K. Tien and J. P. Gordon, Phys. Rev. **129**, 647 (1963).

[489] Yu. Dakhnovskii, Phys. Rev. B **49**, 4649 (1994);
Yu. Dakhnovskii and R. D. Coalson, J. Chem. Phys. **103**, 2908 (1995);
I. A. Goychuk, E. G. Petrov, and V. May, Chem. Phys. Lett. **253**, 428 (1996).
See also A. Lück, M. Winterstetter, U. Weiss, and C.H. Mak, Phys. Rev. E **58**, 5565 (1998).

[490] F. Grossmann and P. Hänggi, Europhys. Lett. **18**, 571 (1992).

[491] J. M. Gomez-Llorente, Phys. Rev. A **45**, R6958 (1992);
erratum Phys. Rev. E **49**, 3547 (1994).

[492] P. Jung, Phys. Rep. **234**, 175 (1993);
Proc. NATO Workshop on *Stochastic Resonance in Physics and Biology*, F. Moss *et al.* (Eds.), J. Stat. Phys. **70**, 1 (1993);
L. Gammaitoni, P. Hänggi, P. Jung, and F. Marchesoni, Rev. Mod. Phys. **70**, 223 (1998).

[493] R. Löfstedt and S. N. Coppersmith, Phys. Rev. Lett. **72**, 1947 (1994);
Phys. Rev. E **49**, 4821 (1994).

[494] M. Grifoni and P. Hänggi, Phys. Rev. Lett. **76**, 1611 (1996); Phys. Rev. E **54**, 1390 (1996).

[495] H. Adam, M. Winterstetter, M. Grifoni, and U. Weiss, Phys. Rev. Lett. **83**, 252 (1999).

[496] *The Photosynthetic Reaction Center*, Vol. 1 and 2, ed. by J. Deisenhofer and J. R. Norris (Academic Press, New York, 1993).

[497] R. Egger, C. H. Mak, and U. Weiss, Phys. Rev. E **50**, R655 (1994).

[498] M. P. A. Fisher and W. Zwerger, Phys. Rev. B **32**, 6190 (1985).

[499] W. Zwerger, Phys. Rev. B **35**, 4737 (1987).

[500] C. L. Kane and M. P. A. Fisher, Phys. Rev. Lett. **68**, 1220 (1992); Phys. Rev. B **46** 15233 (1992).

[501] X. G. Wen, Phys. Rev. B **41**, 12838 (1990); **43** 11025 (1991); **44**, 5708 (1991).

[502] H. J. Schulz, Phys. Rev. Lett. **71**, 1864 (1993).

[503] M. Fabrizio, A. O. Gogolin, and S. Scheidl, Phys. Rev. Lett. **72**, 2235 (1994).

[504] H. Rodenhausen, J. Stat. Phys. **55**, 1065 (1989).

[505] Y.-C. Chen, J. Stat. Phys. **65**, 761 (1991).

[506] A. Lenard, J. Math. Phys. **2**, 682 (1961).

[507] Yu. M. Ivanchenko and L. A. Zil'berman, Sov. Phys.–JETP **28**, 1272 (1969).

[508] U. Weiss and M. Wollensak, Phys. Rev. B **37**, 2729 (1988).

[509] F. Guinea, Phys. Rev. B **32**, 7518 (1985).

[510] U. Weiss, R. Egger, and M. Sassetti, Phys. Rev. B **52**, 16707 (1995).

[511] M. Sassetti, M. Milch, and U. Weiss, Phys. Rev. A **46**, 4615 (1992).

[512] M. Sassetti, H. Schomerus, and U. Weiss, Phys. Rev. B **53**, R2914 (1996).

[513] H. Grabert, G.-L. Ingold, and B. Paul, Europhys. Lett. **44**, 360 (1998).

[514] K. Leung, R. Egger, and C. H. Mak, Phys. Rev. Lett. **75**, 3344 (1995).

[515] L. S. Levitov and G. B. Lesovik, JETP Lett. **58**, 230 (1993).

[516] D. V. Averin, Solid State Comm. **105**, 659 (1998).

[517] I. S. Beloborodov, F. W. J. Hekking and F. Pistolesi, in *New Directions in Mesoscopic Physics*, ed. by R. Fazio et al. (Kluwer Academic Publisher, 2003).

[518] G.-L. Ingold and H. Grabert, Phys. Rev. Lett. **83**, 3721 (1999).

[519] U. Weiss, Solid State Comm. **100**, 281 (1996);
U. Weiss, in *Tunneling and Its Implications*, ed. by D. Mugnai, A. Ranfagni, and L. S. Schulman, p. 134 (World Scientific, Singapore, 1997).

[520] P. Fendley and H. Saleur, Phys. Rev. Lett. **81**, 2518 (1998).

[521] N. Seiberg and E. Witten, Nucl. Phys. B **426**, 19 (1994); B **431**, 484 (1994).

[522] N. M. Temme, *Special Functions* (Wiley, New York, 1996), p. 49.

[523] L. Álvarez-Gaumé and F. Zamora, ArXiv:hep-th/9709180v3 (1997).

[524] K. A. Matveev, D. Yue, and L. I. Glazman, Phys. Rev. Lett. **71**, 3351 (1993).

[525] H. Saleur and U. Weiss, Phys. Rev. B **63**, 201302(R) (2001).

[526] P. Fendley, A. W. W. Ludwig, and H. Saleur, Phys. Rev. B **52**, 8934 (1995).

[527] P. Fendley and H. Saleur, Phys. Rev. B **54**, 10845 (1996).

[528] A. Komnik and H. Saleur, Phys. Rev. Lett. **96**, 216406 (2006).

[529] Y.-C. Chen and J. L. Lebowitz, Phys. Rev. B **46**, 10 743 (1992).

[530] P. Fendley, F. Lesage, and H. Saleur, J. Stat. Phys. **85**, 211 (1996).

[531] J. Honer and U. Weiss, Chem. Phys. **375**, 265 (2010).

[532] R.P. Stanley, Adv. Math. **77**, 76 (1989).

[533] A. O. Gogolin, A. A. Nersesyan, and A. M. Tsvelik, *Bosonization and Strongly Correlated Systems* (Cambridge University Press, Cambridge, 1998).

[534] T. Giamarchi, *Quantum Physics in One Dimension* (Clarendon, Oxford, 2004).

[535] A. Luther and I. Peschel, Phys. Rev. B **9**, 2911 (1974).

[536] P. Fendley, A. W. W. Ludwig, and H. Saleur, Phys. Rev. Lett. **75**, 2196 (1995).

[537] R. Egger and H. Grabert, Phys. Rev. Lett. **77**, 538 (1996), and Ref. 10 therein.

[538] I. Safi and H. Saleur, Phys. Rev. Lett. **93**, 126602 (2004).

[539] M. Kindermann and B. Trauzettel, Phys. Rev. Lett. **94**, 166803 (2005).

[540] A. O. Gogolin and A. Komnik, Phys. Rev. B **73**, 195301 (2006).

Index